D1605483

The Coccidia

The Coccidia

Eimeria, *Isospora*, *Toxoplasma*, and Related Genera

Edited by
Datus M. Hammond
with
Peter L. Long

University Park Press • Baltimore
Butterworths • London

University Park Press

International Publishers in Science and Medicine

Chamber of Commerce Building
Baltimore, Maryland 21202

Printed in the United States of America

Designed by Grant Treaster, Jr.

Published jointly by
University Park Press, Baltimore

and

Butterworth & Co. (Publishers) Ltd., London

ISBN 0-8391-0732-3 (University Park Press)
ISBN 0-408-70462-4 (Butterworths)

Library of Congress Cataloging in Publication Data

Hammond, Datus M.
The Coccidia.

Includes bibliographical references.
1. Coccidia. I. Title. [DNLM: 1. Coccidiosis.
2. Sporozoa—Toxoplasmosis. QX 123 H225c 1972]
QL368.C7H23 593'.19 72-7523
ISBN 0-8391-0732-3

Contents

Preface . . . *v*

1 Introduction, History, and Taxonomy . . . *1*
Norman D. Levine

2 Host and Site Specificity in the Coccidia . . . *23*
William C. Marquardt

3 Life Cycles and Development of Coccidia . . . *45*
Datus M. Hammond

4 Ultrastructure . . . *81*
Erich Scholtyseck

5 Cytochemistry, Physiology, and Biochemistry . . . *145*
John F. Ryley

6 Cultivation of Coccidia in Avian Embryos and Cell Cultures . . . *183*
David J. Doran

7 Pathology and Pathogenicity of Coccidial Infections . . . *253*
P. L. Long

8 Immunity . . . *295*
M. Elaine Rose

9 Toxoplasmosis: Parasite Life Cycle, Pathology, and Immunology . . . *343*
J. K. Frenkel

10 Techniques . . . *411*
Leonard Reid Davis

Index . . . *459*

Preface

The coccidia are the cause of coccidiosis, a disease of considerable importance in domestic animals, especially poultry, cattle, sheep, and goats. Coccidiosis is also important in wild animals, but seldom occurs in man. Coccidia have a complex life cycle and other unusual characteristics which have stimulated investigations by increasing numbers of biologists. During the last seven or eight years, fine-structural studies have indicated that the coccidia are related to *Plasmodium*, which causes malaria, and to other blood Sporozoa. A still closer relationship between coccidia and *Toxoplasma*, the organism which causes toxoplasmosis, and to other similar organisms, such as *Besnoitia*, was suggested by these studies. With the discovery in 1970 that *Toxoplasma gondii* undergoes a cycle of stages in the intestine of cats similar to those of coccidia and is transmissible by oocysts discharged in the feces of cats, the coccidian nature of this organism was proved. The coccidia have now assumed medical as well as veterinary and general biological importance.

Previously, the only book which covered in detail the full range of the biological characteristics and host-parasite relationships of this important group was *Coccidia and coccidiosis of domesticated game and laboratory animals and of man* by E. R. Becker (1934). Recent books in English have emphasized taxonomy, life cycles, or the disease-producing aspects. These include *Coccidiosis* by Davies, Joyner, and Kendall (1963), *Coccidia and coccidiosis* by L. P. Pellérdy (1965), *The coccidian parasites (Protozoa, Sporozoa) of rodents* by N. D. Levine and V. Ivens (1965), *The coccidian parasites (Protozoa, Sporozoa) of ruminants* by N. D. Levine and V. Ivens (1970), *Coccidiosis of farm animals* by N. P. Orlov (1956), English translation by the Israel Program for Scientific Translations (1970), and the *Life cycles of coccidia of domestic animals* by Y. M. Kheysin (1967), English translation edited by K. S. Todd, Jr. (1972). Some books in other languages have also appeared, including *Estudios sobre coccidiosis* by Cordero del Campillo (1962) and *Kotsidii gryzunov SSSR* (Coccidia of rodents of the USSR) by M. A. Musaev and A. M. Veisov (1965).

In September of 1968, during a simultaneous visit of Erich Scholtyseck and Peter L. Long to my laboratory, we discussed the writing of a book dealing with the biology of the coccidia. The plans initiated at that time were elaborated in further discussions among the three of us in Leningrad in July 1969, in Bonn in March 1970, and in Washington, D. C. in September 1970. We were fortunate in obtaining the agreement of two other scientists from Europe and five others from the United States, each to contribute a chapter covering an aspect in which he is a leading investigator. Peter L. Long has assisted with the editing of the manuscripts of the European contributors and with the preparation of the index, as well as with numerous editorial decisions. Erich Scholtyseck, Norman D. Levine, David J. Doran and other contributors have also participated in a number of the planning decisions.

In the present book we have attempted to provide an up-to-date coverage of the general aspects of the coccidia and their parasite-host relationships. We have omitted detailed consideration of chemotherapy, although this is briefly considered in Chapter 5, because we thought that this subject could not be adequately covered within a single chapter. Each author discusses work published by autumn, 1971, and other data through the spring of 1972; he presents interpretations of his own work and of work by others in his field, and a summary of the current status of information in his particular area of interest.

It is our hope that the volume will be useful to advanced undergraduates, graduate students, research workers, and teachers in biology, the health sciences, veterinary medicine and human medicine, as well as to veterinarians and physicians.

I am indebted to Mrs. Madeline J. Codella for her expert assistance with the manuscripts, to each of the contributors for their cooperation in meeting deadlines and in revision of manuscripts, and to Linton M. Vandiver of University Park Press for assistance in publication of the book.

D. M. H.

The Coccidia

1 Introduction, History, and Taxonomy

NORMAN D. LEVINE
College of Veterinary Medicine, Agricultural Experiment Station, and Department of Zoology, University of Illinois, Urbana, Illinois

Contents

I. Introduction . . . *1*
II. History . . . *4*
III. Taxonomy . . . *10*
Literature Cited . . . *19*

I. Introduction

Customarily, the coccidia are thought of as members of the genera *Eimeria, Isospora,* and a few others, primarily in the single family Eimeriidae. This is much too restricted a view. A second view is that they comprise all the members of the subclass Coccidiasina as opposed to the subclass Gregarinasina, which are parasites of invertebrates. This is too broad a view, since it includes not only quite a few poorly known species but also the malaria parasites, which appear to be quite distinct. A middle-of-the-road view, which I prefer, is to include within the group only members of the suborder Eimeriorina. This suborder includes (in my classification—see below) eight families and a great sufficiency of poorly known species. There are so many unanswered questions about many of them that we should have no difficulty in spending many man-years trying to answer them. It is appropriate to review our knowledge of the group, or at least part of it, and to point out some of the problems, both for our own benefit and to provide a baseline for future workers.

We now know that coccidia can cause serious diseases in domestic animals. Indeed, coccidiosis is so important in poultry that all commercial chicken feeds in the United States contain coccidiostats.

This was not always so. Minchin (1903) remarked that the Sporozoa were considered an obscure group of no practical importance, so that for a long time they did not appeal to "common-sense" Englishmen. It took the discovery of the malaria parasite to bring them into prominence.

Even before that, however, it was known that a few coccidia could be pathogenic. Minchin described the effects of *Eimeria stiedai* in rabbits:

> The liver is greatly enlarged, and its blood vessels compressed, leading to functional derangements; the secretion of bile is reduced to a minimum; the blood becomes pale and watery, as in pernicious anemia; the respiration becomes gasping, and the animal finally dies in convulsions. In all these cases the destructive power of the parasite varies directly as its power of multiplying by schizogony, and so overrunning the tissues which it attacks; and it is a very interesting and important fact, that in no case, apparently, can the schizogony continue indefinitely, but has its own natural, intrinsic limit, after which conjugation, with consequent sporogony, is necessary for the recuperation of the parasite and the continuance of its race. If, therefore, the patient can safely pass the acute stage, the disease heals itself through the failing reproductive powers of the parasite on the one hand, and the regenerative capacity of the epithelium on the other.

This was a very accurate description, although it has been modified to some extent by more recent research.

In his section on Protozoa in the *Cambridge Natural History,* Hartog (1906) devoted 14 pages to the Sporozoa. The only coccidia he mentioned by name were *Coccidium schubergi, C. lacazei, C. cuniculi,* and *Adelea ovata.* He also said, "*Coccidium* may also produce a sort of dysentery in cattle on the Alpine pastures of Switzerland; and cases of human coccidiosis are by no means unknown. *Coccidium*-like bodies have been demonstrated in the human disease, 'molluscum contagiosum,' and the 'oriental sore' of Asia; similar bodies have also been recorded in smallpox and vaccinia, malignant tumors and even syphilis, but their nature is not certainly known; some of these are now referred to the Flagellata." Hartog was a good protozoologist, and the fact that coccidia do not cause molluscum contagiosum, oriental sore, smallpox, vaccinia, cancer, or syphilis does not reflect his own deficiencies but the state of our knowledge at the time.

Coccidiosis is essentially a man-made disease, the result of abnormal crowding of single host species in a limited area. Only under these circumstances do animals become infected with enough oocysts to become ill. We now recognize that coccidiosis occurs in many species of domestic animals, including especially chickens, turkeys, rabbits, sheep, and cattle.

The coccidia belong to the subphylum Apicomplexa, all of whose members are parasitic. Electron microscopy has revealed common structures in the group and has enabled taxonomists to arrange them better than before (see Andreassen and Behnke, 1968; Levine, 1969). All members of the subphylum have an apical complex (Fig. 1) at one stage or another, composed of (1) one or more electron-dense *polar rings,* (2) a *conoid* formed by a spirally coiled filament, inside the polar ring, (3) a number of *micronemes,* (sarconemes, rod-shaped granules, convoluted tubules)—elongate, electron-dense organelles extending longitudinally in the anterior part of the body, (4) a number of *rhoptries* (toxonemes, paired organelles, lankesterellonemes, eimerianemes, dense bodies)—electron-dense, tubular or saccular organelles, often enlarged posteriorly, extending back from the anterior region within the conoid, and (5) a number of *subpellicular tubules*—slender, electron-dense hollow structures extending back from the polar ring region beneath the pellicle. In addition, they may have one or more *micropores*—openings in the side visible only with the electron microscope, often used for food intake. Members of the two subphyla, Microsporida and Myxosporida, which were once thrown into the same group with the Apicomplexa, lack all of these structures.

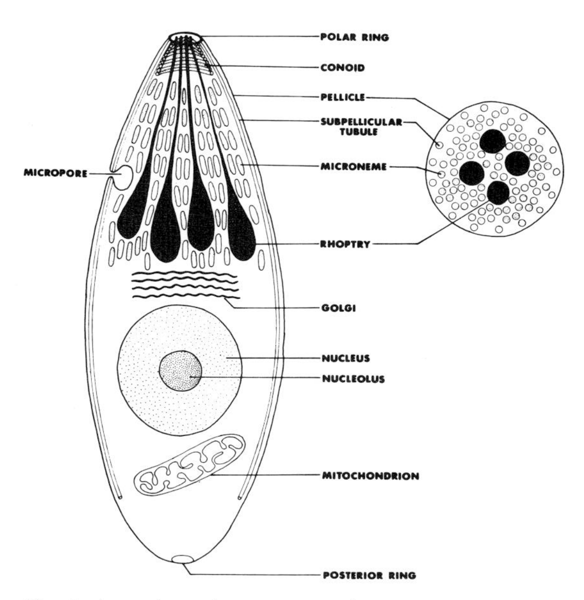

Fig. 1. An apicomplexan sporozoite or merozoite, illustrating the apical complex.

The subphylum Apicomplexa contains two classes, the Sporozoasida and the Piroplasmasida. I shall say nothing about the latter, except that in this group the components of the apical complex are reduced and some of them are absent. I discussed its taxonomy earlier (Levine, 1971).

The class Sporozoasida is divided into two subclasses, the Gregarinasina and the Coccidiasina. In the former the mature gamonts are extracellular and large, and the conoid is modified into a mucron or epimerite. These protozoa are generally monoxenous (i.e., have a single type of host), and are parasites of the digestive tract or body cavity of invertebrates or lower chordates. There are many species and subgroups of gregarines, but they are outside the scope of this book.

In the subclass Coccidiasina, the mature gamonts are typically intracellular and small, and the conoid is not modified into a mucron or epimerite. Some species are monoxenous and others are heteroxenous; they are found in the digestive tract epithelium, blood cells, or other cells, mostly in vertebrates but sometimes in invertebrates.

This subclass is divided into two orders. There are only a few known species in the order Protococcidiorida. They occur in marine invertebrates, and merogony is absent in them. This group is considered primitive.

The main order is the Eucoccidiorida. Most of its members are found in vertebrates, and merogony is present.

The order Eucoccidiorida is divided into three suborders. In the suborder Adeleorina, the macrogamete and microgametocyte are associated in syzygy during development. Because of this, only a few microgametes need to be formed, and actually each microgametocyte produces only from one to four. The sporozoites are enclosed in an envelope (i.e., there are sporocysts). Endodyogeny (internal budding into two daughter organisms) is absent. Members of this suborder are either monoxenous or heteroxenous (i.e., they either do not or do have an intermediate host). I am not discussing this suborder, fascinating as it is, because it lies outside the scope of the present book. It is considered to be the most primitive suborder in the Eucoccidiorida, and most of its members occur in invertebrates.

The second suborder of Eucoccidiorida is the Haemosporina. This group contains the malaria parasites and their relatives. There is no syzygy, the macrogamete and microgametocyte develop independently and the latter produces about eight microgametes, the zygote is motile, there are no sporocysts, and all species are heteroxenous. These are blood cell parasites of vertebrates and are transmitted by Diptera. More has been written about this group than about any other apicomplexans, but again these parasites lie outside the scope of this book.

The third suborder, Eimeriorina, includes the coccidia proper. These are the organisms that come to mind when we think of the coccidia, but they are really only one of three suborders in one of two orders in the subclass Coccidiasina.

The suborder Eimeriorina is currently divided into eight families, but only four of them occur in vertebrates, and we really know little about the others. Further, our knowledge of the four families that occur in vertebrates has many gaps. I shall discuss these below, but first something should be said about the history of the group.

II. History

Practically all the protozoa are so small that it requires a microscope to see them. As a result, it was not until the advent of the microscope that they were discovered. In 1674 Leeuwenhoek saw the first parasitic protozoon, the oocysts of *Eimeria stiedai* in the bile of a rabbit, but it was more than 150 years later that it was described. Hake (1839), who did so, thought that the oocysts were pus globules associated with carcinoma of the liver.

The history of the coccidia is so tangled and so involved in the history of other protozoa (and helminths) that it is impossible to give a straightforward chronological account of it. If we omit Leeuwenhoek, the gregarines were discovered before the coccidia. In 1787 Cavolini was the first to see one. He described and figured what is now known as *Cephaloidophorus conformis* (Diesing, 1851) Léger and Duboscq, 1911 in the glandular appendages of the stomach of the cancroid crab *Pachygrapsus marmoratus* on the Mediterranean coast of Europe. He found gamonts in syzygy and thought that each pair was a kind of tapeworm with two segments.

His was merely an incidental observation, however, and it was really Léon Dufour who discovered the group in the scientific sense. Dufour was interested primarily in insect anatomy, but in studying it he found the gregarines and described many species. He (1828) gave them the generic name *Gregarina,* and thought that they were a peculiar group of worms related to the trematodes. Other workers later described other species, and von Siebold (1839) described the nucleus accurately for the first time. Despite this, he, too, thought they were worms, although he did not say that they had an alimentary canal as one of his predecessors had done (Minchin, 1903). He also described the cysts that he found associated with the gregarines, but he did not realize the connection between them.

In 1845 and 1848, von Kölliker described many species of Sporozoa and said for the first time that gregarines were unicellular and therefore protozoa. Henle (1835) and others had found cysts in the seminal vesicles of earthworms which he called "pseudonavicellae," and Stein (1848) agreed with von Kölliker and showed that these were actually gregarines. However, Henle, Bruch, and Leydig continued to maintain that gregarines were "in some way connected with the embryonic stages of nematodes" (Minchin, 1903), and more especially of *Filaria.* However, as Minchin (1903) said, this theory died a natural death in the course of time.

A name that was once used for coccidia was *Psorospermium.* This name has an interesting history. Johannes Müller (1841) gave the name "Psorospermien" to the cysts of a myxosporidan of pike and other fish. The name comes from the same root as psoriasis—the parasites were associated with a dermatitis of the fish. Remak (1845) remarked on the resemblance between the oocysts of *Eimeria stiedai* of the rabbit and Müller's psorosperms, and Lieberkühn (1854) called the former Psorospermien. Rivolta (1878) named the rabbit species *Psorospermium cuniculi.* Later workers talked of "egg-shaped psorosperms" (coccidia) and "fish-psorosperms" or "Müller's psorosperms" (myxosporidan spores). Later, however, when it was realized that the two were very different, the name *Psorospermium* was dropped for the coccidia.

Another name formerly used was *Coccidium*. This was introduced as a generic name by Leuckart (1879) for the rabbit parasite which he called *Coccidium oviforme*. However, Lindemann (1865) had previously named it *Monocystis stiedae*, thinking that it was a gregarine. Later, when it was realized that *Eimeria* and *Coccidium* were the same genus, the correct name for this species became *Eimeria stiedai* (Lindemann, 1865) Kisskalt and Hartmann, 1907; *Psorospermium cuniculi* and *Coccidium oviforme* were thus synonyms.

The life cycle of coccidia, with its alternation of sexual and asexual generations, was worked out piecemeal. Sporulation inside the oocyst was first described by Kauffmann (1847) and then in more detail by Stieda (1865) in rabbit coccidia. Balbiani (1884) described it more accurately. Kloss (1855) worked out the life cycle of the snail coccidium later named *Klossia helicina* by Aimé Schneider. This was the first finding of coccidia in invertebrates. Eimer (1870) described the endogenous cycle of "*Gregarina falciformis*" in the mouse. This species was later named *Eimeria falciformis* by Schneider (1875) and designated the type of the new genus. Eimer thought that the oocysts spread the infection from one mouse to another, and that the coccidia multiplied in the mouse by schizogony. This is true, but it was denied by Schneider (1892), Labbé (1896) and others, who thought that two different genera were involved. Labbé retained this view as late as 1899, when he accepted both *Eimeria falciformis* and *Coccidium falciforme* as names for the same species; he said that *E. falciformis* might be only a stage in the development of *C. falciforme*, which reversed the correct names.

In the meantime, L. Pfeiffer (1890, 1891) and R. Pfeiffer (1892) said that in the rabbit liver the parasites first multiplied and then produced oocysts. They suggested that there was alternation of generations, but this idea was criticized vigorously—until proved correct.

A great deal of work was done on coccidia in the 1890's because they were thought to be a cause of cancer. Schuberg (1895) described the life cycle of *E. falciformis* (which he called *Coccidium falciforme*) in the mouse, confirming Eimer's work. Then Schaudinn and Siedlecki (1897) studied the life cycle of *Adelea ovata* and *Eimeria lacazei* of the centipede *Lithobius forficatus*, and, in a seminal paper, Schaudinn (1900) worked out the complete life cycle of *E. schubergi* in the same centipede.

There was now no question that *Eimeria* and *Coccidium* were the same, and Stiles (1902) and Lühe (1902) pointed this out, asserting the priority of *Eimeria*. One would think that this would have settled the matter, but it did not, or at least many later authors retained the name *Coccidium*, as will be seen shortly.

The class Sporozoa was established by Leuckart (1879), who included in it only the gregarines and coccidia. These all have spores and thus deserve the designation. However, later writers added some groups that do not have spores, so that finally, in the words of Ball (1960), "probably the only character they all possess is a parasitic mode of life, hardly a distinguishing trait separating them from other kinds of Protozoa."

Balbiani (1882) recognized five orders of Sporozoa: Gregarines, Coccidia, Sarcosporidia, Myxosporidia, and Microsporidia. He said that there was no question as to the relationship of the first two groups, but that the relationship of the last three to the first two was much less manifest.

Bütschli (1882) included the subclasses Gregarinida, Myxosporidia, and Sarcosporidia in the class Sporozoa. He placed the coccidia as a family under the gregarine order

Monocystidea, while he placed the Microsporidia in an appendix to the Sarcosporidia.

Lankester (1885) accepted Leuckart's class Sporozoa, dividing it into four subclasses, Gregarinidea, Coccidiidea, Myxosporidia, and Sarcocystidia. He did not mention malaria parasites, of course, since they had not yet been discovered. He accepted three orders of Coccidiidea: In the order Monosporea A. Schneider, "The whole content of the cyst forms but a single spore"; this order contained a single genus, *Eimeria*. In the order Oligosporea A. Schneider, the "cyst content develops itself into a definite and constant but small number of spores"; this order, too, contained a single genus, *Coccidium*. In the order Polysporea (a name which he presumably introduced), the "cyst-content develops itself into a great number of spores (sixty or more)"; there were two genera, *Klossia* and *Drepanidium*, in this order.

In his monumental review of the Sporozoa, Labbé (1899) used few families. He separated the Sporozoa into the legions Cytosporidia (containing the coccidia) and Myxosporidia, adding the Sarcosporidia under Sporozoa incertae sedis. The Cytosporidia he divided in turn into the orders Gregarinida (containing the family Aggregatidae, now known to be a coccidium), Coccidiida, Haemosporidiida (containing only the genera *Lankesterella, Caryolysus* and *Haemogregarina*), and Gymnosporidiida (containing the genera *Caryophagus, Halteridium, Haemoproteus, Plasmodium, Laverania*, and *Cytamoeba*—i.e., essentially our present suborder Haemosporina).

Labbé (1899) defined the order Coccidiida as spherical or ovoid cytosporidia, never having a free or motile adult stage nor an amoeboid phase nor myophanic fibrils, and sporulating in cysts always formed in the host epithelium. He divided this order into two suborders, Polyplastina and Oligoplastina. The former he defined as having an unlimited number of archispores and the latter as having a limited number. (The term "archispore" has gone out of use; Labbé defined it as a simple nucleated cell which could be transformed directly into a sporozoite but which ordinarily secreted two membranes—an epispore and an endospore—and then turned into a spore in which one or more sporozoites developed). The suborder Polyplastina contained two tribes. In the tribe Polyplastina digenica the sporozoites were formed inside spores, while in the tribe Polyplastina monogenica the sporozoites were formed directly from the archispores. The former contained the genera *Minchinia, Klossia, Hyaloklossia, Adelea, Barrouxia, Echinospora*, and *Gymnospora*, and the latter the genera *Rhabdospora, Eimeria, Gonobia, Pfeifferella, Molybdis*, and *Cretya*.

Labbé's suborder Oligoplastina contained three tribes, based on the number of spores in the oocyst. The tribe Disporea had two, the tribe Trisporea three, and the tribe Tetrasporea four. Again, he used no families. The first tribe contained the genera *Cyclospora, Diplospora*, and *Isospora*, the second tribe the single genus *Bananella*, and the third tribe the genera *Crystallospora, Goussia*, and *Coccidium*. Thus Labbé put *Eimeria* and *Coccidium* in different suborders. He recognized 94 certain and 29 uncertain genera, 239 certain and 259 uncertain species, 18 subspecies and 15 varieties of Sporozoa. Some of his names, of course, have disappeared.

Léger (1900) differed from Labbé, and thought that the coccidia should first be divided upon the basis of the number of sporozoites per oocyst rather than upon the number of archispores. He thus established two families with polyzoic oocysts (Asporocystidae, containing *Eimeria*, without sporocysts; and Polysporocystidae, containing *Barroussia, Adelea, Benedenia*, and *Klossia*, with sporocysts), two families with octozoic oocysts (Disporocystidae, containing *Diplospora*, with two sporocysts each with four sporozoites per oocysts; and Tetrasporocystidae, containing *Coccidium* and *Crystallospora,* with four sporocysts each with two sporozoites per oocyst), and one family (?) with tetrazoic oocysts *(Cyclospora,* with two sporocysts each with two sporozoites per oocyst).

Schaudinn (1900) added the groups now known as the Haemosporidia, Microspora, and Myxospora to the gregarines and coccidia under the class Sporozoa, and divided the class into two subclasses, Telosporidia and Neosporidia. He put the gregarines, haemogregarines, and coccidia in the Telosporidia (from the Greek word *telos,* meaning the end of life and signifying that the spores came at the end of the life cycle; the spores in this group had no polar filaments). He put the Microsporidia, Myxosporidia, Actinomyxidia, Sarcosporidia, and Haplosporidia in the Neosporidia; in this group spores were said to be produced throughout the life cycle. Doflein (1901) established the subclass Cnidosporidia for the first three neosporidian groups because they have polar filaments (*cnida* is the Greek word for nettle), and Hartmann (1923) eventually established the subclass Acnidosporidia for the Sarcosporidia and Haplosporidia.

So much for the major groups. To return to the coccidia proper, Minchin (1903) revised the classifications of Labbé (1899), Léger (1900), and Mesnil (1900), and divided the order simply into four families based on the number of sporocysts (if any) formed in each oocyst. He followed Labbé in retaining the Aggregatidae among the gregarines. His classification is given in Table 1.

Minchin listed 11 other genera as doubtful, including the fungus *Coccidioides.* Minchin said in a footnote that the four families had not been named in accordance with the accepted rules of zoological nomenclature, which require that a family name be based on that of its type genus. For this reason, he said that the Asporocystidae should be Eimeridae (or Legerellidae), the Disporocystidae should be Isosporidae, the Tetrasporocystidae should be Coccididae (or Eimeridae), and the Polysporocystidae should be Klossidae. Later Poche (1913) emended the name Eimeridae to Eimeriidae, which is the presently accepted name of the family.

Minchin retained only one species of *Eimeria.* This was *E. nova* (now *Legerella nova*) from the Malpighian tubules of *Glomeris.* Most of the other previously named species he said belonged in the genus *Coccidium,* since they had been found to be merely a part of the life cycle of members of the latter genus. However, the generic name *Eimeria* dates from 1875 and *Coccidium* from 1879, so the former is the valid name and not the latter, as many subsequent authors have pointed out and as we recognize today. The generic name *Coccidium* has disappeared from the literature, and is commemorated only in the common name of the group.

Minchin's classification was entirely logical for its time, but it has now been superseded.

Doflein (1901) divided the Sporozoa into two orders, the Coccidiomorpha and Gregarinida, and Wenyon (1926) followed him in this, but raised the divisions to subclasses. He divided the Coccidiomorpha into the orders Coccidiida (containing the suborders Eimeriidea, Haemosporidiidea, and Piroplasmidea) and Adeleida (containing the suborders Adeleidea and Haemogregarinidea). The suborder Eimeriidea (which is the one of greatest concern to us in this book) he divided into the families Selenococcidiidae; Cryptosporidiidae; Eimeriidae (with subfamilies Cyclosporinae, Isosporinae, Eimeriinae, Barrouxiinae, Caryosporinae, and Pfeifferellinae); Caryotrophidae; Aggregatidae; and Lankesterellidae (with subfamilies Schellackinae and Lankesterellinae).

Hoare (1933) set up a "periodic table" of coccidian genera, based on the numbers of sporocysts and sporozoites per oocyst, and he later (1957) revised it. I have already (Levine, 1963) commented on this purely ar-

TABLE 1. Minchin's classification of the order Coccidiida

Family Asporocystidae Léger
No sporocysts formed within oocysts; sporozoites naked.

Genus *Eimeria* A. Schneider, 1875
With the characters of the family.

Family Disporocystidae Léger
Oocyst with two spores (i.e., sporocysts).

Genus *Cyclospora* A. Schneider, 1881
Sporocysts each with two sporozoites.

Genus *Diplospora* Labbé, 1893
Sporocysts each with four sporozoites.

Genus *Isospora* A. Schneider, 1881
Sporocysts each with numerous sporozoites.

Family Tetrasporocystidae Léger
Oocyst with four sporocysts.

Genus *Coccidium* Leuckart, 1879
Sporocysts each with two sporozoites; sporocysts spherical or oval.

Genus *Crystallospora* Labbé, 1896
Sporocysts each with two sporozoites; sporocysts in form of a double pyramid.

Family Polysporocystidae Léger
Oocyst with numerous sporocysts.

Genus *Barroussia* A. Schneider, 1885
Sporocysts each with one sporozoite; sporocysts spherical, with smooth bivalved shell.

Genus *Echinospora* Léger, 1897
Sporocysts each with one sporozoite; sporocysts oval, with spiny bivalved shell.

Genus *Diaspora* Léger, 1898
Sporocysts each with one sporozoite; sporocysts oval, not bivalved, with a micropyle at one end.

Genus *Adelea* A. Schneider, 1875
Sporocysts each with two sporozoites; sporocysts spherical or compressed, smooth.

Genus *Minchinia* Labbé, 1896
Sporocysts each with two sporozoites; sporocysts oval, with a long filament at each end.

Genus *"Benedenia"* A. Schneider, 1875
Sporocysts each with three sporozoites; sporocysts spherical; schizogony absent.

Genus *Klossia* A. Schneider, 1875
Sporocysts each with four or many sporozoites; sporocysts spherical.

Genus *Caryotropha* Siedlecki, 1902
Sporocysts each with twelve sporozoites; sporocysts spherical.

Genus *Klossiella* Smith and Johnstone, 1902
Sporocysts each with 30-34 sporozoites; sporocysts subspherical.

tificial classification. The idea is convenient, however, for rapid identification. Figure 2 shows, for each genus, the number of sporocysts and sporozoites in each oocyst.

Becker (1934) wrote a classic review of the coccidia, assembling most of the information then known about them. He was not particularly interested in taxonomy, however, and merely followed Hoare.

Grassé (1953) further revised the classification. He considered the Sporozoa to be a subphylum and to consist of the classes Gregarinomorpha, Coccidiomorpha, and Sarcosporidia. He divided the Coccidiomorpha into two subclasses, Protococcidia Léger and Duboscq, 1910 and Eucoccidia Léger and Duboscq, 1910. The former contained only the family Selenococcidiidae and that family only a single species, *Selenococcidium intermedium* of the lobster. The latter contained the orders Adeleidea Léger, 1911 (with the families Adeleidae, Klossiidae, Legerellidae, Haemogregarinidae, and Dobellidae), Eimeriidea Léger, 1911 (divided into the suborders Holoeimeriidea and Haemosporidiidea), and Sarcosporidia Bütschli, 1882. He put the class Haplosporidia Caullery and Mesnil, 1899 in an appendix to the Sporozoa, and considered members of the superfamily Babesioidea (including *Anaplasma* and its relatives) to be uncertain Sporozoa. The genera *Spirocystis*, *Joyeuxella*, *Rhytidocystis*, *Exoschizon*, and *Piridium* he placed in the Sporozoa incertae sedis, and the genera *Dactylosoma*, *Toxoplasma*, *Bartonella*, etc. among protistan parasites of uncertain affinities. Of these, only members of his new suborder Holoeimeriidea concern us here. He divided it into 20 families, of which eight were new—Cryptosporidiidae; Mantonellidae (new); Cyclosporidae; Pfeifferinellidae (new); Schellackiidae (new); Caryosporidae; Diplosporidae; Eimeriidae; Lankesterellidae; Yakimovellidae; Caryotrophidae; Angeiocystidae; Pseudoklossiidae (new); Myriosporidae (new); Merocystidae (new); Aggregatidae; and Eleutheroschizonidae. This classification is quite modern, but it was made before the electron microscope permitted us to sort out some of the groups and place them where they belong.

The Committee on Taxonomy and Taxonomic Problems of the Society of Protozoologists made a major stride in classification of

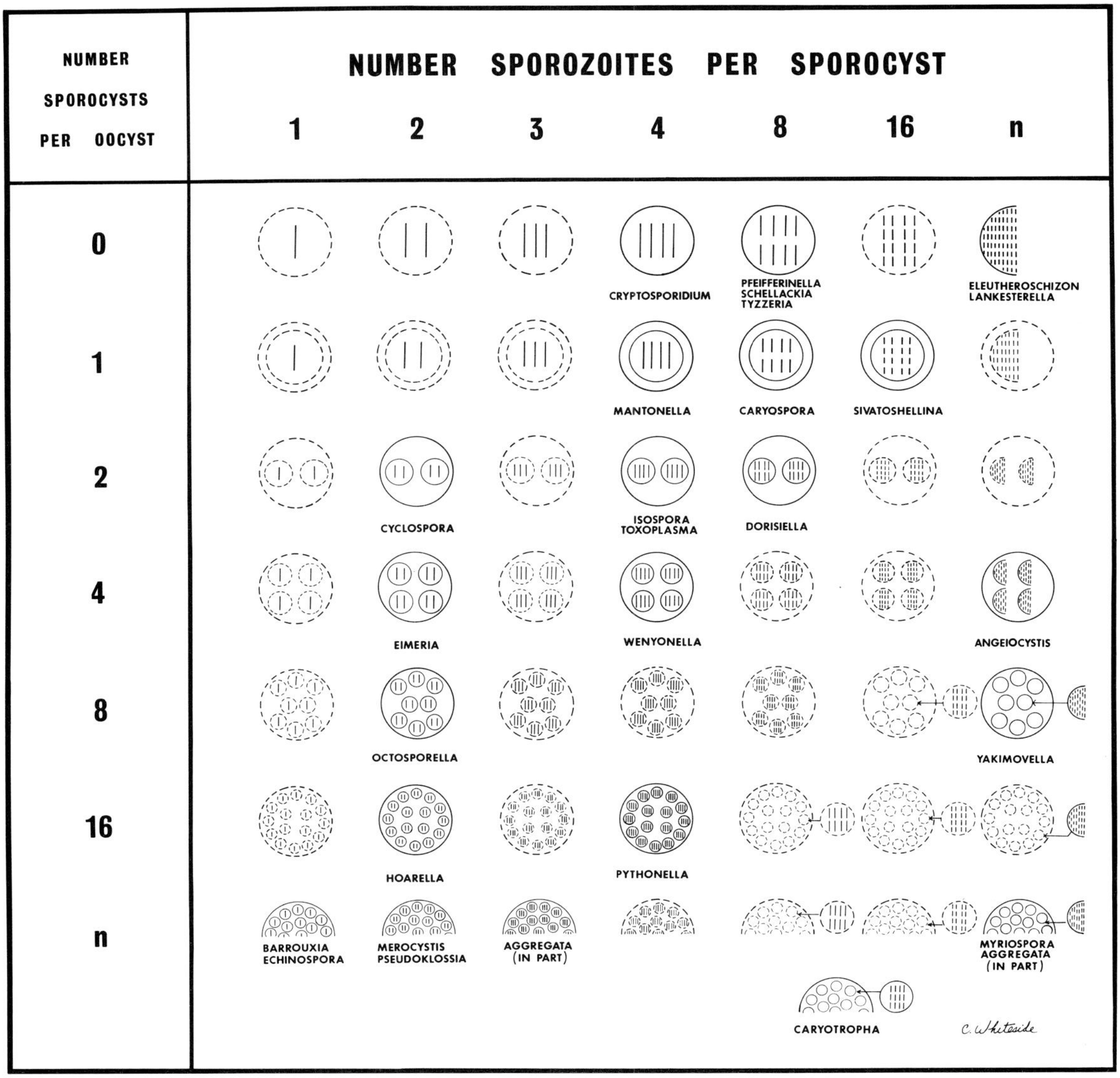

NUMBER SPOROCYSTS PER OOCYST	NUMBER SPOROZOITES PER SPOROCYST 1	2	3	4	8	16	n
0				CRYPTOSPORIDIUM	PFEIFFERINELLA SCHELLACKIA TYZZERIA		ELEUTHEROSCHIZON LANKESTERELLA
1				MANTONELLA	CARYOSPORA	SIVATOSHELLINA	
2		CYCLOSPORA		ISOSPORA TOXOPLASMA	DORISIELLA		
4		EIMERIA		WENYONELLA			ANGEIOCYSTIS
8		OCTOSPORELLA					YAKIMOVELLA
16		HOARELLA		PYTHONELLA			
n	BARROUXIA ECHINOSPORA	MEROCYSTIS PSEUDOKLOSSIA	AGGREGATA (IN PART)				MYRIOSPORA AGGREGATA (IN PART)
					CARYOTROPHA		

C. Whiteside

Fig. 2. Numbers of sporocysts per ocyst and of sporozoites per sporocyst in the genera of the suborder Eimeriorina. (In the genera without sporocysts, the numbers of sporozoites per oocyst are given.)

the Protozoa. It (Honigberg *et al.*, 1964) transferred the class Piroplasmea to the Sarcodina, separated the subphylum Cnidospora from the subphylum Sporozoa, and divided the Sporozoa into three classes, Telosporea, Toxoplasmea, and Haplosporea. This classification went only to suborders and did not have the benefit of the knowledge gained in subsequent years from electron microscope studies. Only eight years later, it is out of date, and a new classification is necessary. The Piroplasmea are more closely related to the Sporozoa than to the Sarcodina, and belong with them in the subphylum Apicomplexa (Levine, 1970); the Toxoplasmea have now been recognized as coccidia; and the Haplospora are simply a class of the subphylum Microspora.

Among recent reviews dealing with the coccidia, Orlov (1956) discussed those of domestic animals but was seriously handicapped by lack of information about non-Russian work. Davies, Joyner and Kendall (1963) and Pellérdy (1965) reviewed all the coccidia, Kheisin* (1967) discussed their life cycles, Cuckler (1970) and Kozar (1970) reviewed immunity in birds and mammals, respectively, and Becker (1956) and Pellérdy (1956, 1957, 1963, 1969) gave checklists of the species. The coccidia of the avian orders Galliformes, Anseriformes, and Charadriiformes were reviewed by Levine (1953), those of the avian orders Galliformes, Anseriformes, and Passeriformes by Todd and Hammond (1971), those of rodents by Levine and Ivens (1965) and Musaev and Veisov (1965), and those of ruminants by Levine and Ivens (1970).

III. Taxonomy

The classification given in Table 2 is my own. It has to do primarily with the coccidia proper (members of the suborder Eimeriorina), and shows their relationship to the other groups in the subphylum Apicomplexa. This classification is certainly not definitive. It is subject to considerable revision, and the discussion below will bring this out. Nevertheless, it is a base on which to build.

Members of the family Aggregatidae are poorly known, and indeed it may be that some species should be placed elsewhere. Grassé (1953), for instance, used five families for the six genera that I have assigned to the family. He placed *Aggregata* and (doubtfully) *Ovivora* in the Aggregatidae, *Angeiocystis* in the Angeocystidae, *Merocystis* in the Merocystidae, *Myriospora* in the Myriosporidae, and *Pseudoklossia* in the Pseudoklossiidae. However, until more information is gained regarding these species, it seems best to retain them in a single family.

The family Caryotrophidae, too, is poorly known. *Caryotropha* is the only genus in it, and it has only a single species, *C. mesnili,* in the spermatogonia in the coelom of the polychaete *Polymnia nebulosa.* Perhaps further research will reveal that this family should be characterized differently, Kudo (1966), for instance, included *Caryotropha* in the Aggregatidae, and he may have been right.

There are two genera, *Schellackia* and *Lankesterella,* in the family Lankesterellidae. The principal differences are in the hosts (reptiles for *Schellackia* and amphibia and birds for *Lankesterella*), and the fact that *Schellackia* has eight naked sporozoites per oocyst and *Lankesterella* has 32 or more. Both have the same type of life cycle, in which all development takes place in the vertebrate host; the sporozoites enter the blood cells and are transferred to a new vertebrate host by a blood-sucking invertebrate; they do not develop in the invertebrate, so that transmission is mechanical. The only known vectors of *Schellackia* are mites, while the vectors of *Lankesterella* are either leeches (for the single frog species known) or mites (for

* Alternative spellings in various publications: Kheysin, Kheisin, Cheissin (ed).

TABLE 2. Classification of Coccidiorina

Subphylum APICOMPLEXA Levine, 1970

Apical complex, generally consisting of polar ring, micronemes, rhoptries, subpellicular tubules, and conoid, present at some stage; micropore(s) generally present; single type of nucleus; cilia and flagella absent, except for flagellated microgametes in some groups; sexuality, when present, syngamy; cysts often present; all species parasitic.

Class SPOROZOASIDA Leuckart, 1879

Apical complex well developed; reproduction generally both sexual and asexual; oocysts (sometimes called "spores" in some forms) present; locomotion by body flexion, gliding or undulation of longitudinal ridges; microgametes flagellated in some groups; pseudopods ordinarily absent; if present used for feeding, not locomotion; monoxenous or heteroxenous.

Subclass GREGARINASINA Dufour, 1828

Mature gamonts extracellular, large; conoid modified into mucron or epimerite in mature organisms; endodyogeny absent; gametes similar (isogamous) or nearly so; equal numbers of male and female gametes produced by gamonts; zygotes form oocysts within gametocysts; parasites of digestive tract or body cavity of invertebrates or lower chordates; generally monoxenous.

Subclass COCCIDIASINA Leuckart, 1879

Mature gamonts small, typically intracellular; conoid not modified into mucron or epimerite; endodyogeny present or absent; mostly in vertebrates, but a few in invertebrates.

Order PROTOCOCCIDIORIDA Kheisin, 1956

Merogony absent; in marine invertebrates.

Order EUCOCCIDIORIDA Léger and Duboscq, 1910

Merogony present; in vertebrates or invertebrates.

Suborder ADELEORINA Léger, 1911

Macrogamete and microgametocyte associated in syzygy during development; microgametocyte usually produces from one to four microgametes; sporozoites enclosed in envelope; endodyogeny absent; monoxenous or heteroxenous.

Suborder EIMERIORINA Léger, 1911

Macrogamete and microgametocyte develop independently; syzygy absent; microgametocyte typically produces many microgametes; zygote not motile; sporozoites typically enclosed in a sporocyst; endodyogeny absent or present; monoxenous or heteroxenous.

Family SELENOCOCCIDIIDAE Poche, 1913

Meronts develop as vermicules in host intestinal lumen; meronts with myonemes and nuclei in a row; in lobster.

Genus *Selenococcidium* Léger and Duboscq, 1910

With the characters of the family.

Family AGGREGATIDAE Labbé, 1899

Development in host cell proper; oocysts typically with many sporocysts; mostly heteroxenous – merogony in one host, gamogony in another.

Genus *Aggregata* Frenzel, 1885

Oocysts large, with many sporocysts; sporocysts with from 3-28 sporozoites; heteroxenous – merogony in a decapod crustacean and gamogony in a cephalopod mollusk.

(Continued)

TABLE 2 *(Continued)*

Genus *Angeiocystis* Brasil, 1909

Oocyst with four sporocysts, each with about 30 sporozoites; gamonts at first in the form of a large sausage; merogony unknown; presumably heteroxenous; known stages in heart of polychaete *Cirriformia* (syn., *Audouinia*).

Genus *Merocystis* Dakin, 1911

Oocyst with many sporocysts, each with two sporozoites; merogony unknown; presumably heteroxenous; known stages in kidney of whelk (prosobranch mollusk) *Buccinum*.

Genus *Myriospora* Lermantoff, 1913

Oocyst with eight to several hundred sporocysts, each with 24 or 32 sporozoites; male gamont helicoid; merogony unknown; presumably heteroxenous; known stages in polychaetes.

Genus *Pseudoklossia* Léger and Duboscq, 1915

Oocyst with no or many sporocysts, each with two sporozoites (if sporocysts occur); merogony unknown; presumably heteroxenous; known stages in lamellibranch mollusks.

Genus *Ovivora* Mackinnon and Ray, 1937

Oocyst with no (?) sporocysts, each presumably with 12 sporozoites; monoxenous; in eggs of echiuroid annelid *Thalassema*.

Family CARYOTROPHIDAE Lühe, 1906

Oocysts large, without definite wall, with about 20 sporocysts each with 12 sporozoites; monoxenous; in polychaete annelids.

Genus *Caryotropha* Siedlecki, 1902

With the characters of the family.

Family LANKESTERELLIDAE Nöller, 1902

Development in host cell proper; oocysts without sporocysts, but with eight or more sporozoites; heteroxenous, with merogony, gametogony, and sporogony in the same vertebrate host; sporozoites in blood cells, transferred without developing by an invertebrate (mite or leech); microgametes with two flagella, so far as is known.

Genus *Lankesterella* Labbé, 1899 (syn., *Atoxoplasma* Garnham, 1950)

Oocyst produces 32 or more sporozoites; in amphibia and birds.

Genus *Schellackia* Reichenow, 1919

Oocyst produces eight sporozoites; in lizards.

Family EIMERIIDAE Minchin, 1903

Development in host cell proper; oocysts and meronts (schizonts) without attachment organelle; oocysts with 0, 1, 2, 4 or many sporocysts, each with one or more sporozoites; monoxenous; merogony in the host, sporogony typically outside; microgametes with two or three flagella.

Genus *Eimeria* Schneider, 1875

Oocysts with four sporocysts, each with two sporozoites.

Genus *Isospora* Schneider, 1881

Oocysts with two sporocysts, each with four sporozoites.

TABLE 2 *(Continued)*

Genus *Barrouxia* Schneider, 1885
Oocysts with *n* sporocysts, each with one sporozoite.

Genus *Caryospora* Léger, 1904
Oocysts with one sporocyst containing eight sporozoites.

Genus *Cyclospora* Schneider, 1881
Oocysts with two sporocysts, each with two sporozoites.

Genus *Dorisiella* Ray, 1930
Oocysts with two sporocysts, each with eight sporozoites.

Genus *Tyzzeria* Allen, 1936
Oocysts with eight naked sporozoites, without sporocysts.

Genus *Wenyonella* Hoare, 1933
Oocysts with four sporocysts, each with four sporozoites.

Genus *Mantonella* Vincent, 1936
Oocysts with one sporocyst containing four sporozoites.

Genus *Octosporella* Ray and Ragavachari, 1942
Oocysts with eight sporocysts, each with two sporozoites.

Genus *Sivatoshellina* Ray and Sarkar, 1968
Oocysts with two sporocysts, each with 16 sporozoites.

Genus *Yakimovella* Gousseff, 1937
Oocysts with eight sporocysts, each with *n* sporozoites.

Genus *Hoarella* Arcay de Peraza, 1963
Oocysts with 16 sporocysts, each with two sporozoites.

Genus *Pythonella* Ray and Das Gupta, 1937
Oocysts with 16 sporocysts, each with four sporozoites.

Family CRYPTOSPORIDIIDAE Léger, 1911
Development just under surface membrane of host cell or within its striated border and not in cell proper; oocysts and meronts with a knob-like attachment organelle at some point on their surface; oocysts without sporocysts, with four naked sporozoites; monoxenous; microgametes without flagella.

Genus *Cryptosporidium* Tyzzer, 1907
With the characters of the family.

Family PFEIFFERINELLIDAE Grassé, 1953
Oocyst without sporocysts, with eight naked sporozoites; fertilization of macrogamete through a "vaginal" tube; monoxenous; in mollusks.

Genus *Pfeifferinella* Wasielewski, 1904
With the characters of the family.

Family SARCOCYSTIDAE Poche, 1913
Syzygy apparently absent; endodyogeny present; cysts or pseudocysts containing zoites in parenteral cells of host; all parasitic in vertebrates; monoxenous.

(Continued)

TABLE 2 *(Continued)*

Subfamily TOXOPLASMATINAE Biocca, 1956

Zoitocysts with thin membranes and pseudocysts present in parenteral cells of host; meronts (schizonts) and gamonts in intestinal cells; gamonts produce oocysts.

Genus *Frenkelia* Biocca, 1968

Cysts in brain lobulate and septate.

Genus *Toxoplasma* Nicolle and Manceaux, 1908

Cysts in brain not lobulate or septate.

Subfamily BESNOITIINAE Garnham, 1966

Zoitocysts with thick, laminated, nucleated walls and pseudocysts present in parenteral cells; sexual reproduction unknown.

Genus *Besnoitia* Henry, 1913

With the characters of the subfamily.

Subfamily SARCOCYSTINAE Poche, 1913

Zoitocysts elongated, often septate, with cytophaneres, in parenteral cells; pseudocysts unknown.

Genus *Sarcocystis* Lankester, 1882

Zoites elongate; septa in cysts, if present, thin.

Genus *Arthrocystis* Levine, Beamer and Simon, 1970

Zoites spherical; septa in cysts thick; cysts jointed like bamboo.

Suborder HAEMOSPORINA Danilewsky, 1885

Macrogamete and microgametocyte develop independently; syzygy absent; microgametocyte produces about eight flagellated microgametes; zygote motile (ookinete); sporozoites naked; endodyogeny absent; heteroxenous; with merogony in vertebrate host and sporogony in invertebrate; pigment (hemozoin) may or may not be formed from host cell hemoglobin.

Class PIROPLASMASIDA Levine, 1961

Small, piriform, round, rod-shaped, or amoeboid; components of apical complex reduced; spores or sporocysts absent; flagella and cilia absent; pseudopods, if present, used for feeding, not locomotion; locomotion by body flexion or gliding; reproduction asexual by binary fission or schizogony; pigment not formed from host cell hemoglobin; heteroxenous; parasitic in vertebrate erythrocytes, leukocytes, other blood system cells or liver parenchymal cells; known vectors ticks.

---

the avian species). The type species of *Lankesterella*, and the only one known for a long time, is *L. minima*, which occurs in the green frog *Rana esculenta* and is transmitted by the leech *Hemiclepsis marginata*. Lainson (1959) said that *Atoxoplasma* Garnham, 1950, which is common in the leukocytes of various birds, is a synonym of *Lankesterella*, and this enlarged the genus considerably. Working in England, Lainson transmitted *L. adiei* (syn., *L. garnhami*) of the English sparrow by means of the common red mite *Dermanyssus gallinae*. However, Box (1967, 1970), working in Galveston, Texas, where sparrows do not have *D. gallinae* and where *Lankesterella* is extremely common in sparrow blood, was unable to transmit the organism with mites. Further, she found that *Lankesterella* exacerbates *Isospora lacazei* infections in the sparrow, and gave evidence that *L. adiei* might really be a parenteral stage of the common English sparrow coccidium *I. la-*

cazei. In another context, Wenyon (1926) described what he considered to be gamonts of *Haemoproteus* in the lungs, liver, and kidneys of sparrows in Baghdad; on this basis Garnham (1966) assigned the species to the genus *Parahaemoproteus* using the name *P. garnhami* for it. (The correct name for this species is *H. passeris* [see Levine and Campbell, 1971].) The question is whether Wenyon was dealing with *Haemoproteus, Isospora, Lankesterella*, or a haemogregarine. This problem requires investigation, and the whole matter has not yet been resolved.

There are several problems in the family Eimeriidae. Members of this family have a single host. Merogony and gametogony take place within the host cells, and sporogony ordinarily occurs outside the host's body. No stage has an attachment organelle. The microgametes have two or three flagella. The oocysts contain from none to many sporocysts, each containing one or more sporozoites. The genera are differentiated by the number of sporocysts in their oocysts and the number of sporozoites in each oocyst.

Figure 3 shows the structure of a typical oocyst, that of *Eimeria*.

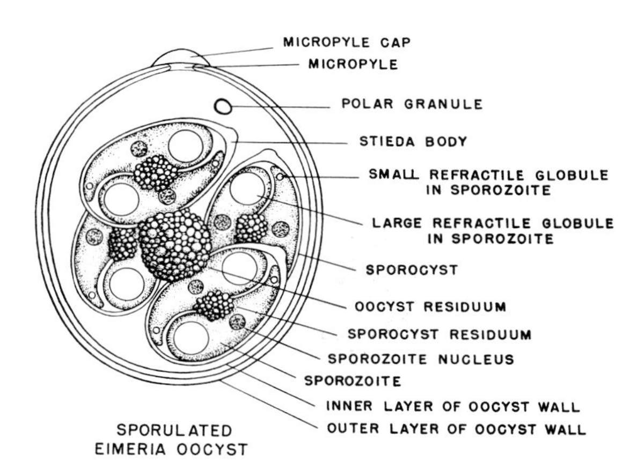

Fig. 3. A coccidian oocyst. (From Levine, 1961.)

I have accepted 14 genera in the family Eimeriidae as valid, but I am far from certain about most of them. The best studied are *Eimeria* and *Isospora*, and a great many species have been named in both of these genera. Nevertheless, there are great gaps in our knowledge. Levine and Ivens (1965) accepted 204 species of *Eimeria* and ten of *Isospora* in rodents, and quite a few more have been named since then. And Levine and Ivens (1970) accepted 95 species of *Eimeria* and four of *Isospora* in ruminants. But they estimated that there must be about 2,700 species of *Eimeria* in rodents alone; the number described was only 7.6% of this. And the location in the host was known for only 45 species of *Eimeria*, the endogenous stages for only 25, and complete life cycles for only four. The situation is little better for any other host group. In ruminants, for example, the location in the host is known for only 17 out of the 95 named species, the endogenous stages are known for only 15, and presumably complete life cycles have been worked out for only two.

I calculated (Levine, 1963) that *Eimeria* had been described from only 1.2% of the world's chordates and 5.7% of the world's mammals. I estimated (Levine, 1962) that, if we examined all the chordates, we might expect to find perhaps 34,000 species of *Eimeria*, and that 3,500 of them might be in mammals. I guessed that there might actually be 45,000 species of *Eimeria*—roughly equivalent to the total number of all living and fossil protozoa heretofore named.

Another problem has to do with the correct name of the coccidia commonly assigned to the genus *Isospora*. This genus was established by Schneider (1881) for *I. rara* from a black slug in France whose scientific name he did not give. Minchin (1903) said that it was *Limax cinereo-niger*. Schneider said that the oocysts of this genus contained two sporocysts, each with quite numerous falciform sporozoites. Labbé (1899) established the genus *Diplospora* for *D. brumpti* of the toad

Bufo viridis. The oocysts of this genus contain two sporocysts, each with four sporozoites. *I. rara* has apparently not been reported again since Schneider's description about 90 years ago, and the question is whether *Isospora* and *Diplospora* are the same. Most authors have thought so, but Minchin (1903) and Grassé (1953) separated them. Grassé said that *I. rara* was too poorly known to be accepted as the type species of the family Diplosporidae Léger, 1911 (which he recognized), and thought that there was a good chance that the parasite of the slug actually belonged to a different genus from the disporocystic coccidia of vertebrates. This may well be, and certainly Schneider was a good microscopist and should have been able to determine whether four or more than four sporozoites were present in each sporocyst. However, until *I. rara* is redescribed, it seems best to retain the customary name for the coccidia of vertebrates with two sporocysts per oocyst, each with four sporozoites. (Incidentally, Grassé divided the genus *Diplospora* into two subgenera; in *Diplospora s. s.* the sporozoites lie head to tail in pairs in the sporocysts, while in *Lucetina* Henry and Leblois, 1925 the sporozoites are all turned the same way; the type species of *Diplospora* is *D. brumpti* and that of *Lucetina* is *L. rivolta* of carnivores; I doubt that this subdivision is justified.)

Eimeria has four sporocysts per oocyst and two sporozoites per sporocyst; *Isospora* (if that is the correct name rather than *Diplospora*) has two sporocysts per oocyst and four sporozoites per sporocyst; both total eight sporozoites per oocyst. These are the two most successful genera of coccidia in terms both of numbers of species and of ubiquity of occurrence. *Eimeria* predominates in gallinaceous birds, rodents, and ruminants, among others. *Isospora* predominates in passerine birds and in carnivores. Primates have few coccidia. Is there some reason for this distribution? If so, I do not know what it is. It seems to be simply a matter of chance. Certainly, rats are not more closely related to chickens than they are to dogs, nor dogs more closely to sparrows than they are to cattle.

We know enough about the genera *Tyzzeria* and *Wenyonella* to feel safe in assigning them to the Eimeriidae. Their endogenous stages are similar to those of *Eimeria* and *Isospora*. There are only a few species of each, and distribution of their hosts—all vertebrates—is spotty. Presumably each could have arisen by a single mutation from either *Eimeria* or *Isospora*. The oocysts of *Tyzzeria* contain eight naked sporozoites, and each species of this genus could have arisen independently by mutations involving the loss of sporocysts. The oocysts of *Wenyonella* contain four sporocysts, each with four sporozoites, and the species of this genus, too, could have arisen independently by mutations involving either a doubling of the number of sporozoites per *Eimeria* sporocyst or a doubling of the number of sporocysts in each *Isospora* oocyst.

Caryospora and *Cyclospora* are well enough known also to be assigned safely to the Eimeriidae. *Caryospora* has a total of eight sporozoites per oocyst (this time in a single sporocyst), and could well have arisen by a single mutation from *Eimeria* or *Isospora*. Indeed abnormal oocysts of various species of *Isospora* or *Eimeria* which resemble those of *Caryospora* have been found; Levine and Ivens (1965a) for example, found *Caryospora*-like oocysts of *I. rivolta* in the dog. *Cyclospora* has a total of four sporozoites per oocyst, each oocyst containing two sporocysts each with two sporozoites.

Perhaps the same could be said of *Mantonella* although it is not as well known. Members of this genus have oocysts with one sporocyst containing four sporozoites. *M. peripati* occurs in the intestine of *Peripatopsis*, while *M. potamobii* occurs in the crayfish *Potamobius leptodactylus*, and it may be that these species really belong to different genera.

I know of only a single paper on *Yakimovella*, and it is not particularly satisfying. Gousseff (1937) established the genus for *Y. erinacei* of the hedgehog *Erinaceus europeus*. Each oocyst contains eight sporocysts, but Gousseff did not try to count the sporozoites in each sporocyst, saying merely that each one contained a very large number. This parasite should be redescribed.

Ray and Das Gupta (1937) found *Pythonella bengalensis* in a snake *Python* sp. in India. Its oocysts contain 16 sporocysts, each with four sporozoites. Duszynski (1969) described *P. scelopori* n. sp. from the colon contents of a Costa Rican lizard *Sceloporus squamosus* on the basis of nine oocysts, only four of which sporulated. Whether this is a valid genus remains to be seen.

Octosporella mabuiae, with eight sporocysts per oocyst, each sporocyst with two sporozoites, was found by Ray and Raghavachari (1942) in the lizard *Mabuia* sp. in India.

Hoarella, too, has been reported only once. Arcay de Peraza (1963) described *H. garnhami* from two of 200 lizards *Cnemidophorus lemniscatus* in Venezuela.

There is only a single paper on *Sivatoshellina* also, but this is recent. Ray and Sarkar (1968) established the genus for *S. lonchurae*, which they described from the intestine of the passeriform birds *Lonchura malabarica* and *L. punctulata* in India. Each oocyst contains two sporocysts, each with 16 sporozoites.

The genus *Barrouxia* deserves special mention. All its members are parasites of insects and myriapods, and all have *n* sporocysts per oocyst, each sporocyst containing one sporozoite. Mesnil (1903) divided the genus into four subgenera: In *Barrouxia s. s.* the sporocysts are ellipsoidal, smooth and bivalved; in *Urobarrouxia* Mesnil, 1903 the sporocysts are similar to those of *Barrouxia s. s.* but have a long terminal "tail"; in *Echinospora* Léger, 1897 the sporocysts are like those of *Barrouxia s. s.*, but the sporocysts are covered with spines; in *Diaspora* Léger, 1898 the sporocysts are ovoid, with a polar micropyle, and have a series of transparent envelopes, and the oocysts are unknown but probably very fragile. Grassé (1953) thought that *Diaspora* was probably a good genus.

Members of the subgenus *Barrouxia s. s.* occur in *Lithobius* and insects, of the subgenera *Urobarrouxia* and *Echinospora* in *Lithobius*, and of the subgenus *Diaspora* in *Polydesmus*. I would be inclined to consider *Diaspora* a separate genus, but it has apparently been described only by Léger (1898), and he found it only once; it should be restudied before a decision is made.

Another genus is *Dorisiella*. It was established by Ray (1930) with *D. scolelepidis* from the marine polychaete *Scolelepis fuliginosa* as the type species. Since then several more species have been named from vertebrates. However, I cannot avoid wondering whether the vertebrate species really belong to this genus or to a separate and different one. They have two sporocysts per oocyst, and each sporocyst contains eight sporozoites. I think that both the polychaete and vertebrate forms will have to be studied more extensively before my question can be answered.

The family Cryptosporidiidae contains only one genus, *Cryptosporidium*. There are several species in mammals and birds. This family is conventionally placed under the Eimeriorina, but I wonder if it really belongs there. It differs in that it does not enter the host intestinal cells proper, but develops on the surface (just under the cell membrane) or within its striated border. The oocysts and meronts have a knob-like attachment organelle at some point on their surface. The microgametes lack flagella. Third, although all the developmental forms known are in the intestine, oocysts have never been found in the feces. The latest paper on this genus is that of Vetterling *et al.* (1971), which described *C. wrairi* n. sp. from the guinea pig and accepted only five species as valid. A number of others have been named, but this was done because their namers mistook free sporocysts of *Isospora* for oocysts of *Cryptosporidium*, since both contain four naked sporozoites.

I approach the family Sarcocystidae with a good deal of trepidation. I am uncertain whether it is really a valid family or whether it should be placed under the Eimeriidae. Further, I am uncertain whether it should be divided into subfamilies as I have done. This group contains the genera *Toxoplasma* and *Sarcocystis*, both of which are common, and three other genera which are not. Sexual reproduction was known for none of them, they occurred commonly in many animals, and they occurred as cysts containing zoites that reproduced by endodyogeny, so it was safe to put them together. But now it has been found independently by Hutchison et al. (1970) in Scotland and Denmark; Frenkel, Dubey, and Miller (1970) in Kansas; Sheffield and Melton (1970) in Maryland; Gordon Wallace (unpublished) in Hawaii; and Overdulve (1970) in the Netherlands, that *Toxoplasma* produces gametes and oocysts in the intestine of the cat (but not in other animals), and that these oocysts are indistinguishable from those of *Isospora bigemina*, also of the cat.

Frenkel, Dubey and Miller (1970) proposed placing the family Toxoplasmatidae in the suborder Eimeriorina, and Overdulve (1970) went so far as to say that *Toxoplasma gondii* was a synonym of *Isospora bigemina*. Either may be correct, or neither.

Parenthetically, I have been unable to satisfy myself as to who introduced the family name Toxoplasmatidae. So far as I can tell, it was first used (without attribution) by Biocca (1956), and I am crediting the name to him, but it may have been used before that.

Sarcocystis was thought to have no sexual stages until Fayer (1972) found macrogametes and biflagellate microgametes in bovine embryonic tissue cultures seeded with zoites from cysts in grackle muscles. No oocysts are known as yet for this genus, but they may well be discovered in the next few years. Perhaps both oocysts and sexual stages will also be discovered eventually in *Frenkelia*, *Besnoitia*, and *Arthrocystis*. Further, recognition of the fact that members of the family Eimeriidae are not necessarily confined to one kind of tissue (usually the intestine) makes one wonder about the true taxonomic position of *Toxoplasma*, *Sarcocystis*, etc.

A related question has to do with the epidemiology of winter coccidiosis, one of the most puzzling types of coccidiosis in cattle. This disease, caused by *Eimeria zuernii*, occurs when it is so cold that oocyst sporulation should be minimal, if it takes place at all. It is believed that the disease, which occurs in calves during the fall and winter in the northern range states and elsewhere, is not caused by exposure of the host to large numbers of infective oocysts but that predisposing factors, such as weaning, severe winter weather and changes in feed, help bring on the condition (Hammond, 1964; Marquardt, 1962; Hammond, Sayin and Miner, 1965; Fitzgerald, 1962). In this connection, the realization that coccidia may have an extraintestinal phase in which merogony may continue for some time if not indefinitely makes it desirable to determine whether this phenomenon may not play a role (see Levine, 1972).

In sum, a mass of uncertainties exists regarding the correct classification of the coccidia. I have not pointed out all of them, but I have mentioned enough to show how limited our knowledge really is. There is work ahead for several generations of protozoologists before a definitive classification can be developed.

Literature Cited

Andreassen, J. and O. Behnke. 1968. Fine structure of merozoites of a rat coccidian *Eimeria miyairii*, with a comparison of the fine structure of other Sporozoa. J. Parasitol. 54:150-163.

Arcay de Peraza, L. 1963. Studies on two new coccidia from the Venezuelan lizard *Cnemidophorus lemniscatus lemniscatus: Hoarella garnhami* gen. nov., sp. nov. and *Eimeria flaviviridis americana* subsp. nov. Parasitology 53:95-107.

Balbiani, E. G. 1882. Les sporozoaires. Seconde partie du cours d'embryogénie comparé professé au Collège de France pendant le second semestre de 1882. J. Microg. 6:281-290, 348-356, 448-457, 514-524, 565-574, 615-627.

Balbiani, E. G. 1884. Lecons sur les sporozoaires. Paris.

Ball, G. H. 1960. Some considerations regarding the Sporozoa. J. Protozool. 7:1-6.

Becker, E. R. 1934. Coccidia and coccidiosis of domesticated, game and laboratory animals and of man. Collegiate Press, Ames, Ia.

Becker, E. R. 1956. Catalog of Eimeriidae in genera occurring in vertebrates and not requiring intermediate hosts. Iowa St. Col. J. Sci. 31:85-139.

Biocca, E. 1956. Schema di classificazione dei protozoi e proposta di una nuova classe. Atti Accad. naz. Lincei., Classe Sci. Fis. Mat. Nat., ser. 8. 21:453-455.

Box, E. D. 1967. Influence of *Isospora* infections on patency of avian *Lankesterella* (*Atoxoplasma*, Garnham, 1950). J. Parasitol. 53:1140-1147.

Box, E. D. 1970. *Atoxoplasma* associated with an isosporan oocyst in canaries. J. Protozool. 17:391-396.

Bütschli, O. 1882. Protozoa. 1 Abt.: Sarkodina und Sporozoa. Bronn's Kl. und Ordn. Thierreichs. 1:321-616.

Cuckler, A. C. 1970. Coccidiosis and histomoniasis in avian hosts. In Jackson, G. J., R. Herman, and I. Singer, eds. Immunity to parasitic animals. Appleton-Century-Crofts, New York. 2:371-397.

Davies, S. F. M., L. P. Joyner, and S. B. Kendall. 1963. Coccidiosis. Oliver and Boyd, Edinburgh.

Diesing, C. M. 1851. Systema helminthium. 2:15. Vindobonae.

Doflein, F. 1901. Die Protozoen als Parasiten und Krankheitserreger. Jena.

Dufour, L. 1828. Note sur la grégarine, nouveau genre de ver qui vit en troupeau dans les intestines de divers insectes. Ann. Sci. Nat. 13:366-368.

Duszynski, D. W. 1969. *Pythonella scelopori* sp. n. (Protozoa: Eimeriidae) from a Costa Rican lizard. J. Parasitol. 55:684-685.

Eimer, T. 1870. Ueber die ei- oder kugelförmigen sogenannten Psorospermien der Wirbeltiere. Würzburg.

Fayer, R. 1972. Gametogony of *Sarcocystis* sp. in cell culture. Science 175:65-67.

Fitzgerald, P. R. 1962. Coccidia in Hereford calves on summer and winter ranges and in feedlots in Utah. J. Parasitol. 48:347-351.

Frenkel, J. K., J. P. Dubey, and N. L. Miller. 1970. Toxoplasma gondii in cats: Fecal stages identified as coccidian oocysts. Science 167:893-896.

Garnham, P. C. C. 1966. Malaria parasites and other Haemosporidia. Blackwell, Oxford.

Gousseff, W. F. 1937. *Yakimovella erinacei* n. gen., n. sp. a coccidium from the hedgehog. J. R. Microsc. Soc. 57:200-202.

Grassé, P.-P. 1953. Classe des coccidiomorphes. In Grassé, P. -P., ed. Traité de Zoologie. Vol. I, Fasc. II. Protozoaires: Rhizopodes, actinopodes, sporozoaires, cnidosporidies. Masson, Paris. 691-906.

Hake, T. G. 1839. A treatise on varicose capillaries, as constituting the structure of carcinoma of the hepatic ducts, and developing the law and treatment of morbid growths. With an account of a new form of pus globule. London.

Hammond, D. M. 1964. Coccidiosis of cattle. Some unsolved problems. Utah St. Univ., Logan.

Hammond, D. M., F. Sayin, and M. L. Miner. 1965. Nitrofurazone as a prophylactic agent against experimental bovine coccidiosis. Am. J. Vet. Res. 26:83-89.

Hartmann, M. 1923. Sporozoa: Amöbosporidia und Gregarina-Coccidia. In Kukenthal, W. and T. Krumbach, eds. Handbuch der Zoologie. De Gruyter, Berlin. 1:186-255.

Hartog, M. 1906. The Protozoa. Cambridge Natural History. 1:1-162.

Henle, J. 1835. Ueber die Gattung *Brachiobdella* and über die Bedeutung der inneren Geschlechtstheile bei den Anneliden und hermaphroditischen Schnecken. Arch. Anat. Physiol. 2:574-608.

Hoare, C. A. 1933. Studies on some new ophidian and avian coccidia from Uganda, with a revision of the classification of the Eimeriidea. Parasitology 25:359-388.

Hoare, C. A. 1957. (1956). Classification of Coccidia Eimeriidae in a "periodic system" of homologous genera. Rev. Brasil. Malariol. 8:197-202.

Honigberg, B. M., W. Balamuth, E. C. Bovee, J. O. Corliss, M. Gojdics, R. P. Hall, R. R. Kudo, N. D. Levine, A. R. Loeblich, Jr., J. Weiser, and D. H. Wenrich. 1964. A revised classification of the phylum Protozoa. J. Protozool. 11:7-20.

Hutchison, W. M., J. F. Dunachie, J. C. Siim and K. Work. 1970. Coccidian-like nature of Toxoplasma gondii. Brit. Med. J. 1970:142-144.

Kaufmann, W. 1847. Analecta ad tuberculorum et entozoorum cognitionem. Diss. Inaug., Berol.

Kheisin, E. M. 1967. Zhiznennye tsikly koktsidii domashnikh zhivotnykh. Izdat. Nauka, Leningrad. (English translation by F. K. Plous, Jr., edited by K. S. Todd, Jr. 1971. Life cycles of coccidia of domestic animals. University Park Press, Baltimore.)

Kloss, H. 1855. Ueber Parasiten in der Niere von *Helix*. Abh. Senckenb. Naturf. Ges. 1:189-213.

von Kölliker, A. 1845. Die Lehre von der thierischen Zelle und den einfacheren thierischen Formelementen, nach den neuesten Fortschritten dargestellt. Z. Wiss. Bot. 1(2):46-102.

von Kölliker, A. 1848. Beiträge zur Kenntniss niederer Thiere. I. Ueber die Gattung *Gregarina*. Z. Wiss. Zool. 1:1-37.

Kozar, Z. 1970. Toxoplasmosis and coccidiosis in mammalian hosts. In Jackson, G. J., R. Herman, and I. Singer, eds. Immunity to parasitic animals. Appleton-Century-Crofts, New York. 2:871-912.

Kudo, R. R. 1966, Protozoology. 5th ed. C. C Thomas, Springfield.

Labbé, A. 1896. Recherches zoologiques, cytologiques et biologiques sur les coccidies. Arch. Zool. Exp. Gener. 14:517-654.

Labbé, A. 1899. Sporozoa. *In* Schulze, F. E., ed. Das Tierreich. 5:xx-180 pp.

Lainson, R. 1959. *Atoxoplasma* Garnham, 1950, as a synonym for *Lankesterella* Labbé, 1899. Its life cycle in the English sparrow (*Passer domesticus domesticus*, Linn.). J. Protozool. 6:360-371.

Lankester, E. R. 1885. Protozoa. Encyclopaedia Britannica, 9th ed. 19:830-866.

Léger, L. 1898. Essai sur la classification des coccidies et description de quelques espèces nouvelles ou peu connues. Ann. Mus. Hist. Nat., Marseille, S. II 1:71-123.

Léger, L. 1900. Le genre *Eimeria* et la classification des coccidies. C. R. Soc. Biol. 52:576-577.

Leuckart, R. 1879. Die Parasiten des Menschen. 2nd ed. Winter, Leipzig.

Levine, N. D. 1953. A review of the coccidia from the avian orders Galliformes, Anseriformes and Charadriiformes with descriptions of three new species. Am. Midl. Nat. 49:696-719.

Levine, N. D. 1961. Protozoan parasites of domestic animals and of man. Burgess, Minneapolis.

Levine, N. D. 1962. Protozoology today. J. Protozool. 9:1-6.

Levine, N. D. 1963. Coccidiosis. Ann. Rev. Microbiol. 17:179-198.

Levine, N. D. 1969. Taxonomy of the Sporozoa. Prog. Protozool. III. Int. Congr. Protozool, Leningrad. 365-366.

Levine, N. D. 1970. Taxonomy of the sporozoa. J. Parasitol. 56(II):208-209.

Levine, N. D. 1971. Taxonomy of the piroplasms. Trans. Am. Microsc. Soc. 90:2-33.

Levine, N. D. 1973. Protozoan parasites of domestic animals and of man. 2nd ed. Burgess, Minneapolis.

Levine, N. D. and G. R. Campbell. 1971. A check-list of the species of the genus *Haemoproteus* (Apicomplexa, Plasmodiidae). J. Protozool. 18:475-484.

Levine, N. D. and V. Ivens. 1965. The coccidian parasites (Protozoa, Sporozoa) of rodents. Ill. Biol. Monogr. #33. Univ. Ill. Press, Urbana.

Levine, N. D. and V. Ivens. 1965a. *Isospora* species in the dog. J. Parasitol. 51:859-864.

Levine, N. D. and V. Ivens. 1970. The coccidian parasites (Protozoa, Sporozoa) of ruminants. Ill. Biol. Monogr. #44. Univ. Ill. Press, Urbana.

Lieberkühn, N. 1854. Ueber die Psorospermien. Arch. Anat. Physiol. 1:1-24, 349-368.

Lindemann, K. 1865. Weiteres über Gregarinen. Bull. Soc. Imp. Nat. Moscow 38(2):381-387.

Lühe, M. 1902. Ueber Geltung und Bedeutung der Gattungsnamen *Eimeria* und *Coccidium*. Centralbl. Bakt. I Abt. 31:771-773.

Marquardt, W. C. 1962. Subclinical infections with coccidia in cattle and their transmission to susceptible calves. J. Parasitol. 48:270-275.

Mesnil, F. 1900. Sur la conservation du nom

générique *Eimeria* et la classification des coccidies. C. R. Soc. Biol. 52:603-604.

Mesnil, F. 1903. Les travaux récents sur les coccidies. Bull. Inst. Pasteur 1:1-14.

Minchin, E. A. 1903. The Protozoa (*continued*). Section K.—The Sporozoa. In Lankester, E. R., ed. A treatise on zoology. Black, London. 1(2):150-360.

Mueller, J. 1841. Ueber eine eigenthümliche krankhafte parasitische Bildung mit specifisch organisierten Samenkörperchen. Arch. Anat. Physiol. 1841:477-496.

Musaev, M. A. and A. M. Veisov. 1965. Koktsidii gryzunov SSSR. Izdat. Akad. Nauk Azerbaid. SSR, Baku.

Orlov, N. P. Koktsidiozy sel'skokhozyaistvennykh zhivotnykh. Moscow (English translation by A. Ferber and A. Storfer, edited by H. Mills. 1970. Israel Program for Scientific Translations, Jerusalem. Publ. TT 70-50040, U. S. Dept. of Commerce, Clearinghouse for Fed. Sci. Tech. Information, Springfield, Va.

Overdulve, J. P. 1970. The identify of *Toxoplasma* Nicolle & Manceaux, 1909 with *Isospora* Schneider, 1881. Proc. Konikl. Nederl. Akad. Wetenschappen, Amsterdam. Ser. C. 73:129-151.

Pellérdy, L. 1956. Catalogue of the genus Eimeria (Protozoa: Eimeriidae). Acta Vet. Acad. Sci. Hung. 6:75-102.

Pellérdy, L. 1957. Catalogue of the genus Isospora (Protozoa: Eimeriidae). Acta Vet. Acad. Sci. Hung. 7:209-220.

Pellérdy, L. 1963. Catalogue of Eimeriidea (Protozoa: Sporozoa). Akad. Kiado, Budapest.

Pellérdy, L. 1965. Coccidia and coccidiosis. Akad. Kiado, Budapest.

Pellérdy, L. 1969. Catalogue. Eimeriidea (Protozoa, Sporozoa). Supplementum I. Akad. Kiado, Budapest.

Pfeiffer, L. 1890. Vergleichende Untersuchungen über Schwärmsporen und Dauersporen bei den Coccidieninfektionen und bei Intermittens. Fortschr. Med. 8:939-951.

Pfeiffer, L. 1891. Die Protozoen als Krankheitserreger. 2nd ed. Fischer, Jena.

Pfeiffer, R. 1892. Beiträge zur Protozoenforschung. I. Die Coccidienkrankheit der Kaninchen. Berlin.

Poche, F. 1913. Das System der Protozoa. Arch. Protist. 30:125-321.

Ray, H. N. 1930. Studies on some Sporozoa in polychaete worms. II. *Dorisiella scolelepidis* n. g., n. sp. Parasitology 22:471-480.

Ray, H. N. and M. Das Gupta. 1937. On a new coccidium from the intestine of *Python* sp. Proc. Ind. Sci. Congr. 24:292.

Ray, H. N. and K. Raghavachari. 1942. Observations on a new coccidium *Octosporella mabuiae* n. gen., n. sp., from the intestine of *Mabuia* sp. Proc. Ind. Sci. Congr. 28:170.

Ray, H. N. and A. C. Sarkar. 1968. A new coccidium *Sivatoshella lonchurae* n. gen., n. sp., from *Lonchura malabarica* and *L. punctulata*. J. Protozool. 15:640-643.

Remak, R. 1845. Diagnostische und pathogenetische Untersuchungen in der Klinik des Herrn Geh. Raths. Dr. Schönlein. Berlin.

Rivolta, S. 1878. Della gregarinosi dei polli e dell' ordinamento delle gregarine e dei psorospermi degli animali domestici. Giorn. Anat. Fisiol. Pat. Anim., Pisa 10:220-235.

Schaudinn, F. 1900. Untersuchungen über den Generationswechsel bei Coccidien. Zool. Jahrb., Abt. Anat. 13:197-292.

Schaudinn, F. and M. Siedlecki. 1897. Beitrag zur Kenntnis der Coccidien. Verhandl. Deutsch. Zool. Ges. 192-204.

Schneider, A. 1875. Note sur la psorospermie oviforme du poulpe. Arch. Zool. Exp. Gen. 4(n. + r.):xl-xliv.

Schneider, A. 1881. Sur les psorospermies oviformes ou coccidies, espèces nouvelles ou peu connues. Arch. Zool. Exp. Gen. 9:387-404.

Schneider, A. 1892. Le cycle èvolutif des coccidies et M. le docteur L. Pfeiffer. Tab. Zool. 2:105-111.

Schuberg, A. 1896. Die Coccidien aus dem Darme der Maus. Verhandl. Naturh.-Med. Ver. Heidelberg (N.F.) 5:369-398.

Sheffield, H. G. and M. L. Melton. 1970. Toxoplasma gondii: The oocysts, sporozoite, and infection of cultured cells. Science 167:892-893.

von Siebold, C. T. 1839. Beiträge zur Naturgeschichte der wirbellosen Thiere. 4. Ueber die aus Gattung Gregarina gehörigen Helminthen. Schrift. Naturf. Ges. Danzig 3:56-71.

Stein, F. 1848. Ueber die Natur der Gregarinen. Arch. Anat. Physiol. Wiss. Med. 182-223.

Stieda, L. 1865. Ueber die Psorospermien der Kaninchenleber und ihre Entwicklung. Arch. Path. Anat. 32:132-139.

Stiles, C. W. 1902. *Eimeria stiedae* (Lindemann, 1865), correct name for the hepatic coccidia of rabbits. USDA Bur. Anim. Ind. Bull. 25:18-19.

Todd, K. S., Jr. and D. M. Hammond. 1971. Coccidia of Anseriformes, Galliformes, and Passeriformes. In Davis, J. W., R. C. Anderson, L. Karstad, and D. O. Trainer, eds. Infectious and parasitic diseases of wild birds. Iowa St. Univ. Press, Ames. 234-281.

Vetterling, J. M., H. R. Jervis, T. G. Merrill, and H. Sprinz. 1971. *Cryptosporidium wrairi* sp. n. from the guinea pig *Cavia porcellus* with an emendation of the genus. J. Protozool. 18:243-247.

Wenyon, C. M. 1926. Protozoology. 2 vols. Wood, New York.

2 Host and Site Specificity in the Coccidia

WILLIAM C. MARQUARDT
Department of Zoology, Colorado State University, Fort Collins, Colorado

---

Contents

I. Introduction . . . *24*
  A. Objectives . . . *24*
  B. General Concepts . . . *24*
    1. General Specificity among Infectious Agents . . . *24*
    2. Specificity among the Coccidia *Sensu Strictu* . . . *25*
  C. Organisms to be Considered . . . *26*
II. Host Specificity . . . *26*
  A. Excystation . . . *26*
  B. Partial Completion of the Life Cycle . . . *27*
  C. Specificity Based on Oocyst Structure . . . *27*
    1. Size . . . *28*
    2. Variability in Outer Oocyst Wall . . . *28*
    3. Shape and Shape Index . . . *29*
    4. Color . . . *29*
  D. Successful Completion of the Life Cycle in a Range of Hosts . . . *29*
  E. Relation of the Classification of the Hosts to their Serving as Adequate Hosts . . . *34*
  F. Completion of the Life Cycle in Chemically Treated Hosts . . . *34*
III. Site Specificity . . . *35*
  A. Normal Life Cycle . . . *35*
  B. Distribution of Parasites along the Intestine . . . *36*
  C. Depth in Mucosa . . . *38*

D. Cell Type . . . *38*
E. Location within the Cell . . . *39*
F. Abnormal Sites . . . *39*
Literature Cited . . . *40*

I. Introduction

A. OBJECTIVES

The purpose of this chapter is to look at various aspects of the host range of species of coccidia and to consider the specificity of location of the parasites within a host. In many instances, it is necessary merely to render a description of things as they are, since it is not possible at this point to state what mechanisms are involved in either host or site specificity.

There is an inordinately large volume of literature to be considered in some areas of investigation, and no attempt has been made to cite every paper. Instead, I have referred to certain recent works which do have an exhaustive review of the literature or have chosen papers which exemplify the principle in question.

B. GENERAL CONCEPTS

1. General Specificity among Infectious Agents

When we consider infectious agents—bacteria, rickettsia, viruses, and fungi, as well as those organisms which we term "parasites," protozoa and helminths,—their host ranges are usually rather broad. In addition, these organisms generally have some latitude in the site within the host which they may inhabit. From an ecologic viewpoint, it could be said that most organisms are capable of invading a number of discrete niches. An agent such as rabies virus has an exceedingly wide host range, being found in a wide range of carnivores, herbivores, and bats, to cite a few groups. *Bacillus anthracis*, the cause of anthrax, also has little host specificity. The causative agent of Rocky Mountain spotted fever, *Rickettsia rickettsi*, is normally found in a broad range of small rodents, among other mammals, and is transmitted by several species of ixodid ticks.

Among helminth parasites, there is frequently a normal host range that encompasses three or more species. For example, one of the digenetic flukes of man, *Schistosoma japonicum*, has a natural reservoir in domestic cattle. *Diphyllobothrium latum*, although considered to be a parasite of man, is normally found in a wide range of wild and domestic carnivores.

Although there are exceptions to the rule, most parasitic protozoa are found in more than one species of host. One can point to *Entamoeba histolytica* which is found both in man and the domestic dog. Members of the genus *Leishmania* are usually found, under natural conditions, in several species of mammals in any one geographic area. *Plasmodium* species, the cause of malaria, which are related to the coccidia, have a relatively high degree of specificity, but studies in recent years have shown that they can be transmitted, at least experimentally, to a number of species of animals in addition to the normal host.

2. Specificity among the Coccidia Sensu Strictu

If we take a broad view of the host specificity of the coccidia, certain general features stand out. Although there are exceptions which we will consider, the coccidia have as high a degree of specificity as any other group of infectious agents. Not only are they naturally limited to a narrow range of host species, but they have an affinity for certain organ systems, portions of the system, specific types of cells and ultimately, even specific locations within the cells which they inhabit.

The general consensus has been that the life cycle of each species of coccidia has been programmed to pass through a series of stages. This program is genetically determined and is one from which little deviation has been observed. For example, the life cycle of one of the agents causing malaria is one in which asexual reproduction, schizogony, may continue almost indefinitely. In contrast, the coccidia, at least in the genus *Eimeria*, have a specific number of schizogenous generations characteristic for each species: *Eimeria bovis* has two schizogenous generations, and *Eimeria nieschulzi* has four. It has generally been considered that once schizogony has been started, it continues through the specific number of asexual generations and then enters gametogony. Such programming was ingeniously demonstrated by Roudabush (1935) who showed that merozoites transferred from an infected animal to an uninfected animal continued their development as if they had remained within the same host. In the view of Fairbairn (1970), the developmental cycle of an organism is closely programmed and is not readily diverted from completing its goal. The programming in the life cycle of a parasite is such that biochemical events appear, disappear, and perhaps reappear later in a closely sequenced series. Recent studies have indicated that this may not apply to some species of coccidia, but variability is to be expected in a group of organisms which have relatively short generation times and are as prolific as the coccidia.

In discussing the specificity of coccidia, there are three principles which most authors accept either explicitly or implicitly.

a. The oocyst morphology is useful for determining the species of coccidia. There is much evidence that the oocyst is a good means of either identifying or describing a new species of coccidia. But there are some instances which should make one careful not to depend too much on oocyst morphology. For one thing, oocysts are likely to change in structure as they age (Pellérdy, 1965). Although Pellérdy pointed out that residua change or may disappear upon storage, it has been observed in some species that the residua are stable for a long period of time (Duszynski and Marquardt, 1969). In another context, despite the fact that their hosts are widely separated taxonomically, it has been assumed that the coccidia of sheep and goats are the same because their oocysts are similar. In some studies, however, it has been shown that the organisms isolated from one host are not readily transmissible to the other species (see review by Levine and Ivens, 1970). The turkey coccidia, *Eimeria meleagridis*, *E. adenoeides* and *E. gallopavonis*, have oocysts which cannot readily be distinguished from one another. They are also found in the same locations in the intestinal tract of the host. The only means by which they have been separated from one another is by cross immunity studies (Moore and Brown, 1952). Lastly, we know that there can be qualitative variability in oocyst structure. The outer wall of *E. auburnensis* may occur either as rough or smooth variants (Christensen and Porter, 1939).

b. The suitability of a host is usually determined only by the completion of the life cycle and the appearance of oocysts. There are a number of processes which take place from the time an oocyst is ingested until fresh oocysts are produced at the completion of the life cycle. For example, if excystation does not occur, no development will take place even though the environment may be otherwise suitable. Some schizogony, but no gametogony may take place in an abnormal host. Or, if the history of a particular animal is unknown, its immune status also remains completely unknown. Some early studies on the cross transmission of the coccidia suffer from this particular shortcoming.

c. Failure to find infection in a potential host where it appears to have ample opportunity to become infected means that the host is not suitable. This assumption is contradicted by Doran (1953) who found that *E. mohavensis* occurred naturally only in *Dipodomys panamintinus mohavensis* at a rate of 8.7% of 251 individuals. *D. m. merriami* is sympatric with *D. p. mohavensis*, but none of 179 individuals examined was found to be infected. In laboratory experiments it was found that *E. mohavensis* could be transmitted experimentally to *D. m. merriami*.

C. ORGANISMS TO BE CONSIDERED

Among the coccidia, three genera, *Eimeria*, *Isospora*, and *Toxoplasma*, have been studied more than any others within the group. This is understandable when one considers that they have members of considerable economic or medical importance and the fact that members of the genus *Eimeria* are probably more prevalent than any of the other genera. Although we will consider other genera, such as *Tyzzeria*, on occasion, most of the studies to be discussed will concern the three genera named above. Listings of coccidia and the hosts in which they have been found are located in the excellent compendia of Pellérdy (1963, 1969b).

II. Host Specificity

A. EXCYSTATION

Nearly all studies on cross transmission of coccidia have involved exposing presumably uninfected animals to infective oocysts and then looking for fresh oocysts at the appropriate time. The principle behind such studies is that the sporozoites will escape from the oocysts, invade the tissue and complete the normal life cycle. However, it should be considered that excystation of sporozoites might not take place within any one host, and although the host might be susceptible to infection, the oocysts would pass through the intestinal tract without change.

Nearly all studies have shown that excystation is a nonspecific phenomenon. Andrews (1930) and Kartchner and Becker (1930) found at nearly the same time that oocysts from such diverse hosts as cats, dogs, guinea pigs, prairie dogs, and pigs all excysted in the small intestine of the laboratory rat. Although other studies (Lotze *et al.*, 1961; Marquardt, 1966) have confirmed that sporozoite excystation is nonspecific, a sequence of stimuli is necessary *in vitro* for sporozoites to become active and to excyst (Jackson, 1962; Nyberg and Hammond, 1964). In poultry coccidia there appears to be a need for mechanical breakage of the oocyst wall (Doran and Farr, 1962, 1965; Farr and Doran, 1962). In contrast, there is evidence indicating that sporozoites of various species of poultry coccidia may undergo excystation without the wall of the oocyst having undergone mechanical breakage (Nyberg *et al.*, 1968; Lotze and Leek, 1968; Bunch and Nyberg, 1970). Disparity between the results of these investigators cannot be resolved at this time. It is probably best to keep in mind that sporozoites will excyst in nearly any host but that there may be some conditions under which they will fail to do so.

B. PARTIAL COMPLETION OF THE LIFE CYCLE

In studies of the susceptibility of mice to the rat coccidium, *E. nieschulzi*, it was found that excystation and invasion of the tissue by sporozoites took place normally in the laboratory mouse (Marquardt, 1966). Sporozoites could be traced from the lumen of the small intestine of the mouse, through epithelial cells of the villi; later they appeared in the cells of the crypts, where they then rounded up. Early stages of schizogony could be recognized, but no endogenous stages were seen later than 36 hr after exposure of the mice to sporulated oocysts. In only a single instance was any significant development seen and this was a single mature first generation schizont found in a crypt of only one mouse. Lotze *et al.* (1961) reported that oocysts derived from an infection in sheep underwent partial development when administered to goats. Schizonts were found, but they generally failed to mature and produce merozoites. More extensive studies than either of the two above have shown that *E. tenella* underwent partial development in the laboratory mouse and that *E. falciformis* of mice developed trophozoites in chickens and second generation schizonts in laboratory rats. An as yet unnamed *Eimeria* isolated from a rat was found to complete its development in laboratory mice, but a large number of second generation schizonts and gametocytes died (Haberkorn, 1970). In my studies with *E. nieschulzi* the disappearance of the early schizonts was not accompanied by any cellular reaction (Marquardt, 1966). Therefore, it was my opinion at the time that lack of essential nutritional substances was instrumental in preventing the development of the endogenous stages. This view has been supported by the lack of success in attempts to infect parenterally foreign hosts with *E. tenella* or *E. stiedai* (Pellérdy, 1969a). Recent work (Rose and Long, 1969; Todd *et al.*, 1971) has indicated that there may be an active response on the part of the host, in the elimination of the organisms or the prevention of their development in some way.

C. Specificity Based on Oocyst Structure

The classification of species of coccidia has been based traditionally on oocyst morphology. Although some descriptions of new species include the location where the coccidia develop in the tissues of the host and sometimes detailed descriptions of endogenous stages are given, by far the greatest proportion of descriptions includes only the morphology of the sporulated oocyst. The crucial position of the oocyst in the description of any species is amply demonstrated by noting that of the 95 named species of *Eimeria* of ruminants, the location in the host is known for only 17 species, and portions of the life cycle are known for only 15 species. Complete life cycles have been described for only two of the 95 species (Levine and Ivens, 1970). If one looks at the coccidia of any group of hosts, with the exception of domestic chickens, the same sort of pattern can be seen: a large number of species described solely on the basis of oocyst structure.

Cross transmission studies have shown that species of coccidia can be transmitted to some extent within the genus of host from which they were isolated, but with only a few exceptions have they been transmitted outside the genus (see Section D of this chapter).

In describing new species, authors have generally held to the principles that oocysts of identical morphology from closely related hosts are members of a single species. As a corollary, it has also been concluded that oocysts of similar structure from hosts that are widely separated taxonomically are different species. Likewise, oocysts of different structure but derived from the same host are different species.

We would do well to look at the validity of these generalizations and determine whether there are pitfalls which might trap the unwary. It has been estimated that with the use of both quantitative and qualitative characteristics there can be at least 2,654,736 different species in the genus *Eimeria* alone (Levine, 1962). Nowhere near that many species of *Eimeria* have been described, and one should not be misled into thinking either that slight morphological differences are justification for erecting a new species, or that similarities in oocyst morphology always mean that only a single species is at hand. The following factors should be kept in mind.

1. Size

The oocysts of *Eimeria bovis* are generally given as 23-34 μ by 17-23 μ (Levine and Ivens, 1970), and this is roughly the size range which one might expect with almost any species. But there are species in which there may be nearly a three-fold difference in either the length or width. The size range of *Eimeria ellipsoidalis*, given by various authors, is from 12-32 μ by 10-29 μ.

Factors which are known to correlate with oocyst size are the time after infection that oocysts are collected and the host from which oocysts are obtained. It has been found that when animals are infected experimentally, the size of oocysts tends to increase through the patent period (see Duszynski, 1971 for review). In studies on laboratory rats infected with *E. separata*, Duszynski (1971) found that the oocyst length increased from 11.7μ to 16.3 μ and that the width increased from 10.1 μ to 14.3 μ during the patent period of four days. *Isospora lacazei* varies in size depending on whether oocysts are obtained from canaries or sparrows (Box, 1967).

In instances where the investigator has access to large numbers of oocysts, it does appear to be possible to differentiate among species of oocysts by statistical means where simple inspection of the data might fail to point up differences. It was found feasible in this way to differentiate among three of four species of *Eimeria* of the cottontail rabbit, *Sylvilagus*. Even though length and width measurements had been made on at least 200 oocysts and sporocysts of each species, it was not evident on simple inspection that there were quantitative differences to match the qualitative differences which had been documented. Application of the New Duncan multiple-range test showed that it was possible to differentiate among *E. poudrei* and *E. neoirresidua* on one hand and *E. environ* and *E. media* from *E. honessi* on the other with at least 95% confidence (Duszynski and Marquardt, 1969). Moore and Brown (1952) found that the oocysts of *E. innocua* and *E. adenoeides* were nearly identical. One of the criteria which they set forth for differentiating between the two species was the immunity which followed experimental infections. They determined that infection of birds with one species left the animals completely susceptible to infection with the other species; challenge with the same species showed complete immunity.

2. Variability in the Outer Oocyst Wall

The description of *E. auburnensis* by Christensen and Porter (1939) shows that there can be striking differences in the oocyst structure of individual specimens of the same species. They found that typical oocysts had a smooth outer wall but that a small percentage had an outer wall which was heavily mammillated. Inoculation of mammillated oocysts into calves resulted in the production of both smooth and mammillated oocysts, although the former predominated. The variability in the structure of the outer wall does not result from the loss of a loosely attached outer wall as may occur in such species as *E. robusta* as described by Supperer and Kutzer (1961).

3. Shape and Shape Index

Under most circumstances, the shape index of a species of coccidia, i.e., the length divided by width, is remarkably constant. Oocyst shape, whether ellipsoid or ovoid, is usually stable, but there are some circumstances in which it has been seen to vary. *E. alabamensis* develops in the nuclei of epithelial cells of the intestine of domestic cattle. Davis *et al.* (1957) noted that the oocysts may be pyriform, elliptical, or spherical. When there were multiple infections of nuclei with macrogametes, the oocysts were distorted, and it appears that the pyriform shape may result from pressure of the oocysts upon one another in such a crowded condition.

4. Color

The color of an oocyst is always indicated in a description of a species. In nearly all instances, color refers to the outer wall or walls of the oocyst. Real variation in the color may occur, but apparent differences in color may come about through biological pigments picked up in the intestinal tract or from the medium in which the oocysts are allowed to sporulate. Potassium dichromate is usually used to inhibit bacterial growth during sporulation of oocysts, and it sometimes stains the outer oocyst walls. It was first pointed out to me some years ago by Dr. L. R. Davis that the type of lens in the microscope determines the apparent color of some oocysts. When comparing species, one should determine whether the original description was given for oocysts examined under an achromatic or apochromatic lens.

Despite the precautions which should be taken in light of the discussion above, it has been possible to use oocysts almost exclusively to set up a satisfactory working hypothesis of the spectrum of organisms found in any particular group of hosts. For example, a large number of species of coccidia have been described from reptiles, and this has been done almost exclusively on the basis of oocyst morphology, the taxonomic position of the hosts, and the geographic distribution of the hosts (Vetterling and Widmer, 1968; Bovee and Telford, 1965; Duszynski, 1969; Bovee, 1971; Cannon, 1967a, b, and c).

The study of coccidia of North American geese and swans by Hanson *et al.* (1957) entailed an examination of 363 individuals of four genera, six species and 13 subspecies of Anseriformes. Based purely on oocyst morphology, the authors concluded that of the species of *Eimeria*, four were found in only a single species of host, three were found in two and one in three species of hosts. *Tyzzeria anseris* was found in all six species of hosts examined. The authors' conclusions are supported by experimental studies such as that of Farr (1953) who found that *E. magnalabia* could be transmitted from wild to domestic geese. A number of other examples could be cited to further document the use of oocyst morphology for determining the species complement of a host or a group of hosts, but the pattern would generally be the same. It is likely that some species will be lumped with others, and that new ones will be erected as we attain additional information, but the bulk of the evidence supports the validity of the conclusions that have been made at this point.

D. SUCCESSFUL COMPLETION OF THE LIFE CYCLE IN A RANGE OF HOSTS

The real crux of the cross transmissibility of any agent lies in exposing a series of experimental animals to it. A typical experiment involves administering infective oocysts of a coccidium by mouth to a species of host in which the organism is not ordinarily found. After the normal prepatent period, the investigator examines fecal material for the presence of fresh oocysts. If none appear, it is concluded that the recipient cannot be infected.

The major difficulty in performing these types of experiments lies in obtaining experimental animals which have not been exposed to any coccidia. For the greatest assurance that experimental animals have not been infected, an investigator should raise his own animals and examine fecal material both of the mother and of the offspring at frequent intervals to determine whether any of them become infected.

Sometimes it is necessary to institute extreme measures in order to insure that the animals are and remain free from infection (Mahrt, 1967), but the need for taking animals by hysterectomy and raising them by hand must be sufficiently great to justify the enormous expenditure of time and money. If a pregnant female is found to be free of coccidial infection by repeated fecal examination, it can be placed in isolation and allowed to rear its offspring normally. This technique was used by Shah (1970) in raising puppies for studies on the transmission of *Isospora felis* from the cat to the dog. Moderate success can at times be achieved by obtaining large animals shortly after birth and removing them to relatively clean isolation facilities (Senger *et al.*, 1959; Marquardt, 1962). If animals are infected with one species of coccidium, they are generally considered to be completely susceptible to infection with the second species, but Long and Horton-Smith (1968) have reviewed previous works and concluded that there may be more cross immunity than we had suspected. Therefore, it is best to note with caution those negative cross infection experiments in which the history of the animals is unknown or in which they have had some light infection with a species of coccidium other than the one being tested.

Recent critical reviews have appeared on the host specificity of the coccidia of various groups of hosts: rodents (Levine and Ivens, 1965), *Isospora* of dogs and cats (Shah, 1970), domestic chickens, and turkeys (McLoughlin, 1969), lagomorphs (Duszynski and Marquardt, 1969), ruminants (Levine and Ivens, 1970), and *Isospora* of passerine birds (Anwar, 1966).

Birds represent ideal experimental animals on which to perform cross transmission experiments since eggs of a number of species of common domesticated birds can be obtained almost year-round, cleaned, incubated, and the hatchlings introduced into clean quarters. The risk of contamination between parent and offspring is not a problem with birds as it clearly is when one is working with mammals.

Quality of accommodations is a crucial factor in maintaining animals free of extraneous infection. An investigator who does not have available to him animal rooms which have at least a moderate degree of isolation and the capability of being thoroughly cleaned would do well to undertake some other line of investigation.

A great proportion of studies which have been done on cross transmission of coccidia have utilized those parasites which are of importance in domestic animals. In light of the discussion above, much of the reliable work has been done on the coccidia of the domestic chicken and turkey. McLoughlin (1969) tabulated the cross transmission studies which have been done with coccidia of chickens and turkeys (Table 1). It becomes evident upon scanning the table that there are remarkable inconsistencies. *E. tenella* was successfully transmitted from quail to chicken by one investigator, but the same organism was unable to complete its life cycle when transmitted from the chicken to the quail. One investigator was successful in transmitting *E. tenella* from the chicken to the pheasant, but another failed in the same attempt. And even more remarkable are three reports of *E. tenella* being transmitted from the chicken to a pigeon; this is successful transmission not only outside the genus but to a completely different order of birds. Most, but by no means all, of the successful reports of transmission listed in the table are rela-

TABLE 1. Summary of results of transfaunation experiments with *Eimeria* of chickens and turkeys (McLoughlin, 1969)

Species[a]	Transmission			Author
	From	To	Successful	
Eimeria sp.	Turkey	Chicken	Yes	Henry, 1931
	Duck	Chicken	No	Tiboldy, 1933[d]
E. acervulina	Chicken	Quail	No	Patterson, 1933
	Chicken	Quail	No	Tyzzer, 1929
	Chicken	Pheasant	No	Tyzzer, 1929
	Chicken	Turkey	No	Steward, 1947
	Quail (?)	Chicken	Yes	Henry, 1931
	Quail (?)	Chicken	No	Venard, 1933
E. maxima	Chicken	Quail	No	Patterson, 1933
E. mitis	Chicken	Quail	No	Patterson, 1933
	Chicken	Turkey	No	Tyzzer, 1929
	Quail (?)	Chicken	Yes	Henry, 1931
E. tenella[b]	Quail (?)	Chicken	Yes	Henry, 1929
	Quail (?)	Chicken	Yes	Venard, 1933
	Chicken	Quail	No	Patterson, 1933
	Chicken	Turkey	No	Patterson, 1933
	Chicken	Duck	No	Patterson, 1933
	Chicken	Pheasant	No	Patterson, 1933
	Chicken	Pheasant	Yes	Haase, 1939
	Chicken	Pigeon	Yes	Hofkamp[e]
	Chicken	Pigeon	Yes	Nieschulz, 1921
	Chicken	Pigeon	Yes	Yakimoff and Iwanoff-Gobzem, 1931
E. avium	Grouse	Chicken	Yes	Fantham, 1910
(= *E. tenella*?)	Chicken	Turkey	No	Johnson, 1923
	Chicken	Duck	No	Johnson, 1923
E. adenoeides	Turkey	Chicken	No	Moore and Brown, 1951
	Turkey	Guinea	No	Moore and Brown, 1951
	Turkey	Pheasant	No	Moore and Brown, 1951
	Turkey	Quail	No	Moore and Brown, 1951
	Turkey	Chicken	No	Clarkson, 1959
E. gallopavonis	Turkey	Partridge	Yes	Hawkins, 1952
	Turkey	Pheasant	No	Hawkins, 1952
	Turkey	Quail	No	Hawkins, 1952
	Turkey	Chicken	Yes	Gill, 1954
E. meleagridis	Turkey	Chicken	Yes	Steward, 1947
	Turkey	Chicken	Yes	Gill, 1954
	Turkey	Chicken	No	Tyzzer, 1929
	Turkey	Chicken	No	Moore *et al.*, 1954[c]
	Turkey	Chicken	No	Clarkson, 1959
	Turkey	Pheasant	No	Tyzzer, 1929
	Turkey	Partridge	No	Hawkins, 1952
	Turkey	Quail	No	Tyzzer, 1929

(Continued)

TABLE 1 *(Continued)*

Species[a]	Transmission			Author
	From	To	Successful	
	Turkey	Quail	No	Hawkins, 1952
E. meleagrimitis	Turkey	Quail	No	Hawkins, 1952
	Turkey	Partridge	No	Hawkins, 1952
	Turkey	Chicken	Yes	Gill, 1954
E. innocua	Turkey	Chicken	No	Moore and Brown, 1952
	Turkey	Guinea	No	Moore and Brown, 1952
	Turkey	Pheasant	No	Moore and Brown, 1952
	Turkey	Quail	No	Moore and Brown, 1952
E. subrotunda	Turkey	Chicken	No	Moore *et al.*, 1954
	Turkey	Guinea	No	Moore *et al.*, 1954
	Turkey	Pheasant	No	Moore *et al.*, 1954
	Turkey	Quail	No	Moore *et al.*, 1954

[a]Natural infections of *E. dispersa* have been reported from a number of hosts. Therefore, studies with this species are not included here.
[b]Haase, 1939, also reported *E. tenella* from quail.
[c]Moore, Brown, and Carter, cited by Moore, 1954.
[d]Cited by Becker, 1934.
[e]Quoted by Yakimoff & Iwanoff-Gobzem, 1931.

tively old. Certain of the older reports should be greeted with some skepticism in view of the negative results obtained by investigators working under tightly controlled conditions. I suspect that many negative experiments have failed to reach publication purely because they are negative.

The recent monograph of Levine and Ivens (1970) includes a tabulation of the cross transmission studies which have been done with all coccidia of ruminants. The genus *Eimeria* is by far the most important one found in ruminants since 95 of the 100 species described in the monograph are *Eimeria*. The table lists 156 attempts to transmit 28 different species of *Eimeria* from one host to another host species. Of the 156 attempts, 11 (7%) were successful and these involved seven species of *Eimeria*. Even if one accepts at face value the positive results which have been obtained, transmission has been obtained only between the domestic sheep, *Ovis aries* and the domestic goat, *Capra hircus*, from the chamois, *Rupricapra rupricapra*, to sheep and goats and from the water buffalo, *Bubalis bubalis*, to the domestic ox, *Bos taurus*. These positive reports represent an exceedingly low success rate. The authors conclude that transmission between members of different tribes probably does not occur, and they leave open the question whether transmission between the domestic goat and the domestic sheep actually does take place. In considering a number of conflicting reports, this is probably the best approach.

Levine and Ivens (1965) have reviewed and tabulated the cross infection experiments performed with *Eimeria* of rodents. Their tabulation shows that completion of the life cycle of all species tested occurred only within the genus from which the coccidium was originally isolated. There is only a single instance in which transmission to another species of a genus, in this case *Peromyscus*, failed. In general, transmission within the genus has been successful in the attempts made in rodents.

Some studies carried out since the completion of the monograph above are of interest. Todd *et al.* (1968) showed that *E. bilamellata* could be transmitted among at least four species of the genus *Spermophilus*. In another study on *E. callospermophili*, it was found that the coccidium was readily transmitted among at least six species of the ground squirrel, *Spermophilus*, and in addition the same parasite isolated from the white tailed prairie dog, *Cynomys leucurus*, could be transmitted to at least two species of the genus *Spermophilus* (Todd and Hammond, 1968). It is of interest that *Cynomys* and *Spermophilus* are generally considered to be closely related. Transmission of *E. callospermophili* to other rodents such as the least chipmunk, *Eutamius minimus*, was not successful. A wider range of hosts was seen in an as yet unnamed species of *Eimeria* isolated from a rat. It completed development in laboratory mice, but with a reduced number of oocysts (Haberkorn, 1970).

E. chinchillae shows an unusual lack of host specificity (de Vos, 1970). The organism was originally isolated from the chinchilla by de Vos and Van der Westhuizen and described as a new species (1968). Although chinchilla are raised commercially in South Africa, they are native to South America. Oocysts originally isolated from the chinchilla were inoculated into a variety of other mammals. Patent infections were observed in the following seven species of native wild rodents: *Praomys natalensis, Rhabdomys tumilio, Otomys irroratus, Mystromys albicaudatus, Arvicanthis niloticus, Saccostomus campestris,* and possibly *Tapara leucogaster.* In addition, laboratory white mice and laboratory white rats were also susceptible to the infection. Rodents which were not susceptible to infection were golden hamsters and guinea pigs. Rabbits, presumably *Oryctolagus cuniculus*, were not susceptible to infection. In a single shrew, *Crocidura*, which was exposed, no oocysts were found in the feces. Oocyst production and pathogenesis were not the same in all animals which were successfully infected. In fact, de Vos did not include *T. leucogaster* among those hosts in which the infection was completed, since only objects resembling damaged oocysts appeared in the feces. de Vos concluded that *E. chinchillae* is most likely a parasite of a number of species of rodents native to South Africa and has only secondarily been transmitted to the chinchilla, in which it has caused severe problems for some raisers. For our purposes here, it is of interest to note that this organism can be successfully transmitted among at least seven genera of rodents coming from two families.

It has been the general consensus that members of the genus *Isospora* have a lower degree of specificity than do members of the genus *Eimeria*. The basis of this view appears to lie largely in the fact that *I. lacazei* has been described from a large number of species of birds (see Anwar, 1966 for review). In addition, the early work on *Isospora* of dogs and cats indicated that they had coccidia in common.

The work with *Isospora* of birds has been hampered by the inability of investigators to obtain many species of birds, particularly small ones, free from infection. Therefore, most identifications of *Isospora* species are based on oocyst structure. The hazards of relying solely on oocyst characteristics have been discussed earlier in this chapter. Anwar (1966) found *I. lacazei* and *I. chloridis* in three species of fringillids and confirmed that both species developed in all of the hosts by superinfecting birds and studying tissue sections of intestines. By contrast, cross infection of *I. lacazei* between the common house sparrow and canaries was not successful (Box, 1967). Obviously, additional studies are required before generalizations can be made.

In the case of *Isospora* of carnivores, recent studies have demonstrated that *I. felis* from the cat will not infect puppies (Shah, 1970) and *I. rivolta* from the dog will not infect kittens (Mahrt, 1967). Studies just now reaching publication (Frenkel and Dubey, 1972; Dubey and Frenkel, 1972) may alter the idea that carnivore *Isospora* species have a high degree of host specificity, but, for now, the specificity appears to be about the same as for *Eimeria*.

The recent finding that *Toxoplasma gondii* is similar to *Isospora bigemina* requires a reassessment of certain aspects of host specificity (see Chapters 3 and 9). At present, it appears that *T. gondii* is an exceedingly adaptable organism with respect to the hosts in which it can develop. While the normal host is the domestic cat, in which oocysts are produced, the schizogonous portions of the life cycle can take place in an enormous range of other host species. Certain species of *Eimeria* undergo schizogony, but show reduced or absent gametogony in abnormal hosts (Haberkorn, 1970).

E. RELATION OF THE CLASSIFICATION OF THE HOSTS TO THEIR SERVING AS ADEQUATE HOSTS

Nearly all cross transmission studies involve exposing a potential host to large numbers of oocysts, waiting a suitable period for the development of the organisms and then examining fecal material for oocysts. Transmission is considered to have taken place if oocysts appear in the fecal material, and not to have taken place if no oocysts appear. Nearly all reports present the results merely as a positive or negative. Doran (1953) examined six species and eleven subspecies in the genus *Dipodomys* for the presence of *E. mohavensis*. Under natural conditions, only *D. panamintinus mohavensis* was found to be infected with *E. mohavensis*. Doran points out specifically that no natural infections were found in any of the rodents locally sympatric with *D. p. mohavensis* or in any of the other subspecies of *D. panamintinus*. When members of the eleven subspecies were exposed to infective oocysts of *E. mohavensis*, it was found that *D. p. mohavensis*, the normal host, was not the best host based on numbers of oocysts produced (Table 2). When oocysts were administered to *D. m. merriami*, 1.8 times more oocysts were produced than when an equivalent number was given to *D. p. mohavensis*. To the writer's knowledge, no other study has been done which is as extensive and as detailed as this one. The results are surprising in that one would predict that the normal host for a coccidium would probably provide conditions allowing maximal reproduction. That is obviously not the case with *E. mohavensis*, and one can only speculate on the conditions which allow better reproduction of a parasite in one subspecies, especially representing an abnormal host, than in another. When it is considered that coccidia often prefer such exotic cells as bovine embryonic tracheal cells when cultivated (see Chapter 6), our view of mechanisms of their requirements and preferences is not at all clear.

TABLE 2. Oocyst production of *E. mohavensis* within the genus *Dipodomys* (Doran, 1953)

Host	Oocysts produced/ Oocyst fed ($\times 10^3$)
D. merriami merriami	125.7
D. panamintinus panamintinus	89.5
D. p. mohavensis[a]	70.5
D. p. leucogenys	64.1
D. p. caudatus	62.8
D. agilis agilis	35.1
D. heermanni tularensis	22.0
D. h. morroensis	8.9
D. h. swarthi	7.0
D. nitratoides brevinasus	2.9
D. deserti deserti	1.1

[a]The normal host for *E. mohavensis*

F. COMPLETION OF THE LIFE CYCLE IN CHEMICALLY TREATED HOSTS

It is known that administration of corticosteroid drugs to a recipient of an infectious agent will generally make it more susceptible to infection. There are two reports of the use of a corticosteroid in increasing the susceptibility of abnormal hosts to infection with coccidia. McLoughlin (1969) found that chicks which received daily intramuscular injections of dexamethasone became infected with and shed oocysts of the turkey coccidium, *E. meleagrimitis*. Those which received dexamethasone in the feed did not become infected as measured by the production of oocysts. In the obverse experiment, turkey poults were medicated with dexamethasone and then inoculated with oocysts of *E. tenella*, but none of them became infected as measured by oocyst production. Some of the questions raised by McLoughlin's work may be answered by studies currently in progress by Todd *et al.* (1971), who have been able to transmit *Eimeria vermiformis* from mice to rats by submitting rats to whole body irradiation and/or administering dexamethasone to them. The complete asexual development takes place in rats but gametogony rarely occurs. A single negative report (Pellérdy and Durr, 1969) indicated that steroid-treated rats and mice were not susceptible to infection with *E. stiedai*. On the basis of these fragmentary reports, it might be predicted that administration of corticosteroids to potential hosts may broaden the host range somewhat, but will not break down the specificity altogether. Completion of the life cycle of a coccidium in an abnormal host treated either with the corticosteroid or by irradiation would appear to result from the suppression of the immune response of the host. Perhaps we can begin to see some of the mechanisms operating in host specificity through the work of Rose and Long (1969) who studied the effect of the corticosteroid betamethasone on serum leakage in chickens. Using pontamine sky blue administered intravenously, they found that upon being exposed to eimerian sporozoites birds treated with betamethasone had a strikingly lower leakage of dye, and presumably serum, into the tissue spaces than untreated birds. The leakage appears to be a response to invasion in birds not previously exposed to sporozoites. If such a response were to appear in animals exposed to oocysts derived from another species of host, the sporozoites would be exposed to a large quantity of "hostile" serum. Although sporozoites of *E. tenella* were lysed in serum from abnormal hosts, it is not clear how the serum would act on the parasites once inside cells. Also it should be remembered that in an abnormal host, sporozoites may invade the tissue and undergo early development, but disappear by 48 hours after infection (Marquardt, 1966). In an inexperienced host, it is generally considered that a longer time than 36 hours is required for the development of a significant immune response (see Chapter 8).

III. Site Specificity

A. Normal Life Cycle

A typical coccidian life cycle involves the excystation of sporozoites in the intestinal tract of the host, a series of asexual generations and a single sexual generation. Certain generalizations can be made about those coccidia which occur in the intestinal tract of their hosts since we have more information on them than on those occurring in other sites or organ systems. Each of the generations will have a typical location along the length of the intestinal tract, usually progressing more posteriorly with successive generations. When looked at more specifically, it will be found that specific types of cells are utilized, and the cell type may change from generation to generation.

Eimeria bovis, a pathogenic coccidium of cattle, provides a good, straightforward example of location and cell type preferences typical for the group. The sporozoites of *E. bovis* excyst in the small intestine of the ox. The sporozoites enter villi of the posterior half of the small intestine, and invade endothelial cells of the lacteals. At the completion of the first schizogonous generation, the merozoites invade epithelial cells in the crypts of the large intestine. At the completion of schizogony, the merozoites invade epithelial cells in the large intestine, both cecum and colon. Most often, the gametocytes are found in the epithelial cells of the intestinal crypts.

B. DISTRIBUTION OF PARASITES ALONG THE INTESTINE

We consider here aspects of site specificity from the standpoints of normal locations and the ways in which specificity may break down under either normal or experimental conditions. The lengthwise distribution of the coccidia in poultry is sufficiently distinct so the lesions at postmortem will allow one to reach a diagnosis of the species involved with a relatively high degree of confidence (Fig. 1). The tendency in the coccidian life cycle is for the successive stages to occur progressively farther down the intestinal tract. *E. debliecki* of swine occurs in the small intestine of the host, and the life cycle consists of two schizogonous generations followed by gametogony. The first generation schizonts are found between the first and sixth feet of the small intestine (Fig. 2), second generation between the first and thirteenth feet and the gametocytes between the first and twenty-first feet of the small intestine (Vetterling, 1966). This is one of the few studies quantifying the occurrence of the various stages along the length of the intestine of the host.

The results of parenteral inoculations indicate that site selection is rather strongly determined. Landers (1960) injected sporulated oocysts of *E. nieschulzi* into the blood stream, skeletal muscle, or the peritoneal cavity of rats and found that the infection would establish in the small intestine of the host in the same location expected if the animals had been exposed by mouth. Additional studies (Sharma and Reid, 1962; Davies and Joyner, 1962; Sharma, 1964) confirmed that infections could be established in poultry by the parenteral administration of sporulated oocysts. In all instances, it was found that the number of developmental stages was considerably reduced when compared to infections established through administration of oocysts by mouth. The mechanism by which the parasites reach their normal site of development has been a matter for conjecture, but preliminary observations have indicated that the organisms are trapped in the liver, make their way to the biliary passages, and finally reach the small intestine via the common bile duct (Haberkorn, 1971).

It is a universal observation that heavy infection by any species of coccidium results in spreading of the organisms over a greater length of intestine. Although there must be some element of population pressure causing the colonization of additional sites, it seems likely that the probability of finding stages is proportionately increased in heavy infections.

Despite the remarkable manner in which the sporozoites seem to find the proper location in the intestine of the host, it has been

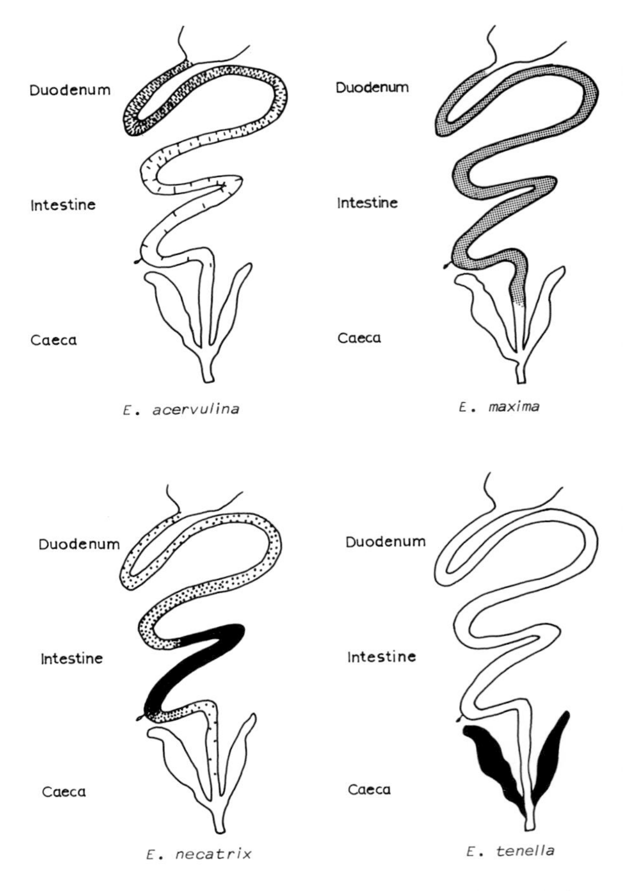

Fig. 1. Sites of infection with four species of *Eimeria* in the fowl. (From Davies, Joyner, and Kendall, 1963.)

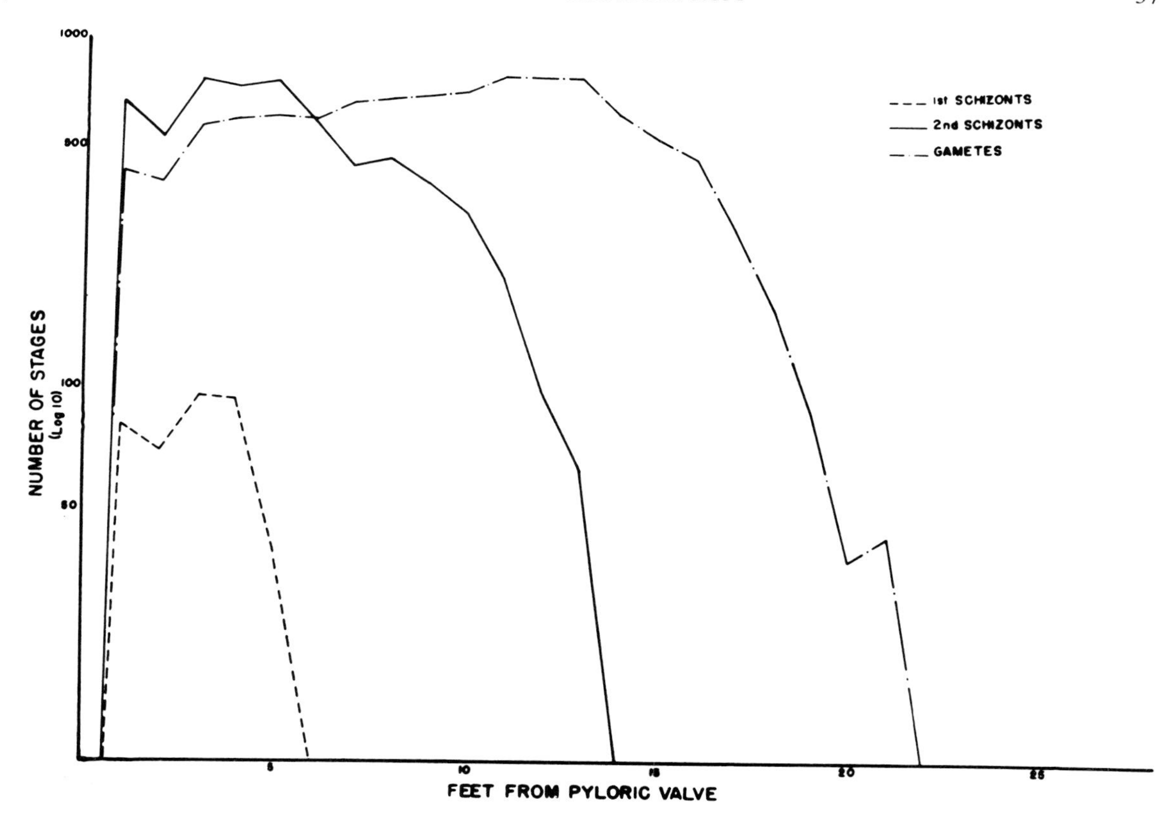

Fig. 2. Distribution of the endogenous stages of *Eimeria debliecki* in the small intestine of three-month-old pigs. (From Vetterling, 1966.)

possible experimentally to alter the location where endogenous stages will undergo development. Certain species of *Eimeria* of the chicken can be induced to develop in the ceca of the chicken if sporozoites are introduced into it. It was found that *E. mivati, E. brunetti,* and *E. necatrix* undergo complete development in the ceca of birds, but *E. acervulina, E. praecox,* and *E. maxima* would not (Table 3). As a matter of interest, *E. praecox*, although introduced into the ceca, was found to develop in the normal site, the intestine (Horton-Smith, 1966; Horton-Smith and Long, 1965, 1966; Long, 1967). *E. tenella* is normally found in the cecum of chickens, but it has been shown that the organism will develop in chickens which have been cecectomized (Leathem, 1969). All stages of the life cycle were found in both small and large intestines of operated chickens, but the number which developed was considerably smaller than in normal birds and the pathogenicity was low. *E. tenella* has also been found to develop normally in the bursa of Fabricius, probably gaining entrance to the bursa through the duct which connects it to the large intestine. Although most of the developmental stages were found in epithelial cells, some were also found in the blood vessels as well (Asdrubali *et al.*, 1967).

It is obvious that the development of a species of coccidium in a particular location provides an adaptive advantage. Under normal conditions, the number of endogenous stages developing at any one time would be relatively small, and the probability of macro-

TABLE 3. Results of injection of seven *Eimeria* species into either the ceca or the crop of the chick or into the embryo allantois (Horton-Smith and Long, 1966; Long, 1967b)

Species	Development of sporozoites in ceca	Development of sporozoites in chorioallantoic membrane	Normal site of development in natural infections
E. acervulina	−	−	Small intestine
E. brunetti	+	+	Small intestine and ceca
E. maxima	−	−	Small intestine
E. mivati	+	+	Small intestine and ceca
E. necatrix	+	(+)	Schizogony in small intestine, gametogony in ceca
E. praecox	−	−	Small intestine
E. tenella	+	+	Restricted to ceca

+ = complete life cycle; − = no development; (+) = development to schizogony only.

gametes being fertilized would be increased if development were restricted to a particular location. Although site specificity can be altered, often resulting in less multiplication of the parasite than in the normal site, we know nothing about the mechanisms in these phenomena.

C. DEPTH IN MUCOSA

Another dimension to site specificity is the location in the tissues with respect to depth. In *E. tenella*, the sporozoites enter the tips of the villi and finally locate in the epithelial cells of the gland fundi. In contrast, *E. debliecki*, which is found in the small intestine of swine, is always located in epithelial cells in the distal portions of the villi (Vetterling, 1966). *E. callospermophili*, of ground squirrels, represents organisms which may develop at any level in the villus, from the cells at the tip clear down into the fundus of the gland (Todd and Hammond, 1968). It would seem to be a disadvantage to the parasite to locate in the distal portions of the villi since the cells are shed off of the tips of the villi at a rapid rate.

D. CELL TYPE

The next aspect of site specificity has to do with the type of cell which the coccidia invade and develop in. As indicated earlier, *E. bovis* has its first generation schizont in endothelial cells, its second generation in epithelial cells and its sexual generation also in epithelial cells. The greatest majority of species of coccidia develop in epithelial cells whether they are located in the intestine or other organs. Only rarely do the stages occur in cells other than epithelial. An interesting exception is *E. auburnensis*, which has its gametocytes in cells of mesodermal origin. The gametocytes develop in the lamina propria of the villi, and it is considered to be possible that the host cells are members of the lymphocyte series (Hammond *et al.*, 1961).

In several instances, as in *E. maxima*, the primary location of the host cell is altered. *E. maxima* has all its endogenous stages in epithelial cells, but during the second schizogenous generation, the epithelial cells are caused to sink into a subepithelial position.

The type of cell which a coccidium prefers at any stage of the life cycle is constant. Were it not for the recent work on the cultivation of coccidia in tissue culture, one would be forced to conclude that there is such an intimate association between the host cell and the parasite that if the parasite were to find itself in another type of cell, it would not develop. The array of cell types which has served as good hosts of coccidia in tissue culture (Chapter 9) destroys certain of our conceptions of the degree of specificity which the coccidia show, but it should be kept in mind that the specificity is maintained under normal conditions. The striking exception to cell type specificity is *Toxoplasma,* which may be found in nearly any cell type in either natural or experimental infections (see Chapter 9). However, such exceptions also occur in *Eimeria* species, as demonstrated by the finding of Long (1971) and Lee and Long (1972) that *E. tenella* will develop in parenchyma and other cells in the liver of chick embryos.

E. LOCATION WITHIN THE CELL

The ultimate in location specificity is seen in the exact location of the coccidium within the host cell. When an author describes the location of an endogenous stage in an epithelial cell, he will typically indicate whether a parasite is located in the proximal or distal portion of the cell, i.e., whether it is below or above the nucleus. Within any one species, the location varies to some degree, and probably is influenced by the route of entry of the parasite. For example, the first generation schizonts of *E. tenella* are spoken of as lying below the nucleus of the host cell. This organism invades the host by the sporozoites entering the tips of the villi and reaching the cells in the fundi of the glands from the lacteal or the basal portion of the cell.

A few coccidia do have a particular location within the host cell, notably those which develop in the host cell nucleus. There are about eight species of coccidia which are known to develop within the nucleus of the host cell, and the hosts are invertebrates as well as lower and higher invertebrates (Davis *et al.*, 1957).

F. ABNORMAL SITES

Until relatively recently, we had been secure in our conception of the high degree of location specificity of the endogenous stages of the coccidia. There were a few reports in which organisms were found in sites which were not normal for the species in question. In studies of the cross transmission of sheep and goat coccidia, it was found that schizogonous stages were sometimes found in the mesenteric lymph nodes (Lotze *et al.*, 1964). Cable and Conway (1953) found gametocytes of a coccidium in the mammary glands of a shrew. These have been considered to be of little importance, but Lotze *et al.* (1964) pointed out correctly that endogenous stages of the life cycle of coccidia might well be disseminated widely throughout the body by entrance into the lymphatic system. They state, "Our findings, together with those of others, raise a question as to whether coccidial parasites commonly occur in the areas of host's body beside their well-known sites of development and, if so, how they get to these locations and what damage is caused."

The recent findings on the life cycle and identity of *Toxoplasma gondii* as well as other members of the genus *Isospora* gives a partial answer to their query. Members of the genus *Isospora* which have been considered to be limited to cells of the intestine, have been found by appropriate experimental manipulation to migrate to organs some distance from the gut, such as liver, lung, and brain (see Chapter 3). The work of Box (1967, 1970) has demonstrated that the organism referred to as *Lankesterella* or *Atoxoplasma*, occurring in the white blood cells of passerine birds, is in actuality a stage in the life cycle of *Isospora lacazei*. Her early studies showed that sparrows, *Passer domesticus domesticus* died from *Atoxoplasma* (*Lankesterella*) infections when they were held under conditions allowing ingestion of large numbers of oocysts of *I. lacazei*. In studies with canaries, *Serinus canarius*, it was demonstrated that transmission of *Atoxoplasma* was not likely to take place by way of mites as had been reported earlier (Lainson, 1959, 1960). The pattern which has emerged (Box, 1970) shows that after ingestion of *I. lacazei* oocysts by canaries, schizogony occurred in the epithelial cells of the intestine, and then at about five days, organisms could be found in liver, spleen, and lung. Schizogony continued in extraintestinal cells, primarily monocytes, but gametogony occurred only in epithelial cells of the intestine. It is theorized that infected monocytes return to the gut perhaps by reaching the lung alveoli and ultimately being swallowed. Infection of gut epithelium would take place from merozoites released from the monocytes. It is interesting to note that schizogony of *Toxoplasma* also occurs in a variety of sites, but gametogony has been found only in epithelial cells of the intestine of the cat. Perhaps more exacting conditions are required for gametocytes than schizonts. This was found to be true for *E. tenella* by Long (1971) and Lee and Long (1972), who reported that gamonts could be detected only in hepatocytes in the liver of chick embryos, whereas schizonts developed in the fibrocytes of the portal tract and in the endothelial cells lining the sinusoids. The liver represents an abnormal site for *E. tenella*; Long (1970) observed schizonts in this organ in chickens treated with dexamethasone, but the complete life cycle was found by Long (1971) and Lee and Long (1972) to occur in the liver of chicken embryos. The development of *Eimeria* species in the chorioallantoic membrane of chicken embryos, as discussed in Chapter 6, represents another abnormal site of development.

Although it has been known for many years that the sporozoites of *E. stiedai* reach the liver of the rabbit by migrating through the hepatic portal system (Smetana, 1933; Horton, 1967), little is known about the migration of other extraintestinal coccidia such as the renal coccidium of the woodcock (Locke *et al.*, 1965), or *E. neitzi* which causes uterine coccidiosis in the impala (McCully *et al.*, 1970).

Literature Cited

Andrews, J. M. 1930. Excystation of coccidial oocysts *in vivo*. Science 71:37.

Anwar, M. 1966. *Isospora lacazei* (Labbé, 1893) and *I. chloridis* sp. n. (Protozoa: Eimeriidae) from the English sparrow (*Passer domesticus*), greenfinch (*Chloris chloris*) and chaffinch (*Fringilla coelebs*). J. Protozool. 13:84-90.

Asdrubali, G., E. Vasconez Lopez, and L. Mughetti. 1967. Sulla presenza di *Eimeria tenella* nella borsa di Fabrizio. Vet. Ital. 18:391-398.

Bovee, E. C. 1971. New species of *Eimeria* from lizards of Japan. Trans. Am. Microsc. Soc. 90:336-343.

Bovee, E. C. and S. R. Telford, Jr. 1965. *Eimeria sceloporis* and *Eimeria molochis* spp. n. from lizards. J. Parasitol. 51:81-94.

Box, E. D. 1967. Influence of *Isospora* infections on patency of avian *Lankesterella* (*Atoxoplasma*, Garnham, 1950). J. Parasitol. 53:1140-1147.

Box, E. D. 1970. *Atoxoplasma* associated with an isosporan oocyst in canaries. J. Protozool. 17:391-396.

Bunch, T. D. and P. A. Nyberg. 1970. Effects of carbon dioxide on coccidian oocysts from 6 host species. J. Protozool. 17:364-370.

Cable, R. M. and C. H. Conway. 1953. Coccidiosis of mammary tissue in the water shrew, *Sorex palustris navigator*. J. Parasitol. 39(4, Sect. 2):30.

Cannon, L. R. G. 1967a. *Caryospora demansiae* sp. nov. (Sporozoa: Eimeriidae) from the Australian snake *Demansia psammophis* (Elapidae). Parasitology 57:221-226.

Cannon, L. R. G. 1967b. New coccidia from Australian lizards I. Isospora. Parasitology 57:227-236.

Cannon, L. R. G., 1967c. New coccidia from Australian lizards. II. *Eimeria*. Parasitology 57:237-250.

Christensen, J. F. and D. A. Porter. 1939. A new species of coccidium from cattle, with observations on its life history. Proc. Helm. Soc. Wash. 6:45-48.

Davies, S. F. M., L. P. Joyner, and S. B. Kendall. 1963. Coccidiosis. Oliver and Boyd Ltd. Edinburgh and London.

Davis, L. R., G. W. Bowman, and D. C. Boughton. 1957. The endogenous development of *Eimeria alabamensis* Christensen, 1941, an intranuclear coccidium of cattle. J. Protozool. 4:219-225.

de Vos, A. J. 1970. Studies on the host range of *Eimeria chinchillae* De Vos & Van der Westhuizen, 1968. Onderst. J. Vet. Res. 37:29-36.

de Vos, A. J. and I. B. Van der Westhuizen. 1968. The occurrence of *Eimeria chinchillae* n. sp. (Eimeriidae) in *Chinchilla laniger* (Molina, 1782) in South Africa. J. S. Afr. Vet. Med. Assoc. 39:81-82.

Doran, D. J. 1953. Coccidiosis in the kangaroo rats of California. Univ. Calif. Publ. Zool. 59:31-60.

Doran, D. J. and M. M. Farr. 1962. Excystation of the poultry coccidium, *Eimeria acervulina*. J. Protozool. 9:154-161.

Doran, D. J. and M. M. Farr. 1965. Susceptibility of 1- and 3-day-old chicks to infection with the coccidium, *Eimeria acervulina*. J. Protozool. 12:160-166.

Dubey, J. P. and J. K. Frenkel. 1972. Extra-intestinal stages of *Isospora felis* and *I. rivolta* (Protozoa: Eimeriidae) in cats. J. Protozool. 19:89-92.

Duszynski, D. W. 1969. Two new coccidia (Protozoa: Eimeriidae) from Costa Rican lizards with a review of the *Eimeria* of lizards. J. Protozool. 16:581-585.

Duszynski, D. W. 1971. Increase in size of *E. separata* oocysts during patent period. J. Parasitol. 57:948-952.

Duszynski, D. W. and W. C. Marquardt. 1969. *Eimeria* (Protozoa:Eimeriidae) of the cottontail rabbit *Sylvilagus audubonii* in northeastern Colorado with descriptions of three new species. J. Protozool. 16:128-137.

Fairbairn, D. 1970. Biochemical adaptation and loss of genetic capacity in helminth parasites. Biol. Rev. 45:29-72.

Farr, M. M. 1953. Three new species of coccidia from the Canada goose, *Branta canadensis* (L.) J. Wash. Acad. Sci. 43:336-340.

Farr, M. M. and D. J. Doran. 1962. Comparative excystation of four species of poultry coccidia. J. Protozool. 9:403-407.

Frenkel, J. K. and J. P. Dubey. 1972. Rodents as transport hosts for cat coccidia, *Isospora felis* and *I. rivolta*. J. Infect. Dis. 125:69-72.

Haberkorn, A. 1970. Zur Empfänglichkeit nicht spezifischer Wirte für Schizogonie-Stadien verschiedener *Eimeria*-Arten. Z. Parasitenk. 35:156-161.

Haberkorn, A. 1971. The problem of host specificity and variability in the pathogenic behavior of coccidia. Vet. Med. Rev. 1971 (2/3):341-348.

Hammond, D. M., W. N. Clark, and M. L. Miner. 1961. Endogenous phase of the life cycle of *Eimeria auburnensis* in calves. J. Parasitol. 47:591-596.

Hanson, H. C., N. D. Levine, and V. Ivens. 1957. Coccidia (Protozoa: Eimeriidae) of North American wild geese and swans. Can. J. Zool. 35:715-733.

Horton, R. J. 1967. The route of migration of *Eimeria stiedae* (Lindemann, 1865) sporozoites between the duodenum and bile ducts of the rabbit. Parasitol. 57:9-17.

Horton-Smith, C. 1966. On the site selection of coccidia in the fowl and the relationship of *Eimeria necatrix* and *Eimeria tenella*. *In* A. Corradetti, ed. Proc. 1st Int. Congr. Parasitol. pp. 278-279. Pergamon Press, London.

Horton-Smith, C. and P. L. Long. 1965. The development of *Eimeria necatrix* Johnson, 1930 and *Eimeria brunetti* Levine, 1942 in the caeca of the fowl *(Gallus domesticus)*. Parasitology 55:401-405.

Horton-Smith, C. and P. L. Long. 1966. The fate of sporozoites of *Eimeria acervulina, E. maxima* and *E. mivati* in the caeca of fowl. Parasitology 56:569-574.

Jackson, A. R. B. 1962. Excystation of *Eimeria arloingi* (Marotel, 1905): stimuli from the host sheep. Nature 194:847-849.

Kartchner, J. A. and E. R. Becker. 1930. Observations on *Eimeria citelli*, a new species of coccidium from the striped ground squirrel. J. Parasitol. 17:90-94.

Lainson, R. 1959. *Atoxoplasma* Garnham, 1950, as a synonym for *Lankesterella* Labbé, 1899. Its life cycle in the English sparrow (*Passer domesticus domesticus*, Linn.) J. Protozool. 6:360-371.

Lainson, R. 1960. The transmission of *Lankesterella* (*Atoxoplasma*) in birds by the mite *Dermanyssus gallinae*. J. Protozool. 7:321-322.

Landers, E. J. 1960. Studies on excystation of coccidial oocysts. J. Parasitol. 46:195-200.

Leathem, W. D. 1969. Tissue and organ specificity of *Eimeria tenella* (Railliet & Lucet, 1891) Fantham, 1909 in cecectomized chickens. J. Protozool. 16:223-226.

Lee, D. L. and P. L. Long. 1972. An electron microscopical study of *Eimeria tenella* grown in the liver of the chick embryo. Int. J. Parasitol. 2:55-58.

Levine, N. D. 1962. Protozoology today. J. Protozool. 9:1-6.

Levine, N. D. and V. Ivens. 1965. The coccidian parasites (Protozoa, Sporozoa) of rodents. Ill. Biol. Monogr. No. 33.

Levine, N. D. and V. Ivens. 1970. The coccidian parasites (Protozoa, Sporozoa) of ruminants. Ill. Biol. Monogr. No. 44.

Locke, L. N., W. H. Stickel, and S. A. Geis. 1965. Some diseases and parasites of captive woodcocks. J. Wildl. Mgmt. 29:156-157.

Long, P. L. 1967. Studies on *Eimeria praecox* Johnson, 1930 in the chicken. Parasitology 57:351-361.

Long, P. L. 1970. Development (schizogony) of *Eimeria tenella* in the liver of chickens treated with corticosteroid. Nature 225:290-291.

Long, P. L. 1971. Schizogony and gametogony of *Eimeria tenella* in the liver of chick embryos. J. Protozool. 18:17-20.

Long, P. L. and C. Horton-Smith. 1968. Coccidia and coccidiosis in the domestic fowl. Adv. Parasitol. 6:313-325.

Lotze, J. C.; R. C. Leek, W. T. Shalkop, and R. Behin. 1961. Coccidial parasites in the "wrong host" animal. J. Parasitol. 47 (Suppl.): 34.

Lotze, J. C., W. T. Shalkop, R. G. Leek, and R. Behin. 1964. Coccidial schizonts in mesenteric lymph nodes of sheep and goats. J. Parasitol. 50:205-208.

Lotze, J. C. and R. G. Leek. 1968. Excystation of the sporozoites of *Eimeria tenella* in apparently unbroken oocysts in the chicken. J. Protozool. 15:693-697.

McCully, R. M., P. A. Basson, V. De Vos, and A. J. De Vos. 1970. Uterine coccidiosis of the impala caused by *Eimeria neitzi* spec. nov. Onderstepoort J. Vet Res. 37:45-58.

McLoughlin, D. K. 1969. The influence of dexamethasone on attempts to transmit *Eimeria meleagrimitis* to chickens and *E. tenella* to turkeys. J. Protozool. 16:145-148.

Mahrt, J. L. 1967. Endogenous stages of the life cycle of *Isospora rivolta* in the dog. J. Protozool. 14:754-759.

Marquardt, W. C. 1962. Subclinical infections with coccidia in cattle and their transmission to susceptible calves. J. Parasitol. 48:270-275.

Marquardt, W. C. 1966. Attempted transmission of the rat coccidium *Eimeria nieschulzi* to mice. J. Parasitol. 52:691-694.

Moore, E. N. and J. A. Brown. 1952. A new coccidium of turkeys, *Eimeria innocua* (Protozoa, Eimeriidae). Cornell Vet. 42:395-402.

Nyberg, P. A. and D. M. Hammond. 1964. Excystation of *Eimeria bovis* and other species of bovine coccidia. J. Protozool. 11:474-480.

Nyberg, P. A., D. H. Bauer and S. E. Knapp. 1968. Carbon dioxide as the initial stimulus for excystation of *Eimeria tenella* oocysts. J. Protozool. 15:144-148.

Pellérdy, L. 1963. Catalogue of Eimeriidea (Protozoa; Sporozoa). Akad. Kiadó, Budapest.

Pellérdy, L. 1965. Coccidia and coccidiosis. Akad. Kiado, Budapest.

Pellérdy, L. 1969a. Attempts to alter the host specificity of eimeriae by parenteral infection experiments. Acta Vet. (Brno) 38:43-46.

Pellérdy, L. 1969b. Catalogue. Eimeriidea (Protozoa, Sporozoa). Supplementum I. Akad. Kiado, Budapest.

Pellérdy, L. and U. Dürr. 1969. Orale und parenterale Übertragungsversuche von Kokzidien auf nicht spezifische Wirte. Acta Vet. Acad. Sci. Hungar. 19:253-268.

Rose, M. E. and P. L. Long. 1969. Immunity to coccidiosis: gut permeability changes in response to sporozoite invasion. Experientia 25:183-184.

Roudabush, R. L. 1935. Merozoite infection in coccidiosis. J. Parasitol. 21:453-454.

Senger, C. M., D. M. Hammond, J. L. Thorne, A. E. Johnson, and M. Wells. 1959. Resistance of calves to reinfection with *Eimeria bovis*. J. Protozool. 6:51-58.

Shah, H. L. 1970. *Isospora* species of the cat and attempted transmission of *I. felis* Wenyon, 1923 from the cat to the dog. J. Protozool. 17:603-609.

Sharma, N. N. and M. W. Reid. 1962. Successful infection of chickens after parenteral inoculation of oocysts of *Eimeria* spp. J. Parasitol. 48(Suppl.):33.

Sharma, N. N. 1964. Response of the fowl (*Gallus domesticus*) to parenteral administration of seven coccidial species. J. Parasitol. 50:509-517.

Smetana, H. 1933. Coccidiosis in the liver of rabbits. II. Experimental study on the mode of infection of the liver by sporozoites of *Eimeria stiedae*. Arch. Pathol. 15:330-339.

Supperer, R. and E. Kutzer. 1961. Die Kokzidien von Reh, Hirsch und Gemse. Jubiläums-Jahrbuch 1960-1961 des Ö. A. W. 128-136.

Todd, K. S. and D. M. Hammond. 1968. Life cycle and host specificity of *Eimeria callospermophili* Henry, 1932 from the Uinta ground squirrel *Spermophilus armatus*. J. Protozool. 15:1-8.

Todd, K. S., D. M. Hammond, and L. C. Anderson. 1968. Observations on the life cycle of *Eimeria bilamellata* Henry, 1932 in the Uinta ground squirrel *Spermophilus armatus*. J. Protozool. 15:732-740.

Todd, K. S., D. L. Lepp, and C. V. Trayser. 1971. Development of the asexual cycle of *Eimeria vermiformis* Ernst, Chobotar and Hammond. 1971, from the mouse, *Mus musculus* in dexamethasone-treated rats, *Rattus norvegicus*. J. Parasitol. 57:1137-1138.

Vetterling, J. M. 1966. Endogenous cycle of the swine coccidium *Eimeria debliecki* Douwes, 1921. J. Protozool. 13:290-300.

Vetterling, J. M. and Widmer, E. A. 1968. *Eimeria cascabeli* sp. n. (Eimeriidae, Sporozoa) from rattlesnakes, with a review of the species of *Eimeria* from snakes. J. Parasitol. 54:569-576.

3 Life Cycles and Development of Coccidia

DATUS M. HAMMOND
Department of Zoology, Utah State University, Logan, Utah

Contents

I. Life Cycles . . . *46*
- A. General Considerations . . . *46*
- B. *Eimeria bovis* . . . *46*
- C. Other Species . . . *51*
- D. *Isospora* Species of Dogs, Cats, and Man . . . *54*
  1. Dogs and Cats . . . *54*
  2. Man . . . *54*

II. Sporogony . . . *56*
- A. Unsporulated Oocysts . . . *56*
- B. Sporulation . . . *56*
- C. Sporulated Oocysts . . . *57*

III. Excystation . . . *57*
- A. Alteration of Oocyst Wall and/or Micropyle . . . *57*
- B. Alteration of Stieda Body and Substiedal Body . . . *58*
- C. Escape of Sporozoites from Sporocysts and Oocysts . . . *58*

IV. Penetration of Cells . . . *59*

V. Intracellular Sporozoites and Trophozoites . . . *60*

VI. Schizogony . . . *64*
- A. Nuclear Division . . . *64*
- B. Growth and Intake of Nutrients . . . *64*
- C. Formation of Merozoites . . . *66*
  1. Ectomerogony . . . *66*
  2. Endomerogony . . . *68*
     - a. Splitting . . . *68*
     - b. Internal Formation of Merozoites in *Eimeria* Species . . . *68*
     - c. Endodyogeny . . . *68*

d. Endopolygeny . . . 68
D. Mature Schizonts and Merozoites . . . 70
E. Termination of Schizogony . . . 71
VII. Gametogony . . . 71
VIII. Fertilization . . . 72
IX. Transmission . . . 73
Literature Cited . . . 74

I. Life Cycles

A. GENERAL CONSIDERATIONS

The life cycles of typical coccidia may be subdivided into three phases: sporogony, merogony, and gametogony. Sporogony usually occurs outside the host and involves the formation of sporozoites, usually eight in number. During the nuclear divisions which are associated with sporogony, meiosis occurs, so that all stages of the life cycle are haploid except for the zygote.

The sporozoites resulting from sporogony may initiate an infection if the oocyst in which they lie is ingested by an appropriate host. The sporozoites excyst from the oocyst in the digestive tract and invade the lining of the intestine. Usually, they enter epithelial cells and undergo merogony (schizogony). The elongate sporozoite usually transforms into a spheroidal trophozoite and then the first nuclear division occurs, after which the parasite is called an immature schizont (meront). The schizont reproduces asexually by multiple fission to form a variable number of merozoites. These escape from the host cell, invade new cells, and undergo one or several more cycles of merogony before the onset of gametogony.

Gametogony involves the formation of two sexually differentiated cells, the macrogamont and the microgamont. The former gives rise directly to a single macrogamete and the latter undergoes many nuclear divisions leading to the formation of numerous microgametes. The mature microgametes, which are flagellated and motile, fertilize the macrogametes. The resulting zygote forms an oocyst wall; it may then be called an oocyst. These are released into the intestinal lumen and then discharged with the feces, after which sporogony (sporulation) occurs.

The details of the life cycles of coccidia vary considerably. *Eimeria bovis*, a pathogenic coccidium of cattle, will be used as an example of a specific life cycle.

B. *EIMERIA BOVIS*

The sporulated oocyst is ingested with the feed or water and completes excystation in the posterior part of the small intestine (Fig. 1). There the sporozoites which have escaped from the oocyst invade the mucosa and enter endothelial cells lining the central lacteals of the villi (Hammond *et al.*, 1946). The intracellular sporozoite begins to grow and transforms into a trophozoite, which is still uninucleate but is spheroidal. As the trophozoite enlarges, the host cell hypertrophies and its nucleus undergoes alterations, becoming larger, with an enlarged nucleolus and with chromatin dispersed into finer particles than normal (Fig. 8); its cytoplasm becomes arranged into two concentric zones, with the outer frequently having radial striations or a vacuolated appearance (Fig. 9) (Hammond,

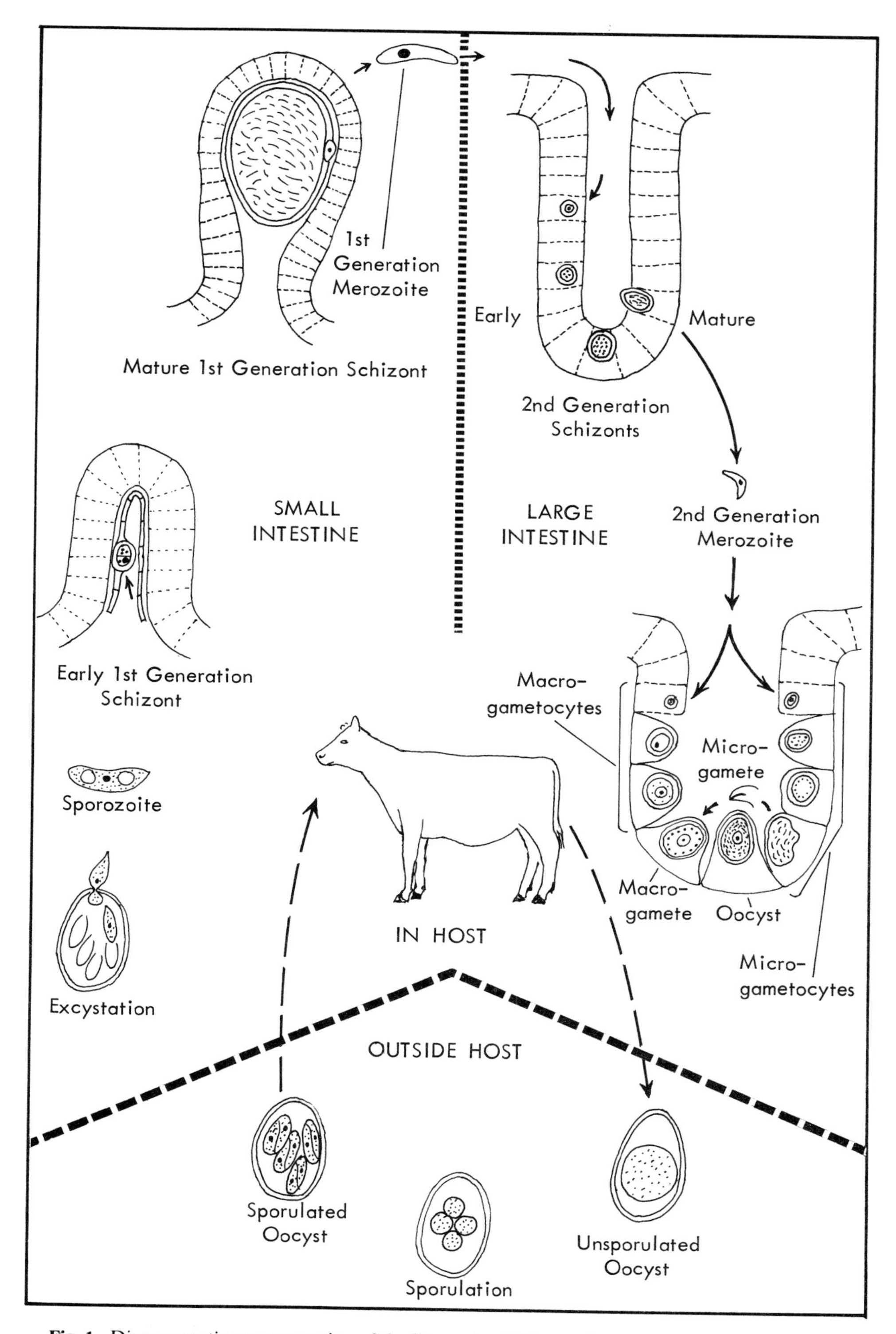

Fig. 1. Diagrammatic representation of the life cycle of *Eimeria bovis*. (From Hammond, 1964.)

Ernst, and Miner, 1966). At first, the nuclei have a random distribution, but beginning about eight days after ingestion of oocysts by the calf, the nuclei become arranged in a single peripheral layer, which becomes infolded (Fig. 10) to form compartments within the schizont (Fig. 11). These undergo subdivision, resulting in the formation of spheroidal or ellipsoidal structures termed blastophores, which have a single peripheral layer of nuclei. A merozoite begins forming at the surface over each nucleus and grows outward radially (Fig. 12). When mature, the merozoites (Fig. 13) become separated from the residual body, representing the remnant of the blastophore. Mature schizonts, which average about 300 μ by 200 μ, each have about 120,000 merozoites (Figs. 14, 15; 9a, Chapter 7). About 14 or 15 days are required for the schizonts to reach maturity. The merozoites escape into the lumen of the small intestine and are carried to the large intestine, where they enter crypts of Lieberkühn. There they penetrate epithelial cells lining the crypt and develop into second-generation schizonts (Figs. 16, 17) (Hammond, Andersen, and Miner, 1963). These require only 1.5 to 2 days to reach maturity, at which time they average about 10μ in diameter and have from 30 to 36 merozoites (Fig. 20). Each infective oocyst ingested by a calf may yield as many as 24 million second-generation merozoites (8 × 100,000 × 30), so that 1,000 such oocysts could result in the destruction of about 24 billion intestinal cells (Hammond, 1964).

The second-generation merozoites escape from the host cell, penetrate adjacent epithelial cells and develop into gamonts (gametocytes) (Fig. 20). At first, these cannot be differentiated into macrogamonts and microgamonts, but the latter soon undergo nuclear division (Fig. 19), whereas the former do not (Fig. 18). Gamonts quickly cause alteration of their host cells, which lose their columnar arrangement and become distorted in shape (Fig. 9b, Chapter 7) (Hammond, Andersen, and Miner, 1963). The pathological changes and clinical signs associated with *E. bovis* infection are caused chiefly by the gamonts. At first, the nuclei of the microgamonts are randomly distributed, but they later assume a peripheral arrangement (Fig.

Figs. 2-34. Photomicrographs of life-cycle stages of various *Eimeria* and *Isospora* species. Abbreviations: AR, anterior refractile body; C, conoid; CE, centriole; CO, compartment formed by infolding of peripheral layer of nuclei; CR, crescent body; DM, developing merozoite; E, nucleus of cell in envelope around host cell; G, Golgi complex; GA, Golgi adjunct; GM, developing gamonts; HC, host cell; HN, host cell nucleus; HNU, host cell nucleolus; IM, inner membrane complex of merozoite anlage; IN, infolding of peripheral layer of nuclei; IZ, inner zone of host cell cytoplasm; M, merozoite; MA macrogamont; MI, microgamont; MP, micropyle; N, nucleus of parasite; NE, nuclear envelope; Nu, nucleolus of parasite, OR, oocyst residuum; OZ, outer zone of host cell cytoplasm; PG, plastic granule in cytoplasm; PR, posterior refractile body; PV, parasitophorous vacuole; RA, rhoptry anlage; RB, residual body; SF, intranuclear spindle fibrils; SMI, microtubules of intranuclear spindle; SP, cone-shaped structure at pole of spindle; SR, sporocyst residuum; ST, Stieda body; YO, young oocyst.

Fig. 2. Living, unsporulated oocyst of *E. magna*, Nomarski interference contrast. × 2000. (Photographed by C. A. Speer.)

Fig. 3. Living, sporulated oocyst of *E. magna*, Nomarski interference contrast. × 2000. (Photographed by C. A. Speer.)

Figs. 4-7. Excystation in sporocysts of *E. utahensis*, fresh preparations; arrow indicates location of substiedal body. × 1300. (From Hammond, Ernst, and Chobotar, 1970.) **4.** Before addition of excysting fluid. **5.** After addition of excysting fluid; Stieda body swollen and protruded, substiedal body in different position than in Fig. 4. **6.** Stieda body no longer present; substiedal body in early stage of protrusion through area formerly occupied by Stieda body. **7.** Sporozoite moving through opening at end of sporocyst; substiedal body still visible adjacent to the opening.

Figs. 8-10. *E. bovis* schizonts fixed in Zenker's and stained with iron hematoxylin. (Figs. 8 and 9 from Hammond, Ernst and Miner, 1966.) **8.** Young schizont in hypertrophied host cell having enlarged nucleus and nucleolus. × 1100. **9.** Intermediate schizont with randomly arranged nuclei; note hypertrophy of host cell cytoplasm and its arrangement in two concentric zones. × 600. **10.** Intermediate schizont with nuclei peripherally arranged except for areas in which infolding of the peripheral nuclear layer has occurred. × 600.

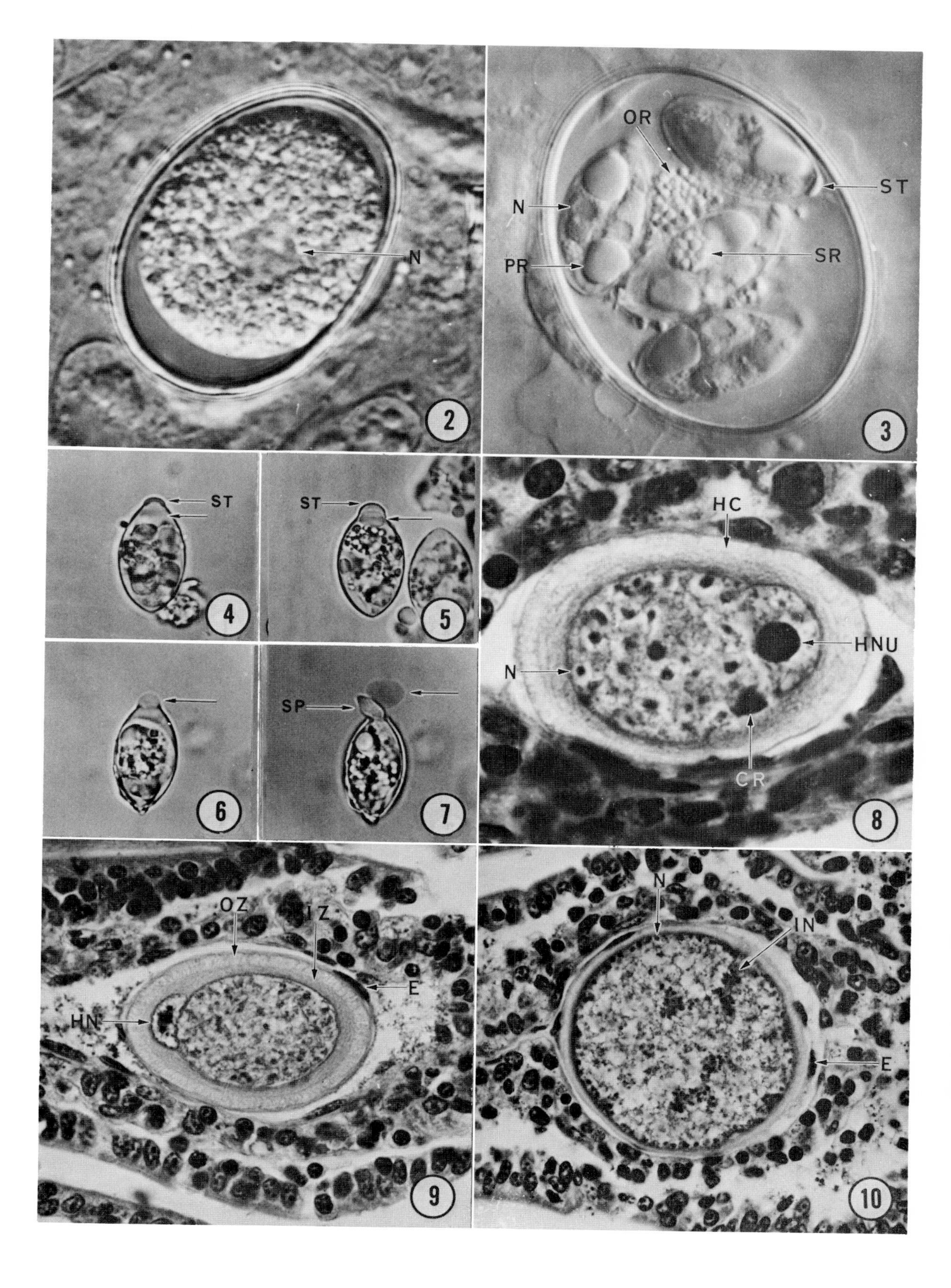

N
2
OR
ST
N
SR
PR
3
ST
4
ST
5
HC
HNU
N
CR
8
6
SP
7
OZ
IZ
E
HN
9
N
IN
E
10

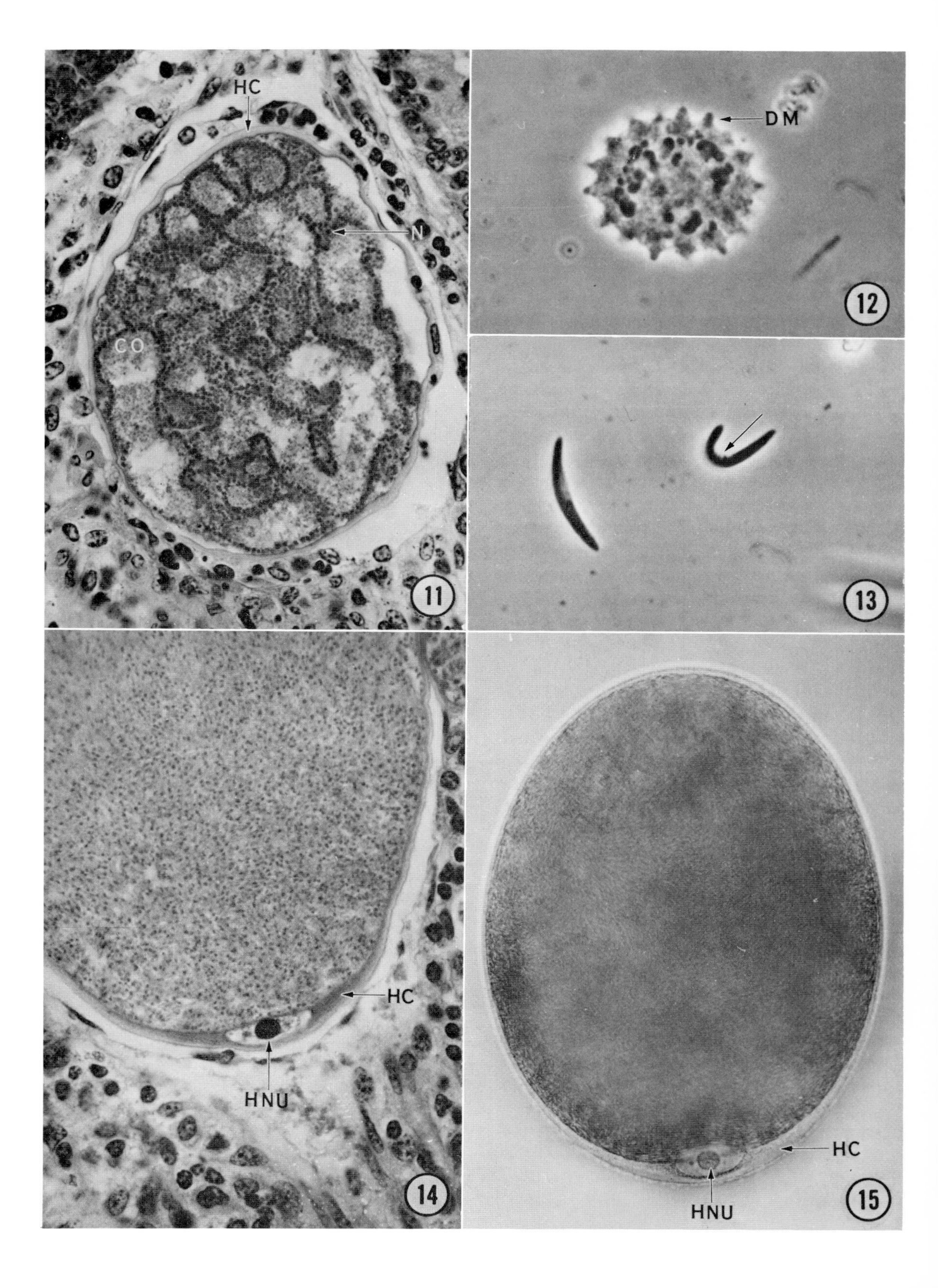
HC
N
CO
11
DM
12
13
HC
HNU
14
HC
HNU
15

21). Microgametes are formed at the surface of the microgamont (Fig. 22) and escape (Fig. 23) to fertilize the macrogametes. Macrogamonts have a relatively large nucleus and nucleolus (Fig. 18); during development, plastic granules (wall-forming bodies) appear in the cytoplasm (Fig. 21). These granules are prominent in the mature macrogamont (macrogamete) (Figs. 22, 24). After fertilization, the plastic granules give rise to the oocyst wall (Figs. 22, 25), which is formed around the zygote. The oocysts are discharged into the lumen of the large intestine and then leave the host with the feces, beginning about 18 days after the ingestion of oocysts by the host, with the peak discharge occurring two or three days later. Oocyst discharge in calves given a single inoculation usually continues for from five to seven days, but in calves given daily inoculations of from 10 to 15,000 oocysts the patent period was extended 7-15 days (Fitzgerald, 1967).

Under favorable conditions of temperature and moisture, the oocyst (Fig. 2) undergoes sporulation. This requires from two to three days at room temperature. The sporulated oocyst has four sporocysts each with two sporozoites (Fig. 3) and is resistant to unfavorable environmental conditions.

C. OTHER SPECIES

Only certain aspects of the life cycles of the large number of species of coccidia will be considered here. For information as to the life cycles of particular species in various hosts, the reader may refer to Becker (1934), Kheysin (1967), Davies, Joyner, and Kendall (1963), Levine (1961), and Pellérdy (1965); for species occurring in rodents, Levine and Ivens (1965a); for species occurring in ruminants, Levine and Ivens (1970); and for species occurring in wild birds, Todd and Hammond (1971).

As discussed in Chapter 2, most coccidia parasitize epithelial cells lining the intestinal tract, but some live in other kinds of cells and/or organs. Each lies within a parasitophorous vacuole in the cytoplasm of the host cell (Fig. 30). The epithelial cells harboring some stages in some species, such as the gamonts of *E. bilamellata* and *E. maxima,* are displaced into the lamina propria. Certain stages of other species invade and develop in cells of mesodermal origin. For example: the first-generation schizonts of *E. bovis* as described above, the first-generation schizonts of *E. ninakohlyakimovae* in reticular connective tissue of the lamina propria (Fig. 10, Chapter 7) (Wacha, Hammond, and Miner, 1971), the second-generation schizonts of *E. tenella* and *E. necatrix* in connective tissue cells between the intestinal glands (Lee and Long, 1972), the gamonts of *E. leuckarti* in lamina propria cells (Barker and Remmler, 1972), the second-generation schizonts and gamonts of *E. auburnensis* (Figs. 18, 19) in lamina propria cells (Chobotar and Hammond, 1969), the endogenous stages of *E. truncata* in the epithelial cells of the kidney tubules (Levine, 1961), and the endogenous stages of *E. neitzi* in the endometrium of the uterus in the impala (McCully *et al.*, 1970). The occurrence of a portion of the life cycle in cells other than those of the epithelial surface of the intestine may indicate that the species in question is in a more advanced evolutionary stage than those which live in cells at the surface.

The number of asexual generations in various species ranges from two to five. In most, the schizonts are comparable in size and in yield of merozoites with those of the second-

Figs. 11-15. *E. bovis* schizonts and merozoites. Figs. 11 and 14 from preparations fixed in Zenker's and stained with iron hematoxylin. Figs. 12, 13, and 15 from living specimens. (Figs. 12 and 14 from Hammond, Ernst and Miner, 1966) **11.** Schizont with compartments formed by infolding of the peripheral nuclear layer. × 600. **12.** Blastophore, with developing merozoites at surface, phase-contrast. × 1200. **13.** Merozoites, with surface wrinkles at concave surface of flexed specimen (arrow), phase-contrast. × 1200. **14.** Portion of mature schizont in host cell with flattened nucleus having enlarged nucleolus. × 600. **15.** Mature schizont within host cell with enlarged nucleus and nucleolus. × 300.

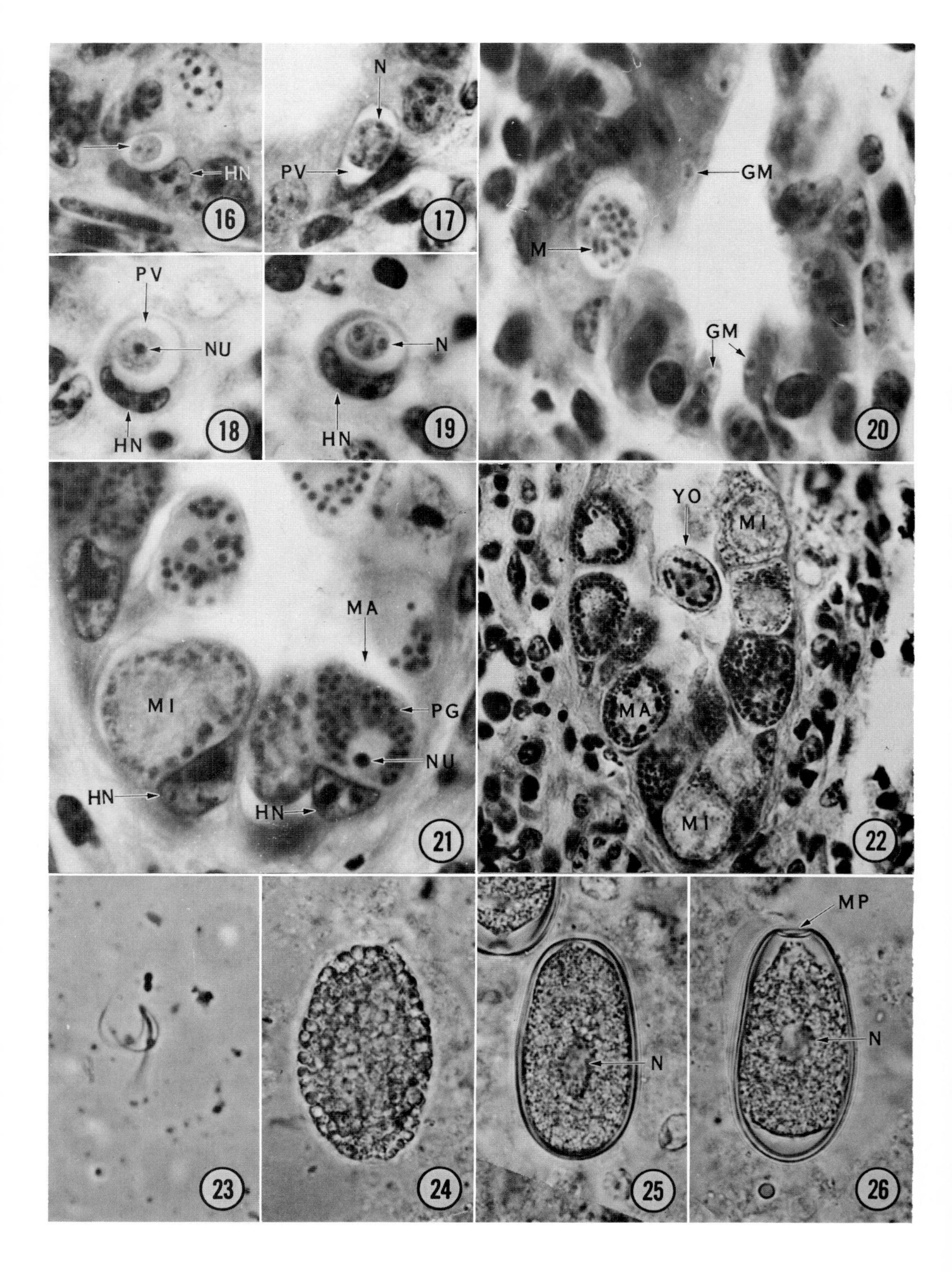
HN
16
N
PV
17
GM
M
GM
20
PV
NU
HN
18
N
HN
19
MA
MI
PG
NU
HN
HN
21
YO
MI
MA
MI
22
23
24
N
25
MP
N
26

generation of *E. bovis*. According to Pellérdy (1970), three schizogonic generations are usual in most of the better known *Eimeria* species parasitizing gallinaceous birds.

A number of *Eimeria* species besides *E. bovis* have unusually large schizonts. In one of these, *E. gilruthi*, life cycle stages other than the large schizont have not been found as yet. This schizont occurs in the abomasum of sheep and a similar or identical one is found in goats. It was at first called *Gastrocystis gilruthi,* but the name was later changed to *Globidium gilruthi* because of its similarity to *Globidium leuckarti,* found in the intestine of the horse. Reichenow and Carini (1937) proposed that *Globidium* be considered a subgenus of *Eimeria,* distinguished by the unusually large size of all of the stages and their occurrence in giant cells in the subepithelial tissue. Later, Reichenow (1953) stated that the demarcation between the subgenus *Globidium* and other *Eimeria* is by no means sharp. He included *E. bovis, E. leuckarti, E. gilruthi, E. cameli* and others in this subgenus. Bhatia and Pande (1966) reported that mature schizonts of *E. gilruthi* in sheep were 677μ-878μ by 584μ-830μ.

Chatton (1910) described the formation of merozoites in schizonts of *E. gilruthi;* it is similar to that of *E. bovis*. Triffitt (1925, 1928) observed that the early development of the schizonts is accompanied, as in *E. bovis*, by considerable hypertrophy and modification of the host cell. Later, Chatton (1938) coined the term "xénoparasitome" for this and similar host-cell and parasite complexes. He described this as consisting of an enormous host cell under the control of the parasite, with the complex of host cell and parasite behaving as an autonomous organism. Such a concept is applicable also to the large schizonts and host cells of *E. bovis* and other *Eimeria* species. According to Sénaud (1969), the cyst wall of *Besnoitia jellisoni* consists of a hypertrophied host cell which corresponds with such a "xénoparasitome," serving both to isolate the parasite from the host and to provide for exchange of materials between them. Sénaud (1969) suggested that the walls of *Toxoplasma* and *Sarcocystis* cysts have a similar origin.

Weissenberg (1922) had proposed the term "xenon" for host cell-parasite complexes in certain microsporidian infections of fishes, such as those caused by *Glugea,* in which the host cell becomes multinucleate and greatly hypertrophied. Weissenberg (1922b) referred to the complexes as gall-like formations, and extended the concept to include a variety of infections in plants and animals, including the coccidian *Caryotropha mesnili.* The term "xenon" was applied to *E. tenella* infections by Scholtyseck (1953) and Péllerdy (1970). Weissenberg (1949), however, later stated his preference for "xenom" or "xenoma" as a name for this complex, and has recently (Weissenberg, 1968) used the latter in a report concerning the intracellular development of *Glugea anomala*. The similarity of the symbiotic complex consisting of the parasite and hypertrophied host cell in microsporidian

Figs. 16-26. Endogenous stages of *Eimeria* species. Figs. 16, 17, 20 and 21 from preparations fixed in Helly's and stained with iron hematoxylin; Figs. 18, 19, and 22 from preparations fixed in Zenker's solution and stained with iron hematoxylin; Figs. 23-26 from fresh preparations. (Figs. 17 and 21 from Hammond, Andersen, and Miner, 1963; Fig. 19 from Hammond, Clark, and Miner, 1961.) **16.** Early second-generation schizont of *E. bovis* (arrow); note the two nuclei. × 1400. **17.** Intermediate second-generation schizont of *E. bovis*, with seven visible nuclei. × 1400. **18.** Early macrogamont of *E. auburnensis*; note relatively large nucleus and prominent nucleolus. × 1400. **19.** Early microgamont of *E. auburnensis*, with three nuclei. × 1400. **20.** Mature second-generation schizont of *E. bovis*. Some merozoites from other mature schizonts have entered cells and are developing into gamonts (GM). × 1400. **21.** Immature macrogamont and microgamont of *E. bovis*. × 1700. **22.** Nearly mature gamonts and young oocyst of *E. bovis*. × 600. **23.** Microgamete of *E. bovis*, phase-contrast. × 1500. **24.** Mature macrogamete of *E. auburnensis*, note prominent plastic granules in cytoplasm. × 1200. **25.** Young oocyst of *E. auburnensis*; note that all the space within the oocyst wall is filled by cytoplasm. × 1200. **26.** Slightly older oocyst of *E. auburnensis*, in which condensation of protoplasm has begun; note spaces at either end of oocyst. × 1200.

and coccidian infections is of considerable interest.

Host cells parasitized by several different *Eimeria* species undergo marked changes, including enlargement of the nucleus, a considerable increase in size of the nucleolus, and rearrangement of the chromatin into smaller aggregations than normal (Fig. 8). These changes are most pronounced in host cells harboring unusually large schizonts (Figs. 14, 15) (Hammond, 1971). The cytoplasm of the host cells harboring schizonts also hypertrophies in *E. bovis* (Hammond, Ernst, and Miner, 1966) and *E. ninakohlyakimovae* (Wacha, Hammond, and Miner, 1967), but this does not occur in *E. auburnensis* (Chobotar, Hammond, and Miner, 1969). This difference may be associated with the type of host cell parasitized; in the former two species, this is mesodermal in origin, whereas in the latter, it is epithelial. The host cells (in the lamina propria of villi) harboring gametocytes of *E. leuckarti* undergo hypertrophy of cytoplasm and nucleus, but do not have a single enlarged nucleolus (Barker and Remmler, 1972).

Thus, parasites of this kind are evidently able to cause modifications in the physiology and growth of the host cell, which presumably results in a more favorable environment for the parasite. Such modifications in the host cell may occur in a manner similar to that hypothesized by Read (1970a, 1970b) for muscle fibers invaded by *Trichinella spiralis*. These fibers undergo reorganization rather than degeneration. Soon after entering the fiber, the larva is stimulated by substances from the host cell to rapidly synthesize RNA, and the worm then furnishes new chemical information to the host cell, so that synthetic activities are directed toward establishment of a cellular "home" for the larva. Thus, the muscle cell becomes part of a highly differentiated host cell-parasite system which develops as a result of a series of reciprocal chemical interactions between the two partners. In *E. bovis* and other similar species, the enlargement of the nucleolus of sporozoites soon after entering host cells, and the ensuing enlargement of the host cell nucleolus, provide indirect evidence supporting the applicability of such a hypothesis to coccidia.

D. *ISOSPORA* SPECIES OF DOGS, CATS, AND MAN

1. *Dogs and cats*

According to Levine and Ivens (1965b), *Isospora canis*, *I. rivolta*, and *I. bigemina* occur in dogs. Traditionally, dogs and cats have been thought to have the same species of coccidia, but this idea has been questioned recently. Nemeséri (1960) was unable to transmit *I. felis* from cats to puppies and *I. canis* from dogs to kittens, but gave no information as to the source or history of the animals. Mahrt (1966) obtained negative results in attempts to transmit *I. rivolta* from the dog to kittens; he was uncertain as to the interpretation of the results because the history of the kittens was unknown. Shah (1970a) described the oocysts of *I. rivolta*, *I. bigemina*, and *I. felis* from cats; he could not transmit *I. felis* from cats to dogs, and concluded that it was host-specific. Dubey, Miller, and Frenkel (1970) suggested that the term *I. cati* Railliet and Lucet, 1891, be used for *I. bigemina*-like oocysts from cats and the term *I. bigemina* be restricted to the species in the dog.

Mahrt (1967) reported that the endogenous stages of *I. rivolta* occur in the posterior half of the small intestine in dogs, predominantly parasitizing subepithelial cells of the lamina propria in the distal third of the villi, although epithelial cells are also invaded. The evidence indicated that two asexual generations were present. The prepatent period was from 142 hr to 146 hr and the patent period averaged 19 days.

Shah (1971) found that the endogenous stages of *I. felis* occurred in cats in the epithelial cells of the distal parts of the villi in the ileum, and occasionally in the duodenum and jejunum. The prepatent period was from seven to eight days and the patent period 10 to 11 days. Three asexual generations were observed. The second-generation schizonts gave rise to from two to ten spindle-shaped second-generation merozoites, which were at first uninucleate. Without leaving the parasitophorous vacuole in which it had developed, each merozoite enlarged and became multinucleate, with as many as six nuclei. Within the spindle-shaped schizonts thus formed, third-generation merozoites were developed asynchronously. This phenomenon is unusual among coccidia, but Long (1970) reported that in many of the second-generation schizonts of *E. tenella* cultivated in the chorioallantois of chick embryos, the merozoites did not leave the parasitophorous vacuole and further merozoite formation occurred *in situ*. Also, Danforth and Hammond (1972) recently found evidence that a similar kind of development may occur in *E. magna* (Fig. 41).

Recently, certain *Isospora* species have been found to have extra-intestinal stages. Frenkel and Dubey (1972) found stages of *I. felis* and *I. rivolta* in the mesenteric lymph nodes of mice from 3 to 14 days after these had been fed oocysts of *I. felis* and *I. rivolta* from cats. Suspensions of lung, liver, and spleen of such mice were infective to kittens from 14 to at least 67 days after inoculation of the mice. Such tissues of rats and hamsters were infective to kittens ten days after they had been fed oocysts of these species. Dubey and Frenkel (1972a) reported that these coccidia also invade the liver, spleen, mesenteric lymph nodes, and other extra-intestinal tissues in cats and that such tissues are infectious to new-born kittens. An *Isospora* species was found by Box (1970) to multiply asexually in the liver, lungs, and spleen of canaries as well as in the intestine, and she suggested that some isosporan life cycles may be quite different from that of the typical coccidian cycle. However, Lee and Long (1972) have found that schizonts of *E. tenella*, grown in the liver of chick embryos given dexamethasone, develop within endothelial cells lining the sinusoids and fibrocytes (mesodermal in origin). Oocysts and gametocytes were found in hepatocytes (endodermal).

According to Nemeséri (1960), *I. canis* oocysts are first discharged on the 11th day after inoculation. Endogenous stages were observed in mucosal scrapings from the small and large intestine. In sections they were found in the villi, in a subepithelial location, as well as in the epithelium. Lepp (1972) found the endogenous stages of *I. canis* directly beneath the epithelium of the distal portion of the villus, predominantly in the lower third of the small intestine. Three asexual generations occurred; some of the second-generation merozoites gave rise to schizonts without leaving the parasitophorous vacuole in which they had developed, in a way similar to that in *I. felis* (Shah, 1971). The prepatent period was from 9 to 11 days.

The life cycle of *I. bigemina* has been studied only by Wenyon and Sheather (1925) and by Wenyon (1926) and then only sketchily. The former authors found that in one dog, the endogenous stages occurred only in the epithelium, especially in that covering the distal portion of villi, along the entire small intestine. Schizonts had eight small merozoites, and oocysts measured from 10 μ to 11 μ in length. In another dog, the stages occurred only in the subepithelial tissues and the oocysts were from 13.5 μ to 15.5 μ long. Wenyon and Sheather (1925) stated that in chronic infections of dogs and cats with *I. bigemina*, mature oocysts occur in the subepithelial tissues. However, Wenyon (1926) found stages, including schizonts with eight or occasionally 16 merozoites, only in the epithelium in a dog which was discharging oocysts, the majority of which were from 10 μ to 11 μ in length. Wenyon observed one infection in a cat in which fully sporulated oocysts of *I. bigemina* occurred in the subepithelial tissues. He concluded that *I. bigemina* occurs in the intestinal epithelium of cats and dogs, but is very liable to invade the subepithelial tissues of the villi, in which case the oocysts attain maturity there. The host specificity of this species is still open to question. Possibly, the form with large oocysts will be found to be a species different from the one with small oocysts.

2. *Man*

Of the three species of *Isospora* infecting man, *I. belli* (Fig. 27) and *I. hominis* are more common than *I. natalensis*, which is rare (Levine, 1961). According to Marcial-Rojas (1971), the oocysts of *I. hominis* are sporulated at the time they are discharged. Little is known about the life cycles of any of these species. Matsubayashi and Nozawa (1948) reported that oocysts first appeared on the 10th or 11th day after ingestion of oocysts, presumably of *I. belli,* by two human volunteers; the discharge of oocysts continued for from 32 to 38 days. Brandborg, Goldberg, and Breidenbach (1970) found that the endogenous stages of *I. belli* occurred in the epithelium of the small intestine, from the crypts to the mucosal surface. They were observed rarely in the lamina propria and submucosa, and occasionally were present within epithelial nuclei. In one case, coccidia were present in all levels of the small intestine examined, with decreasing numbers in the distal portion.

II. Sporogony

A. UNSPORULATED OOCYSTS

When first formed, oocysts have all or most of the space within the oocyst wall filled with the zygote (sporont) (Fig. 25). Later, this becomes condensed to form a spheroidal mass, separated by a fluid-filled space from the oocyst wall (Figs. 2, 26). At the time they are discharged, the oocysts may or may not have undergone contraction of the cytoplasm, but in the latter case this usually takes place within the first day. Ordinarily, condensation of the protoplasm does not occur in a small proportion of the oocysts; the explanation for this is unknown. In some species, such as *Eimeria ahtanumensis* from the gall bladder of lizards (Clark, 1970) and *E. neitzi* from the uterus of the impala (McCully et al., 1970) sporulation occurs in the organ in which the oocysts develop. Whether these may excyst and initiate a new infection without leaving the host is unknown as yet.

The oocyst wall varies in thickness and may consist of one, two, or more layers. At one end of the oocyst, the wall may be thinner or lacking some of the layers; this area is called the micropyle (Fig. 26). The sporozoites escape through the micropyle during excystation. The nucleus may be visible as a light area in the sporont (Figs. 2, 26). Each species has oocysts of characteristic size, shape, and appearance.

B. SPORULATION

This occurs within from one day to three weeks or more, depending on the temperature and the species. Certain *Eimeria* species with thick walls require a relatively long time for completion of sporulation; for example, in *E. lemuris* of the galago, oocysts sporulate in 20 days (Poelma, 1966).

The process of sporulation as it occurs in *E. debliecki* includes two nuclear divisions and splitting off of the polar granule (Vetterling, 1968). The divisions may not be synchronous, so that three nuclei are sometimes observed. After completion of nuclear division, the cytoplasm divides to form four sporoblasts. Usually, the sporoblasts are at first spheroidal, then assume a pyramidal shape, with the nucleus near the tip. Each sporoblast then rounds up, becomes elongate, and forms a surface membrane with a thickening at one end, which later forms the Stieda body. Two sporozoites and a residuum are developed within each sporocyst. In *Isospora rivolta,* sporulation occurs similarly, but the two sporoblasts elongate and form a pyramid at each end, no polar granule is seen, and four sporozoites are formed in each sporocyst (Mahrt, 1968). In the latter, 94% of the oocysts complete sporulation at 20° C within 48 hr, whereas in *E. debliecki,* 100% of the oocysts complete sporulation within 240 hr at 20° C. In *I. felis*, no pyramid stage has been observed and 96% of the oocysts completed sporulation within 40 hr at 20° C (Shah, 1970b).

Meiosis occurs during sporogony, and the polar granule is considered to be a polar body (Levine, 1963). Canning and Anwar (1968) reported that in *E. tenella* and *E. maxima,* at the beginning of meiosis, the nucleus of the zygote moves from a central to a peripheral location, and centriolar granules appear on opposite sides of the nuclear membrane. The nuclear membrane breaks down, the nucleolus disappears, and a large clear spindle appears over the surface of the zygote. Ten rod-like chromosomes, representing homologous pairs lying in two rows of five, align themselves parallel with the spindle. Five chromosomes migrate as beaded filaments to each pole, where they condense. Reduction thus occurs in a single phase without duplication of chromosomes or centromeres. Scholtyseck (1963) also found a diploid number of ten chromosomes in *E. maxima.* The description of the meiotic division by Canning and Anwar (1968) differs from that of schizogonic divisions in *E. callospermophili* (Speer and Hammond, 1970a; Hammond, 1971) in that the nucleolus does not disappear and the nuclear membrane remains intact in the latter.

C. SPORULATED OOCYSTS

There is a characteristic morphology for each species, so that identification of a species is usually possible by examination of the sporulated oocysts when the host is known. As shown in Fig. 2, Chapter 1, the oocyst may include an oocyst residuum, and there may also be a residuum within each sporocyst. The presence or absence of these, their location and appearance if present, the number and relative thickness of layers in the oocyst wall, the presence or absence of a micropyle and micropyle cap, the appearance of the Stieda body if present, the presence or absence of a substiedal body, and the appearance and arrangement of sporozoites within the sporocysts are characteristics which are often useful in distinguishing species. The size and shape of the oocysts and sporocysts are important diagnostic features.

With increasing age oocysts gradually lose their vitality and ability to initiate an infection. Vetterling and Doran (1969) found that oocysts of *E. acervulina* which had been stored for six years were not infective and only a few of a batch of oocysts that had been stored for two years were infective. These authors found a coincidental decrease in the amount of amylopectin in the sporozoites during storage of oocysts (see Chapter 5).

III. Excystation

The physiological and biochemical aspects of excystation are discussed in Chapter 5; in the present chapter, the morphological aspects will be emphasized.

A. ALTERATION OF OOCYST WALL AND/OR MICROPYLE

In ruminants, oocysts probably undergo the first phase of excystation in the rumen (Jackson, 1962). This phase involves alteration of the micropyle as the result of exposure to carbon dioxide. When the oocysts of the ovine species *E. arloingi* were so exposed, the cap over the micropyle was lifted or split off, whereas in *E. bovis* from cattle, the micropyles were thinned or flattened. Although avian coccidia differ somewhat from ruminant coccidia with respect to excystation, Nyberg, Bauer, and Knapp (1968) found similar changes in the micropyles of treated *E. tenella* oocysts and an opening in this region was observed with the scanning electron microscope (Nyberg and Knapp, 1970). However, Vetterling, Madden, and Dittemore (1971) saw micropyles with the same characteristics in untreated oocysts studied with freeze-fracture techniques. Similar changes in the micropyle were found in oocysts which had been in intestinal fistulas for 24 hr or longer (Nyberg and Hammond, 1964). Also, inoculation of calves by injection of *E. bovis* oocysts into the abomasum caused as severe infections as inoculation by nippled flask or by gelatin capsules, showing that passage through the rumen is not essential for excystation of *E. bovis* (Hammond, McCowin, and Shupe, 1954).

In avian coccidia, the first phase of excystation probably consists of mechanical release of the sporocysts from the oocysts in the gizzard (Doran and Farr, 1962). Oocysts which remain intact during passage through this organ apparently do not undergo excystation, although Lotze and Leek (1968, 1969) found that some such oocysts of *E. tenella* had motile sporozoites when recovered from the small intestine, large intestine, or feces in from one to three hours after inoculation. Some of these oocysts were altered in appearance, possibly as a result of exposure to enzymes or carbon dioxide in the upper digestive tract.

B. ALTERATION OF STIEDA BODY AND SUBSTIEDAL BODY

Sporocysts typically have a thickening at one end, called the Stieda body (Figs. 3, 4). In some *Eimeria* and *Isospora* species, an additional plug-like structure, the substiedal body (Fig. 5), is present immediately beneath the Stieda body. In the *Eimeria* species thus far studied, excystation is associated with disintegration or removal of the Stieda body and of the substiedal body, if present. In electron microscope studies of *E. callospermophili* and *E. larimerensis,* the Stieda body was found to occur in a gap in the sporocyst wall; in the latter species it consisted of two layers, but in the former, only one was observed (Roberts, Speer, and Hammond, 1970). The Stieda body of *E. bovis* was found in light microscope studies to consist of two different portions, both of which were protein; the anterior portion had tyrosine, whereas the posterior portion had little or none (Hibbert, 1969). It is uncertain whether the sporocyst wall structures are affected by carbon dioxide treatment, but Hibbert (1969) found that the Stieda body of the sporocysts in carbon dioxide-treated oocysts of *E. bovis* and *E. auburnensis* was more protuberant than those within untreated oocysts.

In *E. bovis* sporocysts freed mechanically from the oocysts, suspended in a mixture of 1% taurocholic acid and 0.25% trypsin buffered at pH 7.5 with tris-maleate, and observed on a warm stage at 39° C, movement of the sporozoites occurred before any change in appearance of the Stieda body could be detected. This held true also for *E. callospermophili* and *E. larimerensis* (Roberts, Speer, and Hammond, 1970); in these species, motility of the sporozoites was stimulated by bile or bile salts. Sporozoites were motile in some free sporocysts observed in saline solution, however. Marquardt (1963) reported that the sporozoites of *E. nieschulzi* became activated when the oocysts were crushed.

During excystation in *E. bovis,* the Stieda body became detached from the sporocyst as an intact structure in some specimens (Hibbert, 1969), but in the majority of specimens, the Stieda body disappeared (Speer and Hammond, unpublished data). In *E. acervulina*, the inner portion of the Stieda body became swollen or enlarged before any movement of the sporozoites could be seen (Doran and Farr, 1962). The inner margin of the

Stieda body then became less distinct and the body disappeared; at this time, the sporozoites began moving. Swelling and disappearance of the Stieda body was also observed in *E. tenella* (Nyberg, Bauer, and Knapp, 1968).

In *E. utahensis,* the Stieda body became swollen and protruded more than normal (Fig. 5), then became indistinct and apparently disappeared (Hammond, Ernst, and Chobotar, 1970). The substiedal body began to protrude, appearing to be pushed outward through the opening earlier occupied by the Stieda body (Fig. 6); later the body suddenly popped out, leaving a relatively wide opening through which the sporozoites immediately passed (Fig. 7). Similar observations were made in *E. callospermophili* and *E. larimerensis* (Roberts, Speer, and Hammond, 1970) and in *E. papillata* (Chobotar and Hammond, 1971), except that in the last species, the substiedal body was sometimes only partially removed and the sporozoites then escaped through a channel in this body.

C. ESCAPE OF SPOROZOITES FROM SPOROCYSTS AND OOCYSTS

As soon as the Stieda body and substiedal body, if present, have disappeared, the sporozoites begin leaving the sporocyst (Fig. 35). In *E. acervulina* the sporozoites moved away from the residuum, straightened out, and circled over each other (Doran and Farr, 1962). Then a sporozoite pushed into the opening left by disappearance of the Stieda body and squeezed through this opening. Usually, the sporozoite leaves with the anterior end first and its body is constricted to about ¾ to ¼ of the normal diameter as it passes through the opening in the sporocyst wall. The degree of constriction varies with the species. Occasionally, two sporozoites escape at the same time. Some *E. larimerensis* oocysts excyst in a U-shaped position, with the convex side coming out first (Roberts, Speer, and Hammond, 1970). Some of these sporozoites straightened out after becoming free, but others did not. In *E. papillata,* the first sporozoite to emerge often did so in a J- or U-shaped position or with the posterior end first (Chobotar and Hammond, 1971). In this species, it appeared that excystation was caused primarily by an increase in pressure within the sporocyst, and the sporozoites seemed to play little or no active role in the process.

After the sporozoites escape from the sporocysts, they usually move about within the oocyst for some time before escaping through the micropyle, in a manner similar to that of the escape from the sporocysts, except that less constriction of the body is required.

This phase of excystation is usually completed in the small intestine (Jackson, 1962). Farr and Doran (1962) found that oocysts of certain intestinal *Eimeria* species from chickens and turkeys excysted more rapidly and farther anteriorly in the small intestine than did certain cecal species from the respective hosts, and suggested that this represented an aspect of adaptation of the species to their hosts.

IV. Penetration of Cells

Observations of sporozoites of various *Eimeria* species in cell cultures have shown that such sporozoites readily and quickly enter and leave cells. Generally, the penetration process is completed within a few seconds (Kelley and Hammond, 1970). The sporozoite is usually undergoing a gliding movement immediately before penetrating a host cell, and this movement is apparently continued with little or no perceptible interruption or change in rate during the invasion.

In cinemicrographic observations of *E. larimerensis*, some sporozoites repeatedly thrust a slender protuberance into the cell and withdrew this before entering (Speer, Davis, and Hammond, 1971). Occasionally, the pathway of penetration was visible after entrance of a sporozoite, indicating that the surface membrane of the host cell had undergone invagination at the site of entrance. Escape of host cell cytoplasm was rarely seen after sporozoites entered cells but was observed frequently after sporozoites left cells, indicating that the exit of sporozoites is more disruptive to the host cell than entrance. Some sporozoites passed through host cell nuclei, which then immediately decreased to approximately one-half of the original size, suggesting the escape of nucleoplasm into the cytoplasm. Sporozoites fixed within host cell nuclei were not separated from the nucleoplasm by any membrane.

In EM studies of *E. larimerensis*, sporozoites fixed in the process of penetrating or leaving host cells were constricted at the site of entrance or exit (Fig. 36) (Roberts, Speer, and Hammond, 1971). The pellicle of the sporozoite was longitudinally folded at the place of the constriction. Such constrictions also occurred in the portion of the sporozoite inside the host cell; the explanation of this is unknown as yet. The findings indicated that the host cell membrane may be interrupted either at the initial site of entry or after becoming invaginated to some extent as penetration begins. In the latter instance, a portion of the membrane lining the parasitophorous vacuole may be derived directly from the limiting host cell membrane. Otherwise, the former membrane probably does not originate in this way; its method of origin is unknown as yet. Some sporozoites, which presumably had been in the cell for some time, were covered by a thin host cell layer as they left the host cell, and such a covering layer was observed around some extracellular sporozoites.

As will be seen in Chapter 5, the conoid and associated organelles may be protruded and retracted and this phenomenon may play a role in penetration through the host cell membrane. Vetterling, Madden, and Dittemore (1971) have used the term "rostrum" for the protruded conoidal complex which they observed in sporozoites of *E. tenella* and *E. adenoeides*. They attributed this protrusion to the activation of the sporozoite by trypsin-bile solution.

Substances secreted by the parasite may also play a role in penetration. Fayer (1971) found that quinine markedly inhibits entrance of *E. meleagrimitis* sporozoites into host cells in cultures, without affecting motility. Alteration of surface phenomena and interference with enzyme actions were suggested as possible explanations of the mechanism of inhibition.

V. Intracellular Sporozoites and Trophozoites

During the period of from one to several days after entrance into a host cell, the sporozoite (Fig. 30) undergoes some changes in appearance. The clear globules (refractile bodies) usually undergo changes in shape, size, and location. They also frequently decrease in number by fusion of some of the globules into

Figs. 27-34. Oocysts of *Isospora belli*; intracellular sporozoites and schizonts of *Eimeria* species in cell cultures; Figs. 28-33, phase-contrast. × 1800; Figs. 28, 29 of *E. callospermophili* (from Speer and Hammond, 1970a); Figs. 30, 32 and 33 of *E. larimerensis* (from Speer and Hammond, 1970b); and Fig. 31 of *E. magna* (from Speer and Hammond, 1971). **27.** Sporulated and unsporulated oocysts of *I. belli*. × 800. (Original photograph kindly supplied by Peter L. Long.) **28.** Sporozoite in an early stage of nuclear division, with elongated nucleolus. **29.** Two sporozoites, one having a dividing nucleus with two nucleoli (arrow). **30.** Sporozoite with two refractile bodies and nucleus with prominent nucleolus. **31.** Sporozoite-shaped schizont with four nuclei, one having two nucleoli (arrow). **32.** Sporozoite-shaped schizont with approximately eight, relatively small, indistinct nuclei. **33.** Spheroidal schizont; note that parasitophorous vacuole still has the outline of a sporozoite, indicating that the transformation to the spheroidal shape has just occurred. **34.** Mature schizont of *E. magna* with about 20 merozoites, some of which have two or three nuclei (arrows). Note binucleate condition of host cell. Nomarski interference-contrast. × 1100. (Original photograph by C. A. Speer, from culture inoculated with merozoites.)

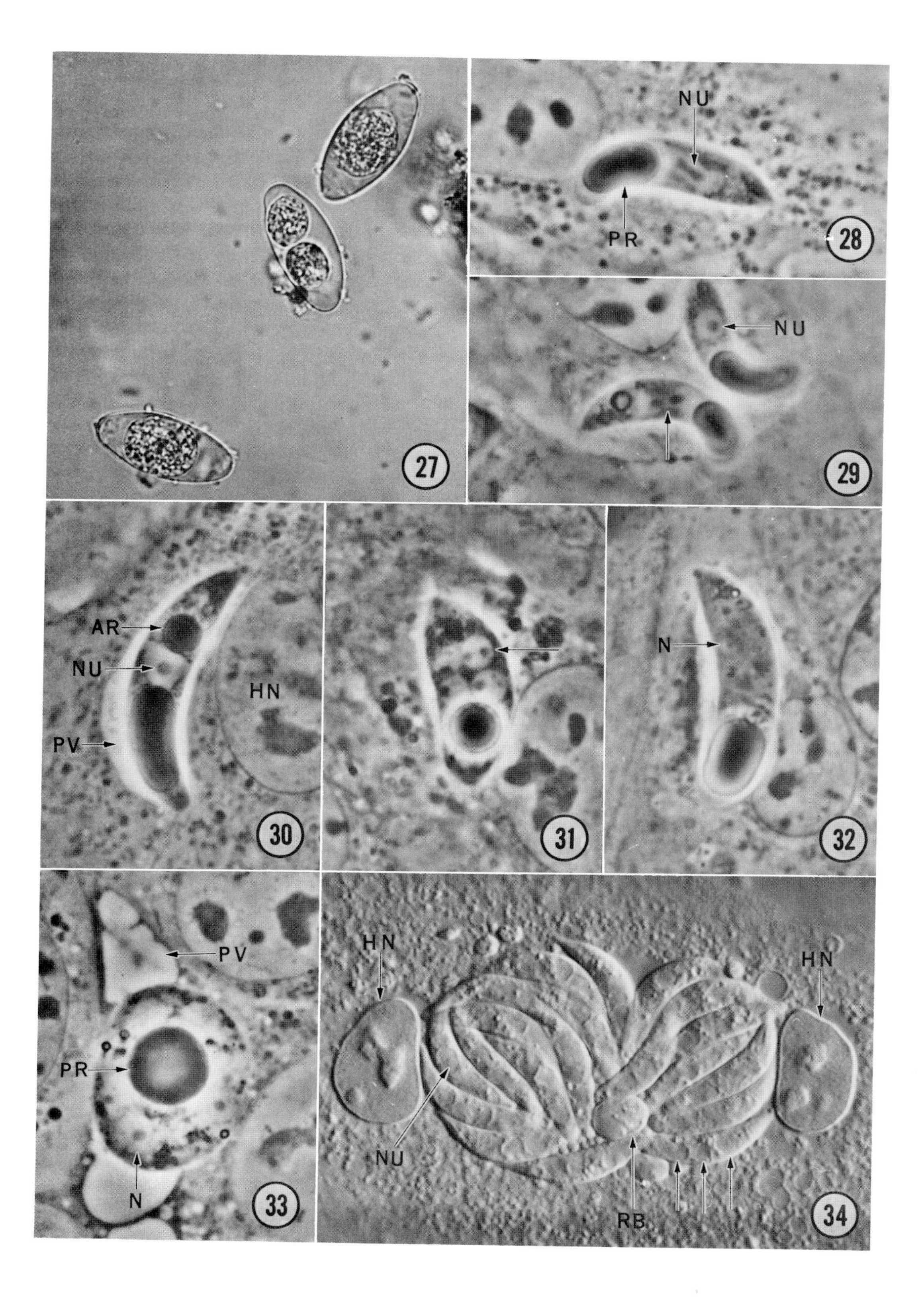

NU
PR
28
NU
29
27
AR
NU
HN
PV
30
31
N
32
PV
PR
N
33
HN
HN
NU
RB
34

a single one. This decrease is greater in *E. crandallis* in cell cultures (deVos, Hammond, and Speer, 1972) than in other species which have been similarly studied. In some species, such as *E. alabamensis,* the clear globules usually increase in number from a single one to two or more during this period, evidently by fusion of small granules pinched off from the original single body (Sampson, Hammond, and Ernst, 1971). These changes in the refractile bodies are evidently associated with the use of their substance in development of the parasite. The composition of the refractile bodies is discussed in Chapter 5 and changes observed *in vitro* are discussed in Chapter 6.

The elongate sporozoite of *E. bovis* transforms in cell cultures into a spheroidal trophozoite (Fig. 30, Chapter 6) during a period which extends from three days until from five to eight days after inoculation (Fayer and Hammond, 1967); in *E. ninakohlyakimovae,* this is completed as early as three days after inoculation (Kelley and Hammond, 1970). The sporozoite changes shape by a gradual widening of the body or by more rapid lateral outpocketings. This change in shape is probably associated with loss of the inner membrane complex, which leaves only the outer unit membrane to limit the cell, resulting in greater lability of the body outline. In *E. ninakohlyakimovae,* this loss occurs gradually in relatively small areas interspersed with areas still retaining a complete pellicle (Kelley and Hammond, 1972). The apical complex also disappears during transformation, but the micronemes and rhoptries persist longer than the other organelles. Remnants of the micronemes and rhoptries are still present in trinucleate schizonts, in which the surface is limited only by the outer membrane, except for small areas (Fig. 38). In *E. intestinalis,* the dedifferentiation process occurs more rapidly and is already completed in trophozoites, but some trophozoites have fragments of inner membrane complex and subpellicular tubules (Snigirevskaya, 1969a). During transformation in *E. bovis,* the nucleus changes from vesicular to compact, approximately doubles in size, and the nucleolus becomes greatly enlarged (Fig. 30, Chapter 6) (Fayer and Hammond, 1967).

In several other species, such as *E. auburnensis, E. callospermophili, E. bilamellata, E. larimerensis,* and *E. magna,* the uninucleate trophozoite stage is usually omitted. The parasite undergoes considerable growth and completes several nuclear divisions while retaining the elongate sporozoite shape (Fig. 31). Such multinucleate forms, called sporozoite-shaped schizonts (Fig. 32), later transform into spheroidal or ellipsoidal schizonts (Fig. 33) by processes similar to those associated with transformation of sporozoites into trophozoites. In *E. callospermophili,* the sporozoite-shaped schizonts retain all of the organelles and the motility characteristic of the sporozoite (Roberts, Hammond, Anderson, and Speer, 1970). They have the ability to leave their host cells and enter new ones (Speer, Hammond, and Anderson, 1970); this also holds true for *E. larimerensis* (Speer and Hammond, 1970b).

VI. Schizogony

A. NUCLEAR DIVISION

Scholtyseck (1953) described the occurrence of an intranuclear spindle and the separation

Figs. 35-37. Electron micrographs of *Eimeria* species. **35.** Sporocyst of *E. larimerensis*, Karnovsky's fixative, with sporozoite beginning to excyst. (From Roberts, Speer and Hammond, 1970.) × 9500. **36.** Sporozoite of *E. larimerensis*, Karnovsky's fixative; about one-third of the body has entered a host cell; note constriction of sporozoite at site of penetration (arrow) (From Roberts, Speer and Hammond, 1971). × 16,000. **37.** Sporozoite-shaped schizont of *E. callospermophili* fixed in 2.5% glutaraldehyde and osmium tetroxide together; merozoites are forming internally in association with nuclei. (From Roberts, Hammond, Anderson and Speer, 1970.) × 9000.

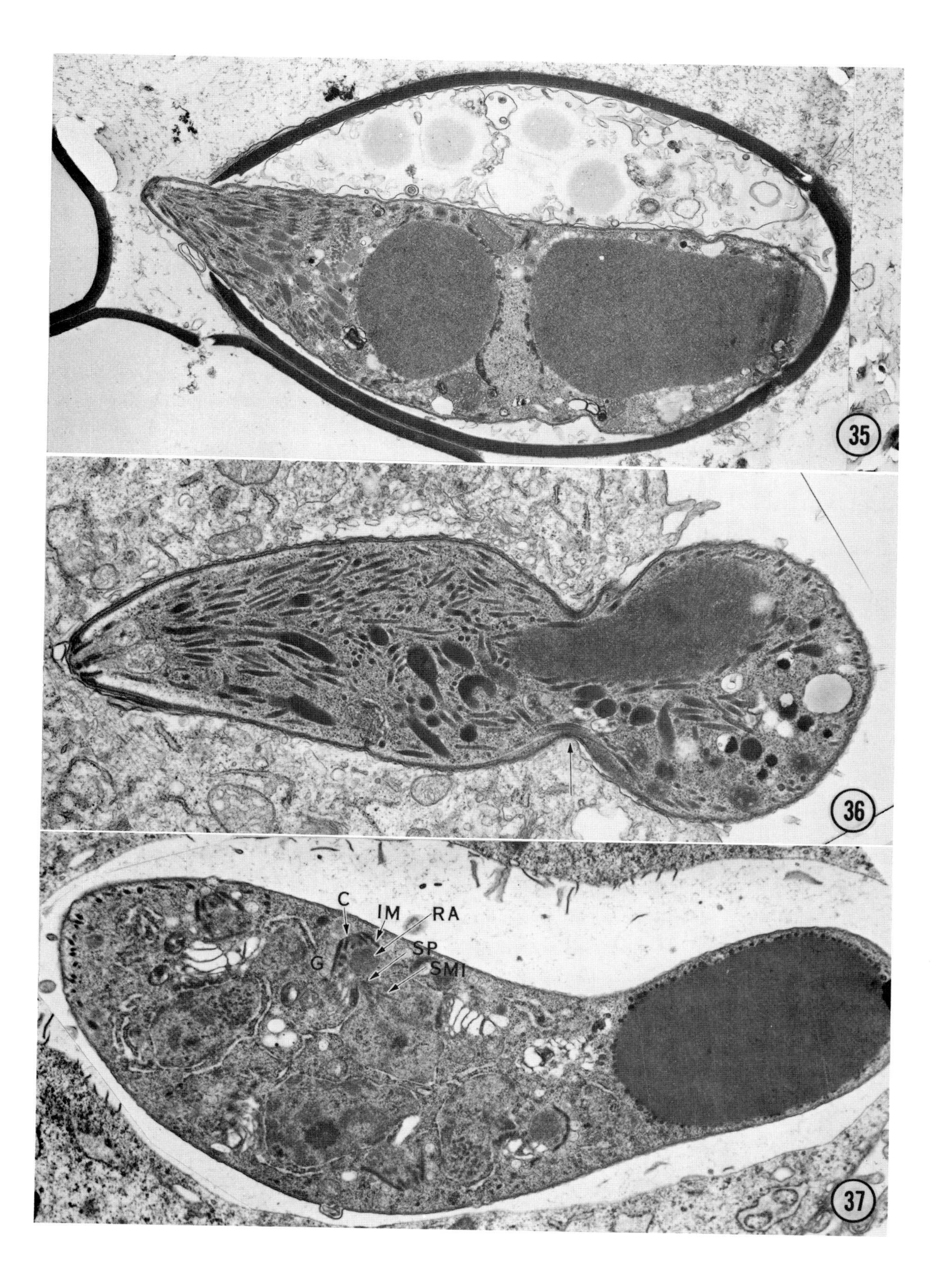
35
36
C
IM
RA
G
SP
SMI
37

of a single central chromatin body into two during nuclear division in the schizonts of *E. tenella*. As determined by the study of living specimens of *E. callospermophili* schizonts in cell cultures (Speer and Hammond, 1970a), the nuclear envelope remains intact during division. The prominent nucleolus elongates and divides into two (Figs. 28, 29). In EM studies, apparent division of the Golgi complex was noted in an early stage of division (Sampson and Hammond, 1972). Centrioles were seen at the poles of the elongated nucleus. The centriole in *Eimeria* species appears to be peculiar in having only single rather than triple peripheral microtubules (Roberts et al., 1970; Dubremetz, 1971). Evidence indicating the occurrence of a centrally located spindle extending between opposite poles of a dividing nucleus having an elongated nucleolus was seen in early schizonts of *E. ninakohlyakimovae* (Fig. 42) (Kelley and Hammond, 1972) and in *E. callospermophili* (Hammond, 1971). Relatively small spindles, lying eccentrically in the nucleus, have been observed in early divisions of the schizont in *E. ninakohlyakimovae* (Fig. 38) (Kelley and Hammond, 1972) and in *E. intestinalis* (Snigirevskaya, 1969a). In the latter species, the nucleolus was reported to disappear during cell division, but this was not observed in *E. callospermophili* or *E. alabamensis*. In *E. alabamensis*, both kinds of spindle occurred in early schizonts, but only eccentric spindles were observed in advanced schizonts (Sampson and Hammond, 1972). Eccentric spindles occur in association with merozoite formation in *E. magna* (Fig. 40) (Danforth and Hammond, 1972) and in *E. callospermophili* (Fig. 37) (Roberts, Hammond, Anderson, and Speer, 1970), as well as with microgamete formation in *E. auburnensis* (Hammond, Scholtyseck, and Chobotar, 1969) and in *E. intestinalis* and *E. magna* (Snigirevskaya, 1969b). The significance of the two kinds of spindle is unknown as yet. Chromosomes have not been observed during schizogony, although they have been reported in zygotes (Scholtyseck, 1963; Canning and Anwar, 1968). Possibly, the chromatin may be distributed to the daughter nuclei by attachment of chromosomes to the nuclear envelope as suggested for *Plasmodium* species by Ladda (1969). The nuclei become somewhat smaller as successive divisions occur in the schizont.

B. GROWTH AND INTAKE OF NUTRIENTS

Schizonts increase considerably in volume. The duration of the growth period and the size attained vary considerably among different species and among different generations in the same species. As an example of the latter, mature first-generation schizonts of *E. ninakohlyakimovae* (Fig. 10, Chapter 7) are about 330 μ by 240 μ and require about ten days to reach maturity, whereas the second-generation schizonts of this species require only from one to two days to mature and are about 9 μ by 12 μ in size (Wacha, Hammond, and Miner, 1971). In the species with relatively large schizonts, these undergo infoldings and/or invaginations of the surface during the later stages of growth. In some instances (*E. bovis*), these result in the formation of numerous daughter individuals, termed blastophores, which give rise to the merozoites. In other instances (*E. auburnensis*), the schizont develops fissures

Figs. 38-40. Electron micrographs of *Eimeria* species, fixed in Karnovsky's solution. **38.** Schizont of *E. ninakohlyakimovae* with three nuclei; six days after inoculation of sporozoites into lamb embryonic kidney culture; note spindle microtubules, cone-like pole of spindle, and centriole (from Kelley and Hammond, 1972.) × 19,000. **39.** Intracellular sporozoite of *E. ninakohlyakimovae*, six days after inoculation into lamb embryonic kidney culture, with crescent body in parasitophorous vacuole (from Kelley and Hammond, 1972). × 7200. **40.** Portion of merozoite of *E. magna*, from rabbit killed four days after inoculation; merozoite with two daughter merozoites forming in association with a dividing nucleus, having an eccentric intranuclear spindle; note Golgi adjunct (from Danforth and Hammond, 1972). × 22,000.

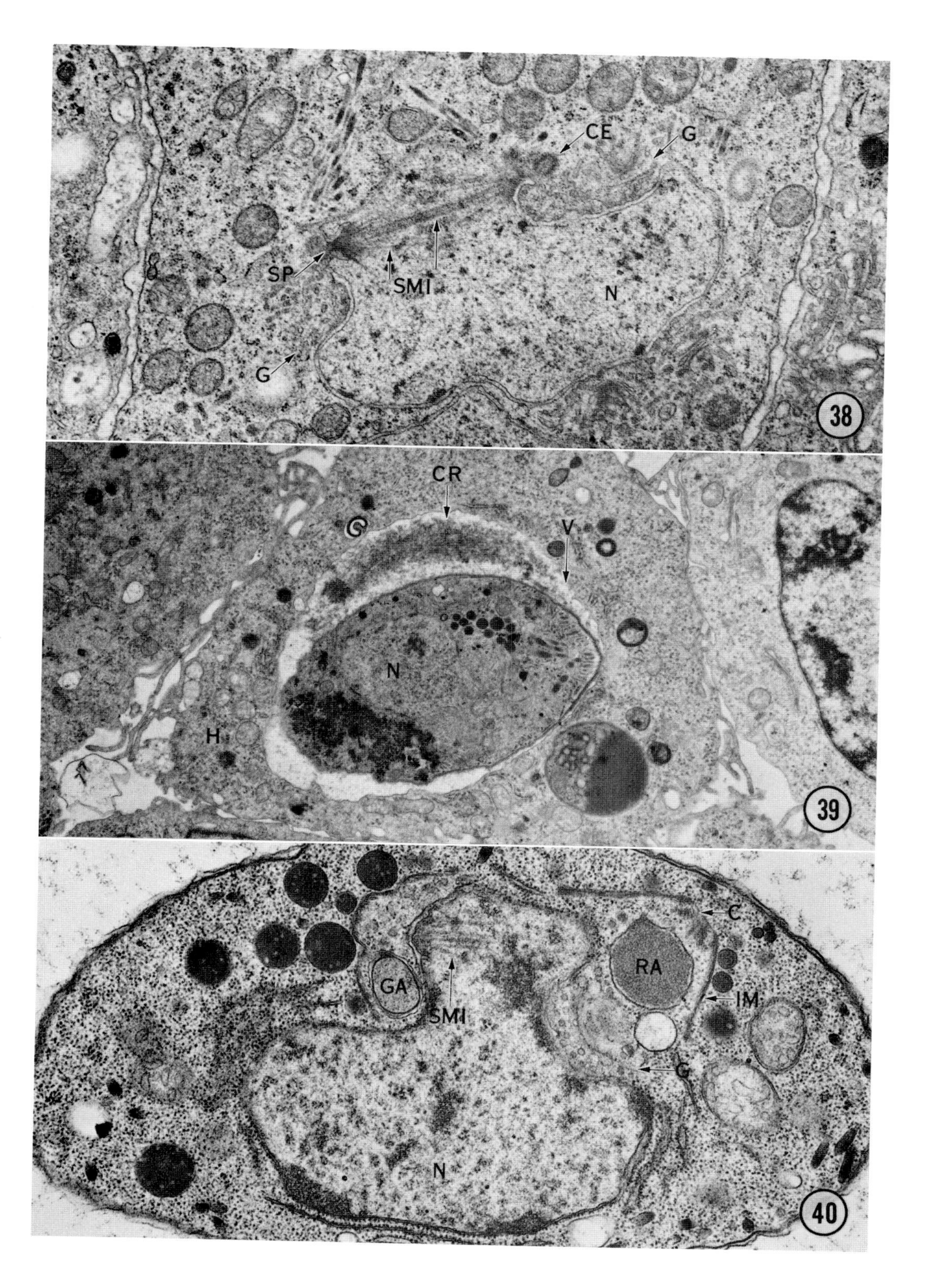
CE
G
SP
SMI
N
G
38
CR
V
N
H
39
C
RA
GA
SMI
IM
G
N
40

or channels lined with a membrane continuous with that at the surface. By these means, the surface area of the schizont is greatly increased. Similar processes occur in unusually large microgamonts, such as those of *E. auburnensis* (Hammond, Scholtyseck, and Chobotar, 1969) and *E. leuckarti* (Barker and Remmler, 1972).

Intake of nutrients may occur by ingestion of substances through the micropores. Evidence indicating such ingestion and the associated formation of food vacuoles has been reported for schizonts and other stages of *E. intestinalis* by Snigirevskaya (1968). In this species and in *E. ninakohlyakimovae* (Kelley and Hammond, 1972), particulate matter similar to that occurring in the parasitophorous vacuole is apparently ingested. This particulate matter evidently consists of fragmented and/or degraded host cell cytoplasm and/or membranes. The occurrence of this material in the parasitophorous vacuole may come about by the separation of finger-like extensions (*E. ninakohlyakimovae*) or of folds into the vacuole (*E. callospermophili*, Roberts, Hammond, Anderson, and Speer, 1970; *E. auburnensis*, Hammond, Scholtyseck, and Chobotar, 1967). In *E. alabamensis*, unaltered host cell cytoplasm is apparently taken into the micropores (Sampson and Hammond, 1971). Vivier *et al.* (1970) described various kinds of micropores occurring in coccidia and related groups and cited observations suggesting a relation with pinocytosis. They concluded, however, that the significance of this organelle is yet to be resolved.

Transport of material from the host cell into the vacuole may occur in *E. bovis* by formation of blebs at the surface of the membrane lining the parasitophorous vacuoles; these blebs then become separated to form free vesicles in the vacuoles (Sheffield and Hammond, 1966). Evidence indicating the occurrence of pinocytosis in association with V-shaped invaginations was observed in macrogametes of *E. auburnensis* (Hammond, Scholtyseck, and Chobotar, 1967). Deep invaginations of the surface apparently served for the intake of nutrients in microgamonts of this species (Hammond, Scholtyseck, and Chobotar, 1969).

C. FORMATION OF MEROZOITES

1. Ectomerogony

In the majority of species which have been studied thus far, merozoites originate at the surface of schizonts, of blastophores, as in *E. bovis*, or of invaginations or infoldings into the schizont, as in *E. auburnensis* (Chobotar, Hammond, and Miner, 1969). The term "ectomerogony" is proposed to distinguish this type of merozoite formation from that in which merozoites originate in the interior of the schizont ("endomerogony").

The first indication of merozoite formation in *E. bovis* is a thickening of the surface membrane of the blastophore in an area overlying a nucleus (Fig. 30, Chapter 4) (Sheffield and Hammond, 1967). At this area, which later becomes elevated into a conical protuberance (Fig. 12), an inner membrane complex representing the anlage of a merozoite appears just beneath the surface membrane. The origin of this inner membrane complex is unknown, but Porchet-Henneré and Richard (1971) hypothesized that in *Aggregata eberthi* it was derived from saccules, possibly originating from ergastoplasm. In the merozoite anlage, conoid, subpellicular tubules and anlagen of rhoptries appear, and the inner membrane extends posteriorly. A Golgi complex and a centriole are associated with the site of merozoite formation. A nucleus and other organelles are incorporated into the forming merozoite before it becomes separated from the residual body (Fig. 41). This occurs by closure of the outer membrane over the opening at the posterior end, leaving a posterior pore. Merozoites develop similarly in *E. nieschulzi* (Colley, 1968), in *E.*

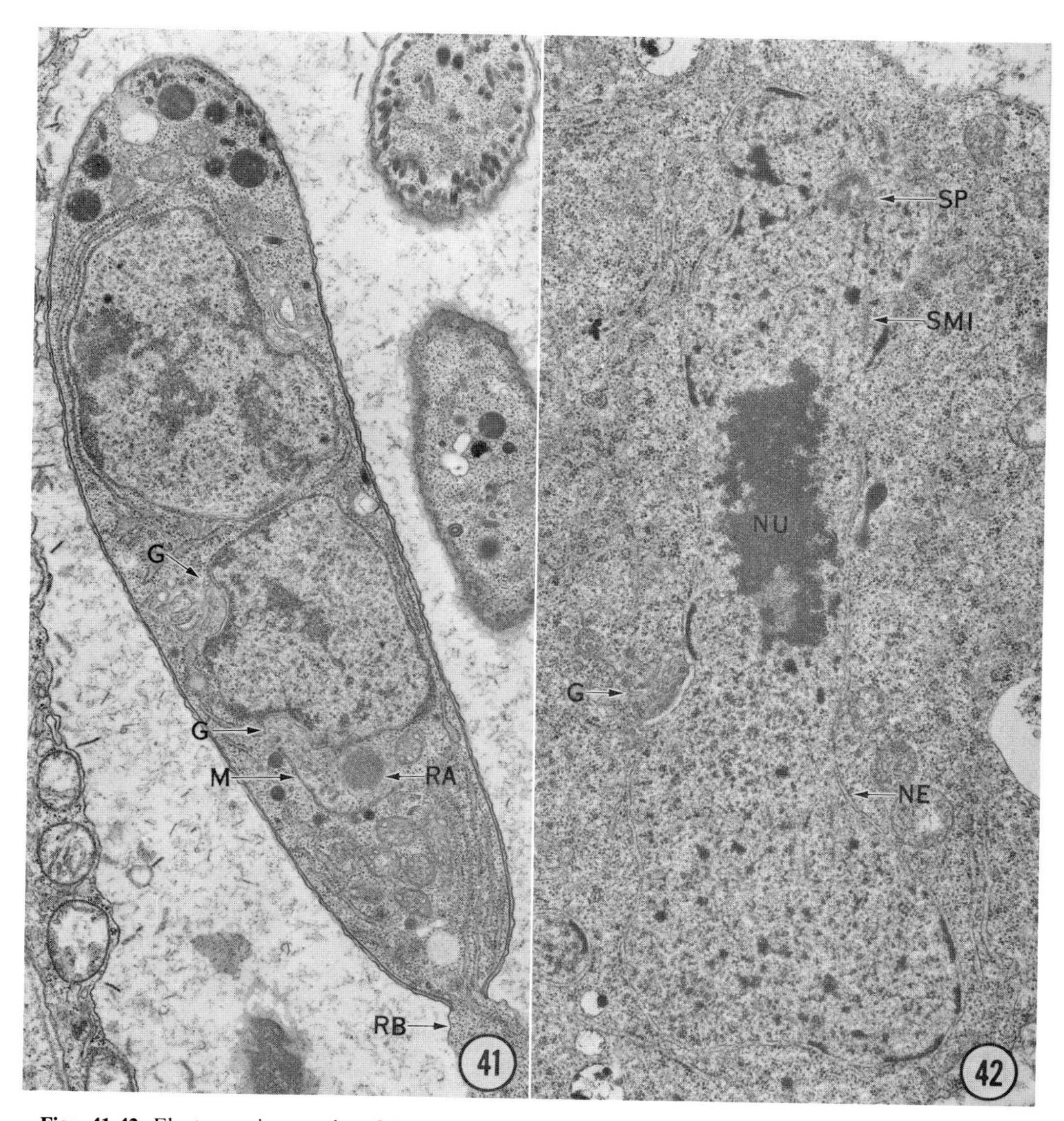

Figs. 41-42. Electron micrographs of *Eimeria* species, fixed in Karnovsky's solution. **41.** Incompletely formed multinucleate merozoite of *E. magna* with a daughter merozoite forming in association with a nucleus. Note that the parent merozoite is still attached to the residual body (from Danforth and Hammond, 1972). × 12,500. **42.** Portion of schizont of *E. ninakohlyakimovae,* 12 days after inoculation into lamb embryonic trachea cell culture; note dividing nucleus, with elongated nucleolus, and with intranuclear spindle pole and microtubules near one nuclear pole (original micrograph by G. L. Kelley). × 15,500.

tenella (McLaren, 1969; Sénaud and Cerná, 1969), in *E. magna* and *E. pragensis* (Sénaud and Cerná, 1968, 1969), in *E. intestinalis* (Snigirevskaya, 1969a), in an *Isospora* species (Schmidt, Johnston, and Stehbens, 1967), and in *Aggregata eberthi* (Porchet-Henneré and Richard, 1971). A conical protuberance of the nucleus, as well as centrioles, and Golgi apparatus were observed in association with the origin of the merozoites in *E. magna* and *E. pragensis* (Sénaud and Cerná, 1969). In certain of the types of *T. gondii* occurring in the intestinal epithelium of cats, many merozoites bud from the surface simultaneously; this probably corresponds with ectomerogony (Dubey and Frenkel, 1972b).

2. *Endomerogony*

a. Splitting

Scholtyseck (1965) described a process of internal merozoite formation in *E. perforans* and *E. stiedai* in which the merozoites were separated from each other and from the remaining cytoplasm of the schizont by fissures formed by confluence of vesicles of the endoplasmic reticulum. These fissures were concentrically arranged around areas of cytoplasm, each having a nucleus. After the merozoites had become free from the schizont, the two membranes at the surface appeared. This process was not described in detail, however, and the relationships between it and the other known methods of merozoite formation are as yet uncertain. Dubey and Frenkel (1972b) used the term "splitting" for separation of single organisms from a multinucleate parent organism in *T. gondii* intestinal stages in cats. This appears to differ from the process described by Scholtyseck (1965).

b. Internal formation of merozoites in *Eimeria* species

In some species, such as *E. callospermophili* (Roberts, Hammond, Anderson, and Speer, 1970), *E. alabamensis* (Sampson and Hammond, 1972) and *E. magna* (Danforth and Hammond, 1972) merozoites originate in the interior of the schizont in association with the nuclei, which are randomly distributed. In *E. callospermophili,* merozoites begin forming in sporozoite-shaped schizonts with 4-6 nuclei (Fig. 29, Chapter 4). The anlage of each merozoite appears near the protruding, conical pole of an eccentric intranuclear spindle (Fig. 37). One or two centrioles, each having one central and nine peripheral single microtubules, a Golgi apparatus and, frequently, a Golgi adjunct, as found in *Toxoplasma* by Ogino and Yoneda (1966) (Figs. 40, 41), are also associated with the site of merozoite formation. The inner membrane complex of the anterior region of the future merozoite, annuli, and an immature conoid constitute each anlage. The subpellicular tubules, polar rings, anlagen of rhoptries, and refractile body (clear globule) are seen soon after the first appearance of the merozoite anlagen. This later elongates and assumes a peripheral location. It then begins to grow outward from the surface of the schizont, and from this point onward, merozoite formation resembles that in the species in which merozoites originate at the surface of schizonts. Nuclear division is completed during the period of merozoite development and the two daughter nuclei are incorporated into the two merozoites formed in association with each parent nucleus. The outer membrane of the merozoite is derived from that of the schizont.

c. Endodyogeny

Endodyogeny is characterized by the formation of two daughter organisms within a mother cell, which is used up during the process (Figs. 26, 27, Chapter 4). According to Sheffield and Melton (1968), two separate Golgi complexes and Golgi adjuncts are present in early stages of endodyogeny in *T. gondii*. In the anterior portion of the cell, the two daughter cells begin to form as a dome-shaped inner membrane complex with an open top. The polar ring, conoid, anlagen of rhoptries, and subpellicular tubules also appear at about this time. Nuclear division occurs during an early stage of daughter cell formation, and the poles of the dividing nucleus are directed into the two daughter cells. At each pole is a conical condensation of nuclear material, the "centrocône" of Sénaud (1967), who stated that this consists of fibrils, apparently having a periodic structure. The anlagen of the daughter organisms undergo elongation, incorporating the nuclei. As they reach the surface of the mother cell, the inner membrane of the latter disappears, and its outer membrane joins the inner membrane complex of the daughter to complete the formation of the pellicle.

According to Dubremetz and Vivier (1971, personal communication*), the centrocône consists of microtubules within an opaque matrix and is associated, at least in one stage of development, with an intranuclear spindle. Similar centrocônes were found in *Eimeria necatrix*. Centrioles associated with the centrocônes were seen in *T. gondii* and *E. necatrix*. The centrioles and centrocônes were observed only as the anterior complex of the daughter individual first appeared, and preceding nuclear division. The centrocône may function in drawing the daughter nucleus toward the developing merozoite (Dubremetz, 1971).

According to Vivier and Petitprez (1968), the inner membrane complex originates from that of the mother cell by a kind of proliferation and infolding process, whereas the outer membrane is formed directly from that of the parent organism. This was found to hold true whether one, two, or more daughter individuals were formed simultaneously (Vivier, 1970). Infolded inner membrane complex was also seen in schizonts of *E. alabamensis* which were in the process of forming daughter cells, but its significance is unknown (Sampson and Hammond, 1972). More recently, however, Vivier (1971, personal communication) has stated that in endodyogeny in *Toxoplasma*, the daughter individuals are formed *de novo*, except for the external membrane. Binary or multiple division may also occur and in these, the daughter organisms are formed in part from intracytoplasmic proliferations of membranes associated with the maternal membrane complex. Variations in these processes were found to occur among different strains examined. Some of these variations appeared to represent intergradations between the two types of reproduction. Also, budding of daughter individuals toward the exterior, similar to that occurring in *Eimeria* species, was sometimes seen.

Recently, *Isospora rivolta* has been found to undergo endodyogeny in cell cultures (Fayer, 1972).

d. Endopolygeny

In a study of the schizogonic stages of *T. gondii* occurring in the intestinal epithelium of cats, Sheffield (1970) found that merozoites may begin forming after the second nuclear division in schizonts. These resemble the sporozoite-shaped schizonts of certain *Eimeria* species, such as *E. callospermophili*, in having an inner membrane as well as an outer membrane at the surface. Twenty-four mature merozoites were observed to form from one schizont. Although this kind of merozoite formation resembles endodyogeny, it differs in simultaneously having more than two offspring formed within the parent organism and in having two or possibly more nuclear divisions before merozoite formation begins (Fig. 28, Chapter 4). Sheffield (1970) suggested that endodyogeny may be a variation of schizogony as it occurs in the cat. Colley and Zaman (1970) reported similar findings, except that they found that approximately 14 merozoites are formed from each schizont. They depicted an early stage of

* J. E. Dubremetz and E. Vivier, Laboratoire de Protistologie et Microscopie électronique, Université des Sciences et Techniques de Lille I, B.P. 36-59 Villeneuve d'Ascq.

merozoite formation in schizonts with four visible nuclei.

Vivier (1970) reported the simultaneous occurrence of four or five daughter individuals in proliferative forms of *T. gondii* in ascitic fluid or peritoneal cells from infected mice, and referred to such multiplication as "endopolygenèse." Piekarski, Pelster, and Witte (1971) found schizonts with a number of daughter individuals in the intestine of cats infected with *T. gondii* and applied the term "Endopolygenie" to this. They also found that merozoite formation begins at the 4-nucleus stage; supposedly 32 merozoites were formed in each schizont. Dubey and Frenkel (1972) also reported the occurrence of endopolygeny in intestinal stages of *T. gondii* occurring in the cat.

The process of merozoite formation in the schizonts of *T. gondii* has not as yet been completely described, but it appears to closely resemble the internal formation of merozoites in *E. callospermophili* (Roberts, Hammond, Anderson, and Speer, 1970) and in *E. magna* (Danforth and Hammond, 1972). This kind of merozoite formation in *Eimeria* species might provisionally be included under the term endopolygeny until more detailed studies of the development of merozoites in *T. gondii* are made.

Vivier (1970) has described the occurrence of a similar kind of multiplication of *T. gondii* in ascitic fluid or peritoneal cells of mice, as well as binary division and formation of a single daughter individual. He proposed the terms "exogenesis" (exogenèse) for formation of daughter individuals at the surface, and "endogenesis" (endogenèse) for internal formation of daughter individuals. These terms are applicable to gamogony and sporogony as well as merogony, whereas the terms "ectomerogony" and "endomerogony" apply only to the formation of merozoites. Vivier (1970) considered that endodyogeny, endopolygeny, and binary division as they occur in *T. gondii* are similar processes, but in later work (see page 69) found some differences among these. Further work must be done before a coherent understanding of asexual reproduction in this group of organisms can be attained.

D. MATURE SCHIZONTS AND MEROZOITES

After completion of merozoite formation, the schizont is no longer present as such, but may be represented by the merozoites and residual bodies, if present. The mature merozoites, after detachment from the residual bodies, lie free within the parasitophorous vacuole. Such merozoites generally show little or no motility, but movement of merozoites within the parasitophorous vacuole was seen in *E. auburnensis* (Clark and Hammond, 1969). Merozoites of *E. callospermophili* (Speer, Hammond, and Anderson, 1970), and *E. alabamensis* (Sampson, Hammond, and Ernst, 1971) sometimes actively left the parasitophorous vacuole. Merozoites of several different species were more motile within the parasitophorous vacuole after bile or bile salts had been added to the preparation than before, and more merozoites left the host cell after exposure to such stimulation (Speer, Hammond, and Kelley, 1970). The mechanism of escape of the merozoites from the parasitophorous vacuole and host cell is not known for most species, but in *E. callospermophili,* the merozoites appear to undergo constriction as they escape (Roberts, Hammond, Anderson, and Speer, 1970).

Multinucleate merozoites have been reported to occur in *E. magna* (Kheysin, 1960; Sénaud and Cerná, 1969), *E. piriformis* (Pellérdy, 1965), *E. callospermophili* (Speer, Hammond, and Anderson, 1970), and *E. stiedai* (Pellérdy and Dürr, 1970), but their significance is unknown. Pellérdy and Dürr (1970) surmised that the multinucleate merozoites of *E. stiedai* may be formed in response to an immune reaction within the host or to an inadequate supply of nutrients which resulted in nuclear degeneration and fragmentation. Speer and Hammond (1971) found that about 30% of the mature first-generation schizonts of *E. magna* observed in cultured cells had multinucleate merozoites (Fig. 34). In rabbits killed three and four days after inoculation, 18% and 23%, respectively, of the schizonts observed had multinucleate merozoites. Recently, young multinucleate merozoites of *E. magna*, free or still attached to the residuum, were observed in tissue from rabbits to have anlagen of daughter merozoites in association with the nuclei, indicating that the multinucleate merozoites in this species represent stages in the formation of new merozoites (Fig. 41) (Danforth and Hammond, 1972). This may correspond with the pattern of development as described by Shah (1971) in *Isospora felis*, in which new merozoites are formed from merozoites in the same parasitophorous vacuole in which they had developed. These merozoites became multinucleate while retaining their elongate shape, and thus resemble sporozoite-shaped schizonts.

Nothing is known about similarities or differences in development within different schizont generations in the same species, although these vary greatly in size, time required to reach maturity, and number of merozoites formed.

E. TERMINATION OF SCHIZOGONY

After a specific number of asexual generations, the merozoites formed develop into gamonts rather than schizonts. Although the factors responsible are unknown as yet, this aspect of the life cycle has generally been considered as genetically determined. Recently, however, Haberkorn (1970) reported that second- to fourth-generation merozoites of *E. falciformis* obtained from infected mice developed from one to three additional asexual generations after injection into the ceca of uninfected mice, whereas normally not more than four generations occur. Long and Rose (1970) have found that schizogony of *E. mivati* was extended beyond the normal duration in chickens treated with betamethasone. These findings indicate that the immune response or other factors associated with the host may play a role in the termination of schizogony.

VII. Gametogony

Only the general features of these stages are to be considered here; the processes are described in detail in Chapter 4. Whether differentiation of trophozoites into microgamonts or macrogamonts is determined phenotypically or genotypically is unknown as yet. Grell (1953, 1968) found that such determination is phenotypic in *Eucoccidium dinophili*, a member of a group of coccidia having no schizogony. The occurrence of microgamonts and macrogamonts in the same host cell (Hammond, Clark, and Miner, 1961) indicates that differentiation is genotypically determined. Rutherford (1943) found that in each of four intestinal species in rabbits, two types of merozoites are produced, and stated that it is possible that one type gives rise to microgamonts, whereas the other type gives rise to macrogamonts. Pellérdy and Dürr (1970) reported that in *E. stiedai*, two types of schizonts could be distinguished, beginning with the second generation. One type gave rise to few, plump merozoites and, after additional schizont generations, to microgamonts. The other type gave rise to numerous slender merozoites and eventually to macrogamonts. Haberkorn (1970) found evidence for phenotypic determination, however. He demonstrated that oocysts were discharged after inoculation of mice with a single merozoite of *E. falciformis*, which could not occur with genotypic determination.

Young macrogamonts are recognizable by the presence of a relatively large nucleus and nucleolus (Fig. 18), and in later stages by the numerous plastic (wall-forming) granules in the cytoplasm (Fig. 21). Microgamonts usually cannot be distinguished from macrogamonts with certainty until after nuclear division has occurred (Fig. 19), and then it is difficult to distinguish them from schizonts. At the time of microgamete formation, eccentric intranuclear spindles, with a conical protuberance at each pole, and an associated centriole or pair of centrioles, occurs at the site of origin of each microgamete anlage in *E. auburnensis* (Hammond, Scholtyseck, and Chobotar, 1969) and in *E. intestinalis* and *E. magna* (Snigirevskaya, 1969b). The poles of the spindle apparatus may play a role in inducing the formation of microgametes, as well as of merozoites in schizonts (Hammond, 1971) and centrioles may also be involved in these processes (Roberts, Hammond, Anderson, and Speer, 1970).

The nucleus of gamonts of coccidia characteristically has a central, deeply staining body (Fig. 18), which is termed a nucleolus by some authors and a karyosome by others (Hutchison *et al.*, 1971). Frequently, a slightly eosinophilic, Feulgen-negative, flattened body occurs adjacent to the nucleolus in macrogametes of *E. auburnensis* (Chobotar and Hammond, 1969) and of *E. ninakohlyakimovae* (Wacha, Hammond, and Miner, 1971). A similar body, which stained blue with bromphenol blue, was reported by Kheysin (1960) in macrogametes of *E. magna*. *E. auburnensis* microgamonts often have an eccentric nucleolus (Chobotar and Hammond, 1969). Recently, Youssef and Hammond (1972) found that the nucleolus of immature *E. magna* gamonts is compact and has no nucleonema. The microgamont has an eccentric nucleolus, which is smaller than the centrally located one of the macrogamont. These nucleoli consist of granules with a diameter of about 20 nm, which were more numerous in the macrogamonts, especially in the peripheral region, and smaller granules, about 5 nm in diameter, randomly distributed in a fibrillar matrix. The large granules, which are similar to cytoplasmic ribosomes, were no longer present in ribonuclease-treated preparations, but were affected only slightly by deoxyribonuclease or deoxyribonuclease and pepsin. Therefore, they appear to be of a ribonucleoprotein nature. The small granules and fibrillar matrix appeared less dense after treatment with deoxyribonuclease or deoxyribonuclease and pepsin and were not affected by ribonuclease, indicating the probable presence of DNA. These findings support the interpretation that the nuclear body in coccidia is a nucleolus.

Crescent-shaped bodies (Fig. 39) have been found associated with gamonts of certain species (*E. auburnensis,* Chobotar and Hammond, 1969; *E. ninakohlyakimovae,* Wacha, Hammond, and Miner, 1971; and *E. leuckarti,* Barker and Remmler, 1972). Such bodies are also found in association with the schizonts of certain *Eimeria* species. Their significance is unknown, but Kelley and Hammond (1972) found that the crescent body associated with the first-generation schizont of *E. ninakohlyakimovae* apparently consists of granular material, which probably arises from degradation and/or fragmentation and subsequent condensation of portions of host cell cytoplasm.

VIII. Fertilization

Little is known concerning the process of fertilization in coccidia. The entrance of the microgamete into the macrogamete is thought to stimulate oocyst wall formation. Few observations of fertilization in *Eimeria* and closely related genera have been made, however. Scholtyseck (1963) reported the occurrence of a microgamete at the nuclear membrane of macrogametes of *E. maxima*. The microgametes increased in volume and five rod-shaped chromosomes appeared in each, representing the haploid set. The oocyst wall had not yet begun forming at this time. Marquardt (1966) observed a living, mature macrogamete of *E. nieschulzi* free of host tissue, surrounded by microgametes which were adhering to it. Scholtyseck and Hammond (1970) found a macrogamete of *E. bovis* with a microgamete in its interior. The wall-forming bodies were altered from the normal appearance, with those of type II undergoing fusion into larger units, indicating that oocyst wall formation might have begun.

The above-mentioned findings suggest that entrance of the microgamete precedes formation of the oocyst wall. However, it is possible that the wall might also form in unfertilized macrogametes; if so, the resulting oocysts may be those in which the protoplasm does not condense into a spheroidal sporont and sporulation does not occur. Long (1972) found that oocysts of *E. mivati* which had developed in chick embryos incubated at 38°-39° C had poor sporulation rates. Histological examinations indicated that the development of gametocytes, especially microgametocytes, was retarded at this temperature. Long postulated that the formation of the oocyst wall occurs without the presence of microgametes, thus giving rise to oocysts which fail to sporulate. Alternatively, the viability of the microgametes may have been affected in such a way that these died soon after entrance into macrogametes.

IX. Transmission

Transmission of infection with coccidia usually involves the discharge of oocysts from infected animals, sporulation outside of the host, and ingestion by another host animal. As discussed in Chapter 9, *Toxoplasma gondii* may be transmitted by ingestion of cysts or tachyzoites in the tissues of infected hosts, and it has recently been shown that *Isospora felis* and *I. rivolta* can be transmitted by ingestion of stages occurring in the lungs, liver, spleen, and mesenteric lymph nodes of mice, rats, and hamsters (Frenkel and Dubey, 1972). Such extra-intestinal stages were also found in cats after inoculation with these species, and the tissues of such cats were infectious to kittens (Dubey and Frenkel, 1972a). Stages occurring in organs other than the intestine may have a longer duration than those in the intestine, as indicated by the finding that the former were still present and infectious in mice 67 days after inoculation (Frenkel and Dubey, 1972). If such stages were found to occur in *Eimeria* species of cattle, it would help to explain the problem of winter coccidiosis, discussed below. If the animals had a reservoir of infection in the form of extra-intestinal stages, these might be able to serve as the source of a clinical infection.

In the western United States and western Canada, clinical coccidiosis is frequently associated with cold weather (Fitzgerald, 1962; Hammond, 1964; Niilo, 1970b). Fitzgerald (1962) found that there was a marked increase in incidence of infections caused by *E. bovis* and *E. zuernii* during the fall and winter in Hereford calves on summer ranges and then on winter ranges or in feedlots in Utah. *Eimeria zuernii* appeared to be responsible for the majority of the clinical outbreaks in the intermountain western United States. Only very few sporulated oocysts were found in the feedlots at the time the calves evidently acquired the infections responsible for the outbreaks, indicating that such oocysts may not be readily available to calves under these conditions. Fitzgerald (1962) suggested that the condition of the animals as affected by weaning, type of feed, environment, and climatic conditions has a bearing on the outbreaks of coccidiosis in calves. Attempts to prevent the outbreaks by weaning the calves slowly and by use of tranquilizers were unsuccessful, however.

Niilo (1970a) observed in an experiment conducted in Lethbridge, Canada, that calves sheltered from cold during the winter and housed with inoculated calves did not develop clinical infections, while two out of three calves maintained similarly, except that they were not sheltered from the cold, had such infections. Niilo (1970a) stated that the results suggest that cold may increase the host's susceptibility to coccidiosis, but may not affect the severity of the signs after the clinical infection is established. More research is needed before the occurrence of winter coccidiosis can be satisfactorily explained.

The oocysts of coccidia are able to withstand exposure to various chemical and physical agents which are deleterious to most organisms (Pellérdy, 1965). Marquardt, Senger, and Seghetti (1960) found that, of various chemical and bactericidal agents tested, only mercuric chloride exhibited a high level of activity against unsporulated oocysts of *E. zuernii.* Sunlight for as short a time as 4 hr and freezing below about -7°C were lethal for *E. zuernii,* and survival was directly proportional to relative humidity. Normal sporulation occurred from about 8° C up to 32.5°C.

It is difficult to explain the high incidence of coccidia in animals such as sheep and cattle in the light of our knowledge that most species are immunogenic under experimental conditions. Pout (1969) has suggested that there could be a large number of strains of sheep coccidia in the field so that oocyst production might be maintained for many weeks, as has been observed, although animals quickly become resistant to low doses (50 oocysts per day) when pure cultures are given. This is a hypothesis which merits further investigation; little is known about the occurrence of such strains in either sheep or cattle.

Recent additions to our knowledge of the life cycles and development of the coccidia provide a basis for a more thorough understanding of these parasites and should enable the development of improved methods of prevention and control of the diseases they cause.

Acknowledgments. I hereby express my appreciation to Clarence A. Speer, Gary L. Kelley, William L. Roberts, Harry D. Danforth, Albertus J. de Vos, and Nabil N. Youssef for assistance in preparation of the illustrations and for suggestions concerning the text; to Peter L. Long, Norman D. Levine, Paul R. Fitzgerald, Leonard Reid Davis, John V. Ernst, Kenneth S. Todd, Jr., Bill Chobotar, and William C. Marquardt for reviewing the manuscript; and to Madeline J. Codella for secretarial assistance.

Some of the research reported herein was supported by PHS Research Grant No. AI-07488 from the National Institutes of Allergy and Infectious Diseases and by Grant No. GB-27423 from the National Science Foundation.

Literature Cited

Barker, I. K. and O. Remmler. 1972. The endogenous development of *Eimeria leuckarti* in ponies. J. Parasitol. 58:112-122.

Becker, E. R. 1934. Coccidia and coccidiosis of domesticated game and laboratory animals and of man. Collegiate Press, Ames, Iowa.

Bhatia, B. B. and B. P. Pande. 1966. Merozoites from megaloschizonts in abomasum and small intestine of sheep. Indian. J. Microbiol. 6:41-44.

Box, E. D. 1970. *Atoxoplasma* associated with an isosporan oocyst in canaries. J. Protozool. 17:391-396.

Brandborg, L. L., S. B. Goldberg, and W. C. Breidenbach. 1970. Human coccidiosis—a possible cause of malabsorption. New England J. Med. 283:1306-1313.

Canning, E. U. and M. Anwar. 1968. Studies on meiotic division in coccidial and malarial parasites. J. Protozool. 15:290-298.

Chatton, E. 1910. Le kyste de Gilruth dans la muqueuse stomacale des ovidés. Arch. Zool. Exp. Gener., 5th Ser. 5:114-124.

Chatton, E. 1938. Un xénoparasitome dans la muqueuse stomacale des ovidés. Titres et Travaux Scientifique, E. Settano ed. (Sète), pp. 181-184.

Chobotar, B. and D. M. Hammond. 1969. Development of gametocytes and second asexual generation stages of *Eimeria auburnensis* in calves. J. Parasitol. 55:1218-1228.

Chobotar, B. and D. M. Hammond. 1971. Cinemicrographic observations on excystation in *Eimeria utahensis* and *E. papillata*. J. Protozool. 18 (Suppl.):12.

Chobotar, B., D. M. Hammond and M. L. Miner. 1969. Development of the first-generation schizonts of *E. auburnensis*. J. Parasitol. 55:385-397.

Clark, G. W. 1970. *Eimeria ahtanumensis* n. sp. from the northwestern fence lizard *Sceloporus occidentalis* in central Washington. J. Protozool. 17:526-530.

Clark, W. N. and D. M. Hammond. 1969. Development of *Eimeria auburnensis* in cell cultures. J. Protozool. 16:646-654.

Colley, F. C. 1968. Fine structure of schizonts and merozoites of *Eimeria nieschulzi*. J. Protozool. 15:374-382.

Colley, F. C. and V. Zaman. 1970. Observations on the endogenous stages of *Toxoplasma gondii* in the cat ileum. II. Electron microscope study. Southeast Asian J. Trop. Med. Public Health 1:465-480.

Danforth, H. D. and D. M. Hammond. 1972. Merogony in multinucleate merozoites of *Eimeria magna* Perard, 1925. J. Protozool., 19:454-457.

Davies, S. F. M., L. P. Joyner and S. B. Kendall. 1963. Coccidiosis. Oliver and Boyd, Edinburgh.

de Vos, A. J., D. M. Hammond, and C. A. Speer. 1972. Development of *Eimeria crandallis* from sheep in cultured cells. J. Protozool. 19: 335-343.

Doran, D. J. and M. M. Farr. 1962. Excystation of the poultry coccidium, *Eimeria acervulina*. J. Protozool. 9:154-161.

Dubey, J. P. and J. K. Frenkel. 1972a. Extra-intestinal stages of *Isospora felis* and *I. rivolta* (Protozoa: Eimeriidae) in cats. J. Protozool. 19:89-92.

Dubey, J. P. and J. K. Frenkel. 1972b. Cyst-induced toxoplasmosis in cats. J. Protozool. 19: 155-177.

Dubey, J. P., N. L. Miller, and J. K. Frenkel. 1970. The *Toxoplasma gondii* oocyst from cat feces. J. Exp. Med. 132:636-662.

Dubremetz, J.-F. 1971. L'ultrastructure du centriole et du centrocône chez la coccidie *Eimeria necatrix*. Étude au cours de la schizogonie. J. Microscop. 12:453-458.

Farr, M. M. and D. J. Doran. 1962. Comparative excystation of 4 species of poultry coccidia. J. Protozool. 9:402-407.

Fayer, R. and D. M. Hammond. 1967. Development of first-generation schizonts of *Eimeria bovis* in cultured bovine cells. J. Protozool. 14:764-772.

Fayer, R. 1971. Quinine inhibition of host cell penetration by eimerian sporozoites *in vitro*. J. Parasitol. 57:901-905.

Fayer, R. 1972. Development of *Isospora rivolta* from cats in cultured mammalian cells. J. Parasitol., in press.

Fitzgerald, P. R. 1962. Coccidia in Hereford calves on summer and winter ranges and in feedlots in Utah. J. Parasitol. 48:347-351.

Fitzgerald, P. R. 1967. Results of continuous low-level inoculations with *Eimeria bovis* in calves. Am. J. Vet. Res. 28:659-665.

Frenkel, J. K. and J. P. Dubey. 1972. Rodents as transport hosts for cat coccidia, *Isospora felis* and *I. rivolta*. J. Infect. Dis. 125:69-72.

Grell, K. G. 1953. Entwicklung and Geschlechtsbestimmung von *Eucoccidium dinophili*. Arch. Protistenk. 99:156-186.

Grell, K. G. 1968. Protozoologie. Second edition. Springer-Verlag, Berlin, Heidelberg, New York, pp. 174-175.

Haberkorn, A. 1970. Die Entwicklung von *Eimeria falciformis* (Eimer 1870) in der weissen Maus (*Mus musculus*). Z. Parasitenk. 34:49-67.

Hammond, D. M. 1964. Coccidiosis of cattle; some unsolved problems. Thirtieth Faculty Honor Lecture, The Faculty Association, Utah State University, Logan.

Hammond, D. M. 1971. The development and ecology of coccidia and related intracellular parasites. *In* A. M. Fallis, Ecology and Physiology of Parasites, a symposium. Toronto University Press, Toronto, pp. 3-19.

Hammond, D. M., F. L. Andersen and M. L. Miner. 1963. The occurrence of a second asexual generation in the life cycle of *Eimeria bovis* in calves. J. Parasitol. 49:428-434.

Hammond, D. M., G. W. Bowman, L. R. Davis, and B. T. Simms. 1946. The endogenous phase of the life cycle of *Eimeria bovis*. J. Parasitol. 32:409-427.

Hammond, D. M., W. N. Clark, and M. L. Miner. 1961. Endogenous phase of the life cycle of *Eimeria auburnensis* in calves. J. Parasitol. 47:591-596.

Hammond, D. M., J. V. Ernst, and B. Chobotar. 1970. Composition and function of the substiedal body in the sporocysts of *Eimeria utahensis*. J. Parasitol . 56:618-619.

Hammond, D. M., J. V. Ernst, and M. L. Miner. 1966. The development of first-generation schizonts of *Eimeria bovis*. J. Protozool. 13:559-564.

Hammond, D. M., T. W. McCowin, and J. L. Shupe. 1954. Effect of site of inoculation and of treatment with sulfathalidine-arsenic on experimental infection with *Eimeria bovis* in calves. Proc. Utah Acad. Sci. 31:161-162.

Hammond, D. M., E. Scholtyseck, and B. Chobotar. 1967. Fine structures associated with nutrition of the intracellular parasite *Eimeria auburnensis*. J. Protozool. 14:678-683.

Hammond, D. M., E. Scholtyseck, and B. Chobotar. 1969. Fine structural study of the microgametogenesis of *Eimeria auburnensis*. Z. Parasitenk. 33:65-84.

Hibbert, L. E. 1969. Excystation of *Eimeria* species: composition and removal of the Stieda body, activation of sporozoites, and alteration in permeability of the micropyle. Dissertation filed in library of Utah State University.

Hutchison, W. M., J. D. Dunachie, K. Work and J. C. Siim. 1971. The life cycle of the coccidian parasite, *Toxoplasma gondii*, in the domestic cat. Trans. Roy. Soc. Trop. Med. Hyg. 65:380-399.

Jackson, A. R. B. 1962. Excystation of *Eimeria arloingi* (Marotel, 1905): Stimuli from the host sheep. Nature 194:847-849.

Kelley, G. L. and D. M. Hammond. 1970. Development of *Eimeria ninakohlyakimovae* from sheep in cell cultures. J. Protozool. 17:340-349.

Kelley, G. L. and D. M. Hammond. 1972. Fine structural aspects of early development of *Eimeria ninakohlyakimovae* in cultured cells. Z. Parasitenk., 38:271-284.

Kheysin, E. M. 1960. Cytological investigation of the life cycle of rabbit coccidia. 2. *Eimeria magna* Perard, 1924. Problems of Cytology and Protistology, USSR Acad. Sci., Inst. Cytology, pp. 311-331.

Kheysin, E. M. 1967. Zhiznennye tsikly koktsidii domashnikh zhivotnykh. Isdat Nauka, Leningrad. (English translation by F. K. Plous, Jr., edited by K. S. Todd, Jr. 1971. Life cycles of coccidia of domestic animals. University Park Press, Baltimore).

Ladda, R. L. 1969. New insights into the fine structure of rodent malarial parasites. Milit. Med. 134:825-865.

Lee, D. L. and P. L. Long. 1972. An electron microscopal study of *Eimeria tenella* grown in the liver of the chick embryo. Int. J. Parasitol. 2:55-58.

Lepp, D. L. 1972. Life cycle of *Isospora canis* Nemeséri, 1959 in the dog. M. S. Thesis, Univ. of Illinois, Urbana.

Levine, N. D. 1961. Protozoan parasites of domestic animals and of man. Burgess, Minneapolis.

Levine, N. D. 1963. Coccidiosis. Ann. Rev. Microbiol. 17:179-198.

Levine, N. D. and V. Ivens. 1965a. The coccidian parasites (Protozoa, Sporozoa) of rodents. Ill. Biol. Monogr. #33. Univ. Ill. Press, Urbana.

Levine, N. D. and V. Ivens. 1965b. *Isospora* species in the dog. J. Parasitol. 51:859-864.

Levine, N. D. and V. Ivens. 1970. The coccidian parasites (Protozoa, Sporozoa) of ruminants. Ill. Biol. Monogr. #44. Univ. Ill. Press, Urbana.

Long, P. L. 1970. *In vitro* culture of *Eimeria tenella*. J. Parasitol. 56 (4, Sect 2):214-215.

Long, P. L. 1972. Observations on the oocyst production and viability of *Eimeria mivati* and *E. tenella* in the chorioallantois of chicken embryos incubated at different temperatures. Z. Parasitenk. 39:27-37.

Long, P. L. and M. E. Rose. 1970. Extended schizogony of *Eimeria mivati* in betamethasone-treated chickens. Parasitology 60:147-155.

Lotze, J. C. and R. G. Leek. 1968. Excystation of the sporozoites of *Eimeria tenella* in apparently unbroken oocysts in the chicken. J. Protozool. 15:693-697.

Lotze, J. C. and R. G. Leek. 1969. Observations on *Eimeria tenella* in feces of inoculated chickens. J. Protozool. 16:496-498.

McCully, R. M., P. A. Basson, V. de Vos, and A. J. de Vos. 1970. Uterine coccidiosis of the impala caused by *Eimeria neitzi* spec. nov. Onderstepoort J. Vet. Res. 37:45-48.

McLaren, D. J. 1969. Observations on the fine structural changes associated with schizogony and gametogony in *Eimeria tenella*. Parasitology 59:563-574.

Mahrt, J. L. 1966. Life cycle of *Isospora rivolta* (Grassi, 1879), Wenyon, 1923 in the dog.

Ph.D. Thesis, Univ. of Illinois, Urbana.
Mahrt, J. L. 1967. Endogenous stages of the life cycle of *Isospora rivolta* in the dog. J. Protozool. 14:754-759.
Mahrt, J. L. 1968. Sporogony of *Isospora rivolta* oocysts from the dog. J. Protozool. 15:308-312.
Marcial-Rojas, R. A. 1971. Coccidiosis (Isosporiosis). *In* R. A. Marcial Rojas, Pathology of protozoal and helminthic diseases with clinical correlations. Williams and Wilkins, Baltimore, pp. 189-194.
Marquardt, W. C. 1963. Observations on living *Eimeria nieschulzi* of the rat. J. Parasitol. 49 (Suppl.):28.
Marquardt, W. C. 1966. The living, endogenous stages of the rat coccidium, *Eimeria nieschulzi*. J. Protozool. 13:509-514.
Marquardt, W. C., C. M. Senger, and L. Seghetti. 1960. The effects of physical and chemical agents on the oocyst of *Eimeria zurnii* (Protozoa, Coccidia). J. Protozool. 7:186-189.
Matsubayashi, H. and T. Nozawa. 1948. Experimental infection of *Isospora hominis* in man. Am. J. Trop. Med. 28:633-637.
Nemeséri, L. 1960. Beiträge zur Ätiologie der Coccidiose der Hunde. I: *Isospora canis* sp. n. Acta Vet. Acad. Sci. Hung. 10:95-99.
Niilo, L. 1970a. Experimental winter coccidiosis in sheltered and unsheltered calves. Can. J. Comp. Med. 34:20-25.
Niilo, L. 1970b. Bovine coccidiosis in Canada. Can. Vet. J. 11:91-98.
Nyberg, P. A., D. H. Bauer, and S. E. Knapp. 1968. Carbon dioxide as the initial stimulus for excystation of *Eimeria tenella* oocysts. J. Protozool. 15:144-148.
Nyberg, P. A. and D. M. Hammond. 1964. Excystation of *Eimeria bovis* and other species of bovine coccidia. J. Protozool. 11:474-480.
Nyberg, P. A. and S. E. Knapp. 1970. Scanning electron microscopy of *Eimeria tenella* oocysts. Proc. Helm. Soc. Wash. 37:29-32.
Ogino, N. and C. Yoneda. 1966. The fine structure and mode of division of *Toxoplasma gondii*. Arch. Ophthalmol. 75:218-227.
Pellérdy, L. P. 1965. Coccidia and coccidiosis. Akad. Kiado, Budapest.
Pellérdy, L. 1970. Life cycles involving sexual and asexual generations of *Eimeria* in gallinaceous birds. Parasitol. Hung. 3:133-146.
Pellérdy, L. and U. Dürr. 1970. Zum endogenen Entwicklungszyklus von *Eimeria stiedai* (Lindemann, 1865) Kisskalt & Hartmann, 1907. Acta Vet. Acad. Sci. Hung. 20:227-244.
Piekarski, G., B. Pelster and H. M. Witte. 1971. Endopolygenie bei *Toxoplasma gondii*. Z. Parasitenk. 36:122-130.
Poelma, F. G. 1966. *Eimeria lemuris* n. sp., *E. galago* n. sp. and *E. otolicni* n. sp. from a galago *Galago senegalis*. J. Protozool. 13:547-549.
Porchet-Henneré, E. and A. Richard. 1971. La schizogonie chez *Aggregata eberthi* étude en microscope électronique. Protistologica 7:227-259.
Pout, D. D. 1969. Coccidiosis of sheep. A critical review of the disease. Vet. Bull. 39:609-618.
Read, C. P. 1970a. Some physiological and biochemical aspects of host-parasite relations. J. Parasitol. 56:643-652.
Read, C. P. 1970b. Chemical pathology of trichinosis. *In* S. E. Gould, ed., Trichinosis in man and animals. Academic Press, New York, pp. 91-101.
Reichenow, E. 1953. Doflein's Lehrbuch der Protozoenkunde. 6th ed. 2 Teil, 2 Hälfte. Jena. 777-1213.
Reichenow, E. and A. Carini. 1937. Über *Eimeria travassosi* und die Gattung *Globidium*. Arch. Protistenk. 88:374-386.
Roberts, W. L., D. M. Hammond, L. C. Anderson, and C. A. Speer. 1970. Ultrastructural study of schizogony in *Eimeria callospermophili*. J. Protozool. 17:584-592.
Roberts, W. L., C. A. Speer, and D. M. Hammond. 1970. Electron and light microscope studies of the oocyst walls, sporocysts and excysting sporozoites of *Eimeria callospermophili*. and *E. larimerensis*. J. Parasitol. 56:918-926.
Roberts, W. L., C. A. Speer, and D. M. Hammond. 1971. Penetration of *Eimeria larimerensis* sporozoites into cultured cells as observed with the light and electron microscopes. J. Parasitol. 57:615-625.
Rutherford, R. L. 1943. The life cycle of four intestinal coccidia of the domestic rabbit. J. Parasitol. 29:10-32.
Sampson, J. R. and D. M. Hammond. 1971. Ingestion of host-cell cytoplasm by micropores in *Eimeria alabamensis*. J. Parasitol. 57:1133-1135.
Sampson, J. R. and D. M. Hammond. 1972. Fine structural aspects of development of *Eimeria*

alabamensis schizonts in cell cultures. J. Parasitol. 58:311-322.

Sampson, J. R., D. M. Hammond, and J. V. Ernst. 1971. Development of *Eimeria alabamensis* from cattle in mammalian cell cultures. J. Protozool. 18:120-128.

Scholtyseck, E. 1953. Beitrag zur Kenntnis des Entwicklungsganges des Hühnercoccids, *Eimeria tenella*. Arch. Protistenk. 98:415-465.

Scholtyseck, E. 1963. Untersuchungen über die Kernverhältnisse und das Wachstum bei Coccidiomorphen unter besonderer Berücksichtigung von *Eimeria maxima*. Z. Parasitenk. 22:428-474.

Scholtyseck, E. 1965. Elektronenmikroskopische Untersuchungen über die Schizogonie bei Coccidien (*Eimeria perforans* und *E. stiedae*). Z. Parasitenk. 26:50-62.

Scholtyseck, E. and D. M. Hammond. 1970. Electron microscope studies of macrogametes and fertilization in *Eimeria bovis*. Z. Parasitenk. 34:310-318.

Schmidt, K., M. R. L. Johnston, and W. E. Stehbens. 1967. Fine structure of the schizont and merozoite of *Isospora* sp. (Sporozoa: Eimeriidae) parasitic in *Gehyra variegata* (Dumeril and Bibron, 1836) (Reptilia: Gekkonidae). J. Protozool. 14:602-608.

Sénaud, J. 1967. Contribution à l'étude des Sarcosporidies et des toxoplasmes (Toxoplasmea). Protistologica 3:167-232.

Sénaud, J. 1969. Ultrastructure des formations kystiques de *Besnoitia jellisoni* (Frenkel 1953) Protozoaire, Toxoplasmea, parasite de la souris (*Mus musculus*). Protistologica 5:413-430.

Sénaud, J. and Z. Cerná. 1968. Etude en microscopie electronique des merozoites et de la merogonie chez *Eimeria pragensis* (Cerná et Sénaud 1968), coccidie parasite de l'intestin de la souris (*Mus musculus*). Ann. Station Biol. de Besse en Chandesse 3:221-242.

Sénaud, J. and Z. Cerná. 1969. Etude ultrastructurale des mérozoites et de la schizogonie des coccidies (Eimeriina): *Eimeria magna* (Perard 1925) de l'intestin des lapins et *E. tenella* (Railliet et Lucet, 1891) des coecums des poulets. J. Protozool. 16:155-165.

Shah, H. L. 1970a. *Isospora* species of the cat and attempted transmission of *I. felis* Wenyon, 1923 from the cat to the dog. J. Protozool. 17: 603-609.

Shah, H. L. 1970b. Sporogony of the oocysts of *Isospora felis* Wenyon, 1923 from the cat. J. Protozool. 17:609-614.

Shah, H. L. 1971. The life cycle of *Isospora felis*. J. Protozool. 18:3-17.

Sheffield, H. G. 1970. Schizogony in *Toxoplasma gondii*: An electron microscope study. Proc. Helm. Soc. Wash. 37:237-242.

Sheffield, H. G. and D. M. Hammond. 1966. Fine structure of first-generation merozoites of *Eimeria bovis*. J. Parasitol. 52:595-606.

Sheffield, H. G. and D. M. Hammond. 1967. Electron microscope observations on the development of first-generation merozoites of *Eimeria bovis*. J. Parasitol. 53:831-840.

Sheffield, H. G. and M. L. Melton. 1968. The fine structure and reproduction of *Toxoplasma gondii*. J. Parasitol. 54:209-226.

Snigirevskaya, E. S. 1968. The occurrence of micropore in schizonts, microgametocytes and macrogametes of *Eimeria intestinalis*. Acta Protozool. 5:381-386.

Snigirevskaya, E. S. 1969a. Electron microscopic study of the schizogony process in *Eimeria intestinalis*. Acta Protozool. 7:57-70.

Snigirevskaya, E. S. 1969b. Changes in some ultrastructures during microgametogenesis in rabbit coccidia. Tsitologiya 3:382-385.

Speer, C. A. and D. M. Hammond. 1970a. Nuclear divisions and refractile-body changes in sporozoites and schizonts of *Eimeria callospermophili* in cultured cells. J. Parasitol. 56:461-467.

Speer, C. A. and D. M. Hammond. 1970b. Development of *Eimeria larimerensis* from the Uinta ground-squirrel in cell cultures. Z. Parasitenk. 5:105-118.

Speer, C. A. and D. M. Hammond. 1971. Development of first- and second- generation schizonts of *Eimeria magna* from rabbits in cell cultures. Z. Parasitenk. 37:336-353.

Speer, C. A., L. R. Davis, and D. M. Hammond. 1971. Cinemicrographic observations on the development of *Eimeria larimerensis* in cultured bovine cells. J. Protozool. 18 (Suppl.):11.

Speer, C. A., D. M. Hammond, and L. C. Anderson. 1970. Development of *Eimeria callospermophili* and *E. bilamellata* from the Uinta ground-squirrel *Spermophilus armatus* in cultured cells. J. Protozool. 17:274-284.

Speer, C. A., D. M. Hammond, and G. L. Kelley. 1970. Stimulation of motility in merozoites of five *Eimeria* species by bile salts. J. Parasitol. 56:927-929.

Todd, K. S., Jr. and D. M. Hammond. 1971. Coccidia of Anseriformes, Galliformes and Passeriformes. *In* Davis, J. W., R. C. Anderson, L. Karstad and D. O. Trainer, eds. Infectious and parasitic diseases of wild birds. Iowa State Univ. Press, Ames. pp. 234-281.

Triffitt, M. J. 1925. Observations on *Gastrocystis gilruthi*, a parasite of sheep in Britain. Protozoology 1:7-18.

Triffitt, M. J. 1928. Further observations on the development of *Globidium gilruthi*. Protozoology 4:83-90.

Vetterling, J. M. 1968. Sporogony of the swine coccidium, *Eimeria debliecki* Douwes, 1921. J. Protozool. 15:167-172.

Vetterling, J. M. and D. J. Doran. 1969. Storage polysaccharide in coccidial sporozoites after excystation and penetration of cells. J. Protozool. 16:772-775.

Vetterling, J. M., P. A. Madden and N. S. Dittemore. 1971. Scanning electron microscopy of poultry coccidia after *in vitro* excystation and penetration of cultured cells. Z. Parasitenk. 37:136-147.

Vivier, E., G. Devauchelle, A. Petitprez, E. Porchet-Henneré, G. Prensier, J. Schrevel, and D. Vinckier. 1970. Observations de cytology comparée chez les sporozoaires. I. Les structures superficielles chez les formes végétatives. Protistologica 6:127-150.

Vivier, E. and A. Petitprez. 1968. Le complexe membranaire superficiel et son evolution lors de l'elaboration des individus-fils chez *Toxoplasma gondii*. J. Cell. Biol. 43:329-342.

Vivier, E. 1970. Observations nouvelles sur la reproduction asexuée de *Toxoplasma gondii* et considérations sur la notion d'endogenèse. C. R. Acad. Sci. Paris 271:2123-2126.

Wacha, R. S., D. M. Hammond, and M. L. Miner. 1971. The development of the endogenous stages of *Eimeria ninakohlyakimovae* (Yakimoff and Rastegaieff, 1930) in domestic sheep. Proc. Helm. Soc. Wash. 38:167-180.

Weissenberg, R. 1922a. Mikrosporidien, Myxosporidien and Chlamydozoen als Zellparasiten von Fischen. Verhandl. Deutsch. Zool. Gesellsch. 27:41-43.

Weissenberg, R. 1922b. Fremddienliche Reaktionen beim intrazellulären Parasitismus, ein Beitrag zur Kenntnis gallenähnlicher Bildungen in Tierkörper. Verhandl. Deutsch. Zool. Gesellsch. 27:96-98.

Weissenberg, R. 1949. Cell growth and cell transformation induced by intracellular parasites. Anat. Rec. 103:101-102.

Weissenberg, R. 1968. Intracellular development of the microsporidian *Glugea anomala* Moniez in hypertrophying migratory cells of the fish *Gasterosteus aculeatus* L., an example of the formation of "Xenoma" tumors. J. Protozool. 15:44-57.

Wenyon, C. M. 1926. Coccidia of the genus *Isospora* in cats, dogs and man. Parasitology 18:253-266.

Wenyon, C. M. and L. Sheather. 1925. Exhibition of specimens illustrating *Isospora* infections of dogs. Trans. Roy. Soc. Trop. Med. Hyg. 19:10.

Youssef, N. N. and D. M. Hammond. 1972. The ultrastructural cytology of the nucleolus in the gamonts of *Eimeria magna* gamonts. J. Parasitol. 58:669-679.

4 Ultrastructure

ERICH SCHOLTYSECK

Zoologisches Institut der Universität Bonn,
Abteilung für Protozoologie, Bonn, Germany

Contents

I. Introduction . . . *82*
II. General Fine-Structural Characteristics of the Motile Stages of Coccidia and Related Organisms . . . *83*
- A. Pellicle . . . *84*
- B. Polar Rings . . . *84*
- C. The Microtubules . . . *84*
- D. Conoid . . . *88*
- E. Paired Organelles or Rhoptries . . . *91*
- F. The Micronemes . . . *92*
- G. The Micropore . . . *92*

III. The Fine Structure of the Developmental Stages of Coccidia . . . *100*
- A. Infective Stages . . . *100*
  - 1. Sporozoites . . . *100*
    - a. *Eimeria* species . . . *100*
    - b. Sporozoites of Other Coccidian Genera . . . *101*
  - 2. Merozoites . . . *102*
    - a. *Eimeria* species . . . *102*
    - b. *Isospora* species . . . *103*
    - c. *Frenkelia* species (M-Organism) . . . *103*
    - d. *Toxoplasma gondii* . . . *104*
    - e. *Besnoitia jellisoni* . . . *104*
    - f. *Sarcocystis* . . . *104*
    - g. Merozoites of Other Sporozoa . . . *104*
- B. Microgamonts and Microgametes . . . *105*
  - 1. Microgamonts . . . *105*
    - a. *Eimeria* species . . . *105*
  - 2. Microgametes . . . *107*
    - a. *Eimeria* species . . . *107*
    - b. *Toxoplasma gondii* . . . *108*
    - c. Microgametes of Other Genera . . . *108*
    - d. Microgametes of Other Sporozoa . . . *108*
- C. Macrogametes . . . *111*

1. Mature Macrogametes of *Eimeria* . . . *111*
2. Mature Macrogametes of *Toxoplasma* . . . *118*
3. Mature Macrogametes of *Aggregata eberthi* . . . *119*
4. Macrogametes of *Klossia helicina (Adeleidea)* . . . *119*
5. Macrogametes of Other Sporozoa . . . *119*
IV. Ultrastructural Aspects of the Development of Coccidia . . . *121*
A. Asexual Reproduction . . . *121*
B. Sexual Reproduction . . . *121*
1. Microgametogenesis . . . *122*
2. Growth and Development of Macrogametes . . . *127*
a. Fertilization . . . *127*
b. Formation of the Oocyst Wall . . . *129*
V. Enzyme Activities in Macrogametes . . . *129*
VI. Host-Parasite Relationships . . . *133*
VII. Conclusions . . . *137*
VIII. Appendix: Abbreviations . . . *138*
Literature Cited . . . *139*

I. Introduction

In the last ten years, the analysis of fine structure has yielded a number of new findings and advances in the knowledge of Sporozoa, and particularly of the coccidia. This new knowledge did not come about by chance, but is directly related to the recent improvements in techniques of electron microscopy. At first, it was possible only to conduct simple fine-structural analyses; later, investigations could be directed toward particular developmental stages, as well as to all of the processes of development in the life cycle. Most recently, comparative studies in functional micromorphology have become feasible.

The fine structure of the coccidia cannot be considered as an isolated topic, but belongs in the more comprehensive realm in common with that of other related Sporozoa such as gregarines and Haemosporidia. This viewpoint is taken in the present chapter, and examples from the latter two groups are cited in order to provide a better understanding of the fine-structural relationships of the coccidia.

Although the coccidian cell is similar in its fine structure to the typical protozoan or metazoan cell, it has certain ultrastructural peculiarities. These are characteristic of the coccidia and distinguish them from all other protozoan cells. The typical fine-structural features are found predominantly in the motile stages, such as the sporozoite and merozoite. These features include the cell boundary, with the underlying microtubules; the structures constituting the apical complex, namely, the polar ring, conoid, rhoptries and micronemes; also the characteristic micropore, which functions as an ultracytostome.

The discovery of these fine structures and organelles resulted in their characteristics being used for the first time as the basis for determining systematic relationships; indeed, these characteristics were proposed as the chief criteria for certain Sporozoa and the coccidia (Honigberg *et al.*, 1964; Garnham, 1969a,b; Levine, 1969a,b). As a result of this, the coccidia became a more inclusive category, through the addition of several protozoan groups of uncertain systematic status in which the typical fine structural characteristics had been found.

Another important area of fine-structural investigation of coccidia is the detailed analysis of their developmental processes, such as endodyogeny, endopolygeny, internal schizogony, and the asexual reproduction known as schizogony. In these developmental processes, which result in the formation of from two (endodyogeny) to thousands (schizogony) of merozoites, relationships are sometimes found which are of significance in the phylogeny of the coccidia. Also, the study of microgametogenesis, including the differentiation and fine structure of the microgametes, has yielded new knowledge, especially in comparative investigations of gregarines, haemogregarines, and Haemosporidia. Although the growth and differentiation of the macrogametes of coccidia have been studied relatively little, these processes are more completely known for the genus *Eimeria*. The macrogametes of *Eimeria* species have a number of typical inclusions, among which the wall-forming bodies and polysaccharide granules are the most prominent. Although the process of oocyst formation is fairly well understood, little is known as yet concerning the fine structural aspects of the phases of sporogony.

Fine structural investigations with physiological aspects concern primarily the processes of locomotion, excystation, penetration of the host cell by sporozoites and merozoites, intake and digestion of nutrients, and demonstration of enzyme activity in organelles and in the fine structures of developmental stages, as well as the cytochemical and EM demonstration of various reserve and inclusion materials.

In the ecology of the coccidia, rapid advances have been made since 1965, especially in the fine-structural studies of developmental stages in the natural host and in cultured cells. The findings are related to the problems of host-specificity as well as to those of parasite-host relationships. The latter are of particular significance, because many fine-structural differentiations occur in the contact zone between host and parasite; these are indicative of distinctive correlations between the two partners. The parasitophorous vacuole, which is formed by the host cell and harbors the parasite, plays an important role in these phenomena. Also fine-structural analyses of parasites and host cells may indicate whether the intracellular parasite may be expected to be pathogenic or nonpathogenic. However, little or no fine-structural knowledge has yet been obtained in the important area of pathology, that is, the degenerative changes caused in the host cell by intracellular parasitism.

The study of the fine structure of coccidia is a relatively new one, which started only about fifteen years ago. It is surprising that in such a short time the area has grown so greatly in depth and breadth by the numerous and outstanding findings that have occurred. Naturally, there are still large gaps in our knowledge in nearly all aspects of the field, and some aspects have scarcely been touched; these are promising topics for future studies.

In this chapter, the entire fine-structural complex of the coccidia and related groups in its morphological, physiological, developmental, and ecological aspects is comprehensively presented for the first time.

II. General Fine-Structural Characteristics of the Motile Stages of Coccidia and Related Organisms

The characteristic structures and organelles of the motile stages of coccidia and related groups are the pellicle, the polar rings, the microtubules, the conoid, the rhoptries (paired organelles), the micronemes, and the micropores.

A. PELLICLE

The cell boundary of the sporozoites and merozoites consists of an outer unit membrane and an inner layer which is composed of two unit membranes, normally closely attached to one another. The outer unit membrane (OM)* and the inner membranous layer (MM, IM) are separated by an intermediate osmiophobic space with a diameter of from 150 Å to 200 Å (Fig. 27). The pellicle varies in width in the different species but usually measures about 400 Å (Vivier *et al.*, 1970; Roberts and Hammond, 1970; Scholtyseck, Mehlhorn, and Friedhoff, 1970). The outer membrane is continuous and encloses the whole cell, whereas the inner membranous layer terminates near the anterior end at the polar ring. In young merozoites and sporozoites of the genus *Eimeria* and in the endodyocytes of *Frenkelia* sp., the inner layer is also interrupted at the posterior end, there forming a posterior polar ring (Scholtyseck and Piekarski, 1965; Andreassen and Behnke, 1968; Roberts, Hammond, and Speer, 1970; Roberts, Speer, and Hammond, 1971). Aikawa (1967) stated that the merozoites of *Plasmodium* were covered with a three-layered pellicular complex, including the continuous outer unit membrane, a labyrinth structure revealed by negative staining, and a layer of microtubules. Nothing comparable to the labyrinth structure, however, was observed in other sporozoa.

B. POLAR RINGS

The polar ring, an osmiophilic thickening, is formed by the inner membranous layer of the pellicle (Figs. 1c, 1d, 1f, 1h, 2a, 2b, 3, 4, 5, 9-16, 18, 27, 30). Usually one polar ring is present. Two of these structures occur in the sporozoites of *Lankesterella hylae*, *E. callospermophili*, *E. larimerensis*, and in the merozoites of *Isospora* sp., whereas three polar rings are observed in merozoites of *Plasmodium* sp. (Schmidt *et al.*, 1967; Aikawa, 1966, 1967; Roberts *et al.*, 1970; Roberts *et al.*, 1971). The function of the polar rings is still unknown. The circular osmiophilic thickenings may serve as a supporting structure around the opening of the inner layer of the pellicle at the anterior end. They probably play a role in the movement of the conoid. Some observations indicate that the polar ring is a structure to which the microtubules are attached (Figs. 2a, 2b, Roberts and Hammond, 1970).

C. THE MICROTUBULES

The microtubules (subpellicular fibrils) are present in many stages of the life cycle of Sporozoa. In sporozoites and merozoites of the coccidia they are regularly distributed around the periphery, running from the anterior end to the posterior region. Definite proof of their origin and attachment was not obtained for a long time. Each microtubule originates at the polar ring and extends posteriorly as far as the nucleus. This was verified by Roberts and Hammond (1970) in a negatively stained sporozoite of *Eimeria ninakohlyakimovae* in which a polar ring and the attached 24 microtubules were isolated from the remainder of the parasite cell (Fig. 2b). These microtubules average 6 μ in length, about half the length of a sporozoite. In cross sections they appear to be composed of an osmiophilic central core about 100 Å in diameter and a densely stained osmiophilic cortex about 60 Å wide. In some cases the microtubules show a transverse periodicity. At a high magnification from 8 to 13 subunits in the form of roughly circular profiles are observed. In negatively stained material these subunits consist of beaded filaments which run nearly parallel to the long axis of the microtubules.

---

*A list of the abbreviations used throughout this chapter is given in the appendix, pp. 138-139.

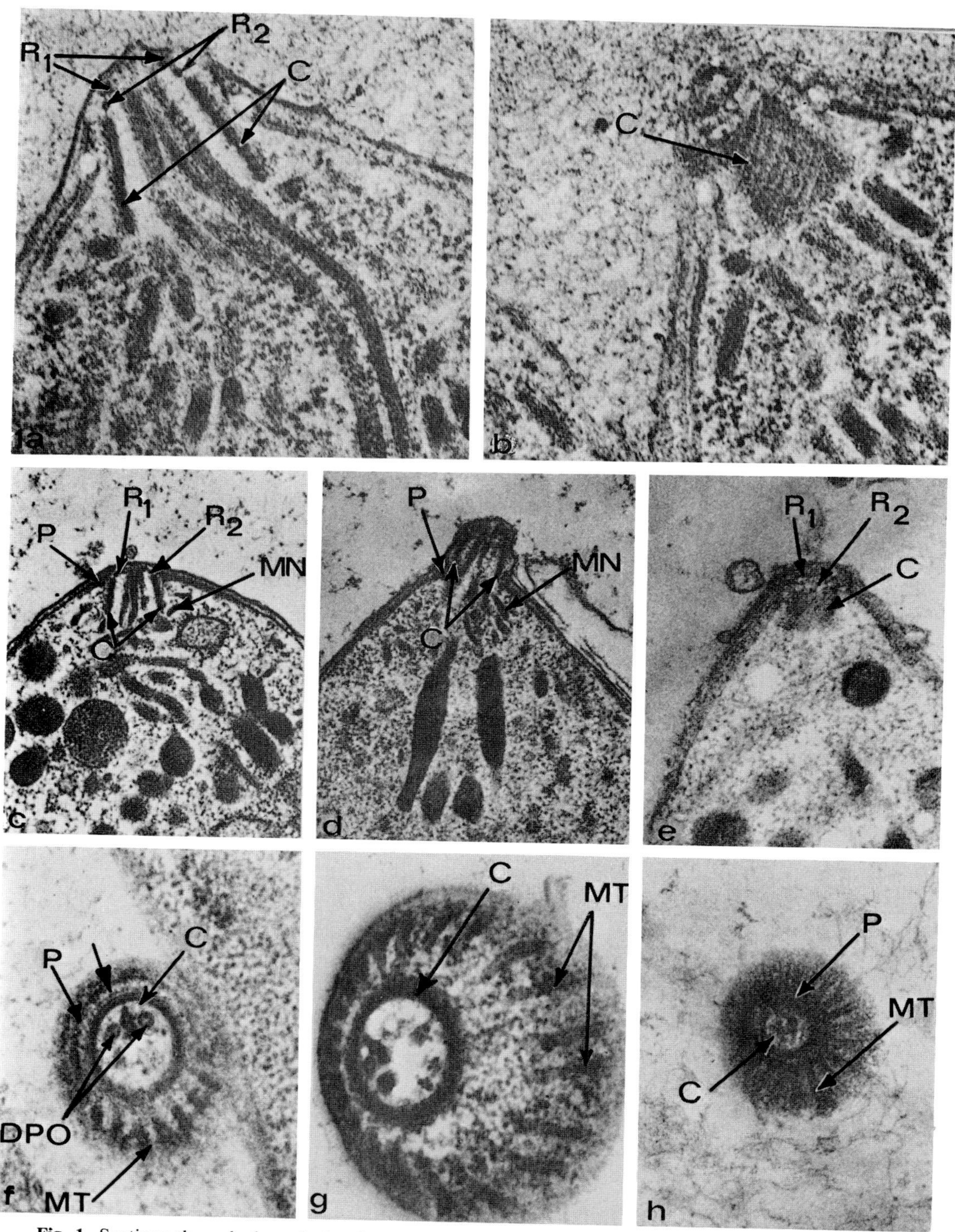

Fig. 1. Sections through the apical pole of merozoites with different aspects of the conoid; *a*, *b*, *f*, *g*, *Sarcocystis tenella*; *c*, *d*, *e*, *h*, *Toxoplasma gondii*. In *a*, *c*, *d* the conoid is sectioned longitudinally, whereas in *b* and *e* it is sectioned tangentially and shows a diagonal arrangement of elements. *f-h*, Cross sections through the anterior end of the cell. *a*, *b*, × 57,000; *c*, *d*, × 27,000; *e*, × 35,000; *f*, × 36,000; *g*, × 50,000; *h*, × 45,000 (from Sénaud, 1967).

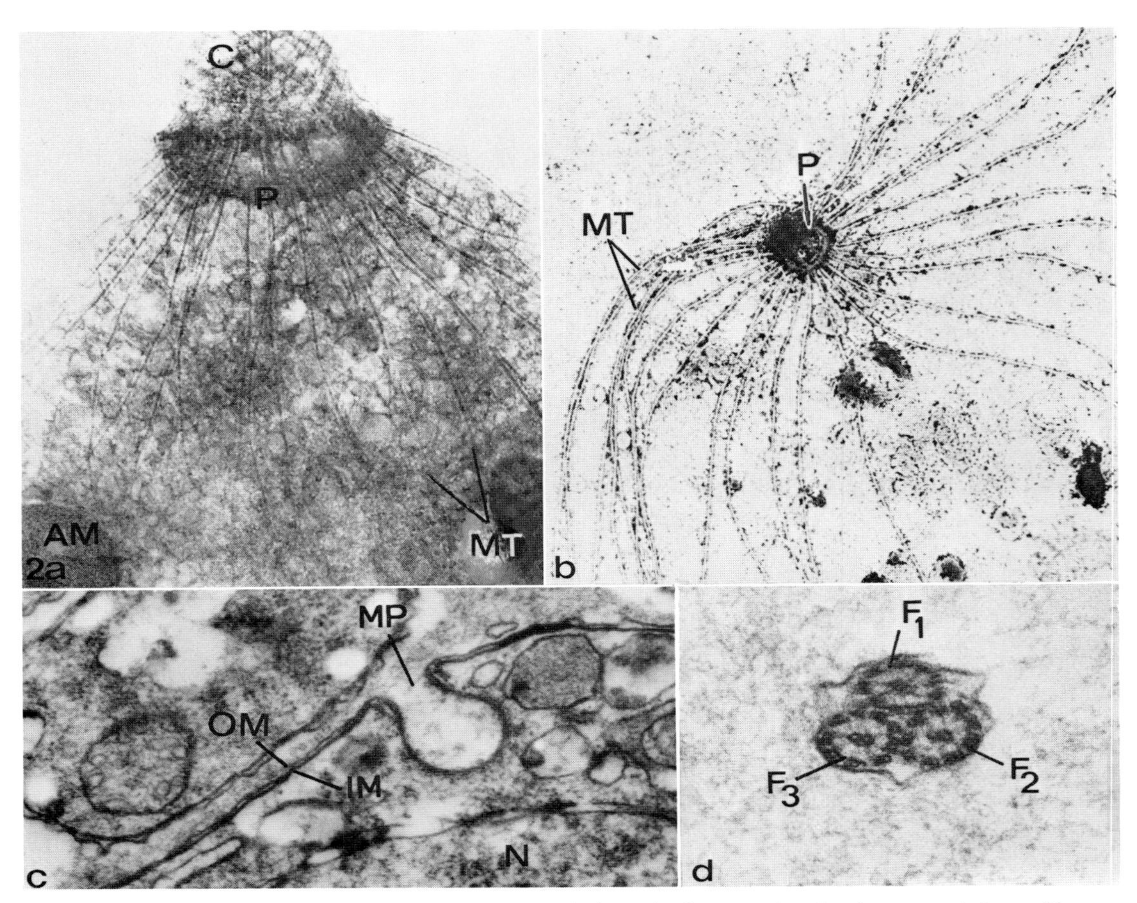

Fig. 2. *a*, *b*, *Eimeria ninakohlyakimovae*. *a*, Negatively stained sporozoite showing protruded conoid with spirally arranged fibrillar structures, polar ring, subpellicular microtubules, and tubules extending through the conoid. *b*, Negatively stained polar ring and subpellicular microtubules (from Roberts and Hammond, 1970). *c*, *Frenkelia* sp., section through a micropore of a metrocyte. *d*, *Aggregata eberthi*. Cross section through the apical pole of a microgamete showing three flagella (from Heller and Scholtyseck, 1969). *a*, × 41,000; *b*, × 10,800; *c*, × 42,000; *d*, × 58,000.

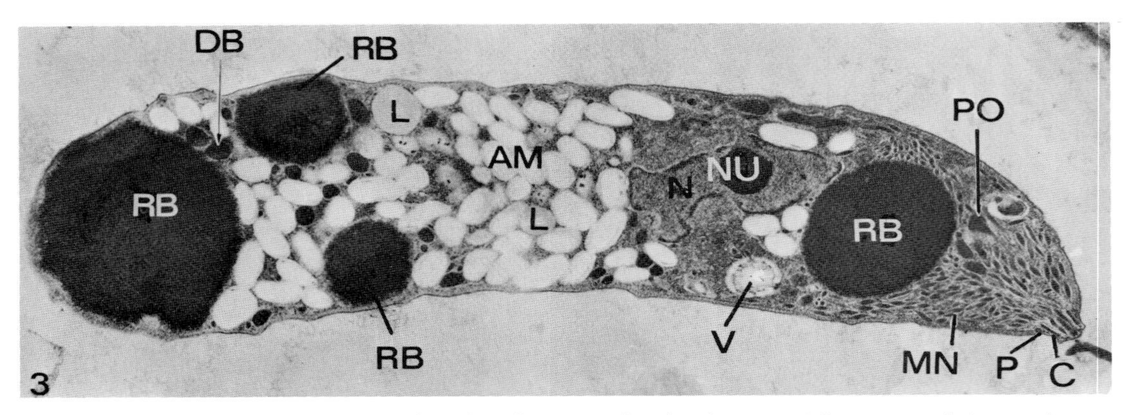

Fig. 3. *Eimeria bovis*. Longitudinal section of a sporozoite, showing general fine structural characteristics. × 12,000 (from Roberts and Hammond, 1970).

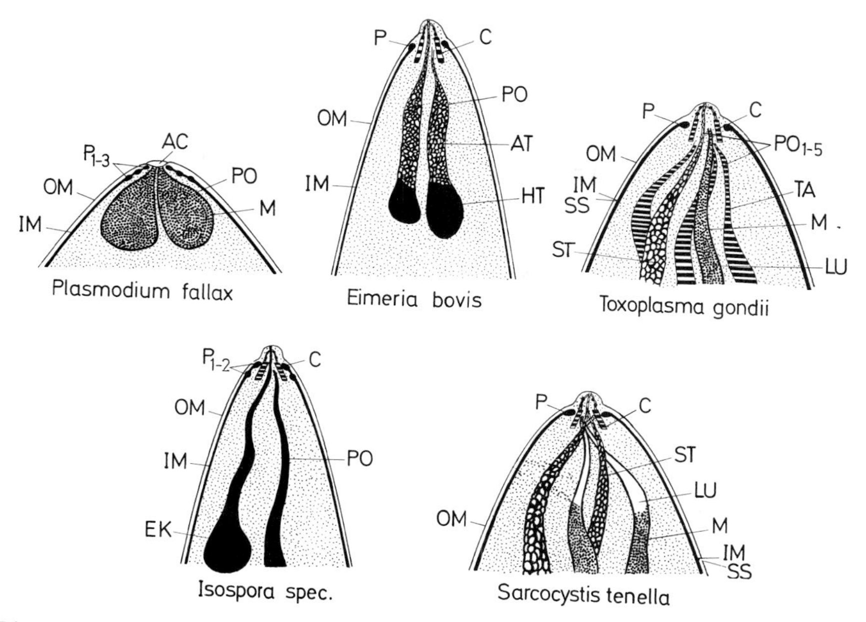

Fig. 4. Diagrammatic representation of some types of rhoptries (from Scholtyseck and Mehlhorn, 1970).

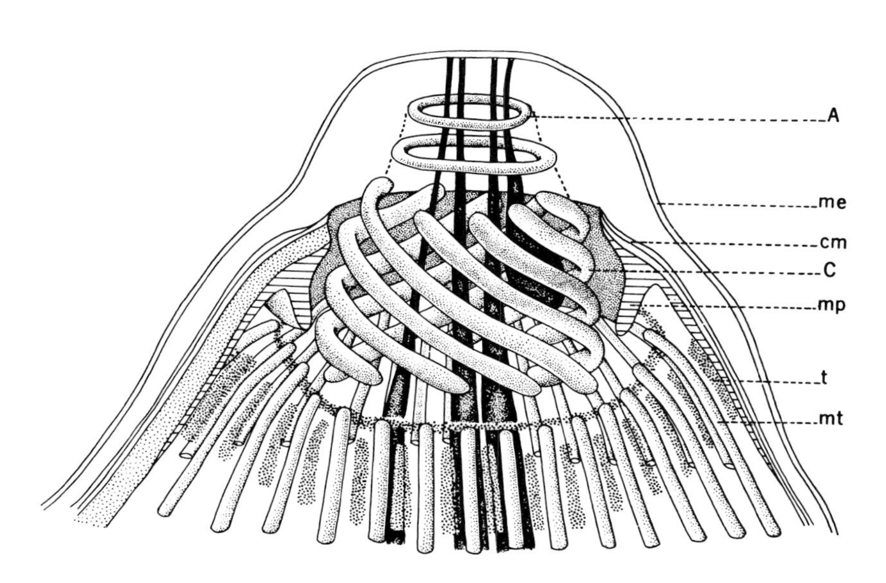

Fig. 5. Diagrammatic representation of the apical pole of a sporozoite of *Stylocephalus* sp. A, conoidal rings; me, outer membrane of the pellicle; mp, polar ring; C, conoid; mt, microtubules; t, intertubular space (from Desportes, 1969).

The number of microtubules may differ in different developmental stages of coccidia (Table 1). In sporozoites and merozoites of many *Eimeria* species there are 24 subpellicular microtubules. In merozoites of *E. stiedai* the microtubules have a structural connection with the inner layer of the pellicle; therefore the two structures may be a single functional unit.

Microtubules are cytoplasmic structures of wide occurrence in the cells of both plants and animals. There are several possible functions of microtubules in cytology. As the subpellicular tubular elements are present mainly in motile stages, their function may be related to cell motility in Sporozoa.

In sporozoites and merozoites of many species two "preconoidal rings" were described anterior to the cone (*Eimeria bovis, E. stiedai, E. callospermophili, E. tenella, Aggregata eberthi, Toxoplasma gondii, Besnoitia jellisoni, Sarcocystis tenella, Frenkelia* sp., and others). These two rings are connected with the hollow cone and with each other by a thin osmiophilic layer. In some *Eimeria* species, in *T. gondii, Frenkelia* sp., and *Lankesterella hylae*, the conoid and the preconoidal rings appear to be covered by a canopy-like membrane (Sheffield and Hammond, 1966; Scholtyseck, Mehlhorn, and Friedhoff, 1970). In some cases a spherical

D. CONOID

In 1954, Gustafson, Agar, and Cramer reported that the trophozoite (endodyocyte, merozoite) of *Toxoplasma gondii* possesses two differently structured poles. At the apical end they described a hollow cone-like structure as a special differentiation of the cytoplasm, and named it conoid. Since then this organelle has been found at the anterior pole running through the polar ring in gregarines, in the developmental stages of *Sarcocystis* sp., of *Besnoitia* sp., of *Frenkelia* sp., and in all sporozoites and merozoites of coccidia. A conoid is not present in the haemosporidians and the piroplasms.

The conoid consists of a truncated hollow cone of spirally arranged fibrillar structures from 260 Å to 300 Å in diameter (Figs. 1, 2a, 5, 6). In the different groups of Sporozoa the conoid is from 0.15 μ to 0.4 μ wide at the anterior end, from 0.2 μ to 0.53 μ at the posterior end, and from 0.08 μ to 4 μ long. The number of the spiral fibrillar structures varies slightly from six to eight. The proportions given in Table 2 are of interest, because the conoids of the different stages are from 1/20 to 1/75 of the length of the cell.

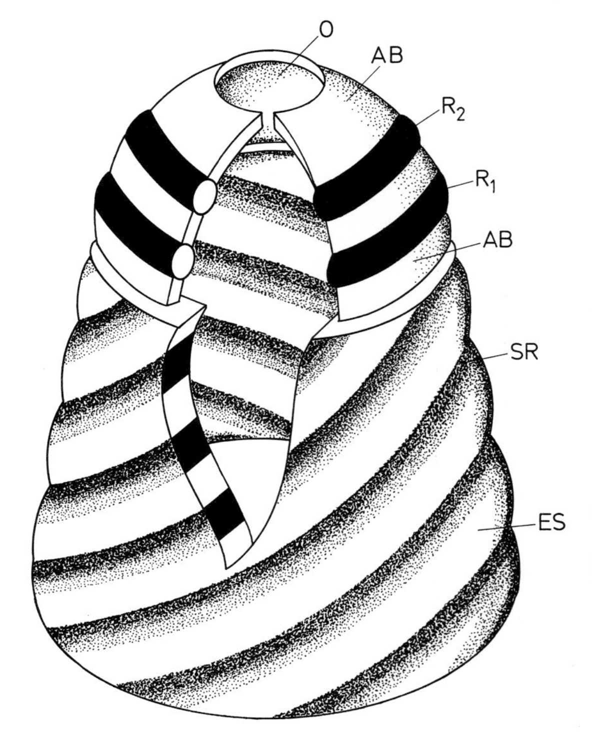

Fig. 6. Diagrammatic representation of a conoid. The conoidal rings (R) are connected with the hollow cone and with each other by fibrillar elements. Note the orifice of the rhoptries (O). The ring-like or spiral elements (SR) in some species seem to be solid, in others tubular (from Scholtyseck, Mehlhorn, and Friedhoff, 1970).

TABLE 1. Number of microtubules

Parasite	Stage	Number of microtubules
Eimeria tenella Ryley (1969)	Sporozoite	24
Eimeria ninakohlyakimovae *Eimeria ellipsoidalis* *Eimeria bovis* *Eimeria auburnensis* Roberts and Hammond (1970)	Sporozoite Sporozoite Sporozoite Sporozoite	24 24 24 24
Eimeria nieschulzi Colley (1967)	Sporozoite	25
Lankesterella hylae Stehbens (1966)	Sporozoite	27
Eucoccidium dinophili Bardele (1966)	Sporozoite	50
Isospora species Schmidt *et al.* (1967)	Merozoite	18-22
Eimeria bovis Sheffield and Hammond (1966)	Merozoite	22
Besnoitia jellisoni Sheffield (1966, 1968)	Merozoite	22
Sarcocystis tenella Sénaud (1967)	Merozoite	22-25
Toxoplasma gondii Sheffield and Melton (1968)	Merozoite	22
Eimeria stiedai *Eimeria perforans* Scholtyseck and Piekarski (1965)	Merozoite Merozoite	24 24
Eimeria tenella McLaren and Paget (1968)	Merozoite	24
Frenkelia species Scholtyseck *et al.* (1969)	Merozoite	24
Plasmodium fallax Aikawa (1967)	Merozoite	24-26
Eimeria nieschulzi Colley (1968)	Merozoite	25

(Continued)

TABLE 1 *(Continued)*

Parasite	Stage	Number of microtubules
Eimeria intestinalis Kheysin and Snigirevskaya (1965)	Merozoite	24-30
Eimeria miyairii Andreassen and Behnke (1968)	Merozoite	26
Eimeria magna Sénaud and Cerná (1969)	Merozoite	26
Babesia bigemina Friedhoff and Scholtyseck (1969)	Merozoite	32

TABLE 2. Morphological features of the conoid

Parasite	Length of the parasites/ Length of the conoid	Width of the parasites/ Width of the conoid	Width of the conoid/ Diameter of its wall	Width of the conoid/ Length of the conoid	Angle of inclination
Haemogregarina sp.	±73.5	±10	±15	±±2.1	72°
Eucoccidium dinophili	±75	±10	± 8.7	± 2	90°
Eimeria bovis	±54	± 7.2	±10	± 1	85°
Eimeria stiedai	±30.6	± 7.8	±10	± 1	80°
Eimeria tenella	±31.3	± 5.8	± 8.0	± 0.9	88°
Eimeria callospermophili	±20.6	± 9.3	±11.2	± 1.1	83°
Eimeria magna	±21.4	±10	± 9.3	± 0.9	85°
Isospora sp.	±55.5	± 8.0	± 7.2	± 1.4	75°
Lankesterella hylae	±37	± 8.0	±10	± 1	81°
Sarcocystis tenella	±30.7	±10.7	± 9.3	± 0.7	80°
Toxoplasma gondii	±20.7	±10	±10	± 1	70°
Besnoitia jellisoni	±22.7	±10	±10	± 1.2	85°
M-Organism	±20.0	± 7.5	±10	± 0.8	82°
Aggregata eberthi	±60	±11	±13	± 2.0	84°

opening was observed in the canopy-like membrane (Scholtyseck and Piekarski, 1965; Sheffield and Hammond, 1966; Andreassen and Behnke, 1968). Figure 6 shows such an organelle consisting of a hollow cone and a canopy-like membrane reinforced by two rings. It is, however, not definitely proved whether the two preconoidal rings are integral components of the conoid or are separate structures helping to maintain the shape of the anterior pole.

No structural connections between the conoid and the polar ring are observed. The location of the conoid with respect to the polar ring varies. Sometimes the polar ring surrounds the conoid at its top, and sometimes at its base, so that the posterior margin of the conoid is opposite or anterior to the polar ring. These observations led to the suggestion that the conoid is an organelle which can be protruded and retracted (McLaren and Paget, 1968). Figure 2a shows such a protruded conoid in a negatively stained sporozoite of *E. ninakohlyakimovae* (Roberts and Hammond, 1970).

There are many suggestions and theories concerning the function of the conoid; however, definite proof has not been obtained by electron microscopy. The conoid probably functions as an organelle of penetration of host cells. The exact mechanism of penetration is not known. Protrusion of the conoid may possibly assist in penetration.

As a conoid does not exist in haemosporidians and piroplasms, Aikawa (1967) proposed to give the name "conoid" to the entire apical complex of a sporozoite or merozoite; however, he did not propose a new term for the structure named by Gustafson *et al.*, (1954).

The origin of the conoid—the term is used here always in the sense of Gustafson *et al.* (1954)—is not known. There are some suggestions that this organelle may originate from a centriole.

E. PAIRED ORGANELLES OR RHOPTRIES

The paired organelles are electron-dense, club-shaped structures located in the anterior region of sporozoites and merozoites of different Sporozoa (Fig. 4, PO). Garnham *et al.* (1960) first used this term to describe a "pair" of these osmiophilic structures found in the anterior portion of the sporozoites of *Haemamoeba* (*Plasmodium*) *gallinacea*. Since then, these organelles have been observed in many genera of coccidians, haemosporidians, and piroplasms.

The paired organelles vary considerably in their shape. They may be club-shaped, drop-shaped, elongate or tortuous in shape. The anterior, neck portions of the organelles are narrow and extend into the conoid area (Fig. 1f). The posterior club-shaped portions, from 250mμ to 400mμ in diameter, appear to be made up of a dense, posterior homogeneous osmiophilic part and a less dense, sometimes alveolar, anterior part about 120mμ wide. Occasionally some portions of the organelles seem to be empty, having a covering membranous layer and showing a transverse periodicity in tangential sections (Sheffield and Melton, 1968; Fig. 4). In some *Eimeria* species a short rod-shaped body has been observed between the neck portions of the paired organelles (Sheffield and Hammond, 1966, and other authors).

A pair of these organelles occurs in the sporozoites of *Plasmodium*, in merozoites of *Isospora* and of many *Eimeria* species and in the erythrocytic merozoites of *Babesia bigemina*. More than two club-shaped structures (6-8) are found in *Eimeria* sporozoites as well as in *Toxoplasma, Besnoitia, Sarcocystis, Frenkelia*, in some *Eimeria* merozoites, and in the merozoites of *Babesia ovis* from the ovary of the tick *Rhipicephalus bursa*. For this reason Sénaud (1967) called these organelles "rhoptries" to describe the shape of the structures observed. The literature on Sporozoa includes a great variety of names used for the paired organelles. Jacobs (1967) called them "club-shaped organelles." The term used by most authors seems to be the name proposed by Sénaud (1967).

The function of the rhoptries is not known. There are, however, some indications that the paired organelles have a gland-like appearance and that they may secrete a proteolytic enzyme to assist in the mechanical function of the conoid. The narrow neck-like portions may serve as secretory ductules. This suggestion is supported by the results of Schrevel (1968), who found activity of acid phosphatase in the rhoptries of the gregarine *Selenidium*.

F. THE MICRONEMES

The micronemes are small, osmiophilic, convoluted, cord-like structures in the anterior region of many Sporozoa (Figs. 1c, 3, 9-18). They may be more tortuous than is indicated by their oval appearance in ultrathin longitudinal sections. In cross sections, they appear to be spherical, with a diameter of from 600 Å to 900 Å. They may vary considerably in length. In most cases these structures appear to be completely dense and osmiophilic. In some organisms, however, differences in their fine structure were described. In *Isospora* sp., *Haemogregarina* sp., *Babesia gibsoni*, *Selenidium* and others they appear to be covered by a membrane (Schmidt *et al.*, 1967; Stehbens and Johnston, 1967; Büttner, 1968; Schrevel, 1968). In some species, they may be surrounded by an electron-lucid space and in others are found to be partly empty. These structures were first described by Gustafson *et al.* (1954) in *Toxoplasma* and called "toxonemes". Since then various names have been given to them, such as "Cytoplasmastränge" in *E. stiedai* and *E. perforans* (Scholtyseck and Piekarski, 1965), "sarconemes" in *Sarcocystis* (Ludvik, 1960), "lankesterellonemes" in *Lankesterella* (Garnham *et al.*, 1962), "convoluted tubules" in *Plasmodium* (Garnham, 1963), "radial clubs" in *Toxoplasma* (Ogino and Yoneda, 1966) and "rod-shaped granules" in *E. miyairii* (Andreassen and Behnke, 1968). The term "microneme," coined by Jacobs (1967), is the name most frequently used.

The function of the micronemes is not known. Some authors suggested a secretory function and others indicated that the micronemes and the rhoptries belong to a common functional system. Structures similar to micronemes were observed in stages of *Babesia ovis*, there assisting in the formation of the inner membranous layer of the pellicle (Scholtyseck and Mehlhorn, 1970).

G. THE MICROPORE

The micropore is an organelle formed by the pellicle. In longitudinal section it consists of a punctate invagination of the outer unit membrane and the inner membranous layer. The outer membrane continues without interruption through the invagination, whereas the inner membranous complex forms a thickened cylindrical structure that surrounds the invagination of the outer membrane (Figs. 7, 8). This structure can also be found in those developmental stages which do not have the inner pellicle layer, such as the microgamonts of *E. auburnensis* (Hammond *et al.*, 1969), *E. intestinalis* (Snigirevskaya, 1968), *E. nieschulzi* (Colley, 1967) and the macrogametes of *E. perforans, E. stiedai, E. auburnensis* and *E. bovis* (Scholtyseck, Hammond, and Ernst, 1966). There the cylindrical structure is present, indicating the invagination of the two pellicle membranes during a previous stage. In transverse section the micropore is composed of two concentric rings. The outer ring is thicker than the inner one and represents the cylindrical structure. The inner ring corresponds with the invagination of the outer pellicle membrane (Figs. 7, 8).

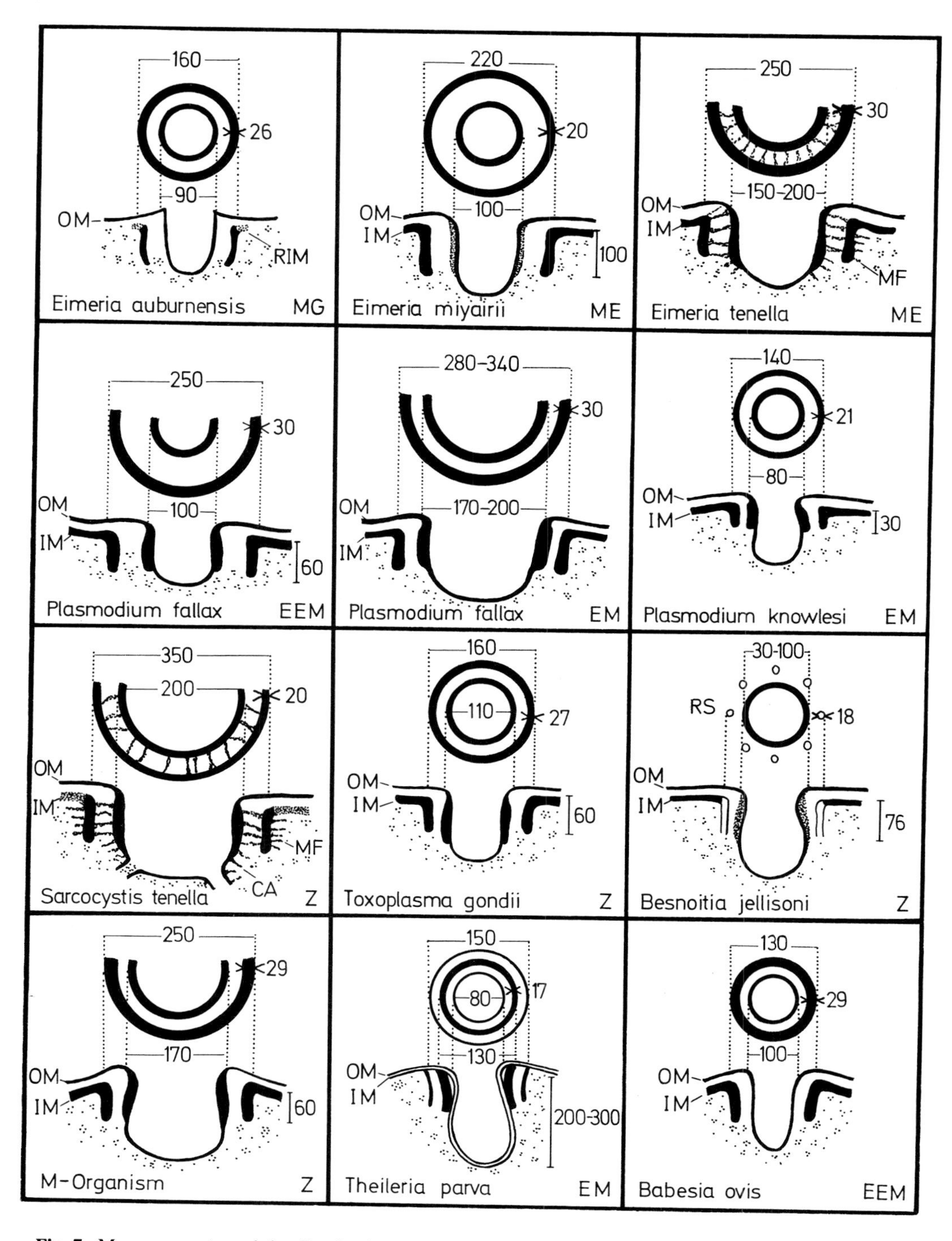

Fig. 7. Measurements and details of micropores in different groups of parasites (from Scholtyseck and Mehlhorn, 1970).

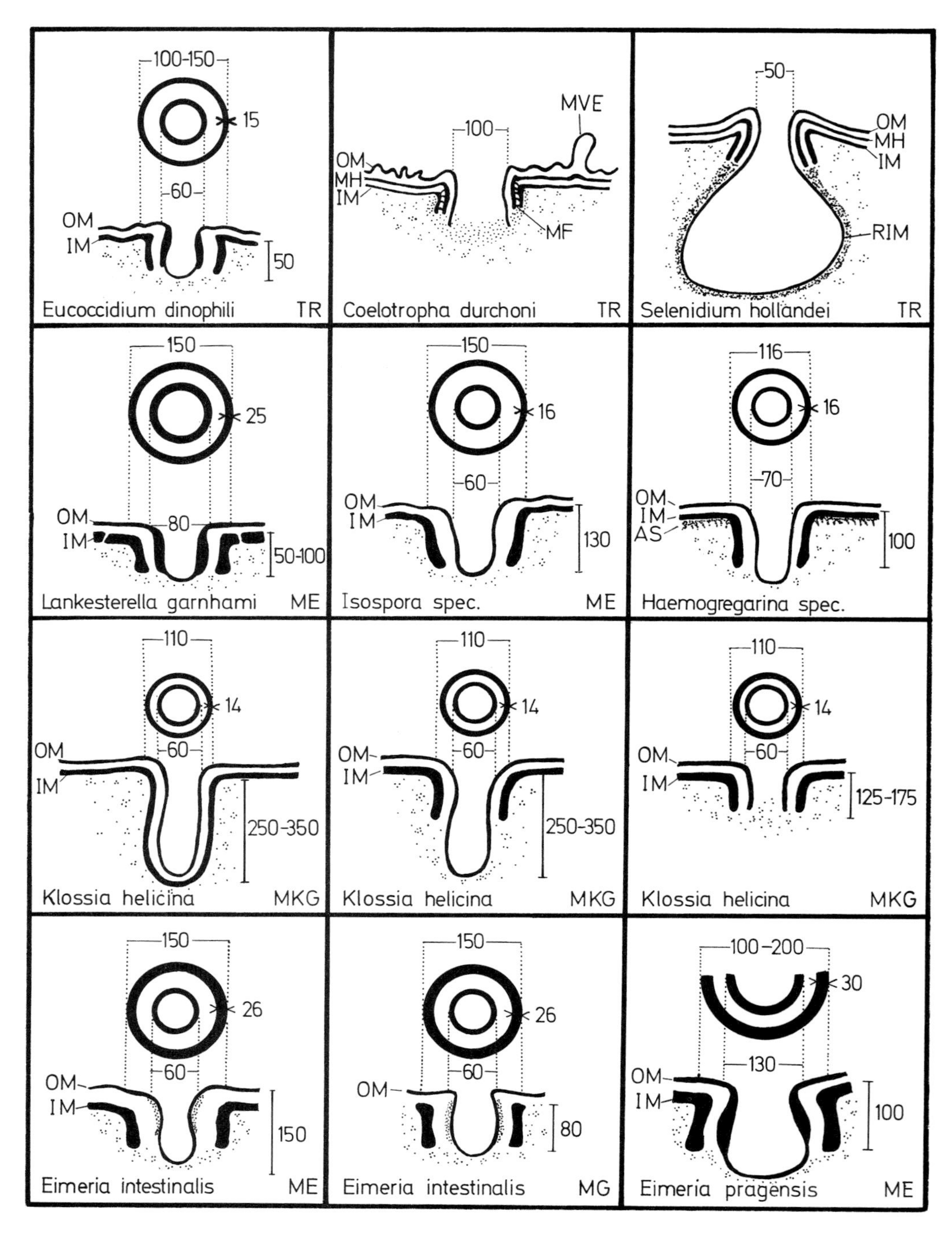

Fig. 8. Measurements and details of micropores in different groups of parasites (from Scholtyseck and Mehlhorn, 1970).

The micropores are 200mμ wide at the surface and 350mμ deep. Their size reported in the different groups of Sporozoa is given in Figures 7 and 8.

A few modifications of the basic structure of this organelle were described by some authors. In *Sarcocystis* from two to three side canals were observed branching off from the invagination of the outer membrane (Sénaud, 1967), and in *E. bovis* and *Klossia helicina* micropores without an interruption in the inner membranous layer (Sheffield and Hammond, 1966; Volkmann, 1967) were seen. Such structures may represent developing micropores.

These organelles are, in accordance with their possible function, not confined to the motile stages, but are present in almost every stage of the life cycle of Sporozoa. In sporozoites and merozoites of *Eimeria* and in zoites of *Sarcocystis, Toxoplasma,* and *Frenkelia,* the number of micropores is limited. In most cases, only one such organelle occurs, but in some species, such as *E. alabamensis,* several have been observed (Sampson and Hammond, 1970). In *Lankesterella, Haemogregarina, Besnoitia* and in *Plasmodia,* two of these organelles were found. In longitudinal sections of metrocytes of *Frenkelia,* even three micropores are present (Fig. 12). In microgamonts of *Eimeria, Eucoccidia* and in the trophozoites of the gregarine *Selenidium* the number of micropores may be great (Scholtyseck, Volkmann, and Hammond, 1966; Snigirevskaya, 1968; Bardele, 1966; Schrevel, 1968). Nine micropores in close proximity to each other, irregularly distributed at the surface, were found in a microgamont of *E. auburnensis*. In sporozoites and merozoites of *Eimeria* and *Isospora*, as well as in the zoites of *Toxoplasma, Besnoitia, Sarcocystis* and *Frenkelia*, a single micropore may occur at a particular place, namely between the anterior end and the region of the nucleus (Figs. 9-11, 13-16). A micropore occurs very early in young developing merozoites.

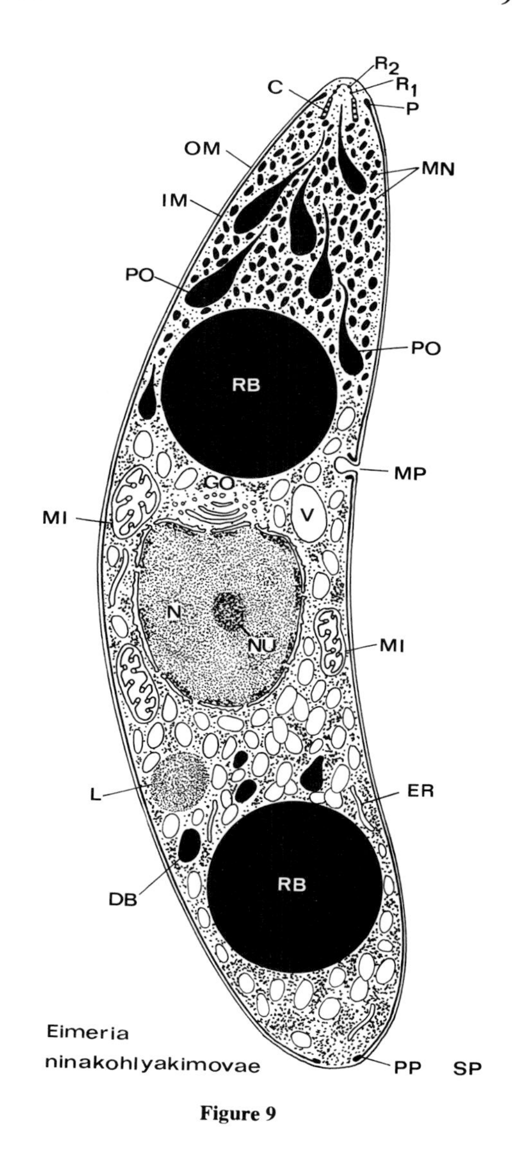

Figure 9

Figs. 9-18. Diagrammatic representations of a sporozoite (SP), a metrocyte (MZ), and merozoites (M) of Coccidia and Piroplasmea, showing the great similarities of the fine structure of the motile stages. (Fig. 9 after Roberts and Hammond, 1970; Fig. 10 after Scholtyseck and Piekarski, 1965; Fig. 11 after Schmidt, Johnston, and Stehbens, 1967; Fig. 12 after Kepka and Scholtyseck, 1970; Fig. 13 after Scholtyseck, Kepka, and Piekarski, 1965; Fig. 14, after Scholtyseck and Piekarski, 1965; Figs. 15, 16 after Sénaud, 1969, 1967; Figs. 17, 18 after Friedhoff and Scholtyseck 1968, 1969.)

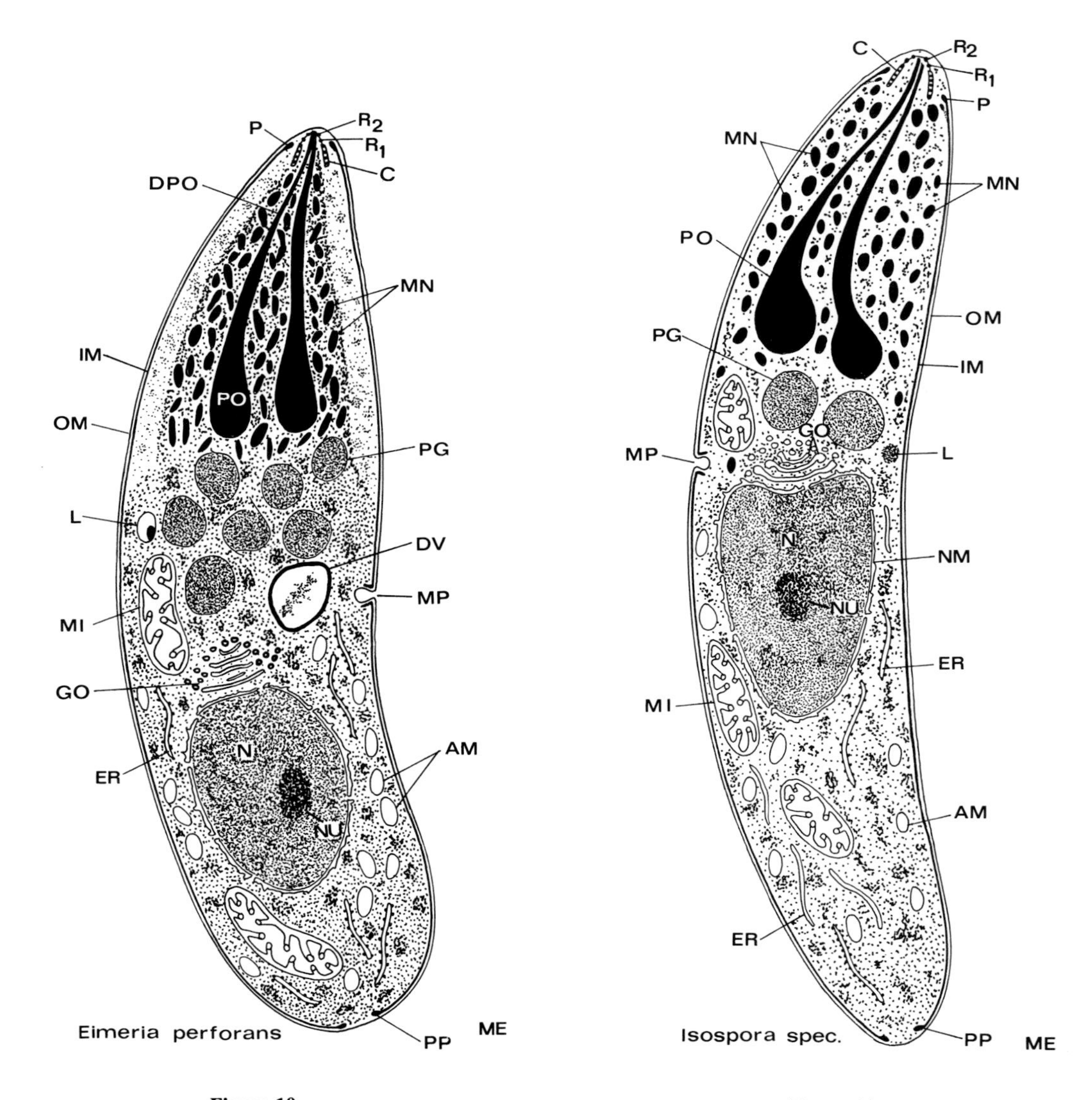

Figure 10

Figure 11

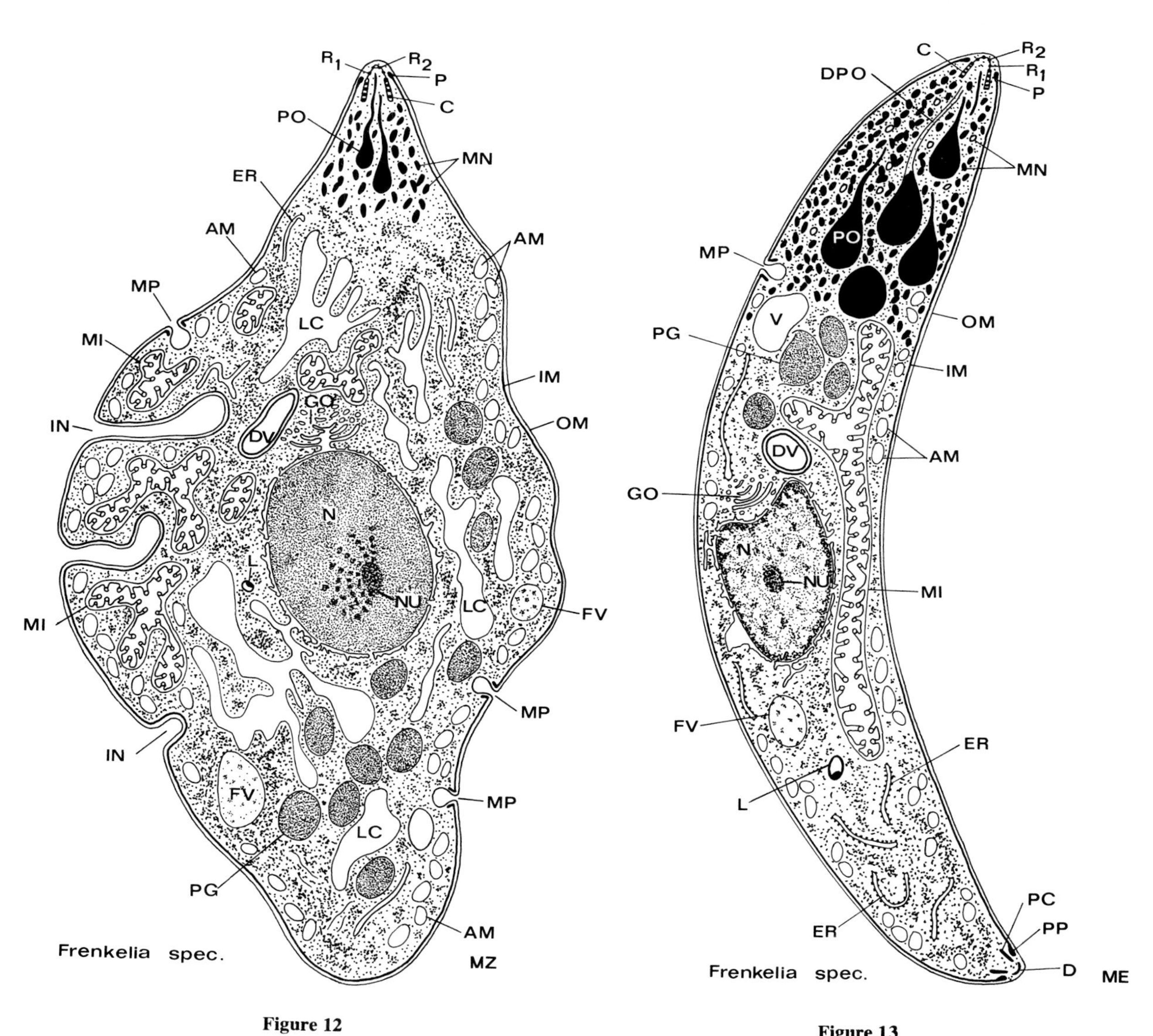

Figure 12

Figure 13

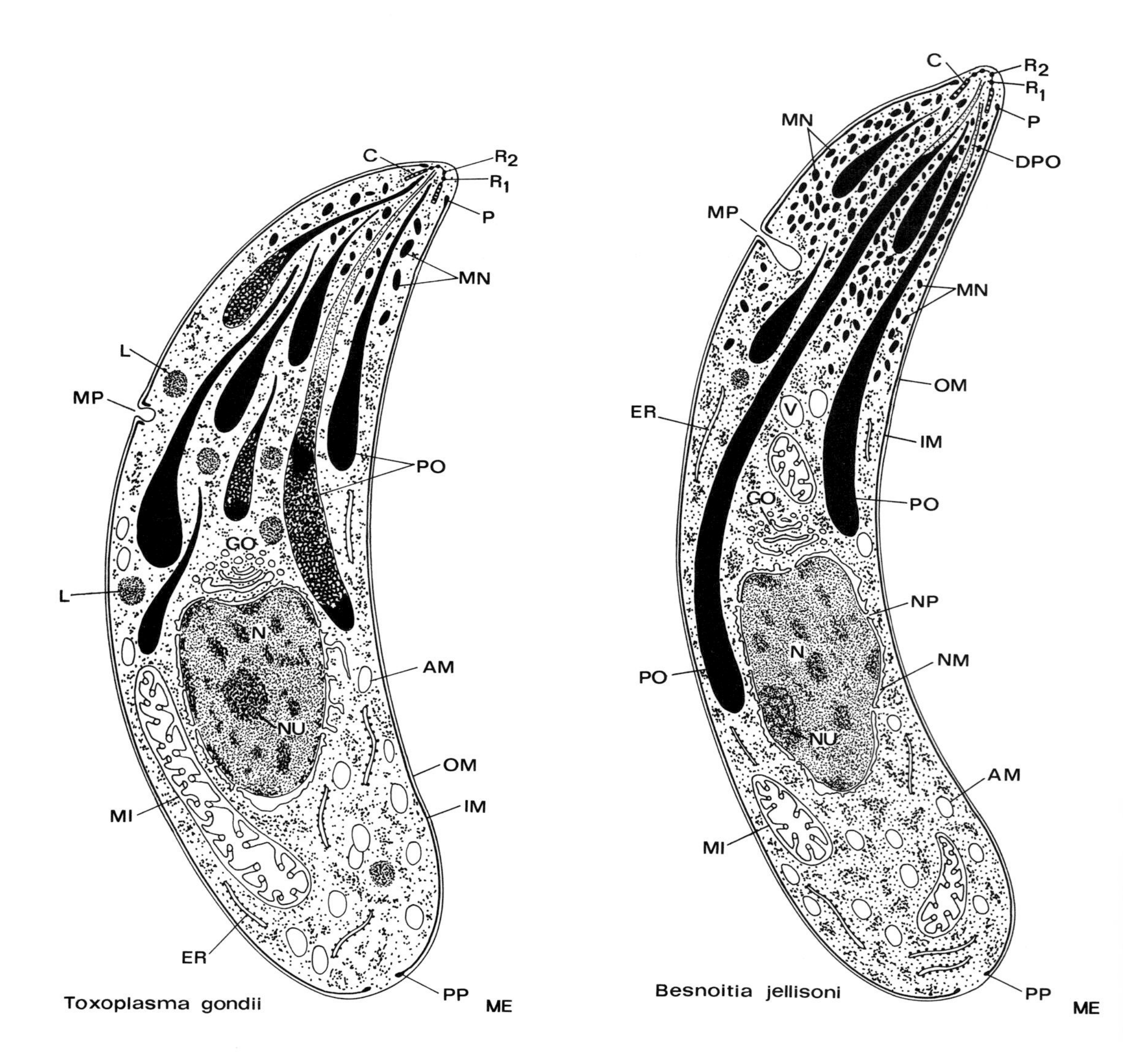

Figure 14

Figure 15

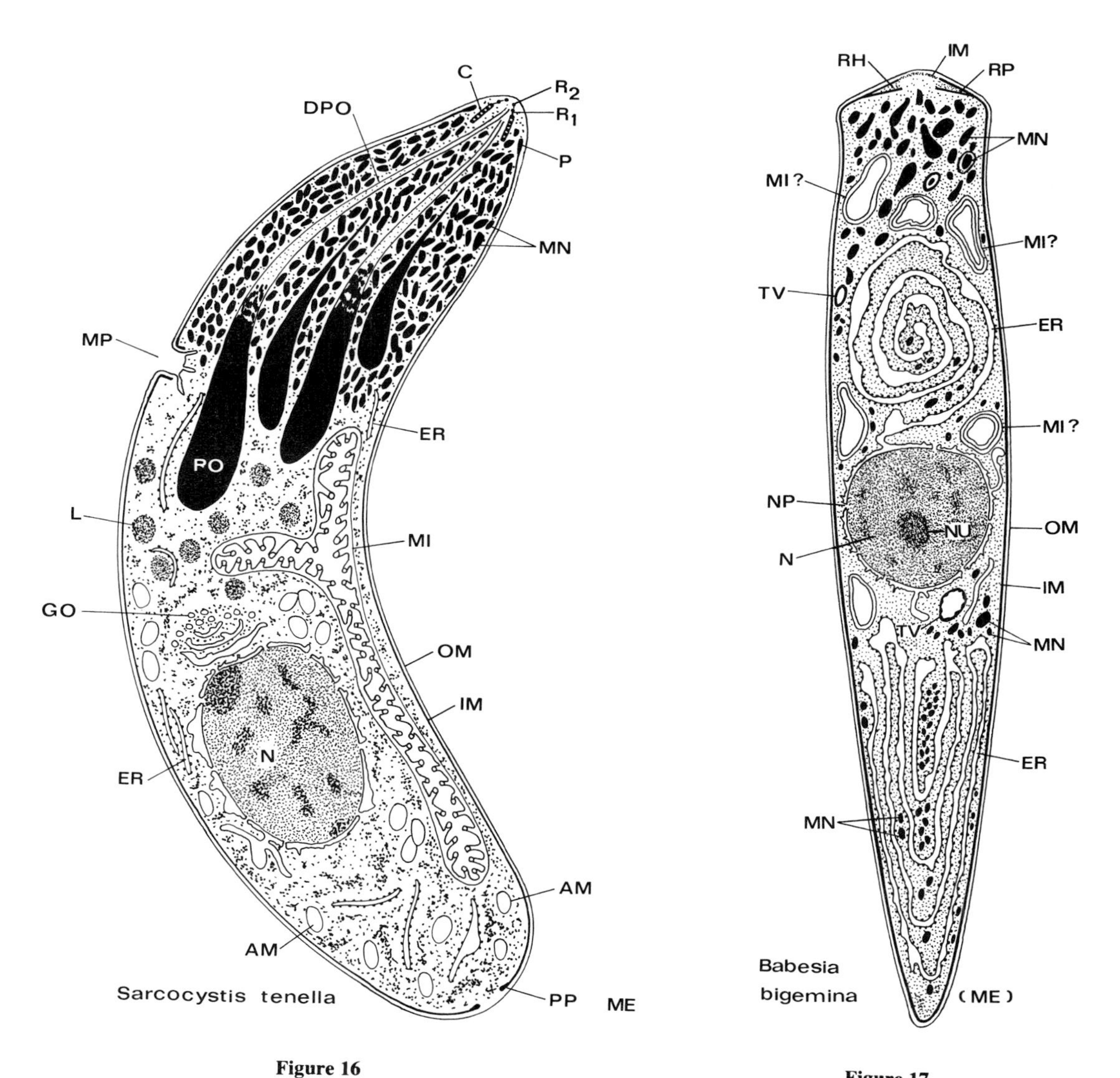

Figure 16

Figure 17

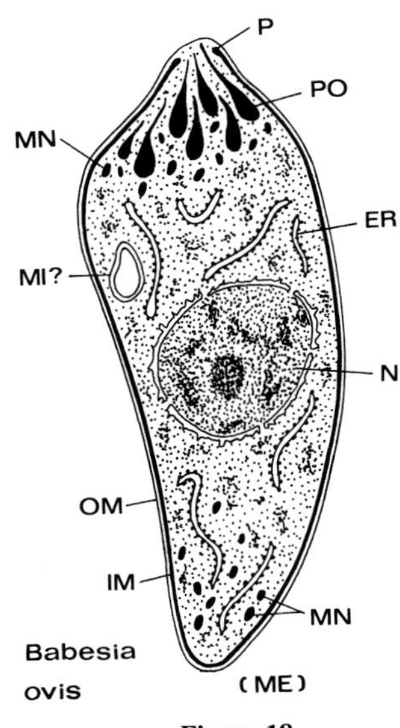

Figure 18

This pellicular organelle was first observed and described by Garnham *et al.* (1961, 1962) in sporozoites of *Haemamoeba* (*Plasmodium*) *gallinacea* and *Lankesterella garnhami* and called "micropyle." Kheysin and Snigirevskaya (1965) proposed the term "ultracytostome," but they found no direct evidence for a feeding mechanism of the organelle. Vivier and Henneré (1965) first used the term "micropore." This name is now the most frequently used one.

The micropore functions as a feeding organelle in *Plasmodium* species. Some evidence for such a function in *Eimeria* was described by Ryley (1969) and by Strout and Scholtyseck (1970). The latter observed micropores with food vacuoles forming at their bases in sporozoites of *E. tenella* within cultured cells. They found that parts of the host cell cytoplasm in the vicinity of the parasite undergoes disintegration soon after invasion of the host cell, and that this material later is included in the parasitophorous vacuoles and ingested by the micropores. The occurrence of active micropores in developmental stages of *E. intestinalis* and in merozoites of *E. pragensis* was reported by Snigirevskaya and Kheysin (1968) and by Sénaud and Černá (1968). Micropores, apparently in the process of ingesting unaltered host cell cytoplasm, were seen in intracellular sporozoites of *E. alabamensis* by Sampson and Hammond (1971). In developmental stages which do not take in nutrients, the micropores may be vestigial or may have an unknown function.

III. The Fine Structure of the Developmental Stages of Coccidia

A. INFECTIVE STAGES

Sporozoites and merozoites are motile infective stages able to penetrate host cells. In the life cycle of coccidia, the mobile vermiform sporozoite is the initial stage, whereas the merozoite derives from schizogony or a similar type of reproduction. The literature on Sporozoa has a confusing variety of terms for the merozoite stage. In the different groups, the names schizozoite, zoite, endodyocyte, vermicule, and others are used for the developmental stage representing a merozoite.

1. Sporozoites

Only a few reports in the electron microscope literature deal with sporozoites, therefore relatively little is known of the ultrastructure of this stage.

a. *Eimeria* species

The sporozoites of the *Eimeria* species which have been studied are relatively uniform in their fine structure. Differences occur mostly in their shape and size. The pellicle, the polar ring, the 24 microtubules, and the conoid show normal structures (Figs. 3-9). According to Colley (1968) there are 25 subpellicular microtubules in the sporozoites of *E. nieschulzi* (Table 1). Several rhoptries are located in the anterior region. In some specimens round and oval bipartite vesicles are found in the anterior half of the cell. They are thought to be portions of the rhoptries. The micronemes appear as elongate bodies. The most prominent and typical structures of the *Eimeria* sporozoites are the refractile bodies, also called paranuclear bodies or clear globules. Usually two of these, one anterior and one posterior, are present. They are osmiophilic, electron dense, and homogeneous. There is no indication of a limiting membrane. The size of these bodies may vary. Occasionally, the anterior body is not present. Until recently, these osmiophilic inclusions were thought to occur only in sporozoites of *Eimeria*. Bodies somewhat resembling these have been reported in the merozoites of *E. nieschulzi*, and refractile bodies have been found in merozoites of *Eimeria* species from ground squirrels (Colley, 1967; Hammond *et al.*, 1970). The anterior refractile bodies may disappear or undergo fragmentation soon after entrance of a sporozoite of *E. bovis* into a host cell (Fayer and Hammond, 1967). The function of the refractile bodies is not known. They may serve as a storage area for certain nutrients.

A single micropore is found in the sporozoites of most *Eimeria* species which have been studied. It is usually located in the anterior portion of the cell, sometimes in the nuclear region. Mitochondria with typical protozoan internal structures are found throughout the cytoplasm. The Golgi apparatus is represented by several lamellar and vesicular structures anterior to the nucleus. An assembly of small vesicles found adjacent to the nucleus may also belong to the Golgi complex. The presence of a rough endoplasmic reticulum (ER) is indicated by the occurrence of ribosomes arranged close to each other in two parallel lines. Numerous ribosomes are scattered throughout the ground substance of the cytoplasm. Spherical bodies, from 0.25 μ to 0.6 μ in diameter, are found in the cytoplasm. They are homogeneous and less electron dense than the refractile bodies and probably represent lipid droplets.

Numerous electron-pale, ovoid granules 0.4 μ by 0.2 μ to 0.3 μ by 0.8 μ, are distributed throughout the cytoplasm. They appear to be concentrated between the nucleus and the posterior refractile body. When treated with lead citrate, these bodies appear darkly stained. They contain the carbohydrate reserve of the parasite cell. Ryley *et al.* (1968) concluded that the coccidian polysaccharide is a typical amylopectin. The vesicular nucleus, sometimes irregular in shape, is usually located just behind the anterior refractile body. It has a perinuclear space between the two nuclear membranes. The nucleoplasm is similar to the cytoplasm in density, and is less osmiophilic than the chromatin clumps and nucleolus which are usually present. The fine structure of the sporozoites here described is based upon the results of investigations with *E. nieschulzi, E. tenella, E. ellipsoidalis, E. auburnensis, E. bovis, E. ninakohlyakimovae, E. callospermophili,* and *E. larimerensis* (Colley, 1967; Ryley, 1969; Strout and Scholtyseck, 1970; Roberts and Hammond, 1970; Roberts *et al.*, 1970, 1971).

b. Sporozoites of other coccidian genera

The fine structural appearance of the sporozoites of the genera *Lankesterella, Aggregata, Eucoccidium, Coelotropha, Myriosporides,* and *Angeiocystis* studied by several authors (Garnham *et al.*, 1962; Bardele, 1966; Porchet-Henneré, 1967, 1969) shows a close affinity to the sporozoites of *Eimeria* species; however, there are some differences. The refractile bodies do not occur in the sporozoites of *Aggregata, Eucoccidium, Coelotropha, Myriosporides,* and *Angeiocystis*, whereas they are present in *Lankesterella* (Emschermann and Staubesand, 1968). In the sporozoites of this genus, numerous microneme-like structures are found not only in the anterior region but are also present in the posterior half of the cell. There are 27 subpellicular microtubules in the sporozoites of *Lankesterella hylae* (Stehbens, 1966), and 50 in the sporozoites of *Eucoccidium dinophili*, according to Bardele (1966).

2. Merozoites

Among the coccidia the developmental stages most thoroughly studied by means of electron microscopy are the merozoites. They originate from parent cells which undergo an asexual reproduction. The general appearance of the merozoites is reminiscent of the infective stages of other Sporozoa, including the piroplasms (Figs. 10 through 18). The similarity in the fine structure of the anterior organelles in merozoites of *Eimeria, Isospora, Frenkelia, Toxoplasma, Besnoitia,* and *Sarcocystis* is too great to be ignored. Moreover it is difficult to recognize the different genera by their ultrastructure. Generally, the merozoites are distinguished from the sporozoites by the absence of inclusions such as the refractile bodies but these are present in some species, such as *E. nieschulzi* (Colley, 1968) and *E. callospermophili* and *E. bilamellata* (Hammond, Speer, and Roberts, 1970).

a. *Eimeria* species

The merozoite of *Eimeria* is usually fusiform. It is enclosed by a pellicle characteristic of the Sporozoa. The inner membranous layer terminates at the anterior pole, forming the polar ring. In young merozoites, a second polar ring is present at the posterior end (Fig. 10). The merozoites of *Eimeria* species have from 22 to 30 microtubules (Table 1). These subpellicular elements extend posteriorly from the anterior polar ring. A conoid of normal fibrillar structure is situated within the polar ring. The rhoptries extend anteriorly through the conoid from the posterior region. Each rhoptry is club-shaped, having a narrow neck within the conoid region and a wider posterior portion. The alveolar appearance of the major portion of the rhoptries, the neck-like anterior part, with an indication of a less dense core, and the occasional occurrence of a vesicle-like structure at the anterior end of the organelle, all are suggestive of a secretory function. A rod-like body of unknown function lying between the neck segments of the rhoptries has been described by some authors (Sheffield and Hammond, 1966; Colley, 1968; Sénaud and Černá, 1968, 1969). In immature and mature merozoites of *E. intestinalis* (Kheysin, 1967), *E. bovis* (Sheffield and Hammond, 1967), *E. magna* (Sénaud and Černá, 1969), *Lankesterella garnhami* (Büttner, 1968), *Toxoplasma gondii* (Sheffield and Melton, 1968), *Besnoitia jellisoni* (Sheffield, 1966), *Babesia ovis* (Friedhoff and Scholtyseck, 1968), peculiar structures resembling centrioles were observed in the vicinity of the nucleus or conoid. Numerous ovoid bodies, averaging 0.15 μ by 0.40 μ, are concentrated in the central one-third of the merozoite. These bodies, having the characteristic properties of glycogen, are seen in unstained and lead-stained ultrathin sections and are presumably amylopectin. Several mitochondria are usually present in or near the region of the polysaccharide granules. They show the typical microtubular structure, having a matrix of medium density and are surrounded by the two usual membranes.

The anterior region of the merozoite contains numerous micronemes. In the central region of the merozoites there are rounded dense bodies, identified by Kheysin and Snigirevskaya (1965) as protein granules, and structures surrounded by several membranes tentatively interpreted by Sheffield and Hammond (1966) as lysosomes. Thick-walled vesicles containing a variety of nondescript structures are also observed in the central region of different *Eimeria* species. Golgi elements consisting of a group of small vesicles and cisternae lie next to the flattened anterior surface of the nucleus. A single micropore seems to occur in the merozoites of all *Eimeria* species. This ultracytostome is usually found near the level of the Golgi apparatus. Some canaliculi and cisternae of the rough surfaced ER are found both anterior and posterior to the nucleus. Scattered through the cytoplasmic matrix are numerous ribosomes, as well as one or two lipid droplets. The nucleus, located in the posterior one-third of the merozoite, has a double membrane and often a central nucleolus. The nucleoplasm, however, may vary considerably in this developmental stage. Irregular clumps of dense nuclear material have been described by some authors. Detailed electron microscope studies of the following species have been reported: *E. intestinalis* (Mossevitch and Kheysin, 1961); *E. perforans, E. stiedai* (Scholtyseck, 1965b); *E. bovis* (Sheffield and Hammond, 1967); *E. nieschulzi* (Colley, 1967); *E. tenella* (McLaren and Paget, 1968; Ryley *et al.*, 1969; Sénaud and Černá, 1969); *E. pragensis, E. magna* (Sénaud and Černá, 1968); and *E. miyairii* (Andreassen and Behnke, 1968).

b. *Isospora* species

Only a single report in the electron microscope literature deals with merozoites of an *Isospora* species parasitic in the gecko, *Gehyra variegeta* (Schmidt, Johnston, and Stehbens, 1967). As this genus is closely related to *Eimeria*, no major differences between the merozoites of the two genera are expected. The merozoites of the *Isospora* species studied are elongate and have a striking similarity to those of *Eimeria* species (Fig. 11). A peculiarity is the presence of two polar rings at the anterior end; only a few *Eimeria* species are known to have these. The 18 to 20 microtubules observed may be attached at the posterior border of one of these rings. In contrast to *T. gondii* merozoites, only two rhoptries occur.

c. *Frenkelia* species (M-Organism)

This sporozoan, forming large cysts in the bank vole (*Clethrionomys glareolus*), belongs to the endodyogeny group of Coccidia. The cysts contain three different types of cells, namely the metrocytes located in the peripheral zone of the cyst, intermediary cells and endodyocytes, each of which has at its anterior end a conoid, rhoptries, and micronemes and a large thick-walled vesicle of unknown function near the well developed Golgi zone. The metrocytes have an irregular, elongate shape and also have irregular outlines in cross section (Fig. 12). Their major features in comparison with the other two types of cells are a deeply folded cell surface, a widely distributed system of vesicles and lacunae in the cytoplasm, and a nucleolus surrounded by a spiral structure, which resembles that of a chromosome. In all three cell types, only a single mitochondrion may occur. The mitochondrion is a ramified network with tubular inner structures. Up to three micropores are found in ultrathin sections of a metrocyte. The intermediary cells are mainly located in close proximity to the marginal metrocytes. They are found randomly, along with endodyocytes, in the cyst chambers. Intermediary cells of *Frenkelia* have an irregular, elliptical shape with a slightly folded pellicle; they lack the deep infoldings of the metrocytes. The nucleus is similar to that of the metrocyte, but it has in addition osmiophilic concentrations of the karyoplasm at the margin. At both poles of the elliptical cells, micronemes and rhoptries are observed. The mitochondrion is similar to that of the metrocytes. The endodyocytes are elongated, with typical anterior organelles (Fig. 13). They have from five to eight rhoptries. At the posterior end is a cone-like structure which is characteristic only of this cell type. The mitochondrion is Y-shaped. The nucleus is elliptical with a flattened anterior surface. A spherical osmiophilic body corresponds to the nucleolus. As both the intermediary cells and the endodyocytes originate from their parent cells by endodyogeny, they may be considered as merozoites.

The account of the ultrastructure of *Frenkelia* is drawn from studies made by Ludvik (1961), Scholtyseck, Kepka, and Piekarski (1970), and Kepka and Scholtyseck (1970).

d. *Toxoplasma gondii*

For obvious reasons, the parasite *T. gondii* has been far more extensively investigated by means of electron microscopy than any other sporozoan. With the finding of sexual forms in *Toxoplasma*, the endodyogeny group of coccidia, including the genera *Toxoplasma*, *Frenkelia*, *Besnoitia*, and *Sarcocystis* is of special interest.

The elongate or crescent-shaped merozoite of *T. gondii* becomes somewhat rounder prior to division. It has a pointed anterior end and a rounded posterior pole (Fig. 14). The inner membranous layer of the pellicle is continuous around the cytoplasm except for an opening at the anterior end and one at the posterior end, each bordered by a polar ring. Subpellicular microtubules, usually numbering 22, originate at the anterior polar ring and may extend posteriorly to the polar ring at the posterior pole (Table 1). Several rhoptries, from four to eight, lie in the cytoplasm with their necks extending into the conoid. All the other organelles and fine structures are identical with those of *Eimeria* merozoites, indicating a close relationship to the Eimeriidae (Scholtyseck and Piekarski, 1965).

Detailed studies of the ultrastructure of *T. gondii* have been made by Gustafson, Agar, and Cramer (1954), Ludvik (1956, 1958), Braunsteiner, Pakesch, and Thalkammer (1957), Wanko, Jacobs, and Gavin (1962), Wildführ (1964), Scholtyseck and Piekarski (1965), Sénaud (1965, 1966, 1967), van der Zypen and Piekarski (1967), Sheffield and Melton (1968), Vivier and Petitprez (1969), Colley and Zaman (1970), and by Sheffield (1970).

e. *Besnoitia jellisoni*

This sporozoan was first described by Frenkel (1953) who recognized its similarity to *Toxoplasma*. Ultrastructural studies of merozoites of *Besnoitia* (Fig. 15) later confirmed the close relationship of this organism to *T. gondii*, *Sarcocystis* and *Frenkelia*. The pellicle, the fine structure of the apical pole, the cytoplasmic inclusions, and the nucleus all are strikingly similar to those of *T. gondii*.

Studies of the ultrastructure of *B. jellisoni* have been made by Sheffield (1966, 1967, 1968) and Sénaud (1969).

f. *Sarcocystis*

Sarcocystis is a sporozoan genus of uncertain systematic position and incompletely known life cycle. It is a parasite in the striated muscles of mammals, birds, and reptiles, where large cysts are found containing thousands of merozoite-like stages. The first report of the ultrastructure was given by Ludvik (1956). A detailed study of the fine structure of *Sarcocystis* has been made by Simpson (1966) and Sénaud (1967).

The merozoites of *S. tenella* are more or less banana-shaped (Fig. 16). Their ultrastructural features are so similar to those of *Frenkelia*, *Toxoplasma*, and *Besnoitia* that it would be difficult to distinguish one from the others in electron micrographs. The nearly identical fine structure indicates a close taxonomic relationship of these four genera.

g. Merozoites of other Sporozoa

Several organelles and structures similar to those in the anterior end of the merozoites of the coccidia have been reported in the merozoites of other Sporozoa; e.g., in schizogregarines (Schrevel and Vivier, 1966), in various malaria parasites (Garnham, 1966), and in piroplasms (Friedhoff and Scholtyseck, 1968). The fine-structural characteristics of the infective stages as described here and modifications of these seem to be unique to the Sporozoa.

The merozoites of *Babesia bigemina* and *B. ovis* (Figs. 17, 18) from tick tissues have most of the typical fine structures and organelles of Sporozoa, except for typical mitochondria with inner tubular structures and except for a conoid, which does not occur in Haemosporidia, either. The fact that the piroplasms do indeed belong to the Sporozoa Apicomplexa, Polannulifera (Levine, 1969b, 1970) has been established in recent years by electron microscopy (Büttner, 1966, 1967, 1968; Simpson *et al.*, 1967; Friedhoff and Scholtyseck, 1968, 1969; Frerichs, 1970; Scholtyseck, Mehlhorn, and Friedhoff, 1970; Scholtyseck and Mehlhorn, 1970).

B. MICROGAMONTS AND MICROGAMETES

The sexual phase in the life cycle of the coccidia is similar in general to that in other Protozoa with differentiated gametes. The microgamete (male gamete) is flagellated, which is remarkable, because the coccidia, except for the Protococcidia, are intracellular parasites and in this situation such a method of locomotion is likely to be disadvantageous. The occurrence of flagella in the microgametes is undoubtedly an indication of phylogenetic relationship of the coccidia with certain groups of Mastigophora.

Detailed studies of the ultrastructure of microgametogenesis in the coccidia and related groups have been made by Kheysin (1965), Scholtyseck (1965a), Bardele (1966), Colley (1967), Snigirevskaya (1968), Hammond *et al.* 1969), Porchet-Henneré (1970), and Colley and Zaman (1970).

1. Microgamonts

The parent cells of the microgametes are designated as microgamonts.

a. *Eimeria* species

The general fine structure of the *Eimeria* microgamonts is relatively uniform. Among various species, these stages differ mostly in size and in the number of developing nuclei or microgametes. In some species the number of developing microgametes is small, whereas in others many thousands of microgametes are produced. The cell boundary of the microgamonts of *Eimeria* consists of a unit membrane. In some species, deep infoldings and fissures increase the surface area.

Micropores occurring singly, doubly, or in numbers as high as nine, are seen in the membrane lining fissures. Although no inner membranous layer is present in this area, the micropores have an osmiophilic cylinder surrounding the inpocketing of the outer membrane. The micropores found in microgamonts of *E. auburnensis* are of special interest because so many occur close together. This observation suggests that micropores may be of a temporary nature and variable location in gamonts (Hammond *et al.*, 1969).

In a stage of development in which the nuclei are no longer dividing, microgamonts of *E. perforans* range from 10 μ to 14 μ in size and the number of nuclei varies considerably. In small microgamonts of some species (Fig. 19), the nuclei are located in the peripheral zone. In large specimens, e.g., in those of *E. auburnensis*, the nuclei are arranged in rows at the periphery and adjacent to the fissures. The nuclei have a conventional double membrane and their karyoplasm seems to be condensed in the marginal regions. Centrioles appear in the immediate vicinity of the nuclei, usually in the space between the nuclei and the surface of the gamont, or between the nuclei and the limiting membrane of the fis-

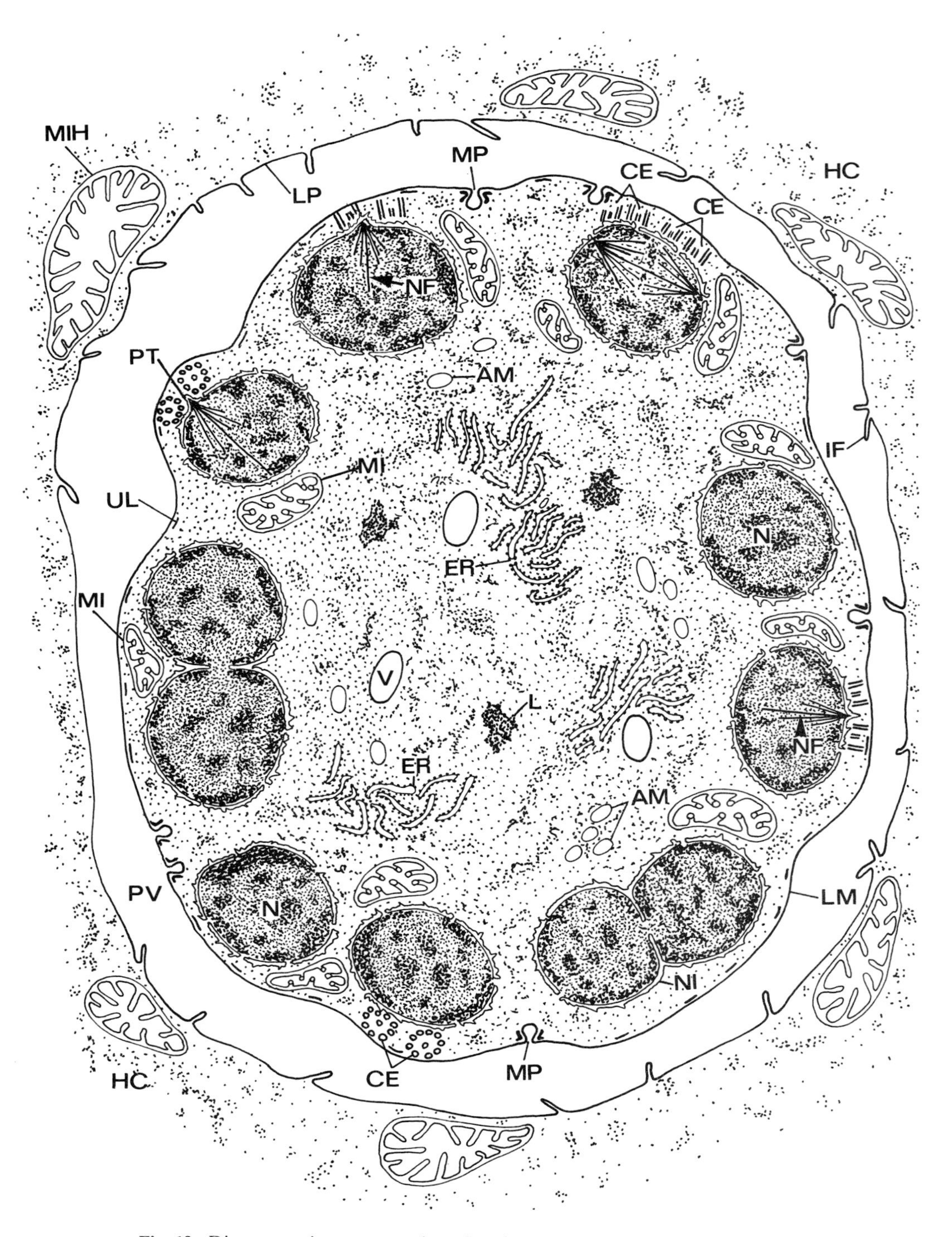

Fig. 19. Diagrammatic representation of a microgamont of the genus *Eimeria*.

sures. Generally, a single pair of centrioles is seen (Fig. 19); however, two pairs are sometimes observed in association with a single nucleus (Hammond *et al.*, 1969). The centrioles are constructed on a plan closely similar to that of kinetosomes. Between the two paired centrioles, a cone-shaped protuberance of the nucleus frequently occurs; from this protuberance an intranuclear spindle apparatus, consisting of microtubular fibrils, extends inward (Fig. 19). The centrioles later give rise to the flagella.

Scattered through the matrix of the cytoplasm are numerous membrane-enclosed compartments ranging from small vesicles and canaliculi to extensive large vesicles. Some of these are surrounded by canals of the ER. Within the cytoplasmic space between the canals, numerous inclusions of light density are observed. These are probably amylopectin granules because they resemble inclusions of this nature found in *Eimeria* sporozoites and merozoites. Many large mitochondria of typical tubular structure are located in the vicinity of the nuclei. The cytoplasmic matrix has a relatively dense appearance, with numerous fine granules clearly differentiating it from the less dense cytoplasm of the host cell.

Detailed reports on the fine structure of *Eimeria* microgamonts have been given by Kheysin (1965), Scholtyseck (1965a), Hammond, Scholtyseck, and Miner (1967), Hammond *et al.* (1969), Colley (1967), and Sénaud and Černá (1970).

2. *Microgametes*

Three different morphological types of microgametes are known to occur in the coccidia:

1. the lens- or cup-shaped microgametes of the Protococcidia (*Eucoccidium, Coelotropha*);
2. the elongate, slender, and slightly curved microgametes of the Eimeriina and of *T. gondii*;
3. the filiform microgametes of the Haemosporina, which are always formed in small numbers and which supposedly have no mitochondria, in contrast to the first two types.

Nothing is known about the fine structure of adeleids. Perhaps they represent still another fundamental type.

a. *Eimeria* species

The micromorphology of *Eimeria* microgametes seems to be more or less uniform. The mature male gamete is an elongate stage averaging from 6 μ to 7 μ in length in *E. perforans*; in *E. maxima* it is from 4 μ to 5 μ long (Figs. 20, 21b). The microgamete consists mainly of nuclear substance. It is bounded by a unit membrane, which is derived from that of the microgamont. At the anterior end is a pointed structure, the perforatorium, in which three basal bodies lie. The basal bodies are located in close proximity to one another at the apical pole. The reinforced tip of the microgamete, termed "perforatorium" by Kheysin (1965), presumably assists in the penetration of the macrogamete by the microgamete. From the basal bodies originate three flagella, two long and free ones, and one relatively short flagellum, the anterior half or two-thirds of which is attached to the surface of the microgamete. The unit membrane which serves as a cell boundary for the microgamete also covers the flagella. These vary somewhat in number and arrangement in the microgametes of *Eimeria*. Three flagella occur in the microgametes of *E. perforans, E. nieschulzi, E. tenella,* and *E. falciformis*. In the microgametes of *E. auburnensis* and *E. maxima*, however, three flagella are found only occasionally. In microgametes with two or three flagella, additional microtubules, from four to eight, are always present (Fig. 20); these originate in the basal apparatus zone at the apical pole and extend to the posterior end of the microgamete (Scholtyseck, 1965a; Colley, 1967; McLaren,

1969; Hammond *et al.*, 1969; Mehlhorn, 1972; Scholtyseck, Mehlhorn, and Hammond, 1972). Typically, an elongate mitochondrion, about 2.5 μ in length, lies in a groove of the nucleus. It has a characteristic structure, with two longitudinal rows of club-shaped tubules. In microgametes of *E. auburnensis* and *E. maxima*, more than two rows of ellipsoidal tubules are present in the mitochondrion, which does not occur in a groove of the nucleus, so that these two species differ from the others with respect to these features.

b. *Toxoplasma gondii*

The microgametes of *T. gondii* appear to be laterally compressed, because in longitudinal sections, narrow, elongated specimens as well as broad ones are seen (Fig. 21a). The microgamete in general has a construction similar to that of *Eimeria* species, with nucleus, mitochondrion, and flagella. The nucleus, which is membrane-bound, occupies most of the space in the body of the microgamete. The mitochondrion differs somewhat from those of *Eimeria* species, in that the ellipsoidal tubules are irregularly arranged, instead of occurring in parallel rows. Two flagella are present. Usually, in the vicinity of the basal apparatus of these, five microtubules originate and extend posteriorly along the nucleus for a relatively short distance. These microtubules may represent the rudiment of the third flagellum. In a few specimens, 15 additional very short microtubules are observed (Fig. 21a); these are arranged in a row, with an underlying osmiophilic layer, at one side of the perforatorium (Colley and Zaman, 1970; Pelster and Piekarski, 1971; Scholtyseck, Mehlhorn, and Hammond, 1972).

c. Microgametes of other genera

i. *Aggregata eberthi*

The microgamete of *A. eberthi* is from 25 μ to 30 μ in length. Its pellicle seems to consist of three unit membranes. Typical micropores are reported to occur. The microgametes have three flagella (Fig. 2d). Numerous subpellicular microtubules are reported (Porchet-Henneré and Richard, 1970).

ii. *Coelotropha*

The more or less spheroidal microgametes of the Protococcidia do not have a polarity as do the *Eimeria* microgametes. In *Eucoccidia* and *Coelotropha* the microgametes seem to be very similar. They are lens- or cup-shaped, as mentioned above. However, Porchet-Henneré (1970) reported an electron-dense area in the microgametes of *C. durchoni*, to which she ascribed a perforating function. The microgamete of *C. durchoni* has two free flagella; one flagellum is from 18 μ to 20 μ in length, while the other is from 13 μ 15 μ long (Vivier and Henneré, 1964). The flagella originate side by side just inside the limiting membrane on the under side of the microgamete. Numerous subpellicular microtubules are present at this place. Typical micropores are found in the microgamete of *C. durchoni*. The mitochondrion in the microgametes of *E. dinophili* and *C. durchoni* has a round form, corresponding to that of the microgamete. It lies in the immediate vicinity of the basal bodies of the flagella, so that the usual close spatial relationship between the energy source and the organelles using energy is observed (Bardele, 1966; Porchet-Henneré, 1970).

d. Microgametes of other Sporozoa

The microgametes of *Leucocytozoon simondi*, *Plasmodium berghei*, *Haemoproteus kochi*, and *L. marchouxi* are relatively long, attaining a length of from 12 μ to 23 μ, double that of the microgametes of eimerians. A unit membrane bounding the microgametes of *L.*

Fig. 20. *Eimeria maxima*, longitudinal section through a microgamete with three flagella and additional microtubules. × 53,000 (from Scholtyseck, Mehlhorn, and Hammond, 1972).

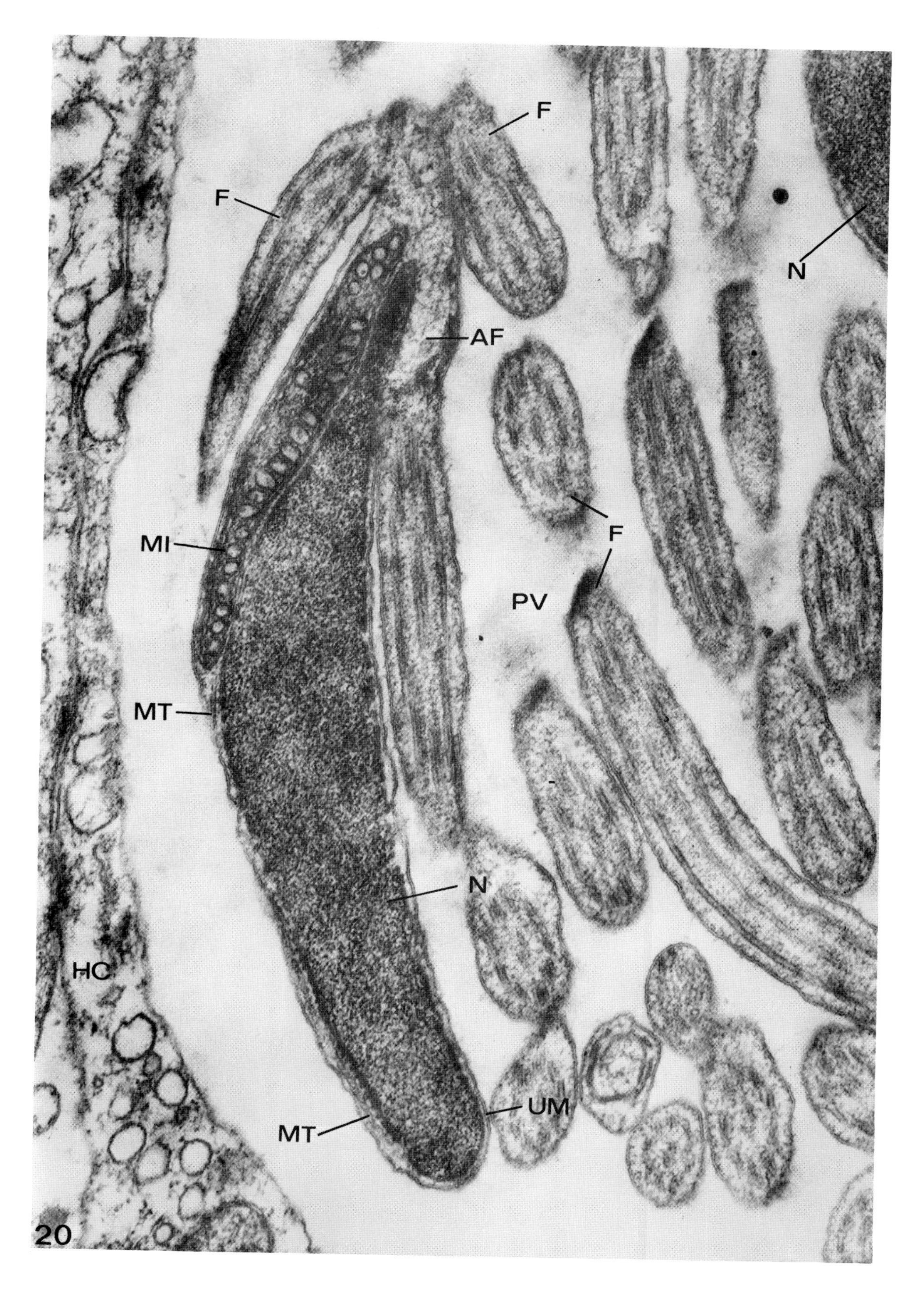

F
F
N
AF
F
F
PV
MI
MT
N
HC
MT
UM
20

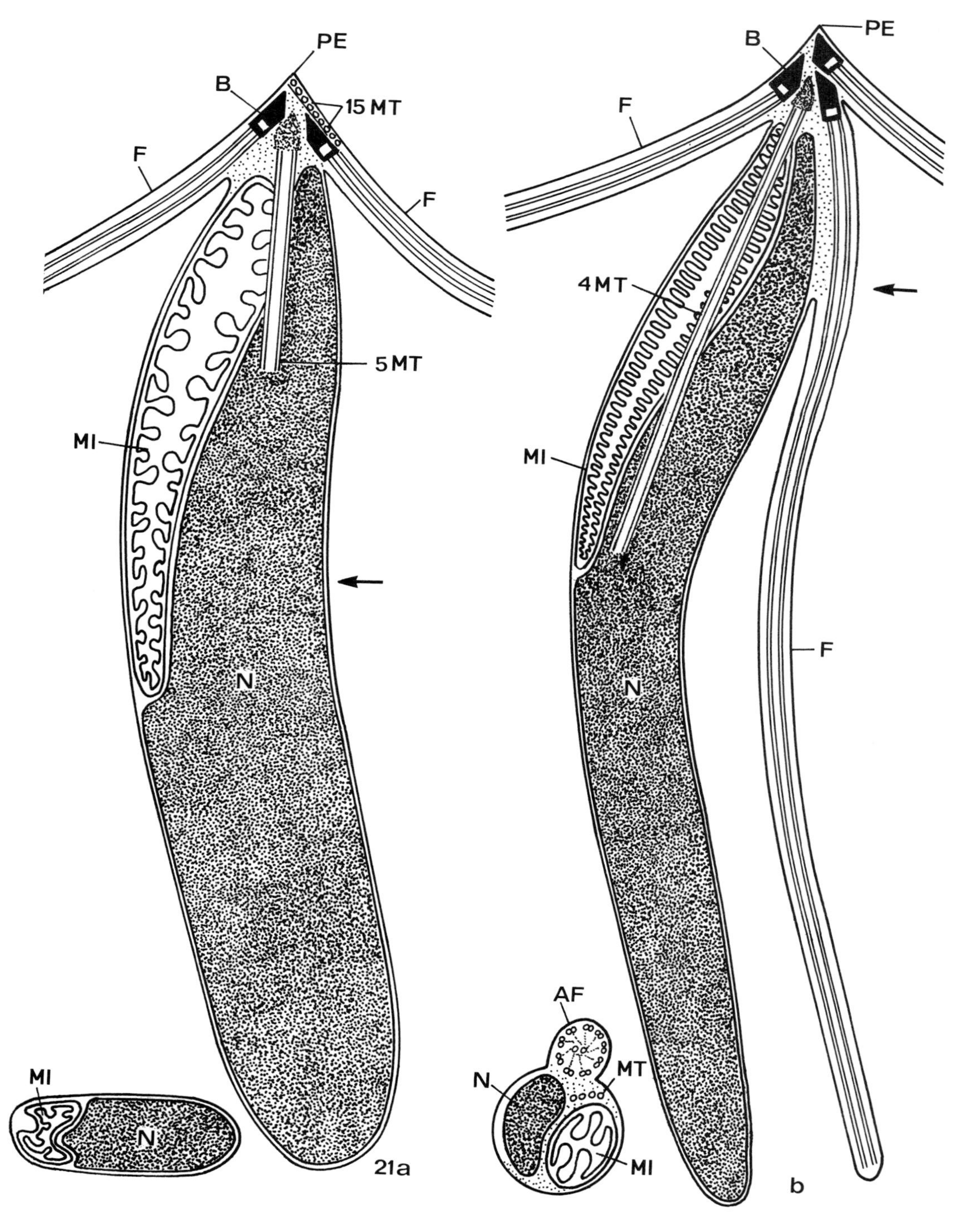

Fig. 21. Diagrammatic representations of microgametes in longitudinal and cross sections. *a, Toxoplasma gondii; b. Eimeria maxima.*

simondi and *H. columbae* has been described by Desser (1970), Aikawa *et al.* (1970), and Bradbury and Trager (1968). In *P. berghei, H. kochi,* and *L. marchouxi,* the microgametes are bounded by two unit membranes (Garnham *et al.*, 1967; Porchet-Henneré, 1970). Numerous subpellicular microtubules are found; these are identical in size and structure with those of merozoites and sporozoites.

The number and arrangement of the flagella in the microgametes of coccidia are of much interest, because of their significance in the systematics and phylogeny of the Sporozoa. These characteristics have been long considered of fundamental importance in determining phylogenetic relationships among the different groups of Protozoa. In this connection, it is important to note the difference between the haemosporidians and the coccidians and eimerians. In the former, the microgametes have only one or two axonemes, similar in structure to a flagellum, but embedded in a cytoplasmic ground substance, whereas in the latter group, free flagella occur (Aikawa *et al.*, 1970; Desser, 1970). Desportes (1970) observed one axoneme in fertile male gametes of the gregarine genus *Stylocephalus,* and two axonemes in sterile male gametes. In *Plasmodium* species, the nucleus lies in the middle of the microgamete, and the axoneme winds spirally around it (Garnham *et al.*, 1967).

C. MACROGAMETES

Relatively little is known about the fine structure of sporozoan macrogametes, with the exception of the *Eimeria* species. The female gametes of the *Eimeria* species studied are relatively uniform in their ultrastructure, but *E. falciformis* shows some differences and the macrogametes of *Toxoplasma gondii* are also different in some respects. In the macrogametes of *Eucoccidium dinophili* and *Klossia helicina* are found a number of structures not seen in the others.

1. Mature macrogametes of Eimeria

The most significant ultrastructures of *Eimeria* macrogametes are the pellicle, the intravacuolar tubules, the wall-forming bodies of the first and second type, and the polysaccharide granules.

a. The pellicle

The cell boundary of macrogametes is derived from the pellicle of the merozoite stage. In some species, the inner layer undergoes dedifferentiation during the development of the macrogamete. Only one unit membrane, with underlying osmiophilic material apparently representing the inner layer of the merozoite pellicle, has been found in *E. tenella, E. stiedai, E. perforans,* and *E. nieschulzi* (Figs. 22, 24, 37). The macrogamete of *E. auburnensis* has a cell boundary of two layers with the three characteristic unit membranes, of which only the middle one is completely retained, the other two being interrupted in places (Fig. 37). In the macrogametes of *E. bovis* and *E. maxima*, the cell boundary consists of two unit membranes, probably representing the two layers of the pellicle (Fig. 37). Deep invaginations of the cell boundary occur in the macrogametes of *E. auburnensis, E. nieschulzi,* and *E. intestinalis* (Fig. 37; Hammond, Scholtyseck, and Chobotar, 1967; Colley, 1967; Snigirevskaya, 1968, 1969; Scholtyseck, 1966-1970). These invaginations, which average 0.1 μ in diameter and from 1.2 μ to 2.2 μ in length in *E. auburnensis,* may serve for the intake of nutrients (Hammond, Scholtyseck, and Chobotar, 1967). In these macrogametes there were vesicles which were apparently being pinched off from the deepest point of the invagination. Numerous, smaller V-shaped invaginations were also observed at the surface of *E. auburnensis* macrogametes, and pinocytotic vesicles were associated with these invaginations.

b. Micropores

In general, intake of nutrients may also occur through the micropores. These pellicular organelles have a similar fine structure to that in other developmental stages of Sporozoa (Figs. 7, 8, 37). Even in the stages in which the inner layer of the pellicle is no longer present, a remnant of this is retained as a cylindrical supporting part of the micropore. These organelles occur in large numbers in the macrogametes of *Eimeria* species. In *E. maxima,* six micropores were found in a surface area of about 2.5 μ^2.

c. Intravacuolar tubules

Characteristic differentiations occur at the cell surface, even in very young macrogametes of *E. perforans, E. nieschulzi, E. tenella, E. auburnensis,* and *E. bovis.* These structures, first described by Scholtyseck and Schäfer (1963), are tubular in form and lie within the parasitophorous vacuole, close to the surface of the macrogamete. In cross section, they resemble vesicles. They have a characteristic fine structure. In longitudinal section, the tubules have transverse striae, each having a width of 75 Å, with an interval of 75 Å between successive striae. The striation apparently is located in the walls of the tubules because the lumen is electron-transparent. The wall is about 125 Å thick and is not a unit membrane. It evidently consists of a number of subunits (Figs. 22, 23, 24, 37). The tubules have an average diameter of from 500 Å to 600 Å in *E. tenella, E. perforans,* and *E. auburnensis,* whereas in *E. nieschulzi, E. falciformis,* and *E. maxima* otherwise identical tubules average from 800 Å to 900 Å in diameter.

The site of origin of these organelles at the cell surface of the parasite is difficult to distinguish, because osmiophilic material lies under the pellicle at these locations (Figs. 23a, 37). The tubules are not evenly distributed along the surface in some species, but occur only in certain areas (Fig. 22). Occasionally, a group of tubules is arranged in a circle around a single tubule with a unit membrane; these structures occur in association with an invagination of the cell surface (Figs. 23b, 37).

Since most longitudinal sections of the tubules are incomplete, their length cannot be determined exactly. There are portions of tubules as long as 4 μ in *E. tenella, E. maxima, E. bovis, E. auburnensis,* and *E. perforans,* indicating that they may have a considerable length.

Little is known about the development of these tubules. They are first found in vacuole-like spaces in the interior of young macrogametes. At a later stage, these areas with the tubules assume a peripheral location. It is obvious from these findings that the tubules originate within the parasite. It is not possible at present to explain how the tubules become established and anchored at the surface.

These organelles may be associated with transport of materials between the parasite and its host cell. This is indicated by the finding of a direct structural connection between the tubules and the host cell cytoplasm. Also, acid phosphatase has been demonstrated in similar tubules at the surface of microgamonts in *Klossia helicina.*

Probably, such tubules are characteristic of all macrogametes of *Eimeria* species. The designation "intravacuolar tubules" was proposed for these organelles by Scholtyseck, Hammond, and Ernst (1966), because of their location within the parasitophorous vacuole.

These tubules do not occur in the macrogametes of Protococcidia, Haemosporidia, and Adeleidae with the exception of *Eucoccidium dinophili.* The Protococcidia are extracellular, and thus do not need such organelles. In

Fig. 22. *Eimeria maxima*, section through a mature macrogamete. Note the numerous intravacuolar tubules (IT) and the Golgi-vesicles near the wall-forming bodies II, × 15,000. (From Mehlhorn, 1971.)

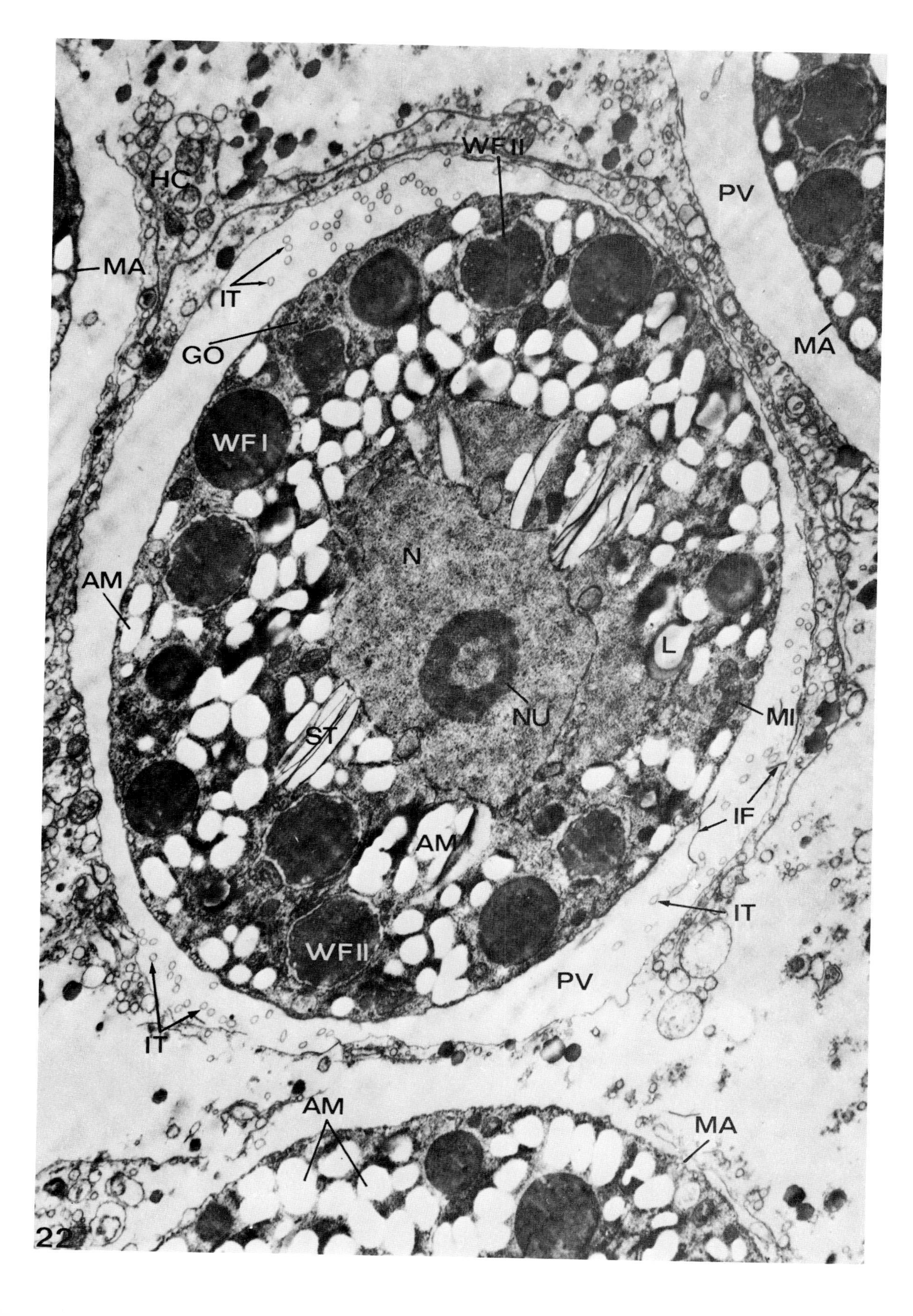
WFII
HC
PV
MA
IT
GO
MA
WFI
N
AM
L
NU
MI
ST
IF
AM
IT
WFII
PV
IT
AM
MA
22

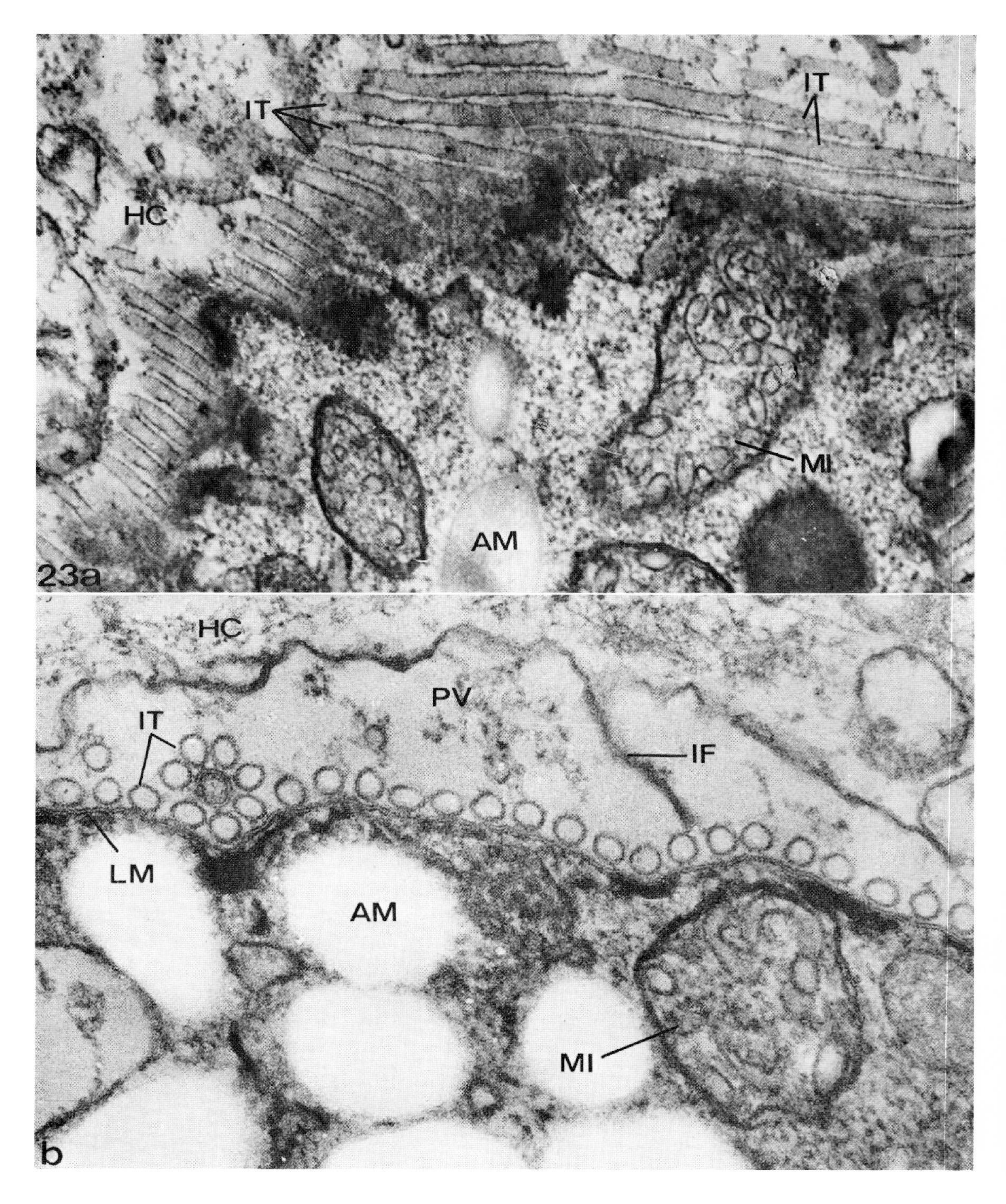

Fig. 23. Longitudinal sections (*a*) and cross sections (*b*) through the intravacuolar tubules (*IT*), which are composed of subunits resulting in a tangential striation. × 85,000. (From Scholtyseck and Schäfer, 1963.)

Plasmodium species, the parasitophorous vacuole develops differently than that in *Eimeria* species, so intravacuolar tubules might not be expected to occur. The absence of such structures in *K. helicina* might be associated with the mucous layer, which may serve as a contact zone between parasite and host cell.

d. The wall-forming bodies

The occurrence of large granules, which later fuse to form the typical two-layered oocyst wall is characteristic of the macrogametes of all *Eimeria* species which have been studied. *Plasmodium* species do not have such inclusions, since their zygote state is motile, and no oocyst wall is formed. The oocyst wall of *Klossia, Aggregata,* and *Protococcidia* is very thin, and such large granules are not present. Usually, *Eimeria* species have two kinds of granules known as wall-forming bodies. Those of the first type are homogeneous and osmiophilic; they usually have a peripheral location. The bodies of the second type have a sponge-like fine structure and are mostly in the central region of the cytoplasm.

The homogeneous bodies were named according to their location and function as wall-forming bodies I, whereas the bodies with a sponge-like structure were given the name wall-forming bodies II (Scholtyseck *et al.*, 1969).

Wall-forming bodies I. These structures appear later than the granules of type II (Figs. 22, 24, 35, 37). Their origin and development are unknown. There are some indications that they develop from thick-walled vesicles. Other observations suggest that mitochondria might give rise to the bodies, since, in some species, similar homogeneous, osmiophilic inclusions occur in mitochondria. No conclusive evidence for such an origin exists, however. Comparative studies with the light and electron microscopes showed that the wall-forming bodies I are identical with the plastinoid granules described by Kheysin (1958) and by Scholtyseck, Rommel, and Heller (1969). They are PAS-positive, and probably contain muco-proteins, among other substances.

The shape of these granules is spherical to ellipsoidal. They are sharply delimited from the ground substance of the cytoplasm. In some cases (*E. tenella, E. maxima, E. intestinalis, E. perforans,* and *E. auburnensis*) they are bounded by a membrane-like layer. Possibly, this layer is present in other species as well, but cannot be demonstrated because the granules are so strongly contrasted with osmium tetroxide. Occasionally, a striation is observed in tangential sections of the bodies in *E. perforans* and *E. auburnensis*. The size of the bodies is a species-specific characteristic. They are larger than the wall-forming bodies II in *E. tenella, E. stiedai, E. perforans,* and *E. maxima;* the reverse is true in *E. bovis* and *E. auburnensis*

Wall-forming bodies II. These bodies have a loose, sponge-like structure (Scholtyseck, 1962). They appear earlier in the development of macrogametes than the wall-forming bodies I (Figs. 22, 24, 35, 37). They occur singly or in groups in membrane-bound spaces, which belong either to the ER or the Golgi complex. The bodies first appear in terminal expansions of lamellar systems as osmiophilic masses or irregular thread-like structures (Scholtyseck, 1964). The wall-forming bodies II have a relatively uniform structure in the different *Eimeria* species. They differ among the various species chiefly with respect to the size of the internal spaces. Such differences, however, may be caused, at least in part, by the techniques used in preparation. In the macrogametes of *T. gondii* and *E. falciformis* the wall-forming bodies of type II are presumably represented by large, dense, homogeneous bodies, which lie in vesicles (Fig. 37).

e. Polysaccharide inclusions

In the macrogametes of all coccidia with the exception of the Haemosporidia, there are many granules, whose appearance varies from electron-pale to electron-dense according to the preparation methods. These granules consist of a reserve carbohydrate, which was demonstrated for the first time in coccidia by Giovannola (1934). On the basis of histochemical studies, this polysaccharide was considered to be genuine glycogen, as in the liver of vertebrates, for example. Electron microscope studies showed no histochemical difference between the coccidian polysaccharide and metazoan glycogen but the fine structure of these are different. This finding led to the proposed term "coccidian glycogen" (Scholtyseck, 1964). Ryley, Bentley, Manners, and Stark (1969) found, with microchemical methods, that the polysaccharide of *E. tenella* and *E. brunetti* consists of amylopectin with a chain length of 20 glucose residues. Probably other coccidia have similar polysaccharides. With the presently available electron microscope techniques, amylopectin cannot be identified.

It is difficult to determine the precise size of the polysaccharide granules, because the amount of shrinkage apparently varies with different preparation methods. Generally, the granules attain an appreciable size (in macrogametes of *Eimeria* species, they average 0.62 μ, about 1.0 μ in maximum length and 0.5 μ in breadth) in the Eimeriidea, Adeleidea, and Protococcidia, considering the relatively small size (from 20mμ to 30mμ) of the inclusions in Metazoa. The micromorphology of the granules is poorly understood. With some methods of preparation, a boundary membrane can be distinguished, but not with others.

f. Mitochondria

The mitochondria always have internal tubular structures (Figs. 23, 24, 34, 37). Their distribution is irregular, but sometimes they are more numerous in the peripheral region. In certain species, such as *E. falciformis, E. perforans,* and *E. maxima,* the mitochondria contain large, opaque bodies, which resemble the wall-forming bodies I (Fig. 37).

g. Golgi complex

In the macrogametes of *Eimeria* species, typical dictyosomes do not occur. It is likely that the Golgi complex is represented only by lamella and vesicles, so that identification is difficult. The large vesicles which contain the wall-forming bodies II may belong to the Golgi complex.

h. Endoplasmic reticulum

This cytoplasmic system is usually well developed and has a lamellar structure in young macrogametes of *Eimeria* species. In older stages, however, only vesicles of different sizes, with walls of variable thickness, represent the ER. In Eimeriidae, rough and smooth canals and spaces of the ER are associated with polysaccharide granules and lipid inclusions. In nearly all species of *Eimeria*, there are prominent structures belonging to the ER. These closely arranged canaliculi occur in groups running in different directions. Frequently, they are closely associated with the nucleus (Fig. 22). The interior of the canaliculi is granule-free. In some cases, their walls are highly osmiophilic. The function of these structures is unknown.

i. Additional cytoplasmic inclusions

Lipid inclusions occur singly throughout the cytoplasm, as well as in groups (Figs. 22, 24). Frequently, such inclusions are associated with channels of the ER. The lipids may be considered as storage material, along with the polysaccharides. Often, the lipids are extracted during preparation.

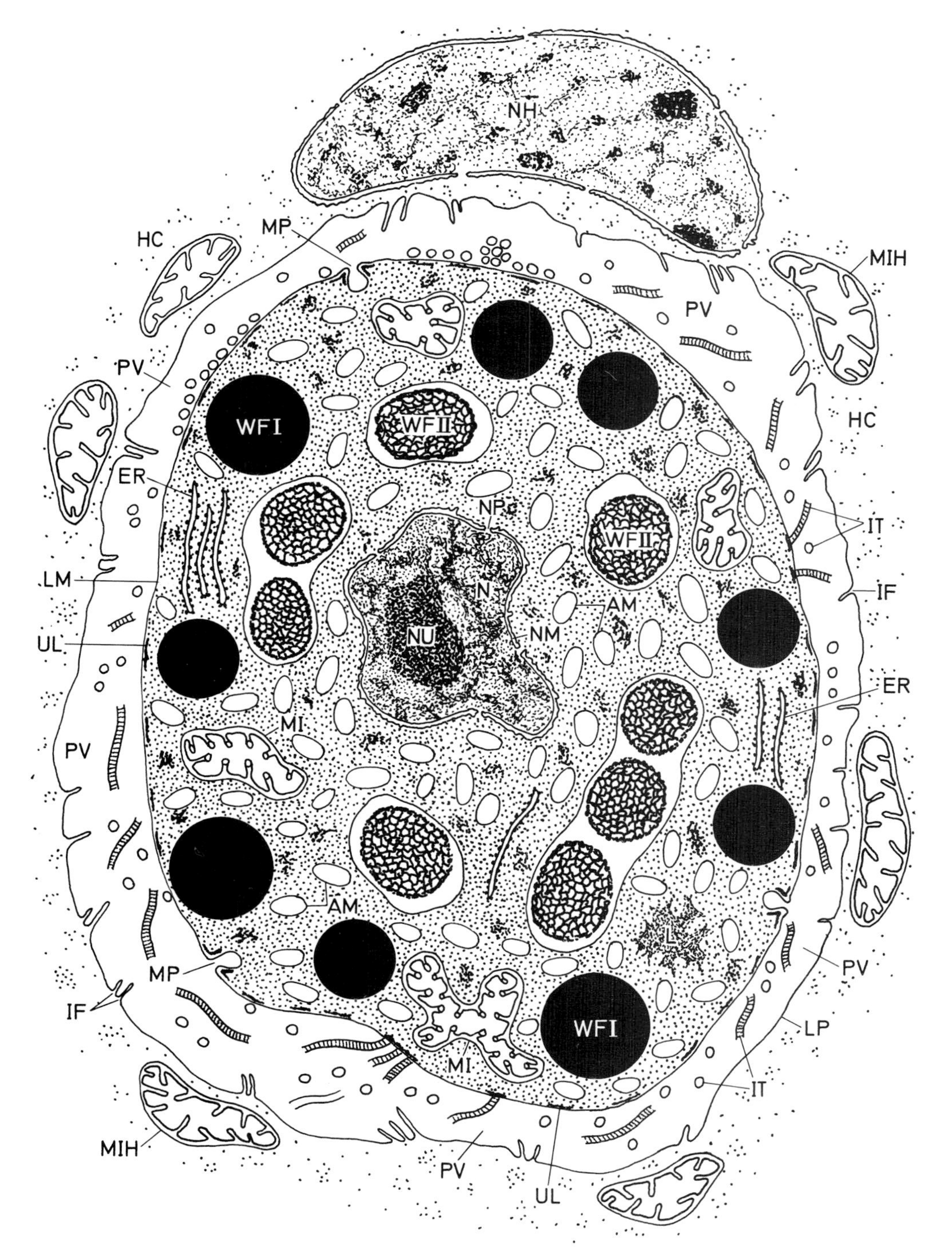

Fig. 24. *Eimeria perforans*, diagrammatic representation of a mature macrogamete.

Small dark bodies occur mainly at the periphery of the macrogamete, and resemble the wall-building bodies I in their fine structure. Their function is unknown. They are no longer present after completion of the oocyst wall.
chondria occur in the macrogametes of
Organelles which resemble metazoan mito-
E. tenella and *E. stiedai*. They are ellipsoidal, about 0.4 μ in size, and bounded by a double membrane, from the inner one of which, structures similar to cristae originate in some specimens.

Another structure of unknown significance is a crescent-shaped or irregular organelle, which has amorphous, osmiophilic masses, and usually lies near the nucleus. Such a structure was found by Reich (1913) in the "Kernsaftzone" of the macrogametes of *E. stiedai*. Also Simond (1897) reported a similar body in macrogametes of the genus *Coccidium*. The function of this organelle is unknown as yet. However, it is possible that it represents the vacuole surrounding the microgamete, which has been observed intact inside the cytoplasm of the macrogamete (Bradbury and Trager, 1968; Scholtyseck and Hammond, 1970; Hammond and Scholtyseck, 1970).

j. Nucleus

The relatively large nucleus is a prominent characteristic of the female gametes of the coccidia. In young stages it is spheroidal, but sometimes it is elongate. It has a normal structure, with a double nuclear membrane, pores, and perinuclear space. In older macrogametes, as for example, in *E. maxima* and *E. perforans*, the nucleus has an irregular shape (Figs. 22, 24). Frequently, a compact or ringlike nucleolus is present (Figs. 22, 24, 34). It appears to correspond with the nucleolus of Metazoa and consists of an accumulation of small osmiophilic granules (ribosomes), which are not membrane-bound.

2. *Mature macrogametes of* Toxoplasma

The macrogamete of *T. gondii* is more or less ellipsoidal, as in the other coccidia. There is a typical pellicle, consisting of three unit membranes; such a pellicle is characteristic of the motile stages of Sporozoa. Immediately under the cell surface are numerous, usually elongate, mitochondria, with a tubular internal structure. The amylopectin granules are conspicuous (Fig. 37, AM), being larger than in the previously described macrogametes of *Eimeria* species. Because of the staining with lead citrate, they appear slightly electron-dense. Wall-forming bodies of type II are presumably represented by large, dense, homogeneous bodies, which lie in vesicles. They are similar in structure to the wall-forming bodies II in *E. falciformis*. The wall-forming bodies of type I are apparently represented by structures which are smaller, more dense, and less numerous than the wall-forming bodies II. Lipid inclusions are observed as large, vacuole-like spaces whose content has mostly or entirely disappeared during the preparation process. The ground cytoplasm has numerous granules and the endoplasmic reticulum is well developed. In the relatively large nucleus is a compact nucleolus; small, irregularly distributed dense granules and clumps of granules are seen in the nucleoplasm.

The macrogamete of *T. gondii* usually occurs immediately beneath the surface of epithelial cells in the small intestine of the domestic cat. It is located in a narrow, electron-pale parasitophorous vacuole, which is bordered by a remarkably thick, dense membrane of the host cell. This membrane, which differs from any associated with previously described *Eimeria* macrogametes, has prominent, irregular folds.

The description of the macrogametes of *T. gondii* is based upon the reports of Colley and Zaman (1970), Scholtyseck, Mehlhorn, and Hammond (1971), and Pelster and Piekarski (1972).

3. Mature macrogametes of Aggregata eberthi

The macrogametes of *A. eberthi* have a surface with deep infoldings and numerous micropores. The cell boundary consists of a typical sporozoan pellicle showing the two layers with three membranes. Intravacuolar tubules, about 25 mμ in diameter, similar to those of *Eimeria* macrogametes, are found mainly in large infoldings. The cytoplasm is arranged into two distinct concentric zones. Mitochondria, dictyosomes, and vacuolar parts of the ER occur chiefly in the outer zone, whereas, polysaccharide granules, lipids, so-called granular bodies, and numerous cisternae of the ER occur in the inner zone. The parasitophorous vacuole appears to be filled with osmiophilic irregular masses. The fine structure is shown diagrammatically in Figure 38 (after Heller, 1969; Porchet-Henneré and Richard, 1970).

4. Macrogametes of Klossia helicina *(Adeleidea)*

The macrogametes of *K. helicina*, from the kidney of the pulmonate snail, *Cepea nemoralis*, have a relatively dense mucous layer, which uniformly covers the entire surface (Figs. 25, 38). The mucous layer is apparently bounded internally and externally by membranes of the parasite cell. The cytoplasm shows three concentrically arranged zones. Mitochondria with internal tubular structures occur in the outer zone. The narrow middle zone contains the so-called gray bodies and parts of the ER, whereas large polysaccharide granules and Golgi systems occur in the inner zone. The Golgi complexes have a characteristic structure, with a variable number of flattened vesicles, which may reach a length of nearly 1 μ. The complex is bipolar, with one pole having spherical vesicles; at the other pole is a large ergastoplasmic space, which belongs to the rough ER. The fine structure of the macrogametes is presented in Figures 25 and 38 (after Scholtyseck, Volkmann, and Hammond, 1966; Volkmann, 1967).

5. Macrogametes of other Sporozoa

The overall appearance of the fine structure of the macrogametes here described is very similar to that of other Sporozoa. The original merozoite pellicle is usually retained in the macrogametes of *Adelina tribolii, Haemogregarina* sp., *Haemoproteus columbae, Plasmodium* sp., *P. floridense, P. coatneyi*, and *P. gallinaceum*. Also, the subpellicular microtubules are often present in these (Fig. 38). The macrogamete of *Leucocytozoon simondi* shows a peculiar ten-layered complex which surrounds mature stages (Fig. 38; Desser *et al.*, 1970). In the macrogametes of *Plasmodium* specific osmiophilic structures are closely associated with the pellicle (Aikawa *et al.*, 1968, 1969). These have an average diameter of from 0.2 μ to 0.3 μ and are connected with the inner layer of the pellicle by a tortuous neck-like portion. They resemble rhoptries, but are smaller. Micropores occur in large numbers in the macrogametes of Protococcidia, Eimeridae, and Adeleidae. In contrast, only a single micropore is usually found in each macrogamete of *Plasmodium* species. Occasionally, however, a second micropore was observed in the macrogametes of *P. lophurae, P. gallinaceum, P. pinotti, P. elongatum*, and *P. floridense* as well as *P. knowlesi* and *P. cynomolgi* (Aikawa, Huff, and Sprinz, 1969). This organelle has a neck portion with an average diameter (from 130mμ to 180mμ) similar to that of micropores in other Sporozoa, but relatively large food vacuoles are formed at its inner end. In *Plasmodium* species, micropores appear to have a characteristic location as well as number, whereas in

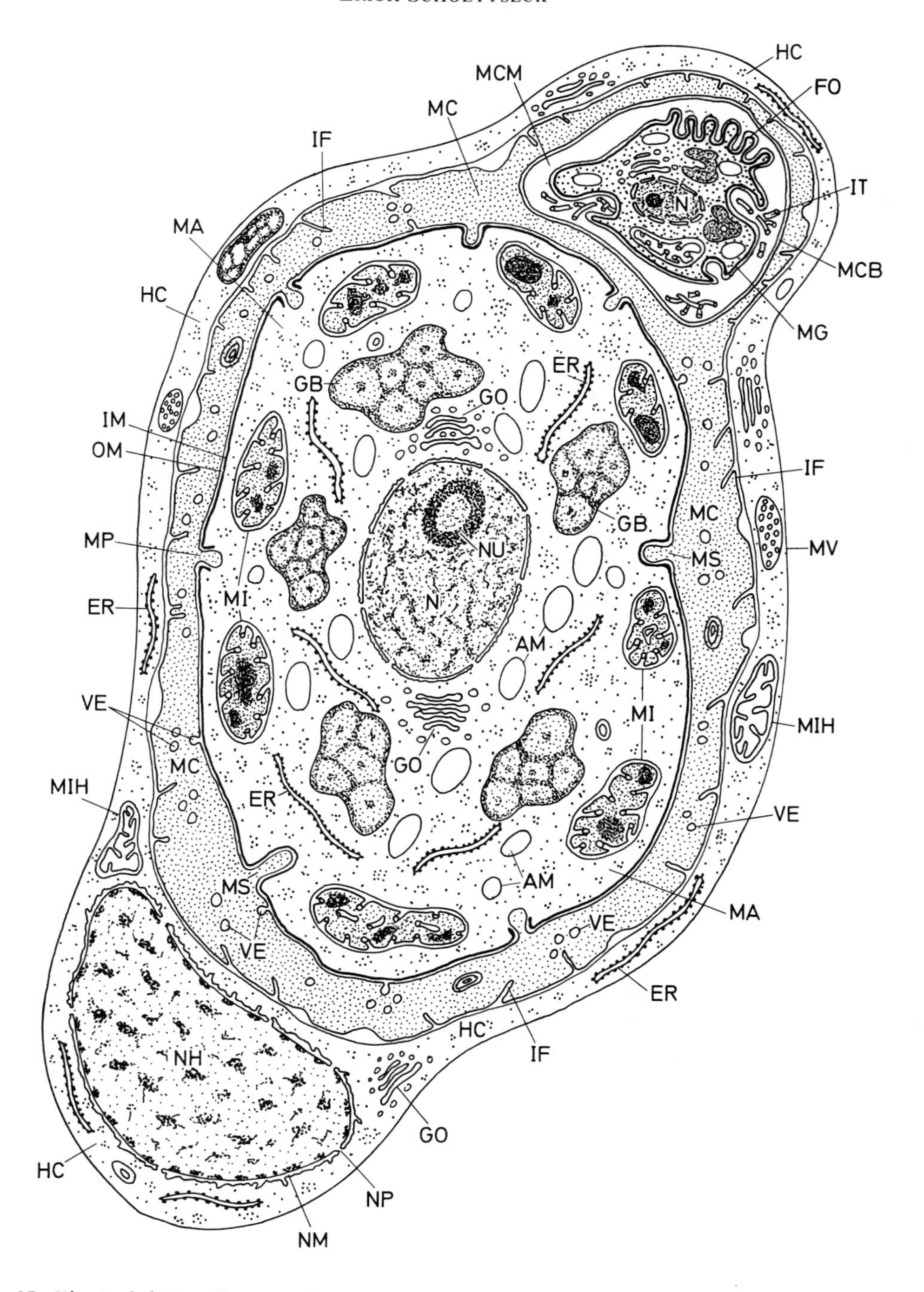

Fig. 25. *Klossia helicina*, diagrammatic representation of a macro- and microgamete. Note that both have a mucous layer (from Scholtyseck, Mehlhorn, and Hammond, 1971).

other Sporozoa, they vary in number and location, apparently in accordance with the level of intake activity.

Whereas the macrogametes of all *Eimeria* species have distinct wall-forming bodies, *Plasmodium* species do not have these. The nuclei of *Plasmodium* occurring in mammals have no nucleolus, whereas those of reptiles and birds do have one. In *Adelina tribolii* the nucleolus sometimes appears ring-shaped.

The account of the ultrastructure of coccidian macrogametocytes is predominantly derived from reports by Bardele (1966), Kheysin (1964), Colley (1967), Hammond *et al.* (1967, 1969, 1970), McLaren (1969), Porchet-Henneré (1967, 1969, 1970), Seliverstova (1969), Snigirevskaya (1968, 1969), and Volkmann (1967).

IV. Ultrastructural Aspects of the Development of Coccidia

The development of coccidia is characterized by a more or less regular alternation of asexual and sexual generations (Pellérdy, 1965). The sexual stages develop by cell growth from vegetative stages; these are merozoites if schizogony precedes the sexual stages, or the gamonts arise by transformation of sporozoites if schizogony is lacking, as in the Protococcidia. The factors responsible for the differentiation of the sexual stages are unknown. In this chapter, major emphasis is laid on the sexual stages of coccidia. The asexual reproduction is described in greater detail in Chapter 3.

A. ASEXUAL REPRODUCTION

Electron microscope studies have facilitated a detailed analysis of the different processes of reproduction in coccidia. One of the most important advances in the knowledge of coccidian development is the recognition that the different types of merogony, including endodyogeny, endopolygeny, endomerogony and ectomerogony are very similar processes. They differ chiefly, insofar as is known, in the number of offspring produced. Figures 26 through 30 depict multiplying stages of *Frenkelia* (endodyogeny), *Toxoplasma* (endopolygeny) and *Eimeria* species (endomerogony and ectomerogony), which illustrate the great cytological similarity of these methods of asexual reproduction. With respect to these processes, endodyogeny may be considered the initial stage of the schizogonic reproduction in the evolution of coccidia. A peculiar pattern of development observed in *E. tenella, T. gondii*, and *Babesia ovis* may be regarded as a more advanced stage in the phylogeny of Sporozoa (Fig. 31); the schizonts subdivide into uninuclear cytomere-like bodies, each of which transforms into a merozoite (Scholtyseck, Friedhoff, and Piekarski, 1970).

Fundamental electron microscope studies on the asexual reproduction of coccidia and related groups have been made by Rudzinska and Trager (1961); by Gavin, Wanko, and Jacobs (1962); Sheffield (1966, 1970); Sheffield and Hammond (1967); Sheffield and Melton (1968); Sénaud (1967, 1970); Roberts, Hammond, Anderson, and Speer (1970); Scholtyseck, Mehlhorn, and Friedhoff (1970); and Piekarski *et al.* (1971).

B. SEXUAL REPRODUCTION

The sexual forms of coccidia, except for the Protococcidia, usually develop from merozoites. In transformation of a merozoite into a gamont, the elongate shape is usually retained for a short time. The organelles typical of the motile stages gradually disappear. After this occurs, the uniformly granulated cytoplasm is relatively free of differentiated structures (Scholtyseck, Rommel, and Heller, 1969). For example, the polysaccharide granules, which represent reserve material, are absent. This material is later newly synthesized in the young macrogametes. In *Eimeria* species, the mitochondria may increase in number, especially in the posterior region of the cell. The young gamete then transforms into an ellipsoidal, subspherical, or irregular form. During this process, one often observes a separation of the two layers of the pellicle, and an appearance of finely granulated cytoplasm between these layers. The young sexual stages can first be distinguished as male or female in electron micrographs when organelles characteristic of either appear.

1. Microgametogenesis

The development of microgametes in coccidia may be devided into two phases:

a. Growth phase, during which nuclear divisions occur.
b. Differentiation phase, during which microgametes develop.

In the growth phase, mitotic nuclear division occurs. Two daughter nuclei are usually formed by each nuclear division in eimerians. In the course of mitosis in *Eimeria* species, intranuclear fibrils appear, which resemble a spindle apparatus (Hammond *et al.*, 1969; Snigirevskaya, 1969). The nuclei later become arranged peripherally, where numerous mitochondria are present (Fig. 19). In some *Eimeria* species, the microgamont undergoes subdivision into several, multinuclear, cytomere-like masses (Scholtyseck, 1963; Kheysin, 1965).

After completion of nuclear division, the second phase of microgametogenesis begins with the appearance of two centrioles in the vicinity of each nucleus. The mechanism of differentiation of the microgametes of *E. perforans* was first reported by Scholtyseck (1965). The following stages occur:

a. The nuclei of the microgamont are arranged in the peripheral zone of cytoplasm; each one is associated with a tubular mitochondrion.
b. An elevation of the surface occurs above each nucleus; in this elevation are located the centrioles, from which the flagella later develop. The nucleus elongates and becomes somewhat pointed and osmiophilic at its peripheral pole. In the cytoplasmic elevation the fibrils of the flagella are visible.
c. The nucleus gives rise to a cylindrical outgrowth, which later incorporates most of the karyoplasm, leaving a saclike structure representing the remainder of the nucleus. The associated mitochondrion becomes elongated.
d. The immature microgamete, which at this stage has a spheroidal or ovoidal form, becomes separated from the microgamont; during this process the nucleus is bent into a circular shape. A short flagellum is attached alongside the nucleus; the two other flagella are free. Later the nucleus straightens out and the microgamete assumes its typical form.

The process of differentiation takes a somewhat different course in *E. auburnensis*, *E. maxima*, *E. falciformis*, and *T. gondii*. In these species, a relatively small electron-pale portion of the nucleus is pinched off as a residual nucleus, while the remaining portion forms the nucleus of the microgamete. This development is diagrammatically shown in Figure 32 (Hammond *et al.*, 1967, 1969; Scholtyseck *et al.*, 1972). It is of interest that there is a thickening of the layer underlying the limiting unit membrane where the pinching off occurs. This thickened layer, which forms a cylinder around the narrowed

Fig. 26. *Frenkelia* sp., metrocyte beginning merozoite formation by endodyogeny. Note the nuclear spindle apparatus (NF) and centrioles (CE). × 30,000.

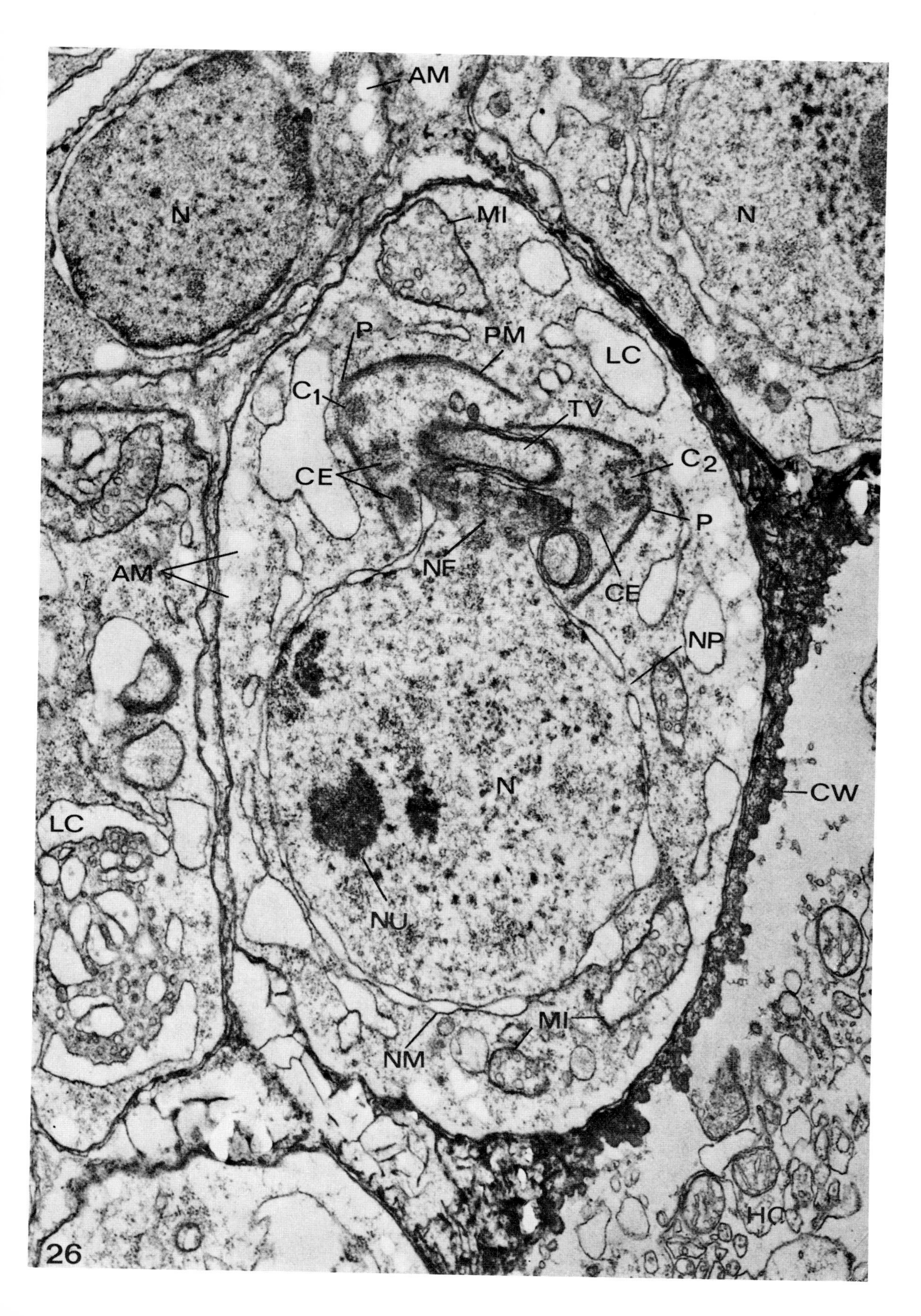

AM
N
MI
N
P
PM
LC
C_1
TV
CE
C_2
P
NF
AM
CE
NP
N
CW
LC
NU
MI
NM
HC
26

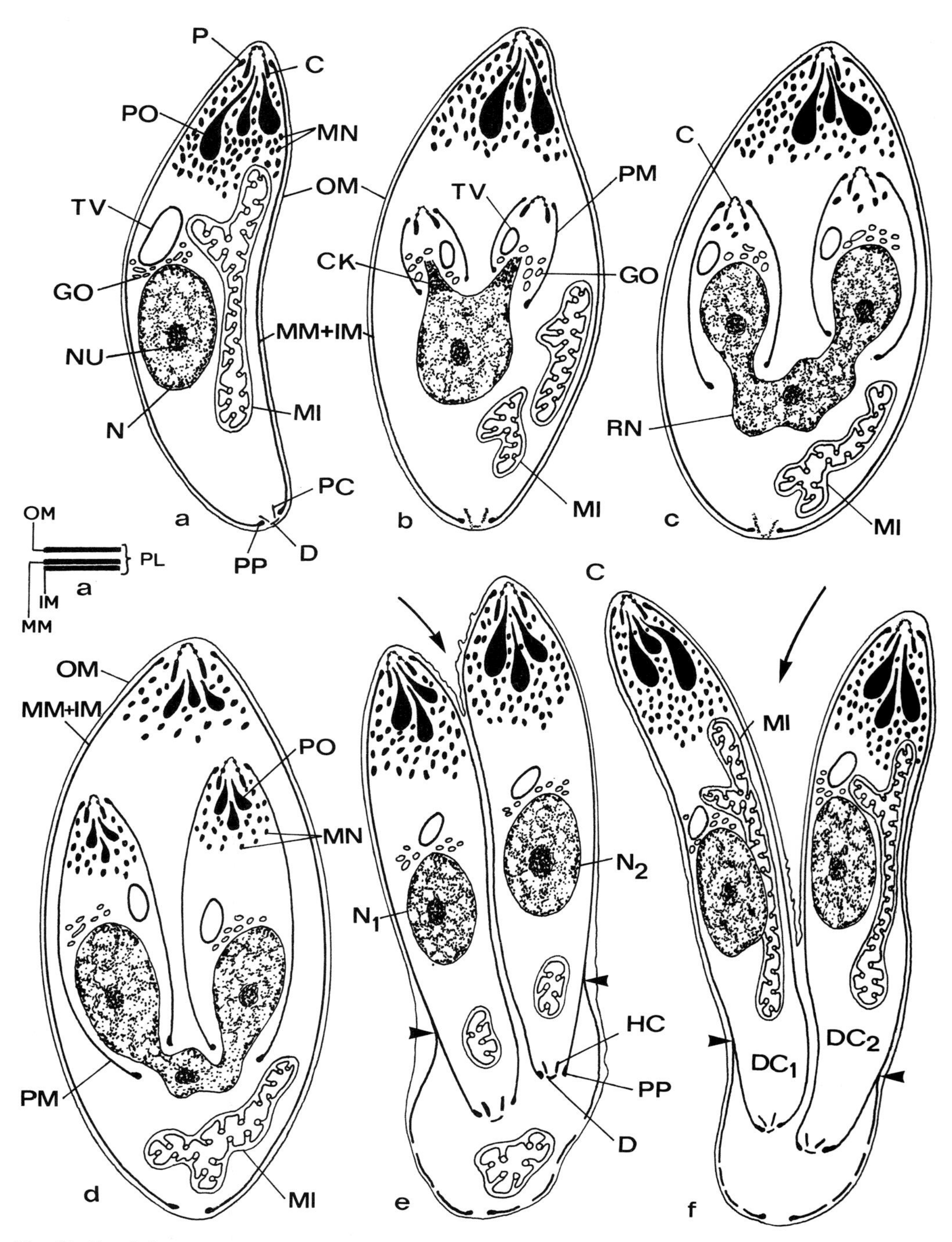

Fig. 27. *Frenkelia* sp., diagrammatic representation of endodyogeny within a mother cell. Note that the pellicle of the mother cell is used for the formation of the pellicle of the daughter cells (after Kepka and Scholtyseck, 1970).

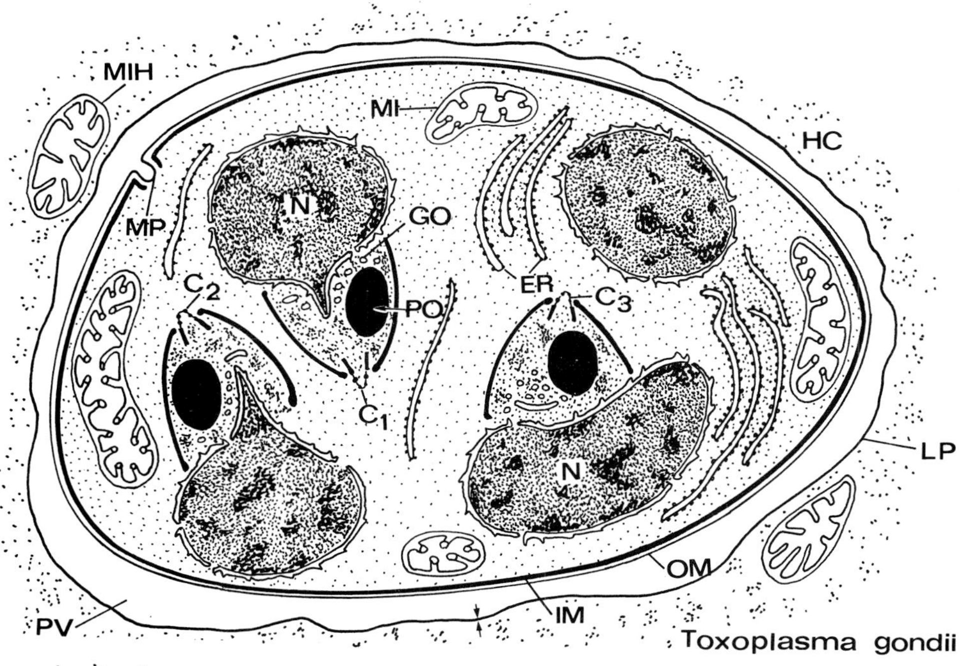

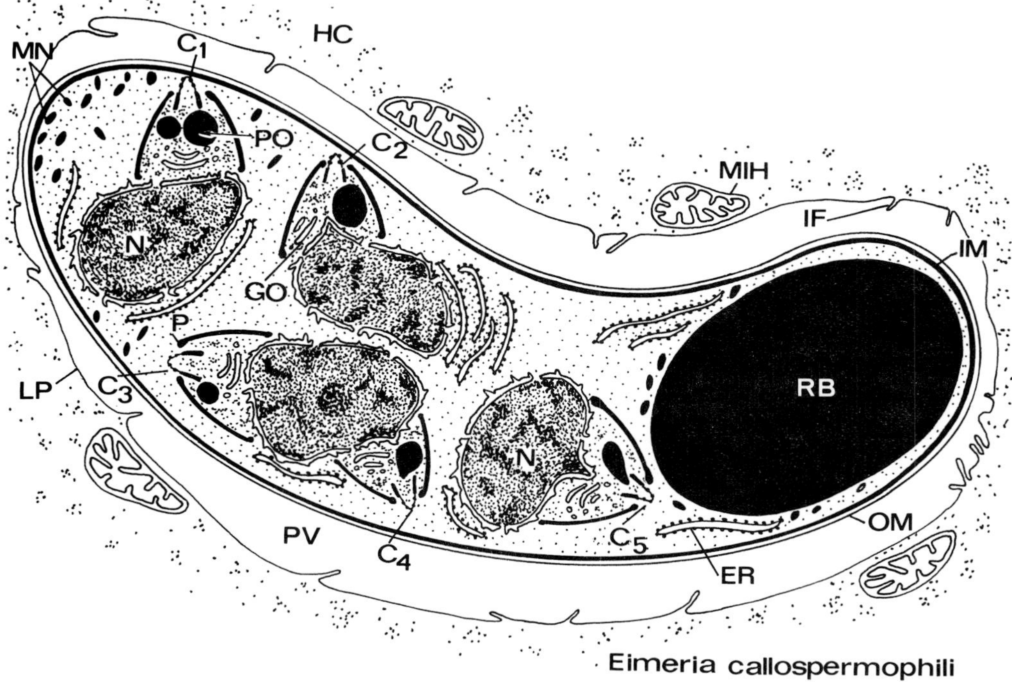

Figs. 28 and 29. Diagrammatic representations of different types of merozoite formation in *Toxoplasma gondii* and *Eimeria callospermophili*. **28.** Dividing schizont of *T. gondii.* Note polar regions of nuclei extending into the forming merozoites (after Sheffield, 1970). This process was called "endopolygeny" by Piekarski, Pelster, and Witte (1970). **29.** *E. callospermophili.* Longitudinal section of a sporozoite 16 hr after penetration. Note the merozoite anlagen with numbered conoids (after Roberts, Hammond, Anderson, and Speer, 1970).

stalk connecting the microgamete and the residual body, was also observed in *E. nieschulzi, E. tenella, E. intestinalis*, and in *T. gondii*; in some cases it has been erroneously interpreted as a micropore (Colley, 1967; McLaren, 1969; Snigirevskaya, 1969; Pelster and Piekarski, 1971).

The development and differentiation of microgametes in coccidia and related groups proceed differently in the various groups. The micromorphologic construction of the microgametes is similar, but characteristic peculiarities occur in certain groups, such as the haemosporidians. These were interpreted by

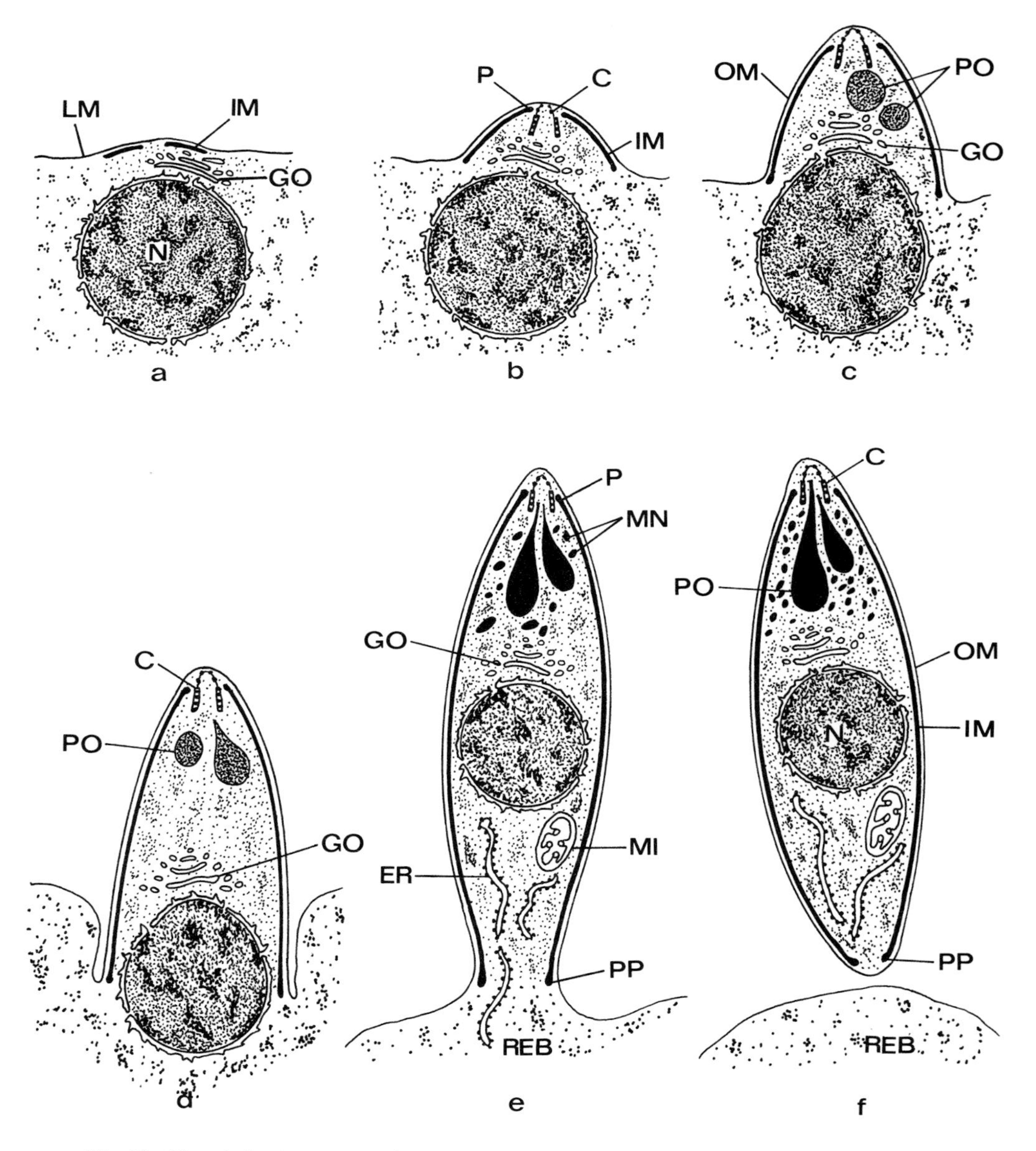

Fig. 30. *Eimeria bovis*, process of merozoite formation (after Sheffield and Hammond, 1967).

Garnham *et al.* (1967) as an indication that the Haemosporina have progressed further in evolution than the Eimeriina. This interpretation is not necessarily in accordance with the account of the ultrastructure of microgametes here presented. The microgametes of some gregarines, considered the simplest members of the Telosporidia, have axonemes, which may be regarded as a primitive character. Axonemes also occur in the Haemosporina, but not in the microgametes of the Protococcidia and the Eimeriina, which have well-developed flagella. With respect to this characteristic, therefore, the Eimeriina and Protoccoccidia can be considered to be more advanced than the Haemosporina.

2. Growth and development of macrogametes

The female gamete is recognizable by the occurrence of the wall-forming bodies in all species with typical oocysts (Fig. 35). In a young macrogamete of *E. perforans*, the intravacuolar tubules first occur within cytoplasmic vacuoles and later at the surface (Fig. 22). The homogeneous wall-forming bodies of type I appear later than the bodies of type II in the development of the macrogametes. They may develop from thick-walled vesicles (Fig. 33). In a more interior region are the wall-forming bodies of type II. These originate from large vesicles apparently derived from the Golgi complex; the vesicles contain irregular, thread-like bodies (Figs. 33, 34). The mitochondria of such young macrogametes are often elongate and dumbbell-shaped. In the cytoplasm lie vesicles with walls of various thickness; the function of these is unknown but they probably originate from the ER.

In young macrogametes, no amylopectin granules are found. These first appear in close association with accumulations of lamellar ER near the nucleus as well as at the periphery in *E. perforans* (Fig. 33). Frequently, polysaccharide granules are observed in the vicinity of lipid inclusions. In young macrogametes of *Klossia helicina*, lipid inclusions may be involved in the synthesis of polysaccharide granules (Volkmann, 1967). This author observed the formation of gray bodies from apical vesicles of the Golgi complex. These bodies transform into lipid

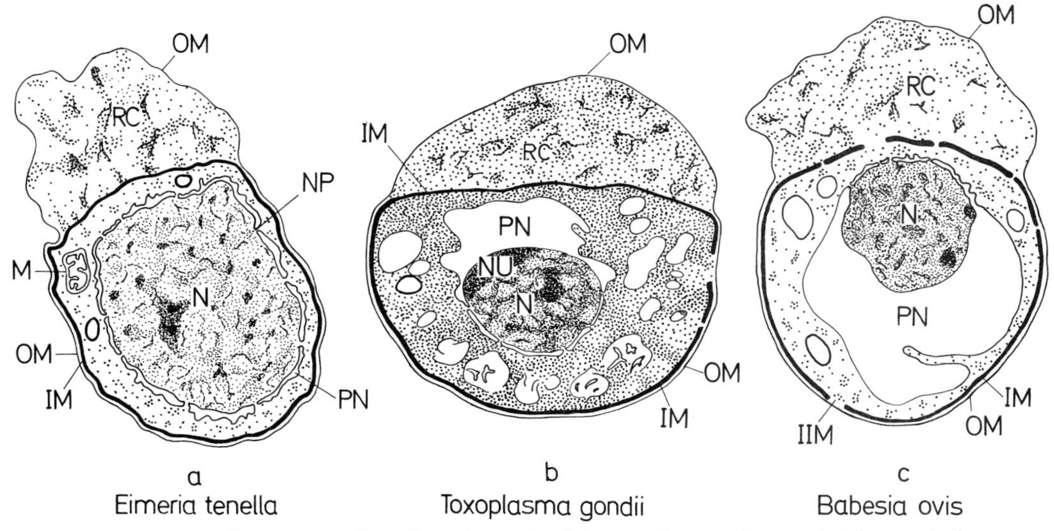

Fig. 31. Diagrammatic representation of transformation into a motile stage (merozoite) from a stationary cell (from Scholtyseck, Friedhoff, and Piekarski, 1970).

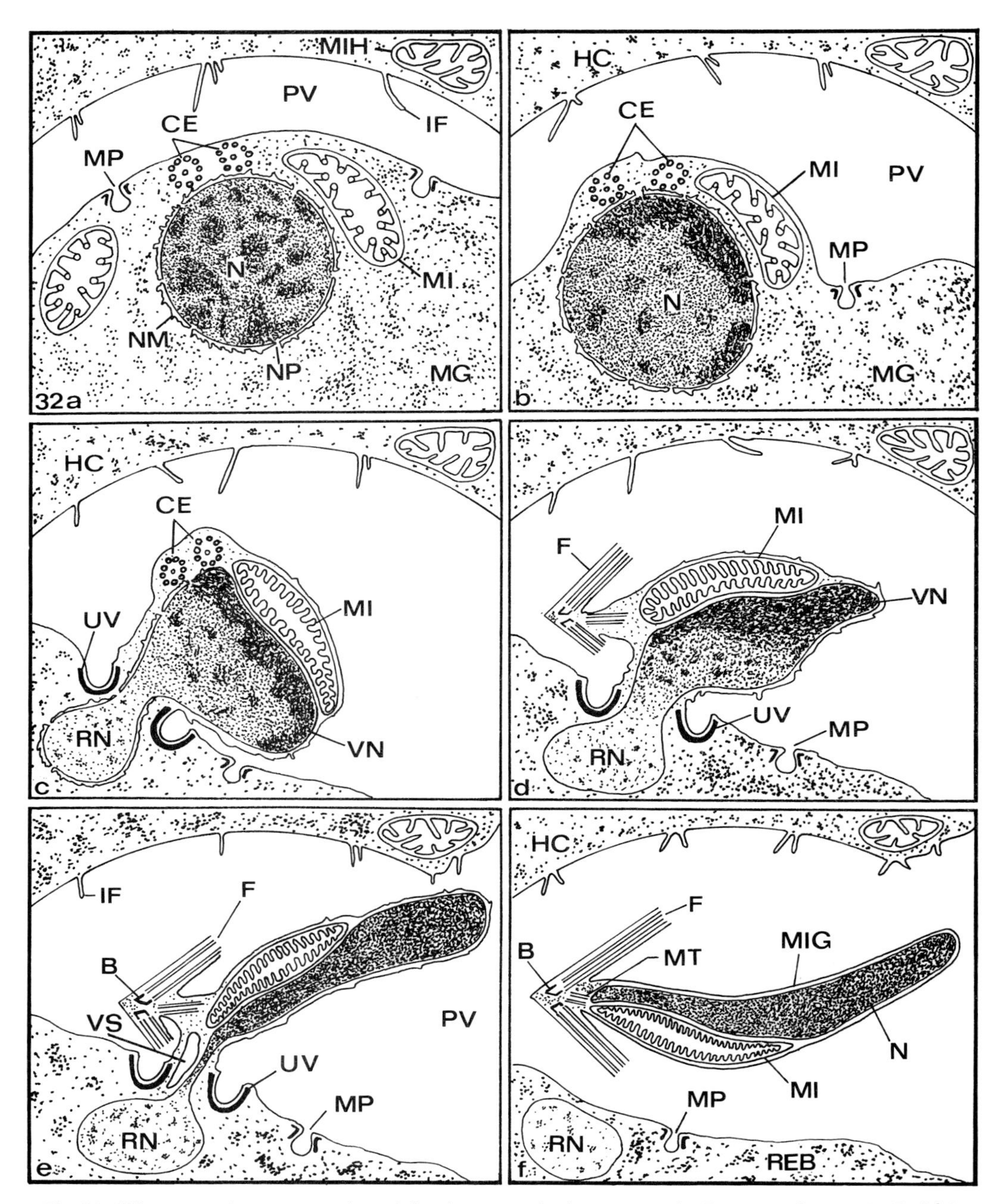

Fig. 32. Diagrammatic representation of development of microgametes in *Eimeria auburnensis*, *E. falciformis*, *E. tenella*, and *E. maxima*. In all four species, specimens with three flagella were found. In each of the micrographs which served as the basis for these figures, only two flagella were seen, because of the orientation of the sections. *a*, The nuclei of the microgamont are arranged peripherally. A mitochondrion is closely associated with the nucleus. *b*, Centrioles (in most micrographs, two) appear between the nucleus and the limiting membrane of the microgamont. Between the two centrioles, a nuclear protuberance (PT) occurs. Peripheral condensations of the karyoplasm become more prominent. *c*, *d*, The nucleus and associated organelles are protruded outward from the surface of the microgamont. The condensation of the karyoplasm progresses further. An osmiophilic layer (UV) supports the cytoplasmic stalk connecting the developing microgamete with the residual body (REB). *e*, The residual nucleus (RN), with electron-pale material being pinched off from the osmiophilic nucleus of the microgamete. *f*, The mature microgamete lies free in the parasitophorous vacuole (PV).

inclusions, at whose margin numerous polysaccharide granules occur; these are small at first, but later become large.

a. Fertilization

The process of fertilization in coccidia is poorly understood. It is usually suggested that in mature macrogametes of coccidia, the nucleus, or a canal-like outgrowth of the nucleus, approaches the surface and the microgamete enters only at this place. The penetration of the microgamete directly into the nucleus of the macrogamete is considered to be a general phenomenon in the coccidia. Only a single report deals with an electron microscope observation of the fertilization process (Hammond and Scholtyseck, 1970). The authors observed an intact microgamete with two flagella within the cytoplasm of a macrogamete of *E. bovis*. The microgamete appeared to lie within a vacuole delimited by a membrane. The nucleus of the macrogamete showed an approximately central location. This finding suggests that the fertilization process in *E. bovis* differs in several respects from the generally accepted concept. In this species, the entire microgamete, including the flagella, evidently enters the cytoplasm rather than the nucleus. The microgamete probably undergoes a change in appearance soon after the penetration; otherwise, findings such as those reported would have been made more frequently.

b. Formation of the oocyst wall

Mature macrogametes cannot be distinguished from young zygotes. Occasionally, in such developmental stages, however, a peculiar, crescent-shaped structure of unknown significance occurs, which has amorphous osmiophilic masses and usually lies near the nucleus. This structure probably represents the cytoplasmic vacuole surrounding the microgamete, which has been observed in the cytoplasm of a macrogamete (Scholtyseck and Hammond, 1970).

In young zygotes of *Eimeria* species, the wall-forming bodies I undergo disaggregation, and in this condition they appear to consist of irregular osmiophilic masses with intervening electron-pale spaces. This material becomes fused together to form a membrane-bound, homogeneous, osmiophilic layer lying at the periphery of the cell (Fig. 35). This process occurs earlier in some places than in others, so that occasionally intact wall-forming bodies I are seen in areas next to those in which fusion of these bodies has occurred. After the formation of the outer layer, which is at first irregular and relatively thick (about 0.7 μ), wall-forming bodies of type I can no longer be seen, because they were completely used up in forming the outer layer of the wall. Later, this layer becomes thinner (0.2 μ), smoother, and more electron-dense. In this stage, the wall-forming bodies II assume a more peripheral location, become fused into large aggregates, and later form the inner layer of the oocyst wall, which is less electron-dense than the outer layer. After completion of this process, wall-forming bodies are no longer present in the young zygote (Scholtyseck, 1963; Scholtyseck and Voigt, 1964; Scholtyseck, Rommel, and Heller, 1969; Scholtyseck, Mehlhorn, and Hammond, 1971). The two layers are not separated by any intervening space; the occurrence of such a space is usually an artifact. The appearance of several membranes external to the oocyst wall is not yet understood. The completed oocyst wall has a thickness of 0.3 μ in *E. perforans*. The formation of the oocyst wall in *Eimeria* species is represented diagramatically in Figure 35.

In *Aggregata eberthi* and *Klossia helicina*, the wall is formed differently. Some covering layers are formed, but a firm oocyst wall is not present.

V. Enzyme Activities in Macrogametes

In certain organelles and fine structures of macrogametes of *E. stiedai, A. eberthi,* and

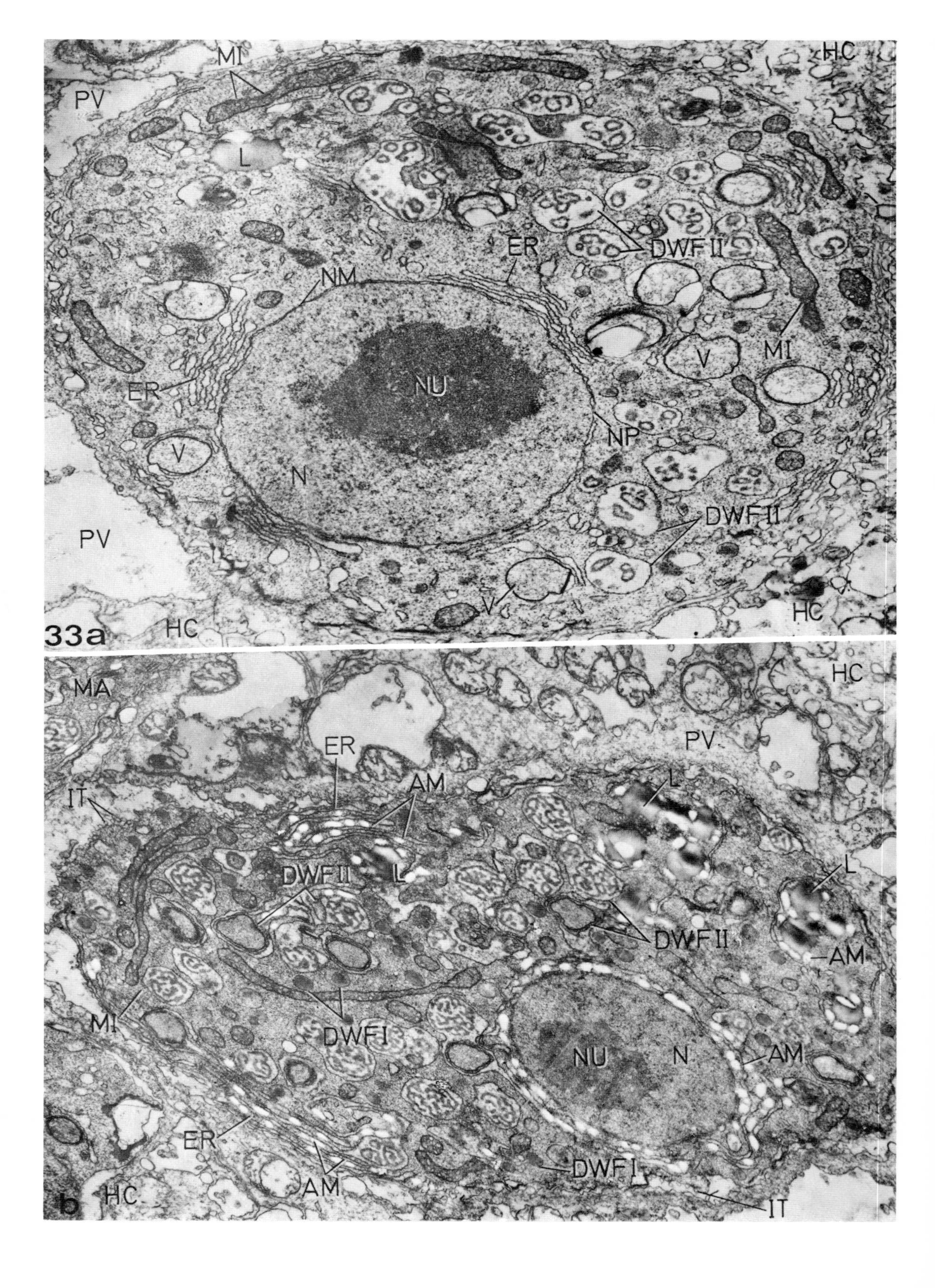

MI
PV
L
HC
ER
DWF II
NM
MI
V
ER
NU
NP
V
N
DWF II
PV
V
HC
HC
33a
MA
HC
ER
PV
IT
AM
L
L
DWF II
L
DWF II
AM
MI
DWF I
DWF I
NU
N
AM
ER
AM
DWF I
b
HC
IT

Fig. 33. *Eimeria perforans*, developing stages of macrogametes. *a*, Early stage. The wall-forming bodies II develop (DWF II). *b*, Later stage. Note the small developing wall-forming bodies I (DWF I) and the arrangement of the polysaccharide granules (AM) along the ER. × 14,000. (From Scholtyseck, 1963, 1964.)

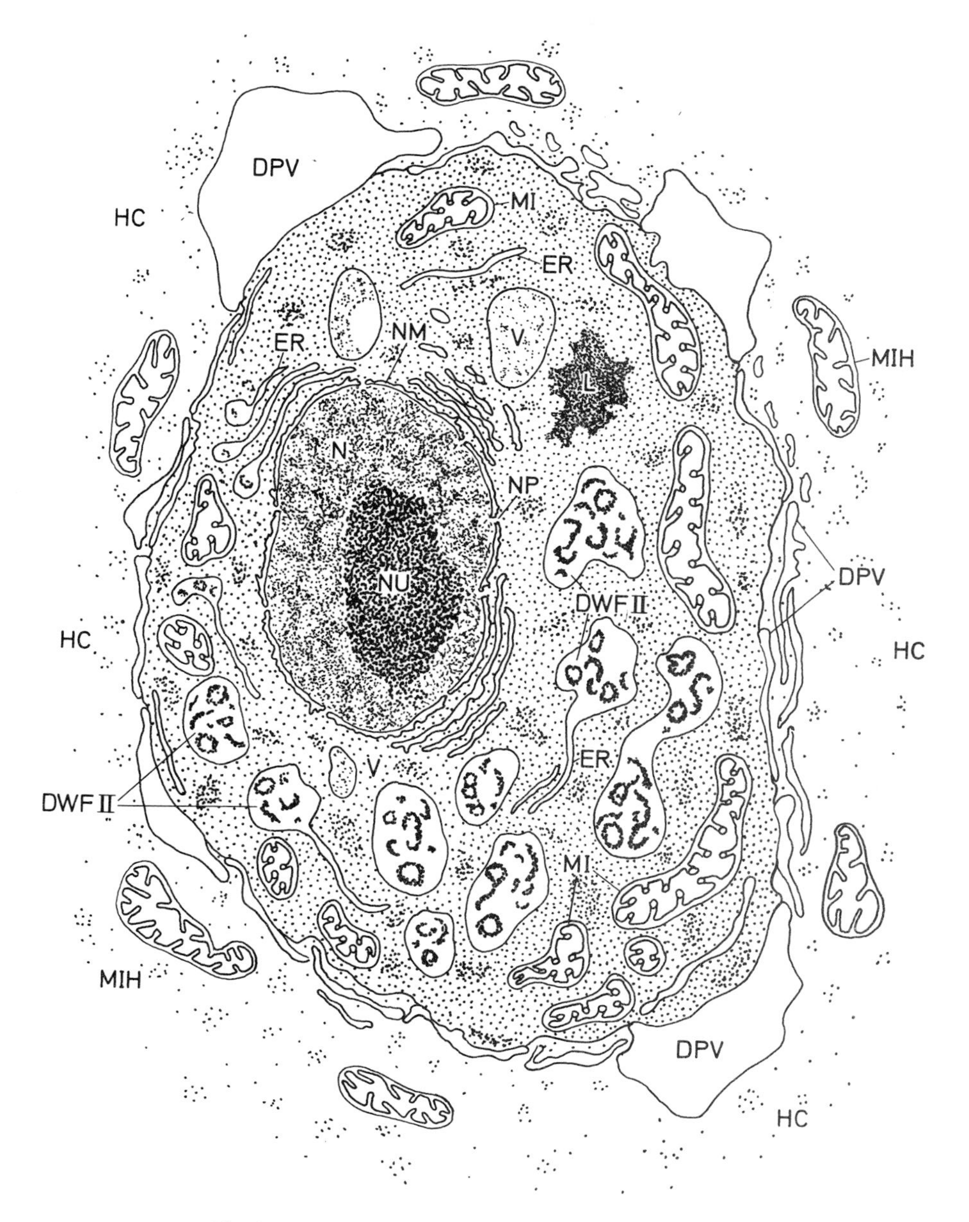

Fig. 34. *Eimeria perforans*, developing macrogamete.

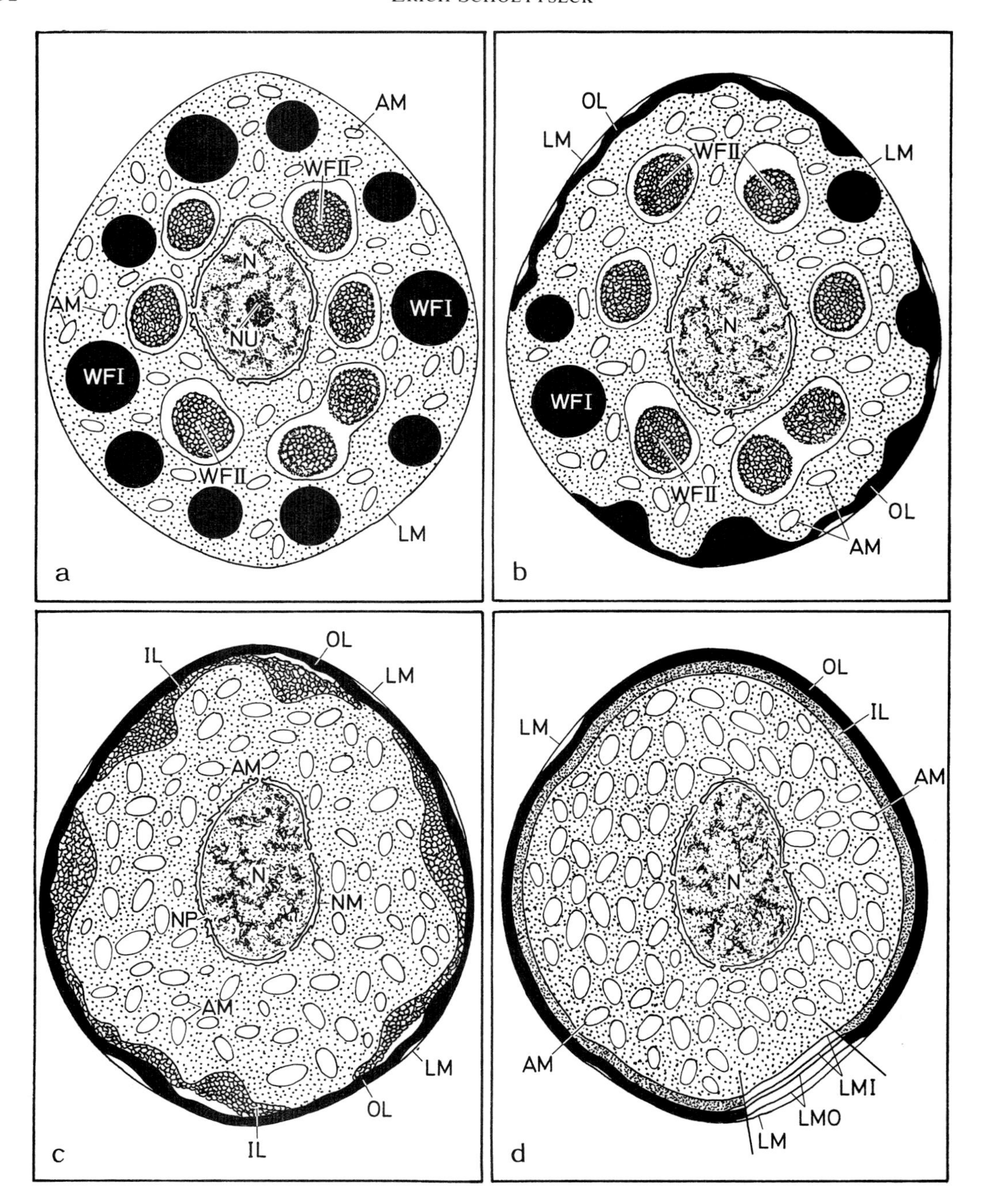

Fig. 35. Diagrammatic representation of the formation of the oocyst-wall in the genus *Eimeria*.

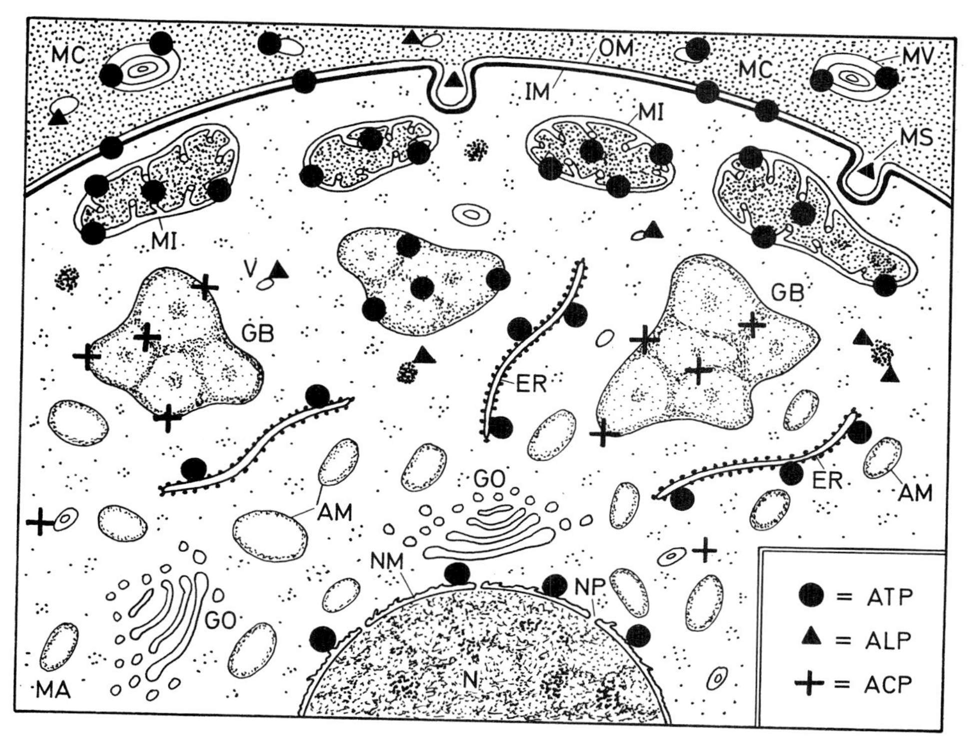

Fig. 36. *Klossia helicina*, diagrammatic representation of enzyme localization in a macrogamete (after Schulte, 1971).

K. helicina, phosphatase activity was demonstrated with cytochemical methods (Heller and Scholtyseck, 1970; Schulte, 1971). Acid phosphatase was found in association with membrane-enclosed compartments in the cytoplasm of the macrogametes, for example, along the vesicles and canaliculi of the ER (Fig. 36). This enzyme was present in the wall-forming bodies II. No activity was seen in the wall-forming bodies I. Acid phosphatase was also demonstrated in the membrane system of *A. eberthi,* especially in the components of the Golgi apparatus and different kinds of vesicles. Since acid phosphatase was found in the gray bodies of the macrogametes of *K. helicina,* these apparently originate from the Golgi complex. Alkaline phosphatase was present in *K. helicina* primarily in the cytoplasmic matrix surrounding the gray bodies and on the membranes of thick-walled vesicles with a diameter of from 0.1 μ to 0.35 μ. This enzyme was also found within the micropores of *K. helicina*. Adenosine triphosphatase occurred within channels and vesicles of the ER, and in the marginal areas of the gray bodies.

Each phosphatase probably has a specific function in the macrogamete. The acid phosphatase may participate in metabolic processes, as for example, in the formation of the wall-forming bodies II and the gray bodies. The alkaline phosphatase probably plays a role in the intake of nutrients, whereas the adenosine triphosphatase may be involved in specific transport. Further studies of all developmental stages of the coccidia should lead to a greater understanding of the function of various organelles.

VI. Host-Parasite Relationships

Each partner in a host-parasite system acts in many ways upon the other partner and is exposed to such actions of that partner. The course of infection, invasion, and development of the parasite is dependent upon this reciprocal relationship, which also influences the defense system of the host and pathogenesis. Major emphasis is given here to the contributions made by modern cytology and functional micromorphology to the analysis of the relationships between parasite and host. These contributions have been markedly influenced and advanced by electron microscopy. Among the different aspects of host-parasite relationships to which ultrastructural studies can make contributions, the most important question is, "Which cell structures are especially concerned with the host-parasite relationships?" By studying various developmental stages of coccidia and their host cells, it is shown that with the help of cell structures recognizable by means of the electron microscrope, the knowledge of the relationship between host cell and intracellular parasite is significantly extended.

The intracellular developmental stages of coccidia lie within the host cell in a vacuolar space, which has been termed "parasitophore Vakuole" (Scholtyseck and Piekarski, 1965) or "parasitophorous vacuole" (Hammond *et al.*, 1967) and "periparasitic vacuole" (Stehbens, 1966). The term most frequently used is that proposed by Hammond, Scholtyseck, and Miner (1967). The cytoplasmic membrane of the host cell bordering the vacuole is often smooth, but in some species it has numerous fine folds or villus-like structures extending into the vacuole. In *E. auburnensis* macrogametes these "intravacuolar folds" (Hammond, Scholtyseck, and Miner, 1967) vary in depth and number in different specimens. In some, the majority of these intravacuolar folds apparently become disconnected from the host cell membrane and evidently disintegrate, forming particulate material, which may be taken into the parasite. Thus, these structures may assist in the transport of nutrients, including membrane material, from the host cell to the parasite. Similar fold-like structures are observed in parasitophorous vacuoles harboring schizonts of *E. callospermophili* (Hammond, 1971), *E. nieschulzi* (Colley, 1968), and *E. miyairii* (Andreassen and Behnke, 1968). Intravacuolar folds of various depth and distribution are also present in the vacuolar membranes surrounding macrogametes of *Eimeria* species and *T. gondii*. These structures are represented diagrammatically in Figure 37. The vacuole membrane surrounding macrogametes of *T. gondii* is slightly folded and of peculiar thickness (Scholtyseck, Mehlhorn, and Hammond, 1971). In the large first-generation schizonts of *E. bovis* as well as in the giant microgametes of *E. auburnensis*, the vacuole surrounding the parasite is filled with amorphous electron-dense masses. This material appears to originate by the pinching-off of numerous vesicles, which evidently disintegrate in the vacuole, thus facilitating the transfer of material from the host cell to the parasite (Scholtyseck, Volkmann, and Hammond, 1966; Sheffield and Hammond, 1966).

The intravacuolar tubules are peculiar structures, making a structural connection with the vacuolar membrane of the host cell (Fig. 37). They may be present at the surface of macrogametes of most *Eimeria* species. Similar tubules were reported from macrogametes of *Aggregata eberthi* and *Klossia helicina* (Fig. 38; Heller, 1969; Schulte, 1971). The function of these structures is unknown. They may serve to transport material from the host cell to the parasite.

The Protococcidia are extracellular parasites (Fig. 38). The *Plasmodium* species lie in a different kind of space; this difference may be associated with the peculiar penetration mechanism of the *Plasmodium* species (Aikawa *et al.*, 1968, 1969; Rudzinska and Trager, 1969; Ladda *et al.*, 1969). The space surrounding these intraerythrocytic parasites

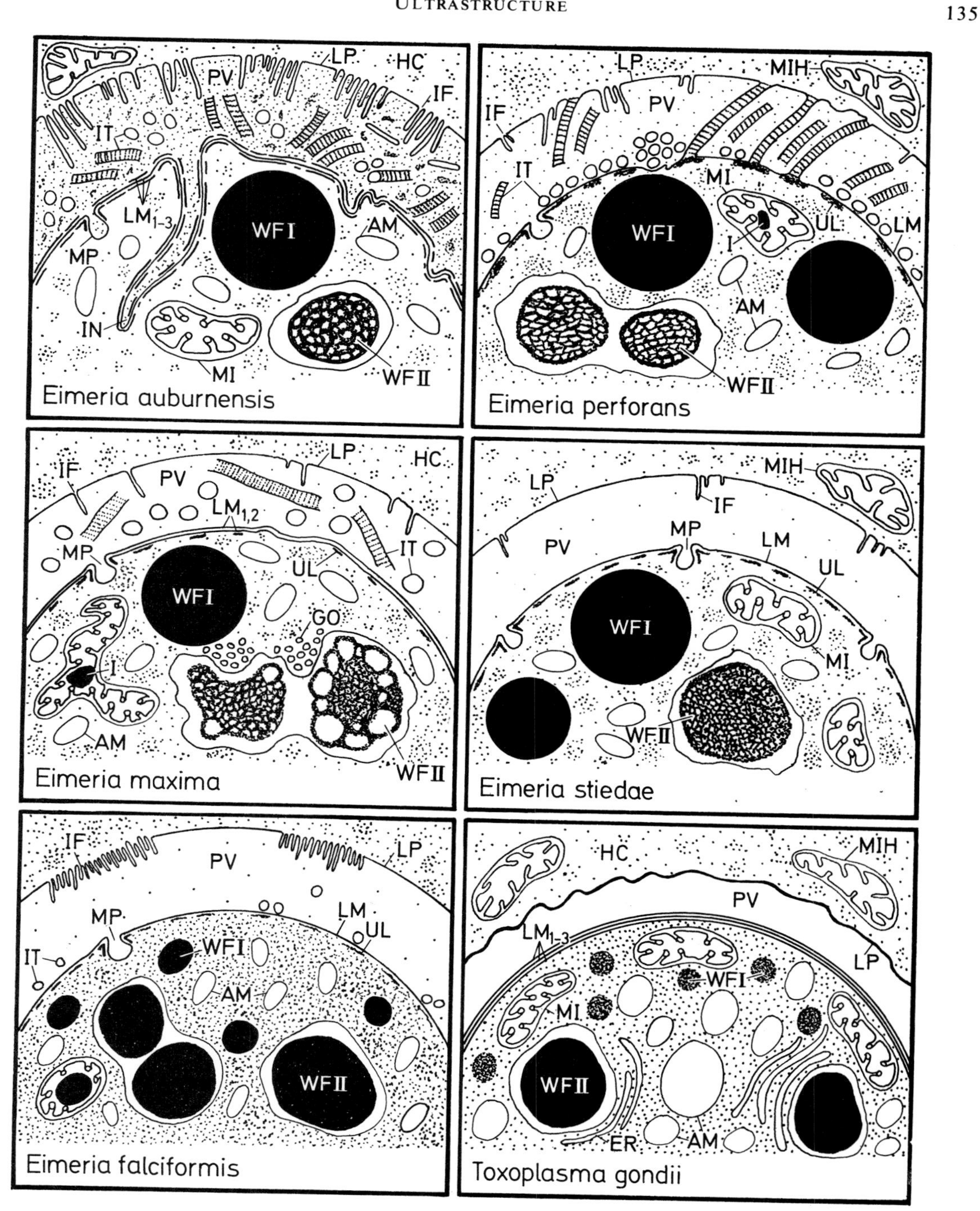

Fig. 37. Diagrammatic representation of macrogametes in the genera *Eimeria* and *Toxoplasma*. Note the limiting membranes of the parasite and the relationship between host cell and parasite (from Scholtyseck, Mehlhorn, and Hammond, 1971).

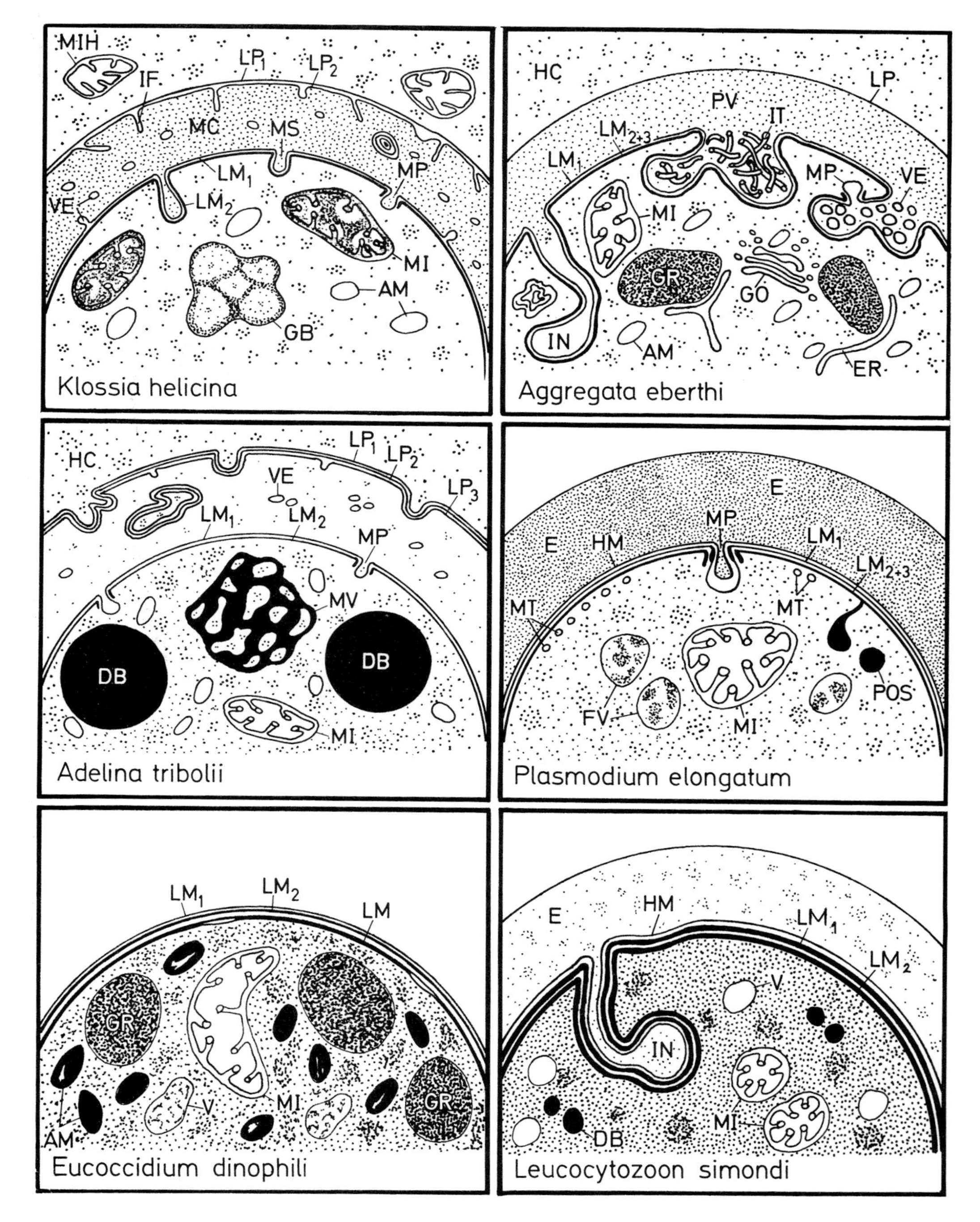

Fig. 38. Diagrammatic representation of macrogametes in different groups of parasites. Note the relationship between host cell and parasite (from Scholtyseck, Mehlhorn, and Hammond, 1971.)

arises as an invagination of the surface during penetration. The spatial relationship between host and parasite is often so close that the host cell membrane limiting this space has frequently been mistaken for a membrane of the parasite (Fig. 38).

In the microgamonts and macrogametes of *K. helicina*, a prominent mucous layer is formed by the parasite at its periphery (Fig. 38). This layer is from 2 μ to 4 μ thick (Volkmann, 1967). Hyaluronic acid was demonstrated in this layer (Schulte, 1971). It is bounded externally by two unit membranes with some septa-like infoldings into the mucous layer. This includes granular and amorphous elements.

The mechanism of ingestion of nutrients in the coccidia is still incompletely known. The intake of nutrients by the intracellular parasites has been discussed in the paragraph dealing with the micromorphology and function of the micropores.

The factors affecting susceptibility of host cells to coccidia are poorly understood. Some studies on the fine structure of parasites developing in cultured cells of various origins revealed that the ultrastructural appearance of the intracellular parasites is identical with that reported from in vivo material (Roberts, Hammond, Anderson, and Speer, 1970; Strout and Scholtyseck, 1970). Scholtyseck, Strout, and Haberkorn (1969) reported developmental stages of *E. tenella*, phagocytized and obviously destroyed by macrophages. While the fine structure of the schizonts appeared morphologically normal, many merozoites were damaged or digested, and were identified only by the typical amylopectin inclusions resisting the digestive process of the host cell.

Ultrastructural changes which have been reported in host cells harboring stages of *Eimeria* species include increases in the number of mitochondria in the cells parasitized as well as in the neighboring cells (Scholtyseck, 1963). This observation indicates a higher than normal rate of metabolism in infected cells. The mitochondria later undergo degeneration.

VII. Conclusions

As shown in this chapter, electron microscope investigations of recent years have yielded an abundance of new results which have had to be integrated into the current framework of science, thereby considerably enriching it. Included among the numerous new descriptions of structures and organelles were many in which homologies among different groups were not recognized, resulting in the assigning of incorrect or duplicate names. Because the discovering of new and surprising structures in the coccidia will undoubtedly continue for some time, additional occurrences of such errors in naming are to be expected. One of the purposes of this chapter has been to provide a basis for a more uniform terminology in this field.

An important consequence of the new information is the considerable disturbance which this has caused in the taxonomic system of the Sporozoa. Systematics and taxonomy are now of current interest and are being investigated with the most modern scientific methods. The "Sporozoa" as a classification concept, which dates back to 1879, will in all likelihood be replaced by new groupings of the organisms in question. A number of changes in grouping and naming of the taxa have already been suggested (Levine, 1969, 1970; also, see Chapter 1). Of particular interest is the use of features recognizable by electron microscopy for systematic characterization and designation of an entire subphylum of protozoa.

VIII. Appendix

ABBREVIATIONS

A	Conoidal rings
AB	Arching canopy-like structure consisting of filamentous elements
AC	Apical cup
ACP	Evidence of acid phosphatase
AF	Attached flagellum
ALP	Evidence of alkaline phosphatase
AM	Amylopectin
AS	Amorphous material
AT	Alveolar aspects of the paired organelles (rhoptries)
ATP	Evidence of ATPase
B	Basal body
BS	Interruptions
C	Conoid
CA	Canals
CE	Centriole
CW	Cyst wall
D	Cover at the posterior end
DB	Dense body
DC	Daughter cell
DPO	Ductules of the rhoptries
DPV	Developing parasitophorous vacuole
DV	Vesicles with thick walls
DWF I	Developing wall-forming bodies of the first type
DWF II	Developing wall-forming bodies of the second type
E	Erythrocyte
EEM	Exoerythrocytic merozoite
EK	Bulbous end (mostly very electron dense)
EM	Erythrocytic merozoite
ER	Endoplasmic reticulum
ES	Space between the ring-like or spiral elements of the conoid (elastic substance?)
F	Flagellum
FO	Folds in the surface of microgamonts
FV	Food vacuole
GB	Gray bodies
GO	Golgi apparatus
GR	Granular bodies
HC	Host cell
HT	Electron-dense posterior end of the PO (rhoptries)
I	Dense inclusions in the mitochondria
IF	Intravacuolar folds
IL	Inner layer of the oocyst wall
IM	Inner layer of the pellicle, consisting of two unit membranes
IIM	Interruption of the inner layer
IN	Invaginations
IT	Intravacuolar tubules
L	Lipid
LC	Lacuna
LM_{1-3}	Limiting membranes of the parasite
LMI	Limiting membranes of the inner layer of the oocyst wall
LMO	Limiting membranes of the outer layer of the oocyst wall
LP	Limiting membranes of the parasitophorous vacuole
LS	Labyrinthine structure
LU	Empty interior of the PO (rhoptries)
M	Membrane
MA	Macrogamete
MC	Mucous layer
MCB	Membranes between the mucous layer of the macro- and micro-gamonts of *Klossia helicina*
MCM	Mucous layer around the microgamont
ME	Merozoite
MF	Microfibrils
MG	Microgamont
MH	Middle membrane
MI	Mitochondrion
MIG	Microgamete
MIH	Mitochondrion of the host cell
MKG	Macrogamont
MM	Middle membrane

MN	Micronemes
MP	Micropore
MS	Structure similar to a micropore
MT	Microtubules
MV	Multivesicular body
MVE	Formation of vesicles
MZ	Metrocyte
N	Nucleus
NF	Intranuclear fibrils (spindle apparatus)
NH	Nucleus of the host cell
NI	Nucleus in division
NM	Nuclear membrane
NP	Nuclear pores
NU	Nucleolus
O	Opening at the top of the conoid; cross section through a ductule of the PO (rhoptries)
OL	Outer layer of the oocyst wall
OM	Outer membrane of the pellicle
P	Polar ring
PC	Posterior cone
PE	Perforatorium
PG	Protein granules
PL	Dense material
PM	Inner membranes of developing merozoites
PN	Perinuclear space
PO	Rhoptries (paired organelles)
POS	Structures similar to the PO
PP	Posterior polar ring
PT	Protrusion of the nucleus
PV	Parasitophorous vacuole
R	Rings belonging to the conoid
RB	Refractile bodies
RC	Residual cytoplasm
REB	Residual body
RH	Hollow space of the ribs
RIM	Remaining part of the inner membrane
RN	Residual nucleus
RP	Ribs
RS	Rod-shaped tubules
SP	Sporozoite
SR	Ring-like or spiral elements, occasionally resembling microtubules
SS	Inner layer of the pellicle
ST	Sponge-like aspect of the paired organelle
TA	Tangential striation
TR	Trophozoite (developing stage)
TV	Vesicle with thick walls
UL	Underlying material
UV	Underlying material at the place where the microgametes are pinched off
V	Vacuole
VE	Vesicle
VN	Condensation of the karyoplasm
VS	Vesicle-like structure, which participates in the pinching off of the microgametes
WF I	Wall-forming bodies of the first type
WF II	Wall-forming bodies of the second type
Z	Zoite

Literature Cited

Aikawa, M. 1966. The fine structure of the erythrocytic stages of three avian malarial parasites, *Plasmodium fallax*, *Pl. lophurae* and *Pl. cathemerium*. Am. J. Trop. Med. Hyg. 15: 449-471.

Aikawa, M. 1967. Ultrastructure of the pellicular complex of *Plasmodium fallax*. J. Cell Biol. 35: 103-113.

Aikawa, M., C.G. Huff and H. Sprinz. 1968. Exoerythrocytic stages of *Plasmodium gallinaceum* in chick-embryo liver as observed electron microscopically. Amer. J. Trop. Med. Hyg. 17: 156-169.

Aikawa, M., C.G. Huff and H. Sprinz. 1969. Comparative fine structure study of the gametocytes of avian, reptilian, and mammalian malarial parasites. J. Ultrastruct. Res. 26: 316-331.

Aikawa, M., C.G. Huff, and C.P.A. Strome. 1970.

Morphological study of microgametogenesis of *Leucocytozoon simondi*. J. Ultrastruct. Res. 32: 43-68.

Andreassen, J., and O. Behnke. 1968. Fine structure of merozoites of a rat coccidian *Eimeria miyairii* with a comparison of the fine structure of other sporozoa. J. Parasitol. 54: 150-163.

Bardele, C.F. 1966. Elektronenmikroskopische Untersuchungen an dem Sporozoon *Eucoccidium dinophili* Grell. Z. Zellforsch. Mikrosk. Anat. 74:559-595.

Bradbury, P.C., and W. Trager. 1968. The fine structure of microgametogenesis in *Haemoproteus columbae* Kruse. J. Protozool. 15: 700-712.

Braunsteiner, H., F. Pakesch, and O. Thalhammer. 1957. Elektronenmikroskopische Untersuchungen über die Morphologie des *Toxoplasma gondii* und das Wesen des Farbtestes nach Sabin-Feldman. Wien. Z. Inn. Med. 38: 16-27.

Büttner, D.W. 1966. Über die Feinstruktur der erythrozytären Formen von *Theileria mutans*. Z. Tropenmed. Parasitol. 17: 397.

Büttner, D.W. 1967. Die Feinstruktur der Merozoiten von *Theileria parva*. Z. Tropenmed. Parasitol. 18: 224-244.

Büttner, D.W. 1968. Vergleichende Untersuchung der Feinstruktur von *Babesia gibsoni* und *Babesia canis*. Z. Tropenmed. Parasitol. 19: 330-342.

Colley, F.C. 1967. Fine structure of sporozoites of *Eimeria nieschulzi*. J. Protozool. 14 (2): 217-220.

Colley, F.C. 1968. Fine structure of schizonts and merozoites of *E. nieschulzi*. J. Protozool. 15 (2): 374-382.

Colley, F.C. and V. Zaman. 1970. Observations on the endogenous stages of *Toxoplasma gondii* in the cat ileum. Southeast Asian J. Trop. Med. Pub. Hlth. 1 (4): 465-480.

Desser, S.S. 1970. The fine structure of *Leucocytozoon simondi*. IV. The microgamete. Can. J. Zool. 48: 647-649.

Desser, S.S., J. R. Baker and P. Lake. 1970. The fine structure of *Leukocytozoon simondi*. I. Gametocytogenesis. Can. J. Zool. 48: 331-341.

Desportes, I. 1970. Ultrastructure des Gregarines du genre *Stylocephalus;* La Phase Enkystee. Ann. Sci. Nat., Zool., 12: 73-170.

Emschermann, P. and J. Staubesand. 1968. Struktureigentümlichkeiten im Feinbau eines Sporozoons in Erythrocyten des Frosches. Anat. Anz. 121: 599-601.

Fayer, R. and D.M. Hammond. 1967. Development of first-generation schizonts of *Eimeria bovis* in cultured bovine cells. J. Protozool. 14: 764-772.

Frenkel, J.K. 1953. Infections with organisms resembling *Toxoplasma* together with the descriptions of a new organism: *Besnoitia jellisoni*. Atti del VI. Congr. Intern. di Microbiol. Rome, 5: 426-434.

Frerichs, W.M. 1970. Fine structure of *Babesia equi*: Electron microscopy and phase-contrast microscopy. (Abstr.) J. Parasitol., 56(No. 4, Sect. II): 108-109.

Friedhoff, K., and E. Scholtyseck. 1968. Feinstrukturen von *Babesia ovis* (Piroplasmidea) in *Rhipicephalus bursa* (Ixodoidea): Transformation sphäroider Formen zu Vermiculaformen. Z. Parasitenk. 30: 347-359.

Friedhoff, K. and E. Scholtyseck. 1969. Feinstrukturen der Merozoiten von *Babesia bigemina* im Ovar von *Boophilus microplus* und *Boophilus decoloratus*. Z. Parasitenk. 32: 266-283.

Garnham, P. C. C. 1963. The ultrastructure of Haemosporidia and allied protozoa with special reference to the motile stages. Proc. 1st Int. Conf. Protozool., Prague, 1961. 427-433.

Garnham, P. C. C. 1966. Locomotion in the parasitic protozoa. Biol. Rev. 41: 561-586.

Garnham, P. C. C. 1969. The structure and function of the cytostome (micropyle) in the Sporozoa. *In* Progress in protozoology, p. 8. (Abstracts of papers read at the IIIrd Int. Congr. on Protozoology). Leningrad: Academy of Sciences of the USSR, Publishing House Nauka.

Garnham, P. C. C., J. R. Baker, and R. G. Bird. 1962. The fine structure of *Lankesterella garnhami*. J. Protozool. 9 (1): 107-114.

Garnham, P. C. C., R. G. Bird, and J. R. Baker. 1960. Electron microscope study of motile stages of malarial parasites. 1. The fine structure of the sporozoites of *Haemamoeba (Plasmodium) gallinacea*. Trans. Roy. Soc. Trop. Med. Hyg. 54 (3): 274-278.

Garnham, P. C. C., R. G. Bird, and J. R. Baker. 1967. Electron microscope studies of motile stages of malaria parasites. V. Exflagellation in *Plasmodium*, *Hepatocystis* and *Leucocytozoon*. Trans. Roy. Soc. Trop. Med. Hyg. 61: 58-68.

Garnham, P. C. C., R. G. Bird, J. R. Baker, and R. S. Bray. 1961. Electron microscope studies of motile stages of malaria parasites. II. The fine structure of the sporozoite of *Laverania (Plasmodium) falcipara*. Trans. Roy. Soc. Trop. Med. Hyg. 55: 98-102.

Gavin, M. A., T. Wanko, and L. Jacobs. 1962. Electron microscope studies of reproducing and interkinetic *Toxoplasma*. J. Protozool. 9 (2): 222-234.

Giovannola, A. 1934. Die Glykogenreaktionen nach Best und Bauer in ihrer Anwendung auf Protozoen. Arch. Protistenk. 83: 270-274.

Gustafson, P. V., H. D. Agar, and D. J. Cramer. 1954. An electron microscope study of *Toxoplasma*. Amer. J. Trop. Med. Hyg. 3: 1008-1021.

Hammond, D. M. 1971. The development and ecology of coccidia and related intracellular parasites. In Ecology and physiology of parasites: a symposium, A. M. Fallis, ed., University of Toronto Press, Toronto, pp. 3-19.

Hammond, D. M. and E. Scholtyseck. 1970. Observations concerning the process of fertilization in *Eimeria bovis*. Naturwissenschaften 8: 399.

Hammond, D. M., E. Scholtyseck, and B. Chobotar. 1967. Fine structures associated with nutrition of the intracellular parasite *Eimeria aubernensis*. J. Protozool. 14: 678-683.

Hammond, D. M., E. Scholtyseck, and B. Chobotar. 1969. Fine structural study of the microgametogenesis of *Eimeria auburnensis*. Z. Parasitenk. 33: 65-84.

Hammond, D. M., E. Scholtyseck, and M. L. Miner. 1967. The fine structure of microgametocytes of *Eimeria perforans*, *E. stiedae*, *E. bovis*, and *E. auburnensis*. J. Protozool. 53 (2): 235-247.

Hammond, D. M., C. A. Speer, and W. Roberts. 1970. Occurrence of refractile bodies in merzoites of *Eimeria* species. J. Parasitol. 56: 189-191.

Heller, G. 1969. Elektronenmikroskopische Untersuchungen an *Aggregata eberthi* aus dem Spiraldarm von *Sepia officinalis* (Sporozoa, Coccidia). I. Die Feinstruktur der Merozoiten, Makrogamonten und Sporen. Z. Parasitenk. 33: 44-46.

Heller, G. and E. Scholtyseck. 1970. Histochemistry of coccidia as demonstrated by means of electron microscopy. J. Parasitol. 56 (No. 4, Sect. II): 142.

Honigberg, B. M., W. Balamuth, E. C. Bovee, J. O. Corliss, M. Gojdics, R. P. Hall, R. R. Kudo, N. D. Levine, A. R. Loeblich, Jr., J. Weiser, and D. H. Wenrich. 1964. A revised classification of the phylum Protozoa. J. Protozool. 11 (1): 7-20.

Jacobs, L. 1967. *Toxoplasma* and toxoplasmosis. Adv. Parasitol. 5: 1-45.

Kepka, O., and E. Scholtyseck. 1970. Weitere Untersuchungen der Feinstruktur von *Frenkelia* spec. (M-Organismus, Sporozoa). Protistologica 6: 249-266.

Kheysin, E. M. (Cheissin, E. M.). 1958. Cytologische Untersuchungen verschiedener Stadien des Lebenszyklus der Kaninchencoccidien. I. *Eimeria intestinalis* E. Cheissin, 1948. Arch. Protist. 102: 265-290.

Kheysin, E. M. (Cheissin, E. M.). 1965. Electron microscopic study of microgametogenesis in two species of coccidia from rabbit (*Eimeria magna* and *E. intestinalis*). Acta Protozool. 3: 215-224.

Kheysin, E. M. 1967. A finding of the centriole at electron microscope studies of merozoites of *E. intestinalis*. Tsitologiya. 9: 1411-1413 [In Russian].

Kheysin, E. M. 1964. Electron microscopic study of microgametes in *Eimeria intestinalis* (Sporozoa, Coccidia). Zool. J. 43: 647-651.

Kheysin, E. M. and E. S. Snigirevskaya. 1965. Some new data on the fine structure of the merozoites of *Eimeria intestinalis* (Sporozoa, Eimeriidea). Protistologica 1: 121-128.

Ladda, R., M. Aikawa, and H. Sprinz. 1969. Penetration of erythrocytes by merozoites of mammalian and avian malarial parasites. J. Parasitol. 55: 633-644.

Levine, N. D. 1969a. Uniform terminology for sporozoan protozoa. *In* Progress in protozoology, p. 340 (Abstracts of papers read at the IIIrd Int. Congr. on Protozoology). Leningrad: Academy of Sciences of the USSR, Publishing House Nauka.

Levine, N. D. 1969b. Taxonomy of the sporozoa. *In* Progress in protozoology, pp. 365-367 (Abstracts of papers read at the IIIrd Int. Cong. on Protozoology). Leningrad: Academy of Sciences of the USSR, Publishing House Nauka.

Levine, N. D. 1970. Taxonomy of the sporozoa. J. Parasitol. 56 (No. 4, Sect. II): 208.

Ludvik, J. 1956. Vergleichende elektronenoptische Untersuchungen an *Toxoplasma gondii* und *Sarcocystis tenella*. Zbl. Bakt. 1. Orig. 166:60.

Ludvik, J. 1958. Morphology of *Toxoplasma gondii* in electron microscope. Zestnik. Ceskoslov. Zool. Spolec. 22:130.

Ludvik, J. 1960. The electron microscopy of *Sarcocystis miescheriana*. (Kuhn 1865). J. Protozool. 7 (2): 128-135.

Ludvik, J. 1961. Electron microscopic study of some parasitic Protozoa. Progress in Protozoology. 1st International Congress on Protozoology, Prague, 387-392.

McLaren, D. J. 1969. Observations on the fine structural changes associated with schizogony and gametogony in *Eimeria tenella*. Parasitology 59: 563-574.

McLaren, D. J. and G. E. Paget. 1968. A fine structural study on the merozoite of *Eimeria tenella* with special reference to the conoid apparatus. Parasitology 58: 561-579.

Mehlhorn, H. 1972. Elektronenmikroskopische Untersuchungen an Entwicklungsstadien von *Eimeria maxima* (Coccidia, Sporozoa) aus dem Haushuhn. Dissertation, Bonn.

Mossevitch, T. N., and E. M. Kheysin. 1961. Certain data on electron microscope study of the merozoites of the *Eimeria intestinalis* from rabbit intestine. Tsitologiya. 3: 34-39. [In Russian]

Ogino, N., and C. Yoneda. 1966. The fine structure and mode of division of *Toxoplasma gondii*. Arch. Ophthal. 75: 218-227.

Pellérdy, L. P. 1965. Coccidia and coccidiosis. Akad. Kiado, Budapest.

Pelster, B. and G. Piekarski. 1971. Elektronenmikroskopische Analyse der Mikrogametenentwicklung von *Toxoplasma gondii*. Z. Parasitenk. 37: 267-277.

Pelster, B. and G. Piekarski. 1972. (in press). Elektronenmikroskopische Untersuchungen der Makrogametenentwicklung von *Toxoplasma gondii*. Z. Parasitenk. 39: 225-232.

Piekarski, G., B. Pelster, and H. M. Witte. 1971. Endopolygenie bei *Toxoplasma gondii*. Z. Parasitenk. 36: 122-130.

Porchet-Henneré, E. 1967. Etude de premiers stades de développement de la coccidie *Coelotropha durchoni*. Z. Zellforsch. 80: 556-569.

Porchet-Henneré, E. 1969. Observations sur la cytologie, l'ultrastructure et la physiologie de quelques coccidies parasites d'Annélides polychètes. Thèse Fac. Sciences Lille.

Porchet-Henneré, E. 1970. La microgamétogenèse chez la coccidie *Coelotropha durchoni* (Vivier-Henneré); étude au microscope électronique. Arch. Protistenk. 112: 21-29.

Porchet-Henneré, E. and A. Richard. 1970. Ultrastructure des stades végétatifs d'*Aggregata eberthi* Labbe: le trophozoite et le schizonte. Z. Zellforsch. Mikrosk. Anat. 103: 179-191.

Reich, F. 1913. Das Kaninchencoccid *Eimeria stiedai* (Lindemann, 1865) nebst einem Beitrage zur Kenntnis von *Eimeria falciformis* (Eimer, 1870). Arch. Protistenk. 28: 1-42.

Roberts, W. L. and D. M. Hammond. 1970. Ultrastructural and cytological studies of sporozoites of four *Eimeria* species. J. Protozool. 17: 76-86.

Roberts, W. L., D. M. Hammond and C. A. Speer. 1970. Ultrastructural study of the intra- and extracellular sporozoites of *Eimeria callospermophili*. J. Parasitol. 56: 907-917.

Roberts, W. L., D. M. Hammond, L. C. Anderson, and C. A. Speer. 1970. Ultrastructural study of schizogony in *Eimeria callospermophili*. J. Protozool. 17: 584-592.

Roberts, W. L., C. A. Speer and D. M. Hammond. 1971. Penetration of *Eimeria larimerensis* sporozoites into cultured cells as observed with the light and electron microscopes. J. Parasitol. 57: 615-625.

Rudzinska, M. A. and W. Trager. 1961. The role of the cytoplasm during reproduction in a malarial parasite (*Plasmodium lophurae*) as revealed by electron microscopy. J. Protozool 8: 307-322.

Rudzinska, M. A. and W. Trager. 1969. The fine structure of trophozoites and gametocytes in *Plasmodium coatneyi*. J. Protozool. 15: 73-88.

Ryley, J. F. 1969. Ultrastructural studies on the sporozoite of *Eimeria tenella*. Parasitology 59: 67-72.

Ryley, J. F., M. Bentley, D. J. Manners, and J. R. Stark. 1969. Amylopectin, the storage polysaccharide of the coccidia *Eimeria brunetti* and *E. tenella*. J. Parasitol. 55: 839-845.

Sampson, J. R., and D. M. Hammond. 1972. Fine structural aspects of development of *Eimeria alabamensis* in cell cultures. J. Parasitol. 58: 311-322.

Sampson, J. R. and D. M. Hammond. 1971. Ingestion of host-cell cytoplasm by means of micropores in *Eimeria alabamensis*. J. Parasitol. 57: 1133-1135.

Schmidt, K., M. R. L. Johnston, and W. E. Stehbens. 1967. The fine structure of the schizont and merozoite of *Isospora* sp. (Sporozoa: Eimeriidae) parasitic in *Gehyra variegata* (Dumeril and Bibron, 1836) (Reptilia: Gekkonidae). J. Protozool. 14: 602-608.

Scholtyseck, E. 1962. Electron microscope studies on *E. perforans* (Sporozoa). J. Protozool. 9: 407-414.

Scholtyseck, E. 1963. Vergleichende Untersuchungen über die Kernverhältnisse und das Wachstum von Coccidiomorphen unter besonderer Berücksichtingung von *Eimeria maxima*. Z. Parasitenk. 22: 428-474.

Scholtyseck, E. 1964. Elektronenmikroskopisch-cytochemischer Nachweis von Glykogen bei *Eimeria perforans*. Z. Zellforsch. Mikrosk. Anat. 64: 688-707.

Scholtyseck, E. 1965a. Die Mikrogametenentwicklung von *Eimeria perforans*. Z. Zellforsch. Mikrosk. Anat. 66: 625-642.

Scholtyseck, E. 1965b. Elektronenmikroskopische Untersuchungen über die Schizogonie bei Coccidien (*Eimeria perforans* und *E. stiedae*). Z. Parasitenk. 26: 50-62.

Scholtyseck, E. 1968. Neue Einblicke in die Parasit-Wirt-Beziehungen mit Hilfe der Elektronenmikroskopie. Z. Parasitenk. 31: 67-84.

Scholtyseck, E. 1969. Electron microscope studies of the effect upon the host cell of various developmental stages of *Eimeria tenella* in the natural chicken host and in tissue cultures. Acta Vet. 38: 153-156.

Scholtyseck, E., K. Friedhoff, and G. Piekarski. 1970. Über mikromorphologische Übereinstimmungen bei Entwicklungsstadien von Coccidien, Toxoplasmen und Piroplasmen. Z. Parasitenk. 35: 119-129.

Scholtyseck, E. and D. M. Hammond. 1970. Electron microscope studies of macrogametes and fertilization in *Eimeria bovis*. Z. Parasitenk. 34: 310-318.

Scholtyseck, E., D. M. Hammond, and J. V. Ernst. 1966. Fine structure of the macrogametes of *Eimeria perforans, E. stiedae, E. bovis,* and *E. auburnensis*. J. Parasitol. 52: 975-987.

Scholtyseck, E., O. Kepka, and G. Piekarski. 1970. Die Feinstruktur der Zoiten des M-Organismus aus reifen Cysten im Gehirn der Rötelmaus (*Clethrionomys glareolus*). Z. Parasitenk. 33: 252-261.

Scholtyseck, E. and H. Mehlhorn. 1970. Ultrastructural study of characteristic organelles (paired organelles, micronemes, micropores) of Sporozoa and related organisms. Z. Parasitenk. 34: 97-127.

Scholtyseck, E., H. Mehlhorn, and K. Friedhoff. 1970. The fine structure of the conoid of Sporozoa and related organisms. Z. Parasitenk. 34: 68-94.

Scholtyseck, E., H. Mehlhorn, and D. M. Hammond. 1971. Fine structure of macrogametes and oocysts of coccidia and related organisms. Z. Parasitenk. 37: 1-43.

Scholtyseck, E., H. Mehlhorn, and D. M. Hammond. 1972. Electron microscope studies of microgametogenesis in coccidia and related groups. Z. Parasitenk. 38: 95-131.

Scholtyseck, E. and G. Piekarski. 1965. Elektronenmikroskopische Untersuchungen über die Merozoiten von *Eimeria stiedae* und *E. perforans* und *Toxoplasma gondii*. Zur systematischen Stellung von *T. gondii*. Z. Parasitenk. 26: 91-115.

Scholtyseck, E., A. Rommel, and G. Heller. 1969. Licht- und elektronenmikroskopische Untersuchungen zur Bildung der Oocystenhülle bei Eimerien (*Eimeria perforans, E. stiedai* und *E. tenella*). Z. Parasitenk. 31: 289-298.

Scholtyseck, E. and D. Schäfer. 1963. Über schlauchförmige Ausstülpungen an der Zellmembran der Makrogametocyten von *Eimeria perforans*. Z. Zellforsch. Mikrosk. Anat. 61: 214-219.

Scholtyseck, E., R. G. Strout, and A. Haberkorn. 1969. Schizonten und Merozoiten von *Eimeria tenella* in Makrophagen. Z. Parasitenk. 32: 284-296.

Scholtyseck, E. and W. H. Voigt. 1964. Die Bildung der Oocystenhülle bei *Eimeria perforans* (Sporozoa). Z. Zellforsch. Mikrosk. Anat. 62: 279-292.

Scholtyseck, E., B. Volkmann and D. M. Hammond. 1966. Spezifische Feinstrukturen bei Parasit und Wirt als Ausdruck ihrer Wechselwirkungen am Beispiel von Coccidien. Z. Parasitenk. 28: 78-94.

Schrevel, J. 1968. L'ultrastructure de la région antérieure de la grégarine *Selenidium* et son intérêt por l'étude de la nutrition chez les sporozoaires. J. Micros. 7 (3): 391-410.

Schrevel, J. 1971. Les polysaccharides de reserve chez les Sporozoaires. Ann. Biol. 10: 31-51.

Schrevel, J. and E. Vivier. 1966. Etude de l'-ultrastructure et du rôle de la région antérieure (mucron et épimérite) des Grégarines parasites d'Annélides polychétes. Protistologica 2(3): 17-28.

Schulte, E. 1971. Cytochemische Untersuchungen an den Feinstrukturen von *Klossia helicina* (Coccidia, Adeleidea). Z. Parasitenk. 36: 140-157.

Seliverstova, V. C. 1969. Some data on the ultrastructure of macrogametes of *Eimeria tenella* (Sporozoa, Coccidia). Acta Protozool. 7: 49-56. [In Russian]

Sénaud, J. 1965. Ultrastructure comparée des endodyozoites de *Sarcocystis tenella* et de *Toxoplasma gondii*. 2nd Intern. Conf. Protozool. Excerpta Medica Foundation, Amsterdam, p. 188-189.

Sénaud, J. 1966. L'ultrastructure du micropyle de Toxoplasmida. C.R. Acad. Sci., Série, D. (262): 119.

Sénaud, J. 1967. Contribution à l'étude des sarcosporidies et des toxoplasmes (Toxoplasmea). Protistologica 3: 170-232.

Sénaud, J. 1969. Sur l'ultrastructure des kystes de *Besnoitia jellisoni* (Frenkel 1953) chez la souris (*Mus musculus*). C. R. Acad. Sci. (Paris) 268: 816-819.

Sénaud, J. 1970. Ultrastructure des formations kystiques de *Besnoitia jellisoni* (Frenkel, 1953) Protozoaire, Toxoplasmea, parasite de la souris (*Mus musculus*). Protistologica 5 (3): 413-430.

Sénaud, J. and Z. Černá. 1968. Etude en microscopie électronique des mérozoites et de la mérogonie chez *Eimeria pragensis* (Černá et Sénaud, 1968), Coccidie parasite de l'intestin de la souris (*Mus musculus*). Ann. St. Biol. Besse-en-Chandesse, 3: 221-242.

Sénaud, J. and Z. Černá. 1969. Etude ultrastructurale des mérozoites et de la schizogonie des Coccidies (*Eimeriina*): *Eimeria magna* (Perard, 1925) de l'intestin des Lapins et *E. tenella* (Raillet et Lucet, 1891) des caecums des poulets. J. Protozool. 16: 155-165.

Sénaud, J. and Z. Černá. 1970. La microgamétogénése chez *Eimeria pragensis* (Černá et Sénaud 1969) *Sporozoa, Telosporea, Coccidia, Eimeriina*, Parasite de l'intestin de la souris: étude au microscope électronique. Protistologica 6 (1): 5-19.

Sheffield, H. G. 1966. Electron microscope study of the proliferative form of *Besnoitia jellisoni*. J. Parasitol., 52: 583.

Sheffield, H. G. 1967. The function of the micropyle in the cyst organisms of *Besnoitia jellisoni*. J. Parasitol. 53: 888.

Sheffield, H. G. 1968. Observations on the fine structure of the "cyst stage" of *Besnoitia jellisoni*. J. Protozool. 15: 685.

Sheffield, H. G. 1970. Schizogony in *Toxoplasma gondii:* An electron microscope study. Proc. Helm. Soc. Wash. 37: 237.

Sheffield, H. G. and D. M. Hammond. 1966. Fine structure of first-generation merozoites of *Eimeria bovis*. J. Parasitol. 52: 595.

Sheffield, H. G. and D. M. Hammond. 1967. Electron microscope observations on the development of first-generation merozoites of *Eimeria bovis*. J. Parasitol. 53: 831.

Sheffield, H. G. and M. L. Melton. 1968. The fine structure and reproduction of *Toxoplasma gondii*. J. Parasitol. 54: 209.

Simond, P. L. 1897. L'évolution des sporozoaires du genre *Coccidium*. Ann. Inst. Pasteur 11: 545-581.

Simpson, C.F. 1966. Electron microscopy of *Sarcocystis fusiformis*. J. Parasitol. 52: 607-613.

Simpson, C. F., W. W. Kirkham, and J. M. Kling. 1967. Comparative morphologic features of *Babesia caballi* and *B. equi*. Amer. J. Vet. Res. 28 (127): 1693-1697.

Snigirevskaya, E. S. 1968. The occurrence of micropore in schizonts, microgametocytes, and macrogametes of *E. intestinalis*. Acta protozool. 5 (25): 381-388.

Snigirevskaya, E. S. 1969. Electron microscope study of the macrogametes of *Eimeria intestinalis* (Coccidia). Tsitologiya. 11: 700-706.

Snigerevskaya, E. S. and E. M. Kheysin. 1968. The role of the micropore in the nutrition of endogenous developmental stages of *Eimeria intestinalis* (Sporozoa, Coccidia). Tsitologiya 10: 940-944.

Stehbens, W. E. 1966. The ultrastructure of *Lankesterella hylae*. J. Protozool. 13: 63-73.

Strout, R. G. and E. Scholtyseck. 1970. The ultrastructure of first generation development of *Eimeria tenella* (Railliet and Lucet 1891) Fantham, 1909 in céll cultures. Z. Parasitenk. 35: 87-96.

Vivier, E. and E. Henneré. 1964. Etude à l'aide du microscope électronique des trophozoites de la Coccidie: *Coelotropha durchoni*. Coll. Soc. Franc. Microsc. électron. Strasbourg. J. Microsc. 3.

Vivier, E. and E. Henneré. 1965. Ultrastructure des stades végétatifs de la Coccidie *Coelotropha durchoni*. Protistologica, I (1): 89-104.

Vivier, E. and A. Petitprez. 1969. Le complexe membranaire superficiel et son évolution lors de l'élaboration des individus-fils chez *Toxoplasma gondii*. J. Cell. Biol. 43: 329-342.

Vivier E., G. Devauchelle, A. Petitprez, E. Porchet-Henneré, G. Prensier, J. Schrevel, and D. Vinckier. 1970. Observations de cytologie comparée chez les sporozoaires. I. Les structures superficielles chez les formes végétatives. Protistologica 6: 127-150.

Volkmann, B. 1967. Vergleichende elektronenmikroskopische und lichtmikroskopische Untersuchungen an verschiedenen Entwicklungsstadien von *Klossia helicina* (Coccidia, Adeleidea). Z. Parasitenk. 29: 159-208.

Wanko, T., L. Jacobs, and M. A. Gavin. 1962. Electron microscope study of *Toxoplasma* cysts in mouse brain. J. Protozool. 9: 235-242.

Wildführ, W. 1964. Elektronenmikroskopische Untersuchungen an *Toxoplasma gondii*. Z. ges. Hyg. 10: 541-546.

Zypen, E. van der, and G. Piekarski. 1967. Die Endodyogenie bei *Toxoplasma gondii*. Eine morphologische Analyse, Z. Parasitenk. 29: 15-35.

5 Cytochemistry, Physiology, and Biochemistry

JOHN F. RYLEY
Imperial Chemical Industries Ltd., Pharmaceuticals Division, Alderley Park, Macclesfield, Cheshire, England

Contents

I. Introduction . . . *146*
II. Cytochemistry . . . *146*
   A. DNA . . . *147*
   B. RNA . . . *147*
   C. Protein . . . *148*
   D. Lipid . . . *148*
   E. Polysaccharide . . . *149*
   F. Mucopolysaccharides . . . *151*
III. Nature of the Oocyst Wall . . . *151*
IV. Nutrition and Chemotherapy . . . *154*
   A. Proteins . . . *154*
   B. Nucleic Acids . . . *155*
   C. Growth Factors and Antagonists . . . *155*
      1. Thiamine . . . *156*
      2. Riboflavin . . . *157*
      3. Biotin . . . *157*
      4. Nicotinic Acid . . . *157*
      5. Pyridoxine . . . *157*
      6. *p*AB-Folic Acid . . . *158*
      7. Vitamins A and K . . . *159*
V. Metabolism . . . *159*
   A. Sporulation . . . *160*
   B. Excystation . . . *162*
      1. Primary Phase . . . *162*
      2. Secondary Phase . . . *163*
      3. Energy Considerations . . . *164*
   C. Penetration . . . *166*

D. Proliferative *Toxoplasma gondii* . . . *166*
E. Sporozoites of *Eimeria tenella* . . . *167*
F. Histochemical Investigations . . . *169*
VI. Host-Parasite Interrelationships . . . *173*
A. Toxins . . . *173*
B. Effects on Intestinal Function . . . *174*
C. Effects on Host Metabolism . . . *175*
Literature Cited . . . *175*

I. Introduction

Since they are intracellular parasites for the greater part of their life cycle, the coccidia have not been particularly amenable to biochemical investigation. In this chapter we discuss cytochemical techniques, which within the limitations of the methods (not always appreciated), give some clues about the constituents from which the organisms are built up and of the materials which form the basis of their metabolism. A combination of cytochemical, chemical, and physical techniques has been used in studies on the coccidial reserve polysaccharide, and particularly in connection with the nature of the oocyst wall, which is such an obstacle to the practical control of coccidiosis by hygienic measures. Although nutritional studies are complicated by the obligate parasitic nature of the coccidia, limited information is available. Attention will be drawn to the interesting and practical interplay between nutrition and chemotherapy. Metabolic studies have been mainly confined to the nonparasitic phases of the life cycle – sporulation of the oocyst and excystation – but histochemical techniques have indicated just a few of the enzyme systems present in the parasitic stages. The effects of the coccidia on the host are to some extent explicable by the disruptive mechanical effects of growth and development. Some effects, however, probably arise as a direct result of the parasites' metabolic requirements and of the end products of their metabolism.

II. Cytochemistry

Classical staining methods have been used in studies aimed at elucidating life cycles, tissue distribution of coccidia, and subcellular structure. If sufficient caution is exercised in interpreting the results, and if undue emphasis is not placed on the specificity of the staining procedures, useful evidence can be gleaned regarding the chemical nature and intracellular location of the substances from which the parasite is built and which form the basis of its metabolism. No attempt will be made here to review exhaustively the numerous cytological studies which have been carried out, but rather attention will be drawn to some of the materials which form the basis of the cell's economy. Among extensive cytological studies are those of Pattillo and Becker (1955) on *Eimeria brunetti* and *E. acervulina*, Kheysin's (1958) investigation of the rabbit parasite *E. intestinalis* and Cross' (1947) investigation of *Toxoplasma gondii*, which although it failed to reveal multiplication by endodyogeny, was nevertheless a notable study. Dasgupta and Kulasiri (1959) give a useful historical review of the cytochemistry of *T. gondii*, while Davis and

Bowman (1963) discuss various microscopic and staining techniques helpful in the location and identification of coccidia.

A. DNA

Although it is difficult to imagine nuclear function without DNA, not all stages of the life cycle give a positive Feulgen reaction. Thus Pattillo and Becker (1955) were unable to find DNA in the macrogametocytes, although it was present in sporozoites, schizonts, merozoites, microgametocytes, and unsporulated oocysts, and Kheysin (1959) experienced similar difficulties in demonstrating DNA in macrogametocytes and oocysts of *E. intestinalis*. Beyer (1968) discusses the possibilities that variations in the Feulgen staining reflect different degrees of DNA polymerization and that a vast increase in nuclear volume such as occurs in the macrogamont may lead to a dilution of DNA to a level below that detectable. In some circumstances, Pattillo and Becker (1955) were able to demonstrate DNA, not by the Feulgen reaction, but by staining in toluidine blue following ribonuclease treatment. Kheysin (1959) showed that the Feulgen negative nucleus of growing macrogametes of rabbit coccidia would stain with methyl green, and confirmed the presence of DNA by a much more sensitive technique utilizing auramine OO - SO_2 and fluorescence micrography with deoxyribonuclease treatment for control (Kheysin, Rozanov and Kudriavtsev, 1968). Quantitative measurements of fluorescence showed that the nuclei of merozoites and of microgametes contain the same amount of DNA, which means that reduction division must take place after fertilization. DNA has similarly been located in the nuclei of *Toxoplasma* from the brain or peritoneal exudate of mice (Lillie, 1947; Dasgupta and Kulasiri, 1959), although Cross (1947) found that doubling the recommended hydrolysis time was necessary for adequate Feulgen staining.

B. RNA

RNA seems to be present in the cytoplasm and karyosome of all stages of coccidia and toxoplasms, and is demonstrable with toluidine blue as an intense basophilia which is not produced after ribonuclease treatment. Beyer and Ovchinnikova (1964, 1966) studied the RNA content of parasites during macrogametogenesis and zygote formation with two rabbit coccidia by quantitative cytophotometry. Tissues were fixed in Zenker and 7μ paraffin sections were stained with gallocyanin-chromalum; sections treated with crystalline ribonuclease were used as control. Scanning was carried out at 579 mμ. In spite of a sharp decrease in the intensity of cytoplasmic basophilia attributable to RNA during growth of the macrogamete, the total quantity of RNA appeared to increase during macrogametogenesis, reaching a peak just before fertilization took place. Once fertilization had taken place, active contact between parasite and host was severed. A slight reduction in RNA was recorded during wall formation, and thereafter the content remained constant.

An interesting aspect of cytoplasmic RNA was illustrated by the investigation of Kulasiri and Dasgupta (1959) into the basis of the Sabin-Feldman dye test for toxoplasmosis. Living toxoplasms lose their affinity for alkaline methylene blue when incubated in the presence of immune serum and an accessory factor present in normal serum. Such a reaction can obviously be made the basis for assay or screening of sera from suspected cases of toxoplasmosis. It was found that the RNA normally present throughout the cytoplasm disappeared or at least became undemonstrable during incubation with immune sera, and thus treated organisms failed to take up the dye in the test. Extraction of cytoplasmic RNA could also apparently be achieved with hot water, and organisms thus treated produced a clearer image in the electron microscope.

C. PROTEIN

It is inconceivable that coccidia should not contain protein. The most conspicuous features of sporozoites when viewed by phase contrast are the refractile granules. They are prominent too – with no indication of structure or function – in electron micrographs. When sporozoites are treated with mercuric bromphenol blue as described by Mazia, Brewer, and Alfert (1953), the refractile bodies show a greater affinity for the stain than any other structures in the cell, indicating their proteinaceous nature. Fayer (1969) made a study of three species of coccidia developing in tissue culture. A posterior refractile body was always present and, although it might decrease slightly in volume, persisted to be a distinctive and diagnostic feature of first generation schizonts. An anterior refractile body was usually present which moved during development to a site alongside or behind the nucleus, often fragmented, and decreased in size and either fused with the posterior body or disappeared entirely. Another phase of the coccidial life cycle giving a particularly strong reaction with mercuric bromphenol blue is the macrogametocyte, where the type II wall-forming bodies, and at a later stage the oocyst walls (Pattillo and Becker, 1955) are intensely stained. In the sporulated oocyst, the Stieda body of the sporocyst also responds to mercuric bromphenol blue; Pattillo and Becker (1955) with *E. brunetti* and *E. acervulina* say that the staining is as intense as that of the refractile globule, while Hammond, Ernst, and Chobotar (1970) with *E. utahensis* describe the reaction as very weak.

D. LIPID

Lipids have not been found with cytochemical techniques in schizonts, merozoites, or microgametocytes (Pattillo and Becker, 1955). Material staining with Sudan black B was observed in the cytoplasm of developing macrogametocytes, and following fertilization, these fat bodies seemed to accumulate around the nucleus. After sporulation they were found as large polar bodies in the sporocyst; the "plastic granules" (wall-forming bodies) of the macrogamete did not react. Similar results were obtained with rabbit coccidia by Dasgupta (1960) and by Kheysin (1958, 1959) who also noted that fat stored by the macrogamete gets into the residual body of the oocyst (when there is one) and into the residual bodies of the sporocysts during sporogony. Although plastic granules were readily removed from macrogametocytes in paraffin sections after fixation in fluids which are poor fixatives for lipids, Long and Rootes (1959) were unable to obtain a definite positive lipid reaction for the granules of *E. maxima*.

Using three different staining methods with *E. stiedai,* Frandsen (1970) was able to demonstrate a few lipid droplets in sporozoites, mostly in the region of the nucleus. A certain amount of stain was also taken up by the refractile globules, but Hammond, Chobotar, and Ernst (1968) note that this stain persists in spite of extraction with hot chloroform-methanol, and so may not indicate lipid. Diffuse staining of fat with Sudan 4 was noted by Cross (1947) for *T. gondii* from ascitic fluid of the mouse, but no reaction with osmic acid was obtained. Only very small amounts were found in asexual stages – structural rather than storage lipids – but heavy accumulations were built up in the central area of female gametocytes, and small droplets were found randomly scattered in the cytoplasm of male gametocytes and gametes.

E. POLYSACCHARIDE

Polysaccharide granules are readily discernible at several stages in the coccidial life cycle and provide a good example of erroneous conclusions drawn from cytochemical techniques whose specificities and limitations are not always fully appreciated. Using iodine as a histochemical stain, Edgar, Herrick, and Fraser (1944) claimed that glycogen was synthesized by the developing macrogametocyte of *E. tenella,* and persisted until the sporozoite stage. The material in question gave a mahogany brown color with iodine, could not be extracted with water, but was destroyed by digestion with saliva. It has since become clear that proteins and, to a lesser extent, fats can interfere with the iodine staining properties of polysaccharides. Numerous subsequent studies (e.g., Gill and Ray, 1954a: Horton-Smith and Long, 1963; Wagner and Foerster, 1964) using staining techniques of the periodic acid - Schiff type (PAS) detected "glycogen" in other stages of the parasites. It has been assumed that because magenta staining material was present in granules which could be destroyed by digestion with amylase, then this material was glycogen. The PAS reaction is in fact given by any polysaccharide or polysaccharide complex containing adjacent diol groups (-CHOH·CHOH-). Thus glycogen, amylose, amylopectin, dextran, 1,4-xylan, or even cellulose may be expected to give a positive stain, although in the latter case the reaction is likely to be extremely slow because of the insoluble nature of the product. An amylase control eliminates substances not having α-1,4-glucosidic linkages, e.g., dextran, xylan, and cellulose; it does not however distinguish between glycogen, amylose, and amylopectin. Scholtyseck (1964) in a thorough ultrastructural study of the dense oval bodies in *E. perforans,* recognized that the material was different from metazoal glycogen, and described it as "Coccidienglykogen." The polysaccharide was visualized in the electron microscope as oval transparent bodies 620 mμ × 500 mμ, which stained intensely with lead, and whose affinity for lead was destroyed by prior digestion of sections with amylase. Schrével (1970) has carried out cytochemical and ultrastructural studies on the "paraglycogen" granules of several gregarines utilizing the periodic acid – TCH – silver proteinate technique, and finds the granules similar to those of *E. tenella*. Ryley, Bentley, Manners, and Stark (1969) extracted polysaccharide from oocysts of *E. brunetti* and oocysts, sporozoites, and merozoites of *E. tenella* with hot 20% KOH and precipitated the material with 1.1 volumes ethanol. A similar procedure has been applied to the gregarine of the cockroach *Gregarina blaberae* (Mercier, Stark, and Schrével, unpublished). Following purification by repeated precipitation with ethanol, the polysaccharides were characterized by their iodine-staining properties both in water and in half-saturated ammonium sulfate, and by the extent of their degradation to maltose by α- and β-amylases. Both approaches enable an assessment to be made of the average chain length of the molecule. Table 1 reproduces data obtained with material from the oocyst of *E. tenella*, a comparison being made with the gregarine polysaccharide and a typical glycogen and amylopectin. Branch points were characterized as α-1, 6-glucosidic linkages by treating β-amylolysis limit dextrins with pullulanase, an enzyme specific for the hydrolysis of such linkages. It will be readily seen that both coccidial and gregarine polysaccharides have properties typical of amylopectin rather than glycogen, and the term amylopectin should, therefore, be used instead of glycogen, Coccidienglykogen, or paraglycogen.

Although glycogen and amylopectin are both composed of linear chains of α-1,4-linked glucose residues which are joined by

TABLE 1. Characterization of reserve polysaccharide

	E. tenella	Gregarine	Glycogen	Amylopectin
Iodine staining				
In water λ_{max}	535	522	445	530
In water ϵ_{max}	1.05	0.90	0.1	1.1
In half-saturated ammonium sulfate λ_{max}	540	528	460	560
In half-saturated ammonium sulfate ϵ_{max}	1.33	1.48	0.6	1.6
Ratio: $\frac{\epsilon_{max}\text{ (amm. sulf.)}}{\epsilon_{max}\text{ (water)}}$	1.27	1.65	6.00	1.45
Chain length	21	19	12	23
Enzymatic degradation				
β-Amylolysis limit (%)	56	54	40-50	50-60
α-Amylolysis limit (%)	92	93	71-84	90-96
Average chain length	20	21	10-14	20-25
Average exterior chain length	14	14	6-9	12-17
Average interior chain length	5	6	3-4	5-8

α-1, 6-glucosidic interchain linkages to form a multiply branched macromolecule, amylopectin has much longer unit chains and hence a more open structure. This architectural difference between the two polysaccharides results in differing physical properties – solubility, viscosity, interaction with proteins and with iodine, physical form of intracellular deposits, etc. – and in tissues so far examined (which unfortunately do not yet include the coccidia) arises from differences in the pathways by which the two polysaccharides are synthesized. An analogy may be made to charcoal and diamond, two rather dissimilar forms of the same basic substance! The coccidia have therefore energy reserves similar to those found in rumen ciliates and plants, rather than the animals in which they live.

Amylopectin was found during the study of Edgar *et al.* (1944) in the macrogametocyte and to a limited extent in the sporozoite of *E. tenella*; none was found in schizonts or merozoites. It was concluded that its function was an energy store to support the organism in its nonparasitic phase. Gill and Ray (1954a) were unable to detect polysaccharide in first generation schizonts or merozoites of *E. tenella*, but did find it in small quantities in second generation parasites, in large amounts as merozoite differentiation took place, and in microgametocytes and gametes. They sought to correlate the amounts of polysaccharide present with the energy requirements of the particular stage of the parasite, and arrived at rather dubious conclusions concerning the origin and synthesis of the material. Wagner and Foerster (1964) went further and demonstrated polysaccharide in first generation merozoites and third generation schizonts as merozoite differentiation took place, but were unable to find the material in microgametocytes or microgametes. Kheysin (1959) drew attention to the reserves of polysaccharide dispersed among fat droplets in the residual bodies of oocysts and sporocysts, and states that the latter as well as the material actually within the sporozoite can be utilized during conditions of prolonged oocyst storage to maintain the vital activity of the sporozoite.

Polysaccharide was noted in *T. gondii* by Dasgupta and Kulasiri (1959); amounts were small and variable in organisms derived from mouse peritoneal fluid, but the cytoplasm of organisms from brain "pseudocysts" was filled with strongly-reacting PAS positive granules. The pseudocyst walls also reacted, but the intensity of this reaction was very variable. Gangi and Manwell (1961) also noted sparse and variable amounts of PAS positive material, sensitive to amylase digestion, in the cytoplasm of free parasites.

In his studies on the "paraglycogen" of gregarines, Schrével (1970) demonstrated that the polysaccharide bodies seen by light microscopy were in fact made up of minute granulations more numerous at the periphery than in the center when viewed by electron microscopy. The fact that trypsin as well as amylase was able to bring about eventual disintegration of the polysaccharide bodies suggests that, if the enzyme was pure, the amylopectin is possibly laid down on a proteinaceous framework.

F. MUCOPOLYSACCHARIDES

During a study of nucleic acids in *E. tenella*, Gill and Ray (1954b) noted that certain stages of the parasite displayed intense metachromasy with toluidine blue which was not affected by ribonuclease treatment. This was further investigated by staining techniques supposedly specific for mucoid sulfate and hyaluronic acid type polysaccharides. Mucoid sulfate granules were found in the cytoplasm of all stages of the parasite, but in varying amounts which were said to reflect the physiological activity of the particular stage. Hyaluronic acid type polysaccharide was more prominent in the nucleus, especially the karyosome, but during schizogony it tended to spread out from the nucleus into the cytoplasm in the form of an irregular colloidal matrix. It was also said to form the protective covering of the oocyst although no significant reaction for hyaluronic acid could be obtained with the oocyst; the logic of this conclusion is derived from the intense staining given by the peripheral globules of the macrogametocyte which subsequent to fertilization coalesced to form the oocyst membrane. An amazing number of virtues were ascribed to the mucopolysaccharides, which, because of their hydrotropic properties were said to prevent cytoplasmic clotting, promote contact between enzymes and substrates, assist in repeated nuclear division, promote osmosis between host and parasite and protect merozoites and oocysts from the unfavorable extracellular environment! Mucopolysaccharides were not found by Pattillo and Becker (1955) or by Kheysin (1959).

III. Nature of the Oocyst Wall

The shell of the cocoidial oocyst is quite remarkable in its ability to protect the oocyst contents from harm. Certain small molecules are apparently able to penetrate the wall, since oocyst sterilization can be achieved with high concentrations of ammonia, methyl bromide or carbon disulfide. On the other hand conventional disinfectants, even in high concentration, are totally without effect, and oocyst cultures are normally stored in solutions such as sulfuric acid or potassium dichromate. Gas exchanges readily take place during the period of sporulation as will be subsequently described, and wall permeability can be markedly altered by exposure to carbon dioxide. Oocysts can be eventually killed by dessication, but although prolonged exposure to hypertonic salts can bring about bizzare distortion of the walls and contents, viability can to a large extent be preserved (Kheysin, 1935). Because of the relatively impervious nature of the wall, cytochemical studies of the wall as well as the oocyst contents are difficult. Strout, Botero, Smith, and Dunlop (1963) find that lipid accounts for one-third of the wall, while Landers (1960) suggests the wall is mainly polysaccharide and protein. Some indications of the nature of the wall can be derived from a study of the wall-forming bodies which are found in the macrogametocyte and which, subsequent to fertilization, coalesce to form the wall. Such projections, coupled with chemical and physicochemical investigations, allow us to form some preliminary conclusions.

Oocysts have a shell consisting of two or sometimes three layers (Henry, 1932). From a study of rabbit coccidia, Kheysin (1935) considered the moderately thick and elastic outer layer to be skeletal, giving protection against mechanical lesions, and to be close to keratin in chemical composition. The inner membrane is much more elastic and protects against chemical as well as physical damage. Kheysin believed this inner membrane to be derived from the plastic granules of the gametocyte and to approximate to nucleoprotein in composition. Soaking oocysts in hypertonic salt solutions resulted in plasmolysis, the inner membrane separating from the outer, invaginating and collapsing onto the oocyst contents. Sporulation occurred in many of the oocysts under these conditions, and shape could be recovered on transferring back to water. This inner layer was more permeable to organic solvents than to inorganic materials, suggesting a lipid barrier. In a delightful study of *E. tenella* and *E. stiedai,* Goodrich (1944) differentiates the shell into ectocyst and endocyst, and considers that it is the ectocyst which is derived from a fusion of the cytoplasmic globules prior to fertilization. Fertilization, she claims, is achieved via the micropyle, after which the endocyst is secreted. (Current opinion – see Scholtyseck and Hammond (1970) – is that wall formation does not begin until after fertilization is complete.) The ectocyst is similar in staining reactions (picric acid, eosin, methylene blue) to keratin, but not identical in view of the absence of birefringence in polarized light. The endocyst is thin and flexible, but Goodrich does not hazard a guess as to its chemical nature.

Scholtyseck and Weissenfels (1956) in an electron microscope study of *E. tenella* oocysts found that the shell is approximately 0.25 μ thick, a figure one-fifth that of estimates made with the light microscope on the highly refractile walls. Two main layers were differentiated; the inner one was less dense, while the more osmiophilic outer layer was covered by a fine overlying membrane. Nyberg and Knapp (1970b) were unable to observe this overlying membrane, although by scanning electron microscopy (Nyberg and Knapp, 1970a) they did note a surface coating of granular debris. Two primary layers were present; the outer layer could be removed by sodium hypochlorite, while the slightly thinner inner layer showed a rather denser region on its outer periphery. Roberts, Speer, and Hammond (1970) noted that the two primary layers of the oocyst wall of *E. callospermophili* had a discernible ultrastructure, but such was not the case with *E. larimerensis.* The inner layer was from 0.1 μ to 0.3 μ thick and showed a greater electron density on its inner aspect, while the outer layer was from 0.7 μ to 1.0 μ thick, was more osmiophilic, and consisted of minute columnar elements arranged in a honeycomb pattern on a dense basement membrane. Granules, presumed to be artifacts associated with oocyst manipulation, were found on the outer surface.

During macrogametocyte development, a number of large granules develop within the cytoplasm. These have been referred to as "plastic" or eosinophilic granules, and in studies such as Pattillo and Becker (1955) and Long and Rootes (1959) have been thought to be of one type, giving reactions for protein and for carbohydrate, but not for lipid; hence it was suggested they consisted of mucoprotein. Nath and Dutta (1962) reviewed ideas then current on the origin of these granules, and suggested that they are derived from lipid spheroids which actively synthesize acid mucopolysaccharides and RNA in their interior, eventually becoming transformed into plastic granules which contain no trace of lipid. The "plastic" granules they claim are not PAS positive. Grassé (1953) and Kheysin (1959) both felt that the oocyst wall had its origin in two distinct types of granule, but it was the electron microscope investigations of Scholtyseck and Voigt (1964) and Scholtyseck, Rommel, and Heller (1969) which first clearly differentiated these two distinct types of wall-forming bodies. Type II granules appear first, and in electron micrographs have only a moderate density. They stain blue with mercuric bromphenol blue, indicating the presence of protein; it is not known whether lipid is present in these bodies or not. Type I granules appear subsequently. They present an extremely dense and homogeneous image in electron micrographs, and give a positive PAS reaction which is not affected by amylase pretreatment. Treatment of gametocyte sections with weak sodium hypochlorite solution destroys the staining ability of these Type I granules, while the Type II granules are not affected. Both types of granule migrate to the periphery of the macrogametocyte during development and, subsequent to fertilization, fuse to form the oocyst wall, the Type I granules giving rise to the outer shell and the Type II granules the inner layer.

In an investigation of the oocyst walls of several species of *Eimeria* and *Isospora*, Monné and Hönig (1954) were able to draw some quite amazing conclusions from a combination of microscopy and a number of potentially destructive treatments. Examination in polarized light shows that the shell is birefringent, and by removal of the outer layer, that this birefringence is chiefly due to the inner layer. The outer part of the inner layer is negatively birefringent in a radial direction, which suggests it consists mainly of protein, while the inner part is positively birefringent and scatters light, suggesting that it consists chiefly of lipid. A radical separation of the two components of the inner layer is not, however, visualized; rather it is regarded as a protein matrix impregnated with lipid, with a particularly high concentration of lipid on the inner aspect.

The two layers of the oocyst wall can be separated by various means. Heating in water or in sodium sulfide or thioglycollate solution, or treatment with sodium hypochlorite or concentrated sulfuric acid produces an initial separation, which may be followed by solution of the outer layer. Blisters first appear and these coalesce so that the outer layer balloons out. This outer layer is either isotropic or weakly birefringent. Because it does not dissolve in sodium sulfide or thioglycollate it was concluded that it cannot be a keratin-type of protein, and because it does dissolve in hypochlorite, and because it will reduce ammoniacal silver nitrate it was concluded that it is a quinone-tanned protein. The presence of lipid in the inner shell and its absence from the outer layer is further suggested by the fact that hypochlorite strips off the outer layer, but has no effect on the inner layer. Such treatment is in fact extremely useful for surface sterilizing oocysts, the contents of the oocyst retaining their viability and being in no way adversely affected. Wilson and Fairbairn (1961) observed that the walls contained considerable amounts of a very stable substance which could be isolated in the same sort of ways as chitin. This substance did not, however, contain N-acetyl glucosamine, nor was it susceptible to chitinase. Monné and Hönig had concluded that chitin was absent since although concentrated sulfuric acid caused a separation of the two layers, neither dissolved, even on heating.

No difference in optical properties was observed between unsporulated and sporulated oocysts, but it is desirable that changes in the wall and its permeability be investigated further by other methods. In a cytological study of sporulation, Canning and Anwar

(1968) found that fixation and subsequent staining were far easier if freshly collected oocysts were incubated with the alkaloid colcemid prior to sporulation; this alkaloid apparently delayed hardening of the oocyst wall although unexpectedly it had no effect on nuclear division.

We have made some preliminary observations on the chemical nature of the oocyst wall of *E. tenella*. Unsporulated oocysts were isolated from feces by sieving and salt flotation, and then cleaned by washing with 5% Tween 80 on a glass bead column followed by sucrose density gradient centrifugation. Oocysts were extracted with 10% sodium hypochlorite for 30 min at room temperature and then well washed with water. Dry weight determinations showed that the ectocyst accounted for from 19.5% to 21.8% of the total weight of the unsporulated oocyst. Stripped oocysts were shaken with grade 7 Ballotini glass beads for five minutes in aqueous suspension. The oocyst contents were reduced to fine particles and soluble material, while the walls persisted as shell fragments. Washing with large volumes of water and slow speed centrifugation gave a preparation of pure endocyst fragments. Chemical investigation of the hypochlorite extract has been hampered by the large amounts of inorganic material present. It does however appear to contain carbohydrate and a protein characterized by high proline content and the absence of basic amino acids. The endocyst fragments contain around 70% protein and 30% lipid with a small amount (about 1.5%) of carbohydrate. The protein shows a fairly normal amino acid make-up, but the lipid material, extractable by chloroform in a Soxhlet, appears to be a mixture of waxes, with chain lengths up to C_{40}, and containing very small amounts of nitrogen and phosphorus.

To determine the location of protein in the oocyst wall, susceptibilities to digestion by three enzymes were determined for three different preparations. As indicated in Table 2, the protein of intact oocysts was virtually inaccessible to enzyme attack. Following stripping of the outer layer with sodium hypochlorite, a small amount of protein could be solubilized, but it was only when the inner surface of the inner layer had been exposed by fragmentation of the walls that real susceptibility to digestion was obtained.

IV. Nutrition and Chemotherapy

Coccidia, being intracellular parasites, depend for their nutrition on the host cell. Nutrients of dietary origin may or may not be modified by the host or by its intestinal flora before becoming available for utilization by the parasite.

A. PROTEINS

Virtually nothing is known concerning the protein requirements of coccidia. Britten, Hill, and Barber (1964), finding that starvation or feeding a diet low in protein to chickens prior to infection reduced the severity of cecal coccidiosis, showed that the

TABLE 2. Susceptibility of oocyst walls to digestion

Enzyme	Per cent amino nitrogen solubilized		
	Intact, complete oocysts	Unbroken, stripped oocysts	Inner wall fragments
Pepsin	0.49	0.4	12.5
Trypsin	0.00	3.8	91.5
Pronase	0.07	10.2	58.0

effect was not caused by depriving the coccidia of essential nutrients, but rather by a lowered intestinal tryptic activity which reduced excystation. Abou-El-Azm (1967) found that coccidia thrived better in chickens fed a commercial diet than in chicks given a synthetic diet containing 20% protein. Tryptophan, he suggested, was essential to the parasite as well as the host, whereas methionine, cystine, and isoleucine were required only by the host.

B. NUCLEIC ACIDS

Although nucleic acid synthesis is undoubtedly also an extremely important sector of coccidial metabolism, here too, pertinent information is scarce.

6-Azauracil was developed by Klimeš (1963) as an anticoccidial, presumably by virtue of its ability to antagonize pyrimidine metabolism. Roberts, Elsner, Shigematsu, and Hammond (1970), unable to obtain incorporation

Uracil   6-aza Uracil

of tritiated thymidine into *E. callospermophili* in cell cultures, concluded "that the parasite does not utilize thymidine directly from the media, from degraded host cell nuclei, or from nucleoside or nucleotide pools of the host cell." A recent abstract by Ouellette, Strout, and McDougald (1972) notes, however, that although *E. tenella* in tissue culture would not incorporate thymidine into DNA or RNA, tritiated cytidine and particularly uridine were taken up by parasites from the medium. Remington, Bloomfield, Russell and Robinson (1970) fractionated the RNA of *T. gondii* by sedimentation and showed the presence of three major components (24S, 19S, and 4-5S). Incubation of free parasites with tritiated uridine *in vitro* resulted in incorporation into all three RNA bands. Perrotto, Keister, and Gelderman (1971) found that while exogenous purines were readily utilized by *T. gondii* for DNA synthesis also, purine precursors such as formate and glycine were utilized only to a very limited extent. Thymidine, cytidine, and orotic acid could be incorporated into DNA by isolated parasites, though at only one tenth the rate of adenine. It is thus not entirely clear to what extent purines and pyrimidines are synthesized *de novo* by *T. gondii*. Cross (1947) notes that although *Toxoplasma* is not an intranuclear parasite, its injurious action seems to be exerted on the nucleus, which is progressively dissolved, presumably to provide a source of nucleic acid precursors.

C. GROWTH FACTORS AND ANTAGONISTS

More information is available in the realm of micronutrients. Warren (1968) gives a useful review of earlier work concerned with requirements for single growth factors, and presents results derived from studies with 19 different vitamins or growth factors whose influence on the course of infection of chickens with *E. tenella* or *E. acervulina* was examined by a deficient diet technique. Chicks were fed diets deficient in a single factor and subgroups were infected with small doses of *E. tenella* or *E. acervulina*; growth (as a measure of host-diet interaction) and total oocyst production (as a measure of coccidial multiplication) were compared with results obtained using a complete diet. Figure 1 presents in diagrammatic form the results obtained with six of these factors; pantothenate, *p*AB, ascorbic acid, and pyridoxine showed small effects, while vitamins A, D, E, and K and lipoic acid, linolenic acid, hesperidin, inositol, or cyanocobalamin could not be shown (with the methods used) to be essential for coccidial nutrition. Rather more detailed experiments were done with thiamine, riboflavin, biotin, and nicotinic acid, histological examination being carried out on infected birds supplied with the vitamin in question

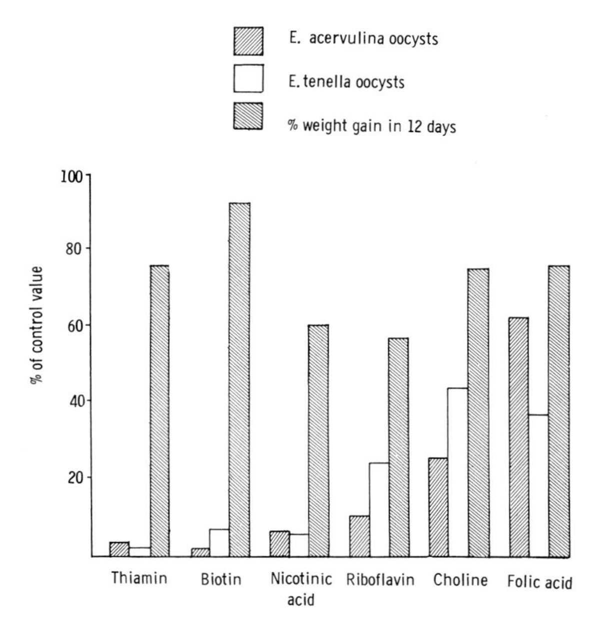

Fig. 1. Effects of vitamin deficiency on coccidial infections.

for but a limited part of the life cycle. Warren concluded, "*E. acervulina* required thiamine for second schizogony and sporulation; riboflavin for first schizogony and possibly gametogony, biotin for the development of the sporozoites/trophozoites and for gametogony, and nicotinic acid for schizogony and gametogony. *E. tenella* required thiamine and riboflavin for gametogony; biotin for first schizogony; and nicotinic acid for second schizogony." In a similar type of study, Abou-El-Azm (1967) concluded that riboflavin, pyridoxine, biotin, and vitamin B_{12} were required by both host and parasite, whereas vitamins A, D, E, and K were required by the host but not by coccidia.

The failure to demonstrate a requirement for a particular factor by this deficient-diet technique does not of course mean the factor is unnecessary for the parasite; the factor may well be synthesized by the host itself or by its gut flora. The recently developed tissue culture techniques (see Chapter 6) should make it possible to study coccidial nutrition in the absence of microbial, but not host cell, interference.

1. Thiamine

Thiamine Amprolium

A practical application of investigations into the vitamin requirements of the coccidia is chemotherapy based on differential vitamin antagonism. Chemotherapeutic studies similarly have the reverse effect of contributing to our meager knowledge of parasite physiology, but it must be remembered that antimetabolites are only one type of chemotherapeutic agent possible. Rogers (1962) gives an excellent account of the history of thiamine antagonists, and in particular of the development of amprolium as an anticoccidial. Chickens on a normal diet are protected against *E. tenella* infection by 125 ppm amprolium, whereas more than 800 ppm dietary amprolium is necessary before the chicks show growth depression and polyneuritis characteristic of thiamine deficiency. Just as the toxic effects on the host can be overcome by increased dietary thiamine, so too the anticoccidial effects of 125 ppm amprolium can be largely nullified by an extra 30 ppm thiamine. Antagonism of the anticoccidial action of amprolium by an equal amount of thiamine has been demonstrated in the chick embryo (Ryley, 1968) and more recently in our laboratories in tissue culture. Because amprolium lacks the hydroxyethyl group of thiamine, it cannot be pyrophosphorylated, and cannot therefore inhibit at the coenzyme level. An early suggestion for the target enzyme system was protozoal thiamine phosphorylase located at the parasite cell wall, but later studies by Polin, Wynosky, and Porter (1963) demonstrated a reduced uptake of thiamine from the intestine in the presence of amprolium; just how this affects the parasite more than the host is not clear. Histological studies (Slater, Hammond, and Miner, 1970) with *E. bovis* show that amprolium causes a retardation in the development of first generation schizonts, indicating that thiamine is important in the economy of the coccidia at an earlier stage than suggested by the report of Warren (1968). Tissue culture studies with *E. tenella* in our laboratories again show that parasite inhibition sets in towards the end of the first asexual generation (Ryley and Wilson, 1972).

2. *Riboflavin*

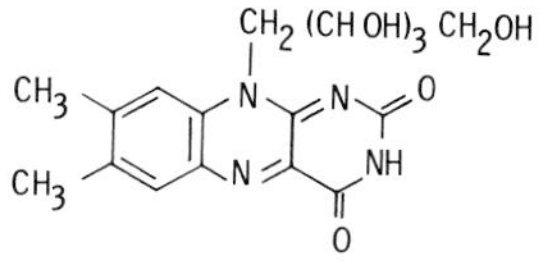

Riboflavine

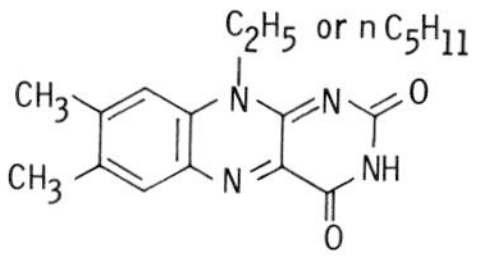

Isoalloxazine analogues (C_2H_5 or nC_5H_{11})

Riboflavin is another vitamin whose requirement (Fig. 1) led to the synthesis of potential antagonists. Ball and Warren (1967) describe two isoalloxazines which at 500 ppm to 1,000 ppm in the diet were able to reduce oocyst output in *E. acervulina* infections. The compounds were somewhat toxic and were inactive against *E. tenella*. Activity and toxicity could be reversed by one-tenth the amount of riboflavin.

3. *Biotin*

It has not so far been possible to control coccidiosis by selective biotin antagonism using synthetic chemicals, but Prasad (1963) and, subsequently, Warren and Ball (1967) were able to inhibit oocyst production with *E. acervulina, E. maxima* and *E. tenella* by feeding diets containing from 30% to 50% dried egg white. These effects could be reversed by daily doses of biotin. By contrast, Abou-El-Azm (1967), who claimed that biotin was essential for *E. acervulina* but not for *E. tenella,* was unable to demonstrate any anticoccidial action with egg white or with desthiobiotin. We have only observed a slight inhibition of growth of *E. tenella* in tissue culture with purified avidin (Ryley and Wilson, 1972).

4. *Nicotinic acid*

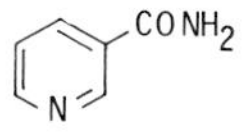

Nicotinamide

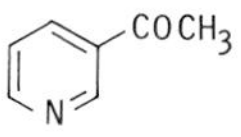

3 - acetyl pyridine

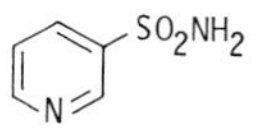

Pyridine - 3 - sulfonamide

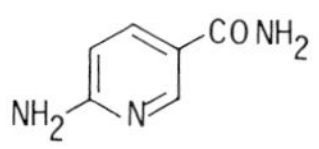

6 - amino - nicotinamide

A further compound shown to be essential for coccidial nutrition is nicotinic acid (Fig. 1). Ball, Warren, and Parnell (1965) tested three known nicotinamide antagonists for anticoccidial activity. 3-Acetyl pyridine was inactive. Pyridine-3-sulfonamide was inactive against *E. tenella* or *E. necatrix*, but markedly reduced oocyst output in *E. acervulina* infections when included in the diet at 60 ppm to 250 ppm; anticoccidial activity was antagonized by equal concentrations of nicotinamide, but not by *p*-amino benzoic acid. 6-Aminonicotinamide on the other hand was inactive against *E. acervulina*, but did show activity against *E. tenella* and *E. necatrix*; there was little margin between activity and toxicity, and both could be nullified by additional nicotinamide.

5. *Pyridoxine*

Although Warren (1968) obtained inconsistent results in efforts to demonstrate a need for pyridoxine, Abou-El-Azm (1967) has shown the factor to be involved in coc-

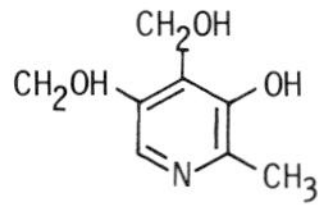

Pyridoxine

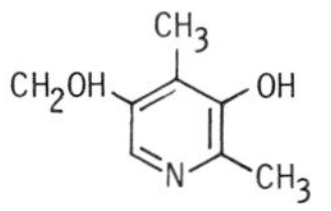

4 - desoxypyridoxine

cidial metabolism; chicks fed 1,000 ppm 4-desoxypyridoxine excreted 99.25% fewer oocysts of *E. acervulina* than unmedicated controls.

6. *pAB—folic acid*

Dietary deficiency of *p*-amino benzoic acid (*p*AB) had no effect, and deficiency in folic acid little effect in the experiments of Warren (1968), although Little (1967) obtained a marked decrease in oocyst production in chicks infected with *E. mivati* and fed *p*AB or folic acid-free diets. Joyner (1963) draws attention to experiments with milk diets in experimental malaria, where the resulting *p*AB deficiency leads to a suppression of parasitemia. A marked reduction in mortality due to *E. tenella* was noted in chicks fed skim milk, but unlike the situation with malaria, this effect could not be overcome by dietary supplementation with *p*AB. Before the advent of synthetic anticoccidials, feeding dried skim milk in the diet was one of the methods used in attempts to control coccidiosis. That *p*AB and folic acid are very much involved in the economy of the coccidia is evident from the susceptibility of coccidia to sulfonamides and diaminopyrimidines or dihydrotriazines. The sulfonamides were the first group of synthetic compounds shown to have definite anticoccidial activity, and Horton-Smith and Boyland (1946) in discussing this activity demonstrated its antagonism by *p*AB. Although pyrimethamine is too toxic to show anticoccidial activity on its own (Arundel, 1959), Lux (1954) was able to show potentiation between this or a number of other diaminopyrimidine or dihydrotriazine folic acid antagonists and sulfonamides. Clarke (1962) showed synergism between sulphaquinoxaline and diaveridine, another diaminopyrimidine which is unusual in that it is an anticoccidial in its own right at nontoxic levels. Kendall and Joyner (1958) showed a similar potentiation of sulfadimidine by a folic acid antagonizing pteridine. Frenkel and Hutchings (1957) discuss the antagonism of

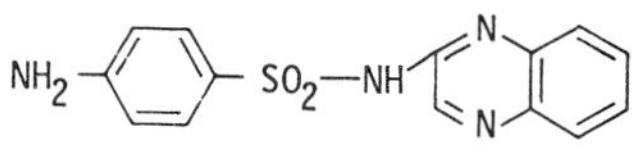

Sulfaquinoxaline

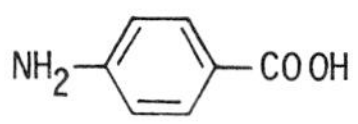

p-aminobenzoic acid

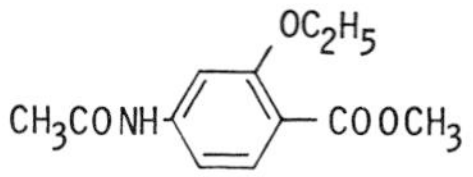

Ethopabate

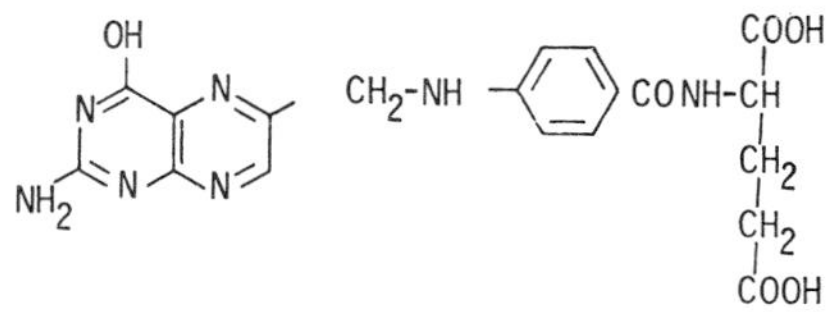

Folic acid

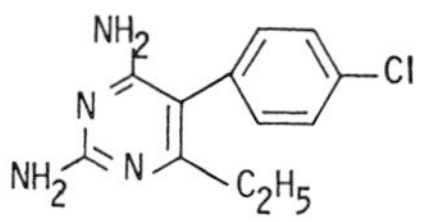

Pyrimethamine

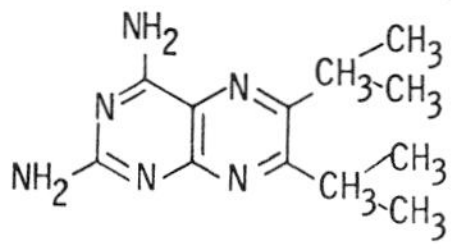

2,4-diamino - 6,7 -di isopropylpteridine

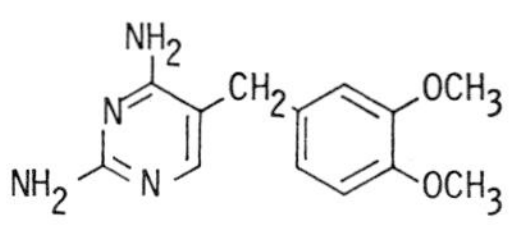

Diaveridine

sulfadiazine and pyrimethamine by *p*AB, folic and folinic acids in *Toxoplasma*, thus implicating these factors in the metabolism of this protozoan.

Rogers *et al.* (1964) describe a series of 2-substituted *p*-amino benzoic acids, which like the sulfonamides are antagonized by *p*AB and potentiate the activity of pyrimethamine. Unlike the sulfonamides, which inhibit the conversion of *p*AB to dihydropteroic acid, this particular class of compound is believed to interfere with the presumed subsequent step in the parasite, the conversion of dihydropteroic acid to dihydrofolic acid. It is interesting to note that while they are active against certain intestinal species in the chick, they are inactive against *E. tenella*; it is not clear whether this is caused by a different *p*AB metabolism in the different coccidial species or whether the drugs have an uneven distribution in various regions of the intestinal epithelium. McManus, Oberdick, and Cuckler (1967) demonstrated considerable *strain* variation in response to ethopabate, one of the more active of the substituted *p*ABs, and made the interesting observation that the strain most sensitive to ethopabate was one of the two strains most refractory to sulfaquinoxaline. Thus although dietary experiments give equivocal results, the development of chemotherapeutics has indicated the importance of the *p*AB → folic acid → folinic acid sequence in the coccidia.

7. *Vitamins A and K*

Although vitamins A and K showed somewhat inconclusive effects in the experiments of Warren (1968), it is interesting to note that they are important in coccidial infections—but by virtue of their protective action on the host rather than as nutritional requirements for the parasite. Gerriets (1961) gives a useful summary of the literature on vitamin A, and demonstrates an inverse relationship between losses from cecal coccidiosis and dietary vitamin A levels. He originally advocated dietary supplementation with vitamin A plus anticoccidials as the most effective prophylactic measure, but later (Gerriets, 1966) suggests that replacement birds can be raised on litter without anticoccidial if 30,000 i.u. of vitamin A per kilogram of food are supplied. Vitamin A is thought to exert its effect by stimulating regeneration of intestinal epithelial tissue. This is not, however, the only effect. Davies (1952) showed that liver reserves of vitamin A are drastically reduced in coccidial infection, and Coles, Biely, and March (1970), in confirming the favorable effect of vitamin A supplementation in *E. acervulina* infections, suggested that deficiency of vitamin A results in a slower immune response and an impaired ability to respond to the stress of infection.

Dietary vitamin K likewise exerts a marked effect on the course of infection, particularly with species where hemorrhage is most apparent—*E. tenella* and *E. necatrix*. Because of its involvement in blood clotting, mortality is particularly severe when the diet does not contain a source of vitamin K. The correlation between the severity of hemorrhage caused by *E. tenella* and the dietary vitamin K level was first shown by Baldwin, Wiswell, and Jankiewicz (1941), while Harms and Tugwell (1956) showed the reverse type of correlation with dicoumarol, a substance which significantly increases the blood clotting time. Experimental studies with *E. tenella* are made easier if one uses diets deficient in vitamin K—diets which contain neither the synthetic vitamin nor supplements such as grass meal or lucerne meal, which are rich in the vitamin. Joyner (1963) discusses further the relation between vitamin K supplementation of diet, blood clotting time, and susceptibility to coccidiosis.

V. Metabolism

Coccidia spend the major part of their active lives as intracellular parasites. Apart from the brief period of sporulation and during the process of excystation, the nonparasitic oocyst is a dormant structure. Metabolic studies by conventional techniques have nevertheless been mostly restricted to this part of the life cycle, which takes place outside the host. Studies on the parasitic phases of the life cycle have mainly been confined to the attempted demonstration and location of individual enzymes or enzyme systems by histochemical methods. Of particular interest would be the correlation of ultrastructure with enzymic activity and cellular function; regrettably such information is very limited.

Smith and Herrick (1944) were the first to attempt metabolic studies with *E. tenella*, when they measured the respiratory rate of unsporulated and sporulated oocysts and of normal and parasitized cecal tissue. Oocysts were isolated from cecal scrapings on the ninth day after infection and prepared as a bacteria-free suspension by repeated washing, salt flotation, and antiformin treatment, a process which took around four hours. The oxygen uptake of such oocysts immediately on isolation averaged 15.6 $\mu l/10^6$ oocysts/hr at 30°C. This respiratory rate fluctuated between 11.2 and 16.6 over a period of ten hours, and then suddenly increased to 26.0 μl/hr during the 11th hour and 35.2 μl/hr during the 12th hour. At 48 hr, when sporulation was complete, the respiratory rate had fallen to 0.4. The respiratory rate was constant over the pH range 1.3 to 8.8; more extreme values of pH were not investigated. (It may be noted that 0.1N H_2SO_4 is a medium sometimes used to carry out sporulation, e.g., Wilson and Fairbairn, 1961). At pH 7.3 and 38°C the oxygen uptake of 13 samples of parasitized cecal tissue taken at an unspecified time in the infection averaged 7.3 μl/mg dry wt/hr compared with a figure of 4.7 μl/hr obtained from ten samples of normal tissue.

Daugherty and Herrick (1952) subsequently found the reverse situation when they studied anaerobic glycolysis by cecal homogenates. Lower rates of metabolism were found for infected tissues when compared with normal, although this may possibly have been the result of the inclusion of inert materials such as blood clots with the tissues from which the homogenates were made. Collier and Swales (1948) on the other hand, who were seeking information on the mode of action of sulfamerazine, were unable to demonstrate any difference between the rates of respiration or of anaerobic glycolysis of normal chick cecal tissue and that parasitized with *E. tenella* at various stages of the infection. Smith and McShan (1949) measured the oxygen uptake of liver homogenates fortified with succinate at various stages of an *E. stiedai* infection. There was a marked decrease in the dry wt / fresh wt ratio as the infection progressed, indicating the development of edema. Oxygen uptake of infected liver when based on dry weight showed an increase over uninfected liver to a peak on the tenth day of infection. Whether this increase was caused by metabolic activity of the parasite or to a stimulation of host cell metabolism could not be determined.

A. SPORULATION

Sporulation is almost certainly a strictly aerobic process. Dürr and Pellérdy (1969) summarize the literature on this topic, and point out that statements to the contrary have in most cases arisen from misunderstanding and misquotation. Duncan (1959) claimed to have obtained limited sporulation of *E. labbeana* of pigeons under anaerobic conditions, but as potassium dichromate was used in the suspending medium, the significance of his findings are doubtful. Dürr and Pellérdy suggest without evidence that anaerobic sporulation may occur in the coccidia of fish. They found, however, no sporulation under anaerobic conditions for *E. stiedai* of rabbits, and at 25°C in a dichromate medium, reported a total oxygen requirement of from 302 μl to

395 μl O_2 per 10^6 oocysts for sporulation to occur.

When the sporulation process in *E. acervulina* was studied by Wilson and Fairbairn (1961), changes in carbohydrate and lipid content were correlated with respiratory rate. Oocysts were isolated from intestinal washings by sedimentation and zinc sulfate flotation at 0°-4°C, and sporulation was carried out in 0.1N H_2SO_4 at 30°C in a water bath with continuous shaking. Respiration was followed manometrically over the whole sporulation period, oocyst samples from additional flasks being taken at intervals for the gravimetric estimation of lipid extractable with 2:1 chloroform:methanol, and the colorimetric estimation of carbohydrate with anthrone. Figure 2 summarizes the changes found over a 50 hr observation period. Microscopic examination showed that the formation of four sporoblasts had not occurred after five hours, was 45% complete after seven hours and was maximal (about 90%) in ten hours; sporulation was virtually complete at 20 hr. Unlike Smith and Herrick who studied *E. tenella*, Wilson and Fairbairn noted with *E. acervulina* a steadily decreasing respiratory rate from an initial value of 3.5 $\mu l/10^6$ oocysts/hr to a level of 0.24 μl after 68 hr. During this period, the respiratory quotient (RQ) changed from a maximal initial value of 1.12 to unity at ten hours, and thereafter lay in the range 0.8 to 0.9. Respiration and sporulation were inhibited by cyanide, but recommenced on removing the cyanide. No sporulation took place under anaerobic conditions, but normal sporulation proceeded on admitting air. Virtually all the alkali-stable carbohydrate was precipitated in 60% ethanol and is now known to be amylopectin. During sporulation, this amylopectin decreased almost linearly from 83 to 46 $\mu g/10^6$ oocysts during the ten hours of sporoblast formation; there was a further decrease to 41 μg over the next five hours, followed by some resynthesis (10 $\mu g/10^6$ oocysts) until 48 hr. During the initial ten-hour period of maximum carbohydrate utilization, lipids were not utilized. Between 10 hr and 25 hr they apparently supplanted carbohydrate almost completely as a source of energy, decreasing from 52 to 36 $\mu g/10^6$ oocysts. Thereafter a further small decrease to 29 μg at 48 hr was noted. The initial high RQ during sporoblast formation is correlated with carbohydrate utilization; the later stages of sporulation, which include some carbohydrate resynthesis, depend on fat utilization with a correspondingly reduced RQ. There was a 19% decrease in oocyst dry wt over a 48-hr period from an initial value of 369 $\mu g/10^6$ oocysts; it is improbable that any loss of cytoplasmic protein occurred, but small amounts of soluble nitrogen and possibly some components of the oocyst wall were lost. Strout, Botero, Smith, and Dunlop (1963) found a halving of oocyst dry weight during sporulation and a 24% decrease in lipids. Phospholipids accounted for one-half of the total lipids present, and cholesterol was detected.

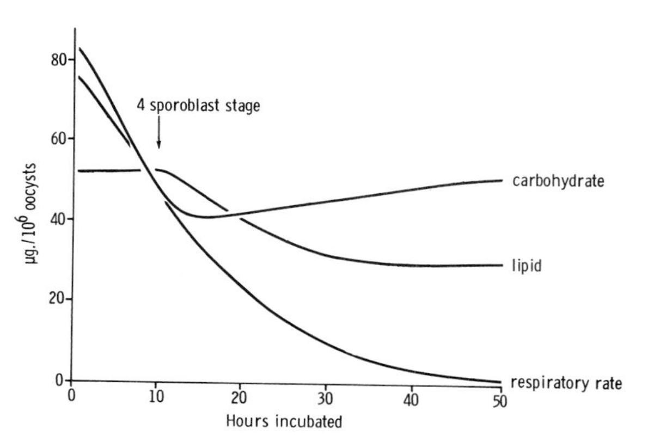

Fig. 2. Metabolism of *E. acervulina* during sporulation (adapted from Wilson and Fairbairn, 1961).

Wagenbach and Burns (1969) have followed the respiration of *E. tenella* and *E. stiedai* oocysts during sporulation by a more sensitive polarographic technique, and have directed much more attention to cytological events. The initial respiratory rate correlated well with that of Wilson and Fairbairn but not with that of Smith and Herrick when compared on a unit surface area basis, but respiratory curves showed a marked temporary depression correlated with the appearance of the early spindle stage. Figure 3 reproduces these respiratory curves and indicates the times when several notable events in oocyst development were taking place.

Once sporulation is complete, oocyst respiration and metabolism fall to a barely detectable level until ingestion by a new host causes excystation to occur.

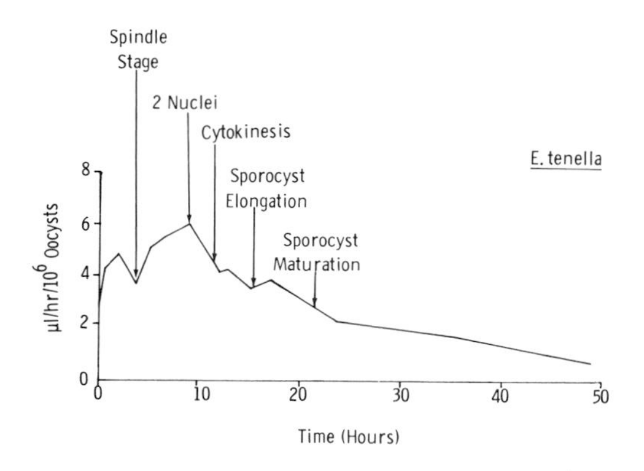

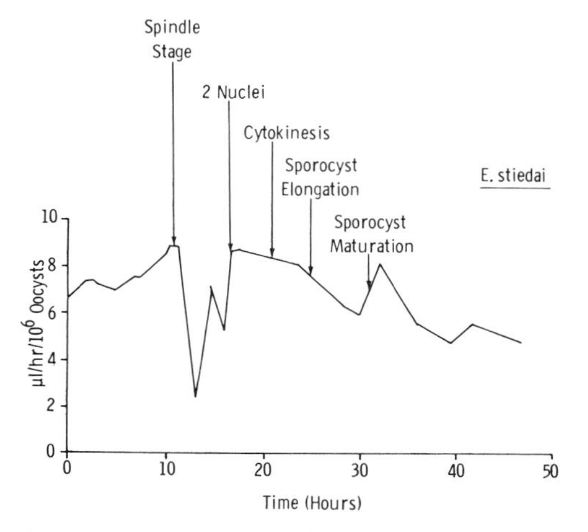

Fig. 3. Respiration of *E. tenella* and *E. stiedae* during sporulation (adapted from Wagenbach and Burns, 1969).

B. EXCYSTATION

Although in Japan excystation has apparently been obtained without the use of enzymes (Itagaki and Tsubokura, 1958) or pretreatment of the oocyst (Ikeda, 1960), it is generally agreed that two separate stimuli are necessary before excystation will take place (Jackson, 1962).

1. Primary phase

Working with *E. arloingi* of sheep, Jackson showed that the primary stimulus, normally provided in the rumen, is caused by carbon dioxide. The secondary stimulus, provided by trypsin and bile in the small intestine, is only effective in producing actual excystation provided sufficient conditioning by CO_2 has already taken place. Under the influence of CO_2, a lifting or splitting of the cap over the micropyle could be seen. That a fundamental change had taken place was apparent from the fact that oocyst contents would now collapse when placed in hypertonic solutions, the inner oocyst wall could be stained with methylene blue, and the contents could be destroyed by sodium hypochlorite. Jackson suggested that the CO_2 stimulus activates an enzyme system within the oocyst which chemically alters both the inner wall of the oocyst and the supposed outer coat of the sporocyst. At least 15% CO_2 in the gas phase during exposure was required and the presence of reducing agents—such as sodium dithionite, cysteine or ascorbic acid—helped. For a reasonable proportion of oocysts to become susceptible to excystation in the presence of trypsin + bile, a temperature of 37° to 40°C and an exposure time of at least four hours was needed. Other species of coccidia are susceptible to "triggering" by CO_2, but the concentration of CO_2 and exposure time necessary for optimal effect varies from species to species. Nyberg and Hammond (1964) found that rather higher CO_2

tensions of 30% to 50% and exposure times of at least ten hours were necessary for good results with *E. bovis* and other bovine coccidia. Nyberg, Bauer, and Knapp (1968) obtained good results with *E. tenella* using pure CO_2 and an exposure time of eight hours at 41°C, while Hibbert and Hammond (1968) obtained some excystation on enzyme treatment following exposure of *E. necatrix* and *E. acervulina* to 90% CO_2. With these poultry coccidia, Hibbert and Hammond obtained much better and more rapid excystation following mechanical breakage of the oocyst, and it would seem that although they are susceptible to CO_2 stimulation in just the same way as ruminant and rodent coccidia, mechanical rupture in the gizzard is more likely to be the normal primary stage of excystation. Hibbert and Hammond (1968) measured the extent of the CO_2 activation of *E. bovis* in terms of subsequent percent excystation under fixed conditions of incubation in trypsin + bile, and investigated the effect of temperature on this activation. A regression analysis using the integrated equation of Arrhenius showed that the CO_2 pretreatment phase of excystation has a linear temperature dependance on the log of velocity in excystation. CO_2 apparently stimulates the activation or production of an enzyme, or an enzymic rate-limiting step, which causes a change in the permeability of the micropyle. Because abnormally sporulated and unsporulated oocysts do not respond to CO_2 pretreatment (also noted by Jackson, 1964), they consider that this activating enzyme occurs in the sporozoite. Nyberg, Bauer, and Knapp (1968) working with *E. tenella* noted that when under the influence of CO_2, a thinning or indentation occurred at the micropylar region of the oocyst wall, the oocyst residual body became prominently located at the peripheral central portion of the thinned area. It could well be that the enzyme activated by CO_2 is present in its masked or inactive form in this residual body (where present) or in the fluid filling the oocyst. Later studies with the scanning electron microscope (Nyberg and Knapp, 1970a) showed the altered micropyle as "a circular depression surrounding a pinpoint opening in a centrally elevated portion." Jackson (1962) noted that oocysts "triggered" by exposure to CO_2 in the presence of isotonic buffer solutions would not liberate free sporozoites on subsequent exposure to trypsin plus bile. Under these conditions, the oocysts had apparently become permeable to enzymes, sporozoites would become motile and escape from the sporocyst, but they were unable to escape from the oocyst. Some form of osmotic (by washing in water) or mechanical (by gently shaking with glass beads) shock was apparently necessary to break the micropylar membrane, thinned and made permeable by enzymic action following CO_2 preconditioning.

2. Secondary phase

The secondary stimulus, which effects the actual escape of the sporozoites, is brought about by trypsin and bile salts. Doran (1966) evaluated previous studies from his laboratory (Doran and Farr, 1962; Farr and Doran, 1962) which had suggested that lipase (steapsin) might also have been involved, but by using pure recrystallized enzymes, he was able to show that trypsin or chymotrypsin were effective, while lipase and carboxypeptidase were not, in bringing about the excystation of *E. acervulina*. Doran, however, doubts whether these enzymes are solely responsible for the enzymic digestion of the sporocystic plug, and suggests that although they initiate the process, the sporozoite also secretes an enzyme which acts on the inner surface of the plug. Jackson (1962), noting that both trypsin and bile were necessary for the second stimulus to be effective, found that bile could be used first followed by washing and then trypsin, but not the other way round. Bile may facilitate the entry of enzymes through the altered micropyle, or may alter the protein or lipoprotein surface of the Stieda body (Doran and Farr, 1962; Hibbert, Hammond, and Simmons, 1969) and seems to be essential for the excystation of ovine, chicken and squirrel but not bovine coccidia (Hibbert and Hammond, 1968). A number of bile acids or surface active agents can replace bile (Doran and Farr, 1962; Jackson, 1962; Hibbert, Hammond, and Simmons, 1969). Bile salts have, however, another interesting effect. Speer, Hammond, and Kelley (1970) report that motility of merozoites, both free and in mature schizonts, was stimulated in tissue cultures of a variety of squirrel, rat, and sheep coccidia by bile or bile salts. This apparently is also the case with sporozoites, since Roberts, Speer, and Hammond (1970) noted that although *E. larimerensis* and *E. callospermophili* could be induced to excyst without bile salts, the sporozoites took longer to escape and had little movement compared with the pivoting, flexing, and gliding sporozoites produced in the presence of bile. Motility could be induced by subsequently adding bile.

This second stage of excystation is pH dependent. Hibbert, Hammond, and Simmons (1969) found maximum excystation of *E. bovis* over the range pH 7.5 to pH 8.5 in tris-maleate buffer. They suggest that this is the best alternative buffering system to bicarbonate for *in vitro* experimental work, since boric acid-borax, ammediol, and glycine-sodium hydroxide buffers brought about disintegration of sporozoites within the sporocysts of intact oocysts, while phosphate buffer has some inhibitory action on both the CO_2 conditioning and the excystation phases (see also Nyberg and Hammond, 1964). We have found that tris-maleate buffer also brings about disintegration of *E. bovis* sporozoites, and prefer to carry out excystation in the absence of any buffer, adjusting the pH from time to time as necessary.

The sporocyst has a thick wall with a small gap at one end. This gap is plugged by the Stieda body, and in certain species of *Eimeria* from rodents and of *Isospora*, by a substiedal body also. The whole is covered by a membrane or membranes. Histochemical tests with *E. utahensis* (Hammond, Ernst, and Chobotar, 1970) showed a weak protein reaction in the Stieda body, and a reaction for PAS positive material (stable to diastase) in addition to protein in the substieda component. On the addition of trypsin plus bile (Hammond, Ernst, and Chobotar, 1970; Nyberg and Hammond, 1964; Roberts, Speer, and Hammond, 1970; Nyberg, Bauer, and Knapp, 1968), sporozoites become motile within the sporocyst; the Stieda body becomes swollen and eventually disappears. Before actually disappearing, it may be actively forced out of the gap it was plugging. The sporozoites then escape through the small hole, a process which is remarkably quick and which involves a constriction passing along the sporozoite body because of the small diameter of the hole. Although one can visualize natural movements of the sporozoite being actively responsible for its escape, the rapidity of this escape and the popping out of the partly degraded Stieda body suggest the possibility of expulsion under pressure generated within the sporocyst (Chobotar and Hammond, 1971). It is interesting to speculate whether this may arise from osmotic forces generated by say the hydrolysis of the amylopectin abundantly present in the sporocyst residual body; a functional explanation of these membrane-bound residual bodies, which often disintegrate during excystation, is more attractive than imagining they are merely the consequence of bad budgeting on the part of the developing oocyst.

3. Energy considerations

Figure 4 summarizes observations made by Vetterling (1968) on respiration during excys-

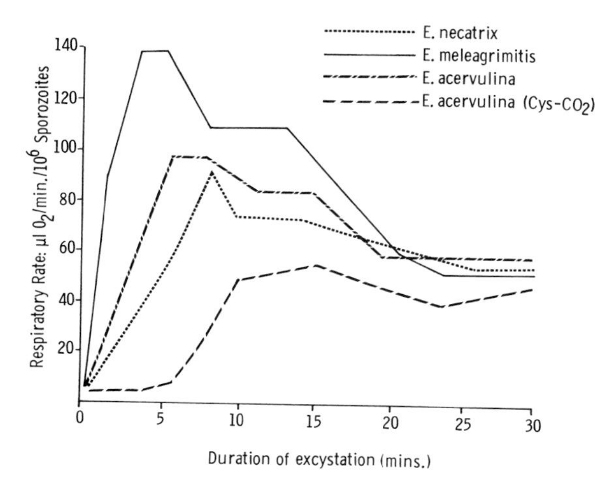

Fig. 4. Respiration of coccidia during excystation (adapted from Vetterling, 1968).

tation of *E. acervulina, E. necatrix,* and *E. meleagrimitis*. Sporulated oocysts were mechanically ground and then incubated in trypsin-bile at pH 7.5 and 41.5°C for 30 min. oxygen consumption being followed with an oxygen electrode. Oxygen consumption increased rapidly to a peak at 1.5 min to 6 min depending on the species, remained steady for a few minutes, decreased to a lower rate for another few minutes and then decreased to a yet lower rate for the remainder of the determination. This latter rate is presumably maintained while the sporozoite seeks out and penetrates a host cell. A lower, delayed respiratory curve was obtained with equivalent numbers of *E. acervulina* oocysts which had been pretreated for 15 hr with CO_2-cysteine instead of mechanical breakage, and this correlated with a final excystation of 47% compared with 95% in the mechanically broken *E. acervulina*. Vetterling noted that respiration was completely suppressed by 10^{-3}M KCN, but did not state whether this influenced excystation; Wagenbach and Burns (1969) reported that excystation of *E. tenella* proceeded normally in the presence of 10^{-2} M NaCN. Although cyanide does not apparently kill oocysts or prevent excystation, the enforced anaerobic metabolism which results is much more wasteful of stored nutrients; Wagenbach and Burns (1969) noted that oocysts would survive for four days at 41.5°C in air, but only for one day at this temperature when incubated under nitrogen.

During excystation and cell penetration, the sporozoite utilizes most of its remaining amylopectin stores to meet energy requirements. Vetterling and Doran (1969) observed that during a 30 min period of excystation at 42.9°C, carbohydrate reserves in the sporozoites of three species of avian coccidia fell by about two-thirds; this utilization was correlated with oxygen consumption values given by Vetterling (1968) to give the ratios quoted in Table 3, which suggest that some other substrate, possibly lipid, is additionally being oxidized. PAS treatment of dormant sporozoites (as observed in sections of frozen oocysts) showed carbohydrate granules around the nucleus, in front of the anterior refractile globule, and behind the posterior globule. During excystation, these granules became fewer in number and more dispersed, and by the time cell invasion had taken place,

TABLE 3. Amylopectin utilization during excystation

	µg Glucose per 10^6 oocysts		
	E. acervulina	*E. necatrix*	*E. meleagrimitis*
Before excystation	33.3	30.1	36.7
After excystation	11.0	9.4	13.3
Moles O_2/moles glucose used during excystation	6.35	5.87	7.08

only traces remained in the area around the nucleus. In spite of the indications from the work of Wilson and Fairbairn (1961) that the dormant oocyst used lipid to meet its basal requirements, Vetterling and Doran found that the amylopectin content of *E. acervulina* fell from 33.3μg glucose/10^6 oocysts at three months to 21.3μg at one year, 7.8μg at two years, and 1.5 μg at six years; this fall in carbohydrate was correlated with a decline in infectivity, which may have been caused by an inability of the sporozoite actively to excyst and penetrate a host cell. Similar observations on carbohydrate utilization during storage were made by Kheysin (1959) who noted that reserves were longer lasting in species of *Eimeria* than in *Isospora*. Kheysin also observed utilization of the reserves in the residual body of the sporocyst as well as in the sporozoites themselves.

C. PENETRATION

The conoid probably serves as an organ of penetration for both sporozoites and merozoites into new host cells. A number of studies (e.g., Ryley, 1969; Roberts and Hammond, 1970) show the conoid in an extruded position; it is possible that a reversible protrusion of the conoid may be connected with membrane penetration. The club-shaped organelles have a gland-like appearance with ducts extending into the conoid, but so far no demonstration of lytic enzymes in these structures has been achieved. Similarly the demonstration of a secretory function for the micronemes is lacking. In view of the rapidity of cell penetration, it is conceivable that the conoid could act like a captive-bolt pistol to achieve mechanical puncturing of the membrane of the prospective host cell. In the case of *Toxoplasma gondii*, Lycke, Lund, and Strannegård (1965) showed that penetration into HeLa cells in tissue culture was enhanced by adding lysozyme or hyaluronidase; these two enzymes would act together synergistically. Although hyaluronic acid did not inhibit penetration at the normal rate, it was found to antagonize the penetration-enhancing effects of hyaluronidase and lysozyme. Lycke and Norrby (1966) proceeded to demonstrate factors in parasite extracts which would similarly enhance cell penetration, but also found another factor which inhibited this stimulation. Although coccidia contain analogous structures to *T. gondii*, Fayer, Romanowski, and Vetterling (1970) were unable to demonstrate any stimulation by hyaluronidase of the penetration into cultured kidney cells of *E. adenoeides* sporozoites, nor any antagonism of penetration with hyaluronic acid or chondroitin sulfate other than that attributable to increases in viscosity at high concentrations. No hyaluronidase activity could be detected in intact or fragmented *E. adenoeides* sporozoites or fragmented *E. tenella* oocysts.

D. PROLIFERATIVE *TOXOPLASMA GONDII*

Fulton and Spooner (1957, 1960) isolated toxoplasms in quantity from the peritoneal cavity of cotton rats, and found they respired at a steady rate for at least two hours, with an average Qo_2 of 30 and an RQ of 0.9 (range 0.83 to 1.14). There was no detectable endogenous respiration, exogenous glucose being the favored substrate. It should be remembered that cytochemical techniques have revealed but little storage polysaccharide in the proliferative stages of *Toxoplasma*, although rich deposits are present in pseudocyst forms. Glutamine also supported active respiration, while mannose, galactose, ribose, glucose-6-phosphate, and fructose gave 36% to 85% of the oxygen uptake found with glucose; the toxoplasms were apparently able to use extracellular glycogen to a limited extent, a substance often available in host cells. No respiratory stimulation was obtained with Krebs cycle intermediates. Cell homogenates were shown to contain hexokinase, and to oxidize glucose, glucose-6-phosphate, fructose-6-phosphate, fructose-1, 6-diphosphate, and DPNH. Under aerobic conditions, about one-third the glucose carbon used was recovered as lactic acid and about one-half as CO_2; acetic acid and traces of propionic, butyric, and valeric acids were also found. An increased yield of lactic acid was obtained under anaerobic conditions.

Respiration was inhibited 85% by 0.46×10^{-3} M cyanide, but not by azide, and respiratory inhibition produced by carbon monoxide was partially reversible by light. Spectroscopic examination revealed cytochrome bands with maxima at 548, 560, and 599 mμ. Hydrogen peroxide had no damaging effect on the parasite, and this was correlated with the presence of an active catalase. No respiratory inhibition was produced by sulfapyridine, but a 63% inhibition was recorded in the presence of 10^{-3} M proguanil. Sera from different animals affected respiration to various extents; some inhibiting effects were produced by sera from patients with toxoplasmosis, but there was little correlation with titers in the Sabin-Feldman dye test. Dumas (1970) has also tried to assay immune sera by their respiratory inhibition effects. Lund, Hansson, Lycke, and Sourander (1966) using the Cartesian diver technique found a Qo_2 about four times that recorded by Fulton and Spooner, and having showed this respiration to be sensitive to cyanide (66% inhibition by 10^{-4} M KCN), demonstrated the localization of cytochrome oxidase in three to seven distinct cytoplasmic granules by histochemical methods. Using diaminobenzidine and peroxide, Akao (1971b) obtained a similar granular reaction which electron microscopy revealed was limited to the mitochondrial cristae (published electron micrographs are not very convincing). The reaction was thought to indicate peroxidase, or possibly the cytochrome system.

E. SPOROZOITES OF *EIMERIA TENELLA*

The sporozoite is an attractive stage for biochemical studies, since it is metabolically active and can readily be obtained free from host tissues and contaminating microorganisms. Preliminary studies in our laboratories show that its metabolism is similar in many respects to that of *T. gondii* and malaria parasites. Oocysts obtained from infected feces by sieving and salt flotation were further purified and surface sterilized by hypochlorite treatment, and then broken by shaking for 20 sec on a mechanical shaker with grade 7 Ballotini glass beads. Sporozoites were obtained by excystation in trypsin + 5% chick bile and separated from unbroken oocysts, membranes, etc., on a glass bead column 9.5 cm in diameter and 8.5 cm deep by the method of Wagenbach (1969).

Unlike *T. gondii*, the coccidial sporozoite has an appreciable endogenous respiration due to the presence of amylopectin reserves. An 18% respiratory stimulation was observed with glucose and a 46% stimulation with fructose; oxygen uptake in the presence of fructose was 330 μl/hr/10^9 sporozoites. Respiration was sensitive to cyanide; 73% inhibition was observed with 10^{-4} M KCN and 93% inhibition with 0.46×10^{-3} M KCN. Homogenates of sporozoites prepared by shaking with glass beads oxidized *p*-phenylene diamine, and this oxidation was stimulated 30% by added mammalian cytochrome *c*. Spectroscopic examination of a thick sporozoite suspension showed cytochrome bands from 556 mμ to 562 mμ and from 600 mμ to 610 mμ; the cytochrome *a* band from 600 mμ to 610 mμ was particularly prominent. Using the diaminobenzidine method of Seligman, Karnovsky, Wasserkrug, and Hanker (1968), we were able to locate cytochrome

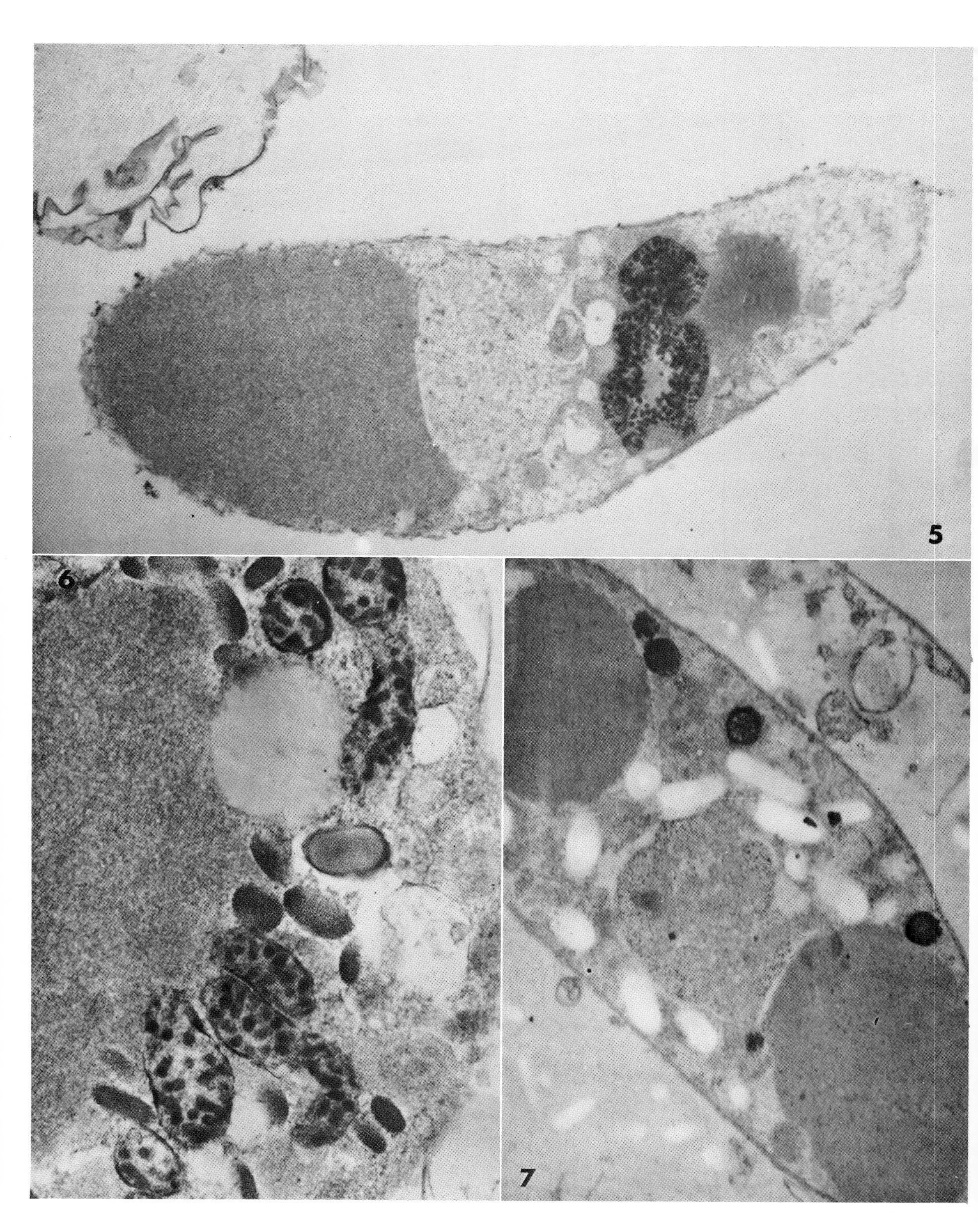
5
6
7

Fig. 5 and 6. Sporozoites of *E. tenella* treated with diaminobenzidine to show mitochondrial location of cytochrome oxidase. **5,** × 20,000; **6,** × 40,000.

Fig. 7. Sporozoite of *E. tenella* treated with DPNH and tetrazolium salt. × 26,000.

oxidase at the EM level on the tubular cristae of the mitochondria (Figs. 5 and 6). Incubation was carried out without prior fixation, and this resulted in rather swollen mitochondria and poor preservation of other cellular structures. Incubation of unfixed sporozoites with DPNH and the tetrazolium salt MTT in the presence of cobalt followed by treatment with ammonium sulfide, showed oxidation had taken place and formazan was deposited in bodies which may be mitochondria or possibly structures analagous to the microbodies of the trypanosomes (Fig. 7).

Under anaerobic conditions, metabolism was maintained at the expense of amylopectin reserves with the production of organic acid and CO_2. From Table 4 it will be seen that the major metabolite was lactic acid with lesser amounts of carbon dioxide and glycerol; 97% of the amylopectin used was accounted for in this experiment. Anaerobic acid production (Table 5) was stimulated by glucose and lactose, and to a lesser extent by fructose and mannose; amylopectin usage was spared by glucose, fructose and mannose, but not by lactose.

F. HISTOCHEMICAL INVESTIGATIONS

In Table 6 an attempt has been made to summarize the histochemical reactions carried out by a number of authors on a variety of coccidia. The presence or absence of an enzyme is indicated by ✓ or –, respectively; in some cases, from the intensities of the reactions reported, an attempt has been made to grade these reactions by a series of +s. In most cases the reactions have been examined with the light microscope, and only in a few cases has there been any definite structural localization of activity.

Capella and Kaufman (1964), in demonstrating the presence of a number of enzymes in *T. gondii* by the use of the tetrazolium salt NBT, noted that they were present both in proliferative organisms from the peritoneal fluid and in cysts from the brain of mice. Six to eight discrete formazan granules per cell were produced in each case, and as these areas could be supra-vitally stained with Janus green B, it was felt that the enzymes were limited to the mitochondria. Niebrój and Wojdala (1967a, b) found both the degree and localization of succinic dehydrogenase activity were pH-dependent. Granules produced in response to different substrates had differing but ill-defined distribution in the cell, and no correlation with subcellular structures was made. They considered that the relatively high activities of lactic dehydrogenase and NADH and NADPH reductases indicated that anaerobic oxidations were the main source of the cell's energy. It should be remembered that although Fulton and Spooner showed that glucose breakdown by toxoplasms was not carried to completion,

TABLE 4. Anaerobic endogenous metabolism of *E. tenella* sporozoites[a]

	μmoles	Moles/mole glucose
Amylopectin (as glucose)	−30.05	−1.00
Acid (from bicarbonate)	+50.25	+1.67
CO_2	+14.95	+0.50
Lactic acid	+45.50	+1.49
Pyruvic acid	0.00	0.00
Succinic acid	0.00	0.00
Glycerol	+8.60	+0.28

[a]Sporozoites of *E. tenella* (16 mg N in a total volume of 6 ml) were incubated in Ringer-bicarbonate medium for 90 min in Warburg flasks at 41°C with gas phase of 5% CO_2 - 95% N_2. Fermentation was stopped by tipping acid from a side bulb. Initial and final values are based on the combined contents of four manometer flasks. Values given are total amounts of substrate used and metabolites recovered, and also moles of metabolite formed per mole of substrate used.

TABLE 5. Anaerobic carbohydrate utilization by *E. tenella* sporozoites [a]

	Acid (bicarbonate) (μequivalents)	CO_2 (μmoles)	Amylopectin (μmoles of glucose)
Blank	19.56	1.66	−8.05
Glucose	32.01	3.48	−0.69
Glycerol	19.38	2.15	−6.78
Maltose	19.60	2.64	−7.28
Sucrose	19.40	2.54	−4.99
Blank	16.25	2.55	−10.31
Fructose	25.45	4.38	−5.40
Lactose	30.59	2.95	−8.38
Galactose	15.28	3.26	−8.50
Mannose	23.93	4.61	−3.04

[a]Sporozoites of *E. tenella* (3.7 and 2.7 mg N in two experiments) in total volume of 3 ml were incubated in Ringer-bicarbonate medium in Warburg flasks at 41°C for 90 min with gas phase of 5% CO_2 - 95% N_2. Amylopectin was determined in initial and final samples after digestion with KOH, precipitation with ethanol and hydrolysis.

the organisms readily consumed oxygen by the cytochrome pathway.

Aspartate aminotransferase, which in rat liver cells is localized in mitochondria or on the limiting membranes of microbodies and nuclei, was shown by Akao (1971a) to be limited in distribution in *Toxoplasma* to the inner surface of the cell membrane; Akao (1969) notes that adenosine triphosphatase has a similar distribution. Although Fulton and Spooner found that metabolism was entirely dependent on a supply of exogenous substrate, Akao and Matsubayashi (1966) state that "the cytoplasm of *T. gondii* is rich in glycogen vacuoles." Tritiated glucose was injected into *Toxoplasma*-infected mice, and after a 30 min to 60 min incorporation period, organisms were harvested and processed for autoradiographic examination in the electron microscope. Dark grains were apparently scattered throughout the cytoplasm, but in some cases were concentrated in vacuoles above the Golgi apparatus or found in the toxonemes. They concluded that their experiments revealed the function of the toxonemes as the primary site of "glycogen" accumulation. In a fuller publication, Akao (1969) suggested that newly synthesized polysaccharide was added to the deposits in the PAS positive granules, and that the toxonemes probably utilize glucose as an energy source for their physiological activities – whatever these may be.

Phosphatases are a group of enzymes whose physiological importance is considerably exaggerated by virtue of the simple methods which make their detection possible. Alkaline phosphatase in the studies of Ray and Gill (1954) appeared to be concentrated in the area of the nucleus of *E. tenella*. An intense reaction was also obtained in the wall-forming bodies of the female gametocyte; this disappeared after fertilization and gave way to a conspicuous reaction at the micropylar end of the oocyst and a weak reaction in the oocyst membrane. Gill and Ray (1954c), in drawing attention to a similar distribution of phosphatases, RNA, and hyaluronic acid in the karyosome and in the wall-forming bodies of the macrogametocyte of *E. tenella*, suggested a correlation between these substances, protein synthesis, and oocyst wall formation.

In the case of *E. stiedai*, Frandsen (1968, 1970) noted that acid phosphatase corresponded in its random distribution to intralysosomal sites; in no instance was it observed in nuclei or in association with plastic gran-

TABLE 6. Histochemical reactions of various coccidia

Organism	Stage	acid phosphatase	alkaline phosphatase	5-nucleotidase	ATP-ase	carboxylic ester hydrolase	leucine aminopeptidase	β-glucosidase	β-galactosidase	α-glucan phosphorylase	sulfatase	aspartate amino transferase	aldolase	cytochrome oxidase	succinic dehydrogenase	malic dehydrogenase	isocitric dehydrogenase	glutamic dehydrogenase	β-hydroxybutyric dehydrogenase	glucose-6-phosphate dehydrogenase	glucose-1-phosphate dehydrogenase	fructose diphosphate dehydrogenase	α-glycerophosphate dehydrogenase	lactic dehydrogenase	alcohol dehydrogenase	NADH diaphorase	NADPH diaphorase	Reference
T. gondii	proliferative														+		+++			+++			++	+++		+++	+++	Capella & Kaufman (1964)
T. gondii	proliferative													√	+						+	+		+		+++	+++	Lund, Hansson, Lycke & Sourander (1966)
T. gondii	proliferative														+		+	+++	+	++			+	++++	+	++++	++	Niebrój & Wojdała (1967a, b)
T. gondii	proliferative				√							√		?														Akao (1969, 1971a, b)
E. tenella	all		√																									Ray & Gill (1954)
E. tenella	all	√		√																								Gill & Ray (1954, c)
E. tenella	sporozoite	-	++																									Tsunoda & Itikawa (1958)
	schizont	+	++																									
	microgamete	-	+																									
	macrogamete	++	++																									
E. stiedai	all		√																									Dasgupta (1961)
E. stiedai	sporozoite	√	-			-	-	-	-											√				√				Frandsen (1968, 1970)
	schizont	√	?			√									√					√				√				
	merozoite	√	√			√	√	-	+	+	-		√		√					√				√				
	male gametocyte	√	√			√	√	-	+	+	-		√		-					√				√				
	female gametocyte	√	√			√	√	-	++++	+++	-		-		-					√				√				
	unsporulated oocyst	√	√			√	√	-	+++	++++	-		√		-					√				√				
E. stiedae	various	√																										
A. eberthi	macrogamete	√																										Heller & Scholtyseck (1970)
K. helicina	macro- & micro-gamete	√	√		√																							
Eimeria	various														√								√					Beyer (1970)
E. tenella	sporozoite													√												√		Ryley (this chapter)

ules of the macrogametocyte. Heller and Scholtyseck (1970), who investigated the phosphatases of *E. stiedai* and two noneimerian coccidia at the electron microscope rather than the light microscope level found acid phosphatase was always associated with membranes, and occurred in vacuoles and vesicles with metabolic function. Although wall-forming bodies I (which give rise to the outer layer of the oocyst wall) in macrogametocytes did not show any reaction, the enzyme was found inside and on the surface of wall-forming bodies II (which give rise to the inner layer of the oocyst wall) as well as on membranes of the endoplasmic reticulum and in vacuoles. Schulte (1971) gives further details of the distribution of phosphatases in *Klossia helicina* at the ultrastructural level. Frandsen found alkaline phosphatase in all stages except the sporozoite; it showed a random distribution in the cytoplasm of unsporulated oocysts, and was not closely associated with the wall. In developing macrogametocytes, much of it was however located in the peripheral plastic granules. Carboxylic ester hydrolases were similarly present in all stages except sporozoites, but no correlation with subcellular structures was possible.

The pattern of random cytoplasmic distribution of formazan was consistent with a mitochondrial occurrence of succinic dehydrogenase, but the enzyme was found regularly only in schizonts and merozoites; most sexual stages and unsporulated oocysts were negative. Glucose-6-phosphate gave the most intense deposits of formazan, but although glucose-6-phosphate and lactic dehydrogenases could be demonstrated in all stages of *E. stiedai,* the distribution was random and not associated with any particular cellular organelles. Lactic dehydrogenase in the sporozoite was particularly prominent in a zone just anterior to the refractile globule. β-galactosidase was particularly active in macrogametocytes, most of the activity being located in the plastic granules. Although small punctate areas of activity were found in the cytoplasm of unsporulated oocysts, the strongest reaction was given by the oocyst wall and the space between the cytoplasm and the wall. Mounting of histochemical preparations in synthetic resin caused differential extraction of indigo, leaving only the shell of the oocyst blue. Although little is known of galactose metabolism in the coccidia or the presence of galactose in the oocyst shell, β-galactosidase is obviously implicated in oocyst wall formation. It is interesting to note that several enzymes demonstrated in endogenous stages of *E. stiedai* were apparently absent from the sporozoite, and Frandsen's studies have so far failed to throw any light on chemical factors which may be associated with penetration by sporozoites and merozoites into host cells.

In a series of studies on rabbit coccidia and *E. tenella,* Beyer (1965, 1970) has drawn some very interesting, though perhaps not entirely warranted, conclusions based apparently on the intensities of reactions for α-glycerophosphate and succinate dehydrogenases at various stages of the life cycle. She considers that under the conditions of ample nutrition provided by the host cell, glycolysis can easily cover the energy requirements of the developing schizont; this is indicated by the absence of succinic dehydrogenase. In the later stages of growth, as the host cell substrate becomes exhausted and as the merozoites differentiate and prepare for a temporary extracellular period of existence, transition to aerobic metabolism with the Krebs cycle takes place; this is indicated by the appearance of succinic dehydrogenase. Similarly the growing macrogamete obtains its energy by glycolysis — as indicated by a high level of α-glycerophosphate dehydrogenase and the absence of succinic dehydrogenase. Immediately after fertilization, high levels of both enzymes can be found. Beyer considers this to be an indication of change to oxidative metabolism, which would be a more economical way of substrate utilization in an independent organism which has to provide energy to support sporulation and, later, excystation and cell penetration. Such an alternation of metabolic pathways with changing environment is reminiscent of the trypanosomes; what factor could bring about the necessary transformations is a mystery.

VI. Host-Parasite Interrelationships

Although the pathology of coccidial infections and the effects of the organism on the host will be dealt with in a subsequent chapter, certain aspects of the host-parasite relationship are pertinent for consideration here. That such a relationship exists implies some defect in the parasite's make-up and physiology which can be remedied by a suitable host, and the fact that disease or death of the host can result from such a relationship suggests that the parasite's biochemical activities are not confined to the parasite itself. As yet, no specific biochemical lesion has been discovered in any of these parasites which would explain their obligate parasitic mode of life.

A. TOXINS

Three different types of "toxic" substance have been found associated with various coccidia. None of them, however, seems to have any bearing on the pathogenesis of the infections concerned, and they are little more than scientific curiosities.

Varela, Vázquez, and Torroella (1956) found a substance pharmacologically similar to LSD in the peritoneal exudate of *Toxoplasma*-infected mice; this finding does not seem to have been followed up.

Weinman and Klatchko (1950) described a substance "toxotoxin" found in the peritoneal exudate of mice infected with toxoplasms; it was lethal on intravenous injection in normal mice, producing convulsions and rapid death. Mice which are ill and moribund from overwhelming infections with toxoplasms, however, show none of the dramatic signs associated with intravenous administration of "toxotoxin." Fulton (1965) studied the symptoms produced in greater detail, and suggested that toxotoxin is a protein-hyaluronic acid complex. A disagreement exists between Woodworth and Weinman (1960) and Nozik and O'Connor (1969) concerning the reasons for the toxicity. The material cannot be extracted from isolated toxoplasms (Weinman, 1952), nor is it produced by toxoplasms growing in chick embryos (Woodworth and Weinman, 1960) or tissue culture (Nozik and O'Connor, 1969). It seems most likely that "toxotoxin" is not a product of parasite metabolism, but rather a substance produced under certain conditions by the host in response to infection, and that it owes its toxicity on intravenous injection into mice to nothing more than its viscosity.

Parasite extracts, proteinaceous in nature, show toxicity also, but as this toxicity is limited to the rabbit and chick embryo, the worthwhileness of the great deal of work done on these "toxins" is in doubt. Thus Burns (1959) found that homogenates of cecal contents or scrapings taken from birds infected with *E. tenella* or homogenates of oocysts were toxic when given intravenously or intraperitoneally to the rabbit. Harmful effects were not, however, produced when the substance was administered to healthy chicks, let alone the symptoms characteristic of coccidiosis. Rikimaru, Galysh, and Shumard (1961) investigated the pharmacology of these extracts in the rabbit, while Sharma and Foster (1964) showed that the toxic material in the oocyst is not located in the oocyst wall, sporocyst, or sporozoite, and must therefore be in the fluid filling the sporulated oocyst. Lunde and Jacobs (1964) showed that *Toxoplasma* lysates were lethal to rabbits on intravenous injection, while Lutz-Ostertag and Sénaud (1966) found that the extract "toxoplasmine" would produce embryonic malformations if injected into the chick embryo at a very early stage of development. "Sarcocystin" is a similar material present in extracts of cysts of *Sarcocystis tenella*, but once again, the toxicity to the rabbit of this material first demonstrated by Pfeiffer in 1890 and recently studied by Sénaud, Vendrely, and Tronche (1968a, b) would seem to be irrelevant to the host-parasite relationship in the sheep.

B. EFFECTS ON INTESTINAL FUNCTION

Coccidia during their metabolism produce considerable quantities of acid, mostly lactic acid. One feature of infection noted by a number of authors is a markedly lowered intestinal pH. With an infection of *E. necatrix* (Stephens, 1965) the pH of the jejunum and ileum anterior to the caeca was 5.55 on day 6 and 6.02 on day 9 compared with 6.62 in noninfected chicks. With *E. maxima*, pH's of 5.31 to 6.07 were observed over days 6 to 12 (Stephens, Kowalski, and Borst, 1967) compared with control values averaging 6.47. With *Eimeria acervulina* (Kouwenhoven and Horst, 1969; Horst and Kouwenhoven, 1970), values as low as 3.1 have been found in the upper small intestine; this lowered pH was discernible by day 2, showed a minimum at days 5 or 6 and returned to normal values by days 7 to 9. Lowered intestinal pH causes a denaturation of proteins; a caseous substance, loosely attached to the mucosa, was observed 4 to 6 days after infection, and a similar effect (Kouwenhoven and Horst, 1970) could be produced by administration of HCl to normal chicks. Whether this lowered intestinal pH is the direct result of acid production by the parasites, or whether it is an indirect effect brought about by an interaction of the parasite with the host's metabolism is not known. It is, however, one of the factors which results in disturbances in absorption of nutrients and leakage of materials from the tissues.

Stephens (1965) noted that lowered intestinal pH during an *E. necatrix* infection was correlated with an increase in light transmittance of plasma at 500 mμ, and suggested this indicated a reduced ability to absorb carotenoid pigments among other materials from the intestine during infection. Similar effects were noted with *E. maxima* (Stephens, Kowalski, and Borst, 1967) and *E. acervulina* (Kouwenhoven and Horst, 1969, 1970; Horst and Kouwenhoven, 1970). Normally xanthophyll and vitamin A are bound by proteins present in the epithelial cells. Protein denaturation caused by a lowered pH upsets this binding and absorption. Blood vitamin-A levels do not show quite the same depression as carotene, since they can be maintained at the expense of liver vitamin A stores, while carotene levels depend on absorption from the intestine. Consequent, however, on the reduced intestinal pH and damage to the epithelium is a leakage of serum protein into the intestine, which makes it more difficult to transport liver vitamin A into the blood.

Preston-Mafham and Sykes (1967) noted this change in gut permeability during infection with *E. acervulina*, and quantified the change in permeability in terms of the extent of leakage of intravenously administered Pontamine sky blue into the intestine. Leakage was maximal 90 hr to 120 hr after infection. Concurrent with the period of maximum dye loss, they noted a depression in absorption of histidine, glucose, and fluid. Turk and Stephens (1966, 1967a, b, c, 1969, 1970) in a series of studies found that coccidial infection depressed the uptake of Zn^{65} - and I^{131}-labeled oleic acid from the intestine, but that the extent of this effect depended on the species of coccidia by virtue of differences in the particular part of the intestine parasitized. Thus absorption of both substances was depressed by *E. acervulina* in the upper small intestine, zinc absorption was more severely affected than oleic acid by *E. necatrix* in the mid intestine, while effects were only mild with *E. tenella* in the ceca or virtually undetectable with *E. brunetti* in the lower intestine.

C. EFFECTS ON HOST METABOLISM

Studies by Daugherty and Herrick (1952) drew attention to a factor in cecal homogenates from birds infected with *E. tenella* which would reduce the ability of chick brain homogenates to carry out anaerobic phosphorylative glycolysis *in vitro*. Whether depressed glycolysis normally occurs in the brains of infected chicks is not known; anaerobic glycolysis rates *in vitro* were the same for brain homogenates prepared from normal and infected chicks. RQ determinations on muscle tissue taken from birds on the fifth or sixth day of infection yielded a value of 0.70 compared with an average value of 0.89 found with uninfected birds; such a lowering of muscle RQ could not be produced by starving birds for 72 hr.

Coccidia, including *Toxoplasma*, during their metabolism consume considerable quantities of carbohydrate, some of which is degraded to lactic acid. Janssen (1970) found that mice infected with *T. gondii* showed lowered blood glucose levels and a large decrease in liver glycogen and glucose. In spite of lactic acid production by the parasites, there was a slight decrease in blood and liver lactate levels. Blood pyruvate and ATP were normal, while small decreases were noted in the liver. Pratt (1940, 1941) had shown with *E. tenella* an *increase* in blood sugar levels during the acute stages of the disease which coincided with a reduced muscle glycogen content, and he attempted to explain these changes in terms of hemorrhage. Early and pronounced fatigue was characteristic of the muscles of infected birds (Levine and Herrick, 1954) and muscular activity of the crop and ceca was reduced (Schildt and Herrick, 1955). This might be a reflection of reduced muscle glycogen.

Freeman (1970) on the other hand did not observe a depression of plasma glucose or liver glycogen in *E. tenella* infections, but he did obtain a significant fall in plasma lactate early in the infection and a transient increase in cardiac glycogen at the fifth day. Reduction of glycogen levels in voluntary muscles on days 5 and 6 were thought to be caused by a reduction in muscular activity rather than an impairment of glycogenesis. Liver changes, particularly appreciable reductions in levels of glucose and inosine monophosphate, were however observed by Stoll, Enigk and Dey-Hazra (1970) in *E. necatrix* infections. Determinations of the liver glutathione quotient (ratio of reduced to oxidized glutathione) showed that this parameter was markedly reduced within 30 min of inoculation of chicks with *E. necatrix*, and that this reduction was maintained for 6-7 days (Harisch, Dey-Hazra, Enigk, and Schole, 1971). In a subsequent paper (Schole, Dey-Hazra, Harisch, and Enigk, 1972) they suggested that the catabolism resulting from this hypothetical stress reaction brought about by the process of infection could be abolished by massive doses of 5-hydroxytryptamine (50 mg/Kg IP), and claimed that with such treatment, mortality due to *E. necatrix* could be markedly reduced. They came to the amazing conclusion that "death following a single inoculation of a large number of oocysts is due to an alarm reaction and not to a specific pathogenic action of the parasites"! We have been unable to confirm the protective action of 5-hydroxytryptamine in *E. necatrix* infections, but have noted that such treatment results in a marked hyperemia of the intestine, particularly the duodenum, and a virtual hold-up in the passage of material through the gut for anything up to 6 hr.

Literature Cited

Abou-El-Azm, I.M. 1967. Relationships between chicken coccidia and certain vitamins, amino acids and antimetabolites. Diss.Abstr. 27B: 2591.

Akao, S. 1969. Ultramicroscopic studies of the localization of adenosine triphosphate activity and H^3-glucose transport in *Toxoplasma gondii.* Jap. J. Parasitol. 18: 488-497.

Akao, S. 1971a. *Toxoplasma gondii*: Aspartate aminotransferase in cell membrane. Exp. Parasitol. 29: 26-29.

Akao, S. 1971b. *Toxoplasma gondii*: Localization of peroxidase activity. Exp. Parasitol. 29: 250-254.

Akao, S. and H. Matsubayashi. 1966. Ultramicroautoradiographic studies on *Toxoplasma gondii.* p. 243 Abstr. 6th. Int. Cong. Electron Microscopy, Kyoto.

Arundel, J. H. 1959. The efficiency and toxicity of pyrimethamine in the control of caecal coccidiosis of chickens. Aust. Vet. J. 35: 7-12.

Baldwin, F. M., O. B. Wiswell and H. A. Jankiewicz. 1941. Hemorrhage control in *Eimeria tenella* infected chicks when protected by antihemorrhagic factor, vitamin K. Proc. Soc. Exp. Biol. Med. 48: 278-280.

Ball, S. J. and E. W. Warren.1967. Activity of riboflavin analogues against *Eimeria acervulina.* Vet. Rec. 80: 581-582.

Ball, S. J., E. W. Warren and E. W. Parnell.1965. Anticoccidial activity of nicotinamide antagonists. Nature, London. 208: 397.

Beyer, T. V. 1965. Analysis of the metabolism of different developmental stages of rabbit intestinal coccidia. Progress in Protozoology. p.159. 2nd. Int. Cong. Protozool., London,. Int. Cong. Series No. 91. Excerpta Medica, The Hague.

Beyer, T. V. 1968. Cytochemical studies on the haemogregarines of Armenian reptiles. I. DNA in the nuclei of some *Karyolysus* species from the rock-lizards *Lacerta armenica* and *L. saxicola.* Acta Protozool., Warsaw, 6: 79-86.

Beyer, T. V. 1970. Coccidia of domestic animals. Some metabolic peculiarities of particular stages of the life cycle. J. Parasit. (No. 4, Sect. 11) 56: 28-29.

Beyer, T. V. and L. P. Ovchinnikova.1964. Cytophotometric study of RNA content in the macrogametogenesis of two rabbit intestinal coccidia *Eimeria magna* and *E. intestinalis.* Acta Protozool., Warsaw, 2: 329-337.

Beyer, T. V. and L. P. Ovchinnikova.1966. A cytophotometrical investigation of the cytoplasmic RNA content in the course of oocyst formation in the intestinal rabbit coccidia *Eimeria intestinalis* Cheissin, 1948.Acta Protozool., Warsaw, 4: 75-80.

Britton, W. M., C. H. Hill and C. W. Barber.1964. A mechanism of interaction between dietary protein levels and coccidiosis in chicks. J. Nutr. 82: 306-310.

Burns, W. C. 1959. The lethal effect of *Eimeria tenella* extracts on rabbits. J. Parasitol. 45: 38-46.

Canning, E. U. and M. Anwar. 1968. Studies on meiotic division in coccidial and malarial parasites. J. Protozool. 15: 290-298.

Capella, J. A. and H. E. Kaufman. 1964. Enzyme histochemistry of *Toxoplasma gondii.* Amer. J. Trop. Med. Hyg. 13: 664-666.

Chobotar, B. and D. M. Hammond. 1971. Cinemicrographic observations on excystation in *Eimeria utahensis* and *E. papillata.* J. Protozool. 18 (Suppl.): 12.

Clarke, M. L. 1962. A mixture of diaveridine and sulphaquinoxaline as a coccidiostat for poultry. I. Preliminary studies on efficiency against *Eimeria tenella* and *E. necatrix* infections, and on toxicity in poultry. Vet. Rec. 74: 845-848.

Coles, B., J. Biely and B. E. March. 1970. Vitamin A deficiency and *Eimeria acervulina* infection in the chick. Poult. Sci. 49: 1295-1301.

Collier, H. B. and W. E. Swales. 1948. On the chemotherapy of caecal coccidiosis (*Eimeria tenella*) of chickens. VI. A note on the metabolism of caecal epithelium, normal and parasitized. Can. J. Res. Ser.D. 26: 77-81.

Cross, J. B. 1947. A cytologic study of *Toxoplasma* with special reference to its effect on the host's cell. J. Infect.Dis. 80: 278-296.

Dasgupta, B. 1960. Lipids in different stages of the life cycles of malaria parasites and some other sporozoa. Parasitology 50: 501-508.

Dasgupta, B. 1961. Alkaline phosphatase reaction in the different stages of the life cycles of certain sporozoa. Věst.Čsl.Zool.Spol. 25: 16-21.

Dasgupta, B. and C. Kulasiri. 1959. Some cytochemical observations on *Toxoplasma gondii.* Parasitology 49: 594-600.

Daugherty, J. W. and C. A. Herrick. 1952. Cecal coccidiosis and carbohydrate metabolism in chickens. J. Parasitol. 38: 298-304.

Davies, A. W. 1952. Lowered vitamin A reserves in avian coccidiosis. Nature, London, 170: 849.

Davis, L. R. and G. W. Bowman. 1963. Diagnosis of coccidiosis of cattle and sheep by histochemical and other techniques. Proc. 67th. Ann. Meet. U.S. Livestock Sanit. Ass., New Mexico, 516-522.

Doran, D. J. 1966. Pancreatic enzymes initiating excystation of *Eimeria acervulina* sporozoites. Proc.Helminthol.Soc.Wash. 33: 42-43.

Doran, D. J. and M. M. Farr. 1962. Excystation of the poultry coccidium, *Eimeria acervulina*. J. Protozool. 9: 154-161.

Dumas, N. 1970. Influence de divers facteurs sur la respiration des toxoplasmes étudiée par l'oxygraphe G.M.E. Bull. Soc. Pathol. Exot. 63: 215-227.

Duncan, S. 1959. The effects of some chemical and physical agents on the oocysts of the pigeon coccidium, *Eimeria labbeana* (Pinto, 1928). J. Parasitol. 45: 193-197.

Dürr, U and L. Pellérdy. 1969. Zum Sauerstoffverbrauch der Kokzidienoocysten während der Sporulation. Acta Vet.Hung. 19: 307-310.

Edgar, S. A., C. A. Herrick and L. A. Fraser. 1944. Glycogen in the life cycle of the coccidium, *Eimeria tenella*. Trans. Amer. Microsc. Soc. 63: 199-202.

Farr, M. M. and D. J. Doran. 1962. Comparative excystation of four species of poultry coccidia. J.Protozool. 9: 403-407.

Fayer, R. 1969. Refractile body changes in sporozoites of poultry coccidia in cell culture. Proc.Helminthol.Soc.Wash. 36: 224-231.

Fayer, R., R. D. Romanowski and J. M. Vetterling. 1970. The influence of hyaluronidase and hyaluronidase substrates on penetration of cultured cells by eimerian sporozoites. J.Protozool. 17: 432-436.

Frandsen, J. C. 1968. *Eimeria stiedae*: Cytochemical identification of acid and alkaline phosphatases, carboxylic ester hydrolases, and succinate, lactate, and glucose-6-phosphate dehydrogenases in endogenous stages from rabbit tissues. Exp.Parasitol. 23. 398-411.

Frandsen, J. C. 1970. *Eimeria stiedae*: Cytochemical identification of enzymes and lipids in sporozoites and endogenous stages. Exp.Parasitol. 27: 100-115.

Freeman, B. M. 1970. Carbohydrate stores in chickens infected with *Eimeria tenella*. Parasitology 61: 245-251.

Frenkel, J. K. and C. H. Hutchings. 1957. Relative reversal by vitamins (p-aminobenzoic, folic and folinic acids) of the effects of sulfadiazine and pyrimethamine on toxoplasma, mouse and man. Antibiot.Chemother. 7: 630-638.

Fulton, J. D. 1965. Toxic exudates in *Toxoplasma gondii* infections. Exp.Parasitol. 17: 252-260.

Fulton, J. D. and D. F. Spooner. 1957. Preliminary observations on the metabolism of *Toxoplasma gondii*. Trans. Roy. Soc. Trop. Med. Hyg. 51: 123-124.

Fulton, J. D. and D. F. Spooner. 1960. Metabolic studies on *Toxoplasma gondii*. Exp. Parasitol. 9: 293-301.

Gangi, D. P. and R. D. Manwell. 1961. Some aspects of the cytochemical anatomy of *Toxoplasma gondii*. J.Parasitol. 47: 291-296.

Gerriets, E. 1961. The prophylactic action of vitamin A in caecal coccidiosis by protection of the epithelium. Brit. Vet.J. 117: 507-515.

Gerriets, E. 1966. Die coccidiostaticafreie Massenaufzucht von Junghennen unter Vitamin A-Schutz. Berl.Münch.Tierärztl. Wschr. 79: 271-275.

Gill, B. S. and H. N. Ray. 1954a. Glycogen and its probable significance in *Eimeria tenella* Railliet and Lucet, 1891. Ind.J.Vet.Sci. 24: 223-228.

Gill, B. S. and H. N. Ray. 1954b. On the occurrence of mucopolysaccharides in *Eimeria tenella* Railliet and Lucet, 1891. Ind. J. Vet. Sci. 24: 229-237.

Gill, B. S. and H. N. Ray. 1954c. Phosphatases and their significance in *Eimeria tenella* Railliet and Lucet, 1891. Ind.J.Vet.Sci. 24: 239-244.

Goodrich, H. P. 1944. Coccidian oocysts. Parasitology 36: 72-79.

Grassé, P. P., ed. 1953. Traité de Zoologie, Vol. 1, Pt. 2. Masson, Paris.

Hammond, D. M., B. Chobotar and J. V. Ernst. 1968. Cytological observations on sporozoites of *Eimeria bovis* and *E. auburnensis*, and an *Eimeria* species from the Ord kangaroo rat. J.Parasitol. 54: 550-558.

Hammond, D. M., J. V. Ernst and B. Chobotar. 1970. Composition and function of the substiedal body in the sporocysts of *Eimeria utahensis*. J.Parasitol. 56: 618-619.

Harms, R. H. and R. L. Tugwell. 1956. The effect of experimentally induced prolonged blood clotting time on cecal coccidiosis of chicks. Poult.Sci. 35: 937-938.

Harisch, G., A. Dey-Hazra, K. Enigk, and J. Schole. 1971. Glutathionquotient und Konzentration einiger Metabolite des Kohlenhydratstoffwechsels in der Leber von

Hühnerküken während einer *Eimeria necatrix*-Infektion. Zbl. Vet.-Med., B, 18:211-220.
Heller, G. and E. Scholtyseck. 1970. Histochemistry of coccidia as demonstrated by means of electron microscopy. J.Parasitol. 56 (No. 4, Sect. 11): 142.
Henry, D. P. 1932. The oocyst wall in the genus *Eimeria*. Univ.Calif.Publ.Zool. 37: 269-279.
Hibbert, L. E. and D. M. Hammond. 1968. Effects of temperature on *in vitro* excystation of various *Eimeria* species. Exp.Parasitol. 23: 161-170.
Hibbert, L. E., D. M. Hammond and J. R. Simmons. 1969. The effects of pH, buffers, bile and bile acids on excystation of sporozoites of various *Eimeria* species. J. Protozool. 16: 441-444.
Horst, C. J. G. van der and B. Kouwenhoven. 1970. Disturbed intestinal resorption and transport of vitamin A and carotenes during *E. acervulina* infection in the fowl. J.Parasitol. 56: 351.
Horton-Smith, C. and E. Boyland. 1946. Sulphonamides in the treatment of caecal coccidiosis of chickens. Brit.J.Pharmacol. 1: 139-152.
Horton-Smith, C. and P. L. Long. 1963. Coccidia and coccidiosis in the domestic fowl and turkey. In B. Dawes, ed., Advances in Parasitology. Vol. 1. Academic Press, New York, 67-107.
Ikeda, M. 1960. Factors necessary for *E. tenella* infection of the chicken. VI. Excystation of oocyst *in vitro*. Jap.J.Vet.Sci. 22: 39-41.
Itagaki, K. and M. Tsubokura. 1958. Studies on the infectious process of coccidium in fowl. V. Further investigations on the liberation of sporozoites. Jap.J.Vet.Sci. 20: 105-110.
Jackson, A. R. B. 1962. Excystation of *Eimeria arloingi* (Marotel, 1905): Stimuli from the host sheep. Nature, London, 194: 847-849.
Jackson, A. R. B. 1964. The isolation of viable coccidial sporozoites. Parasitology 54: 87-93.
Janssen, P. 1970. Zur Pathophysiologie der *Toxoplasma gondii*-Infektion. II. Der Kohlenhydratstoffwechsel bei experimentell infizierten Mäusen. Z. Parasitenk. 35: 97-104.
Joyner, L. P. 1963. Some metabolic relationships between host and parasite with particular reference to the *Eimeriae* of domestic poultry. Proc.Nutr.Soc. 22: 26-32.
Kendall, S. B. and L. P. Joyner. 1958. Potentiation of the coccidiostatic effects of sulphadimidine by five different folic acid antagonists. Vet.Rec. 70: 632-634.
Kheysin, E. M. (Cheissin, E. M.). 1958. Cytologische Untersuchungen verschiedener Stadien des Lebenszyklus der Kaninchencoccidien. I. *Eimeria intestinalis* E. Cheissin 1948. Arch Protistenk. 102: 265-290.
Kheysin, E. M. (Cheissin, E. M.). 1959. Cytochemical investigations of different stages of the life cycle of coccidia of the rabbit. Proc. XVth.Int.Cong.Zool, London 1958, 713-716.
Kheysin, E. M., J. M. Rozanov, and B.N. Kudriavtsev. 1968. Fluorescence microscopic study on the content of nucleic acids in merozoites and microgametes of *Eimeria intestinalis* from the intestine of rabbit. Acta Protozool., Warsaw, 5: 389-393.
Kheysin, E. M. (Cheyssin, E.). 1935. Structure de l'oocyste et perméabilité de ses membranes chez les coccidies du lapin. Ann. Parasitol. 13: 133-164.
Klimeš, B. 1963. Ein neues Kokzidiostaticum 6-Azauracil. Berl.Münch.Tierärztl.Wschr. 76: 298-299.
Kouwenhoven, B. and C. J. G. van der Horst. 1969. Strongly acid intestinal content and lowered protein, carotene and vitamin A blood levels in *Eimeria acervulina* infected chickens. Z. Parasitenk. 32: 347-353.
Kouwenhoven, B. and C. J. G. van der Horst. 1970. Significance and possible cause of the lowered intestinal pH during *Eimeria acervulina* infection in the fowl. J.Parasitol. 56: 191-192.
Kulasiri, C. and B. Dasgupta. 1959. A cytochemical investigation of the Sabin-Feldman phenomenon in *Toxoplasma gondii* and an explanation of its mechanism on this basis. Parasitology 49: 586-593.
Landers, E. J. 1960. Studies on excystation of coccidial oocysts. J.Parasitol. 46: 195-200.
Levine, L. and C. A. Herrick. 1954. The effects of the protozoan parasite *Eimeria tenella* on the ability of the chicken to do muscular work when its muscles are stimulated directly and indirectly. J.Parasitol. 40: 525-531.
Lillie, R. D. 1947. Reactions of various parasitic organisms in tissues to the Bauer, Feulgen, Gram and Gram-Weigert methods. J. Lab. Clin. Med. 32: 76-88.
Little, P. L. 1967. The effect of certain vitamins and their analogues on coccidia and coccidiosis of chickens. Diss.Abstr. 27B: 3363.
Long, P. L. and D. G. Rootes. 1959. Cytochemical studies of *Eimeria* in the fowl. Trans.Roy.Soc.Trop.Med.Hyg. 53: 308-309.
Lunde, M. N. and L. Jacobs. 1964. Properties of toxoplasma lysates toxic to rabbits on intravenous injection. J.Parasitol. 50: 49-51.

Lund, E., H-A. Hansson, E. Lycke, and P. Sourander. 1966. Enzymatic activities of *Toxoplasma gondii.* Acta Pathol.Microbiol.Scand. 68: 59-67.

Lutz-Ostertag, Y. and J. Sénaud. 1966. Action de l'extrait de toxoplasme (Toxoplasmine) sur le développement embryonnaire du poulet. Arch.Anat.Microsc.Morphol.Exp. 55: 363-386.

Lux, R. E. 1954. The chemotherapy of *Eimeria tenella.* I. Diaminopyrimidines and dihydrotriazines. Antibiot.Chemother. 4: 971-977.

Lycke, E., E. Lund, and Ö. Strannegård. 1965. Enhancement by lysozyme and hyaluronidase of the penetration by *Toxoplasma gondii* into cultured host cells. Brit.J.Exp.Pathol. 46: 189-199.

Lycke, E. and R. Norrby. 1966. Demonstration of a factor of *Toxoplasma gondii* enhancing the penetration of toxoplasma parasites into cultured host cells. Brit.J.Exp.Pathol. 47: 248-256.

McManus, E. C., M. T. Oberdick, and A. C. Cuckler. 1967. Response of six strains of *Eimeria brunetti* to two antagonists of para-aminobenzoic acid. J.Protozool. 14: 379-381.

Mazia, D., P. A. Brewer, and M. Alfert. 1953. The cytochemical staining and measurement of protein with mercuric bromphenol blue. Biol.Bull. 104: 57-67.

Monné, L. and G. Hönig. 1954. On the properties of the shells of the coccidian oocysts. Ark.Zool., Stockholm, 7: 251-256.

Nath, V. and G. P. Dutta. 1962. Cytochemistry of protozoa, with particular reference to the Golgi apparatus and the mitochondria. Int.Rev.Cytol. 13: 323-355.

Niebrój, T. K. and Z. Wojdała. 1967a. Studies on *Toxoplasma gondii.* IV. Dehydrogenases. Acta Parasitol.Polon. 15: 51-55.

Niebrój. T. K. and Z. Wojdała. 1967b. Studies on *Toxoplasma gondii.* V. Tetrasolium reductases. Acta Parasitol.Polon. 15: 57-59.

Nozik, R. A. and G. R. O'Connor. 1969. The so-called toxin of *Toxoplasma.* Amer. J. Trop. Med. Hyg. 18: 511-515.

Nyberg, P. A., D. H. Bauer, and S. E. Knapp. 1968. Carbon dioxide as the initial stimulus for excystation of *Eimeria tenella* oocysts. J.Protozool. 15: 144-148.

Nyberg, P. A. and D. M. Hammond. 1964. Excystation of *Eimeria bovis* and other species of bovine coccidia. J.Protozool. 11: 474-480.

Nyberg, P. A. and S. E. Knapp. 1970a. Scanning electron microscopy of *Eimeria tenella* oocysts. Proc.Helminthol.Soc.Wash. 37: 29-32.

Nyberg, P. A. and S. E. Knapp. 1970b. Effect of sodium hypochlorite on the oocyst wall of *Eimeria tenella* as shown by electron microscopy. Proc.Helminthol.Soc.Wash. 37: 32-36.

Ouellette, C. A., R. G. Strout, and L. R. McDougald. 1972. Incorporation of pyrimidine nucleosides into *Eimeridia tenella* cultured in vitro. Abstr. 47th Ann. Meet. Soc. Parasitol. 26.

Pattillo, W. H. and E. R. Becker. 1955. Cytochemistry of *Eimeria brunetti* and *E. acervulina* of the chicken. J.Morphol. 96: 61-95.

Perrotto, J., D. B. Keister, and A. H. Gelderman. 1971. Incorporation of precursors into *Toxoplasma* DNA. J. Protozool. 18: 470-473.

Polin, D., E. R. Wynosky, and C. C. Porter. 1963. *In vivo* adsorption of amprolium and its competition with thiamine. Proc. Soc. Exp. Biol. Med. 114: 273-277.

Prasad, H. 1963. The role of some vitamin B deficient diets in coccidiosis of the domestic fowl. Ind.Vet.J. 40: 478-489.

Pratt, I. 1940. Effect of *E. tenella* (coccidia) upon blood sugar of chickens. Trans. Am. Microsc. Soc. 69: 31-37.

Pratt, I. 1941. Effect of *Eimeria tenella* (coccidia) upon the glycogen stores of the chicken. Amer.J.Hyg.Sect.C, 34: 54-61.

Preston-Mafham, R. A. and A. H. Sykes. 1967. Changes in permeability of the mucosa during intestinal coccidiosis infections in the fowl. Experientia 23: 972-973.

Ray, H. N. and B. S. Gill. 1954. Preliminary observations on alkaline phosphatase in experimental *Eimeria tenella* infection in chicks. Ann.Trop.Med.Parasitol. 48: 8-10.

Remington, J. S., M. M. Bloomfield, E. Russell, and W. S. Robinson. 1970. The RNA of *Toxoplasma gondii.* Proc. Soc. Exp. Biol. Med. 133: 623-626.

Rikimaru, M. T., F. T. Galysh, and R. F. Shumard. 1961. Some pharmacological aspects of a toxic substance from oocysts of the coccidium *Eimeria tenella.* J.Parasitol. 47: 407-412.

Roberts, W. L., Y. Y. Elsner, A. Shigematsu, and D. M. Hammond. 1970. Lack of incorporation of H^3-thymidine into *Eimeria callospermophili* in cell cultures. J.Parasitol. 56: 833-834.

Roberts, W. L. and D. M. Hammond. 1970. Ultrastructural and cytologic studies of the sporozoites of four *Eimeria* species. J.Protozool. 17: 76-86.

Roberts, W. L., C. A. Speer, and D. M. Hammond. 1970. Electron and light microscope studies of the oocyst walls, sporocysts, and excysting sporozoites of *Eimeria callospermophili* and *E. larimerensis*. J.Parasitol. 56: 918-926.

Rogers, E. F. 1962. Thiamine antagonists. Ann.N.Y.Acad.Sci. 98: 412-429.

Rogers, E. F., R. L. Clark, H. J. Becker, A. A. Pessolano, W. J. Leanza, E. C. McManus, F. J. Andriuli, and A. C. Cuckler. 1964. Antiparasitic drugs. V. Anticoccidial activity of 4-amino-2-ethoxybenzioc acid and related compounds. Proc.Soc.Exp.Biol.Med. 117: 488-492.

Ryley, J. F. 1968. Chick embryo infections for the evaluation of anticoccidial drugs. Parasitology 58: 215-220.

Ryley, J. F. 1969. Ultrastructural studies on the sporozoite of *Eimeria tenella*. Parasitology 59: 67-72.

Ryley, J. F., M. Bentley., D. J. Manners, and J. R. Stark. 1969. Amylopectin, the storage polysaccharide of the coccidia *Eimeria brunetti* and *E. tenella*. J.Parasitol. 55: 839-845.

Ryley, J. F. and R. G. Wilson. 1972. Growth factor antagonism studies with coccidia in tissue culture. Z.Parasitenk. 40: 31-34.

Schildt, C. S. and C. A. Herrick. 1955. The effect of cecal coccidiosis on the motility of the digestive tract of the domestic fowl. J.Parasitol. 41 (Suppl.): 18-19.

Schole, J., A. Dey-Hazra, G. Harisch, and K. Enigk. 1972. Zur Pathogenität der Coccidien des Huhnes. Z.Parasitenk. 38: 3-13.

Scholtyseck, E. 1964. Elektronenmikroskopisch-cytochemischer Nachweis von Glykogen bei *Eimeria perforans*. Z. Zellforsch. 64: 688-707.

Scholtyseck, E. and D. M. Hammond. 1970. Electron microscope studies of macrogametes and fertilization in *Eimeria bovis*. Z. Parasitenk. 34: 310-318.

Scholtyseck, E., A. Rommel, and G. Heller. 1969. Licht-und elektronenmikroskopische Untersuchungen zur Bildung der Oocystenhülle bei Eimerien (*Eimeria perforans*, *E. stiedae* und *E. tenella*). Z. Parasitenk. 31: 289-298.

Scholtyseck, E. and W.-H. Voigt. 1964. Die Bildung der Oocystenhülle bei *Eimeria perforans* (Sporozoa). Z.Zellforsch. 62: 279-292.

Scholtyseck, E. and N. Weissenfels. 1956. Elektronenmikroskopische Untersuchungen von Sporozoen. I. Die Oocystenmembran des Hühnercoccids *Eimeria tenella*. Arch. Protistenk. 101: 215-222.

Schrével, J. 1970. Recherches ultrastructurales et cytochimiques sur le paraglycogène, réserve glucidique des grégarines et coccidies. J.Microsc. 9: 593-610.

Schulte, E. 1971. Cytochemische Untersuchungen an den Feinstrukturen von *Klossia helicina* (Coccidia, Adeleidea). II. Nachweis dreier Phosphatasen in *Klossia helicina*. Z. Parasitenk. 35: 188-204.

Seligman, A.M., M. J. Karnovsky, H. L. Wasserkrug, and J. S. Hanker. 1968. Nondroplet ultrastructural demonstration of cytochrome oxidase activity with a polymerizing osmiophilic reagent, diaminobenzidine (DAB). J.Cell.Biol. 38: 1-14.

Sénaud, J., R. Vendrely, and P. Tronche. 1968a. Sur la nature de la substance toxique des kystes de Sarcosporidies du mouton (Toxoplasmea), active sur le lapin. C. R. Acad. Sci. Paris, 266: Ser. D, 1137-1138.

Sénaud, J., R. Vendrely, and P. Tronche. 1968b. Obtention, à partir d'un extrait de kystes de *Sarcocystis tenella* (Rail.), d'une fraction protéique toxique pour le lapin. C.R.Soc.Biol. 162: 1184-1189.

Sharma, N. N. and J. W. Foster. 1964. Toxic substance in various constituents of *Eimeria tenella* oocysts. Am.J.Vet.Res. 25: 211-215.

Slater, R. L., D. M. Hammond, and M. L. Miner. 1970. *Eimeria bovis*: development in calves treated with thiamine metabolic antagonist (amprolium) in feed. Trans.Am.Microsc.Soc. 89: 55-65.

Smith, B. F. and C. A. Herrick. 1944. The respiration of the protozoan parasite, *Eimeria tenella*. J.Parasitol. 30: 295-302.

Smith, B. F. and W. H. McShan. 1949. The effect of the protozoan parasite, *Eimeria stiedae*, on the succinic dehydrogenase activity of liver tissue of rabbits. Ann.N.Y.Acad.Sci. 52: 496-500.

Speer, C. A., D. M. Hammond, and G. L. Kelley. 1970. Stimulation of motility in merozoites of five *Eimeria* species by bile salts. J.Parasitol. 56: 927-929.

Stephens, J. F. 1965. Some physiological effects of coccidiosis caused by *Eimeria necatrix* in the chicken. J.Parasitol. 51: 331-335.

Stephens, J. F., L. M. Kowalski, and W. J. Borst. 1967. Some physiological effects of coccidiosis caused by *Eimeria maxima* in young chickens. J.Parasitol. 53: 176-179.

Stoll, U., K. Enigk, and A. Dey-Hazra. 1970. Der Einfluss einer Coccidieninfektion auf Kohlenhydrate und Ribonucleotide der Leber bei Hühnerküken. Z.Parasitenk. 34: 356-361.

Strout, R. G., H. Botero, S. C. Smith, and W. R. Dunlop. 1963. The lipids of coccidial oocysts. J.Parasitol. 49 (Suppl.): 20.

Tsunoda, K. and O. Itikawa. 1958. Histochemical studies on chicken coccidia (*Eimeria tenella*). II. Action of sulfamerazine, nitrophenide and furacin on *E. tenella*. Bull.Nat.Inst.Animal Health, Tokyo, 34: 181-192.

Turk, D. E. and J. F. Stephens. 1966. Effect of intestinal damage produced by *Eimeria necatrix* infection in chicks upon absorption of orally administered zinc-65. J.Nutr. 88: 261-266.

Turk, D. E. and J. F. Stephens. 1967a. *Eimeria necatrix* infections and oleic acid absorption in broilers. Poult.Sci. 46: 775-777.

Turk, D. E. and J. F. Stephens. 1967b. Gastrointestinal tract disease and zinc absorption: cecal coccidiosis. Poult.Sci. 46: 939-943.

Turk, D. E. and J. F. Stephens. 1967c. Upper intestinal tract infection produced by *E. acervulina* and absorption of zinc-65 and ^{131}I-labelled oleic acid. J.Nutr. 93: 161-165.

Turk, D. E. and J. F. Stephens. 1969. Coccidial infections of the ileum, colon, and ceca of the chick and nutrient absorption. Poult.Sci. 48: 586-589.

Turk, D. E. and J. F. Stephens. 1970. Effects of serial inoculations with *Eimeria acervulina* or *Eimeria necatrix* upon zinc and oleic acid absorption in chicks. Poult.Sci. 49: 523-526.

Varela, G., A. Vázquez, and J. Torroella. 1956. Probable existencia de la dietilamida del ácido D-lisérgico en la infección por *Toxoplasma gondii*. Rev.Inst.Salub.Enferm.Trop. 16: 29-32. (Trop.Dis.Bull. 54, 1359, 1957.)

Vetterling, J. M. 1968. Oxygen consumption of coccidial sporozoites during *in vitro* excystation. J.Protozool. 15: 520-522.

Vetterling, J. M. and D. J. Doran. 1969. Storage polysaccharide in coccidial sporozoites after excystation and penetration of cells. J.Protozool. 16: 772-775.

Wagenbach, G. E. 1969. Purification of *Eimeria tenella* sporozoites with glass bead columns. J.Parasitol. 55: 833-838.

Wagenbach, G. E. and W. C. Burns. 1969. Structure and respiration of sporulating *Eimeria stiedae* and *E. tenella* oocysts. J.Protozool. 16: 257-263.

Wagner, W. H. and O. Foerster. 1964. Die PAS-AO-Methode, eine Spezialfärbung für Coccidien im Gewebe. Z.Parasitenk. 25: 28-48.

Warren, E. W. 1968. Vitamin requirements of the *Coccidia* of the chicken. Parasitology 58: 137-148.

Warren, E. W. and S. J. Ball. 1967. Anticoccidial activity of egg white and its counteraction by biotin. Vet.Rec. 80: 578-579.

Weinman, D. 1952. *Toxoplasma* and toxoplasmosis. Ann.Rev.Microbiol. 6, 281-298.

Weinman, D. and H. J. Klatchko. 1950. Description of toxin in toxoplasmosis. Yale J.Biol.Med. 22: 323-326.

Wilson, P. A. G. and D. Fairbairn. 1961. Biochemistry of sporulation in oocysts of *Eimeria acervulina*. J. Protozool. 8: 410-416.

Woodworth, H. C. and D. Weinman. 1960. Studies on the toxin of toxoplasma (toxotoxin). J.Infect.Dis. 107: 318-324.

6 Cultivation of Coccidia in Avian Embryos and Cell Culture

D. J. DORAN
National Animal Parasite Laboratory, Veterinary Sciences Research Division, Agricultural Research Service, Beltsville, Maryland

---

Contents

I. Introduction . . . *185*
II. Cultivation in Avian Embryos . . . *186*
 A. Development and Pathological Effects . . . *186*
  1. *Toxoplasma* . . . *186*
  2. *Besnoitia* . . . *186*
  3. *Eimeria* . . . *186*
 B. Factors Influencing Development and Severity of Pathological Effects . . . *189*
  1. Strain, Size, and Age of the Inoculum . . . *189*
  2. Temperature of Incubation . . . *190*
  3. Age, Strain, and Sex of the Embryo . . . *190*
III. Cultivation in Cell Culture . . . *191*
 A. Behavior of Organisms Prior to Development . . . *191*
  1. Extracellular Movement . . . *191*
  2. Mode of Entering and Leaving Cells . . . *191*
   a. Active Penetration . . . *191*
   b. Passive Ingestion (Phagocytosis) . . . *196*
  3. Time and Rate of Entry . . . *196*
  4. Intracellular Movement and Location . . . *197*

5. Refractile Bodies in *Eimeria* Sporozoites . . . *198*

B. Development . . . *200*

1. *Toxoplasma* . . . *200*
   - a. Cells Supporting Development . . . *200*
   - b. Mode and Rate of Multiplication . . . *200*
   - c. Aggregates and Cytopathological Effects . . . *205*
   - d. Release of Toxoplasmas, Reinvasion of Cells, and "Histopathological" Effects . . . *206*
   - e. Cysts . . . *207*
2. *Besnoitia* . . . *207*
   - a. Cells Supporting Development . . . *207*
   - b. Mode and Rate of Multiplication . . . *208*
   - c. Aggregates and Cytopathological Effects . . . *208*
   - d. Release of Organisms, Reinvasion of Cells, and "Histopathological" Effects . . . *209*
3. *Eimeria* and *Isospora* . . . *209*
   - a. Cells Used and Extent of Development . . . *209*
   - b. Asexual Development . . . *215*
     - i. The First and Second Generations . . . *215*
     - ii. The Third and Other Generations . . . *221*
     - iii. "Activating" Factors . . . *222*
   - c. Sexual Development . . . *222*
     - i. Gametocytes . . . *222*
     - ii. Oocysts . . . *223*
   - d. Survival in Culture . . . *223*
   - e. Cyto- and "Histopathological" Effects . . . *225*
   - f. Cell Culture and Host Comparisons . . . *226*
4. *Sarcocystis* . . . *227*
   - a. Cells Used in Attempts to Obtain Development . . . *227*
   - b. Nuclear Changes and Stages Found in Culture . . . *227*

C. Factors Other Than Strain Differences Influencing the Number of Parasites that Penetrate Cells and Develop . . . *228*

1. Parasitological Factors . . . *228*
   - a. Size of the Inoculum . . . *228*
   - b. Age of the Parasite and Time When Obtained from the Host . . . *228*
   - c. Treatment of Parasites Prior to Inoculation . . . *229*
2. Environmental Factors . . . *230*
   - a. Culture Chamber . . . *230*
   - b. Temperature . . . *230*
   - c. Debris (Oocyst and Sporocyst Walls) and Toxins . . . *231*
   - d. Media and Media Changes . . . *231*
3. Cellular Factors . . . *232*
   - a. Cell Type and Passage Number . . . *232*
   - b. Age and Number of Cells . . . *234*

IV. Embryo and Cell Culture Applications . . . *234*

A. Isolating Organisms . . . *234*

B. Maintaining Organisms . . . *235*

C. Preparing Antigens . . . *236*

D. Determining the Effect or Mode of Action of Antimicrobial Agents and Other Substances . . . *237*
1. Adenine and Other Purines, Pyrimethamine, and Sulfonamides . . . *237*
2. Aminopterin and Vitamins . . . *238*
3. Antibiotics . . . *238*
4. "Coccidiostats" . . . *239*
5. Quinine Compounds . . . *240*
E. Studying Factors Pertaining to Immunity and Resistance . . . *240*
F. Other Applications . . . *241*
Literature Cited . . . *242*

I. Introduction

Many coccidia are of medical and veterinary importance. The main reason for studying them is to learn more about their behavior in hopes of subsequently discovering a method for controlling or eradicating the disease they cause. Because coccidia are intracellular parasites and are hidden from view during most of their life cycle, they are difficult to study. Avian embryos and cell culture have become useful tools in coccidian research. Although the parasite is still hidden from view, embryos are valuable in that they offer an environment at the tissue level. In cell culture, the parasite is visible and the immediate environment can be altered. Host-parasite relationships can be observed at the cellular level, and behavior of the parasite can be studied under a variety of conditions.

This chapter reviews work on cultivation of species belonging to the genera *Eimeria, Isospora, Toxoplasma, Besnoitia,* and *Sarcocystis.* Species in the genera *Eimeria, Toxoplasma,* and *Besnoitia* have been cultivated in avian embryos and cell culture, whereas *Isospora* and *Sarcocystis* have been grown in cell culture only.

Cultivation of *Toxoplasma gondii* in both avian embryo and cell culture was first reported by Levaditi *et al.* (1929). Until 1953, there were only a few scattered reports of work with embryonated eggs (Barkley and Edwards, 1952; Buttitta, 1951; Crema and Ambrek, 1952; Warren and Russ, 1948; Weinman, 1944; Wolf *et al.*, 1940; Wolfson, 1941, 1942). With cell culture there was even less activity. Cook and Jacobs (1958a) summarized the highlights up to 1956. Their list includes six reports up to 1953 (Bland, 1934; Guimarães and Meyer, 1942; Levaditi *et al.*, 1929; Meyer and de Oliveira, 1942, 1943; Sabin and Olitsky, 1937). Beginning in 1953, research involving *T. gondii* in avian embryos and in cell culture, primarily the latter, increased greatly. This chapter includes data obtained from 192 publications of original research. Most of the cultivation activity has been in Brazil (Meyer and co-workers), Czechoslovakia (Jíra, Schuhová and co-workers), Japan (Matsubayashi, Shimizu, and co-workers), Germany (Bommer and co-workers), Sweden (Lund, Lycke, and co-workers), USSR (Akinshina and co-workers), and USA at the National Institutes of Health (Jacobs, Sheffield, and co-workers). Galuszka (1962a) and Zuckerman (1966) have briefly reviewed some of the work concerning *T. gondii* in avian embryos. Work in cell culture has been reviewed by Akinshina (1963b,c),

Cook and Jacobs (1958a), Galuszka (1962a), Jacobs (1953, 1956), Taylor and Baker (1968), Trager and Krassner (1967), and Zuckerman (1966).

An eimerian species (*E. tenella*) was first cultivated in the embryonated chicken egg by Long (1965). In the same year, Patton (1965) reported development of this species in cell culture. In 1970, Turner and Box first reported development of an isosporan (*I. lacazei* and/or *chloridis*) in culture. Patton's report has been considered the first regarding development of an eimerian species. However, as will be mentioned later, it is possible that Gavrilov and Cowez (1941) obtained development of the rabbit coccidium, *E. perforans*. Since 1965, cultivation work with *Eimeria* and *Isospora*, primarily the former, has been most intense. Data were obtained from 82 publications dealing with these two genera. Although cultivation work has primarily involved species from livestock (cattle and sheep) and poultry, species from ground squirrels, rabbits, rats, cats, dogs, and sparrows have also been studied. All work with species from livestock and most of it with species from rodents has been carried out by Dr. Hammond and his students at Utah State University. Most work with species from poultry has been done in Czechoslovakia (Bedrnik), USSR (Shibalova and coworkers), England (Long and Ryley), and in the USA at the University of New Hampshire (Strout and co-workers) and the National Animal Parasite Laboratory (Doran, Fayer, and co-workers). Hammond and Fayer (1968), Taylor and Baker (1968), and Trager and Krassner (1967) have reviewed earlier work with *Eimeria* in cell culture.

Frenkel (1953) first reported that a besnoitian species (*B. jellisoni*) could be cultivated in embryonated chicken eggs. Bigalke (1962) first grew a besnoitian (*B. besnoiti*) in cell culture. Recently, Fayer (1970) reported cultivation of *Sarcocystis* sp. in cell culture.

II. Cultivation in Avian Embryos

A. Development and Pathological Effects

1. Toxoplasma

There are no reported failures to obtain infection in embryonated eggs. The parasite is not always demonstrable after one or more early passages, but by repeated subinoculation it can easily be demonstrated (Buttitta, 1951; Jacobs and Melton, 1954). Toxoplasmas (trophozoites, proliferative forms) can be inoculated intravenously, into the yolk sac, or into the embryonic membranes or cavities. No comparative study has been made of all of these routes, but Abbas (1967b) compared the yolk sac and allantoic cavity routes and found little difference in intensity of the resultant infections.

Macroscopic yellow-white lesions (abscess-like elevations or islands, pocks, nodules, colonies, plaques) appear on the chorio-allantoic membrane (Anwar and Oureshi, 1954; Jögiste and Ukhov, 1969; Lund *et al.*, 1963b; Weinman, 1944; Wolfson, 1941, 1942) and amniotic membrane (Jögiste and Ukhov, 1969; MacFarlane and Ruckman, 1948). These lesions, which can usually be found one week after inoculation or earlier (Jacobs, 1953), range in diameter from 0.5 mm (MacFarlane and Ruckman, 1948) to 20 mm (Anwar and Oureshi, 1954). They are frequently visible through the unbroken shell on transillumination (Warren and Russ, 1948). The lesion, or the area around it, is necrotic (Anwar and Oureshi, 1954; Jögiste and Ukhov, 1969; Kuwahara, 1959; Warren and Russ, 1948) or thickened (MacFarlane and Ruckman, 1948) and contains numerous toxoplasmas (Jögiste and Ukhov, 1969; MacFarlane and Ruckman, 1948; Warren and Russ, 1948; Wolfson, 1941). According to Anwar and Oureshi (1954), the lesion is surrounded by a red or black line, and the number of toxoplasmas outside of the line is proportional to the size of the lesion. Jögiste and Ukhov (1969) reported (a) extracellular colonies of the parasite on the chorioallantoic membrane (CAM), (b) transformation of proliferating mesenchymal cells into macrophages, and (c) the cystic form of the parasite within the macrophages.

In addition to the CAM and amniotic membrane, toxoplasmas and cysts can be found in tissues within the embryo. The parasite is disseminated through the blood stream (Jacobs, 1953; Jögiste and Ukhov, 1969; Kuwahara, 1956; Weinman, 1944) and can reproduce within blood cells (Galuzo *et al.*, 1969; Novinskaya, 1965; Vasina, 1958; Zassuchin, 1963). Wolfson (1941) found toxoplasmas in erythrocytes, monocytes, and plasma of embryonic duck blood, but not in other tissues of the embryo. In descending order of prevalence, Anwar and Oureshi (1954) found (a) large lesions containing toxoplasmas in the CAM, kidney, bone marrow, and brain and (b) "micro-lesions" in the cerebellum, heart, liver, spleen, and intestine. Cysts can also be found in a variety of tissues (Abbas, 1967a; Galuzo *et al.*, 1969; Jögiste and Ukhov, 1969). Galuzo *et al.* (1969) found them in blood on the 11th day after inoculation. The latter workers also reported "cysts with budding daughter cysts as well as little cysts around one large cyst."

2. *Besnoitia*

There are only a few reports concerning infection of embryonated eggs with besnoitian parasites. The pathology is similar to that produced by *T. gondii*. Frenkel (1953) inoculated chick embryos with *B. jellisoni* organisms obtained from cysts found in white-footed (deer) mice. Acute infections developed that were fatal in from 4 to 12 days. Frenkel later (1965) stated that *B. jellisoni* produces generalized, usually fatal infections in chick embryos. Bigalke (1962) inoculated the proliferative and cyst forms of *B. besnoitia* into the chorioallantoic cavity and yolk sac of chick embryos. Circumscribed, yellow-white foci measuring from 0.5 mm to 4 mm in diameter were found on the CAM and yolk sac membrane. These lesions consisted of necrotic material with cellular infiltration and numerous intra- and extracellular parasites in the surrounding area.

3. *Eimeria*

Results of attempts to obtain complete endogenous development of *Eimeria* when either sporozoites or second-generation merozoites were inoculated into the allantoic cavity are summarized in Table 1. Several other workers (Jeffers and Wagenbach, 1969, 1970; Long, 1970a,d) obtained infection with *E. tenella* in eggs, but their experiments were terminated before the cycle could be completed. There were no failures with *E. tenella* and no successes with *E. acervulina* and *E. maxima*. As will be pointed out later, these results are similar to those obtained in cell culture. Long (1966) did not obtain development of either *E. stiedai* or *E. praecox* and observed only late schizonts of *E. necatrix*. However, Fitzgerald (1970) obtained gametocytes of *E. stiedai* and Shibalova (1970) obtained oocysts with both *E. praecox* and *E. necatrix*. If it were not for the fact that Ryley (1968) failed to obtain development of *E. brunetti* using the same strain with which Long (1966) obtained oocysts, the differences with *E. stiedai* and *E. praecox*, and perhaps even *E. necatrix*, might be attributed to strain differences.

Attempts were also made to obtain development of *E. tenella* using different inocula, routes of inoculation, and different species of embryonated eggs. Intact oocysts fail to produce infection (Long, 1966; Shibalova, 1970) and sporocysts mechanically released from oocysts produce light infections (Long, 1966). Some sporozoites are released from

TABLE 1. Development of *Eimeria* in embryonated chicken eggs after inoculation of sporozoites or second-generation merozoites into the allantoic cavity[a]

Inoculum	*E. acervulina*	*E. brunetti*	*E. maxima*	*E. mitis*	*E. mivati*	*E. necatrix*	*E. praecox*	*E. stiedai*	*E. tenella*	Reference
Sporozoites									oo[b]	Long (1965)
Sporozoites	nd				oo[b]					Long and Tanielian (1965)
Sporozoites	nd	oo[b]	nd		oo[b]	sch	nd	nd	oo[b]	Long (1966)
Sporozoites		nd							gam	Ryley (1968)
Sporozoites									oo	Baldelli *et al.* (1968)
Sporozoites									oo	Shibalova (1968)
Sporozoites	nd		nd						oo	Shibalova *et al.* (1969)
Sporozoites		oo[b]							oo[b]	Shibalova (1969a)
Sporozoites	nd	oo[b]		oo[b]	nd	oo[b]	oo[b]		oo[b]	Shibalova (1970)
Sporozoites								gam		Fitzgerald (1970)
Merozoites						nd			oo	Long (1966)
Merozoites									oo	Shibalova (1968)
Merozoites		oo[b]							oo[b]	Shibalova (1969a)

[a]Abbreviations: gam = gametocytes; oo = oocysts; sch = schizonts; nd = no development.
[b]Sporulated and produced infection in the natural host.

sporocysts while the latter are being released from oocysts (Doran and Farr, 1962, 1965). The infections with a sporocyst inoculum obtained by Long were probably caused by free sporozoites rather than by sporozoites excysting from sporocysts after inoculation into the egg. No infection is obtained when sporozoites are inoculated into the amniotic cavity of chick embryos or into the allantoic cavity of embryonated quail and turkey eggs (Long, 1965). In his initial work, Long (1965) did not obtain development after inoculating chick embryos by the intravenous route; Long (1971a) later reported that developmental stages were present in the liver and CAM. However, he thought presence in the CAM was caused by a leak during the inoculation procedure. Long (1971a) also inoculated sporozoites intravenously into chick embryos and into the allantoic cavity of goose embryos that had been treated with 0.01 mg of dexamethasone prior to inoculation. Gamogony occurred more freely in chick embryos given immunosuppressants than in those untreated. In goose embryos, second-generation schizonts only were found in treated eggs; in untreated, there was no development. As Long mentioned, these results coupled with those of McLoughlin (1969) and Long (1970e) suggest that host- and site-specificity may be determined by immune mechanisms in the host.

In *E. tenella* infections, focal lesions are produced on the membrane where the schizonts and sexual stages develop (Long, 1970a). Many of these lesions, which measure up to 0.5 mm in diameter, are clearly visible macroscopically (Fig. 11). Lesions have not been reported for the other species that develop, but there is no reason to believe that they do not occur. Long (1966) observed that *E. necatrix* schizonts develop in groups well below the epithelium and appear to be surrounded by a membranous layer. He also found merozoites of *E. brunetti, E. mivati*, and *E. tenella*, but not *E. necatrix*, in the allantoic cavity. There is a greater variation in size of *E. tenella* schizonts in the CAM than in the cecum of chicks, and many second-generation merozoites are not released from the CAM (Long, 1966). Long suggests that the presence of trypsin and/or bile, which are present in the intestine of the host and absent in the allantoic cavity, is needed in order to stimulate a greater release of merozoites from schizonts. In *E. tenella* infections, hemorrhaging occurs (Baldelli *et al.*, 1968; Jeffers and Wagenbach, 1970; Long, 1966; Ryley, 1968) and, if the dosage is large enough, death of the embryo results (Jeffers and Wagenbach, 1970; Long, 1970c; Ryley, 1968). Jeffers and Wagenbach (1970) found that infection did not affect egg weight or the weight of the hatched chick. They did find that the mortality rate following the feeding of *E. tenella* oocysts is much higher among chicks hatched from eggs that are inoculated with sporozoites than it is among chicks hatched from noninoculated eggs. These workers believe that their observations suggest that the egg inoculation induced partial immunological tolerance to antigens associated with the parasite.

B. Factors Influencing Development and Severity of Pathological Effects

1. *Strain, Size, and Age of the Inoculum*

Jacobs and Melton (1954) inoculated seven-day-old embryos with 100, 1,000, 10,000, or 100,000 toxoplasmas of strain 113-CE. All embryos died, but the survival time was longer with the lower dosages. The survival time with 100 trophozoites was about 14 days, whereas with 100,000 it was only eight days. Size of lesions and the time of appearance depend on size of the inoculum (Weinman, 1944; Wolfson, 1942). Jõgiste and Ukhov (1969) claimed that the highly virulent RH strain produces changes on the CAM similar to those produced by a strain of low virulence, but no necrotic foci. Their work, however, was carried out using 1,500 of the RH strain and 65,000 of an avirulent strain.

Dosages of 11,000 to 56,000 *Eimeria necatrix* sporozoites and 17,000 to 70,000 *E. mivati* sporozoites produce good infections in embryos (Long, 1966). Dosages of 17,000 to 85,000 *E. tenella* give good infections, but dosages larger than 85,000 produce early deaths in which the parasites cannot be found (Long, 1966). Ryley (1968), also working with *E. tenella*, found that 10,000 sporozoites kill 30% of the embryos and 100,000 kill all embryos on the 5th day. Long (1970c) compared hemorrhagic mortality using different dosages of the "W" (Weybridge) and "H" (Houghton) strains of *E. tenella*. With dosages of 5,000, 15,000, and 45,000 sporozoites of the virulent "W" strain, the mortality rate was 70%, 92%, and 100%, respectively; with similar dosages of the less virulent "H" strain, the rate is only 29%, 68%, and 92%. With both strains, mortalities caused by a dosage of 45,000 are first found at 89 hr, but with the other dosages they are not found until from 92 to 95 hr. Long (1970c) also reported first finding mature second-generation schizonts of the "W" strain on the CAM at 92 hr, whereas those of the "H" strain did not appear until 95 hr. He suggests that this earlier time of appearance might explain, at least in part, the greater virulence of the strain. Jeffers and Wagenbach (1970) also found that embryonic death during the hemorrhagic phase of the infection (4th to 8th days) was highly dependent on dosage level. Mortality occurs earlier and at a slightly higher rate among embryonated eggs inoculated with large doses of sporozoites. Although mortality during the hatching period (18th through 21st days) exhibits a lesser degree of dose dependency than does hemorrhagic mortality, inoculations even at a low level have an adverse effect on the hatchability of embryos. Long (1970d) inoculated embryos with similar numbers of *E. tenella* sporozoites obtained from oocysts 15, 65, 122, and 162 days old. The percentages of mortality were 83.3, 16.6, 5.5, and 7.6, respectively. The mean numbers of foci per CAM were 140.1, 15.4, 3.2, and 3.5, respectively.

2. Temperature of Incubation

Multiplication of *T. gondii* occurs equally well when eggs are incubated at 35°C as at 37.5°C (Warren and Russ, 1958).

Development of *E. tenella* is more rapid at the temperature of the chicken (41°C) than at slightly lower temperatures (Table 2). There is also greater and more rapid mortality at 41°C (Long, 1970c).

3. Age, Strain, and Sex of the Embryo

The RH strain of *T. gondii* multiplies equally as well in eggs from six to 12 days old (MacFarlane and Ruckman, 1948; Warren and Russ, 1948), but deaths in the younger embryos are spread over a longer period, i.e. from four to eight days (MacFarlane and Ruckman, 1948). Calcified plaques in the CAM are only occasionally seen in seven-day-old embryos, but are found more often in 12-day-old embryos (Jacobs, 1956). Jõgiste and Ukhov (1969) obtained some interesting results when comparing infection of the RH strain in embryonated eggs 5, 8, 11, and 14 days old. If inoculated before the 14th day, embryos die from seven to nine days later with focal lesions in various organs. However, if inoculated on the 14th day, infection takes a milder course. There are no focal lesions, but an intense proliferation of polyblasts which "actively phagocytize toxoplasmas."

Mortality that is caused by *E. tenella* occurs earlier and in greater amounts in embryos nine days old than in those 11 days old (Long, 1970c). Long (1970c) compared *E. tenella* infections in three strains of embryos ("S," "C," and "K") that had been selectively bred for susceptibility or resistance to Marek's disease. The "S" strain, which was bred for susceptibility, is markedly more susceptible to infection than the other two strains. However, three-week-old chicks of the "S" strain are less susceptible than either "C" or "K" embryos. Long suggests that the difference in susceptibility of chicks and embryos of the "S" strain is due to host resistance factors that develop within the first few weeks after hatching. Using nine different mating types and different dosage levels, Jeffers and Wagenbach (1969) found that mortality during the hemorrhagic period in infections with *E. tenella* is significantly greater in female than in male embryos.

TABLE 2. Earliest time (in days) at which developmental stages of *E. tenella* are found in embryonated chicken eggs incubated at different temperatures following inoculation with sporozoites

Temperature (°C)	Schizonts	Gametocytes or oocysts	Reference
38	4.0	7.0	Long (1965)
	4.0	7.0	Ryley (1968)
	4.2 (second generation)		Long (1970c)
39	4.0 (first generation)	7.0	Long (1966)
	6.0 (second generation)		Long (1966)
41	3.0	5.0	Shibalova (1970)
	2.3 (first generation)		Long (1970c)
	3.4 (second generation)		Long (1970c)

III. Cultivation in Cell Culture

A. BEHAVIOR OF ORGANISMS PRIOR TO DEVELOPMENT

1. Extracellular Movement

Movement of *Toxoplasma gondii* within the medium has been variously described. Guimarães and Meyer (1942), who first demonstrated an extracellular motile stage in cell culture, described movement as twisting, oscillating, and undulating. Since then, it has been described as circular, gliding, and spiral (Bommer, Hofling, and Heunert, 1969), gliding (Jacobs, 1953), rotating (Hirai *et al.*, 1966), rotating and drilling (Schmidt-Hoensdorf and Holz, 1953), and rotating and boring (Bommer, Heunert, and Milthaler, 1969). Toxoplasmas have also been said to frequently turn over (Schmidt-Hoensdorf and Holz, 1953), turn somersaults (Hirai *et al.*, 1966), perform slow, jumping movements (Hansson and Sourander, 1965), move with porpoise-like undulations (Pulvertaft *et al.*, 1954), and swim like a seal (Bommer, Heunert, and Milthaler, 1969). The principal movements are in the direction of the long axis (Bommer, Heunert, and Milthaler, 1969; Hirai *et al.*, 1966; Lund *et al.*, 1961b) with most of the activity at the anterior end (Jacobs, 1953; Mühlpfordt, 1952; Pulvertaft *et al.*, 1954). Movement is most vigorous immediately following release from a cell (Bommer, Hofling, and Heunert, 1969; Jacobs, 1953; Sourander *et al.*, 1960), intermittent (Mühlpfordt, 1952) with progression in any one direction up to 200 μ (Pulvertaft *et al.*, 1954), with periods of rest becoming progressively longer until, after 15 min (Pulvertaft *et al.*, 1954) or three hours (Hirai *et al.*, 1966), movement stops. Speeds of 400 times the body length/sec (Lund *et al.*, 1961b) and from one to two body lengths/sec (Bommer, Heunert, and Milthaler, 1969) have been reported.

Descriptions of movements by *Eimeria* are less diverse. Sporozoites glide (Clark and Hammond, 1969), flex (Sampson *et al.*, 1971), glide and flex (Fayer and Hammond, 1967; Speer and Hammond, 1969, 1970b; Speer, Hammond, and Anderson, 1970), glide, flex, and pivot (Speer *et al.*, 1971), glide and undergo helical movements (Roberts *et al.*, 1971), and flex. They extend their anterior ends to form stylet-like protuberances which are later retracted (Kelley and Hammond, 1970; Sampson *et al.*, 1971), and they probe (Fayer and Hammond, 1967; Speer, Hammond, and Anderson, 1970) or make intermittent contact (Sampson *et al.*, 1971) with cells. They undulate when attached to cells by their posterior ends (Doran and Vetterling, 1967a) and pivot when attached by their posterior ends (Fayer and Hammond, 1967) or anterior ends (Sampson *et al.*, 1971).

Besnoitia jellisoni organisms obtained from cysts flex to the extent of becoming U-shaped, pivot on their posterior ends, glide for short distances, and extend their anterior ends to form stylet-like protuberances (Fayer *et al.*, 1969). *Sarcocystis* organisms obtained from cysts also flex, glide, and pivot (Fayer, 1970).

2. Mode of Entering and Leaving Cells

a. Active penetration

Toxoplasmas have been observed in the process of actively penetrating cells (Bommer, 1969; Bommer, Hofling, and Heunert, 1969; Bommer, Heunert, and Milthaler, 1969; Guimarães and Meyer, 1942; Hirai *et al.*, 1966; Jacobs, 1953; Pulvertaft *et al.*, 1954; Schmidt-Hoensdorf and Holz, 1953) and leaving cells shortly after entering (Bommer, Heunert, and Milthaler, 1969; Hirai *et al.*, 1966). The process of leaving is the same as entering (Bommer, Heunert, and Milthaler, 1969), and the host cell is not harmed (Hirai *et al.*, 1966). However, active entry has been reported to kill cells (Pulvertaft *et al.*, 1954);

a point that could not be supported by later work (Cook and Jacobs, 1958a). The more detailed descriptions of entry into cells are given by Bommer (1969), Bommer, Heunert, and Milthaler (1969), Hirai *et al.* (1966), and Pulvertaft *et al.* (1954). According to Pulvertaft *et al.* (1954), if a host cell lies within 50 μ of a parasite it is approached in one movement; if further away, in a series of bounds. They also mentioned that the parasite (a) selects a point on the "circumference" of the cell for invasion, applies itself to the surface, and then slides around until it reaches the selected point and (b) always enters a cell from below at the selected point which is that area of the cytoplasm underlying the "concavity near the nucleus." Hirai *et al.* (1966) observed that free and attached parasites behave differently. Those attached never penetrate through the cell membrane in the immediate area of attachment, but move to another part of the cell or even to another cell before penetrating. Those moving freely in the medium show no one particular movement. They suddenly "take their course" and penetrate directly into cells. The entry process is shown in Figures 1 through 4. The anterior pole of the parasite is extended and a small hole is made in the cell membrane (Bommer, 1969; Bommer, Heunert, and Milthaler, 1970). This hole, through which the parasite enters, is 1.5 μ wide (Hirai *et al.*, 1966) and appears within a few seconds after initial contact (Bommer, Heunert, and Milthaler, 1969; Hirai *et al.*, 1966). The parasite constricts and becomes gourd-shaped (Hirai *et al.*, 1966) and undergoes amoeboid deformation (Bommer, 1969) while pulling its way through the hole (Bommer, Heunert, and Milthaler, 1969). Propelling (Schmidt-Hoensdorf and Holz, 1953) and boring, drilling (Holz and Albrecht, 1953) motions have been observed while the parasite is entering. Complete entry has been reported to take from 15 to 30 sec (Bommer, Heunert, and Milthaler, 1969), an average of 40 sec (Hirai *et al.*, 1966), and from 30 to 60 min (Schmidt-Hoensdorf and Holz, 1953). The extreme length of the latter interval may be caused by movement on the cell surface after initial contact. Most workers believe that the parasite enters a cell anterior end first. However, Hirai *et al.* (1966) are not sure. They state that a toxoplasma generally penetrates with the obtuse end, but also mention that differentiation between the acute and obtuse ends cannot be made with any degree of accuracy.

Eimeria sporozoites have also been observed entering and leaving cells (Clark and Hammond, 1969; Fayer and Hammond, 1967; Hammond and Fayer, 1967; Kelley and Hammond, 1970; Roberts, Hammond and Speer, 1970; Sampson *et al.*, 1971; Speer and Hammond, 1970b; Speer, Hammond, and Anderson, 1970; Speer *et al.*, 1971). The process of leaving is similar to that of entering (Figs. 5-10) and both appear similar to

Figs. 1-4. Penetration of the host cell membrane by *Toxoplasma gondii*. Phase-contrast, cinematographic study. × 1200. (From Bommer, Heunert, and Milthaler, 1969.) **1.** The anterior part of the parasite is pushed forward penetrating the cell membrane. **2, 3.** The toxoplasma is pulling its body through the narrow opening. **4.** The toxoplasma is completely within a cell.

Figs. 5-10. *Eimeria larimerensis* sporozoite penetrating and then leaving a sixth passage embryonic bovine kidney cell immediately after inoculation; site of entrance or exit indicated by arrow. Living specimens, with phase-contrast microscopy. × 700. (From Speer and Hammond, 1970b.) **5.** Sporozoite penetrating cell, showing constriction in the posterior third of sporozoite body at site of entrance. **6.** Constriction in the posterior fifth of sporozoite body. **7.** Intracellular sporozoite immediately after completion of penetration. **8.** Early stage in exit from the host cell, with constriction in the anterior quarter of sporozoite body at the site of exit. **9.** Constriction in the posterior third of sporozoite body. **10.** Extracellular sporozoite immediately after leaving host cell.

Fig. 11. Chorioallantoic membrane (fresh preparation) six days after infection showing numerous focal lesions associated with *E. tenella* schizogony. Note the frequency of lesions along the walls of blood vessels. Approximate magnification, × 5. (From Long, 1970a.)

Fig. 12. *Toxoplasma* arranged in rosettes in human amnion cells 48 hr after inoculation (Csóka and Kulcsár, original.)

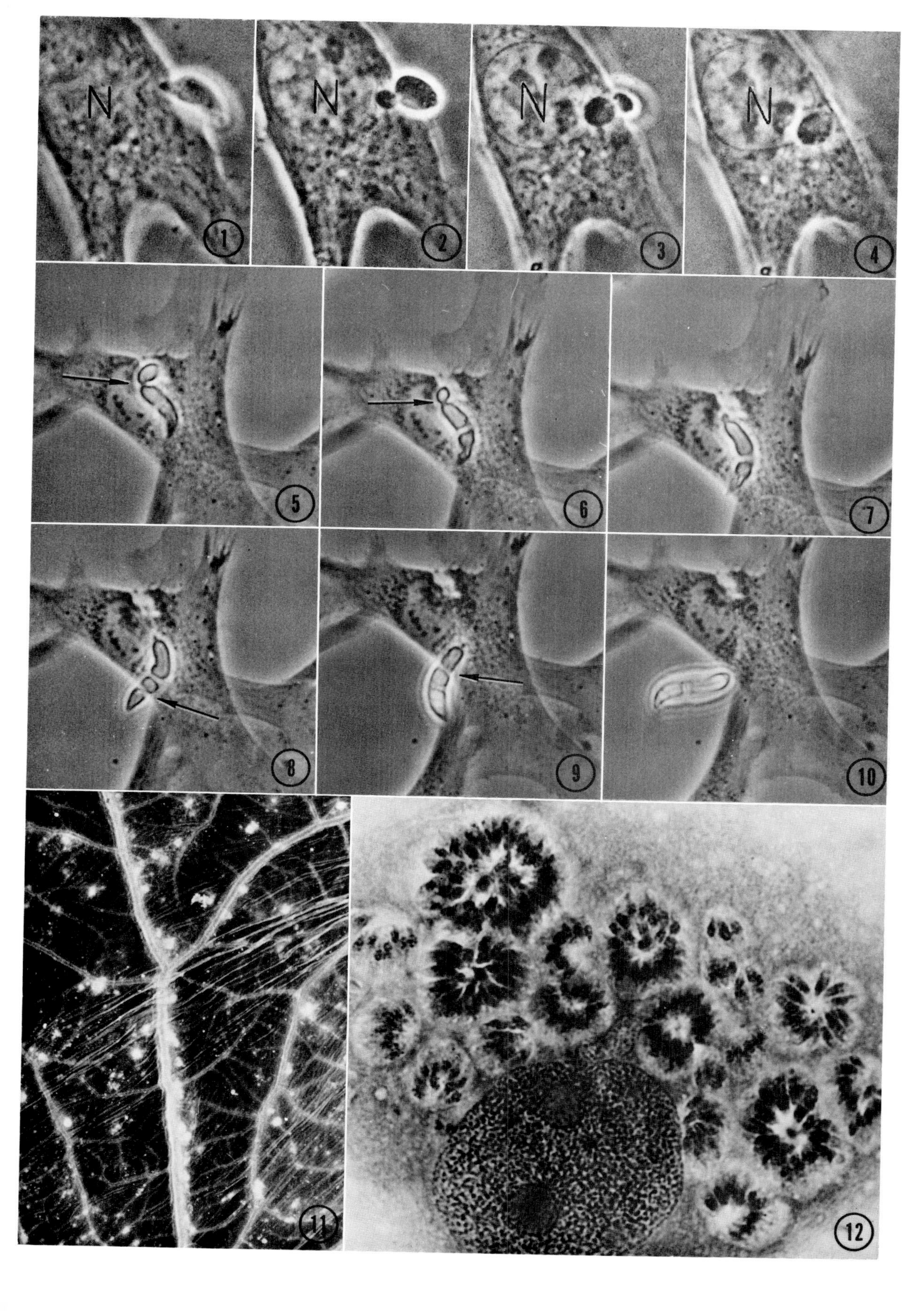
N
N
N
N
1
2
3
4
5
6
7
8
9
10
11
12

the entry process described for *T. gondii*. However, *E. bovis* sometimes leaves a cell in a backward direction (Hammond and Fayer, 1967). According to Roberts *et al.* (1971), a sporozoite first thrusts a slender anterior protuberance into a host cell to a depth of from 1 μ to 2 μ. Occasionally, this protuberance is immediately retracted and inserted at different locations within the immediate vicinity of the original insertion. Usually, the protuberance moves laterally for from 1 to 2 sec within the cell cytoplasm, then appears to

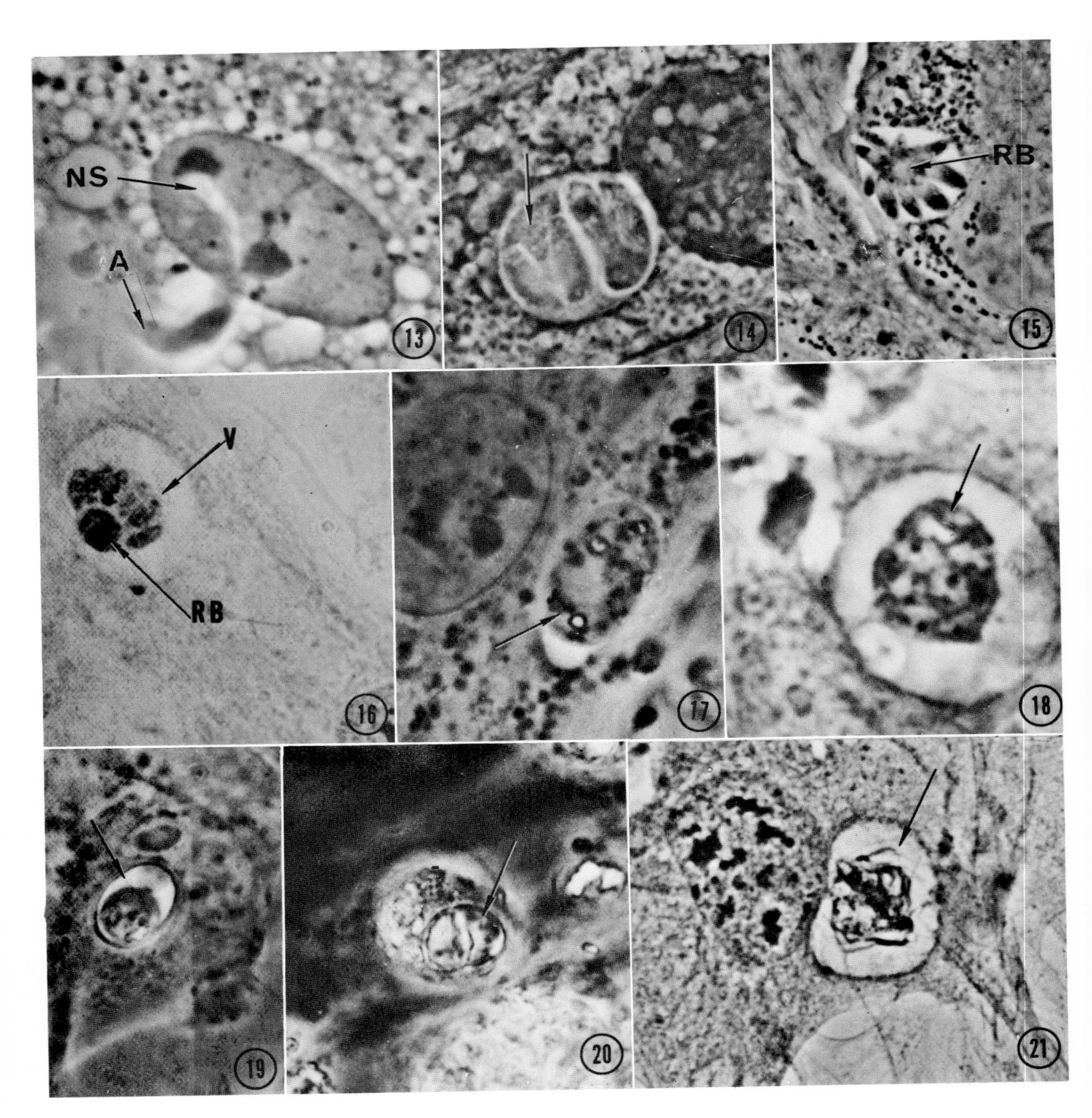

swell, increase in width, and decrease in length. Within three seconds, the remainder of the sporozoite enters the cell. Other time intervals up to one minute have also been reported for complete entry into a cell (Clark and Hammond, 1969; Fayer and Hammond, 1967; Kelley and Hammond, 1970; Speer and Hammond, 1970b; Speer, Hammond, and Anderson, 1970). *Eimeria alabamensis* sporozoites, which normally enter nuclei of host cells, take from 20 to 30 min to enter cells in culture (Sampson *et al.*, 1971). Sporozoites may leave cells either immediately following entry (Clark and Hammond, 1969; Kelley and Hammond, 1970; Roberts, Hammond, and Speer, 1970; Speer and Hammond, 1970b) or after as long as from five to seven days from the time of inoculation (Clark and Hammond, 1969; Hammond and Fayer, 1967). Escape of host cell cytoplasm is frequently seen after a sporozoite leaves a cell, but rarely after it enters (Speer *et al.*, 1971). The act of leaving a cell, which may or may not occur in the host, could either be associated with the artificial environmental conditions or the type of host cell in culture or both (Fayer and Hammond, 1967).

There has been considerable interest since 1965 by workers in Sweden in whether *T. gondii* secretes an enzyme that assists the organism in entering cells. Their work using cell cultures has mainly dealt with (a) the penetration promoting effects produced by the addition of lysozyme, hyaluronidase, and lysed parasites (Lycke, Lund, and Stranegärd, 1965) (b) extraction of a penetration enhancing factor (PEF) from lysed parasites and the effect it produces in cell culture (Lycke and Norrby, 1966; Lycke *et al.*, 1968; Norrby and Lycke, 1967), (c) demonstration of lysosomes in the parasite before and after inoculation into culture (Norrby, 1970b), and (d) fractionation, purification, and localization of PEF (Norrby, 1970a). Results of this series of studies indicate that *T. gondii* contains a substance that is proteinaceous, associated with lysosomes at the anterior end of the parasite, and, when collected and tested after lysis of the organisms, enhances penetration of cells by other organisms. These findings constitute strong circumstantial evidence that a substance is secreted by the intact parasite that aids in penetration.

There is only one report concerning enzymes and penetration by *Eimeria*. Hyaluronidase (1 and 10 mg/ml) does not increase the number of intracellular *E. adenoeides* sporozoites and both chondroitin sulfate (1 and 10 mg/ml) and hyaluronic acid (1 mg/ml) did not reduce the number that were intracellular when these agents were included in the

Fig. 13. *Besnoitia jellisoni* organism leaving a cell nucleus. NS, area in nucleus where nucleoplasm has been displaced; A, anterior tip of organism. Embryonic bovine tracheal cells, 5 days. × 2500. (From Fayer *et al.*, 1969.)

Fig. 14. A pair of *B. jellisoni* organisms in a single vacuole, each organism containing two daughters. Embryonic bovine spleen cells, 4 days. × 2500. (From Fayer *et al.*, 1969.)

Fig. 15. A rosette-shaped aggregate with eight *B. jellisoni* organisms and a round body (RB). Embryonic bovine spleen cells, 4 days. × 1200. (From Fayer *et al.*, 1969.)

Fig. 16. A banana-shaped aggregate with six *B. jellisoni* organisms and a round body (RB) within a vacuole (V) in an embryonic bovine spleen cell, 3 days. × 1000. (From Fayer *et al.*, 1969.)

Fig. 17. Ovoid stage of *Sarcocystis* sp. Embryonic bovine tracheal cell, 24 hr. × 2800. (From Fayer, 1970; copyright 1970 by the American Association for the Advancement of Science.)

Fig. 18. Multinucleate stage of *Sarcocystis* sp. Embryonic chicken kidney cell, 48 hr. × 2800. (From Fayer, 1970; copyright 1970 by the American Association for the Advancement of Science.)

Fig. 19. Cyst-like body of *Sarcocystis* sp. containing sphere of granular cytoplasm. Embryonic bovine tracheal cells, 48 hr. × 1500. (From Fayer, 1970; copyright 1970 by the American Association for the Advancement of Science.)

Fig. 20. Cyst-like body of *Sarcocystis* sp. containing two elongate forms. Embryonic bovine tracheal cells, 48 hr. × 1500. (From Fayer, 1970; copyright 1970 by the American Association for the Advancement of Science.)

Fig. 21. Microgametocyte of *Sarcocystis* sp. with microgametes. Embryonic bovine tracheal cells, 30 hr. × 1500. (From Fayer, 1972; copyright 1972 by the American Association for the Advancement of Science.)

medium in which sporozoites were suspended before inoculation into cell cultures, and the cultures were fixed one hour after inoculation (Fayer *et al.*, 1970). With *T. gondii*, hyaluronidase has a penetration-promoting effect and hyaluronic acid inhibits the effect of hyaluronidase (Lycke, Lund, and Strannegärd, 1965).

b. Passive ingestion (phagocytosis)

Pulvertaft *et al.* (1954) reported that they observed phagocytosis of *T. gondii* by murine macrophages. It has also been reported (Holz and Albrecht, 1953) that macrophages show no tendency to flow around the parasite and take it up, but move in the opposite direction from the parasite. In a study of infected macrophages coated with formvar, Vischer and Suter (1954) concluded that phagocytosis had taken place. However, their conclusion was based on observations of fixed preparations only. Parasites pretreated with formalin have been found intracellular (Lycke, Lund, and Strannegärd, 1965; Norrby and Lycke, 1967). This suggests that phagocytosis occurs. However, Hirai *et al.* (1966) never observed this phenomenon in a cinematographic study when using formalin-killed parasites.

Although phagocytosis of *Eimeria* has not been observed, it has been suggested (Doran and Vetterling, 1967a; Strout *et al.*, 1965). Doran and Vetterling (1967a) found 28 sporozoites within one cell, whereas cells nearby were devoid of the parasite. Since they thought it unlikely that such a large number of parasites within a cell was fortuitous, they attributed it to phagocytosis. Other explanations may be host cell selection by the organism and random entry of the closest cell by clumped organisms.

3. *Time and Rate of Entry*

Toxoplasmas are usually intracellular within one hour after inoculation. Occasionally, they enter as early as 15 min (Akinshina and Doby, 1969b; Hansson and Sourander, 1965; Hirai *et al.*, 1966; Lund *et al.*, 1961b; Sourander *et al.*, 1960). The rate of entry, which is most frequently expressed as percentage of infected cells, increases in proportion to the time of incubation from three to eight hours (Cook and Jacobs, 1958a; Kaufman *et al.*, 1958; Lycke and Lund, 1964a) before decreasing. No increase after 20 min (Matsubayashi and Akao, 1963) and an increase up to 24 hr (Arai *et al.*, 1958) have also been reported. The decrease in penetration rate is thought to be caused by loss of viability of parasites which remain extracellular and a decrease in the total number in the supernatant fluid (Kaufman *et al.*, 1958). Three investigations utilizing similar numbers of organisms have shown that the more virulent strains are more invasive: (a) at 20 min, 40% of cultured cells are infected with the highly virulent RH strain, whereas only 28% are infected with the less virulent Beverley strain after five hours (Matsubayashi and Akao, 1963); (b) at three hours, five times as many cells are infected with the RH strain than are infected with the avirulent 113-CE strain (Kaufman *et al.*, 1958); and (c) an inoculum containing 20,000/ml of the less virulent MF, JQ, and 113-CE strains are necessary to produce the same number of infected cells at 24 hrs as about 5,000/ml of the RH and HI strains (Hogan *et al.*, 1961).

Eimeria sporozoites are also found intracellularly within one hour after inoculation. Many *E. alabamensis* enter within five minutes (Sampson *et al.*, 1971). The time of entry may depend on the cell type. *Eimeria bovis* were found in two cell types after three minutes; in others, only after 30 min (Fayer and Hammond, 1967). Entry rates have not been quantitatively determined. However, it has been mentioned that (a) the number of intracellular sporozoites increases up to three hours, remains constant up to six hours, and then decreases (Strout *et al.*, 1965), (b) all that enter do so by 2.5 to 3 hr (Doran and Vetterling, 1967b), (c) entry continues for 24 hr (Strout and Ouellette, 1968, 1970; Strout *et al.*, 1969b), and (d) the maximum number is reached in different cell types between 1 to 8 hr (*E. callospermophili*) and from 24 to 48 hr (*E. bilamellata*) (Speer, Hammond, and Anderson, 1970). Although there has been no comparative study between species on entry into a single cell type, the latter findings concerning the two species from ground squirrels suggest that species of *Eimeria*, like strains of *T. gondii*, differ in their entry capabilities.

Besnoitia jellisoni (proliferative forms obtained from exudate) are found intracellularly within from 20 to 30 min after inoculation (Doby and Akinshina, 1968; Akinshina and Doby, 1969b). It takes 45 min for all to penetrate, providing the exudate is used within one hour after it is obtained (Doby and Akinshina, 1968).

4. *Intracellular Movement and Location*

Toxoplasmas pass through the cytoplasm and become situated close to the host cell nucleus (Akinshina and Doby, 1969b; Bommer, 1969; Hansson and Sourander, 1965; Hirai *et al.*, 1966; Lund *et al.*, 1961b; Sourander *et al.*, 1960). Hirai *et al.* (1966) observed that the parasite stops immediately after passing through the cell membrane and that, when movement is resumed, it takes from 10 to 90 sec from the time of entry, not always by the shortest route, to reach the nucleus.

Eimeria sporozoites also lie in close proximity to the cell nucleus (Doran and Vetterling, 1967a; Kelley and Hammond, 1970; Roberts *et al.*, 1971; Sampson *et al.*, 1971; Speer and Hammond, 1970b; Strout and Ouellette, 1970). Some even indent the nuclear membrane (Doran and Vetterling, 1967a; Roberts *et al.*, 1971; Sampson *et al.*, 1971; Strout *et al.*, 1965). Naturally, when multiple infection of a cell occurs all are not next to the nucleus. Sporozoites are usually oriented in the same direction when from two to four are in a cell and each is less than one body length from another (Doran and Vetterling, 1967a,b). While moving to the cell nucleus, *E. alabamensis* and *E. larimerensis* sporozoites leave "trails" or pathways in the cytoplasm (Roberts *et al.*, 1971; Sampson *et al.*, 1971; Speer *et al.*, 1971). Intracellular sporozoites flex (Clark and Hammond, 1969; Kelley and Hammond, 1970; Speer and Hammond, 1970b), probe (Kelley and Hammond, 1970), and undergo lateral movements (Speer and Hammond, 1970b) or changes in direction (Roberts *et al.*, 1971). Roberts *et al.* (1971) observed that *E. larimerensis* sporozoites have from one to four constrictions similar in appearance to those during penetration. Each constriction appears to move along the body of the sporozoite as the sporozoite moves. These workers state that such structures may be associated with endoplasmic reticulum or other membranous structures in the cytoplasm. When movement ceases, sporozoites immediately become blunt or rounded at the anterior end and assume the form characteristic of intracellular forms (Clark and Hammond, 1969).

Besnoitia and *Sarcocystis* are also located close to the cell nucleus (Akinshina and Doby, 1969a,b; Fayer, 1970). From one to five *Sarcocystis* organisms are often found in close proximity to the nucleus, and before coming to rest they can move backwards and forwards for about 1 μm in each direction (Fayer, 1970).

Eimeria, Isospora, Toxoplasma, Besnoitia, and *Sarcocystis* organisms become situated within a cytoplasmic vacuole ("halo," parasitophorous vacuole, intracytoplasmic space). In the case of *T. gondii*, this vacuole is formed when the toxoplasma starts to increase in size (Hansson and Sourander, 1965;

Lund *et al.*, 1961b). Several investigators have reported that the vacuole is not always present (Akinshina, 1959; Mühlpfordt, 1952; Sourander *et al.*, 1960; Vischer and Suter, 1954). Doran and Vetterling (1967a) found a few *Eimeria* sporozoites not within vacuoles and suggested that the organisms might merely be lying on the cell surface. It is more likely that they, and perhaps the toxoplasmas not within a vacuole, had just entered and the vacuole had not formed at the time interval that cultures were fixed or examined.

Eimeria alabamensis normally develops within nuclei of host cells. However, in cell culture only one percent of those which are intracellular are also intranuclear (Sampson *et al.*, 1971). Other species have also been observed to pass through nuclei (Speer and Hammond, 1970b; Speer, Hammond, and Anderson, 1970; Speer *et al.*, 1971) or to be situated in fixed preparations in such a manner as to give the impression that they are intranuclear (Doran and Vetterling, 1967a). Roberts *et al.* (1971) described the process by which *E. larimerensis* sporozoites enter and leave cells. Their description indicates that passage through the nuclear membrane occurs in a manner similar to passage through the cell wall. Roberts *et al.* (1971) also mention that (a) sporozoites usually leave a nucleus immediately after entering, but some remain for as long as 30 min and occasionally move around within the nucleus, and (b) sporozoites leaving a nucleus usually leave the cell. Occasionally, *T. gondii* have been found that appeared to be intranuclear (Remington, Earle, and Yagura, 1970; Sourander *et al.*, 1960). However, rather than being attributed to active penetration, it was explained as being caused by a folding of the nucleus about the cytoplasmic vacuole containing the parasite (Sourander *et al.*, 1960) or to a "layering" of the parasite upon the nuclear membrane (Remington, Earle, and Yagura, 1970). *Besnoitia jellisoni* has also been found within nuclei (Akinshina and Doby, 1969a,b; Fayer *et al.*, 1969). Fig. 13 shows an organism that has just left a nucleus.

5. *Refractile Bodies in* Eimeria *Sporozoites*

In addition to the large refractile body (eosinophilic or clear globule) located posterior to the nucleus, freshly excysted sporozoites of some species of *Eimeria* contain several other smaller refractile bodies. The smaller bodies which are located anterior and/or posterior to the nucleus, gradually disappear after a sporozoite enters a cell (Table 3).

de Vos and Hammond (1971) found from two to ten spherical refractile bodies in *E. crandallis* sporozoites shortly after penetration. These randomly located bodies varied in size from approximately 0.5 μ to 2.5 μ, with an average of 2.0 μ. Most of the sporozoites in cultures examined 24 hr after inoculation had only one or two refractile bodies, including a posterior one averaging 2.5 μ in diameter; an anterior body, if present, averaged 1.5 μ. This reduction in number of refractile bodies occurred, in part, by a fusion of small bodies with larger ones (de Vos *et al.*, 1972).

Anterior refractile bodies of *E. bovis* become smaller and move posteriorly (Fayer and Hammond, 1969); those in *E. adenoeides, E. meleagrimitis*, and *E. tenella* merge with the larger posterior body (Fayer, 1969). Anterior bodies can also "disappear" without moving posteriorly by either releasing small amounts of refractile material into the cytoplasm or by forming several smaller bodies that decrease in size and eventually vanish (Fayer, 1969). Prior to and during the merger, which takes 15 min or less, finger-like projections appear randomly along the margin of either or both refractile bodies. These sometimes become detached and are found free in the cytoplasm. Small, spherical granules, similar in appearance to the anterior refractile body, are also observed on its surface either from three to eight hours

TABLE 3. Refractile body changes in *Eimeria* sporozoites

Species	Freshly excysted sporozoites (%)					Intracellular sporozoites (%)					Time of observation (hr)	Reference
	a	b	c	d	e	a	b	c	d	e		
E. adenoeides	58	35	2	5	0	43	34	4	20	0	1	Fayer (1969)
						35	6	6	53	0	3	
						28	6	6	60	0	8	
E. bilamellata	90	0	0	8	2	anterior usually absent					24	Speer, Hammond, and Anderson (1970)
E. bovis	89	0	3	8	0	42	0	40	18	0	1	Fayer and Hammond (1969)
						15	0	18	67	0	3	
						6	0	4	90	0	8	
E. callospermophili	100	0	0	0	0	most with only posterior					6-10	Speer, Hammond, and Anderson (1970)
	93	0	0	4	3	anterior not visible					after 10	Speer and Hammond (1970a)
E. larimerensis	82	0	0	18	0	most with only posterior					24	Speer and Hammond (1970b)
E. meleagrimitis	72	20	2	6	0	21	58	12	10	0	1	Fayer (1969)
						15	40	11	34	0	3	
						11	12	8	69	0	8	
E. ninakohlyakimovae	90	0	0	10	0	16	0	0	84	0	24	Kelley and Hammond (1970)

a = 2 r.b.'s, 1 anterior and 1 posterior to nucleus.
b = 2 or 3 r.b.'s anterior to nucleus, 1 posterior to nucleus.
c = 2 r.b.'s posterior to nucleus.
d = 1 r.b. posterior to nucleus.
e = 3 r.b.'s, 1 anterior and 2 posterior to nucleus.

after inoculation (Speer and Hammond, 1970a) or when the anterior body is decreasing in size (Clark and Hammond, 1969). In *E. alabamensis* at 24 hr, similar granules are found separated from the posterior refractile body; at 48 hr, granules appear at the anterior end and, in some instances, coalesce into one or more refractile bodies (Sampson *et al.*, 1971).

The function and significance of refractile bodies are unknown. Changes in location of the anterior refractile body and its usual disappearance during the first 24 hr in *E. bovis* in culture suggest that it is involved in the early stages of cellular existence (Fayer and Hammond, 1969); changes in the refractile bodies of *E. callospermophili* indicate that they may be used in the metabolism or growth of the parasite (Speer and Hammond, 1970a). Some sporozoites of *E. bovis* do not have an anterior refractile body; possibly such sporozoites do not enter cells or develop into schizonts (Fayer and Hammond, 1969). The large posterior refractile body becomes smaller as the schizont increases in size, but may persist until the schizont is mature or nearly so (Hammond and Fayer, 1968). Fayer and Hammond (1969) state that the large refractile body could be a source of reserve material which is gradually used during development of the schizont. Anterior and posterior refractile bodies occur in the first-generation merozoites of several *Eimeria* species (Hammond *et al.*, 1970).

B. DEVELOPMENT

1. *Toxoplasma*

a. Cells supporting development

Cells from eight different animal species grown in almost all types of culture vessels known have been used for cultivating *T. gondii* (Table 4). Probably because of the medical importance of the parasite, human cells were most frequently used. The parasite reproduces better in cell culture than any other intracellular protozoan species. Not one case was encountered where there was doubt that development occurred.

Cells that have been used to cultivate *T. gondii* are listed in Table 4 as they were reported with two exceptions: (a) reports were consolidated when it was obvious that several investigations were carried out with the same cell line; (b) the American Cell Type Collection was consulted to make information as complete as possible concerning several of the cell lines that were reported only by number and/or the isolator's name. Although most frequently not stated, it was evident from the techniques that most of the cell types not obviously cell lines were primary cultures.

All of the work listed was carried out with toxoplasmas (trophozoites, proliferative forms) obtained from either peritoneal exudate of infected animals (usually mice) or from embryonated eggs or other cell cultures used for maintaining the parasite. In addition, Sheffield and Melton (1970a,b) inoculated monkey kidney cells with sporozoites excysted from *Isospora*-like oocysts obtained from the droppings of cats fed mouse brain containing *T. gondii* cysts. The sporozoites developed in a manner similar to that occurring when toxoplasmas are introduced into culture.

b. Mode and rate of multiplication

Toxoplasmas multiply within from two to five hours (Bommer, 1969; Bommer, Hofling, and Heunert, 1969; Lund *et al.*, 1961b; Sourander *et al.*, 1960) or from eight to ten hours (Burnstein and Bickford, 1968; Kaufman and Maloney, 1962) after entering a cell. Before multiplication commences, toxoplasmas become ovoid (Sheffield and Melton, 1968) or round (Schmidt-Hoensdorf and Holz, 1953) and increase in size (Hansson and Sourander, 1965; Schmidt-Hoensdorf and Holz, 1953; Sourander *et al.*, 1960). Recent time-lapse photography studies (Bommer, 1969; Bommer, Hofling, and Heunert, 1969; Hansson and Sourander, 1965) and electron microscopy studies (Sheffield and Melton, 1968) of organisms in cell culture have confirmed an earlier finding from silver-stained smears (Goldman *et al.*, 1958) that toxoplasmas multiply by a process called endodyogeny. However, endodyogeny may not be the only method of reproduction; longitudinal fission and "some stages" of endodyogeny have been reported to occur in culture (Akinshina and Zassuchin, 1965). According to Bommer (1969) and Bommer, Hofling, and Heunert (1969), the two daughter trophozoites that are formed leave together or in quick succession by active penetration of the maternal membrane. The pair of new parasites remain for a long time connected to the remnants of the membrane, forming a V-shaped "butterfly" figure. Later, the membrane is sloughed off. Lund *et al.* (1961b), although not referring to multiplication as endodyogeny, found that it takes 15 min for the nucleus to divide and from 30 to 45 min for the parasite to divide. Division of the parasite has also been reported to take from 60 to 90 min (Bommer, 1970; Hansson and Sourander, 1965).

The two immediate descendants from one toxoplasma constitute a generation. The generation time, which is the interval between two generations, is frequently used to express rate of multiplication. The more virulent strains multiply at faster rates than those less virulent (Kaufman and Maloney, 1962; Kaufman *et al.*, 1958; Matsubayashi and Akao, 1963). When similar dosages were used, the generation time of the highly viru-

TABLE 4. Cells and culture vessels used for cultivating *Toxoplasma gondii*

Host	Cell type	Culture vessel	Reference
Human	amnion	tubes (roller)	Cook and Jacobs (1958a); Kaufman *et al.* (1958)
		tubes (unspecified)	Akinshina (1963a); Csóka and Kulcsár (1968a, b; 1970)
		undetermined	Andreoni *et al.* (1965)
	amnion (FL)	tubes (stationary)	Kishida and Kato (1965); Osaki et al. (1964)
	bone marrow, sternal (Detroit-6 of Berman and Stulberg)	Gey chamber	Lund *et al.* (1961a)
		Rose chamber	Sourander *et al.* (1960)
	bone marrow (Detroit-98 of Berman and Stulberg)	tubes (roller)	Cook and Jacobs (1958a)
	carcinoma, cervix (HeLa)	bottles (milk dilution)	Lund *et al.* (1963a)
		bottles (Roux)	Pierreaux (1963)
		bottles (other)	Abbas (1967b); Balducci and Tyrrell (1956); Eichenwald (1956); Norrby (1970b); Shimizu (1961a, b; 1963); Shimizu and Shiokawa (1965)
		"coverslip" cultures (vessel not stated)	Bommer, Heunert, and Milthaler (1969); Hogan (1961); Hogan *et al.* (1960b).
		flask (Carrel)	Ebringer *et al.* (1965)
		flask (Spinneret)	Norrby (1970a, b)
		Gey chamber	Abbas (1967b); Lund *et al.* (1961a, b; 1963a, b); Lycke and Lund (1964a, b; 1966); Lycke, Lund, and Strannegärd (1965); Lycke, Lund, Strannegärd, and Falsen (1965); Lycke and Norrby (1966); Lycke *et al.* (1968); Norrby (1970a, b); Norrby and Lycke (1967); Strannegärd and Lycke (1966)
		slide cultures	Bommer (1969)
		tubes (roller)	Balducci and Tyrrell (1956); Chernin and Weller (1954a); Lund *et al.* (1963a)
		tubes (stationary)	Arai *et al.* (1958); de Castro and Amaral (1964); Hogan *et al.* (1960a); Hogan (1961); Matsubayashi and Akao (1962; 1965); Norrby *et al.* (1968); Schuhová (1957; 1960); Schuhová *et al.* (1956; 1957)
		tubes (unspecified)	Akinshina (1963a); Csóka and Kulcsár (1968); Pierreaux (1963)
		not stated	Abbas (1967a); Arai *et al.* (1958); Bell (1961); Bommer (1970); Frenkel (1961)
		undetermined	Cook and Jacobs (1958c)
	carcinoma, larynx (HE-p 2 of Moore)	bottles (milk dilution)	Bickford and Burnstein (1966)
		tubes (roller)	Stewart and Feldman (1965)
		tubes (stationary)	Abbas (1967b); Burnstein and Bickford (1968)
		tubes (unspecified)	Akinshina (1963a)
		not stated	Abbas (1967a); Akinshina and Zassuchin (1965)
	carcinoma, larynx (?) (HE-p 1)	tubes (unspecified)	Akinshina (1963a)

(Continued)

TABLE 4 *(Continued)*

Host	Cell type	Culture vessel	Reference
Human—*Cont'd*		not stated	Akinshina and Zassuchin (1965)
	carcinoma, mouth (KB of Eagle)	bottles (Roux)	Pierreaux (1963)
		tubes (roller)	Cook and Jacobs (1958a)
		tubes (unspecified)	Pierreaux (1963)
		undetermined	Cook and Jacobs (1958c)
	conjunctiva (Chang)	tubes (roller)	Cook and Jacobs (1958a)
		undetermined	Cook and Jacobs (1958c)
	epithelium	tubes (unspecified)	Jacobs and Melton (1965)
	epithelium, nasal	tubes (stationary)	Hogan *et al.* (1961)
	epithelium, pharyngeal	tubes (unspecified)	Akinshina (1963a)
	epithelium, testicle	tubes (roller)	Cook and Jacobs (1958a)
	fibroblasts	tubes (stationary)	Hogan *et al.* (1960a)
		"coverslip" cultures (vessel not stated)	Hogan *et al.* (1960b)
	fibroblasts (MRF)	tubes (roller)	Cook and Jacobs (1958a)
	foreskin, malignant (Detroit-189 of L)	tubes (roller)	Cook and Jacobs (1958a)
	intestine (Henle)	tubes (roller)	Cook and Jacobs (1958a)
		undetermined	Cook and Jacobs (1958c)
	kidney	Gey chamber	Lund *et al.* (1961b)
		Rose chamber	Sourander *et al.* (1960)
		not stated	Vermeil *et al.* (1962)
	leukemia cells, monocytes (J-111 of Osgood)	tubes (roller)	Cook and Jacobs (1958a)
	leukocytes	undetermined	Stadtsbaeder (1965)
	liver (Chang)	tubes (roller)	Cook and Jacobs (1958a)
		tubes (unspecified)	Akinshina (1963a)
	lung, fetal (RU-1)	bottles (prescription)	Suggs *et al.* (1968)
	macrophages	tubes (roller)	Chernin and Weller (1954a)
	myometrium	tubes (roller)	Chernin and Weller (1954a)
	muscle	not stated	Varela *et al.* (1955)
	pleura (Detroit-116 of Berman and Stulberg)	tubes (roller)	Cook and Jacobs (1958a)
	retinoblastoma cells	tubes (stationary)	Sourander *et al.* (1960)
		"coverslip" cultures (vessel not stated)	Hogan (1961); Hogan *et al.* (1960b)
	retinoblastoma (RB)	tubes (stationary)	Hogan *et al.* (1961)
	skin-muscle	tubes (roller)	Chernin and Weller (1954a)
	skin	tubes (roller)	Chernin and Weller (1954a)
	tonsil, epithelial-like cells	not stated	Varela *et al.* (1958a)
	uterus	tubes (roller)	Chernin and Weller (1957)
		tubes (unspecified)	Akinshina (1963a)
chicken (embryo)	blood monocytes	slides ("hanging drop")	Guimarães and Meyer (1942)
	cartilage	slides	Mühlpfordt (1952)
	embryo[a]	flasks (Carrel ?)	Akinshina and Gracheva (1964)
		flasks (other)	Akinshina and Zasuchina (1966)
		Maitland-type suspension culture	Sabin and Olitsky (1932)

TABLE 4 *(Continued)*

Host	Cell type	Culture vessel	Reference
chicken—*Cont'd*		tubes (stationary)	Ma and Mun (1961)
		tubes (unspecified)	Akinshina (1963a)
		not stated	Levaditi *et al.* (1929)
	fibroblasts	bottles	Akinshina and Zassuchin (1969)
		flasks	Akinshina (1964); Akinshina and Zasuchina (1966; 1967)
		petri dishes	Chaparas and Schlesinger (1959); Foley and Remington (1969)
		tubes (unspecified)	Akinshina (1964); Akinshina and Doby (1969b)
		not stated	Freshman *et al.* (1966); Remington and Merigan (1968)
		undetermined	Giroud and Dumas (1957); Tos-Luty (1967)
	heart	flasks	Akinshina (1959)
		slides ("hanging drop")	Guimarães and Meyer (1942); Meyer and de Oliveira (1943; 1945)
		tubes (roller)	Cook and Jacobs (1958a); Jacobs *et al.* (1954)
		undetermined	Cook and Jacobs (1958c)
	intestine	tubes (roller)	Cook and Jacobs (1958a)
	liver	slides ("hanging drop")	Guimarães and Meyer (1942); Meyer and de Oliveira (1943; 1945)
		tubes (unspecified)	Jacobs and Melton (1965)
	lung	tubes (roller)	Cook and Jacobs (1958a)
	muscle	flasks	Akinshina (1959)
		slides ("hanging drop")	Meyer and de Andrade Mendonca (1957)
		slides	Mühlpfordt (1952)
	muscle (leg)	tubes (roller)	Cook and Jacobs (1958a)
	nerve ganglion	slides ("hanging drop")	Guimarães and Meyer (1942); Meyer and de Oliveira (1945)
	spleen	slides ("hanging drop")	Guimarães and Meyer (1942); Meyer and de Oliveira (1943; 1945)
		not stated	Meyer and de Oliveira (1942)
	subcutaneous tissue	slides ("hanging drop")	Meyer and de Andrade Mendonca (1957)
mouse	embryo[a]	bottles (Roux)	Schuhová *et al.* (1963a, b)
		flasks (Carrel)	Akinshina and Gracheva (1964)
		tubes (roller)	Chernin and Weller (1954a, b; 1957); Cook (1958a); Cook and Jacobs (1958b); Jacobs *et al.* (1954)
		tubes (stationary)	Ma and Mun (1961)
		tubes (unspecified)	Akinshina (1963a)
		undetermined	Cook and Jacobs (1958c)
	fibroblasts	tubes (roller)	Jacobs (1956)
		not stated	Frenkel (1961)
	fibroblasts (L-cell of Earle)	bottles (Roux)	Pierreaux (1963)
		bottles (other)	Matsubayashi and Akao (1963); Shimizu 1961a, b); Shimizu and Shiokawa (1965)
		"coverslip" cultures (vessel not stated)	Bommer, Hofling, and Heunert (1969); Remington, Earle, and Yagura (1970)
		slides	Bommer (1969)
		tubes (roller)	Cook and Jacobs (1958a)

(Continued)

TABLE 4 *(Continued)*

Host	Cell type	Culture vessel	Reference
mouse—*Cont'd*		tubes (stationary)	Remington, Yagura, and Robinson (1970; Shimizu and Takagaki (1959)
		tubes (unspecified)	Pierreaux (1963)
		not stated	Bommer (1970); Freshman *et al.* (1966); Remington and Merigan (1968)
	fibroblasts (L-929)	petri dishes	Rytel and Jones (1966)
	heart	tubes (roller)	Cook and Jacobs (1958a)
	intestine	tubes (roller)	Cook and Jacobs (1958a)
	kidney	tubes (unspecified)	Jacobs and Melton (1965)
		undetermined	Virat (1967)
	lung	tubes (roller)	Cook and Jacobs (1958a)
		tubes (unspecified)	Jacobs and Melton (1965)
	macrophages	slides	Pulvertaft *et al.* (1954)
		tubes (unspecified)	Vischer and Suter (1954)
	muscle (leg)	tubes (roller)	Cook and Jacobs (1958a)
	sarcoma (S-180 of Foley)	tubes (roller)	Cook and Jacobs (1958a)
swine	kidney	flasks	Akinshina *et al.* (1967)
		slides	Hirai *et al.* (1966)
		tubes (stationary)	Ma and Mun (1961)
		tubes (unspecified)	Akinshina and Doby (1969b)
		not stated	Yü and In (1966)
monkey	heart	undetermined	de Castro and Amaral (1964)
	kidney	bottles	Balducci and Tyrrell (1956); Chaparas *et al.* (1966); Jírovec and Jíra (1962); Kaufman et al. (1959)
		"coverslip" cultures (vessel not stated)	Bommer, Hofling and Heunert (1969); Hogan (1961)
		slides	Bommer (1969)
		tubes (roller)	Balducci and Tyrrell (1956); Chernin and Weller (1954a); Cook (1958a, b); Cook and Jacobs (1958a, b); Kaufman *et al.* (1958)
		tubes (stationary)	Cook and Jacobs (1958c); Kaufman and Maloney (1962); Kaufman *et al.* (1959); Maloney and Kaufman (1964); Sheffield and Melton (1968; 1969; 1970a, b)
		tubes (unspecified)	Akinshina (1963a)
		not stated	Bommer (1970); Frenkel (1961); Yü and In (1966)
		undetermined	Babudieri and Castellani (1968); Chaparas *et al.* (1966); Schuhová and Jírovec (1962)
	spleen	tubes (roller)	Cook and Jacobs (1958a)
	testicle, epithelium	tubes (roller)	Chernin and Weller (1957)
rat	embryo[a]	tubes (stationary)	Hogan *et al.* (1961)
	heart	flasks (Carrel)	Lock (1953)
		tubes (roller)	Cook and Jacobs (1958a)
	kidney	tubes (roller)	Cook and Jacobs (1958a)
		tubes (unspecified)	Jacobs and Melton (1965)
	leukocytes (buffy coat)	tubes (unspecified)	Yanagawa and Hirato (1963)
	macrophages	tubes (unspecified)	Vischer and Suter (1954)

(Continued)

TABLE 4 *(Continued)*

Host	Cell type	Culture vessel	Reference
rat—*Cont'd*			
	retinal cells (ganglion cells, bipolar nerve cells, rods, neuroglial cells, retinal pigment epithelium, retinal blood vessel)	Gey chamber	Hansson and Sourander (1965)
cattle	kidney (MDBK)	bottles (prescription)	Paine and Meyer (1969)
		tubes (stationary)	Paine and Meyer (1969)
	kidney	tubes (roller)	Cook and Jacobs (1958a)
		tubes (stationary)	da Silva (1960)
		undetermined	Virat (1967)
hamster	fibroblasts	not stated	Frenkel (1961)
dog	kidney	bottles	Shimizu (1961a, b; 1964)
rabbit	kidney	bottles	Balducci and Tyrrell (1956)
		tubes (roller)	Balducci and Tyrrell (1956); Cook and Jacobs (1958a)
		tubes (stationary)	Hogan *et al.* (1961)
		tubes (unspecified)	Akinshina and Doby (1969b)
		undetermined	Virat (1967)
	liver	tubes (roller)	Cook and Jacobs (1958a)
	macrophages	tubes (roller)	Cook and Jacobs (1958a)
		tubes	Vischer and Suter (1954)
	spleen	tubes (roller)	Cook and Jacobs (1958a)
guinea pig	bone marrow	flasks (Carrel)	Holz and Albrecht (1953)
	embryo[a]	not stated	Galuszka (1962a, b)
	kidney	tubes (roller)	Cook and Jacobs (1958a)
		tubes (stationary)	Sourander *et al.* (1960)
	macrophages	tubes (unspecified)	Vischer and Suter (1954)

[a]Minced or freshly trypsinized whole embryo.

lent RH strain was 4.85 hr, whereas for less virulent S-7 and M-7741 strains it was 6.35 hr and 7.46 hr, respectively (Kaufman and Maloney, 1962). In a similar study, the generation time of the RH strain was 7.00 hr, whereas for the less virulent S-5 and 113-CE strains it was 7.42 and 15.67 hr, respectively (Kaufman *et al.*, 1958).

c. Aggregates and cytopathological effects

By repeated synchronous multiplication (Akinshina and Zassuchin, 1965; Bommer, 1969; Bommer, Hofling, and Heunert, 1969; Hansson and Sourander, 1965; Lund *et al.*, 1961b, 1963b), an aggregate of organisms forms. In cell culture work, this aggregation, which is frequently rosette-shaped (Fig. 12), has also been referred to as a clone or terminal colony. Sometimes rosettes do not develop. Hogan *et al.* (1961) obtained rosettes with the RH strain in six different cell types, but with the MF strain they developed in only five of the cell types and in none with the 113-CE strain. Lund *et al.* (1961b), who repeatedly observed rosettes, said that the "characteristic" rosette-like appearances often disappear as multiplication proceeds, but these are reestablished by newly formed toxoplasmas. Rosettes can appear as early as four hours after inoculation (Balducci and Tyrrell, 1956). Thirty hours after inoculation with the RH strain, the cytoplasm of the cell is completely filled with growing rosettes (Lund *et al.*, 1963b). There is apparently no limit to multiplication as long as the host cell remains intact. A single cell can contain 200 toxoplasmas (Lund *et al.*, 1961b) and as many as 32 toxoplasmas can be in a rosette (Beverley, 1969). Single rosettes of the RH strain have more individuals than those of 113-CE (Kaufman *et al.*, 1958).

A cell undergoes normal division in spite of infection (Bommer, 1969; Bommer, Hofling, and Heunert, 1969; Lund *et al.*, 1961b; Lund *et al.*, 1963b; Meyer and de Oliveira, 1942) or it may become multinucleate (Akinshina, 1959; Pulvertaft *et al.*, 1954). Prior to cell rupture, the nucleus becomes pyknotic (Akinshina, 1959; Bickford and Burnstein, 1966) and frequently is pushed toward the periphery of the cell by a developing rosette (Akinshina, 1959; Vischer and Suter, 1954). The cytoplasm has been reported to become vacuolated and "frothy" and then to disappear (Akinshina, 1959). However, it has also been reported that there are no appreciable cytopathological changes prior to rupture of the cell membrane (Chernin and Weller, 1957; Varela *et al.*, 1955).

d. Release of toxoplasmas, reinvasion of cells, and "histopathological" effects

Exit of toxoplasmas from a cell by active screw-like movements leaving an intact cell membrane with multiple perforations has been suggested by Schmidt-Hoensdorf and Holz (1953). A more generally accepted view is that they are passively expelled after the cell membrane ruptures (Bommer, 1969; Bommer, Heunert, and Milthaler, 1969; Hansson and Sourander, 1965; Lund *et al.*, 1961b, 1963b; Lycke and Lund, 1964a; Pulvertaft *et al.*, 1954). Rupture of a cell does not depend on any fixed number of parasites (Lund *et al.*, 1961b, 1963b). It occurs when the elastic capacity of the cell membrane is exhausted as a result of parasite multiplication (Lund *et al.*, 1961b).

Toxoplasmas invade adjoining cells within a few seconds after destroying the cell in which they developed (Lund *et al.*, 1961b). Cells remote from the original zones of infection remain uninfected (Hogan *et al.*, 1961). Repetition of the developmental "cycle" results in degenerative changes in the layer of cells. Plaques appear as white, irregular (Foley and Remington, 1969) or circular (Akinshina and Zassuchin, 1969; Bickford and Burnstein, 1966; Chaparas and Schlesinger, 1959) areas against a pink background of viable cells (Foley and Remington, 1969). A plaque consists of a peripheral ring of infected cells surrounding a central mass of degenerated cells with pyknotic nuclei (Bickford and Burnstein, 1966). The central necrotic area is later sloughed off leaving a clear area (Akinshina, 1965; Bickford and Burnstein, 1966).

Degenerative changes in the cell layer coincide with the liberation of large numbers of parasites (Chernin and Weller, 1954a). The more virulent strains multiply faster than those less virulent and, consequently, are released into the medium earlier. According to Lund *et al.* (1963a), toxoplasmas of the RH strain are released from HeLa cells beginning at 24 hr. The quantity in the medium reaches a maximum at five days (Fig. 22). This corresponds with the time at which plaques or necrotic lesions appear (Akinshina and Zassuchin, 1966; Chaparas and Schlesinger, 1959; Foley and Remington, 1969; Lycke and Lund, 1964a). The decline after five days shown in Fig. 22 is probably caused by a lack of cells and death of organisms in the medium. The speed with which cell layers are

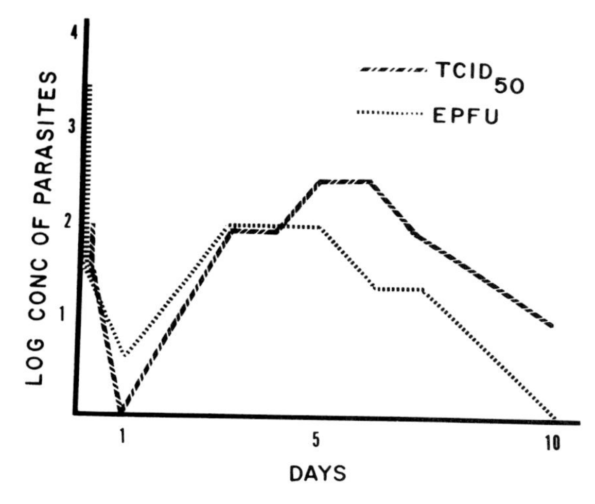

Fig. 22. Rate of liberation of infective *T. gondii* from bottle cultures of HeLa cells infected with 3.6×10^6 parasites per bottle. The titrations of infectious units were carried out in tube cultures and registered as $TCID_{50}$ and on egg membranes as plaque-forming units (EPFU). The log concentration of infectious units is plotted against time. (Redrawn from Lund *et al.*, 1963a.)

destroyed depends on inoculum size and virulence of the strain. The RH strain destroys cells more rapidly and more consistently than the avirulent 113-CE strain (Kaufman, Melton, Remington, and Jacobs, 1959). Cultures infected with avirulent strains sometimes are not destroyed, and a "chronic" or prolonged infection of the cell culture seems to take place (Kaufman, Melton, Remington, and Jacobs, 1959).

The number of toxoplasmas present at any one time in the medium is not great compared to the numbers that can be recovered from exudates of animals (Jacobs, 1956). Also, there is a high proportion of nonviable forms (Jacobs, 1956). However, under optimal growth conditions, yields of from 10 to 12 million/ml nutrient fluid (Akinshina and Gracheva, 1964; Eichenwald, 1956) and from 10 to 18 million/ml of fluid (Shimizu, 1961a) have been obtained. By treating the monolayer with trypsin, the yield can be substantially increased above that normally released (Csóka and Kulcsár, 1968b, 1970).

e. Cysts

Cysts have been frequently found in culture (Akinshina and Zassuchin, 1965; Bickford and Burnstein, 1966; Bommer, 1969; Galuzo *et al.*, 1969; Hansson and Sourander, 1965; Hogan, 1961; Hogan *et al.*, 1960b, 1961; Jacobs and Melton, 1965; Matsubayashi and Akao, 1962, 1963, 1965; Mühlpfordt, 1952; Osaki and Oka, 1963; Pierreaux, 1963). Work carried out utilizing cell cultures indicates that the thick, rigid cyst wall is either derived from only the parasite (Matsubayashi and Akao, 1962), only the host cell (Bommer, 1969; Hansson and Sourander, 1965), or both the host cell and parasite (Matsubayashi and Akao, 1963, 1965).

Cyst formation depends on the rate of multiplication (Hogan *et al.*, 1961; Matsubayashi and Akao, 1963). Cysts are found earlier (Hogan, 1961) and more frequently (Hogan *et al.*, 1961; Matsubayashi and Akao, 1963) with slow-growing, less virulent strains than with faster growing, more virulent strains. The MF and 113-CE strains produce cysts by 72 hr, whereas with the RH strain it sometimes takes seven days and usually from two to three weeks (Hogan *et al.*, 1961).

2. *Besnoitia*

a. Cells supporting development

Cells and culture vessels used for cultivating *Besnoitia* are listed in Table 5. There are probably additional cells in which development took place. Frenkel (1965) states that *B. jellisoni* can be grown in many tissue culture cell lines, but does not mention what they are.

In the work listed, proliferative and "cyst" forms were used as the inoculum. Bigalke (1962) used proliferative forms obtained from the testes of rabbits and cyst forms obtained from bovine skin. Fayer *et al.* (1969) used cyst forms from either the fascia beneath the skin of kangaroo rats or from the peritoneum of white mice. Akinshina and Doby (1969a,b)

TABLE 5. Cells and culture vessels used for cultivating *Besnoitia* and *Sarcocystis*

Species	Cells	Culture type	Culture vessel	Reference
Besnoitia jellisoni	bovine kidney	established cell line (MDBK)	tubes (stationary)	Fayer *et al.* (1969)
	bovine spleen (E)[a]	cell line	tubes (stationary)	Fayer *et al.* (1969)
	bovine trachea (E)	cell line	tubes (stationary)	Fayer *et al.* (1969)
	chicken kidney (E)	primary	not stated	Akinshina and Doby (1969b)
	chicken fibroblasts (E)	primary	tubes (stationary)	Doby and Akinshina (1968)
			flasks	Doby and Akinshina (1968)
			not stated	Akinshina and Doby (1969b)
	human lung (E)	cell line	tubes (stationary)	Suggs and Walls (1967)
	rabbit kidney	cell line	not stated	Akinshina and Doby (1969a, b)
	swine kidney	cell line	not stated	Akinshina and Doby (1969a, b)
B. besnoiti	bovine kidney	primary	tubes (stationary) tubes (roller) flasks (Roux)	Bigalke (1962)
	lamb kidney	primary	tubes (stationary) tubes (roller) flasks (Roux)	Bigalke (1962)
Sarcocystis sp.	bovine kidney (E)	cell line	tubes (stationary)	Fayer (1970; 1972)
	bovine trachea (E)	cell line	tubes (stationary)	Fayer (1970; 1972)
	canine kidney	established cell line (MDCK)	tubes (stationary)	Fayer (1970)
	chicken kidney	primary	tubes (stationary)	Fayer (1970)
	chicken muscle	primary	tubes (stationary)	Fayer (1970)
	turkey kidney	primary	tubes (stationary)	Fayer (1970)

[a]E = embryonic

and Doby and Akinshina (1968) used the proliferative form from peritoneal exudate of mice. Suggs and Walls (1967) state only that they used organisms from other infected cell cultures.

b. Mode and rate of multiplication

According to Doby and Akinshina (1968), *B. jellisoni* enters cells, increases in size, becomes ovoid, and multiplies within from three to five hours after inoculation. Bigalke (1962) reports that multiplication of *B. besnoiti* in cell culture is by synchronous binary fission. In more detailed studies of *B. jellisoni* in culture, Akinshina and Doby (1969a) and Fayer *et al.* (1969) found stages indicating that endodyogeny takes place (Fig. 14) and stages suggesting that binary fission and even schizogony might take place.

Nothing is known with regard to generation times. Doby and Akinshina (1968) state that "divisions" after the first are carried out with an increasingly rapid rhythm.

c. Aggregates and cytopathological effects

Aggregates of organisms, which are formed by repeated multiplication, are either rosette-shaped (Bigalke, 1962; Doby and Akinshina, 1969a; Fayer *et al.*, 1969), shaped like a bunch of bananas (Bigalke, 1962; Fayer *et al.*, 1969), or have no obvious organization (Fayer *et al.*, 1969). Some rosette-shaped aggregates (Fig. 15) and aggregates shaped like a bunch of bananas (Fig. 16) contain sizeable structures interpreted as residual bodies (Fayer *et al.*, 1969). Rosette-shaped aggregates containing from three to eight organisms can be found two days after inoculation and a vacuole within an embryonic bovine spleen cell nine days after inoculation can contain 164 organisms (Fayer *et al.*, 1969).

The nucleus of an infected cell enlarges (Akinshina and Doby, 1969a), contains an enlarged nucleolus and relatively small particles of chromatin (Fayer *et al.*, 1969), and eventually becomes pyknotic (Akinshina and Doby, 1969a). The cytoplasm becomes vacuolated, giving the monolayer a foamy appearance (Fayer *et al.*, 1969).

d. Release of organisms, reinvasion of cells, and "histopathological" effects

Doby and Akinshina (1968) state that extracellular organisms can be found from 30 to 40 hr after inoculation. Fayer *et al.* (1969) found maximum numbers of intracellular organisms after three and four days and large numbers of extracellular organisms after five and six days. In two of their nine experiments, they found a second peak in intracellular organisms at eight and nine days. This indicates that organisms can leave cells in which they develop, invade new cells, and go through a second "cycle" of development. Since most experiments showed declining numbers of intracellular organisms after the initial peak, the authors state that it was evident that free organisms were usually unsuccessful in reentering cells. They offer two explanations for this. First, pathological changes may have made the cells unsuitable as potential host cells; second, destruction of large numbers of cells by infection may have reduced the number of host cells below a level which would support further development.

Like *Toxoplasma*, infection with *Besnoitia* results in a loss of most of the cell layer (Bigalke, 1962; Fayer *et al.*, 1969). *Besnoitia* may be different from *Toxoplasma* in the manner in which organisms exit from cells. Fayer *et al.* (1969) frequently observed organisms passing through the cell membrane. Rupture of the membrane, enabling all organisms within the cell to be passively expelled, was not reported.

3. Eimeria and Isospora

a. Cells used and extent of development

Cells from 17 different animal species were used in attempts to obtain development of *Eimeria* and *Isospora*. Results of these attempts with excysted sporozoites and merozoites as the inoculum are summarized in Tables 6 (pp. 210-215) and 7 (pp. 216-217), respectively. With the exception of various bottles (Roberts, Elsner, Shigematsu, and Hammond, 1969, 1970; Roberts *et al.*, 1971), Sykes-Moore chambers (Strout and Ouellette, 1970), Rose perfusion chambers (Fayer, 1969), and petri dishes (Ryley and Wilson, 1971), all workers grew their cells upon cover glasses in stationary Leighton tubes.

Most of the attempts were in cell line cultures from animals that were not the natural host. However, development of seven of the species (*E. acervulina, E. alabamensis, E. bovis, E. callospermophili, E. larimerensis, E. necatrix,* and *E. tenella*) progressed farther into the life cycle when cultures from the natural host were used. The farthest development with sporozoites as the inoculum was obtained with a species from rabbits (*E. magna*) and certain avian species (*E. tenella, E. necatrix, E. brunetti,* and *Isospora lacazei* and/or *chloridis*). The reason may be the shorter life cycles of these species and the ability to maintain cells in a condition suitable for development for at least a part of the life cycle interval. It is interesting that the four species that did not develop (*E. gallopavonis, E. mivati, E. praecox,* and *E. maxima*) and the species that developed the least (*E.

TABLE 6. Development of *Eimeria* and *Isospora* in cell cultures when sporozoites were used as the inoculum

Species	Cells	Type of culture	Extent of development	Reference
E. acervulina	bovine kidney (E)[a]	primary	no development	Doran and Vetterling (1967a)
		cell line	no development	Doran and Vetterling (1967a)
	chicken fibroblasts (E)	primary	trophozoites (1)[b]	Strout *et al.* (1965)
	chicken kidney	primary	no development	Doran (1971b)
	chicken kidney (E)	primary	trophozoites (1)	Strout *et al.* (1965)
			immature schizonts (1)	(*p.c.*)[d]
	human amnion	cell line (FL)	trophozoites (1)	Strout *et al.* (1965)
	human cervical carcinoma	cell line (HeLa)	trophozoites (1)	Strout *et al.* (1965)
	human kidney (E)	cell line	no development	Doran and Vetterling (1967a)
	mouse fibroblasts	cell line (clone 2445)	trophozoites (1)	Strout *et al.* (1965)
	ovine kidney	primary	no development	Doran and Vetterling (1967a)
		cell line	no development	Doran and Vetterling (1967a)
	porcine kidney	primary	no development	Doran and Vetterling (1967a)
		cell line	no development	Doran and Vetterling (1967a)
E. adenoeides	bovine kidney (E)	cell line	mature schizonts (1)	Doran (1969; 1970a, b)
E. alabamensis	bovine intestine (E)	primary	mature schizonts (1)	Sampson *et al.* (1971)
		cell line	mature schizonts (1)	Sampson *et al.* (1971)
	bovine kidney	established cell line (MDBK)	trophozoites (2)	Sampson *et al.* (1971)
	bovine kidney (E)	primary	mature schizonts (1)	Sampson *et al.* (1971)
	bovine spleen (E)	primary	mature schizonts (1)	Sampson *et al.* (1971)
	bovine synovia (E)	cell line	mature schizonts (1)	Sampson *et al.* (1971)
	bovine thyroid (E)	primary	mature schizonts (1)	Sampson *et al.* (1971)
	bovine trachea (E)	cell line	mature schizonts (1)	Sampson *et al.* (1971)
	hamster kidney	cell line	mature schizonts (1)	Sampson *et al.* (1971)
	human intestine (E)	established cell line (Int 407)	mature schizonts (1)	Sampson *et al.* (1971)
E. auburnensis	bovine kidney	established cell line (MDBK)	mature schizonts (1)	Clark and Hammond (1969)
	bovine spleen (E)	cell line	mature schizonts (1)	Clark and Hammond (1969)
	bovine trachea (E)	cell line	mature schizonts (1)	Clark and Hammond (1969)
	human intestine (E)	established cell line (Int 407)	trophozoites (1)	Clark and Hammond (1969)
E. bilamellata	bovine intestine (E)	cell line	mature schizonts (1)	Speer, Hammond, and Kelley (1970)
	bovine kidney	established cell line (MDBK)	mature schizonts (1)	Speer, Hammond, and Anderson (1970)
	bovine kidney (E)	cell line	mature schizonts (1)	Speer, Hammond, and Kelley (1970); Speer, Hammond, and Anderson (1970)

TABLE 6 *(Continued)*

Species	Cells	Type of culture	Extent of development	Reference
E. bilamellata–Cont'd	bovine liver (E)	cell line	mature schizonts (1)	Speer, Hammond, and Kelley (1970)
	bovine spleen (E)	cell line	immature schizonts (1)	Speer, Hammond, and Anderson (1970)
	bovine trachea (E)	cell line	mature schizonts (1)	Speer, Hammond, and Anderson (1970)
	bovine synovia (E)	cell line	mature schizonts (1)	Speer, Hammond, and Anderson (1970)
	hamster kidney	cell line	mature schizonts (1)	Speer, Hammond, and Anderson (1970)
	human intestine (E)	established cell line (Int 407)	immature schizonts (1)	Speer, Hammond, and Anderson (1970)
E. bovis	bovine intestine (E)	primary	trophozoites (1)	Fayer and Hammond (1967); Hammond and Fayer (1967)
		cell line	immature schizonts (1)	Fayer and Hammond (1967); Hammond and Fayer (1967)
	bovine kidney	established cell line (MDBK)	mature schizonts (1)	Hammond and Fayer (1967)
	bovine kidney (E)	primary	immature schizonts (1)	Fayer and Hammond (1967); Hammond and Fayer (1967)
		cell line	immature schizonts (1)	Fayer and Hammond (1966)
			mature schizonts (1)	Fayer and Hammond (1967); Hammond and Fayer (1967)
	bovine spleen (E)	primary	no development	Fayer and Hammond (1967); Hammond and Fayer (1967)
		cell line	immature schizonts (1)	Fayer and Hammond (1966)
			mature schizonts (1)	Fayer and Hammond (1967; Hammond and Fayer (1967)
	bovine testicle (E)	primary	no development	Fayer and Hammond (1967); Hammond and Fayer (1967)
		cell line	immature schizonts (1)	Fayer and Hammond (1967); Hammond and Fayer (1967)
	bovine thymus (E)	primary	no development	Fayer and Hammond (1967); Hammond and Fayer (1967)
		cell line	mature schizonts (1)	Fayer and Hammond (1967); Hammond and Fayer (1967)
	bovine trachea (E)	cell line	immature schizonts (2)	Hammond and Fayer (1968)
	human intestine (E)	established cell line (Int 407)	mature schizonts (1)	Hammond and Fayer (1968)
	mouse fibroblasts (from connective tissue)	cell line (L-cells)	immature schizonts (1)	Hammond and Fayer (1968)
E. brunetti	chicken fibroblasts (E)	primary	mature schizonts (2)	Shibalova (1970)
	quail fibroblasts (E)	primary	mature schizonts (2)	Shibalova (1970)
E. callospermophili	bovine kidney	established cell line (MDBK)	mature schizonts (1)	Speer, Hammond, and Anderson (1970)
	bovine intestine (E)	cell line	mature schizonts (1)	Roberts, Hammond, and Anderson (1969); Roberts, Hammond, and Speer (1970); Speer and Hammond

(Continued)

TABLE 6 *(Continued)*

Species	Cells	Type of culture	Extent of development	Reference
E. callospermophili—Cont'd				(1969); Speer, Hammond, and Kelley (1970)
	bovine kidney (E)	cell line	trophozoites (2)	Speer, Hammond, and Anderson (1970)
			mature schizonts (1)	Speer, Hammond, and Kelley (1970)
	bovine liver (E)	cell line	mature schizonts (1)	Speer, Hammond, and Kelley (1970)
	bovine spleen (E)	cell line	mature schizonts (1)	Speer, Hammond, and Anderson (1970)
	bovine synovia (E)	cell line	mature schizonts (1)	Speer, Hammond, and Anderson (1970)
	bovine thymus (E)	cell line	mature schizonts (1)	Speer, Hammond, and Anderson (1970)
	bovine thyroid (E)	cell line	mature schizonts (1)	Speer, Hammond, and Anderson (1970)
	bovine trachea (E)	cell line	mature schizonts (1)	Speer, Hammond, and Anderson (1970)
	ground squirrel embryo[c]	primary	immature schizonts (2)	Speer, Hammond, and Anderson (1970)
	hamster kidney	cell line	mature schizonts (1)	Speer, Hammond, and Anderson (1970)
	human intestine (E)	established cell line (Int 407)	mature schizonts (1)	Speer, Hammond, and Anderson (1970)
E. crandallis	bovine kidney (E)	cell line	immature schizonts (1)	de Vos and Hammond (1971); de Vos *et al.* (1972)
	bovine liver (E)	cell line	mature schizonts (1)	de Vos and Hammond (1971); de Vos *et al.* (1972)
	ovine thyroid (E)	cell line	mature schizonts (1)	de Vos and Hammond (1971); de Vos *et al.* (1872)
	ovine trachea (E)	cell line	mature schizonts (1)	de Vos and Hammond (1971); de Vos *et al.* (1972)
E. ellipsoidalis	bovine kidney	established cell line (MDBK)	mature schizonts (1)	Speer and Hammond (1971a)
	bovine spleen (E)	cell line	mature schizonts (1)	Speer and Hammond (1971a)
	bovine synovia (E)	cell line	no development	Speer and Hammond (1971a)
	bovine trachea (E)	cell line	mature schizonts (1)	Speer and Hammond (1971a)
E. gallopavonis	bovine kidney (E)	primary	no development	Doran and Vetterling (1967a)
		cell line	no development	Doran and Vetterling (1967a)
	human kidney (E)	cell line	no development	Doran and Vetterling (1967a)
	ovine kidney	primary	no development	Doran and Vetterling (1967a)
		cell line	no development	Doran and Vetterling (1967a)
	porcine kidney	primary	no development	Doran and Vetterling (1967a)
		cell line	no development	Doran and Vetterling (1967a)
E. larimerensis	bovine intestine (E)	cell line	mature schizonts (1)	Speer, Hammond, and Kelley (1970)
	bovine kidney	established cell line (MDBK)	immature schizonts (2)	Speer and Hammond (1970b)
	bovine kidney (E)	cell line	mature schizonts (1)	Speer, Hammond, and Kelley (1970)
	bovine liver (E)	cell line	mature schizonts (1)	Speer, Hammond, and Kelley (1970)
			immature schizonts (2)	Speer and Hammond (1970b)

TABLE 6 *(Continued)*

Species	Cells	Type of culture	Extent of development	Reference
E. larimerensis—Cont'd	ground squirrel embryo[c]	cell line	immature schizonts (2)	Speer and Hammond (1970b)
	ovine kidney (E)	cell line	immature schizonts (2)	Speer and Hammond (1970b)
E. magna	bovine kidney (E)	primary	no development	Speer and Hammond (1971b)
	bovine kidney	established cell line (MDBK)	mature schizonts (2)	Speer and Hammond (1971b)
	bovine liver (epithelial-like cells)(E)	not stated	mature schizonts (2)	Speer and Hammond (1971b)
	bovine liver (fibroblast-like cells)(E)	not stated	no development	Speer and Hammond (1971b)
	ovine trachea (E)	cell line	no development	Speer and Hammond (1971b)
E. maxima	chicken kidney	primary	no development	Doran (1971b)
E. meleagrimitis	bovine kidney (E)	primary	no development	Doran and Vetterling (1967a)
		cell line	mature schizonts (1)	Doran (1969); Doran and Vetterling (1967a, 1968b, 1969)
			trophozoites (2)	Doran and Vetterling (1967b; 1968a)
	human kidney (E)	cell line	no development	Doran and Vetterling (1967a)
	ovine kidney	primary	no development	Doran and Vetterling (1967a)
		cell line	no development	Doran and Vetterling (1967a)
	porcine kidney	primary	no development	Doran and Vetterling (1967a, b)
		cell line	no development	Doran and Vetterling (1967a, b)
	turkey intestine (E)	primary	mature schizonts (1)	Doran and Vetterling (1968a)
E. mivati	bovine kidney (E)	cell line	no development	Doran (1969)
	chicken kidney	primary	no development	Doran (1971b)
E. necatrix	bovine kidney (E)	primary	no development	Doran and Vetterling (1967a)
		cell line	mature schizonts (1)	Doran (1969); Doran and Vetterling (1967a)
	chicken kidney	primary	mature schizonts (2)	Doran (1971b)
	human kidney (E)	cell line	no development	Doran and Vetterling (1967a)
	ovine kidney	primary	no development	Doran and Vetterling (1967a)
		cell line	no development	Doran and Vetterling (1967a)
	porcine kidney	primary	no development	Doran and Vetterling (1967a)
		cell line	immature schizonts (1)	Doran and Vetterling (1967a)
		cloned cell line (CC1-33)	no development	Doran and Vetterling (1967a)
E. nieschulzi	bovine intestine (E)	cell line	mature schizonts (1)	Speer, Hammond, and Kelley (1970)
	bovine kidney (E)	cell line	mature schizonts (1)	Speer, Hammond, and Kelley (1970)
	bovine liver (E)	cell line	mature schizonts (1)	Speer, Hammond, and Kelley (1970)
E. ninakohlyakimovae	bovine intestine (E)	cell line	mature schizonts (1)	Speer, Hammond, and Kelley (1970)
	bovine kidney	established cell line	mature schizonts (1)	Kelley and Hammond (1970)

(Continued)

TABLE 6 *(Continued)*

Species	Cells	Type of culture	Extent of development	Reference
E. ninakohlyak-imovae—Cont'd		(MDBK)		
	bovine kidney (E)	cell line	mature schizonts (1)	Speer, Hammond, and Kelley (1970)
	bovine liver (E)	cell line	mature schizonts (1)	Speer, Hammond, and Kelley (1970)
	ovine kidney (E)	cell line	mature schizonts (1)	Kelley and Hammond (1970)
	ovine thyroid (E)	cell line	mature schizonts (1)	Kelley and Hammond (1970)
	ovine trachea (E)	cell line	mature schizonts (1)	Kelley and Hammond (1970)
	ovine thymus (E)	cell line	mature schizonts (1)	Kelley and Hammond (1970)
E. praecox	chicken kidney	primary	no development	Doran (1971b)
E. tenella	bovine kidney	primary	mature schizonts (1)	Ryley (1968)
		cell line (MDBK-p.c.)	mature schizonts (1)	Patton (1965)
	bovine kidney (E)	cell line	mature schizonts (1)	Doran (1969; 1970a)
	bovine trachea (E)	cell line	mature schizonts (1)	Matsuoka *et al.* (1969)
	chicken fibroblasts (E)	primary	immature schizonts (1)	Bedrnik (1969a); Patton (1965)
			mature schizonts (2)	Shibalova (1968; 1969a); Shibalova *et al.* (1969)
			mature schizonts (3)	Shibalova (1969b)
		primary (p.c.)	gametocytes	Shibalova (1970)
	chicken kidney	primary	oocysts	Doran (1970c; 1971a, c)
	chicken kidney (E)	primary	mature schizonts (?)	Shibalova (1969a)
			mature schizonts (2)	Shibalova (1968)
			mature schizonts (3)	Shibalova et al. (1969)
			gametocytes	Shibalova (1970); Strout and Ouellette (1968)
			oocysts	Doran (1970c); Strout and Ouellette (1970)
	chorioallantoic membrane (chicken)	primary	no development	Long (1969)
	guinea pig kidney	cell line	mature schizonts (?)	Shibalova (1969a)
	human fibroblasts (E)	cell line	mature schizonts (?)	Shibalova (1969a)
			mature schizonts (2)	Shibalova (1968; 1969b; 1970)
	human kidney (E)	cell line	mature schizonts (?)	Shibalova (1969a)
	human cervical carcinoma	cell line (HeLa)	mature schizonts (?)	Shibalova (1969a)
			mature schizonts (2)	Shibalova (1969b; 1970)
	monkey kidney	cell line	mature schizonts (?)	Shibalova (1970)
			mature schizonts (2)	Shibalova (1969b)
	mouse fibroblasts	cell line (L-cells)	immature schizonts (1)	Patton (1965)
	partridge kidney	primary	gametocytes	Doran (1971c)
	pheasant kidney	primary	oocysts	Doran (1971c)
	quail fibroblasts (E)	primary (p.c.)	gametocytes	Shibalova (1970)
		cell line	mature schizonts (1)	Patton (1965)
	turkey kidney	primary	mature schizonts (2)	Doran (1971c)
I. canis	canine intestine (E)	primary	paired organisms	Fayer and Mahrt (1972)
		established cell line	paired organisms	Fayer and Mahrt (1972)

TABLE 6 *(Continued)*

Species	Cells	Type of culture	Extent of development	Reference
		(MDCK)		
	bovine kidney (E)	cell line	paired organisms	Fayer and Mahrt (1972)
	bovine trachea (E)	cell line	paired organisms	Fayer and Mahrt (1972)
I. felis and *I. rivolta*	monkey kidney	primary	no development	Sheffield and Melton (1970a)
I. lacazei and/or *I. chloridis*	chicken fibroblasts (E)	primary	mature schizonts (2)	Turner and Box (1970)
	canary fibroblasts (E)	primary	mature schizonts (2)	Turner and Box (1970)

[a]E = Embryonic.
[b]Numbers in parentheses indicate generation; question mark indicates generation not stated.
[c]Freshly trypsinized whole embryo.
[d]p.c. = personal communication with senior author.

acervulina) are also avian species. The fact that three of the nondeveloping species failed to develop in the same cell type in which *E. tenella* developed to oocysts (Doran, 1971b) suggests that different organisms have different requirements which cannot be supplied by the same cells.

In addition to the work listed in Tables 6 and 7, there are two instances where parasitized tissue was used. Long (1969) obtained oocysts of *E. tenella* after establishing monolayer cultures with cells from the CAM of embryonated chicken eggs that contained second-generation merozoites and gametocytes. Gavrilov and Cowez (1941) established cultures with explants from rabbit intestine parasitized with various developmental stages of *E. perforans*. Twelve days later, gametocytes were found in cells away from the explant. If these were new cells as claimed, and not merely cells from the explant that had moved, development must have taken place.

b. Asexual development

i. The first and second generations

Fayer and Hammond (1967) give a good description of nuclear changes in *E. bovis* sporozoites before they begin to assume the round shape characteristic of trophozoites (Fig. 23a-d). Types A and B sporozoites of Doran and Vetterling (1968a) correspond to Figures 23a and 23d. Change from a vesicular to a compact type nucleus has also been described for other species (Clark and Hammond, 1969; Sampson *et al.*, 1971; Speer,

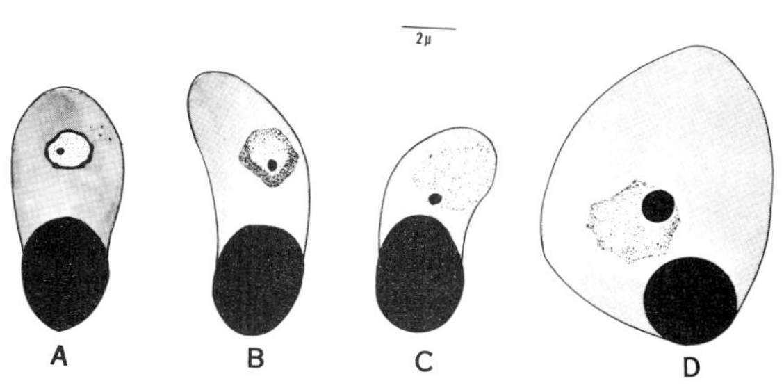

Fig. 23. Semidiagrammatic representation of nuclear changes associated with transformation of *E. bovis* sporozoites into trophozoites; sketches of intracellular specimens from spleen cell culture eight days after inoculation; fixed in Schaudinn's fluid, stained with iron hematoxylin (from Fayer and Hammond, 1967). *a.* Sporozoite with vesicular nucleus; well-defined layer of peripheral chromatin; small, somewhat eccentric nucleolus; and deeply stained, ellipsoidal, posterior refractile body. *b.* Sporozoite with thicker, paler, and less distinct peripheral layer of chromatin; nucleus and nucleolus somewhat enlarged. *c.* Sporozoite with relatively large nucleolus; peripheral chromatin has become dispersed and nucleus is no longer vesicular. *d.* Trophozoite with nucleus similar to that of Fig. 23c, except that nucleolus is larger; posterior refractile body has become spherical.

TABLE 7. Development of *Eimeria* in cell cultures when merozoites were used as the inoculum

Species	Cells	Type of culture	Source of merozoites	Extent of development	Reference
E. bovis	bovine cecum (E)[a]	primary	host (1)[b]	not intracellular	Hammond *et al.* (1969)
	bovine ileum (E)	cell line	host (1)[c]	intracellular, no development	Hammond *et al.* (1969)
			host (1)	intracellular, no development	Hammond *et al.* (1969)
	bovine kidney	established cell line (MDBK)	culture (1)	not intracellular	Hammond *et al.* (1969)
			culture (1)	not intracellular	Hammond *et al.* (1969)
	bovine kidney (E)	primary	host (1)	not intracellular	Hammond *et al.* (1969)
		cell line	host (1)	not intracellular	Hammond *et al.* (1969)
	bovine spleen (E)	cell line	culture (1)	not intracellular	Hammond *et al.* (1969)
	bovine trachea (E)	cell line	culture (1)	intracellular, no development	Hammond *et al.* (1969)
			culture (1)	not intracellular	Hammond *et al.* (1969)
	chicken fibroblasts (E)	primary	host (1)	not intracellular	Hammond *et al.* (1969)
	human intestine (E)	established cell line (Int 407)	culture (1)	not intracellular	Hammond *et al.* (1969)
	mouse macrophages (from peritoneal cavity)	primary	culture (1)	intracellular, no development	Hammond *et al.* (1969)
	ovine kidney (E)	cell line	culture (1)	not intracellular	Hammond *et al.* (1969)
	ovine thyroid (E)	cell line		not intracellular	Hammond *et al.* (1969)
E. brunetti	chicken fibro blasts (E)	primary	host (2)	gametocytes	Shibalova (1970)
	quail fibroblasts (E)	primary	host (2)	gametocytes	Shibalova (1970)
E. magna	bovine kidney	established cell line (MDBK)	host (?)	oocysts	Speer and Hammond (1972)
E. tenella	chicken cecum	primary	host (2)[d]	oocysts	Bedrnik (1969b)
	chicken cecum and liver	primary	host (2)	oocysts	Bedrnik (1969b; 1970)
	chicken embryo[e]	primary	host (2)	oocysts	Bedrnik (1969b; 1970)
	chicken embryo[f]	primary	host (2)	oocysts	Bedrnik (1969b; 1970)
			host (3)	extracellular merozoites (4)	Bedrnik (1969b; 1970)
	chicken fibro-blasts (E)	primary	host (2)	mature schi-zonts (3)	Bedrnik (1967a)
				oocysts	Bedrnik (1967b; 1969a)
		primary (p.c.)	host (2)	no development	Shibalova (1969b)
				oocysts	Shibalova *et al.* (1969)
	chicken kidney (E)	primary	host (2)	no oocysts	Shibalova (1970)
				no development	Shibalova (1969b)
	human fibroblasts (E)	cell line	host (2)	no oocysts	Shibalova (1970)
	human cervical	cell line	host (2)	no oocysts	Shibalova (1970)

TABLE 7 *(Continued)*

Species	Cells	Type of culture	Source of merozoites	Extent of development	Reference
E. tenella—Cont'd	carcinoma	(HeLa)	host (2)	no development	Shibalova (1969b)
			host (2)	mature schizonts	Bedrnik (1967a; 1969b; 1970)
	monkey kidney	cell line		no development	Shibalova (1969b)
	quail fibroblasts (E)	primary (p.c.)		oocysts	·Shibalova (1970)

[a]E = embryonic.
[b]Numbers in parentheses indicate generation; question mark indicates generation not stated.
[c]Pretreated with fluid from bovine cecum or extracts of the large and small intestine.
[d]Along with cells at time of inoculation.
[e]Freshly trypsinized whole embryo.
[f]Freshly trypsinized whole embryo with no cecum.
[g]p.c. = personal communication with senior author.

Hammond, and Anderson, 1970). Nuclear changes are thought to be associated with preparation for marked growth characteristic of the trophozoite and schizont; increase in size of the nucleolus is presumably linked with increased production of RNA (Fayer and Hammond, 1967). Transformation of *E. ninakohlyakimovae* to a trophozoite follows one of two courses (Kelley and Hammond, 1970). The most prevalent course is a gradual increase in width of the sporozoite, which is more pronounced at the anterior end. The other course involves a lateral outpocketing of the sporozoite, usually in the posterior portion of the body. In both methods of transformation, the body increases in size as it becomes spherical. *Eimeria crandallis* sporozoites transform to spherical trophozoites by an increase in width of organisms and/or by formation of lateral outpocketing (de Vos and Hammond, 1971). Also, some sporozoites enlarge in a lengthwise direction, resulting in elongate immature schizonts that later assume an ellipsoidal shape. In *E. alabamensis*, outpocketing occurs at the anterior end and extends posteriorly along one side (Sampson *et al.*, 1971). Doran and Vetterling (1968a) found type B sporozoites and binucleate schizonts of *E. meleagrimitis* with lateral outpocketings from 120 to 144 hr. They thought that the sporozoites were probably undergoing transformation to trophozoites, but considered the outpocketing to be a form of abnormal development.

A detailed description of nuclear division is given by Speer and Hammond (1970a). The first nuclear division in *E. callospermophili* usually occurs from eight to ten hours after inoculation (Speer and Hammond, 1970a); with *E. larimerensis*, it occurs after 15 hr and takes two hours to complete (Speer and Hammond, 1970b). Among various species, immature first-generation schizonts develop from either spherical trophozoites or from elongated sporozoites. Sporozoite-shaped schizonts, which later transform into spherical schizonts following lateral outpocketing and increase in size, have been reported for *E. alabamensis* (Sampson, *et al.*, 1971). *E. auburnensis* (Clark and Hammond, 1967), *E. bilamellata* and *E. callospermophili* (Speer, Hammond, and Anderson, 1970), *E. larimerensis* (Speer and Hammond, 1970b), and *E. magna* (Speer and Hammond, 1971b). Transformation of *E. magna* sporozoite-shaped schizonts to spherical schizonts takes 41 min at 37°C and occurs after from two to eight nuclei are formed. Sporozoite-shaped schizonts flex while intracellular (Speer and Hammond, 1969) and, when freed from the monolayer, undergo gliding and flexing (Speer and Hammond, 1969, 1970b; Speer, Hammond, and Anderson, 1970; Speer *et al.*, 1971). Some even penetrate new cells (Speer and Hammond, 1969; Speer, Hammond, and Anderson, 1970) in a manner similar to that of sporozoites (Speer, Hammond, and Anderson, 1970).

Merozoite formation in ten species of *Eimeria* has been studied in cell culture. Merozoites are formed by a radial budding process along the periphery of either the intact immature schizont (Fig. 24) or from subdivisions of the schizont called blastophores (Fig. 25). All of the species studied—*E. alabamensis* (Sampson *et al.*, 1971), *E. auburnensis* (Clark and Hammond, 1969), *E. bilamellata* and *E. callospermophili* (Speer, Hammond, and Anderson, 1970), *E. bovis* (Fayer and Hammond, 1967; Hammond and Fayer, 1968), *E. crandallis* (de Vos and Hammond,1971; de Vos *et al.*, 1972), *E. larimerensis* (Speer and Hammond, 1969), *E. magna* (Speer and Hammond, 1971), *E. ninakohlyakimovae* (Kelley and Hammond, 1970), and *E. tenella* (Strout and Ouellette, 1970)—form merozoites from intact schizonts; five of them (*E. auburnensis, E. bovis, E. crandallis, E. ninakohlyakimovae,* and *E. tenella*) also form merozoites from blastophores. According to de Vos and Hammond (1971) and de Vos *et al.* (1972), some *E. crandallis* schizonts become lobulated. The lobules eventually separate to form individual schizonts. Merozoites form from these in the same manner as from schizonts which have not undergone subdivision. *Eimeria tenella* merozoites of both the first and second generation that develop from intact schizonts are usually arranged in the shape of a rosette. On other occasions, they are simply detached from the central mass in an irregular manner without evidence of symmetry or pattern (Strout and Ouellette, 1970). Long (1969) found four different types of mature *E. tenella* schizonts and some of his rosette-shaped Type 2 schizonts (second generation?) had two rows of merozoites. It takes from 60 to 96 min for *E. larimerensis* to complete merozoite formation, beginning with the appearance of the conical elevations on the periphery of the schizont and ending with the formation of the posterior refractile body of each merozoite (Speer and Hammond, 1970b). The process in *E. magna* takes 56 min (Speer and Hammond, 1971b).

A merozoite usually contains only one nucleus. However, Speer and Hammond (1971b) found *E. magna* merozoites that were multinucleate. Seventy-two hours after inoculation, 30% of the mature schizonts had four (ranging from 2 to 8) multinucleate merozoites; these take an average of 68 min for complete formation. Multinucleate merozoites were not observed to undergo any further development.

Isospora lacazei and/or *chloridis* develop to mature schizonts by schizogony (Turner and Box, 1970). However, Fayer and Mahrt (1972) found no evidence of schizogony with *I. canis*. Instead, they found evidence that

Fig. 24. *Eimeria bovis* schizont in which budding of merozoites has begun. Dark bodies (db) at tips of some merozoites are probably developing paired organelles. A nucleus is visible at the base of merozoites in a portion of the schizont. Embryonic bovine tracheal cells, 9 days. × 1,000. (From Hammond and Fayer, 1968.)

Fig. 25. Large, ruptured *E. ninakohlyakimovae* schizont, with released merozoites and residual bodies of blastophores (arrows) surrounded by almost completely formed merozoites. Embryonic lamb tracheal cells, 16 days. × 175. (From Kelley and Hammond, 1970.)

Fig. 26. Paired *Isospora canis* organisms in embryonic canine kidney cells. A sporozoite is in an adjacent cell, 9 days. × 1000. (From Fayer and Mahrt, 1972.)

Fig. 27. Cluster of mature second-generation *E. tenella* schizonts in chicken kidney cells, 4 days. × 380. (Doran, original.)

Fig. 28. A macrogamete (MAC), microgametocyte (MIC), and a developing oocyst of *E. tenella* in chicken kidney cells, 6 days. × 380. (From Doran, 1970c.)

Fig. 29. Young microgametocytes of *E. tenella* in chicken kidney cells, 6 days. × 380. (From Doran, 1970c.)

Fig. 30. *E. tenella* oocysts in chicken kidney cells, 7 days. × 380. (Doran, original.)

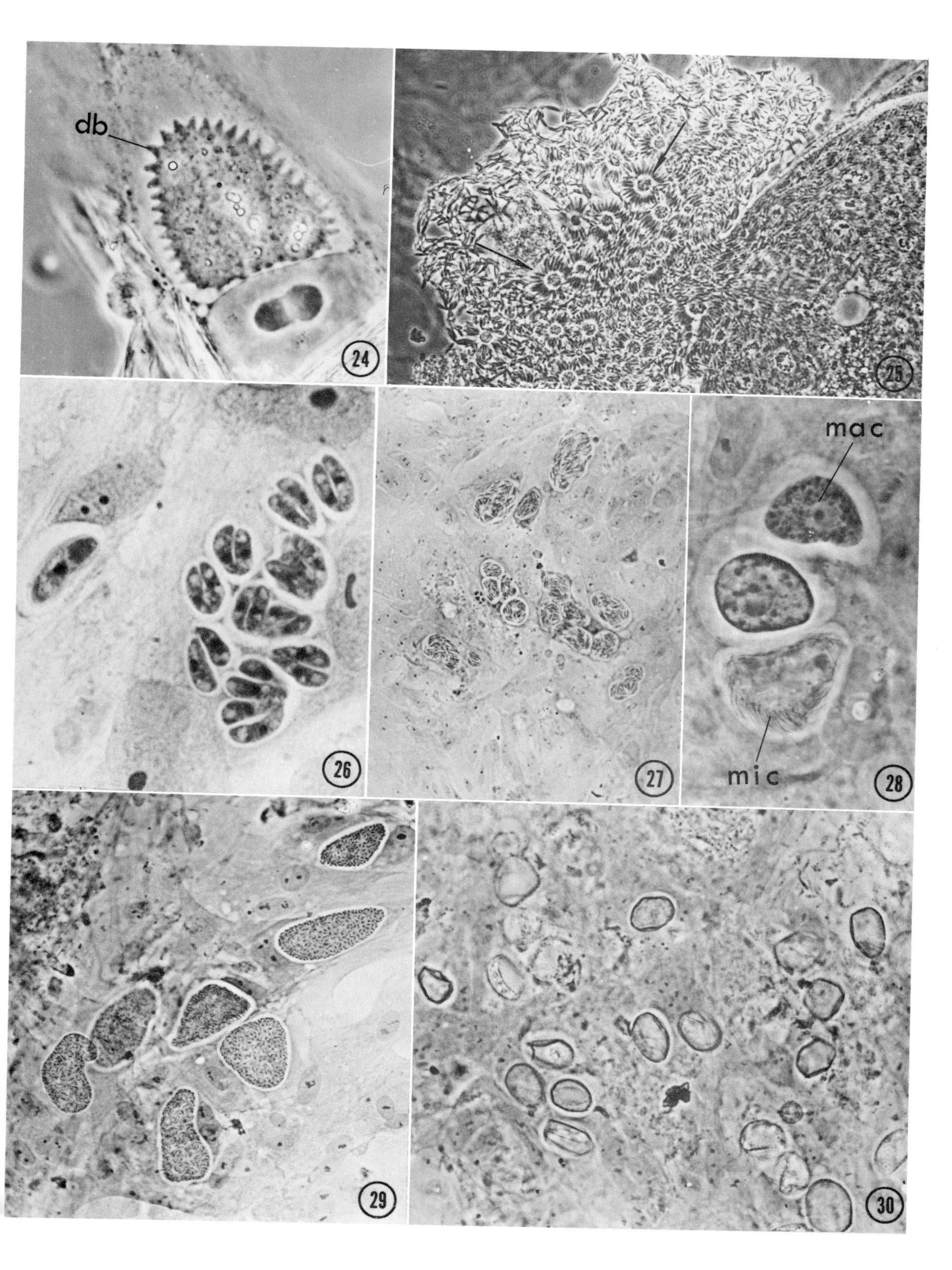
db
24
25
mac
mic
26
27
28
29
30

endodyogeny, a form of development present in *Toxoplasma, Besnoitia,* and *Sarcocystis,* probably takes place. Organisms are arranged in pairs within one parasitophorous vacuole and frequently appear to be attached at one end (Fig. 26). These organisms, which were first found three days after inoculation, were shorter and narrower than sporozoites, stained more intensely, and contained a proportionately larger nucleus.

After development of *E. tenella* first- and second-generation schizonts is complete, the merozoites within become spontaneously activated (Strout and Quellette, 1970). The relatively flat schizont assumes a spherical shape and becomes nearly detached from the cell. The release of activated merozoites from the schizont does not occur simultaneously. Some escape by penetrating the extremely elastic membrane; after from two to three hours of activity, others are released when the schizont membrane eventually tears and disintegrates. Occasionally, a few merozoites do not escape and become inactive within the collapsed membrane. According to Sampson *et al.* (1971), when a first-generation *E. alabamensis* merozoite leaves a cell another immediately follows. There is apparently no constriction of the body while leaving and 10 to 20 min elapse before other merozoites leave the schizont, but one to four merozoites remain in the vacuole after others have escaped. Merozoites of *E. callospermophili* do undergo constriction as they leave (Roberts, Hammond, and Speer, 1970).

Hammond *et al.* (1969) give the only account concerning numbers of merozoites produced in culture. They inoculated one group of 14 bovine tracheal cell cultures with 270,000 *E. bovis* sporozoites and another group of 14 with 390,000 sporozoites. At daily intervals from the 10th through the 30th days after inoculation, they changed the media and counted the number of free merozoites in the media removed. Merozoites first appeared on the 12th or 13th day. The number increased daily until a peak was reached on the 21st day. The number then progressively declined until the 30th day, at which time very few were found. The average total number per tube over the sampling period in the group of cultures inoculated with 270,000 sporozoites was approximateley 15 million; in the group given 390,000 sporozoites it was only two million. The authors consider these to be large yields. Actually, only approximately 55 merozoites were recovered for every sporozoite in the smaller inoculum. In the host, the number of merozoites in a mature first-generation schizont is estimated to be 120,000 (Hammond *et al.*, 1946). The small number of merozoites recovered in culture may be due to small numbers of sporozoites entering cells and developing, to a loss of merozoites in the 24 hr period between counts, or to the development of schizonts in cell culture which contain fewer merozoites than schizonts that develop in the host.

Extracellular first-generation merozoites undergo flexing (Clark and Hammond, 1969; Sampson *et al.*, 1971) and gliding (Sampson *et al.*, 1971; Speer, Hammond, and Anderson, 1970; Speer *et al.*, 1971). *Eimeria callospermophili* occasionally undergoes rapid flexions and glides forward in abrupt movements usually lasting from one to two seconds.

Hammond *et al.*(1969) observed that *E. bovis* merozoites produced *in vitro* entered only mouse macrophages or, if inoculated along with sterile fluid from the cecum of a calf, they also entered embryonic bovine tracheal cells. Merozoites from calves entered bovine ileal cells when untreated or treated with large and small intestine tissue extracts, but no development occurred and intracellular merozoites degenerated within two days. The investigators suggest that (a) merozoites produced in culture may be incapable of further development, (b) an activating substance is needed for penetration and development, and (c) merozoites may be selective with regard to host cells. They disproved the first of these possibilities when *in vitro* merozoites produced gametocytes in a calf, and their experiments are inconclusive with regard to the other two possibilities.

Bile, bile salts, and trypsin have also been investigated as possible stimulators of merozoite activity. *Eimeria bovis* merozoites do not enter cells when preincubated for ten minutes in a medium that contains 0.2% trypsin and 5% bile (Hammond *et al.*, 1969). A medium containing 0.5% trypsin and 2% bovine bile increases the motility of intra- and extracellular *E. callospermophili* merozoites and increases the number of merozoites leaving a schizont (Speer, Hammond, and Anderson, 1970). In media containing 4% bile or 0.5% of a bile salt or one of these in combination with 0.2% trypsin, extracellular merozoites of *E. bilamellata, E. callospermophili, E. larimerensis, E. nieschulzi,* and *E. ninakohlyakimovae* are markedly motile and merozoites within schizonts, except for those of *E. ninakohlyakimovae*, become active and leave the cell (Speer, Hammond, and Kelley, 1970). In such media, the merozoites move for longer distances than normal and those of *E. callospermophili* penetrate new host cells (Speer, Hammond, and Anderson, 1970). Speer *et al.* (1971) later found that 1% bovine bile or sodium taurocholate alone increased the activity of *E. larimerensis* merozoites that were extracellular or within schizonts.

First-generation *E. alabamensis* merozoites penetrate new cells within one hour after leaving a schizont (Sampson *et al.*, 1971). Those of *E. callospermophili* take from five to ten seconds to enter and then move toward the host cell nucleus (Speer, Hammond, and Anderson, 1970). Trophozoites of *E. meleagrimitis* (Doran and Vetterling, 1968a) and trophozoites and schizonts of *E. necatrix* (Doran, 1971b) and *E. tenella* (Doran, 1970c; Strout and Ouellette, 1970) belonging to the second generation are most frequently found in clusters (Fig. 27). Consequently, it is assumed that first-generation merozoites invade cells in the immediate vicinity of the schizont in which they develop.

After leaving a cell, second-generation *E. tenella* merozoites occasionally form a complete circle either by the bending of one or by the apparent fusion of two merozoites (Strout and Ouellette, 1970). Intracellular, amoeboid T-shaped merozoites capable of movement in either direction are also present (Strout and Ouellette, 1970). The present author has frequently encountered bent and nearly circular merozoites. However, since internal structures other than a nucleus were not visible, they were considered degenerate forms. The author has also observed that intracellular merozoites, particularly those of the second generation, change shape and move slowly giving the appearance of amoeboid movement.

Bedrnik (1969a) compared penetration by sporozoites and second-generation merozoites. With a dosage of 1.8 million merozoites, only 58 and 10 were found in 100 microscopic fields after 7 and 24 hr, respectively; with 1.5 million sporozoites, 709 and 625 in 50 fields after 24 and 48 hr, respectively.

ii. The third and other generations

Small, mature *E. tenella* schizonts containing merozoites 10 μ long consistently appear at the time of gametogony in embryonic chick kidney cells (Strout and Ouellette, 1970). Schizonts from 12 μ to 20 μ in diameter with from 5 to 17 merozoites the same size as those of the second generation appeared at six and seven days in one of two experiments with nonembryonic chick kidney cells (Doran, 1970c). These schizonts, which are usually situated close to a cluster of immature or mature second-generation schizonts (Doran, 1970c), were thought to belong to a third generation (Doran, 1970c; Strout and Ouellette, 1970). A few small *E. necatrix* schizonts, similar to the *E. tenella* schizonts possibly belonging to a third generation, were first found five and six days after inoculation. However, since they appeared on the same day as second-generation, mature schizonts, it is unlikely that they belonged to a third generation (Doran, 1971b).

Bedrnik (1967a,b; 1969b) inoculated cell cultures with second-generation *E. tenella* merozoites obtained from the cecum of chickens. Merozoites entered cells within five minutes. Trophozoites appeared at from one to two hours (Bedrnik, 1969b), multinucleate immature schizonts appeared at from two to four hours (Bedrnik, 1967a,b), and mature third-generation schizonts appeared at 14 hr (Bedrnik, 1969b) and 24 hr (Bedrnik, 1967a,b). Merozoites invaded cells in cultures prepared from trypsinized whole chick embryo rather unevenly, sometimes accumulating two or more within a cell, giving the impression that merozoites sought out cells (Bedrnik, 1969b). These areas of accumulation are probably the epithelial-like islets of cells in which most mature schizonts were later found. Schizonts develop in HeLa cells, but the merozoites are not released (Bedrnik, 1967a). In chick embryo cell cultures, merozoites are released and quickly invade new cells (Bedrnik, 1969b). Additional small, mature schizonts appeared from 24 to 48 hr later. On the basis of size and time of appearance, these were assumed to constitute a fourth generation. In other work (Bedrnik, 1969a), a fifth generation is mentioned. Only three generations are known to exist in the host.

iii. "Activating" factors

Various cell types grown under similar conditions differ in their ability to support development of a parasite and parasites differ in their ability to develop within the same cell type. The presence of either intrinsic mechanisms within the parasite or cellular activating factors or both have been suggested to account for these differences (Bedrnik, 1970a; Doran, 1971a; Doran and Vetterling, 1967a,b; Fayer and Hammond, 1967; Hammond *et al.*, 1969). Bedrnik (1970a) obtained greater development of *E. tenella* merozoites after using inocula containing "misintroduced" cells from the cecum of chickens. Following a series of mostly nonquantitated experiments, he claimed that there is an unknown stimulatory factor that (a) is not present in the cecal cells of uninfected chickens, (b) is bound on intact cell structures and can be destroyed by lysis, (c) is "serially" transferred from second- to third-generation merozoites and (d) is probably introduced into the cecum of uninfected chickens from oocysts during excystation. In initial experiments, Bedrnik counted the number of merozoites in a variable number of microscopic fields and then calculated the assumed number of intracellular merozoites at a uniform concentration of 1 million merozoites/1 ml of inoculum. In four of these experiments, the values for the "N" inoculum (merozoites + "misintroduced" cecal cells) were 10, 22.6, 32, and 38; corresponding values for the "P_2" inoculum (merozoites only) were 10, 17.1, 16, and 26. However, he used inocula ranging in size from 0.5 to 1.7 million without determining whether the number entering cells was proportional to dosage. He also did not take into consideration the confluency of the cell layer. On the basis of these two shortcomings, which must be considered before an accurate comparison of infectivity can be made (Doran, 1971a), the present author believes that the limited quantitative data presented are insufficient evidence for supporting the presence of a cellular growth factor.

c. Sexual development

i. Gametocytes

Macrogametes and microgametocytes of *E. tenella* (Fig. 28) are first found six days after inoculation with sporozoites (Doran, 1970c; Strout and Ouellette, 1968, 1970), and one to three days after inoculation with second-generation merozoites (Bedrnik, 1967b; 1969b).

After inoculation with *E. magna* merozoites (generation undetermined), immature gametocytes are present at 12 hr and mature macrogametes and microgametocytes are present at about 60 hr and 72 hr, respectively (Speer and Hammond, 1972).

The number of *E. tenella* gametocytes developing in relation to asexual schizonts is extremely small (Bedrnik, 1967b; Strout and Ouellette, 1968, 1970). They are not found in embryonic chick kidney (ECK) or in nonembryonic chick kidney (CK) cell cultures that are prepared from cell suspensions containing only a few cell aggregates (Doran, 1970c). Sexual stages are found in cultures prepared from cell suspensions with an abundance of aggregates. More are found in CK than ECK at seven days. A single gametocyte is rarely seen. They usually occur in clusters, macrogametes and microgametocytes together, within islets or patches of cells (Bedrnik, 1967b, 1969b; Doran, 1970c; Strout and Ouellette, 1968, 1970). Occasionally, each type of gametocyte occurs by itself in clusters of from 5 to 18 (Fig. 29). Bedrnik (1967a,b; 1969b) states that gametocytes do not appear in fibroblast-like cells in primary cultures of chick embryo cells, but only in islets of epithelial-like cells. He further states (1967b) that the islets of epithelial-like cells occur infrequently and not in all experiments. By this, he may be implying that a certain type of cell is required for gametocyte development.

ii. Oocysts

Eimeria tenella oocysts are produced only sporadically from 1 to 2 days after inoculation with second-generation merozoites (Bedrnik, 1967b). When cultures are established with parasitized tissue from the chorioallantoic membrane, oocysts are produced, but the number is small in comparison with the number of gametocytes (Long, 1969). With sporozoites as the inoculum, oocysts are consistently produced and yields ranging from approximately 1 to 900/50 mm² of coverslip area have been obtained (Doran, 1970c; 1971a). Oocysts produced in culture (Fig. 30), which are found intra- and extracellularly (Bedrnik, 1969b; Doran, 1971a), sporulate when removed to room temperature and produce infection when fed to their natural host (Doran, 1970c; Long, 1969).

The number of oocysts that develop in two or more different culture tubes held under similar conditions vary more than counts of earlier stages in the life cycle. The percentage deviation between oocyst counts ranges from 34 to 120, whereas at 4 hr and 24 hr the deviation between numbers of sporozoites/10 mm² of cells is from 9 to 24 and from 17 to 21, respectively (Doran, 1971a). Counts of *E. meleagrimitis* first-generation schizonts vary by from 5% to 31% (Doran and Vetterling, 1967a,b; 1968a). Greater variation is to be expected as development progresses beyond the first generation. A difference between cultures of merely one developing first-generation schizont could lead to a wide variation between oocyst counts.

d. Survival in culture

Other than the fact that only limited asexual development has been obtained for most species, perhaps the most disappointing thing about cultivation in cell culture is the loss of intracellular parasites that occurs shortly after inoculation. The loss occurs after inoculation with *Isospora* sporozoites (Turner and Box, 1970), *Eimeria* sporozoites (Doran, 1971a,b; Doran and Vetterling, 1967a,b; 1968a; Hammond *et al.*, 1969), and *Eimeria* merozoites (Bedrnik, 1967a). With an inoculum of sporozoites, the amount of the loss depends on several factors: (a) *Cell type.* Between 5 and 48 hr, 41% of *E. meleagrim-*

itis are lost from embryonic bovine kidney (EBK) cells, whereas from 80% to 81% are lost from embryonic turkey intestine (ETI) cells (Doran and Vetterling, 1968a). Most of the loss in ETI is from the fibroblast-like cells and not from the patches of epithelial-like cells. With *E. bovis*, between one and six days there are losses of 95% in mouse fibroblasts and only 30% in an established cell line of bovine kidney (Hammond and Fayer, 1968). Between 3 and 48 hr, the number of *Isospora lacazei* and/or *chloridis* in embryonic chicken fibroblasts and in embryonic canary fibroblasts decreases by 77% and 42%, respectively (Turner and Box, 1970). (b) *Species of parasite*. When development of different species was compared in primary chicken kidney (CK) cell cultures maintained under similar conditions (Doran, 1971b), losses between 5 and 48 hr were as follows: *E. acervulina* and *E. praecox*, 90% to 96%; *E. mivati*, 70% to 92%; *E. maxima*, 62% to 83%; and *E. necatrix*, 11% to 27%. (c) *Age of sporozoites*. Sporozoites from young *E. meleagrimitis* oocysts survive better than those from older oocysts. In EBK cultures, between 5 and 48 hr, there are 18% and 12% losses with sporozoites 5 and 12 weeks old; with those 53 and 60 weeks old, 78% and 72% losses (Doran and Vetterling, 1969). (d) *Debris (oocyst and sporocyst walls)*. The loss of *E. tenella* sporozoites between 4 and 24 hr from CK cell cultures inoculated with sporozoite suspensions cleaned of debris is from 50% to 60%, whereas with those uncleaned it is from 74% to 85% (Doran, 1971a). The loss after inoculation of cultures with *E. adenoeides* sporozoites from uncleaned suspensions increases with dosage, whereas with sporozoites from cleaned suspensions it does not increase (Fig. 31). This difference clearly indicates that debris hinders survival. A number of reasons have been suggested to explain why a loss of parasites occurs (Bedrnik, 1967a, 1969b; Doran and Vetterling, 1967a,b; Patton, 1965; Strout and Ouellette, 1970; Strout *et al.*, 1965). Perhaps the more logical of these suggestions could be summed up by saying that the loss is the result of inadequate or deteriorating living conditions within the cell culture.

Intracellular sporozoites can survive for a long time after inoculation. Sporozoites of *E. bovis* are found on the 17th day (Fayer and Hammond, 1967); those of *E. auburnensis* on the 18th day (Clark and Hammond, 1969). Fairly large numbers of *E. necatrix* are present through the 9th day, but those of *E. acervulina* and *E. praecox* are absent from cultures by the 3rd and 4th days (Doran, 1971b). Development is not synchronous. Immature, "spherical-type" *E. auburnensis* schizonts are present from the 5th to 18th days (Clark and Hammond, 1969). Mature first-generation schizonts are found from day 2 through day 6 (Doran and Vetterling, 1968a), 8 through 18 (Hammond and Fayer, 1968), 2 through 9 (Doran, 1971c) and 2 through 5 (Doran, 1970c). Mature second-

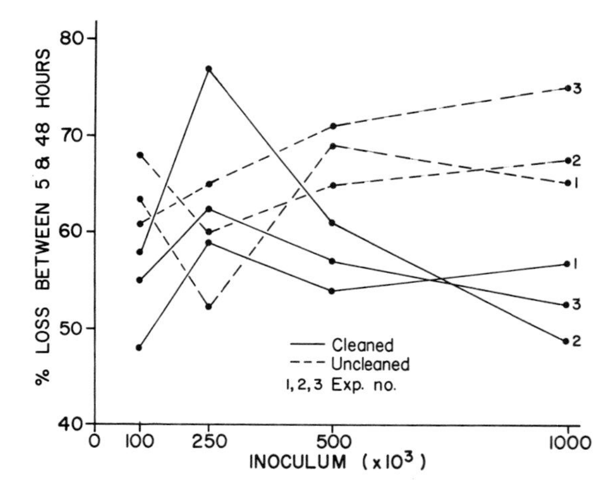

Fig. 31. Percentage loss of intracellular *E. adenoeides* developmental stages in embryonic bovine kidney cell cultures inoculated with sporozoites from cleaned and uncleaned suspensions. (From Doran, 1970b.)

generation schizonts persist from days 4 through 9 (Doran, 1970c).

e. Cyto- and "histopathological" effects

The nucleus of the host cell either increases in size (Clark and Hammond, 1969; Fayer and Hammond, 1967; Sampson *et al.*, 1971; de Vos and Hammond, 1971; de Vos *et al.*, 1972) or decreases in size (Speer, Hammond, and Anderson, 1970). It slowly enlarges early in development of *E. auburnensis* sporozoites with the effect becoming more pronounced in tracheal cells than in an established cell line of bovine kidney (Clark and Hammond, 1969). When *E. alabamensis* develops intranuclearly, the nucleus becomes larger than if development had been in the cytoplasm (Sampson *et al.*, 1971). The nucleus may also become vacuolated and irregular-shaped (Speer, Hammond, and Anderson, 1970) and a cell may become binucleate (Clark and Hammond, 1967; Sampson *et al.*, 1971; Speer, Hammond, and Anderson, 1970) or multinucleate (Clark and Hammond, 1969). The nucleolus enlarges (Fayer and Hammond, 1967; Hammond and Fayer, 1968; Kelley and Hammond, 1970; de Vos and Hammond, 1971; de Vos *et al.*, 1972) and the chromatin either remains unchanged (Clark and Hammond, 1969), or the chromatin clumps become finely granular or are no longer visible (Fayer and Hammond, 1967; Hammond and Fayer, 1968; Kelley and Hammond, 1970).

Cytoplasm of the host cell becomes vacuolated (Bedrnik, 1969b; Doran and Vetterling, 1967a; de Vos and Hammond, 1971; de Vos *et al.*, 1972), vacuolated and granular (Clark and Hammond, 1969; Kelley and Hammond, 1970; Speer, Hammond, and Anderson, 1970), or only granular (Fayer and Hammond, 1967; Hammond and Fayer, 1968). Cells are greatly vacuolated when parasitized by more than four sporozoites, more than one small type schizont, or one large type schizont of *E. necatrix* (Doran and Vetterling, 1967a). Cells enlarge (Clark and Hammond, 1968; Doran and Vetterling, 1967a; Patton, 1965; Speer, Hammond, and Anderson, 1970). Cells infected with large *E. meleagrimitis* schizonts become from two to three times larger than normal (Doran and Vetterling, 1967a) and cells with *E. tenella* schizonts become from eight to ten times larger (Patton, 1965).

Cytopathological changes are most noticeable either soon after mature schizonts are formed or when merozoites are leaving a cell. Clark and Hammond (1969) noticed that vacuolization and granulation are first limited to a cell parasitized with a schizont, but then spread to neighboring nonparasitized cells and eventually affect the entire monolayer. They state that the timing of degenerative changes in relation to time of appearance of mature schizonts suggests that this stage may have a toxic effect on cells.

Like *Toxoplasma, Eimeria* also destroy cells. The host cell membrane is destroyed as merozoites of *E. alabamensis* escape from nuclei (Sampson *et al.*, 1971). Cells containing free merozoites or schizonts with merozoites become detached from the cell layer (Doran and Vetterling, 1967b, 1968a). Unlike *Toxoplasma*, there is very little reinvasion of cells by *Eimeria* merozoites. Consequently, nothing similar to circular foci or plaques is produced. However, *Eimeria* do produce an effect on the monolayer. "Breaks" appear in the monolayer after inoculation with one million or more second-generation merozoites (Bedrnik, 1969b). With sporozoites, the effect on the monolayer depends on size of the inoculum. Doran (1971a) inoculated cell cultures with dosages up to and including one million sporozoites that had been freed of debris (oocyst and sporocyst walls). A progressive retardation of cell growth occurred with increased dosage. Cultures inoculated with 250,000 sporozoites or less were from 95% to 100% confluent at seven days, whereas those that received a higher number were progressively less confluent. In cultures that were given one million sporozoites, the cell layers, which were only 40% confluent, contained a large amount of cellular debris and many vacuolated cells.

f. Cell culture and host comparisons

There is considerable variation between species in how closely development in culture resembles development in the host. Mature first-generation schizonts of most species in culture are smaller and contain fewer merozoites (Clark and Hammond, 1969; Doran, 1970c; Doran and Vetterling, 1967a; Fayer and Hammond, 1967; Kelley and Hammond, 1970; Matsuoka *et al.*, 1969; Patton, 1965; Sampson *et al.*, 1971; Speer and Hammond, 1970b; Speer, Hammond, and Anderson, 1970; Strout and Ouellette, 1970). *Eimeria meleagrimitis* schizonts vary in size depending on the cell type and time after inoculation (Doran and Vetterling, 1968a). In embryonic turkey intestine (ETI) cells they are extremely small; in embryonic bovine kidney (EBK) cells at 48 hr they are larger than in ETI, but still smaller than in the host. However, at 72 hr in EBK cells they measure from 50 μ to 115 μ by from 20 μ to 70 μ and contain far more merozoites than are present in schizonts that are smaller. Occasionally, *E. meleagrimitis* schizonts reach sizes larger than in the host and contain well over 200 merozoites. Mature *E. ellipsoidalis* and *E. magna* schizonts contain nearly the same number of merozoites in culture and in the host (Speer and Hammond, 1971a,b). The merozoites of *E. ellipsoidalis* are similar in size to those in the host (Speer and Hammond, 1971a), but those of *E. auburnensis, E. meleagrimitis,* and *E. necatrix* are smaller than in the host (Clark and Hammond, 1969; Doran and Vetterling, 1967a,b) and those of *E. tenella* are larger (Doran, 1970c; Matsuoka *et al.*, 1969; Strout and Ouellette, 1970; Strout *et al.*, 1969a). Oocysts of *E. magna* are similar in size, shape, and appearance to those obtained from rabbits (Speer and Hammond, 1972). Those of *E. tenella* are mostly similar in size to those from chickens, but a few are larger (Doran, 1970c).

Mature first-generation schizonts of most species first appear in cell culture and in the natural host at the same time (Clark and Hammond, 1969; Doran, 1970b; Doran and Vetterling, 1967a and b, 1968a; Kelley and Hammond, 1970; Sampson *et al.*, 1971). However, with *E. bovis* the time varies in different cell types; in most of the cells tried, it was about the same as in the host, but in some it was several days less (Fayer and Hammond, 1967; Hammond and Fayer, 1968). Schizonts of *E. ellipsoidalis* developed to maturity in from five to six days in cell cultures (Speer and Hammond, 1971a), whereas mature schizonts were seen in calves from 10 to 14 days after inoculation (Hammond *et al.*, 1963). Because of this great time difference, the former workers believe that the schizonts in calves, which were described as belonging to an unknown generation (Hammond *et al.*, 1963), are those of a generation later than the first. Mature schizonts of two ground squirrel species, *E. callospermophili* and *E. larimerensis*, also appear at a time about one-half that in the host (Speer and Hammond, 1970b; Speer, Hammond, and Anderson, 1970). The times when *E. tenella* schizonts, gametocytes, and oocysts first appear in cell cultures and in chickens are shown in Table 8. Development is apparently faster in cell culture at the beginning of the life cycle, but gradually becomes slower. At the time gametocytes appear, the timing of the cycle in the host and in chicken kidney cells is similar, but longer in cells from birds that are not the natural host. By the time oocysts appear, the cycle in cell culture is longer in both embryonic chicken and nonhost kidney cells.

TABLE 8. Timing of first appearance of *E. tenella* schizonts, gametocytes, and oocysts in the chicken and in cell culture

Chicken			Cell culture		
Developmental stage	Time (days)	Reference	Time (days)	Cell type[a]	Reference
Schizont, mature (first generation)	2.5-3	Levine (1961)	2	BK	Patton (1965)
			2.75	EBTr	Matsuoka *et al.* (1969)
			2	CEK	Doran (1970c); Strout and Ouellette (1968; 1970)
			2	CK	Doran (1970c; 1971c)
			2	PhK	Doran (1971c)
			2	PK	Doran (1971c)
			2	TK	Doran (1971c)
Schizont, mature (second generation)	5	Levine (1961)	4	CEF	Shibalova (1969a); Shibalova and Koralev (1969)
			4	CEK	Doran (1970c); Strout and Ouellette (1968; 1970)
			4	CK	Doran (1970c; 1971c)
			5	PhK	Doran (1971c)
			5	PK	Doran (1971c)
			5	TK	Doran (1971c)
Gametocyte	6	Scholtyseck (1953)	6	CEK	Doran (1970c); Strout and Ouellette (1968; 1970)
			6	CK	Doran (1970c; 1971c)
			7	PK	Doran (1971c)
			8	PhK	Doran (1971c)
Oocyst	7	Levine (1961)	6	CK	Doran (1970c; 1971a,c)
			8	CEK	Doran (1970c)
			9	PhK	Doran (1971c)

[a]Abbreviations: BK = bovine kidney; EBTr = embryonic bovine trachea; CEK = embryonic chick kidney; CK = chick kidney; CEF = embryonic chick fibroblasts; PhK = pheasant kidney; PK = partridge kidney; TK = turkey kidney. (All are primary cultures except EBTr.)

4. *Sarcocystis*

a. Cells used in attempts to obtain development

The only reports concerning *Sarcocystis* in cell culture are those of Fayer (1970; 1972). The cells that he used are listed in Table 5. The inoculum always consisted of banana-shaped organisms obtained from cysts found in leg or breast muscle of purple grackles (*Quiscalis quiscula*).

b. Nuclear changes and stages found in culture

In his first report, Fayer described nuclear changes during transformation of banana-shaped forms to ellipsoid-oblong forms. The changes are nearly identical to those that he observed in *Eimeria bovis* sporozoites (Fig. 23a-d). The cytoplasm of the ellipsoid-oblong forms (Fig. 17) contained many granules not observed in other stages. From two- to nine-nucleate stages (Fig. 18) were present in chicken kidney cells at 30 hr and in chicken, turkey, and bovine kidney cells at 48 and 72 hr. Cyst-like bodies, that were usually ovoid or lemon-shaped, were also present at 48 and 72 hr. At one end of the body they contained a sphere of what appeared to be granular cytoplasm that occupied more than one-half the volume (Fig. 19). A few of the cyst-like bodies contained two inclusions, some of which resembled banana-shaped organisms (Fig. 20). Fayer states that the finding of nuclear and cytoplasmic transformation and multinucleate stages similar to *Eimeria* schizonts suggests that *Sarcocystis* may be related to this genus.

In his second study, Fayer found the ellipsoid-oblong forms and multinucleate forms at 24 hr and structures obviously mature microgametocytes (Fig. 21) at 30 hr. The ellipsoid-oblong forms were then considered to be macrogametes and the multinucleate forms, previously reported as resembling *Eimeria* schizonts, were considered to be immature microgametocytes. At 48 and 72 hr, the cyst-like bodies were the only stage present. Attempts to obtain sporulation at room temperature were unsuccessful. Fayer states that the occurrence of sexual stages preceding formation of cyst-like bodies suggests that these bodies may be equivalent to the oocyst stage of coccidia. He also states that the motile banana-shaped organisms obtained from cysts in muscle represent a merozoite stage that gives rise directly to sexual stages in cell culture.

C. FACTORS OTHER THAN STRAIN DIFFERENCES INFLUENCING THE NUMBER OF PARASITES THAT PENETRATE CELLS AND DEVELOP

1. *Parasitological Factors*

a. Size of the inoculum

The size of a *T. gondii* inoculum is most frequently expressed as a dilution of pure peritoneal exudate and not as an actual parasite count. Regardless of the method, size of the inoculum influences the number of parasites entering cells. The number entering, expressed as either the number of infected cells (Holz and Albrecht, 1953) or RNIU's (relative number of infective units) (Lycke and Lund, 1964a; Norrby, 1970b), is proportional to the size of the inoculum. The optimal dosage of the RH strain for maximal invasion is one that results in five parasites/cell (Hirai *et al.*, 1966). Size of the inoculum also influences the amount (Hogan *et al.*, 1961) and time (Balducci and Tyrrell, 1956) of initial cell destruction, plaque size (Akinshina and Zasuchina, 1966), the time when large numbers appear in the culture media (Balducci and Tyrrell, 1956; Chernin and Weller, 1957; Ma and Mun, 1961; Stewart and Feldman, 1965; Vischer and Suter, 1954), and the time when complete degeneration of the cell layer occurs (Cook and Jacobs, 1958a).

All work concerning the effect of inoculum size on *Eimeria* has been carried out with *E. tenella*. Strout *et al.* (1969b) inoculated cultures with different quantities of sporozoites ranging from 10,000 to one million and compared infection and development 96 hr later. Their data represent only what was present at 96 hr and give no indication of the relationship between dosage and the number entering cells. However, they did find that more asexual development occurred in cultures inoculated with 10,000 sporozoites while less occurred in cultures inoculated with one million sporozoites. Doran (1971a) compared infection at four hours and oocyst production at six and seven days in cultures inoculated with from 25,000 to 1 million sporozoites freed of debris (oocyst and sporocyst walls). The number of intracellular sporozoites at four hours was roughly proportional to the size of the inoculum up to 500,000; with 750,000 and 1 million, the number was similar to that with 500,000. The number of oocysts produced was greatest with a dosage of 100,000 sporozoites, but rapidly declined with dosages of 250,000 and greater.

b. Age of the parasite and time when obtained from the host

Age of *Eimeria* oocysts is an important factor and it is one that is frequently overlooked. There is greater development of *E. meleagrimitis* 48 and 72 hr after inoculation with sporozoites from 5 to 34 weeks old than there is with sporozoites 53 and 60 weeks old (Doran and Vetterling, 1969). In experiments using previously frozen sporozoites, development of *E. adenoeides* and *E. tenella* is greater at 48 hr with sporozoites frozen when three or four weeks old than it is with sporozoites 36 or 54 weeks old (Doran, 1970a). The effect is probably entirely the result of age, but might, at least in part, be the result of storage at low temperature.

Toxoplasmas taken from peritoneal exudate earlier than four days after infection of the animal produce a greater infection in culture than those taken from exudate at four days or later (Hirai *et al.*, 1966; Shimizu, 1961a). The difference might be caused by more toxin in exudate on the later days. Cultures are frequently rinsed several times following inoculation and several investigators (Hirai *et al.*, 1966; Lund *et al.*, 1961b; Shimizu, 1961a) mention that the reason for rinsing is to remove toxin. Shimizu (1961a) compared rinsed and unrinsed cultures and obtained greater infection in those that were rinsed. Second-generation *E. tenella* merozoites taken from the cecal mucosa of chickens six days after infection always produce sexual stages when inoculated into cell culture, whereas those taken on the 5th day produce sexual stages only in a few instances (Bedrnik, 1969a,b; 1970a). This is probably caused not by any toxic effect, but rather by a greater number of mature merozoites at six days.

c. Treatment of parasites prior to inoculation

Freeze-thawing and cleaning sporozoite suspensions of oocyst and sporocyst walls by passage over glass beads affect the number of *Eimeria* sporozoites in an inoculum that enter cells. When subjected to freeze-thawing, a smaller number are intracellular at five hours as compared with previously unfrozen controls (Doran, 1969; 1970a). After passage over glass beads, fewer sporozoites are intracellular at four or five hours than when they are untreated (Doran, 1970b; 1971a). The difference between treated and untreated increases with dosage (Fig. 32). Freeze-thawing and passage over beads might alter the sporozoite in a way that would affect its ability to penetrate cells or survive up to four or five hours in a cell. However, it is more probable that some of the sporozoites in the inoculum, judged as living on the basis of their refractiveness, are really dead.

Freeze-thawing has no effect on development. The percentages of development at 48 hr with frozen and unfrozen sporozoites are similar (Doran, 1969, 1970a). However, cleaning by passage over beads has a decided effect on development. With *E. adenoeides*, the number of schizonts in cultures inoculated with cleaned suspensions is directly proportional to size of the inoculum; in cultures inoculated with uncleaned suspensions, this relationship does not occur (Fig. 33). The decreased yield with uncleaned inocula greater than 250,000 is probably caused by

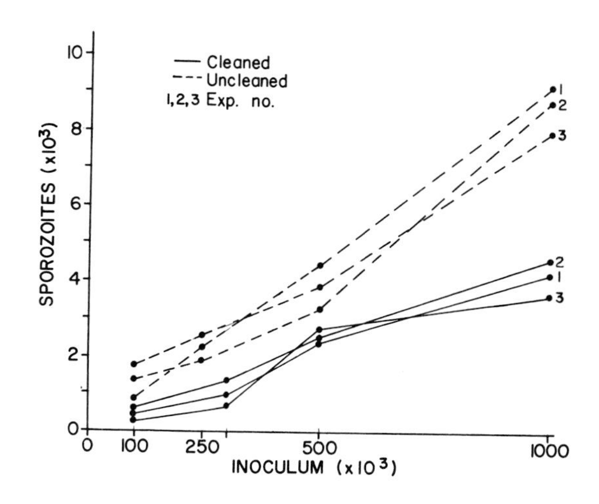

Fig. 32. Number of intracellular *E. adenoeides* sporozoites 5 hr after inoculation of embryonic bovine kidney cell cultures with sporozoites from cleaned and uncleaned suspensions. (From Doran, 1970b.)

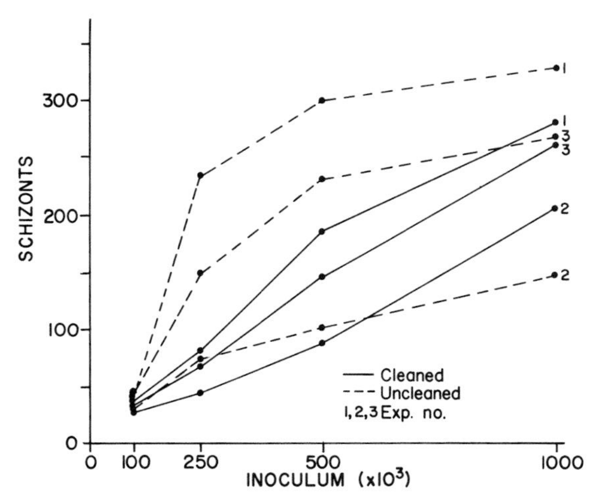

Fig. 33. Number of *E. adenoeides* schizonts 48 hr after inoculation of embryonic bovine kidney cell cultures with sporozoites from cleaned and uncleaned suspensions. (From Doran, 1970b.)

the effect that debris itself has on the monolayer. The yield of *E. tenella* oocysts at seven days is three times greater with cleaned suspensions (Doran, 1971a).

Results with *E. auburnensis* suggest that pretreatment of oocysts with CO_2 before grinding to release sporocysts may have an adverse effect on development (Clark and Hammond, 1969).

2. *Environmental Factors*

a. Culture chamber

Toxoplasmas inoculated into roller tubes (tubes in a revolving drum) infect fewer cells than when inoculated into stationary tube cultures (Kaufman, Melton, Remington, and Jacobs, 1959; Kaufman *et al.*, 1958). With the RH strain, the percentage of infected cells in roller tubes is about one-fourth that in stationary cultures; with strain 113-CE, about one-third as many (Kaufman, Melton, Remington, and Jacobs, 1959). The latter investigators indicate that the differences are probably the result of a more prolonged contact between the organisms and cells in stationary cultures. As judged by lysis time instead of direct counts, differences between infection in roller and stationary tubes are not detectable (Cook and Jacobs, 1958a).

b. Temperature

The optimal temperature for multiplication of *T. gondii* is 37°C (Akinshina, 1965). Plaque formation occurs at 42°C, but the number of plaques is not as great as at 37°C (Akinshina and Zasuchina, 1966, 1967). The rate of multiplication decreases as the temperature is lowered (Cook and Jacobs, 1958a; Kaufman, 1961; Maloney and Kaufman, 1964; Meyer and de Oliveira, 1943; Stewart and Feldman, 1965; Yü and In, 1966). The number of toxoplasmas/cell after 24 hr at 38°C, 29°C, and 4°C is 8.80, 2.23, and 1.44, respectively (Kaufman, 1961). The mean generation time during the exponential growth phase at 37°C is 4.85 hr; at 31°C, 11.1 hr (Maloney and Kaufman, 1964). Balducci and Tyrrell (1956) placed heavily inoculated cultures at 4°C and found that organisms were gradually released into the medium and could be detected there for four weeks. However, Cook and Jacobs (1958a) held cultures at 4°C for 30 days and found no degeneration caused by proliferation of organisms. They also found that the culture material was noninfective to mice upon intraperitoneal inoculation.

Mature *Isospora lacazei* and/or *chloridis* schizonts develop at 37°C, 41°C, and 43°C, but fewer developing parasites are found in cells grown at 37°C than in those grown at the higher temperatures (Turner and Box, 1970). Patton (1965) found mature *Eimeria tenella* schizonts at 41°C, but not at 37°C. In a detailed study involving temperatures from 35°C to 45°C, Strout *et al.* (1969a) later found that *E. tenella* undergoes schizogony at temperatures from 35°C to 43°C, that mature schizonts are present at 36°C to 43°C, and that the optimum temperature for asexual development is 41°C. With second-generation merozoites as the inoculum, mature third-generation schizonts develop at 37°C and from 39°C to 40°C (Bedrnik, 1970a).

c. Debris (oocyst and sporocyst walls) and toxins

After inoculation of cell cultures with either *T. gondii* or *Eimeria,* most investigators rinse their cells one or more times and then add fresh media. As previously mentioned in the section on age of the parasite and time when obtained from the host, rinsing of cultures infected with *Toxoplasma* was performed to remove toxins carried over from the exudate. However, opinions differ as to whether a "toxotoxin" produces an effect in culture. Some claim there is an effect (Hirai *et al.*, 1966; Lund *et al.*, 1961b, 1963a; Lycke and Lund, 1964b; Lycke and Norrby, 1966; Shimizu, 1961a; Vischer and Suter, 1954); others say there is no evidence for a toxin causing degenerative changes in culture (Balducci and Tyrrell, 1956; Chernin and Weller, 1954a, 1957; Cook and Jacobs, 1958a; Hansson and Sourander, 1965; Varela *et al.*, 1955).

With *Eimeria*, rinsing after inoculation was assumed to be for the purpose of removing debris. After Patton (1965) stated that debris was toxic to cells, rinsing became common practice. According to Fayer and Hammond (1967), debris sometimes hinders observation, but does not seem to have any appreciable toxic effect on cells. However, as shown in Figure 33, debris has a definite effect on the numbers of schizonts found after inoculation with uncleaned suspensions. Fayer and Hammond (1967) prepared an extract of *E. bovis* oocysts (debris removed) and found that it killed cells within one hour. The lack of proportionality between size of uncleaned inoculum and numbers of schizonts shown in Figure 33 can hardly be caused by toxins carried over from the oocysts because the uncleaned suspensions were thoroughly washed before inoculation.

d. Media and media changes

The size of *T. gondii* plaques is not influenced by the type of serum in the maintenance medium (Akinshina and Zasuchina, 1966, 1967). However, the amount of serum does have an effect. Lund *et al.* (1963a) compared penetration and development of *T. gondii* in media without serum and in media supplemented with 1%, 10%, and 20% human serum. The greatest penetration occurs when a medium containing no serum is used. With 1% serum, about one-half as many parasites are intracellular at three hours as when there is no serum; with 10% and 20% serum, a further decrease occurs. They also found that with all concentrations of serum there is a two-day delay in the liberation of parasites into the medium. Somewhat different results were obtained by Shimizu (1961a, 1963). Better penetration was obtained when the medium contained no serum (Shimizu, 1963), but the rate of multiplication increased rather than decreased when calf serum was added (Shimizu, 1961a, 1963). Composition of media also influences the number of *Eimeria* and *Isospora* sporozoites that penetrate cells. Greater penetration by *Isospora lacazei* and/or *chloridis* sporozoites is obtained when serum is omitted from the inoculation medium (Turner and Box, 1970). With *E. tenella*, a medium consisting of 90% Hanks' balanced salt solution (HBSS), 5% lactalbumin hydrolysate (LAH; 2.5% solution in HBSS), and 5% fetal calf serum gives better penetration than one composed of Medium 199 and 5% chicken serum (Doran, 1971a). Development of *E. tenella* is also affected, but the effects are apparent only during the latter part of the life cycle and only when using sporozoites as the inoculum. With second-generation merozoites as the inoculum, Bedrnik (1970a) found no differences in development of mature third-generation schizonts when cells were maintained in phosphate buffered saline (PBS), PBS and 10% calf serum, Medium 199 and 10% calf serum, or in PBS and various components of Medium 199. With sporozoites as the inoculum, medium does not influence development of mature first-generation schizonts (Patton, 1965). However, after four or five days, differences with this species become apparent (Doran, 1971a). In a medium (HBME) consisting of Basal Medium Eagle (with HBSS) and 5% or 10% fetal calf serum, very few mature second-generation schizonts are found, macrogametes contain either no plastin granules or a few that are small, and most of the small number of oocysts are small and incompletely formed. On the other hand, in a medium composed of 90% HBSS, 5% LAH, and 5% fetal calf serum there are more sexual stages, macrogametes contain larger plastin granules, and most of the larger number of oocysts have completely formed walls. These differences are probably mainly caused by cell maintenance. The monolayers in cultures maintained in both media were in good condition at four days, but after that, those in HBME and calf serum were not in good condition and contained few of the islets or patches of epithelial-like cells essential for development of sexual stages.

Good penetration of cells by *T. gondii* can be obtained when the pH of the medium is 6.8–8.0 (Shimizu, 1963). Values of pH 6.5 and 8.0 result in inferior cells and poor multiplication (Shimizu, 1963). The optimal pH is reported to be 7.0 (Shimizu, 1961a, 1963) or 7.5 (Akinshina and Zasuchina, 1967). Although the work was carried out axenically, Lund *et al.* (1964b) reported that 7.9 is optimal for stability and that below 7.4 and above 8.4 there is a loss of about 40% of the parasites found at the optimum pH. Penetration and development of *Eimeria* at different pH values have not been compared. A pH of 7.4 is beneficial for development of *Isospora lacazei* and/or *chloridis* (Turner and Box, 1970).

Routine media changes are necessary in order to obtain high yields of *T. gondii* (Chernin and Weller, 1954b, 1957; Shimizu, 1961a, 1963). Changes every three or four days give yields as high as 2.7 million/ml nutrient media; with no media changes about one-fifth that amount are obtained (Chernin and Weller, 1954b, 1957). More *E. tenella* oocysts develop if a total volume of 10 ml rather than 2 ml is used for maintaining cells (Doran, 1971a). With 2 ml, media changes are necessary on days 4, 5, and 6 after inoculation because of a drop in pH from 7.0 to 7.2 to approximately 6.4. If 10 ml are used, the pH changes very little and a medium change is unnecessary. The higher oocyst yield with 10 ml is probably not entirely the result of better regulation of pH, but also because second-generation merozoites free in the medium on days 4 through 6 are retained and not removed from culture by a medium change. Shimizu (1963) states that he held cultures infected with *T. gondii* for 15 days in 2 ml, 5 ml, and 10 ml of media and noticed no difference in multiplication. However, he mentions nothing about the actual comparative yields of toxoplasmas under these three conditions.

3. *Cellular Factors*

a. Cell type and passage number

When two or more cell types are used for cultivating *T. gondii*, one is usually found to be better than the other(s). There are differences in the rates of penetration (Csóka and Kulcsár, 1968a; Vischer and Suter, 1954), rates of multiplication (Hogan *et al.*, 1960b; Lund *et al.*, 1963a; Ma and Mun, 1961; Matsubayashi and Akao, 1963; Vischer and Suter, 1954; Yü and In, 1966) and timing of cytopathogenic effects (Akinshina and Zassuchin, 1965; Cook and Jacobs, 1958a; Hansson and Sourander, 1965; Hogan *et al.*,

1960b). Cook and Jacobs (1958a) found delayed lysis in conjunctival epithelium (Chang) and human intestine (Henle) as compared to primary monkey kidney cells. The longer intervals, which were found not to be due to differences in rates of penetration, are thought to be caused either by metabolic factors, by physical differences such as continuity of cell sheets, or by the resistance of fibroblast-like cells that continued to proliferate after epithelial-like cells were destroyed. In the only comparative study of infection in a single cell type derived from different hosts, Vischer and Suter (1954) observed slower intracellular development of *T. gondii* in macrophages from rabbits, guinea pigs, and rats than in macrophages from mice. The authors state that the pattern of development is correlated with the susceptibility of the four species of animals to infection. Human cervical carcinoma (HeLa) cells are the most frequently used cell type (Table 4), yet in six instances where HeLa was compared with other cell types (Bommer, 1969; Csóka and Kulcsár, 1968a; Hogan, 1961; Hogan *et al.*, 1960b; Lund *et al.*, 1963a; Matsubayashi and Akao, 1963) it was found to be the inferior cell type in three of them (Csóka and Kulcsár, 1968a; Hogan, 1961; Lund *et al.*, 1963a). The best indication that cell preference is not a matter of differences in cell maintenance or some other cultural factor is in the work of Csóka and Kulcsár (1968a). They grew HeLa cells and primary human amnion cells together and found that far more parasites entered the amnion cells.

Some cell types are also better than others for the cultivation of *Eimeria* and *Isospora*. *Isospora lacazei* and/or *chloridis* progresses to mature second-generation schizonts in primary cultures of both canary and chick kidney, but the numbers developing are much greater in the canary cells (Turner and Box, 1970). With the species of *Eimeria* that do develop, the best cell type might be considered the one in which the life cycle progresses the farthest. As mentioned in an earlier section, these are usually cells from the definitive host. Some species develop to mature first-generation schizonts in several different cell types. In these instances, differences exist in time of appearance (Clark and Hammond, 1969; Doran, 1971a; Doran and Vetterling, 1967a; Hammond and Fayer, 1968; Kelley and Hammond, 1970) and percentage development (Clark and Hammond, 1969; Hammond and Fayer, 1968; Sampson *et al.*, 1971; Speer and Hammond, 1970b; Speer, Hammond, and Anderson, 1970). Doran (1971c) compared development of *E. tenella* in primary cultures of kidney cells from the chicken, pheasant, partridge, and turkey (Table 8). Although mature first-generation schizonts first appear at the same time and are present through similar time intervals in cultures from all four birds, development of later stages is delayed in cultures from the pheasant, partridge, and turkey. Oocysts, which do not develop in partridge and turkey cells, appear on the 6th day in chicken cells, but not until the 9th day in pheasant cells. Nonembryonic chicken kidney cells are better for development of *E. tenella* oocysts than those obtained from the embryo (Doran, 1970c).

Apparently, cell type also influences penetration and development of *Besnoitia jellisoni* and *Sarcocystis*. Although confluency of cells at the time of inoculation was not considered, the data of Fayer *et al.* (1969) do suggest that embryonic bovine spleen and tracheal cells are more favorable for development of *B. jellisoni* than an established cell line of bovine kidney (MDBK). Although quantitative data are not presented, Akinshina and Doby (1969a) reported that (1) fibroblast-like cells are better for development of *B. jellisoni* than epithelial-like cells, (2) fibroblast-like cells from swine and rabbit kidney are better than embryonic chick fibroblasts, and that (3) chick embryo fibroblasts are better than epithelial-like cells from swine and rabbit kidney. Fayer (1970) found that *Sarcocystis* penetrated and developed in five of the cell types listed in Table 5, but did not penetrate embryonic chicken muscle cells. This is an interesting observation considering that *Sarcocystis*, at least from what we know of its life cycle, is a parasite of muscle tissue.

The number of times cells are serially passed influences development of *Eimeria meleagrimitis* and *E. necatrix*. Neither species develops in primary cultures or in the 1st and 2nd passage of embryonic bovine kidney; development commences in cells of the 3rd (*E. necatrix*) or 4th (*E. meleagrimitis*) passages and becomes progressively greater through the 20th passage (Doran and Vetterling, 1967a). The reason for this phenomenon is unknown, but it is thought to be caused by a cellular activating factor becoming more pronounced through serial passage (Doran and Vetterling, 1967a).

b. Age and number of cells

Reports differ concerning the effect age of cells (interval in culture before inoculation) has on multiplication of *T. gondii*. Human amnion cells 7 and 41 days old are equally susceptible (Csóka and Kulcsár, 1968a) and there is no variation in timing of lysis in monkey kidney cells 6, 14, 21, and 28 days old (Cook and Jacobs, 1958a). With HeLa cells, approximately four times more organisms are produced eight days after inoculation in cultures two days old than in those five days old (Shimizu, 1961a).

The number of cells in a culture influences development of both *T. gondii* and *E. tenella*. In cultures seeded with 25,000 and 50,000 cells, one week elapses before one-half of the cells infected with *T. gondii* burst; in cultures with 5,000 cells, lysis occurs at 24 hr (Lund *et al.*, 1963a). In order to maintain cultures of chicken kidney cells long enough for *E. tenella* oocysts to develop, it is necessary to use a cell suspension that will result in less than 35% confluency of cells at the time of inoculation with sporozoites (Doran, 1971).

IV. Embryo and Cell Culture Applications

A. ISOLATING ORGANISMS

Toxoplasma gondii can be isolated from macerated or excised bits of tissue by inoculation into embryonated eggs (Anwar and Oureshi, 1954; Gibb *et al.*, 1966; Jacobs, 1956; Jacobs *et al.*, 1954; Warren and Russ, 1948) and cell cultures (Akinshina, 1965; Jacobs *et al.*, 1954; Osaki *et al.*, 1964).

Embryos are not as sensitive to infection with small numbers of organisms as mice (Abbas, 1967b; Buttitta, 1951; Eichenwald, 1956; Weinman, 1944). Eichenwald (1956) states that of the three systems for isolation (chick embryo, cell culture, and animals), chick embryos are the least sensitive. However, Abbas (1967b) compared the three methods using the RH strain in acutely, subacutely, and chronically infected tissue. For isolation from acutely infected tissue, mice intraperitoneally inoculated are at least 16 times more sensitive than embryos and, for subacutely and chronically infected tissues, mice are 50 and 10 times, respectively, more sensitive than embryos. Cell cultures are 316 times less sensitive than mice for isolation from acutely infected tissue and of no value for isolation from subacutely and chronically infected tissue. Abbas reported that infections can be detected sooner with embryos than with mice, although the latter are more sensitive. He states that toxoplasmas can be found within embryos in ten days, whereas with all except the highly virulent strains it takes six weeks in mice.

B. MAINTAINING ORGANISMS

A strain of *T. gondii* can be maintained in chick embryos at the level of virulence characteristic of it at the time of isolation (Jacobs, 1953; Jacobs and Melton, 1954). Occasionally, a strain is found that is difficult to establish in chick embryos, but, once established, it can be maintained readily by weekly passages (Jacobs, 1956). Wolfson (1942) showed for the first time that *T. gondii* can be maintained by serial passage in fertile eggs. MacFarlane and Ruckman (1948) passed the RH strain 27 times in 10- to 11-day-old embryos. The LD-50 titers for the passages and between embryo passages and mice intracerebrally inoculated were about the same. Jacobs (1953) stated that the RH strain had been maintained for from two to three years without loss of virulence. Jacobs and Melton (1954) maintained the 113-CE strain by weekly passages in embryos for one year.

Long-term maintenance of *T. gondii* in stationary culture without serial transfer has been reported (Akinshina, 1963a; Bell, 1961; Bickford and Burnstein, 1966; Csóka and Kulcsár, 1968a; Kaufman *et al.*, 1959; Lund *et al.*, 1963a; Schuhová, 1957, 1960; Schuhová *et al.*, 1957; Stewart and Feldman, 1965). From 26°C to 28°C, organisms survived 44 days with no media change and 80 days with a change every 21 days (Bell, 1961). At 4°C with no media change, organisms are viable after 90 days and, when glycerin is added to the medium before refrigeration, 120 days (Stewart and Feldman, 1965). However, longer intervals were usually obtained when cultures were kept at or close to 37°C. With a media change every three or four days, viable organisms can be maintained for as long as 33 months (Schuhová, 1960). A "chronic" or prolonged infection occurs (Bickford and Burnstein, 1966; Kaufman *et al.*, 1959; Schuhová 1957, 1960; Schuhová *et al.*, 1957). A few islets of cells remain after initial cell destruction (Bickford and Burnstein, 1966; Schuhová, 1960). These grow and, after 17 days (Schuhová, 1960) or from 26 to 30 days (Bickford and Burnstein, 1966), a new sheet of cells is formed. Cell destruction again takes place (Bickford and Burnstein, 1966; Schuhová, 1960), but not as severe as before (Bickford and Burnstein, 1966). Eventually, an equilibrium is established between cell destruction and cell growth (Bickford and Burnstein, 1966; Kaufman *et al.*, 1959; Lund *et al.*, 1963a; Schuhová, 1960). Lund *et al.* (1963a) believed that, in order for this phenomenon to occur, less than one-half of the cells should be initially infected.

Toxoplasma gondii has also been serially propagated in cell culture for long intervals of time. This has been carried out in hanging drop (Carrel) cultures (Meyer and de Oliveira, 1943, 1945), roller tubes (Chernin and Weller, 1954a,b, 1957), and stationary tubes or other culture vessels (Akinshina, 1964, 1965; Balducci and Tyrrell, 1956; Cook and Jacobs, 1958a; Csóka and Kulcsár, 1968b; Galuszka, 1962b,c; Lycke and Lund, 1964b; Pierreaux, 1963; Sabin and Olitsky, 1937; Schuhová *et al.*, 1956; Stewart and Feldman, 1965). Meyer and de Oliveira (1943, 1945) incubated freshly inoculated cultures from 38°C to 39°C for 24 hr and then kept them at ambient temperatures (close to 28°C). After the cultures had become adapted to ambient temperatures, transfers and addition of new tissue were performed every 7 to 13 days; viable parasites were maintained for three years (Meyer and de Oliveira, 1945). Cultures have also been kept at 22°C (Akinshina, 1965; Cook and Jacobs, 1958a) and 4°C (Akinshina, 1964, 1965). Storage at these lower temperatures apparently prolongs the time necessary for a transfer. At 22°C, only three transfers were made during three months (Cook and Jacobs, 1958a); at 4°C, every three or four weeks (Akinshina, 1964).

Csóka and Kulcsár (1968b) used two different temperatures between transfers of the RH strain. They incubated newly inoculated cultures at 37°C for three days and then placed them at 4°C for two days. Forty-five transfers were made during the eight months the parasites were maintained.

In this case, placing cultures at 4°C did not prolong time of transfer. In roller tube cultures of the RH strain (Chernin and Weller, 1954a and b, 1957), transfers were made every five or six days (calculated, based on reported numbers of passages and duration of maintenance).

There is no loss of virulence during maintenance of *T. gondii* in a single culture (Cook and Jacobs, 1958a; Galuszka, 1962b; Schuhová, 1957) and during serial propagation (Chernin and Weller, 1954a and b, 1957; Cook and Jacobs, 1958a; Galuszka, 1962c; Meyer and de Oliveira, 1945; Pierreaux, 1963; Sabin and Olitsky, 1937). However, the parasites kept in culture are less virulent than those from the peritoneal exudate of mice (Cook and Jacobs, 1958a; Galuszka, 1962b; Schuhová, 1957). According to Balducci and Tyrrell (1956), those grown in cell culture are 100-fold less infective for mice than those from the exudate.

Bigalke (1962) maintained *Besnoitia besnoiti* by serial passage in both avian embryos and cell culture. Organisms were transferred six times in embryos. During these passages the virulence of the organisms apparently increased. Mortality began to occur on the 19th day after initial transfer. However, after transfer numbers 2, 3 and 4, and 5 and 6, the mortality began on days 11, 7, and 5, respectively. In cell culture, organisms were lost after four transfers. The loss was probably due to decreased numbers of viable parasites. Bigalke mentioned that multiplication was more profuse after the first rather than the second transfer.

Long (1971b) excysted *Eimeria tenella* sporozoites from sporocysts released from ocysts recovered from the CAM of chick embryos and inoculated them into the chorioallantoic cavity of uninfected embryonated eggs. By this method, he was able to maintain the culture through 16 passages. Bedrnik (1969a) transferred *E. tenella* merozoites from one culture to another. This work was probably initiated with the idea of serial propagation. He claims to have obtained not only the third generation, but also a fourth and fifth generation. However, the numbers present after each transfer gradually diminished. This is probably also the case in Long's work with embryonated eggs.

C. PREPARING ANTIGENS

Antigens (toxoplasmins) have been prepared from (a) the chorioallantoic membrane (CAM) of chicken eggs (Afzelius-Alm and Hahn, 1953; Balducci and Tyrrell, 1956; Chaparas *et al.*, 1966; Eichenwald, 1956; Feldman and Sabin, 1949; Jíra and Bozděch, 1963; Jíra *et al.*, 1961; Jírovec and Jíra, 1962; Morris *et al.*, 1957; Shimizu, 1961b; Shimizu and Shiokawa, 1965; Shimizu and Takagaki, 1959; Stroczyńska, 1966; Warren and Russ, 1948) or duck eggs (Heyl and Gispen, 1955), (b) the allantoic fluid of chicken eggs (Morris *et al.*, 1957), and (c) cultural fluid from infected cell cultures (Akinshina and Gracheva, 1964; Eichenwald, 1956; Ma and Mun, 1961; Pierreaux, 1963; Schuhová and Jírovec, 1962; Schuhová *et al.*, 1963a,b; Shimizu, 1961b; Shimizu and Shiokawa, 1965; Štumpa, 1966; Suggs *et al.*, 1968; Virat, 1967). Antigens attain a maximum level in cell culture fluid at the time of maximum multiplication or from one to two days earlier (Shimizu, 1961b). Antigens prepared from cell culture are reported to be pure (Eichenwald, 1956; Schuhová *et al.*, 1963a,b); the nitrogen content is 0.388 mg N/ml as compared to 0.597 mg to 0.741 mg N/ml for antigens from mouse exudate (Schuhová *et al.*, 1963b). Compared to the antigen from the CAM, the cell culture antigen has a more specific reaction, more stable nature, and is easier to prepare (Shimizu and Shiokawa, 1965). Cell culture antigens are more sensitive detectors of specific antibodies than mouse exudate antigens (Akinshina and Gracheva, 1964; Schuhovaá *et al.*, 1963a,b). Cell culture antigens react with antibodies in from two to eight times higher serum dilutions (Schuhová *et al.*, 1963b). However, Tŏs-Luty (1967) found exudate antigen more sensitive (1:280) than cell culture antigen (1:30-1:240). Compared to mouse exudate antigens, CAM antigens have been reported to be either less reactive (Chaparas *et al.*, 1966; Jíra *et al.*, 1961; Jírovec and Jíra, 1962) or to show similar activity (Jíra and Bozdĕch, 1963; Jírovec and Jíra, 1961).

Baldelli *et al.* (1968) obtained a preparation from the CAM infected with *E. tenella* and injected it intramuscularly into chickens. There was no resistance to challenge dosages of oocysts fed to the chickens 5, 10, and 20 days later.

D. Determining the Effect or Mode of Action of Antimicrobial Agents and Other Substances

1. Adenine and Other Purines, Pyrimethamine, and Sulfonamides

Cook (1958b) studied the effect of adenine and other purines on *T. gondii* in cell culture. Adenine does not inhibit penetration of cells and adenine, adenine sulfate, adenosine, AMP, ATP, and xanthene delay the cytopathogenic effects produced by the parasites. At a concentration of 1.0 mg%, adenine and adenine sulfate cause approximately two-fold increases in the time required for lysis of cells. The effect of adenine, which is on intracellular proliferating forms rather than extracellular forms, occurs even when administered as late as four days after inoculation of cultures. Two purine antagonists (benezimidizole and 2,6-diaminopurine), rather than reversing the effect of adenine, produce effects similar to it. Cook suggests that the action of the two antagonists may be caused by a "poisoning" effect rather than an interference with specific metabolic activities of the cells or parasites. Dimethyladenine and the aminonucleoside of stylomycin also inhibit multiplication of the parasite (da Silva, 1960).

The effect of pyrimethamine on *T. gondii* in cell culture is similar to that of the purine compounds. Penetration of cells is not inhibited, but multiplication of the parasite is inhibited (Cook and Jacobs, 1958b; Lycke and Lund, 1966; Sheffield and Melton, 1969). Although no data are given, it has also been stated (Akinshina and Zassuchin, 1969) that the drug acts on extracellular forms and not on those that are multiplying. Addition of the drug six hours after inoculation results in 50% inhibition of multiplication (Lycke and Lund, 1966). When added at the time of inoculation, as little as 0.005 mg% for 12 hr causes delay in lysis of cells and 0.025 mg% for three days prevents lysis of the cells and renders cultures free of the parasite (Cook and Jacobs, 1958b). Temperature influences action of pyrimethamine (Maloney and Kaufman, 1964). At 37°C, 0.05 mg% for four days kills all organisms; all mice survive when injected with the cultural material. At 34°C, 0.6 mg% is needed for some mice to survive, and at 31°C, even 0.6 mg% is only slightly effective when given for four days. Kaufman *et al.* (1959) report that (a) pyrimethamine is less effective against less rapidly multiplying forms (113-CE strain) than more rapidly multiplying forms (RH strain) and that (b) the concentration of the drug that can be safely reached in serum of man is not greater than the minimal concentrations required to kill organisms of both strains under *in vitro* conditions.

Cook (1958b) treated cells with 8 mg%

adenine for one week, removed the chemical by washing, and then inoculated the cultures with 200,000 toxoplasmas. Both adenine-treated and control cells lysed after the same interval of time, indicating that the effect is produced only when the chemical is present in the medium and that the chemical cannot be "stored" within the cell. Cook and Jacobs (1958b) found no cytopathogenic effects after 21 days in cultures previously exposed to 0.5 mg% pyrimethamine for 1.5 hr. They attributed the failure to lyse cells to an accumulation and storage of the drug in concentrations sufficient to inhibit the parasite. They later (Cook and Jacobs, 1958c) compared seven different cell types for storage of the drug and found that it could be stored in chick embryo heart and HeLa cells at levels sufficient to inhibit the parasite.

Strains more resistant to pyrimethamine have been developed in cell culture. Cook (1958a) initially treated cultures with 0.001 mg% and, with each serial passage, doubled the concentration. After 16 passages, the parasite could live in 16 mg% of the drug; thus, this strain was 640 times more resistant than the parent strain, for which the minimal effective dose was 0.025 mg% (Cook and Jacobs, 1958b). Akinshina and Zassuchin (1969) reported nearly identical results. After from 6 to 11 passages, the parasite could stand 18 mg%; thus, the strain was 600 times more resistant than the original strain for which the minimal effective dose was 0.03 mg%. The reason for the range in passages is that they also found that clones establish resistance with different speeds; some reached 18 mg% in six passages, others in 11 passages.

Lycke and Lund (1966) studied the action of sulfonamides (sulfamerazine, sulfaproxylene, sulfadiazine, and sulfathiazole) and pyrimethamine in culture. All four sulfonamide derivatives have no influence on penetration. Using only sulfathiazole for testing effects on multiplication, they found that, in order to obtain marked inhibition, the drug must be present when parasites are extracellular. It reduces the capacity for reproduction and the effect depends on time of exposure. Pretreatment for 30 min gives no inhibition, but treatment for four hours results in almost 50% inhibition. However, there is no inhibition of multiplication if sulfathiazole is added at the time of inoculation. This is in direct contrast to what they found with pyrimethamine. The synergistic effect of sulfonamide and pyrimethamine has been confirmed in cell culture (Cook and Jacobs, 1958b; Lycke and Lund, 1966) and electron microscopically with infected cells from culture (Sheffield and Melton, 1969).

2. *Aminopterin and Vitamins*

The action of aminopterin, folic and folinic acids, and *p*-aminobenzoic acid (PABA) on *T. gondii* has also been studied in cell culture (Lycke and Lund, 1966). None affect penetration. Aminopterin has no effect when parasites are either pretreated with the drug or when it is added at the time of inoculation. PABA increased the number of multiplying parasites. Folic and folinic acids do not increase the number of dividing parasites; sometimes the use of high concentrations produces a negative effect. PABA eliminates the inhibition due to sulfathiazole and folinic acid reduces the inhibition by pyrimethamine. PABA and folic acid together abolish the inhibition of multiplication caused by the synergistic effect of a mixture of sulfathiazole and pyrimethamine.

3. *Antibiotics*

In embryonated eggs, terramycin either has little or no effect (Crema and Ambrek, 1952; Grönross, 1954) or delays and prolongs infection with *T. gondii* (Anwar and Oureshi, 1954). Other antibiotics having no effect on survival of infected eggs are (a) polymyxin-B-sulfate, bacitracin, and aureomycin

(Grönross, 1954); (b) magnamycin, neomycin, tebezon, and erythromycin (Grönross, 1955); and (c) streptomycin, penicillin, and acetarsone (Adams *et al.*, 1949).

Remington *et al.* (1970) studied the effect of rifampin (rifampin, rifamycin AMP) on *T. gondii* in cell culture. It inhibited multiplication at concentrations of 50 μg/ml, 75 μg/ml, and 100 μg/ml. They also found that the mode of action was not on RNA synthesis. Since the toxicity of the drug for cells and its inhibition of multiplication occurred at the same concentration, the authors state that it is possible that the drug inhibits multiplication by its toxic effect on the cells. Trypacidin is effective against *T. gondii* in culture (Ebringer *et al.*, 1965). There is no effect when cells are treated with 10 μg/ml for two hours before inoculation. However, there is complete "cure" of cells when from 40 μg/ml to 50 μg/ml is added two hours after infection and when 10 μg/ml is added at the time of inoculation and removed 48 hr later. Another antibiotic, spiramycin, is also apparently effective against the parasite. It was mentioned in abstract only (Kien-truong-Thai *et al.*, 1970) that the minimal inhibitory dose for two strains was the same as the effective dosage usually used in man.

Bedrnik (1970b) found that amphotericin B was effective against *E. tenella* sporozoites. A concentration of 1 μg/ml decreased the number of extra- and intracellular sporozoites; with 10 μg/ml there was almost a complete disappearance of sporozoites during 24 hr in culture.

4. "Coccidiostats"

Infections with *E. tenella* in embryonated eggs have been used for testing coccidiostats (Long, 1970a; Ryley, 1968). Ryley (1968) tested 19 drugs using mortality as the criterion of effect and found that some of the drugs active *in vivo* were not active in the embryo. Because of the large quantities of sporozoites and oocysts needed and inability to obtain infection with another species, *E. brunetti*, Ryley concluded that the embryo system was unsuitable for general screening. Long (1970a) used a technique for testing that is not based on mortality and, consequently, very small numbers of sporozoites are needed. The technique involves counting the number of macroscopic lesions visible with a simple viewing box with indirect light on the CAM beginning five days after inoculation with only 2,500-5,000 sporozoites. His results with methyl benzoquate on two strains of *E. tenella* are shown in Table 9. This macroscopic focal count technique appears to be superior in several respects to the one involving mortality. It is not only more quantitative and less wasteful of oocysts, but allows the worker to follow the effect of an antimicrobial agent through the course of infection by merely candling the egg and counting the lesions. This technique might also be applicable to work with *Toxoplasma*. Plaques or lesions have been counted, but only after harvesting the eggs (Lund *et al.*, 1960).

There have been three studies on the effect of "coccidiostats" on *Eimeria* in cell culture. Fayer and McLoughlin (1970) found that amprolium does not inhibit penetration of cells by *E. adenoeides* sporozoites, but does inhibit the formation of trophozoites and first-generation schizonts. The addition of 0.096% of the drug at either the time of inoculation or 24 hr later decreases the number of developmental stages and the total number of intracellular parasites. In turkeys, amprolium acts on the second- and first-generation schizonts, but not on developmental stages earlier than these. Ryley and Wilson (1971) found that (a) robenidene, added to cultures at either the time of inoculation or 24 hr later, completely inhibits development of *E. tenella* schizonts at concentrations from 0.01 ppm to 0.003 ppm, and that (b) methyl benzoquate

TABLE 9. Number of CAM foci occurring in embryos given 2,000 "W" or "H" strain of *E. tenella* sporozoites and treated with different amounts of methyl benzoquate. (From Long, 1970a)

Amount of drug administered (mg) on day 0 + 10	Embryos inoculated with "W" strain on day 0 + 11 mean number foci/CAM			Mean foci count for 3 days	Embryos inoculated with "H" strain on day 0 + 11 mean number foci/CAM			Mean foci count for 3 days
	5th day	6th day	7th day		5th day	6th day	7th day	
0	136.4	75.6	15.8	80.2	117.2	81.2	7.3	72.9
0.001	106.6	102	74	97.3 NS[a]	4.4	33.3	2.2	13.2[b]
0.002	32.8	93.8	61.6	62.6 NS	1	0.4	4	1.8[b]
0.004	0.4	0.6	5.5	1.92[b]	0.4	0	0	0.13[b]

[a]NS = not significantly different from 0 treatment. [b]Significantly different from 0 treatment ($p<0.001$).

is completely effective against intracellular sporozoites at concentrations ten times weaker than that of robenidene. Strout and Ouellette (1971) state that the cell culture system is an accurate, rapid method for screening anti-coccidial activity, which requires only microgram quantities of materials to be tested. They tested ten compounds known to have *in vivo* activity against *E. tenella.* All were active *in vitro* within a 72 hr period of development, whereas two compounds shown to be inactive *in vivo* were ineffective. The effective compounds reduced or eliminated schizogony and resulted in abnormal parasites.

5. *Quinine compounds*

Quinine hydrochloride delays and prolongs infection with *T. gondii* in embryonated eggs (Anwar and Oureshi, 1954). In cell culture, 0.1 mg/ml of quinine hydrochloride, quinine alkaloid, or quinine sulfate added to the media at the time of inoculation reduces the number of *E. meleagrimitis* sporozoites entering cells by more than 80%; a similar concentration of quinine sulfate reduces the number of *E. tenella* sporozoites entering by 97% (Fayer, 1971). Results obtained by Fayer *et al.* (1972) with *T. gondii*, *B. jellisoni* and *Sarcocystis* sp. were similar to those with *Eimeria* and suggest that quinine interferes with a mechanism of host cell penetration common to all four genera.

E. STUDYING FACTORS PERTAINING TO IMMUNITY AND RESISTANCE

Vischer and Suter (1954) studied development of *T. gondii* in cultured macrophages from immune and normal animals. Macrophages from immunized animals have an inhibitory effect on intracellular development. This effect increases in the presence of immune serum. There is less inhibition when either immunized cells and normal serum or normal cells and immune serum are used. However, Yanagawa and Hirato (1963) in a similar study with leukocytes (buffy coat) from pigs, found that the parasites multiplied equally as well in cells from both immune and normal animals and that immune serum inhibited multiplication regardless of whether the cells were from immunized or normal animals. Lycke *et al.* (1963a) found that antiserum affects extracellular parasites and not those already intracellular.

Lycke, Lund, Strannegärd, and Falsen (1965) exposed the RH strain to various preparations of immune serum and activator, and then tested infectivity in cell culture. Penetration of cells was inhibited when the parasites were exposed to either immune serum and activator or immune serum in which heat-labile components were destroyed. Properdin, complement, and Mg^{++} enhanced the effect of *T. gondii* antibodies. Some of their results indicate that properdin might cause inhibition of penetration in the absence of antibodies. Strannegärd and Lycke (1966) later reported that properdin, rather than exerting an antitoxoplasma effect alone, combines with the parasites and sensitizes them to the action of antibody.

Suggs *et al.* (1968) compared the activity of antigens from five different isolates of *T. gondii* and two of *Besnoitia jellisoni* in cell culture. Not all *T. gondii* isolates have the same degree of related antigenicity to *B. jellisoni.* The *T. gondii* RH antigens showed no reaction with *B. jellisoni* antiserum, but cross-reactions were observed between *T. gondii* RH antiserum and *B. jellisoni* antigen.

The effect of interferon (virus inhibitor) on *T. gondii* and *Eimeria* has been tested in cell culture. Remington and Merigan (1968) treated L-cells with mouse interferon before inoculation with *T. gondii* and found that it significantly protects cells from destruction by the parasite. They also obtained a 50% reduction in plaques when chick embryo fibroblast cultures were treated with chick interferon. These authors believe the wide variation in virulence among different strains of *T. gondii* may be related to their ability to induce interferon production. However, Rytel and Jones (1966) found that the parasite does not induce formation of detectable interferon. When the cell culture fluids were assayed for interferon, there was no inhibition of the vesicular stomatitis virus plaque size or number even at the lowest (1/4) fluid dilution. Burnstein and Bickford (1968) also found no evidence for the interferon phenomenon. *Toxoplasma gondii* and each of five different viruses can cohabit the same cell and the parasite can multiply in cells severely damaged by the virus. Fayer and Baron (1971) treated cells with either an interferon inducer, polyinosinic-polycytidylic acid, plus neomycin or chicken interferon and then challenged with either Sinbis virus or *E. tenella.* Penetration of cells was not inhibited, but both the inducer and interferon reduced the number of developmental stages by as much as 74% and 33%, respectively, and reduced the yield of Sinbis virus.

F. OTHER APPLICATIONS

The embryo and cell culture, primarily the latter, have been shown to have several other uses. In addition to its use for comparing virulence of different *T. gondii* strains (Hogan *et al.*, 1961; Jacobs and Melton, 1954; Kaufman *et al.*, 1958; Vermeil *et al.*, 1962), cultures can be used for comparing infectivity of the same strain after serial passage (Chernin and Weller, 1954b; Galuszka, 1962b,c) and determining infectivity or viability after storage at different temperatures (Lund *et al.*, 1960; Stewart and Feldman, 1965) or other axenic experimentation (Lund *et al.*, 1963a; Lycke and Lund, 1964a,b). With *Eimeria*, the effects of storage period and freeze-thawing have been tested in embryos (Long, 1970d) and the effect of storage period (Doran, 1970a; Doran and Vetterling, 1969), freeze-thawing (Doran, 1969, 1970a; Doran and Vetterling, 1968b), and treatment prior to inoculation (Doran, 1970b) have been tested in cell culture.

Plaques in cultures infected with *T. gondii* have been used for isolating and studying genetically pure strains (Akinshina, 1968; Akinshina and Zasuchina, 1966, 1967). Parasites grown in culture have been used in cytochemical studies (Vetterling and Doran, 1969), and in studies concerning uptake of DNA precursors and other nutrients (Perroto *et al.*, 1971; Roberts, Elsner, Shigematsu, and Hammond, 1969, 1970). The ultrastructure of *T. gondii* (Hansson and Sourander, 1968; Hogan, 1961; Matsubayashi and Akao,

1963, 1964; Meyer and de Andrade Mendonca, 1955; Sheffield and Melton, 1969) and various species of *Eimeria* (Roberts, Hammond, and Anderson, 1969; Roberts, Hammond, and Speer, 1970; Scholtyseck, 1969; Scholtyseck and Strout, 1968; Sheffield and Hammond, 1967; Sheffield *et al.*, 1968; Strout and Scholtyseck, 1970) have been studied by using parasites grown in culture.

Literature Cited

Abbas, A.M.A. 1967a. Toxoplasmosis of chickens experimentally infected during embryonic life. Roy. Soc. Trop. Med. Hyg., Trans. 61:514-516.

Abbas, A.M.A. 1967b. Comparative study of methods used for isolation of *Toxoplasma gondii*. Bull. World Health Org. 36:344-346.

Adams, F.H., M. Cooney, J.M. Adams, and P. Kabler. 1949. Experimental toxoplasmosis. Proc. Soc. Exp. Biol. Med. 70:258-260.

Afzelius-Alm, L., and E. Hahn. 1953. Toxoplasmose und Komplement bindungsreaktion. Aerztl. Wochenschr. 8:1100-1103.

Akinshina, G.T. 1959. Cultivation of *Toxoplasma gondii* in tissue cultures. Bull. Exp. Biol. Med. 47:47-50.

Akinshina, G.T. 1963a. Employment of different types of cells for culturing the *Toxoplasma gondii* [In Russian, English summary]. Byull. Eksp. Biol. Med. 56:69-71.

Akinshina, G.T. 1963b. Growth of *Toxoplasma* in tissue cultures [In Russian]. Tr. Inst. Zool. Akad. Nauk Kazakh. SSR 19:5-15.

Akinshina, G.T. 1963c. The cultivation of *Toxoplasma* in tissue cultures [In Russian, English summary]. Med. Parazitar. Bolezni 32:344-345.

Akinshina, G.T. 1964. Prolonged preservation of *Toxoplasma* in tissue cultures [In Russian, English summary]. Byull. Eksp. Biol. Med. 58:98-100.

Akinshina, G.T. 1965. Use of the methods of tissue cultures for diagnostics of toxoplasmosis and for maintenance of *Toxoplasma*'s strains [In Russian], p. 386-390. *In* I.G. Galuzo, Toxoplasmosis of Animals, Academy of Science of the Kazakh SSR, Alma-ata.

Akinshina, G.T. 1968. Materials on the study of variability of *Toxoplasma*. I. Investigation of the composition of the population of an old laboratory strain and some spontaneous mutants isolated from it [In Russian, English summary]. Med. Parazitar. Bolezni 37:657-660.

Akinshina, G.T., and J.M. Doby. 1969a. Multiplication de *Besnoitia jellisoni* Frenkel, 1953 (Protozoaires Toxoplasmatea) en cultures de cellules de tissus d'origines differentes. Protistologica 5:249-253.

Akinshina, G.T., and J.M. Doby. 1969b. Étude comparée de la multiplication de *Toxoplasma gondii* et de *Besnoitia jellisoni* dans les cultures de cellules. Progr. Protozool. (Proc. III Int. Cong. Protozool.), p. 222.

Akinshina, G.T., and L.I. Gracheva. 1964. Production of toxoplasmosis antigen by tissue culture methods [In Russian, English summary]. Med. Parazitar. Bolezni 33:661-665.

Akinshina, G.T., and D.N. Zassuchin. 1965. Multiplication of *Toxoplasma* in tissue culture. Progr. Protozool. (Proc. II Int. Cong. Protozool.), Int. Cong. Series 91, Excerpta Med. Found., The Hague, p. 99-100.

Akinshina, G.T., and D.N. Zassuchin. 1969. Contribution a l'etude de la variabilite de *Toxoplasma gondii*. II. Obtention de clônes resistants aux medicaments in vitro. Progr. Protozool. (Proc. III Int. Cong. Protozool.), p. 223.

Akinshina, G.T., D.N. Zassuchin, V.N. Bliumkin, D.S. Levina, and V.I. Gavrelov. 1967. Certain peculiarities of cultivating toxoplasmas in cultures of re-inoculated renal cells of pig embryos RES and RES-la [In Russian, English summary]. Byull. Eksp. Biol. Med. 64:69-71.

Akinshina, G.T., and G.D. Zasuchina. 1966. Method of investigating mutations in protozoa (*Toxoplasma gondii*) [In Russian]. Genetika 2:71-75.

Akinshina, G.T., and G.D. Zasuchina. 1967. A new method of indicating the presence of *Toxoplasma gondii* in tissue cultures. Bull. Exp. Biol. Med. 63:443-445.

Andreoni, G., D. Curatolo, and G. Rocchi. 1965. Moltiplicazione del *Toxoplasma gondii* ceppo RH in colture cellulari di amnios umono e di rene discimmia. Minerva Ginec. 17:785-787.

Anwar, A., and M. Oureshi. 1954. Experimental toxoplasmosis in chick embryos. Proc. 6 Pakistan Sci. Conf. (Karachi, 1954) III. (Abstracts), p. 225-226.

Arai, H., H. Saito, and T. Nomura. 1958. Invasion and multiplication of *Toxoplasma gondii* in Hela cells. Jap. J. Med. Progr. 45:663-669.

Babudieri, B., and M. Castellani. 1968. Un nouvel antigéne pour la diagnostic de la toxoplasmose par la réaction de fixation du complèment. Arch. Inst. Pasteur. Tunis. 43:5-8.

Baldelli, B., T. Frescura, and G. Asdrubali. 1968. Colutra di *Eimeria tenella* in embrione di pollo. Vet. Ital. 19:91-100.

Balducci, D., and D. Tyrrell. 1956. Quantitative studies of *Toxoplasma gondii* in culture of trypsin-dispersed mammalian cells. Brit. J. Exp. Pathol. 37:168-175.

Barkley, F.A., and R.W. Edwards. 1952. Ethyl vanillate studies on embryonated eggs. V. Lloydia 15:53-54.

Bedrnik, P. 1967a. Further development of the second generation of *Eimeria tenella* merozoites in tissue cultures. Folia Parasitol. 14:361-363.

Bedrnik, P. 1967b. Development of sexual stages and oocysts from the 2nd generation of *Eimeria tenella* merozoites in tissue cultures. Folia Parasitol. 14:364.

Bedrnik, P. 1969a. Some results and problems of cultivation of *Eimeria tenella* in tissue cultures. Acta Vet. (Brno) 38:31-35.

Bedrnik, P. 1969b. Cultivation of *Eimeria tenella* in tissue cultures. I. Further development of second generation merozoites in tissue cultures. Acta Protozool., Warszawa 7:87-98.

Bedrnik, P. 1970a. Cultivation of *Eimeria tenella* in tissue cultures. II. Factors influencing a further development of second generation merozoites in tissue culture. Acta Protozool., Warszawa 7:253-261.

Bedrnik, P. 1970b. The sensitivity of *Eimeria tenella* sporozoites, maintained in tissue culture, to fungizone. Acta Vet. (Brno) 39:473-476.

Bell, J.B. 1961. *Toxoplasma gondii* infection of tissue culture cells and animals. Virginia J. Sci. 12:163.

Beverley, J.K.A. 1969. The biology of *Toxoplasma* infections. 7th Symp. Brit. Soc. Parasitol., London, p. 43-49.

Bickford, A.A., and T. Burnstein. 1966. Maintenance of *Toxoplasma gondii* in monolayer cultures of human epithelial (H. Ep. 2) cells. Am. J. Vet. Res. 27:319-325.

Bigalke, R.D. 1962. Preliminary communication on the cultivation of *Besnoitia besnoiti* (Marotel, 1912) in tissue culture and embryonated eggs. J. So. Afr. Vet. Med. Assoc. 33:523-532.

Bland, J.O.W. 1934. Cultivation of a protozoan (*Toxoplasma*) in tissue cultures. Arch. Exp. Zellforsch. 14:345.

Bommer, W. 1969. The life cycle of virulent *Toxoplasma* in cell culture. Aust. J. Exp. Biol. Med. Sci. 47:505-512.

Bommer, W. 1970. The life cycle of *Toxoplasma gondii*. J. Parasitol. 56(4; Sect. II, Part I):31.

Bommer, W., H.H. Heunert, and B. Milthaler. 1969. Kinematographische Studien über die Eigenbewegung von *Toxoplasma gondii*. Z. Tropenmed. Parasitol. 20:450-458.

Bommer, W., K.H. Hofling, and H.H. Heunert. 1969. Multiplication of *Toxoplasma gondii* in cell cultures. Ger. Med. Monthly 14:399-405.

Burnstein, T., and A.A. Bickford. 1968. Double infections of cultured cells. Nature 220:299-300.

Buttitta, P.L. 1951. Sulla biologia del *Toxoplasma hominis*, nella cavia, nel topolino e nell 'uovo embreonato. Boll. Soc. Ital. Biol. Sper. 27:83-84.

deCastro, M.P., and V.D.O. Amaral. 1964. Divisão multipla do *Toxoplasma gondii* em cultura de célulos. Proc. 7th Int. Cong. Trop. Med. and Malaria 2:352.

Chaparas, S.D., V.J. Fuller, and R.W. Kolb. 1966. A laboratory procedure for determining the potency of toxoplasmins for skin testing. Proc. Soc. Exp. Biol. Med. 121:734-739.

Chaparas, S.D., and R.W. Schlesinger. 1959. Plaque assay of *Toxoplasma* on monolayers of chick embryo fibroblasts. Proc. Soc. Exp. Biol. Med. 102:431-437.

Chernin, E., and T.H. Weller. 1954a. Serial propagation of *Toxoplasma gondii* in roller tube cultures of mouse and of human tissues. Proc. Soc. Exp. Biol. Med. 85:68-72.

Chernin, E., and T.H. Weller. 1954b. Further observations on the growth of *Toxoplasma gondii* in roller tube tissue cultures. J. Parasitol. 40(No. 5, Sect. 2):21.

Chernin, E., and T.H. Weller. 1957. Further observations on the growth of *Toxoplasma gondii* in roller tube cultures of mouse and primate tissues. J. Parasitol. 43:33-39.

Clark, W.N., and D.M. Hammond. 1969. Development of *Eimeria auburnensis* in cell cultures. J. Protozool. 16:646-654.

Cook, M.K. 1958a. The development of a pyrimethamine-resistant line of *Toxoplasma* under *in vitro* conditions. Am. J. Trop. Med. Hyg. 7:400-402.

Cook, M.K. 1958b. The inhibitory effect of adenine and related compounds on proliferation of *Toxoplasma gondii* in tissue culture. J. Parasitol. 44:274-279.

Cook, M.K., and L. Jacobs. 1958a. Cultivation of *Toxoplasma gondii* in tissue cultures of various derivations. J. Parasitol. 44:172-182.

Cook, M.K., and L. Jacobs. 1958b. *In vitro* investigations on the action of pyrimethamine against *Toxoplasma gondii*. J. Parasitol. 44:280-288.

Cook, M.K., and L. Jacobs. 1958c. Storage of pyrimethamine in cells *in vitro*. Proc. Soc. Exp. Biol. Med. 98:195-198.

Crema, A., and A. Ambrek. 1952. Contributo allo studio dell 'azione esercetata doll'aureomycina e terramicina *in vitro* e *in vivo* dolla tirotricina *in vitro* e dolla proteinoterapia aspecifica sullo toxoplasmosi sperimentale. Boll. Soc. Ital. Biol. Sper. 28: 650-654.

Csóka, R., and G. Kulcsár. 1968a. Cultivation of *Toxoplasma gondii* in primary human amnion cell cultures. Acta Microbiol., Acad. Sci., Hungary 15:11-15.

Csóka, R., and G. Kulcsár. 1968b. Maintenance of *Toxoplasma gondii* in primary human amniotic cell culture. Acta Microbiol., Acad. Sci., Hungary 15:357-360.

Csóka, R., and G. Kulcsár. 1970. Comparative study on sensitivity to *Toxoplasma gondii* of human primary amniotic cell culture and of mice. Acta Microbiol., Acad. Sci., Hungary 17:85-89.

Doby, J.M., and F.T. Akinshina. 1968. Possibilités de dévelopment de *Besnoitia jellisoni* (Protozoaire parasite Toxoplasmatea) en culture de cellules. Quelques aspects de son comportement en fibroblasts d'embryon de poulet. Compt. Rend. Séances Soc. Biologie. 162:1207-1210.

Doran, D.J. 1969. Cultivation and freezing of poultry coccidia. Acta Vet. (Brno) 38:25-30.

Doran, D.J. 1970a. Effect of age and freezing on development of *Eimeria adenoeides* and *E. tenella* sporozoites in cell culture. J. Parasitol. 56:27-29.

Doran, D.J. 1970b. Survival and development of *Eimeria adenoeides* in cell cultures inoculated with sporozoites from cleaned and uncleaned suspensions. Proc. Helm. Soc. Wash. 38:45-48.

Doran, D.J. 1970c. *Eimeria tenella*: From sporozoites to oocysts in cell culture. Proc. Helm. Soc. Wash. 70:84-92.

Doran, D.J. 1971a. Increasing the yield of *Eimeria tenella* oocysts in cell culture. J. Parasitol. 57:891-900.

Doran, D.J. 1971b. Survival and development of 5 species of chicken coccidia in primary kidney cell cultures. J. Parasitol. 57:1135-1137.

Doran, D.J. 1971c. Comparative development of *Eimeria tenella* in primary cultures of kidney cells from the chicken, pheasant, partridge, and turkey. J. Parasitol. 57:1376-1377.

Doran, D.J., and M.M. Farr. 1962. Excystation of the poultry coccidium, *Eimeria acervulina*. J. Protozool. 9:154-161.

Doran, D.J., and M.M. Farr. 1965. Susceptibility of 1- and 3-day-old chicks to infection with the coccidium, *Eimeria acervulina*. J. Protozool. 12:160-166.

Doran, D.J., and J.M. Vetterling. 1967a. Comparative cultivation of poultry coccidia in mammalian kidney cell cultures. J. Protozool. 14:657-662.

Doran, D.J., and J.M. Vetterling. 1967b. Cultivation of the turkey coccidium, *Eimeria meleagrimitis* Tyzzer, 1929, in mammalian kidney cell cultures. Proc. Helm. Soc. Wash. 34:59-65.

Doran, D.J., and J.M. Vetterling. 1968a. Survival and development of *Eimeria meleagrimitis* Tyzzer, 1929 in bovine kidney and turkey intestine cell cultures. J. Protozool. 15:796-802.

Doran, D.J., and J.M. Vetterling. 1968b. Preservation of coccidial sporozoites by freezing. Nature 217:1262.

Doran, D.J., and J.M. Vetterling. 1969. Influence of storage period on excystation and development in cell culture of sporozoites of *Eimeria meleagrimitis* Tyzzer, 1929. Proc. Helm. Soc. Wash. 36:33-35.

Ebringer, L., J. Bâlon, G. Cătar, K. Horákova, and J. Ebringerova. 1965. Effect of trypacidin on *Toxoplasma gondii* in tissue culture and in mice. Exp. Parasitol. 16:182-189.

Eichenwald, H. 1956. The laboratory diagnosis of toxoplasmosis. Ann. N.Y. Acad. Sci. 64:207-214.

Fayer, R. 1969. Refractile body changes in sporozoites of poultry coccidia in cell culture. Proc. Helm. Soc. Wash. 36:224-231.

Fayer, R. 1970. *Sarcocystis*: Development in cultured avian and mammalian cells. Science 168:1104-1105.

Fayer, R. 1971. Quinine inhibition of host cell penetration by eimerian sporozoites *in vitro*. J. Parasitol. 57:901-905.

Fayer, R. 1972. Gametogony of *Sarcocystis* sp. in cell culture. Science 175:65-66.

Fayer, R., and S. Baron. 1971. Activity of interferon and its inducers against development of *Eimeria tenella* in cell culture. J. Protozool. 18(Suppl.):12.

Fayer, R., and D.M. Hammond. 1966. In vitro cultivation of first-generation schizonts of *Eimeria bovis*. Am. Zool. 6:22.

Fayer, R., and D.M. Hammond. 1967. Development of first-generation schizonts of *Eimeria bovis* in cultured bovine cells. J. Protozool. 14:764-772.

Fayer, R., and D.M. Hammond. 1969. Morphological changes in *Eimeria bovis* sporozoites during their first day in cultured mammalian cells. J. Parasitol. 55:398-401.

Fayer, R., D.M. Hammond, B. Chobotar, and Y.Y. Elsner. 1969. Cultivation of *Besnoitia jellisoni* in bovine cell cultures. J. Parasitol. 55:645-653.

Fayer, R., and J.L. Mahrt. 1972. Development of *Isospora canis* (Protozoa; Sporozoa) in cell culture. Z. Parasitenk. 38:313-318.

Fayer, R., and D.K. McLoughlin. 1970. Effect of amprolium on *Eimeria adenoeides* in cell culture. J. Parasitol. 56:388-389.

Fayer, R., M.L. Melton, and H.G. Sheffield. 1972. Quinine inhibition of host cell penetration by *Toxoplasma gondii, Besnoitia jellisoni,* and *Sarcocystis* sp. in vitro. J. Parasitol 58:595-599.

Fayer, R., R.D. Romanowski, and J.M. Vetterling. 1970. The influence of hyaluronidase and hyaluronidase substrates on penetration of cultured cells by eimerian sporozoites. J. Protozool. 17:432-436.

Feldman, H.A., and A.B. Sabin. 1949. Skin reaction to toxoplasmic antigen in people of different ages without known history of infection. Pediatrics, Am. Acad. Pediat. 4:798-804.

Fitzgerald, P.R. 1970. Development of *Eimeria stiedae* in avian embryos. J. Parasitol. 56:1252-1253.

Foley, V.L., and J.S. Remington. 1969. Plaquing of *Toxoplasma gondii* in secondary cultures of chick embryo fibroblasts. J. Bacteriol. 98:1-3.

Frenkel, J.K. 1953. Infections with organisms resembling *Toxoplasma*, together with the description of a new organism: *Besnoitia jellisoni.* Atti del VI Congresso Intern'l di Microbiol. 5:426-434.

Frenkel, J.K. 1961. Discussion in the symposium on toxoplasmosis. Surv. Opthalmol. 6:734.

Frenkel, J.K. 1965. The development of the cyst of *Besnoitia jellisoni*: usefulness of this infection as a biologic model. Progr. Protozool. (Proc. II Int. Cong. Protozool.), Int. Cong. Series 91, Excerpta Med. Found., The Hague, p. 187-188.

Freshman, M.M., T.C. Merigan, J.S. Remington, and I.E. Brownlee. 1966. *In vitro* and *in vivo* antiviral action of an interferon-like substance induced by *Toxoplasma gondii.* Proc. Soc. Exp. Biol. Med. 123:862-869.

Galuszka, J. 1962a. Survey of literature concerning the cultivation of *Toxoplasma gondii* in vitro in conditions in which tissues are cultivated [In Polish, English summary]. Wiad. Parazytol. 8:307-313.

Galuszka, J. 1962b. Observations on virulence of *Toxoplasma gondii* in cultures of trypsinized embryonic cells of guinea pigs. Acta Parasitol., Polon. 10:265-269.

Galuszka, J. 1962c. Attempt to use trypsinized embryonic cells of guinea pig for the culture of *Toxoplasma gondii* [In Polish, English summary]. Medycyna Wet. 18:746-747.

Galuzo, I.G., S.I. Konovalova, and A.M. Krivkova. 1969. The behavior of avirulent strains of *Toxoplasma* in tissue cultures and in chick embryos. Progr. Protozool. (Proc. III Inst. Cong. Protozool.), p. 229-230.

Gavrilov, W., and S. Cowez. 1941. Essai de culture in vitro de tissus de moustiques et d'intestins de lapins adultes infectés. Ann. Parasitol. 18:180-186.

Gibb, D.G.A., B.A. Kakulas, D.H. Herret, and D.J. Jenkyn. 1966. Toxoplasmosis in the rottnest quokka (*Setonix brachyurus*). Aust. J. Exp. Biol. Med. Sci. 44:665-671.

Giroud, P., and N. Dumas. 1957. Essai pour la mise en évidence des anticorps dans la toxoplasmose, pouvoir cytotoxique des toxoplasmes lysés. Compt. Rend. Acad. Sci., Paris 245:1185-1186.

Goldman, M., R.K. Carver, and A.J. Sulzer. 1958. Reproduction of *Toxoplasma gondii* by internal budding. J. Parasitol. 44:161-171.

Grönroos, P. 1954. Antibiotics and experimental toxoplasmosis; polymyxin B-sulfate, bacitracin, terramycin, aureomycin and sulfa in experimental toxoplasmosis. Ann. Med. Exp. Fenn. 31:374-377.

Grönroos, P. 1955. Antibiotics and experimental toxoplasmosis. II. Erythromycin, magnamycin, neomycin, and tebezon in experimental toxoplasmosis. Ann. Med. Exp. Fenn. 32:257-259.

Guimarães, F.N., and H. Meyer. 1942. Cultivo de "*Toxoplasma*" Nicolle e Manceaux, 1909, em culturas de tecidos. Rev. Brasil. Biol. 2:123-129.

Hammond, D.M., G.W. Bowman, L.R. Davis, and B.T. Simms. 1946. The endogenous phase of the life cycle of *Eimeria bovis*. J. Parasitol. 32:409-427.

Hammond, D.M., and R. Fayer. 1967. In vitro cultivation of *Eimeria bovis*. J. Protozool. 14 (Suppl):22.

Hammond, D. M., and R. Fayer. 1968. Cultivation of *Eimeria bovis* in three established cell lines and in bovine tracheal cell line cultures. J. Parasitol. 54:559-568.

Hammond, D.M., R. Fayer, and M.L. Miner. 1969. Further studies on *in vitro* development of *Eimeria bovis* and attempts to obtain second-generation schizonts. J. Protozool. 16:298-302.

Hammond, D.M., F. Sayin, and M.L. Miner. 1963. Über den Entwicklungszyklus und die Pathogenität von *Eimeria ellipsoidalis* Becker and Frye, 1929, in Kälbern. Berl. Münch. Tierärztl. Wschr. 76:331-333.

Hammond, D.M., C.A. Speer, and W. Roberts. 1970. Occurrence of refractile bodies in merozoites of *Eimeria* species. J. Parasitol. 56:189-191.

Hansson, H.A., and P. Sourander. 1965. *Toxoplasma gondii* in cell cultures from rat retina. Virchows Arch. Path. Anat. 338:224-236.

Hansson, H.A., and P. Sourander. 1968. Ultrastructural demonstration of lysosomes in *Toxoplasma gondii*. Acta. Path. Microbiol., Scand. 74:431-444.

Heyl, J.G., and R. Gispen. 1955. A complement fixing *Toxoplasma* antigen prepared from duck eggs and enhanced by phenol. Antonie von Leeuwenhock 21:157-160.

Hirai, K., K. Hirato, and R. Yanagawa. 1966. A cinematographic study of the penetration of cultured cells by *Toxoplasma gondii*. Jap. J. Vet. Res. 14:81-90.

Hogan, M.J. 1961. Discussion in the symposium on toxoplasmosis. Surv. Ophthalmol. 6:734.

Hogan, M.J., C. Yoneda, L. Feeney, P. Zweigart, and A. Lewis. 1960a. Morphology and culture of *Toxoplasma*. Am. Ophthalmol. Soc., Trans. 58:167-187.

Hogan, M.J., C. Yoneda, L. Feeney, P. Zweigart, and A. Lewis. 1960b. Morphology and culture of *Toxoplasma*. Arch. Ophthalmol. 67:655-667.

Hogan, M.J., C. Yoneda, and O. Zweigart. 1961. Growth of toxoplasma strains in tissue culture. Am. J. Ophthalmol. 51:920-930.

Holz, A., and M. Albrecht. 1953. Die Züchtung von *Toxoplasma gondii* in Zellkulturen. Z. Hyg. Infekt. 136:605-609.

Jacobs, L. 1953. The biology of *Toxoplasma*. Am. J. Trop. Med. Hyg. 2:365-389.

Jacobs, L. 1956. Propagation, morphology, and biology of *Toxoplasma*. Ann. N.Y. Acad. Sci. 64:154-179.

Jacobs, L., J.R. Fair, and J.H. Bickerton. 1954. Adult ocular toxplasmosis. Report of a parasitologically proved case. Arch. Ophthalmol. 52:63-71.

Jacobs, L., and M.L. Melton. 1954. Modifications in virulence of a strain of *Toxoplasma gondii* by passage in various hosts. Am. J. Trop. Med. Hyg. 3:447-457.

Jacobs, L., and M.L. Melton. 1965. *Toxoplasma* cysts in tissue culture. Progr. in Protozool. (Proc. II Int. Cong. Protozool.), Int. Cong. Series 91, Excerpta Med Found., The Hague, p. 187-188.

Jeffers, T.K., and G.E. Wagenbach. 1969. Sex differences in embryonic response to *Eimeria tenella* infection. J. Parasitol. 55:949-951.

Jeffers, T.K., and G.E. Wagenbach. 1970. Embryonic response to *Eimeria tenella* infection. J. Parasitol. 56:656-662.

Jíra, J., and V. Bozděch. 1963. Complement fixing reaction, its technique and importance for diagnosis of toxoplasmosis [In Czech, English summary]. Česk. Epidemiol., Microbiol., Immunol. 12:118-125.

Jíra, J., V. Bozděch, and K. Heyberger. 1961. Complement fixing reaction in toxoplasmosis. III. Comparison of antigen from mouse ascites with antigen from the chick chorioallantoic. Časop. Lék. Česk. 100:1291-1296.

Jírovec, O., and J. Jíra. 1961. A contribution to the technique of intracutaneous testing with toxoplasmin. J. Clin. Pathol. 14:522-524.

Jírovec, O., and J. Jíra. 1962. Studies on intracutaneous test with the toxoplasmin. Česk. Parasitol. 9:281-297.

Jögiste, A.K., and J.I. Ukhov. 1969. Host-parasite relationships in chick embryos infected with *Toxoplasma gondii*. Progr. in Protozool. (Proc. III Int. Cong. Protozool.), p. 230-231.

Kaufman, H.E. 1961. Discussion in the symposium on toxoplasmosis. Surv. Ophthalmol. 6:734.

Kaufman, H.E., and E.D. Maloney. 1962. Multiplication of *Toxoplasma gondii* in tissue culture. J. Parasitol. 48:358-361.

Kaufman, H.E., M.L. Melton, J.S. Remington, and L. Jacobs. 1959. Strain differences of *Toxoplasma gondii*. J. Parasitol. 45:189-190.

Kaufman, H.E., J.S. Remington, and L. Jacobs. 1958. Toxoplasmosis: The nature of virulence. Am. J. Ophthalmol. 46 (Nov., Part 2):255-261.

Kaufman, H.E., J.S. Remington, M.L. Melton, and L. Jacobs. 1959. Relative resistance of slow-growing strains of *Toxoplasma gondii* to pyrimethane (daraprim). Arch. Ophthalmol. 62:611-615.

Kelley, G.L., and D.M. Hammond. 1970. Development of *Eimeria ninakohlyakimovae* from sheep in cell cultures. J. Protozool. 17:340-349.

Kien-truong-Thai, J.P. Garin, P. Ambroise-Thomas, and J. Despeignes. 1970. Concentration minimale inhibitrice de spiramycine sur 2 souches de toxoplasmes (RH de Sabin et DC Lyon) entretenues sur systeme cellulaire. J. Parasitol. 56(4; Sect. II, Part I):186.

Kishida, T., and S. Kato. 1965. Autoradiographic studies on intracytoplasmic multiplication of *Toxoplasma gondii* in FL cells. Biken J. 8:107-113.

Kuwahara, T. 1956. Studies on the culture of *Toxoplasma* in embryonated eggs. I. Observation on inoculations region and days. Jap. J. Sanit. Zool. 7:138.

Kuwahara, Ch. 1959. Observations on the culture of *Toxoplasma* in embryonated chick eggs and a histological study on chick embryo [In Japanese, English summary]. J. Osaka City Med. Center 8:907-925.

Levaditi, C., V. Sanchis-Bayarri, P. Lepine, and R. Schoen. 1929. Étude sur l'encephalomyélite provoquee par le *Toxoplasma cuniculi*. Ann. Inst. Pasteur, Paris 43:673-736.

Levine, N.D. 1961. Protozoan parasites of domestic animals and of man. Burgess, Minneapolis.

Lock, J.A. 1953. Cultivation of *Toxoplasma gondii* in tissue culture in mammalian cells. Lancet 264:324-325.

Long, P. L. 1965. Development of *Eimeria tenella* in avian embryos. Nature 208:509-510.

Long, P.L. 1966. The growth of some species of *Eimeria* in avian embryos. J. Parasitol. 56:575-581.

Long, P.L. 1969. Observations on the growth of *Eimeria tenella* in cultured cells from the parasitized chorioallantoic membranes of the developing chick embryo. Parasitology 59:757-765.

Long, P.L. 1970a. *Eimeria tenella:* chemotherapeutic studies in chick embryos with a description of a new method (Chorioallantoic membrane foci counts) for evaluating infections. Z. Parasitenk. 33:329-338.

Long, P.L. 1970b. *In vitro* culture of *Eimeria tenella*. J. Parasitol. 56(4; Sect. II, Part I):214-215.

Long, P.L. 1970c. Some factors affecting the severity of infection with *Eimeria tenella* in chicken embryos. Parasitology. 60:435-447.

Long, P.L. 1970d. Studies on the viability of sporozoites of *Eimeria tenella*. Z. Parasitenk. 35:1-6.

Long, P.L. 1970e. Development (schizogony) of *Eimeria tenella* in the liver of chickens treated with corticosteroid. Nature 225:290-291.

Long, P.L. 1971a. Schizogony and gametogony of *Eimeria tenella* in the liver of chick embryos. J. Protozool. 18:17-20.

Long, P.L. 1971b. Maintenance of intestinal protozoa *in vivo* with particular reference to *Eimeria* and *Histomonas*, 65-75. *In* Taylor, A.E.R. and R. Muller, Isolation and maintenance of parasites in vivo. Blackwell Scientific Publications, Oxford and Edinburgh.

Long, P.L., and Z. Tanielian. 1965. The isolation of *Eimeria mivati* in Lebanon during the course of a survey of *Eimeria* spp. in chickens. "Magon" Scientific Series, Lebanon Agr. Res. Inst., 6:1-18.

Lund, E., E. Lycke, and E. Hahn. 1960. Stability of *Toxoplasma gondii* in liquid media. Acta Pathol. Microbiol., Scand. 48:99-104.

Lund, E., E. Lycke, and P. Sourander. 1961a. Studies of *Toxoplasma gondii* in cell cultures by means of irradiation experiments. Brit. J. Exp. Pathol. 42:404-407.

Lund, E., E. Lycke, and P. Sourander. 1961b. A cinematographic study of *Toxoplasma gondii* in cell cultures. Brit. J. Exp. Pathol. 42:357-362.

Lund, E., E. Lycke, and P. Sourander. 1963a. Some aspects of cultivation of *Toxoplasma gondii* in cell cultures. Acta Pathol. Microbiol., Scand. 57:199-210.

Lund, E., E. Lycke, and P. Sourander. 1963b. Study on cultured cells infected wtih *Toxoplasma gondii*. Progr. Protozool. (Proc. I Int. Cong. Protozool.), p. 365.

Lycke, E., and E. Lund. 1964a. A tissue culture method for titration of infectivity and determination of growth rate of *Toxoplasma gondii*. Acta Pathol. Microbiol., Scand. 60:221-233.

Lycke, E., and E. Lund. 1964b. A tissue culture method for titration of infectivity and determination of growth rate of *Toxoplasma gondii*. Acta Pathol. Microbiol., Scand. 60:209-220.

Lycke, E., and E. Lund. 1966. Studies on the reproduction of *Toxoplasma gondii* in a cell culture system. Inhibition or stimulation of growth by changes in the *P*-aminobenzoic acid and the folic acid metabolism. Acta Pathol. Microbiol., Scand. 67:276-290.

Lycke, E., E. Lund, and Ö. Strannegärd. 1965. Enhancement by lysosome and hyaluronidase of the penetration by *Toxoplasma gondii* into cultured host cells. Brit. J. Exp. Pathol. 46:189-199.

Lycke, E., E. Lund, Ö. Strannegärd, and E. Falsen. 1965. The effect of immune serum and activator on the infectivity of *Toxoplasma gondii* for cell culture. Acta Pathol. Microbiol., Scand. 63:206-220.

Lycke, E., and R. Norrby. 1966. Demonstration of a factor of *Toxoplasma gondii* enhancing the penetration of *Toxoplasma* parasites into cultured host cells. Brit. J. Exp. Pathol. 47:248-256.

Lycke, E., R. Norrby, and J. Remington. 1968. Penetration-enhancing factor extracted from *Toxoplasma gondii* which increases its virulence for mice. J. Bacteriol. 96:785-788.

McLoughlin, D.K. 1969. The influence of dexamethasone on attempts to transmit *Eimeria meleagrimitis* to chickens and *E. tenella* to turkeys. J. Protozool. 16:145-148.

Ma, J.S., and J.B. Mun. 1961. Propagation of *Toxoplasma* in tissue culture. Rep. Vet. Lab., Inst. Agr., Korea 7:1-11.

MacFarlane, J.O., and I. Ruckman. 1948. Cultivation of *Toxoplasma gondii* in the developing chick embryo. Proc. Soc. Exp. Biol. Med. 67:1-4.

Maloney, E.D., and H.E. Kaufman. 1964. Multiplication and therapy of *Toxoplasma gondii* in tissue culture. J. Bacteriol. 88:319-321.

Matsubayashi, H., and S. Akao. 1962. Morphological studies on the development of *Toxoplasma* cysts and a comment on the mechanism of cyst production [In Japanese]. Jap. J. Parasitol. 11(4):13.

Matsubayashi, H., and S. Akao. 1963. Morphological studies on the development of the *Toxoplasma* cyst. Am. J. Trop. Med. Hyg. 12:321-333.

Matsubayashi, H., and S. Akao. 1964. Electron microscopical studies on the multiplication of *Toxoplasma gondii*. [In Japanese]. Jap. J. Parasitol. 12(4):285.

Matsubayashi, H., and S. Akao. 1965. The application of immunoelectron microscopy to the study of cyst development in toxoplasmosis. [In Japanese] Jap. J. Parasitol. 14(4):55.

Matsuoka, T., M.E. Callender, and R.F. Shumard. 1969. Embryonic bovine tracheal cell line for *in vitro* cultivation of *Eimeria tenella*. Am. J. Vet. Res. 30:1119-1122.

Meyer, H., and I. de Andrade Mendonca. 1955. Electron microscopic observations of *Toxoplasma* "Nicolle et Manceaux" grown in tissue cultures. Parasitology 45:449-451.

Meyer, H., and I. de Andrade Mendonca. 1957. Electron microscopic observations of *Toxoplasma* "Nicolle et Manceaux" in thin sections of tissue cultures. Parasitology 47:66-69.

Meyer, H., and M.X. de Oliveira. 1942. Observações sobre divisões mitoticas em células parasitadas. Ann. Acad. Brasil. Cienc. 14:289-292.

Meyer, H., and M.X. de Oliveira. 1943. Conservação de protozoáires em culteras de tecido mantidas a temperátura ambiente. Rev. Brasil. Biol. 3:341-343.

Meyer, H., and M.X. de Oliveira. 1945. Resultados de 3 anos de observação de cultivo de "Toxoplasma" (Nicolle e Manceaux, 1909) em cultura de tecido. Rev. Brasil. Biol. 5:145-146.

Morris, J.A., C.G. Aulisio, and J.M. McCown. 1957. Serological evidence of toxoplasmosis in animals. J. Infect. Dis. 98:52-54.

Mühlpfordt, H. 1952. Das Verhalten von *Toxoplasma gondii* (Stamm BK) in der Gewebekulturen. Z. Tropenmed. Parasitol. 4:53-64.

Norrby, R. 1970a. An immunological study on the host cell penetration factor of *Toxoplasma gondii*. In Studies on the host cell penetration of *Toxoplasma gondii*. Elanders Boktryckeri Aktiebolag, Göteborg, p. 79.

Norrby, R. 1970b. Host cell penetration of *Toxoplasma gondii*. Infect. Immunity 3:250-255.

Norrby, R., L. Lindholm, and E. Lycke. 1968. Lysosomes of *Toxoplasma gondii* and their possible relation to the host-cell penetration of toxoplasma parasites. J. Bacteriol. 96:916-919.

Norrby, R., and E. Lycke. 1967. Factors enhancing the host cell penetration of *Toxoplasma gondii*. J. Bacteriol. 93:53-58.

Novinskaya, V.F. 1965. Use of the chick embryos in diagnostics of toxoplasmosis [In Russian], p. 391-393. *In* I.G. Galuzo, Toxoplasmosis of animals. Academy of Science of the Kazakh SSR, Alma-ata.

Osaki, H., and Y. Oka. 1963. Studies on the minority inoculation and preservation of *Toxoplasma gondii* in tissue culture [In Japanese]. Jap. J. Parasitol. 12(4):286-287.

Osaki, H., Y. Oka, K. Yamamoto, and N. Matsuo. 1964. Isolation of *Toxoplasma* by tissue culture [In Japanese]. Jap. J. Parasitol. 13(4):284.

Paine, G.D., and R.C. Meyer. 1969. *Toxoplasma gondii* propagation in cell cultures and preservation at liquid nitrogen temperatures. Cryobiology 5:270-272.

Patton, W.H. 1965. *Eimeria tenella*: Cultivation of the asexual stages in cultured animal cells. Science 150:767-769.

Perrotto, J., D.B. Keister, and A.H. Gelderman. 1971. Incorporation of precursors into *Toxoplasma* DNA. J. Protozool. 18:470-473.

Pierreaux, G. 1963. Culture de *Toxoplasma gondii* sur tissus. Ann. Soc. Belge Med. Trop. 213:241-246.

Pulvertaft, R.J., J.C. Valentine, and W.F. Lane. 1954. The behavior of *Toxoplasma gondii* on serum-agar culture. Parasitology 44:478-485.

Remington, J.S., P. Earle, and T. Yagura. 1970. *Toxoplasma* in nucleus. J. Parasitol. 56:390-391.

Remington, J.S., and T.C. Merigan. 1968. Interferon: protection of cells infected with an intracellular protozoan (*Toxoplasma gondii*). Science 161:804-806.

Remington, J.S., T. Yagura, and W.S. Robinson. 1970. The effect of rifampin on *Toxoplasma gondii*. Proc. Soc. Exp. Biol. Med. 135:167-172.

Roberts, W.L., Y.Y. Elsner, A. Shigematsu, and D.M. Hammond. 1969. Autoradiographic study of the incorporation of H^3-thymidine into *Eimeria callospermophili* in cell cultures. J. Protozool. 16(Suppl.):16.

Roberts, W. L., Y. Y. Elsner, A. Shigematsu and D. M. Hammond. 1970. Lack of incorporation of H^3-Thymidine into *Eimeria callospermophili* in cell cultures. J. Parasitol. 56:833-834.

Roberts, W.L., D.M. Hammond, and L. Anderson. 1969. Electron microscope studies of the early endogenous development of *Eimeria callospermophili* from the Uinta ground squirrel *Spermophilis armatus*. J. Protozool. 16(Suppl.):16.

Roberts, W.L., D.M. Hammond, and C.A. Speer. 1970. Ultrastructural study of the intra- and extracellular sporozoites of *Eimeria callospermophili*. J. Parasitol. 56:907-917.

Roberts, W.L., C.A. Speer, and D.M. Hammond. 1971. Penetration of *Eimeria larimerensis* sporozoites into cultured cells as observed with the light and electron microscopes. J. Parasitol. 57:615-625.

Ryley, J.F. 1968. Chick embryo infections for the evaluation of anticoccidial drugs. Parasitology 58:215-220.

Ryley, J.F., and R.G. Wilson. 1971. Studies on the mode of action of the coccidiostat robenidene. Z. Parasitenk. 37:85-93.

Rytel, M.W., and T.C. Jones. 1966. Induction of interferon in mice infected with *Toxoplasma gondii*. Proc. Soc. Exp. Biol. Med. 123:859-862.

Sabin, A.B., and P.K. Olitsky. 1937. *Toxoplasma* and obligate intracellular parasitism. Science 85:336-338.

Sampson, J.R., D.M. Hammond, and J.V. Ernst. 1971. Development of *Eimeria alabamensis* from cattle in mammalian cell cultures. J. Protozool. 18:120-128.

Scholtyseck, E. 1953. Beitrag zur Kenntnis des Entwicklungsganges des Hühnercoccids, *Eimeria tenella*. Arch. Protistenk. 98:415-465.

Scholtyseck, E. 1969. Electron microscope studies of the effect upon the host cell of various developmental stages of *Eimeria tenella* in the natural chicken host and in tissue cultures. Acta Vet. (Brno) 38:153-156.

Scholtyseck, E., and R.G. Strout. 1968. Feinstrukturuntersuchungen über die Nahrungsaufnahme bei Coccidien in Gewebekulturen (*Eimeria tenella*). Z. Parasitenk. 30:291-300.

Schmidt-Hoensdorf, F., and J. Holz. 1953. Zur Biologie und Morphologie des *Toxoplasma gondii*. Z. Hyg. Infekt. 136:601-604.

Schuhová, V. 1957. Langfristige Kulturen des *Toxoplasma gondii* in He-La-Zellen. Zentralbl. Bakteriol., I Abt. Orig. 168:631-636.

Schuhová, V. 1960. Long-term culture of *Toxoplasma gondii* in Hela cells. J. Hyg. Epidemiol., Microbiol., Immunol. 4:131-132.

Schuhová, V., E. Bonnové, J. Hübnera, and Z. Šaškové. 1956. Serial propagation of *Toxoplasma gondii* in stationary tube cultures of

HeLa cells. Česk. Epidemiol., Mikrobiol., Imunol., 5:161-163.
Schuhová, V., and O. Jírovec. 1962. Frequency of positive toxoplasmin test in people in contact with animals [In Czech, English summary.] Časep. Lěk. Čes. 100:964-966.
Schuhová, V., M. Splítkové, and Z. Šaškové. 1957. Long-term tissue cultures of *Toxoplasma gondii* in HeLa Cells [In Czech, English summary.] Česk. Epidemiol., Mikrobiol., Imunol. 6:9-11.
Schuhová, V., M. Zavádová, and G. Štumpa. 1963a. Complement fixation *Toxoplasma* antigen prepared in tissue culture. J. Hyg. Epidemiol., Microbiol., Immunol. 7:65-73.
Schuhová, V., H. Zavádová, and G. Štumpa. 1963b. Compliment fixation *Toxoplasma* antigen prepared in tissue culture. Progr. Protozool. (Proc. I Int. Cong. Protozool.) p. 371-378.
Sheffield, H.G., R. Fayer, and D.M. Hammond. 1968. Electron microscope observations on sporozoites of *Eimeria bovis* in cultured bovine kidney cells. J. Protozool. 15(Suppl.):18.
Sheffield, H.G., and D.M. Hammond. 1967. Electron microscope observations on the development of first-generation merozoites of *Eimeria bovis*. J. Parasitol. 53:831-840.
Sheffield, H.G., and M.L. Melton. 1968. The fine structure and reproduction of *Toxoplasma gondii*. J. Parasitol. 54:209-226.
Sheffield, H.G., and M.L. Melton. 1969. Ultrastructural changes in *Toxoplasma gondii* after in vitro treatment with chemotherapeutic agents. Progr. Protozool. (III Int. Cong. Protozool.), p. 243.
Sheffield, H.G., and M.L. Melton. 1970a. *Toxoplasma gondii*: The oocyst, sporozoite and infection of cultured cells. Science 167:892-893.
Sheffield, H.G., and M.L. Melton. 1970b. Observations on the sporozoites of *Toxoplasma gondii* and their behavior in cultured cells. J. Parasitol. 56(4; Sect. II, Part I):315.
Shibalova, T.A. 1968. Cultivation of *Eimeria tenella* on chicken embryos and in tissue cultures [In Russian]. Parazitologiya, Leningrad 2:483-484.
Shibalova, T.A. 1969a. Cultivation of coccidian endogenous stages in chicken embryos and tissue cultures. Progr. Protozool. (III Int. Cong. Protozool.), p. 355-356.
Shibalova, T.A. 1969b. Cultivation of the asexual stages of *Eimeria tenella* in cultured tissue cells [In Russian]. Tsitologiya, 11:707-713.
Shibalova, T.A. 1970. Cultivation of the endogenous stages of chicken coccidia in embryos and tissue cultures. J. Parasitol. 56(4; Sect. II, Part I):315-316.
Shibalova, T.A., and A.M. Korolev. 1969. Excystation of sporozoites of *Eimeria tenella in vitro* and their cultivation in cell cultures. Progr. Protozool.(III Int. Cong. Protozool)., p. 387.
Shibalova, T.A., A.M. Korolev, and I.A. Sobchak. 1969. Cultivation of chicken coccidia in chick embryos [In Russian]. Veterinariya, Moskva 11:68-71.
Shimizu, K. 1961a. Studies on toxoplasmosis. III. Observations on the tissue culture method of *Toxoplasma gondii* [In Japanese]. Jap. J. Vet. Sci. 23:33-44.
Shimizu, K. 1961b. Studies on toxoplasmosis. IV. Complement fixation antigen from the tissue culture fluid [In Japanese]. Jap. J. Vet. Sci. 23:167-180.
Shimizu, K. 1963. Studies on toxoplasmosis. V. Complemental observations on the tissue culture method, especially the effect of the nutrient fluid upon the invasion and multiplication of the organisms. Jap. J. Vet. Res. 11:1-11.
Shimizu, K., and H. Shiokawa. 1965. Complement-fixing antigen of *Toxoplasma gondii* derived from tissue-cultured organisms and fluid. Jap. J. Vet. Sci. 27:295-304.
Shimizu, K., and Y. Takagaki. 1959. Studies on toxoplasmosis. II. Some observations on strain "HT" which was isolated from a hare (*Lepus timidus aimu*) in Sapporo. Jap. J. Vet. Res. 7:95-103.
da Silva, L.H.P. 1960. *In vitro* effect of the aminonucleoside of stylomycin and dimethyladenine against *Toxoplasma gondii*. Rev. Inst. Med. Trop., São Paulo 2:155-162.
Sourander, P., E. Lycke, and E. Lund. 1960. Observations on living cells infected with *Toxoplasma gondii*. Brit. J. Exp. Pathol. 41:176-178.
Speer, C.A., L.R. Davis, and D.M. Hammond. 1971. Cinemicrographic observations on the development of *Eimeria larimerensis* in cultured bovine cells. J. Protozool. 18,Suppl.):11.
Speer, C.A., and D.M. Hammond. 1969. Cinemicrographic observations on the development of *Eimeria callospermophili* in cultured cells. J. Protozool. 16(Suppl.):16.
Speer, C.A., and D.M. Hammond. 1970a. Nuclear divisions and refractile body changes in sporozoites and schizonts of *Eimeria callosper-*

mophili in cultured cells. J. Parasitol. 56:461-467.

Speer, C.A., and D.M. Hammond. 1970b. Development of *Eimeria larimerensis* from the Uinta ground-squirrel in cell culture. Z. Parasitenk. 35:105-118.

Speer, C.A., and D.M. Hammond. 1971a. Development of *Eimeria ellipsoidalis* from cattle in cultured bovine cells. J. Parasitol. 57:675-677.

Speer, C.A., and D.M. Hammond. 1971b. Development of first-and second-generation schizonts of *Eimeria magna* from rabbits in cell cultures. Z. Parasitenk. 37:336-353.

Speer, C.A., and D.M. Hammond. 1972. Development of gametocytes and oocysts of *Eimeria magna* from rabbits in cell culture. Proc. Helm. Soc. Wash. 39:114-118.

Speer, C.A., D.M. Hammond, and L.C. Anderson. 1970. Development of *Eimeria callospermophili* and *E. bilamellata* from the Uinta ground squirrel *Spermophilus armatus* in cultured cells. J. Protozool. 17:274-284.

Speer, C.A., D.M. Hammond, and G.L. Kelley. 1970. Stimulation of motility in merozoites of five *Eimeria* species by bile salts. J. Parasitol. 56:927-929.

Stadtsbaeder, S. 1965. Multiplication des toxoplasmes dans les globules blancs humains mis en culture. Rev. Belge Pathol. 31:280-284.

Stewart, G.L., and H.A. Feldman. 1965. Use of tissue culture cultivated *Toxoplasma* in the dye test and for storage. Proc. Soc. Exp. Biol. Med. 188:542-546.

Strannegärd, Ö., and E. Lycke. 1966. Properdin and the antibody-effect on *Toxoplasma gondii.* Acta Pathol. Microbiol., Scand. 66:227-238.

Stroczyńska, M. 1966. Investigations on the complement fixation test in toxoplasmosis. Acta Parasitol., Polon. 13:117-126.

Strout, R.G., and C.A. Ouellette. 1968. Gametogony of *Eimeria tenella* (Coccidia) in cell cultures. Science 163:695-696.

Strout, R.G., and C.A. Ouellette. 1970. Schizogony and gametogony of *Eimeria tenella* in cell culture. Am. J. Vet. Res. 31:911-918.

Strout, R.G., and C.A. Ouellette. 1971. Detecting the activity of anticoccidial compounds *in vitro*. J. Protozool. 18(Suppl.):12.

Strout, R.G., C.A. Ouellette, and D.P. Gangi. 1969a. *Eimeria tenella:* Temperature and asexual development in cell culture. Exp. Parasitol. 25:324-328.

Strout, R.G., C.A. Ouellette, and D.P. Gangi. 1969b. Effect of inoculum size on development of *Eimeria tenella* in cell cultures. J. Parasitol. 55:406-411.

Strout, R.G., and E. Scholtyseck. 1970. The ultrastructure of first generation development of *Eimeria tenella* (Railliet and Lucet, 1891) Fantham, 1909 in cell culture. Z. Parasitenk. 35:87-96.

Strout, R.G., J. Solis, S.C. Smith, and W.R. Dunlop. 1965. In vitro cultivation of *Eimeria acervulina* (Coccidia). Exp. Parasitol. 17:241-246.

Štumpa, 1966. The use of haemagglutination test in the serologic diagnosis of toxoplasmosis [In Czech, English summary]. Česk. Epidemiol., Mikrobiol., Imunol. 15:328-333.

Suggs, M.T., and K.W. Walls. 1967. Growth comparison of *Toxoplasma gondii* and *Besnoitia jellisoni* in human fetal lung cell culture. Proc. 2nd Joint Meet. of Clinic. Soc. and Comm. Officers Assoc. of USPHS., Atlanta, p.70.

Suggs, M.T., K.W. Walls, and I.G. Kagan. 1968. Comparative antigenic study of *Besnoitia jellisoni, B. panamenis* and five *Toxoplasma gondii* isolates. J. Immunol. 101:166-175.

Taylor, A.E.R., and J.R. Baker. 1968. The cultivation of parasites *in vitro.* Blackwell Scientific Publications, Oxford.

Toš-Luty, S. 1967. Investigation on the *Toxoplasma* antigen, prepared from tissue culture, in the complement fixation test [In Polish, English summary]. Wiad. Parazytol. 13:41-48.

Trager, W., and S.M. Krassner. 1967. Growth of parasitic protozoa in tissue cultures, p. 357-382. *In* Tze-Tuan Chen, Research in Protozoology, Vol. 2. Pergamon Press, Oxford.

Turner, M.B.F., and E.D. Box. 1970. Cell culture of *Isospora* from the English sparrow, *Passer domesticus domesticus*. J. Parasitol. 56:1218-1223.

Varela, G., E. Roch, and A. Vásques. 1955. Virulencia, cultivo, polisacaridos, toxinas y la prueba del colorante, estudiodas con una cepa de *Toxoplasma gondii.* Rev. Inst. Salub. Enfer. Trop. 15:73-80.

Vasina, S.G. 1958. Development of *Toxoplasma gondii* in chick's embryo [In Russian, English summary]. Med. Parazitar. Bolezni, Moskva 27:79-82.

Vermeil, C., J. Lavillaureix, and S. Heitz. 1962. Sur l'utilisation de la culture de tissus pour l'identification d'une souche de *Toxoplasma gondii.* Cobiose ultra-virus-*Toxoplasma* in culture de tissus. Bull. Soc. Path Pathol.t. 55:1078-1084.

Vetterling, J.M., and D.J. Doran. 1969. Storage polysaccharide in coccidial sporozoites after excystation and penetration of cells. J. Protozool. 16:772-775.

Virat, J. 1967. Le diagnostic sérologuique de la toxoplasmose. Pathol. et Biol. 15:60-64.

Vischer, W.A., and E. Suter. 1954. Intracellular multiplication of *Toxoplasma* in adult mammalian macrophages cultivated *in vitro*. Proc. Soc. Exp. Biol. Med. 86:413-419.

de Vos, A.J., and D.M. Hammond. 1971. Development of *Eimeria crandallis* from sheep in cell cultures. J. Protozool. 18(Suppl.):11.

de Vos, A.J., D.M. Hammond, and C.A. Speer. 1972. Development of *Eimeria crandallis* from sheep in cultured cells. J. Protozool. 19: 335-343.

Warren, J., and S.B. Russ. 1948. Cultivation of *Toxoplasma* in embryonated eggs. An antigen derived from chorioallantoic membrane. Proc. Soc. Exp. Biol. Med. 67:85-89.

Weinman, D. 1944. Human toxoplasma. Puerto Rico J. Pub. Health Trop. Med. 20:125-193.

Wolf, A., D. Cowen, and B.H. Paige. 1940. Toxoplasmic encephalomyelitis. IV. Experimental transfer of the infection to animals from a human infant. J. Exp. Med. 71:187-214.

Wolfson, F. 1941. Mammalian toxoplasma in erythrocytes of canaries, ducks, and duck embryos. Am. J. Trop. Med. 21:653-658.

Wolfson, F. 1942. Maintenance of human *Toxoplasma* in chicken embryos. J. Parasitol. 28(Suppl.):16-17.

Yanagawa, R., and K. Hirato. 1963. Antitoxoplasmic effect of immune swine serum revealed in the culture of swine leucocytes. Jap. J. Vet. Res. 11:135-142.

Yü, E.S., and F.K. In. 1966. Cultivation of *Toxoplasma* in monkey kidney and hog kidney monolayer tissue cells. K'O Hsüeh t'ung pao: wai wen pa (Foreign Lang. Edition), Peking 17:41-42.

Zassuchin, D.N. 1963. Observations on the biology of *Toxoplasma*. Progr. Protozool. (Proc. I Int. Cong. Protozool.), p. 366-368.

Zuckerman, A. 1966. Propagation of parasitic protozoa in tissue culture and avian embryos. Ann. N.Y. Acad. Sci. 139:24-38.

7 Pathology and Pathogenicity of Coccidial Infections

P. L. LONG
Houghton Poultry Research Station, Houghton, Huntingdon, England

Contents

I. Introduction . . . *254*
II. Factors Affecting the Pathogenicity of Coccidia . . . *254*
  A. Site of Development . . . *254*
  B. Dose of Oocysts . . . *256*
  C. Viability and Virulence of Oocyst Cultures . . . *258*
  D. Effect of Age of Host on Susceptibility to Infection . . . *259*
  E. Effect of Breed or Strain of Host . . . *262*
III. Gross Pathology of Infection . . . *262*
  A. Infections of Poultry . . . *262*
    1. The Domestic Fowl . . . *263*
    2. Ducks, Turkeys, and Geese . . . *264*
  B. Infection of Mammals . . . *265*
    1. Cattle . . . *265*
    2. Sheep . . . *266*
    3. Pigs . . . *267*
    4. Rodents . . . *267*
    5. Dogs and Cats . . . *267*
    6. Man . . . *268*
IV. Histopathology of Coccidial Infections . . . *268*
V. Experimental Pathology . . . *277*
  A. Effect of Coccidiosis on Food and Water Consumption, Body Weight Gain, and Egg Production . . . *277*
  B. Effects of Coccidiosis on Carbohydrate Metabolism and Blood Loss of the Host . . . *278*
  C. Effect of Coccidiosis on Gut Permeability and the Uptake of Nutrients . . . *279*
  D. The Possible Role of Toxins in Coccidial Infections . . . *280*

VI. Pathogenesis of Coccidial Infections . . . *281*
VII. Concluding Remarks . . . *286*
Literature Cited . . . *287*

I. Introduction

It is not the intention of this chapter to describe the gross lesions and histopathology of the numerous species of *Eimeria, Isospora*, and *Tyzzeria* which cause pathological changes in their hosts. This information is, to some extent, available in other books on pathogenic protozoa (Levine, 1961) and coccidia (Pellérdy, 1965; Davies, Joyner, and Kendall, 1963; Levine and Ivens, 1965, 1970). Some general observations on the interaction between coccidia and their hosts will be made and some of the pathological changes induced by parasitism will be described. Factors affecting the pathogenicity of coccidial infection will also be considered, including age and strain of host, dose of oocysts, and viability of sporulated oocysts. In addition, the opportunity will be taken to review recent work on the pathology of experimental infections.

II. Factors Affecting the Pathogenicity of Coccidia

A. SITE OF DEVELOPMENT

The degree of damage caused by coccidia on their hosts depends upon the numbers of parasites occurring at any particular site. With coccidia, the numbers present depend upon the number of sporulated oocysts ingested and, within limits, the reproduction can be predetermined. This situation is different from that of many parasitic protozoa which reproduce by binary fission until either the numbers of parasites overwhelm the host or immune mechanisms intervene to reduce or stop replication. Thus, the degree of damage to the host caused by coccidia might be considered directly proportional to the degree of destruction of host cells. However, this is an oversimplification of the events, as will be shown below. There appears to be some relation between the pathogenicity of species and the depth with which they penetrate the intestinal mucosa. With few exceptions, endogenous stages of coccidia occur in epithelial cells. In order to avoid any confusion with regard to the particular sites in which coccidia develop and cause disruption of the host tissues, diagrams of transverse sections of the small intestine are shown in Figures 1 and 2.

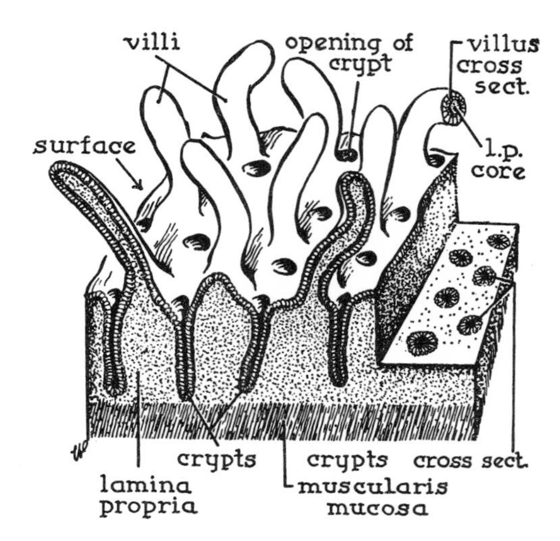

Fig. 1. Three-dimensional drawing of the lining of the small intestine. Observe that villi are finger-like processes, with cores of lamina propria that extend into the lumen. Note also that crypts of Lieberkühn are glands that dip down into the tunica propria. Observe particularly the difference in the cross-section appearance of villi and crypts. (From A. W. Ham, Histology, 6th ed., p. 692. Lippincott Company, Phila. Reproduced by kind permission of the author and the publishers.)

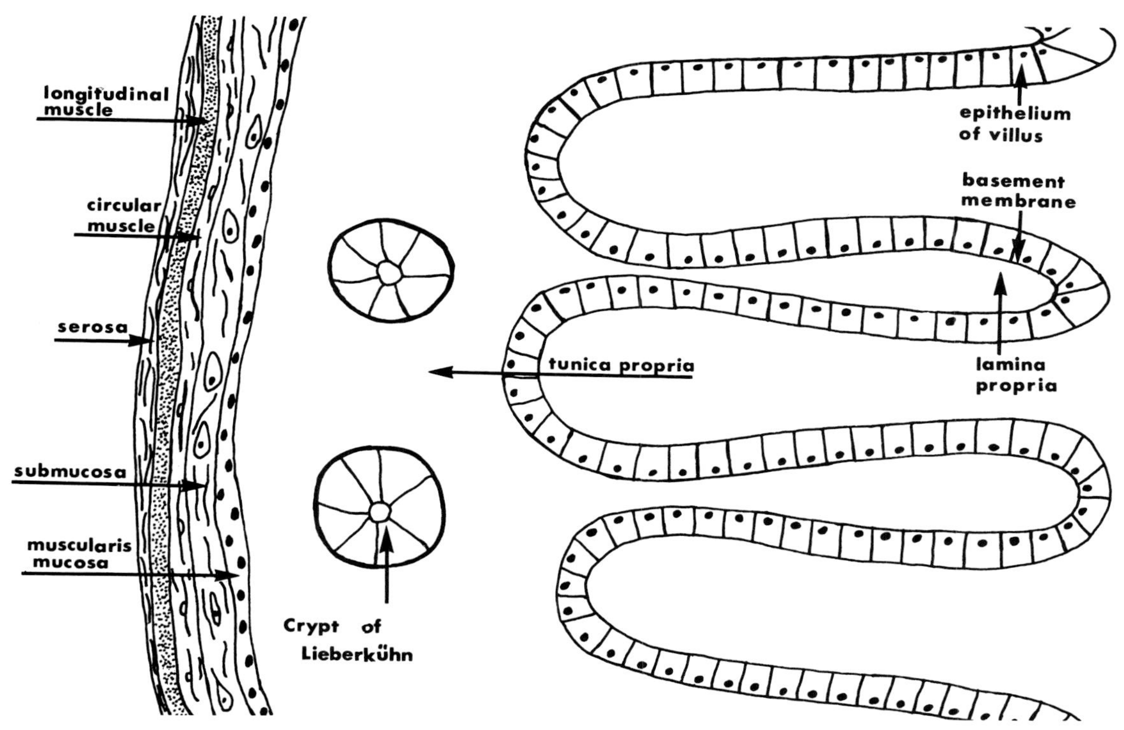

Fig. 2. Diagram of transverse section of small intestine showing the various parts which comprise the intestinal wall. A distinction has been made between the lamina propria and tunica propria since only a few species of *Eimeria* develop in the tunica propria.

It should be noted that the relative thickness of the muscular layers and the length of the villi vary from the small intestine to the rectum. Hill (1971) has indicated the important differences in the structure of the avian intestine and that of mammals. In the chicken there are no central lacteals in the villi of the small intestine, the villous core being occupied by a capillary bed. Hill also illustrates differences in the villus height in different parts of the ceca. Brunner's glands are absent from the duodenum, and mucous secretion is performed by numerous goblet cells in both the surface epithelium and glands. Argentaffin cells are present in large numbers in both the deep glands and surface epithelium of the duodenum.

In order to be specific about the particular parts of the intestinal wall in which the different stages of coccidia develop, a distinction is made between lamina propria and tunica propria; restricting the term "lamina propria" to the central portion of the superficial villi and the term "tunica propria" to the tissue beneath the superficial villi extending to the submucosa, but excluding the crypts of Lieberkühn.

Some *Eimeria* develop only in the superficial villi (*E. praecox*), while others develop first in the crypts of Lieberkühn and later in the superficial villi (*E. acervulina*). Others develop in the lamina propria and crypts of Lieberkühn (*E. bovis*). In *E. tenella* and *E. necatrix* infections, first-generation schizogony occurs in the crypts of Lieberkühn; second-generation schizonts develop away from the epithelium in fibrocytes of the tunica propria, where they cause the disruption of blood vessels resulting in severe hemorrhage. In experimental infections of embryos, the second-generation schizonts of *E. tenella* appear to select sites near the blood vessels of the chorioallantois of chicken embryos (see Figure 11, Chapter 6).

Another factor affecting pathogenicity is the size of the endogenous stages. Stages which do not grow beyond from 1 μ to 15 μ in diameter and develop in epithelial cells of the superficial villi in positions above the host cell nuclei may cause the destruction of only those epithelial cells which are coming to the end of their useful life. Only slight or moderate cellular responses occur with this kind of infection, a good example being *E. praecox* infection in chickens (Long, 1967a). When endogenous stages develop below the host cell nucleus in positions close to the basement membrane (e.g., *E. maxima* and *E. brunetti*) and give rise to large schizonts or gametocytes they tend to cause greater effects on the host. The first-generation schizonts of *E. bovis* and *E. ovina* (the name now preferred by Levine and Ivens, 1970 for *E. arloingi* of sheep) both have giant schizonts which occur in the lamina propria and stimulate severe cellular responses. The first-generation schizont of *E. ninakohlyakimovae* is also extremely large (about 290 μ in diameter) and develops within connective tissue cells near the crypts of Lieberkühn. The second-generation schizonts of this species are small and occur in epithelial cells of the crypts (Wacha, Hammond, and Miner, 1971). These species evoke greater host reactions than those species which parasitize the epithelial cells of the superficial villi.

The response of the host to parasitism by coccidia within sites other than the intestine appears to be very great. With infection of the rabbit liver by *E. stiedai* there appears to be a proliferation of the capillary bile ducts disproportional to the number of coccidia present (Kotlán and Pellérdy, 1937). Jaundice is also a feature of this disease (Bachman and Menendez, 1930). In infections of the goose kidney by *E. truncata*, inflammatory cells are numerous and necrosis of the kidney tubules occurs (Fig. 3). The pathogenesis of coccidial infections varies according to the site parasitized. Intestinal function is obviously greatly affected by intestinal coccidia, as is the nutrition of the host, since the uptake of nutrients is also affected. Where other organs are parasitized (e.g., liver or kidney) the major effects are on the functions of these organs.

B. DOSE OF OOCYSTS

It is now well established that the reproduction of *Eimeria* within the host is dependent upon the dose of sporulated oocysts ingested, and in the absence of reinfection, the infections of the host terminate when oocysts are formed. Provided that sensitive methods are used to assess the degree of infection, it can be shown that the effects on the host are dose related.

Most studies on the relationship between dose of oocysts and effects on the host have been carried out in chickens, with only a limited amount of work in other animals. In general, increased effects on the host are produced by administering increased numbers of oocysts (Long and Horton-Smith, 1968).

Krassner (1963) showed that with *E. acervulina*, fewer oocysts were produced as the dose of oocysts was increased, because of a "crowding factor," in which insufficient numbers of epithelial cells were available when exceptionally large doses of oocysts were administered. This may not be the sole reason for the lack of development in heavy infections. It is possible that interferon or an interferon-like substance is produced in response to coccidia. Long and Milne (1971) showed that chickens given an artificial interferon inducer were protected against *E. maxima* infection and suggested that *E.*

maxima and *E. tenella* may stimulate the production of interferon by the infected chickens. Fayer and Baron (1971) briefly reported that interferon did not inhibit the penetration of cultured cells by sporozoites of *E. tenella* but inhibited the development of the parasites within these cells. This effect is similar to the action of interferon on viruses, virus replication being affected and not invasion of the cells by the virus particles.

Hein (1968) clearly showed that despite the early loss of parasites, as judged by oocyst production in those chickens which received large doses of oocysts of *E. acervulina*, the effects on the host, judged by body weight gain, increased with the oocyst dose. However, Kouwenhoven (1970) was of the opinion that with this parasite, doses in excess of 600,000 oocysts failed to produce proportional effects on the host.

Resistance to *Eimeria* may commence during the primary infection, and exposure of the host to massive numbers of early stages of the life cycle may stimulate immune mechanisms which are effective against later stages in the life cycle. This effect may be most important in infection of hosts with *Eimeria* having long prepatent periods. Long and Rose (1970), working with *E. mivati*, which has a short prepatent period, demonstrated extended schizogony as a result of corticosteroid treatment of the host. These findings

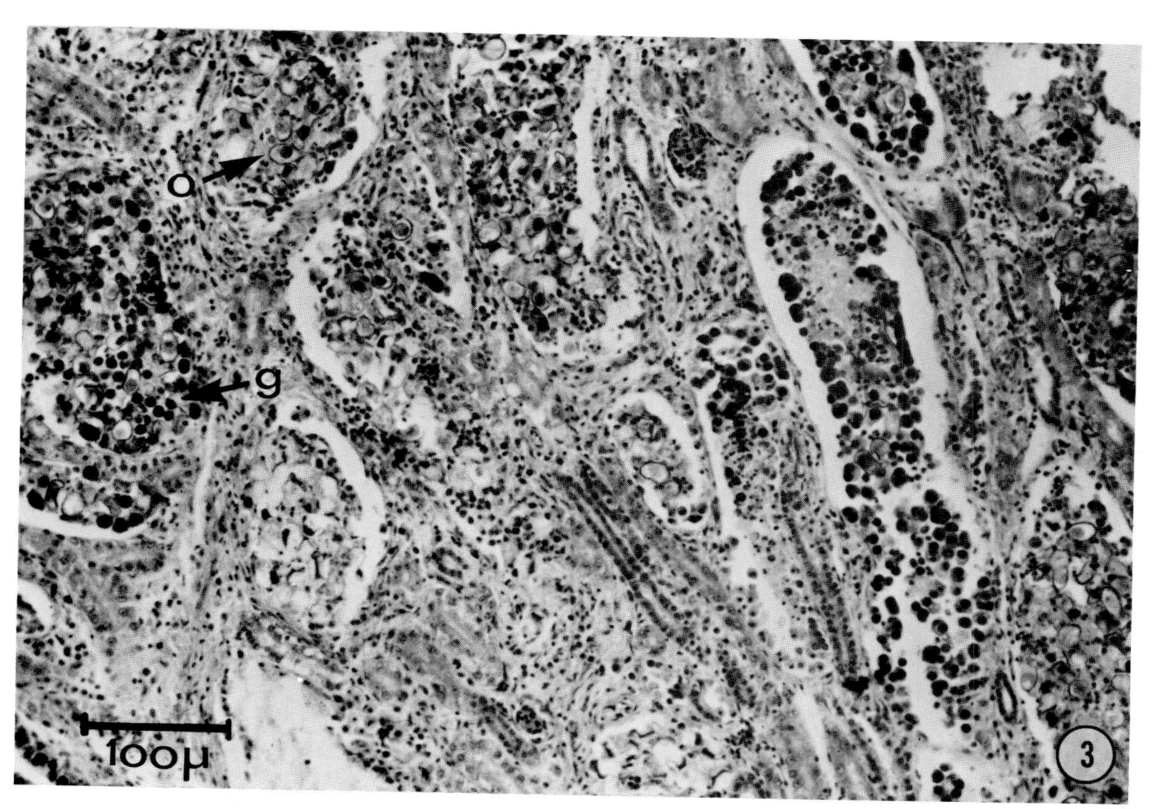

Fig. 3. Photomicrograph of a paraffin section of goose kidney infected with *E. truncata*; tissues were fixed in formol-sublimate and stained with Ehrlich's hematoxylin and eosin. Oocysts (o) and gametocytes (g) have almost completely obstructed the kidney tubules and the normal architecture of the kidney has been lost. Numerous host defense cells have infiltrated the connective tissue.

suggest that extended schizogony in untreated birds was inhibited by host responses to the parasite.

Leathem and Burns (1968) found that mortality as a result of cecal coccidiosis was higher in birds given 50,000 or 100,000 oocysts than in birds given ten million oocysts.

Hein (1969) demonstrated that increasing doses of *E. adenoeides* and *E. meleagrimitis* in turkey poults produced proportional effects on the host judged by mortality and body weight. Christensen (1941) showed that the severity of outbreaks of coccidiosis caused by *E. ninakohlyakimovae* in feeder lambs was proportional to the numbers of sporulated oocysts ingested. In a study of *E. bovis* infection in calves, Hammond, Davis, and Bowman (1944) showed that effects on the host, judged by body weight, oocyst discharge, and abnormalities of fecal discharges, were proportional to oocyst dose. Lotze (1952), working with *E. ovina* in young sheep reared free from coccidia, showed that clinical signs of the disease in the sheep were obtained when doses between two and five million oocysts were given.

Lund (1954) found that the effects of *E. stiedai* infection on rabbit hosts increased as the dose of oocysts administered was increased. Lotze and Leek (1970), working with *E. intricata* in sheep, believed that host reactions in sheep to massive numbers of asexual stages result in the inhibition of sexual stages.

It appears that some interference with the normal endogenous cycle may occur if excessively large doses of oocysts are given. This appears to depend upon the species of coccidium employed; the "crowding" effect is most noticeable when highly immunogenic species are used. The "crowding" effect may, under certain circumstances, be an expression of immunity mediated by antibody or antibody-producing cells (see Chapter 8). Alternatively, interferon or an interferon-like substance may be involved. These mechanisms are most likely to operate when large numbers of sporozoites are present within the epithelial cells several days after the administration of oocysts *per os*. The "crowding" factor may operate in a much simpler way when the tissue disruption caused by large numbers of early stages of a parasite results in the loss of both host cells and parasites into the lumen of the intestine, and the merozoites fail to invade new cells.

C. VIABILITY AND VIRULENCE OF OOCYST CULTURES

It is not valid to assume that batches of oocysts of the same species, obtained from field cases of coccidiosis are similarly pathogenic when administered to susceptible animals. Although some oocysts are able to survive for long periods under unfavorable conditions, most of them are probably destroyed and the remainder probably have low virulence. Horton-Smith and Long (1954) showed that *Eimeria* oocysts from the chicken survive only about three weeks in deep litter upon which chickens are kept. The numbers of oocysts in litter are kept more or less constant by the discharge of new oocysts from the hosts. Horton-Smith (1957) reviewed factors affecting the transmission of coccidia and pointed out that, although the action of bacteria and fungi on oocysts was detrimental, the major factors affecting the viability of oocysts were the presence of oxygen and water. Oocysts die rapidly in the absence of oxygen or water. Environmental temperature is of primary importance in the sporulation of oocysts, the optimum temperature being 29°C. A small number of oocysts held at 8°C for eight weeks sporulated when moved to room temperature, but oocysts placed at 45°C were destroyed within 24 hr (Edgar, 1954). Our unpublished observations showed that of unsporulated oocysts held at

4°C for 14 weeks, 46% sporulated when removed to 28°C for 48 hr, but oocysts kept at 4°C for 26 weeks failed to sporulate when placed at 28°C.

Horton-Smith (1957) discussed the probable influence of environmental temperatures on sporulation and survival of oocysts under field conditions and the effect this might have on the course of epidemics of coccidiosis in the field. During winter conditions, sporulation would not occur and the animals would remain free of (but susceptible to) coccidiosis. In the spring, sporulation of oocysts on the ground would occur and infection of highly susceptible hosts would result.

Some workers have observed low sporulation rates and abnormal sporulation in oocysts obtained from partially resistant animals (see Chapter 8).

Kheysin (1959) noted that sporozoites stored within oocysts for from 18 to 20 months lost their glycogen completely. Vetterling and Doran (1969) confirmed this observation by demonstrating a significant quantitative loss in amylopectin in oocysts of *E. acervulina* stored for long periods at 4°C. Long (1970a) found that the infectivity of sporozoites obtained from oocyst cultures stored for 65 days was significantly lower than that of sporozoites from fresh oocyst cultures.

There is some evidence that some strains of the same species differ in their pathogenicity to their hosts. Joyner (1969) demonstrated differences in the pathogenicity of two strains of *E. acervulina*; Joyner and Norton (1969) and Long (1970b) showed that two strains of *E. tenella* maintained in the laboratory for more than 20 years differed in their pathogenicity to chickens and chick embryos. Long (1970b) suggested that the greater pathogenicity of one of the *E. tenella* strains might be the result of the earlier maturation of the second-generation schizonts. More work is needed in order to confirm whether the rapid passage of coccidia in groups of animals increases the virulence of the parasite. It should now be possible to obtain this information with the aid of freeze-preservation techniques. The use of these techniques is also essential if we are to determine whether seasonal variation in the susceptibility of the host occurs.

D. THE EFFECT OF AGE OF HOST ON SUSCEPTIBILITY TO INFECTION

Mayhew (1934), Jones (1932) and Horton-Smith (1947), working with *Eimeria* of chickens, concluded that age resistance did not occur. In the earlier publications several workers had considered that age resistance did occur, but our present view is that in much of the early work insufficient attention was paid to the rearing of stock free from infection. Recent studies indicate that susceptibity to infection, as judged by the reproduction of the parasite, increases with the age of the host. This was found to be true with *E. maxima* infection (Long, 1959), with *E. praecox* infection (Long, 1967a), and with *E. tenella* infection (Rose, 1967).

Hein (1968) showed that the oocyst production of six-week-old chickens infected with *E. acervulina* was almost ten times greater than in two-week-old chickens given similar numbers of oocysts. However, she concluded, by comparing weight gains at 4 and 14 days after infection, that the pathogenic effects of this parasite were more marked in the two-week-old birds. If, however, the body weight depression of both age groups are compared at the sixth day after the infection with 20 million oocysts, then the older birds were more susceptible, since they lost body weight, whereas the body weight of two-week-old chicks was 23% higher than their starting weight. Kouwenhoven (1970) showed that with chickens aged 5, 10, 15, and 20 weeks given similar numbers of oocysts (5×10^6) there was a progressive increase in the pathogenic effects of *E. acervu-*

lina judged by body weight gain. The birds, aged five and ten weeks when given oocysts, recovered more of their body weight gain depression than the birds of 15 and 20 weeks of age (Fig. 4). It may be that with some infections, pathogenic effects are more easily produced in the younger hosts. This view is supported by work with the turkey coccidia *E. adenoeides* and *E. meleagrimitis*. Clarkson (1958, 1959) showed that young turkeys were more susceptible to the pathogenic effects of these parasites as judged by mortality, but he increased the dose of oocysts with age, giving 100,000 or 200,000 oocysts to three-week-old poults and from two to three million oocysts to eleven-week-old birds. Warren, Ball, and Fagg (1963) attempted to resolve the problem of age resistance of turkeys to *E. meleagrimitis* by inoculating turkeys of different ages with a dose of oocysts proportional to their weight, giving them 100 oocysts per gram body weight. The oldest birds (64 days) produced more oocysts than the other groups (from 8 to 50 days), but the reproductive potential (number of oocysts produced per oocyst fed) was lowest in the oldest birds.

If the true effect of age on the reproduction of coccidia is to be known, the dose of oocysts should be kept constant for each age group, and the dose should be kept small, from 50 to 500 oocysts. It is also preferable to differentiate between age susceptibility judged by parasite reproduction and susceptibility as determined by pathogenic effects. It is possible that with the hemorrhagic forms of coccidiosis in chickens (*E. tenella* and *E. necatrix*), young hosts are more susceptible. But it is likely that with the other infections of the fowl, older birds are more susceptible. The reports where the reverse is true merely show that it is easier to "crowd" the intestine of a small bird with parasites than it is with an older bird. With adult hens, reared free from coccidia, we have obtained mortality with low doses (1,000 oocysts) of *E. maxima* and small numbers of *E. brunetti* or *E. maxima* have caused clinical disease and cessation of egg production. Severe disease can also occur in turkeys from 10 to 12 weeks old (Ford, 1956).

In mammals, Lotze (1954) showed that *E. ninakohlyakimovae* killed lambs given five million oocysts, and one million oocysts given to a two-year-old sheep caused dysentery but not death of the animal. Pout, Ostler, Joyner and Norton (1966), in a survey of the incidence of sheep coccidia in Britain, indicated that the largest numbers of oocysts are recovered from young animals but this may be because older animals have developed some acquired resistance to the infections. From other reports, it appears that young calves are more susceptible to infections than older animals. However, Davis, Boughton, and Bowman (1955) showed that *E. alabamensis* was less pathogenic to calves under three weeks of age than to calves over this age and they were able to produce more severe disease in older than in younger calves. They considered that *E. ellipsoidalis* was found most frequently in calves from 3 to 12 weeks old but *E. alabamensis* was more common in calves from three to nine months of age. Gill and Katiyar (1961) reported the deaths of 11 of 12 lambs from one to two months old, which had been kept with adult sheep. Presumably, the lambs died as a result of coccidiosis.

It seems clear that susceptibility to coccidiosis increases with age and may be caused by a variety of physiological changes in the host. The observation made by Davis, Boughton, and Bowman (1955) that calves three weeks of age were more resistant to *E. alabamensis* than calves over this age is in line with the observations made by Rose (1967) which showed that fewer oocysts released their sporozoites in the intestine of chickens under three weeks of age than in older ones (i.e., those four to six weeks old). It appears, therefore, that very young animals

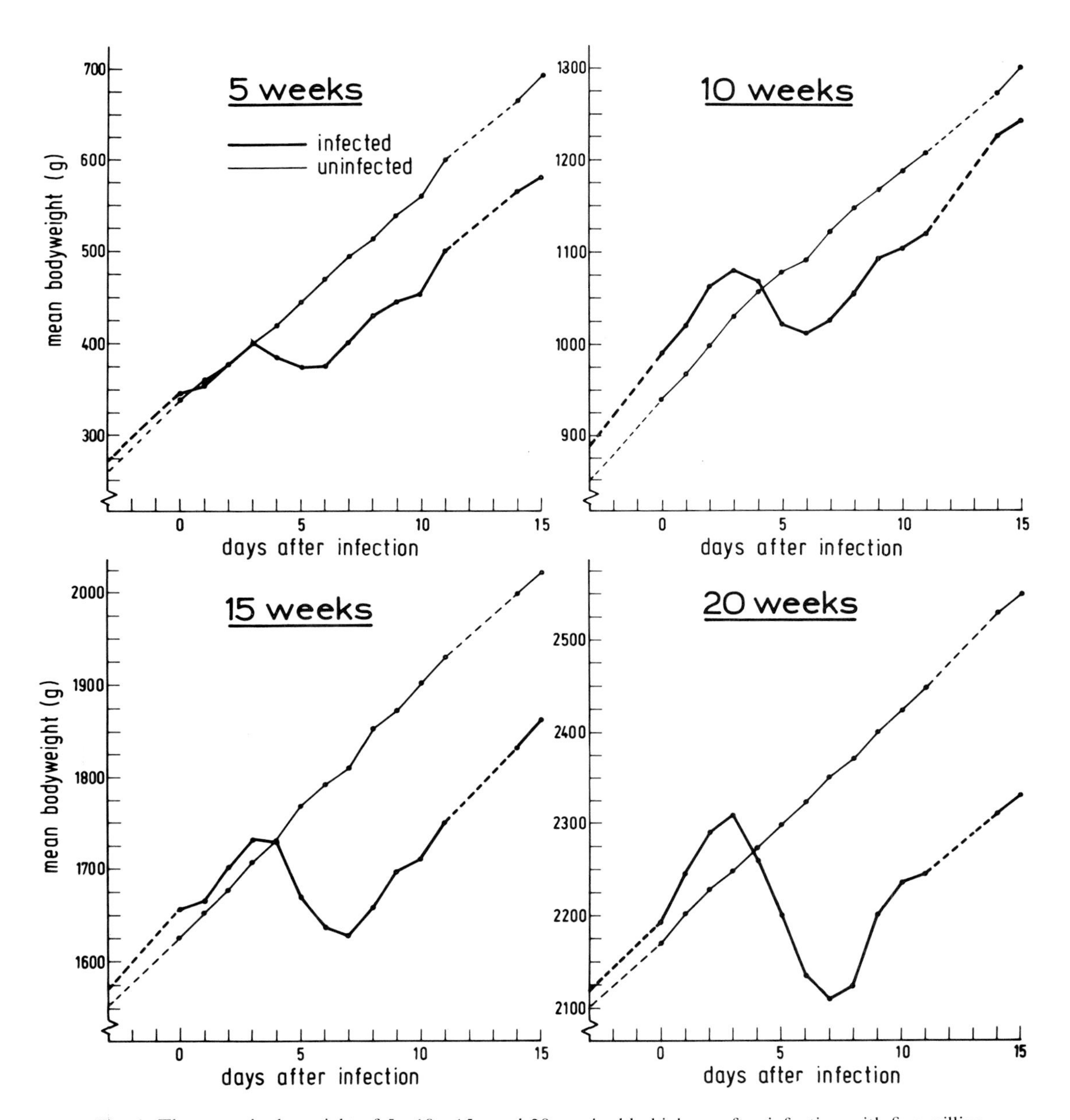

Fig. 4. The mean body weight of 5-, 10-, 15-, and 20-week-old chickens after infection with five million oocysts of *E. acervulina*. (From B. Kouwenhoven, "*Eimeria acervulina* infectie bij de kip een parasitologisch en biochemisch onderzoek," Doctoral Thesis, Rijks Universiteit, Utrecht, 1970. Reproduced with kind permission of the author.)

may not in fact receive the same challenge, in terms of sporozoites invading the intestine. The young of different host animals may vary in this respect, but this aspect has not been fully investigated.

Judged by parasite reproduction, older animals which have been kept free from the parasites during the rearing stage are more susceptible than younger ones, provided the animals are given a similar number of oocysts. Judged by clinical disease, mortality, or weight gain, older animals are probably still more susceptible than young animals, provided that they have been successfully reared free from infection and allowance is made for their size by increasing the oocyst dose in proportion to their body weight. In the field situation, infected older animals produce large numbers of oocysts and contact-susceptible animals have the opportunity of picking up proportionally larger numbers of infective oocysts. In the modern poultry industry, mixed age groups are not reared together and severe disease occurs in any age group if a coccidial species, to which they have little or no previous contact, develops in a particular site. With mammals (with the exception of man) it would be unusual if they were not exposed to coccidiosis before they reached maturity.

E. THE EFFECT OF BREED OR STRAIN OF HOST

The opportunity to study genetic resistance to coccidia has been possible so far only with chickens. Rosenberg (1941) found significant differences between five breeds of young chickens in their susceptibility to *E. tenella* infection and concluded that resistance was partially controlled by hereditary factors. It was demonstrated by Rosenberg, Alicata, and Palafox (1954) that sex linkage and passive immunity were not involved. However, recent work on the passive transfer of protection from immune hens to their progeny has shown that maternal antibody is important (see Chapter 8). Champion (1954), working with *E. tenella*, considered that selective breeding was effective in establishing lines of chickens resistant to the parasite and confirmed that sex linkage did not play a significant part in resistance to infection. Long (1968a) examined five strains of chickens and found that they differed in their resistance to *E. acervulina*, *E. brunetti*, *E. maxima*, and *E. mivati*. Later, Long (1970b) found that the resistance of some strains of white leghorns to *E. tenella* was different when they were examined as embryos rather than as hatched chicks (see Table 1). Jeffers and Wagenbach (1970) confirmed that the susceptibility of different strains of host was different in embryos as compared with hatched chicks, but still found differences in the susceptibility of different strains of embryos.

It now appears that if true genetic differences are to be examined, then embryos or very young chickens may not be suitable subjects, because it has been recently shown that protective maternal antibodies can be transferred from *E. tenella*-infected hens to their embryos (Rose and Long, 1971). Also, if the dams are being treated with anti-coccidial drugs, protective amounts may occur in eggs laid by these birds (Long, 1971a).

However, the results obtained using progeny over three weeks of age indicates that genetic resistance is an important factor and is in part responsible for some of the differences in the results of different investigators working with the same parasite.

III. Gross Pathology of Infection

A. INFECTIONS OF POULTRY

It is difficult to discuss or describe all of the observations made by workers on the large number of species of *Eimeria* occurring in poultry. In chickens, we distinguish two main

TABLE 1. The susceptibility of three strains of White Leghorns to *E. tenella* and *E. brunetti*

Strain	Challenged with *E. brunetti* at 3 weeks of age with 50,000 oocysts		Challenged with *E. tenella* at 3 weeks of age with 200,000 oocysts		Challenged with 40,000 *E. tenella* sporozoites *in ovo* when embryos were 12 days old	
	Number of birds	Number of deaths	Number of birds	Number of deaths	Number of embryos	Number of deaths
"S"	20	6 (30%)[a]	20	1 (5%)	29	23 (79%)
"K"	20	7 (35%)	20	4 (20%)	24	11 (46%)
"C"	20	14 (70%)	20	5 (25%)	29	8 (27%)

[a] = percentage mortality

types of pathology associated with the hemorrhagic and the nonhemorrhagic forms of coccidiosis. With two parasites of chickens (*E. tenella* and *E. necatrix*), severe hemorrhage is associated with the development of large second-generation schizonts which occur in colonies in the interglandular tissue. Although some blood loss can be observed in severe infection with other species, it is relatively small compared with that caused by *E. tenella* or *E. necatrix*. With the other species in the fowl, the damage to the intestine is caused by excessive numbers of gametocytes or a combination of gametocytes and schizonts parasitizing the villi.

1. The Domestic Fowl

With *E. tenella* infection, the onset of hemorrhage is sudden; little or no free blood can be seen in the cecal lumen on the fourth day after infection, but on the fifth day, profuse hemorrhage occurs and numerous hemorrhagic or whitish lesions appear on the cecal wall; these are usually from 1 mm to 2 mm in diameter, but occasionally larger necrotic lesions may be seen (about 5 mm diameter). A large number of deaths may occur at this time. The blood in the cecal lumen may be fluid despite the fact that death may have occurred some hours earlier. If the affected animals survive until the 6th day, the ceca have a more grayish, convoluted appearance and the lesions are often less distinct than those seen on the fifth day. The cecal contents are often semisolid and contain clots of blood. On the fifth and sixth days the ceca become shorter and thicker, and may remain in this condition for several days. The contents of the ceca on the 7th day are solid, and often consist of cores of necrosed blood cells and epithelial cells packed with oocysts. These cecal cores may persist for two weeks or more, but in most birds they are discharged, and the ceca have a more normal appearance after the 14th day of infection. It is possible that the retention of these necrotic cores may be one of the causes of typhlitis of the ceca, sometimes seen in adult chickens.

There is a marked depression in body weight gain from five to seven days after infection, but this is not as severe as that which occurs with infections of the small intestine, perhaps because the ceca are not markedly involved in the absorption of nutrients. There is a marked depression in the number of circulating erythrocytes, as determined by the microhematocrit technique (Joyner and Davies, 1960).

The most severe form of coccidiosis occurring in chickens is caused by *E. necatrix*. Tyzzer, Theiler, and Jones (1932) and Davies (1956) have given descriptions of the clinical signs and pathology of infections with this parasite. In both of these papers the authors mention that gross changes became evident at the end of the fourth day of infection or early on the fifth day, and are greatest at from 24 to 48 hr after this. As with *E. tenella* infection, the pathological changes are associated with the maturation of colonies of large second-generation schizonts. The lesions are distributed throughout the length of the small intestine, being most numerous in the middle portion. In our experiments we have noted that in young birds (one week old) the lesions are more numerous in the lower half of the intestine. This may be caused by delayed excystation of sporozoites in the younger birds, in which sporulated oocysts, given *via* the crop do not release sporozoites in the duodenum, but do so about half way down the intestine; the initial parasitism then occurs further down the intestine. This view is supported by the work of Horton-Smith and Long (1965), who demonstrated that *E. necatrix* was capable of developing the whole of its endogenous cycle in the ceca if the sporozoites were inoculated into this site.

The lesions in the intestinal wall consist of discrete hemorrhagic and whitish spots. The whitish spots may be about 1 mm in diameter, but sometimes they are larger where several colonies of schizonts have coalesced. The intestine is swollen and easily broken when handled with forceps. The intestinal contents contain thick mucoid material mixed with blood, which adheres to the intestinal wall. Blood and mucus from the small intestine are discharged with the feces and some of the material is aspirated into the ceca; merozoites present in this material penetrate the cecal epithelium and oocysts are formed. The cecal contents become dry and hard and cecal walls thin and dry looking. If birds survive the acute phase of the disease, the ceca regain their normal appearance but lesions in the intestine persist and schizonts can be found even after 18 days.

Gross lesions are produced by most of the other species in the fowl. Levine (1942) and Davies (1963) have described the gross pathology of *E. brunetti*. The lesions produced by the parasite occur mainly in the lower third of the intestine; the intestinal wall is inflamed, pin-point hemorrhages are present on the serous surface, and there is a gross mucoid enteritis, sometimes with blood. Later, this discharge becomes necrotic and adheres to the intestinal wall, which at this time is often somewhat thin. In *E. maxima* infection the lesions can occur in any part of the upper two-thirds of the intestine. The intestinal wall is often thickened and the thick mucoid discharge is often flecked with blood and salmon-colored streaks. Petechial lesions often occur. Old chickens are more susceptible to the effects of the parasite.

Tyzzer (1929) concluded that the pathogenicity of *E. acervulina* was not proved. Subsequently, most workers have attributed watery enteritis, suppression of body weight gain, and effects on egg production to this parasite. Dickinson and Schofield (1939) produced 20% mortality in chickens given this parasite; Morehouse and McGuire (1958), who produced up to 75% mortality with this parasite, have described the gross pathology. Edgar and Seibold (1964) have described the gross pathology of *E. mivati*, which appeared to be similar to that of *E. acervulina*.

2. *Ducks, Turkeys, and Geese*

Allen (1936) and Davies (1957) produced clinical disease in ducks with *Tyzzeria perniciosa*, and Becker (1934) included a description of duck coccidiosis in his book. *T. perniciosa* appears to be highly pathogenic and produces lesions and pathology not unlike that of *E. necatrix* in the chicken. Dubey and Pande (1963) described the occurrence of giant coccidian schizonts in the lamina propria of the villi of domestic ducks. The schizonts measured 277 $\mu \times 203$ μ and caused atrophy of the villi. Oocysts were not found, so the species concerned could not be determined.

Of the seven species of *Eimeria* occurring in the turkey, two species are known to be highly pathogenic (*E. adenoiedes* and *E. meleagrimitis*). *Eimeria adenoiedes* produces severe enteritis of the lower third of the turkey intestine, including the ceca. A thick grayish-white mucoid exudate, which contains millions of gametocytes and oocysts, is often present. Moore and Brown (1951), Clarkson and Gentles (1958), and Clarkson (1960) describe the pathology and pathogenicity of this species in detail. Equally pathogenic to turkey poults is *E. meleagrimitis* which produces an inflammatory condition in the upper intestine. Flecks of blood and a blood-tinged mucoid exudate are common. High mortality occurs in young birds; Hawkins (1952) and Clarkson and Gentles (1958) thought that age resistance occurred with this species. Farr, Wehr, and Shalkop (1962) produced gross pathological changes with *E. gallopavonis* in the lower intestine and ceca of young poults. It differed from *E. adenoiedes* in that the pathological lesions and deaths occurred later with *E. gallopavonis*.

An unusual species deserving further study is *E. truncata*, which develops in the epithelium of the goose kidney. The life cycle of this species is still unknown. However, it produces severe disease in goslings (Farr and Wehr, 1952); schizonts, and more commonly gametocytes and oocysts, cause destruction of the uriniferous tubules in the kidney. The disease is acute and usually fatal in goslings.

Pellérdy (1965) described the pathological lesions produced by this species; these include kidney enlargement and typical lesions of yellowish-white streaks and spots. The tubules contain massive numbers of characteristically shaped oocysts, which possess a prominent micropyle. Versényi and Pellérdy (1970) showed that *E. anseris* is a highly pathogenic parasite in both young and old geese, causing severe enteritis with death in a high proportion of the birds.

B. Infections of Mammals

According to published information the pathogenicity of most of the reported species is unknown. This is mainly because experimental infections have been carried out in only a small proportion of the studies. It is obvious that most of the work on mammalian coccidiosis has been carried out with domestic and farm animals.

In their book on coccidia of ruminants, Levine and Ivens (1970) reported that most of the coccidia of sheep and cattle have been shown to produce some pathogenic effects on their hosts. This was true also for *E. wassilewsky* and *E. ponderosa* in deer and *E. impalae* in impala, the latter species causing severe infection of the uterus. Depending on the coccidial species involved, infected animals appear to show loss of appetite, grinding of the teeth; sometimes, tenesmus, evidence of dehydration, and weakness are also present. The fecal discharges may be watery, blood tinged, mucoid, or they may contain blood clots.

1. Cattle

Smith and Graybill (1918), working with mixed infections in which *E. zuernii* (the name *E. zuernii* is now preferred to *E. zürnii* by Levine and Ivens, 1970) was predominant, observed foci on the walls of the large intestine, which grew and coalesced. These lesions were produced by schizonts and gametocytes in the crypts of Lieberkühn; hemorrhage is a feature of the disease, and damaged areas of intestine, devoid of glandular tissue, show extensive necrosis. Hammond (1964) considers that with *E. bovis* the gametocyte is the most pathogenic stage because they are numerous and, even at early stages, they alter the host cells. In severe infections, most of the glandular epithelium is destroyed. With *E. zuernii* infection, anemia, body weakness and emaciation are signs of the disease and deaths often occur within from seven to ten days of infection. The lesions are described by Davis and Bowman (1952) and Niilo (1970a); they consist of a catarrhal enteritis, with blood and desquamative epithelium. The mucous membrane is thickened and diffuse, and petechial hemorrhages may be seen.

In calves infected with *E. alabamensis*, Davis, Boughton, and Bowman (1955) observed that the lower half of the intestine was hyperemic, contained bloody mucus and/or liquid intestinal contents with massive numbers of merozoites and gametocytes. Boughton (1943) produced clinical disease in five young calves by giving them 200×10^6 oocysts of *E. alabamensis* and other species; one calf died.

Hammond, Davis, and Bowman (1944) considered *E. bovis* to be one of the most common and pathogenic of species affecting cattle. A dose of only 125×10^3 oocysts caused marked signs of the disease. A high body temperature (106.6°F) was associated with severe infection and calves given from 250×10^3 to 250×10^6 oocysts died. There was appreciable blood loss, petechial lesions and loss of mucosa. Semi-fluid bloody material was found in the ileum, colon and cecum, the walls of which were thickened. Small numbers of macroscopic schizonts may also be seen (Boughton, 1942) but the main changes were associated with the sexual phases of this parasite and tubular pieces of bloody tissue were sometimes discharged with the feces. Although Niilo (1970b) and Fitzgerald (1962a) considered *E. zuernii* to be the predominant species involved in clinical coccidiosis in the winter it would be difficult to say that it is the most pathogenic. It is possible that *E. bovis* and *E. zuernii* are more or less equally pathogenic.

2. Sheep

Lotze (1952) administered large numbers of oocysts of *E. ovina* (*E. arloingi*) to lambs and found no visible signs of disease in those given 10^6 oocysts or less. Severe diarrhea occurred in lambs given several million oocysts. Lotze considered that the asexual stages were the most damaging. The first lesions in the small intestine occurred about the 13th day and consisted of small hemorrhagic areas; later, the small intestine was thickened and whitish opaque patches were seen. In sheep affected by *E. ovina*, Lotze (1953) and Gill and Katiyar (1961) observed that the mucous membrane of the ileum was raised into numerous pea-sized papilliform elevations and the walls of the intestine were edematous. The intestinal contents may also be blood tinged.

Smith, Davis, and Bowman (1960) considered *E. ahsata* to be the most pathogenic of sheep coccidia. A dose of 1×10^5 oocysts proved fatal to four of nine lambs from one to three-months-old. Edema of the small intestine occurred and Peyer's patches were found in the last part of the small intestine. Later, Smith and Davis (1965) found that about 30×10^3 oocysts would produce the death of lambs.

E. ninakohlyakimovae appears to be highly pathogenic in lambs and adult sheep. Lotze (1954) administered 50×10^3 oocysts, which caused severe diarrhea, and a dose of 5×10^5 oocysts caused deaths in three-month-old lambs. Svanbaev (1967) caused clinical signs in six-week-old lambs by giving them 1×10^4 oocysts. Diarrhea, fever, anorexia, body weight loss, and poor wool quality were features of the disease. Smith and Davis (1965) found that the oocyst dose caused greater effects if given in dry food rather than in water. Hammond, Kuta, and Miner (1967) inoculated lambs from 10 to 14 weeks old with about 97,000 oocysts and caused six deaths in a group of ten lambs; the deaths occurring between two and three weeks after inoculation. Shumard (1957) used 50-day-old lambs and gave them 7×10^6 oocysts of this species. Only one of four lambs died. Levine and Ivens (1970) considered that Shumard had used a less pathogenic strain, but it is possible the dose of oocysts was excessive, and that with a parasite which has a prepatent period of two weeks, the large number of early stages in the host resulting from an excessive dose may stimulate immunity to the later endogenous stages.

3. *Pigs*

In coccidiosis of the pig, diarrhea, emaciation, and thickening of the wall of the large intestine occur. Intestinal contents are mucoid and necrotic enteritis occurs (Biester and Murray, 1929). Rommel (1970) found that as few as 200 oocysts of *E. scabra* or *E. polita* could cause diarrhea and affect the body weight of weaner pigs. Swanson and Kates (1940) investigated an outbreak in 18-week-old pigs, in which there was severe diarrhea.

4. *Rodents*

Levine and Ivens (1965) in their monograph on the coccidia of rodents noted that, apart from infections of rabbits, pathogenicity of most species is unknown. However, the following species have been reported to produce pathogenic effects on their hosts: *E. nieschulzi* and *E. separata* (rat), *E. falciformis* (mouse), *E. ondatrazibethicae* (muskrat), *E. sciurorum* (squirrel), *E. caviae* (guinea pig), *E. seideli* (nutria). *Klossiella muris* also produces lesions in the kidneys of the house mouse, causing destruction of the epithelium and distention of the tubules by sporocysts. Several of the species of *Eimeria* occurring in rabbits have been shown to have marked pathogenicity. In severe infections of *E. stiedai*, the animals become emaciated and have diarrhea. The liver is enlarged and white nodules occur on the surface and throughout the biliary system. The liver may be grossly enlarged. Lund (1954) found that the liver might reach a weight up to 20% of the body weight. The gall bladder is often distended and jaundice occurs. The bile ducts are grossly enlarged and their epithelial cells are filled with gametocytes and oocysts. Smetana (1933) described the gross changes in detail; in old infections, the bile ducts, which were formerly infected, become surrounded with fibrous tissue, which might resemble that of cirrhosis.

Of the species of *Eimeria* which develop in the intestine, *E. magna* appears to be highly pathogenic, causing deaths in rabbits given relatively small numbers of oocysts. According to Levine (1961) affected rabbits discharge large amounts of mucus with their feces. *E. irresidua* is also considered to be pathogenic, and with this disease blood may be discharged and the intestinal epithelium sloughed off.

5. *Dogs and cats*

Of the coccidia occurring in dogs and cats, the species of *Isospora* appear to be more common than *Eimeria* and several species of *Isospora* have been reported to produce pathogenic effects. The exact situation in the cat is now confused by the finding that *Toxoplasma gondii* has oocysts morphologically identical with *Isospora bigemina*.

Lee (1934) and Mahrt (1966) thought that *I. rivolta* had only mild pathogenicity in the dog, but Nemeséri (1960) thought that *I. canis* was somewhat pathogenic, causing catarrhal inflammation or mucosal ulcers. Hitchcock (1955) gave 10^5 oocysts of *I. felis* to nine-week-old kittens without producing clinical disease. Levine (1961) considered this species to be only slightly pathogenic. According to Pellérdy (1965) the signs of canine and feline coccidiosis are not specific and the coccidial species involved cannot be determined by the signs or the lesions. The general signs are diarrhea (sometimes hemorrhagic), anorexia, elevation in body temperature and, later, muscular tremor of the hind legs and anemia. Because of the paucity of experimental work with the coccidia of dogs and cats it would be wrong to attribute such changes to coccidia alone, since these animals suffer from enteritis caused by other pathogens and indeed there may be interactions between coccidia and these disease producing agents.

6. *Man*

According to Levine (1961) most *Isospora* infections of man (*I. hominis* and *I. belli*) appear to be subclinical. However, Barksdale and Routh (1948) examined 50 cases in U.S. Army personnel. Fifteen cases were examined in detail and anorexia, nausea, abdominal pain and diarrhea accompanied many of the infections. Matsubayashi and Nozawa (1948) reported on experimental infections in two human volunteers given about 3,000 *I. hominis* oocysts and found that symptoms commenced one week after exposure and that oocysts were discharged in the feces after ten days; this discharge continued for a month. Brandborg, Goldberg, and Breidenbach (1970) have reviewed the present information on human coccidiosis, which indicates that *Isospora belli* may cause pathogenic effects in man.

IV. Histopathology of Coccidial Infections

Most studies on progressive pathological changes with infection and recovery from infection have been carried out with *Eimeria* infections in chickens. In general, the host's reactions are relatively slight at the time of sporozoite invasion. Sporozoites have been observed invading the intestinal epithelium of chickens within from one to two hours of feeding oocysts *via* the crop. With some species (e.g., *E. praecox*) they appear to invade cells of the superficial villi and do not migrate, but round up shortly after invasion. The whole development then occurs in this site (Long, 1967a). There was little evidence of cell reactions to this parasite, apart from moderate numbers of heterophil polymorphonuclear cells between the epithelial cells; these appeared from 16 to 48 hr after oocyst inoculation. There was little cell reaction during the production of between three and four generations of schizonts and gametogony. Tyzzer, Theiler, and Jones (1932) thought that some sporozoites were taken up by macrophages and transported to the crypts of Lieberkühn, where they were destroyed. In infections with most of the other species in the chicken, the sporozoites migrate to the crypts, where the first-generation schizonts are formed. Whether they are all actively phagocytosed and transported by macrophages, as suggested by Pattillo (1959), Van Doorninck and Becker (1957), and Challey and Burns (1959), or whether only some sporozoites are phagocytosed and the remainder make their way independently has not been fully established. However, since sporozoites are known to be capable of invading a wide range of cells *in vitro,* it is possible that some sporozoites actively invade macrophages. However, it does appear that the sporozoites which become involved with macrophages are those which penetrate the basement membrane of the villi and enter the lamina propria.

While the sporozoites are developing into first-generation schizonts in the crypts of Lieberkühn, there is a progressive increase in the number of heterophil polymorphonuclear leukocytes in the submucosa surrounding the glandular tissue, but few, if any, between or within epithelial cells (Fig. 5b). During schizogony, there is a progressive increase in the number of granulocyte leukocytes and pyroninophilic cells in the submucosa and lamina propria. With *E. tenella* and *E. necatrix*, the granulocyte-cell response becomes most obvious when the first-generation merozoites leave the glandular tissues and migrate into the tunica propria, and begin to develop into the large second-generation schizonts characteristic of this species. Massive numbers of heterophil leukocytes, lymphocytes, and pyroninophilic cells infiltrate the tunica propria, lamina propria, and muscularis mucosae (Figs. 5 and 6). The intestinal wall becomes thickened, erythrocytes and second-generation merozoites are released in large numbers from the blood capillaries of the submucosa and muscularis mucosae, and are discharged, along with desquamative epithelium, into the intestinal lumen. In *E. tenella* infections, the small amount of undamaged epithelium which remains is invaded by merozoites, and gametocytes are formed. There is little or no cellular infiltration of the epithelium and there is a progressive decline in the number of inflammatory cells, but an increase in the number of discrete lymphoid foci in the submucosa. The epithelium is regenerated at a rapid rate, but oocysts may become trapped beneath the epithelium in the submucosa. Lymphocytes and pyroninophilic cells accumulate around the oocysts, and later, giant cells are frequently formed (Fig. 5b). The muscular layers, thickened on days 5 and 6, become more normal and contain fewer inflammatory cells. After about the 14th day, the cecal wall becomes more normal, but may contain numbers of large discrete lymphoid foci (Fig. 7b) and occasional patches of oocysts may still be seen trapped in the submucosa (Pierce, Long, and Horton-Smith, 1962). In *E. necatrix* infection, colonies of schizonts may persist for up to 18 days after primary infection (Davies, 1956). With other species of *Eimeria*, the endogenous development usually occurs entirely within the epithelium and the cellular reactions are often milder compared with those of the hemorrhagic species just described. The main changes depend upon the degree of destruction of the villi and the loss of inflammatory cells and lymphocytes into the intestinal lumen. The cell reactions occur mostly around the parasitized villi and in the submucosa, where both diffuse and discrete lymphoid areas develop; diffuse lymphoid tissue is shown in Figure 6b and discrete lymphoid foci in Figures 7a and 7b. Long and Rose (1970) showed that there was a quantitative increase in numbers of discrete lymphoid foci in birds infected with *E. mivati*. A summary of the findings are given in Figure 8. The villus height is shortened and the villi tend to become broader; this effect has been noted by Pout (1967a,b) and will be discussed later. Pierce, Long, and Horton-Smith (1962) noted a progressive increase in numbers of globule leukocytes in the glands of chickens infected with *E. tenella*. These cells (see Fig. 7c) are believed to be involved in resistance to nematode infections in sheep (Dobson, 1966) and in rats (Whur, 1967).

With infections of mammals, most species develop in the glandular epithelium, schizonts, gametocytes, and oocysts being produced there. With *E. bovis*, large schizonts develop in the endothelial cells of the lacteals (Hammond, Bowman, Davis, and Simms, 1946). They appear to measure up to 416 $\mu \times$ 364 μ, have more than 10^5 merozoites, and are readily observed macroscopically (Fig. 9a). Second-generation schizonts of normal size occur in the crypts of Lieberkühn in the large intestine (Hammond, Anderson, and Miner, 1963b).

Although schizonts obviously cause some tissue damage, Hammond (1964) considered that the gametocytes were more directly responsible for the clinical signs. The parasitized host cells in the villi show enlarged nuclei with prominent nucleoli and hypertrophy of the cytoplasm (Hammond, Ernst, and

Miner, 1966). Cells harboring young gametocytes (Fig. 9b), undergo shrinkage and change of shape. The infected cells lose contact with the neighboring cells except at the base and the typical columnar arrangement is disturbed. In severe infections, most of the glandular epithelium is destroyed by gametocytes and oocysts (Fig. 9c).

Large schizonts similar in size to those of *E. bovis* occur in the lamina propria of sheep infected with *E. ovina* (Lotze, 1953) and also *E. ninakohlyakimovae* and *E. parva*. The tips of the villi are ruptured by the growing macroscopic schizonts. However, it appears that with most species of *Eimeria*, severe damage is caused by the sexual stages or a combination of asexual and sexual stages. Levine, Ivens, and Fritz (1962), working with a coccidium of goats (probably *E. arloingi*), mention the cellular reactions to the parasites, which involved macrophages, lymphocytes, plasma cells, eosinophils. In addition, some small areas of inflammation with infiltration by polymophonuclear cells occurred and these areas showed necrosis of the epithelium and lamina propria.

Hammond (1965), working with *E. bovis*, described the cellular reactions to the giant schizonts of this species. With some schizonts the host cell disintegrates and the schizont is invaded by eosinophils, macrophages, and other cells, which eventually destroy the merozoites. Wacha, Hammond, and Miner (1971) described a similar process in sheep experimentally infected with *E. ninakohlyakimovae* and photomicrographs showing the invasion of schizonts by leukocytes are shown in Figure 10. In infections with *E. bovis*, numerous neutrophil polymorphonuclear cells migrate into the lumen of the infected crypts as this becomes filled with oocysts. After most of the oocysts have been discharged into the intestinal lumen the remaining oocysts become surrounded by macrophages or giant cells. Oocysts remained in this site for several weeks.

Smith and Graybill (1918), Boughton (1945), and Davis and Bowman (1952) all point out that with *E. zuernii* infection, schizonts and gametocytes occur together in the crypts. Sloughing of the epithelium and hemorrhage occur. Leukocytes and erythrocytes adhere to the damaged areas, which later become necrosed. Lotze, Shalkop, Leek, and Behin (1964) reported the interesting observation of the development of schizonts in the mesenteric lymph nodes of sheep and goats.

Fig. 5. Photomicrographs of transverse sections of chicken cecal tissue. Tissues were fixed in formol-sublimate and stained with Ehrlich's hematoxylin and eosin. *A*. Tissue from uninfected chicken. Note flat villi and the small amount of connective tissue. *B*. Three days after inoculation with *E. tenella*. Note first-generation schizonts in the epithelium of a deep gland; lymphocytes and granulocytes have infiltrated the tunica propria. *C*. Six days after inoculation with *E. tenella*. Note colonies of schizonts (s) in the tunica propria and gametocytes (g) in the glandular tissue. The muscularis mucosae and muscle layers are thickened and infiltrated by host defense cells. *D*. Fourteen days after inoculation with *E. tenella*. Note degenerating oocysts (o) in the muscularis mucosae within giant cells.

Fig. 6. Photomicrographs of transverse paraffin sections of the small intestine of chickens inoculated with *E. necatrix* oocysts. Tissues were fixed in formol-sublimate and stained with Ehrlich's hematoxylin and eosin. *A*. Six days after inoculation showing a colony of second-generation schizonts in the tunica propria. Some host defense cells have infiltrated this site and some blood vessels have been ruptured. There is some damage to the muscularis mucosae but little thickening of the muscular layers. *B*. Nine days after infection. Note colony of second generation schizonts in the tunica propria and gross infiltration of lymphoid cells (L) between the glands. *C*. Nine days after infection. Note mature second-generation schizonts and heterophile leukocytes (HL) infiltrating between and into the schizonts.

Fig. 7. Photomicrographs of transverse paraffin sections of chicken intestine. Tissues were fixed in formol-sublimate and stained with Ehrlich's hematoxylin and eosin. *A*. Duodenum ten days after inoculation of *E. mivati* oocysts showing numerous discrete lymphoid foci (L) in the tunica propria and lamina propria. *B*. Ceca 20 days after an acute infection with *E. tenella* showing complete recovery of the glandular tissues. Two large discrete lymphoid foci are situated beneath the glandular tissue. *C*. As in *B*, but high power view of part of deep gland showing globular leukocytes (G) between the epithelial cells. One of the globular leukocytes (lower right corner) has released one of its globules.

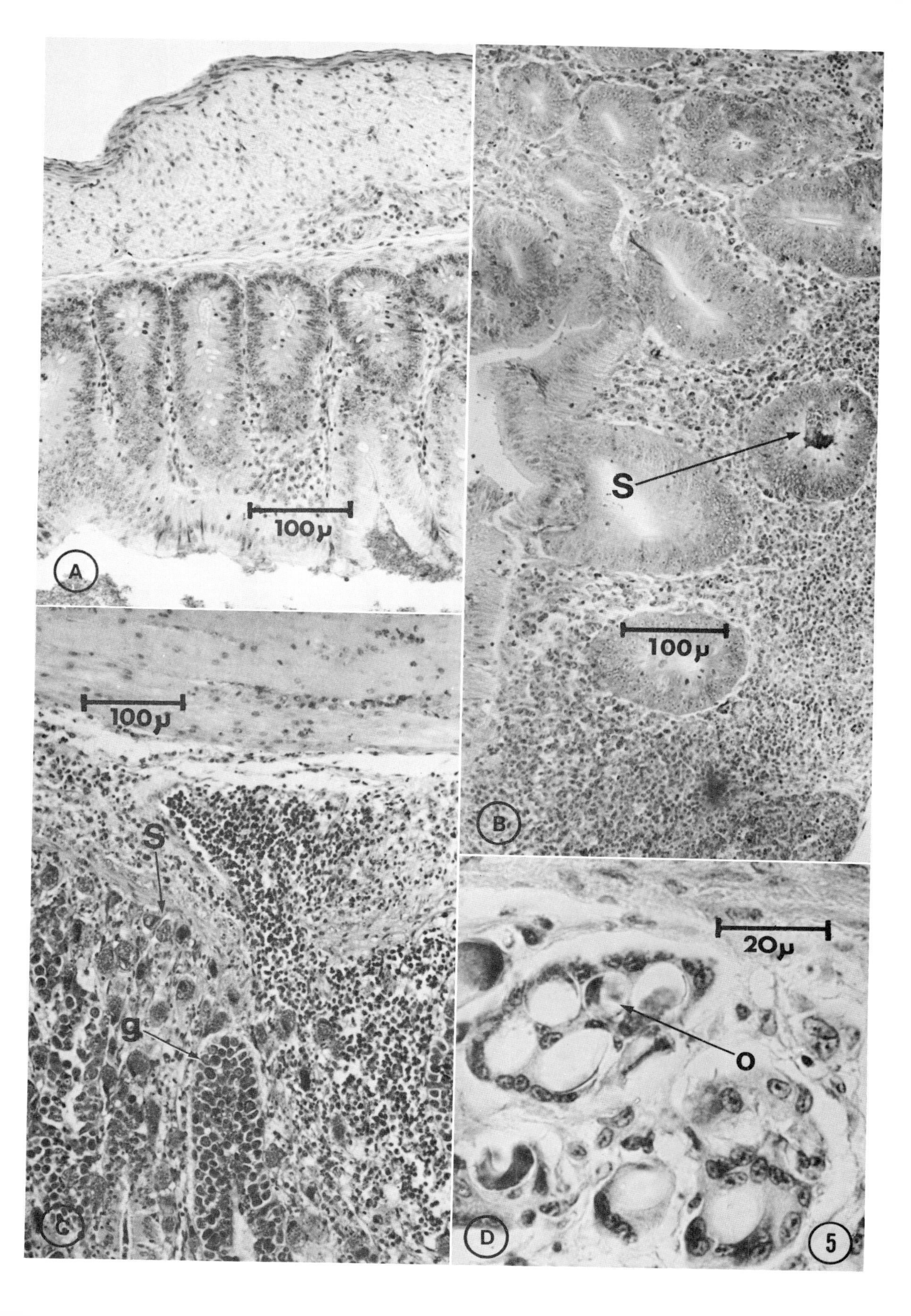
100μ
A
S
100μ
B
100μ
S
g
C
20μ
O
D
5

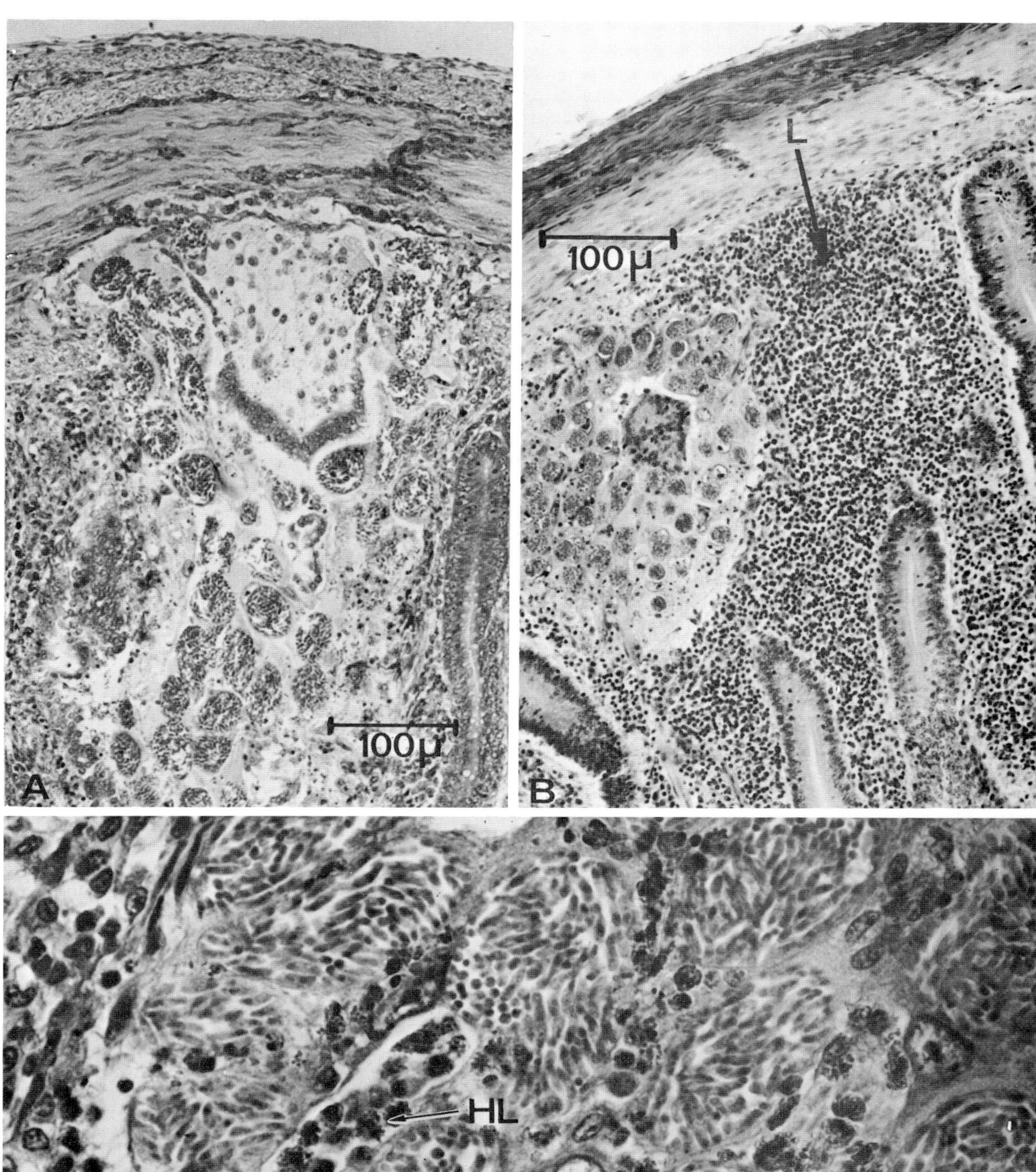
100µ
A
100µ
L
B

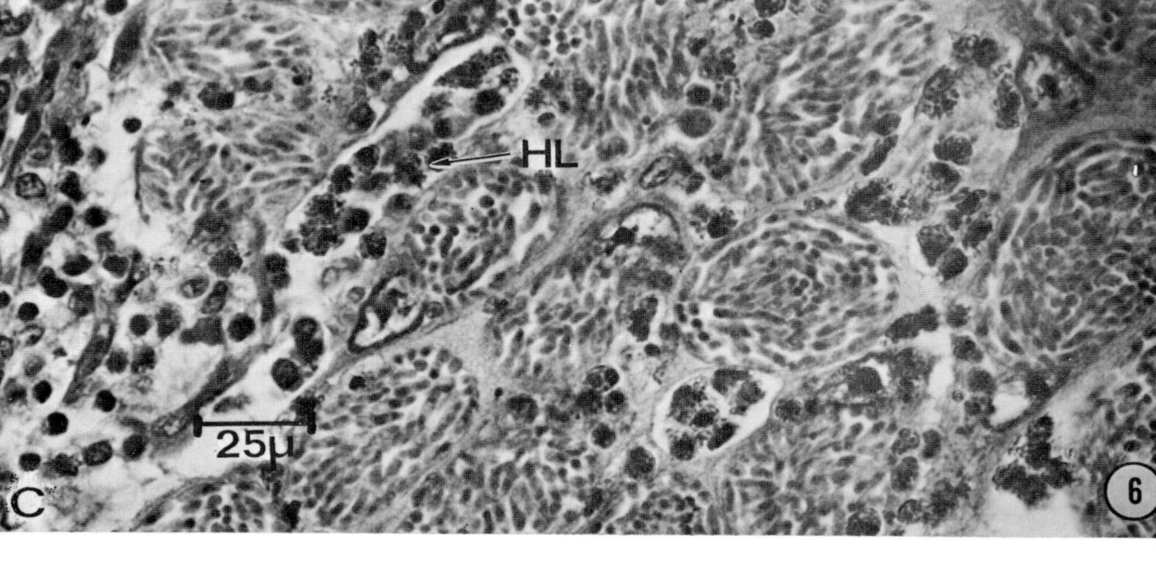
HL
25µ
C
6

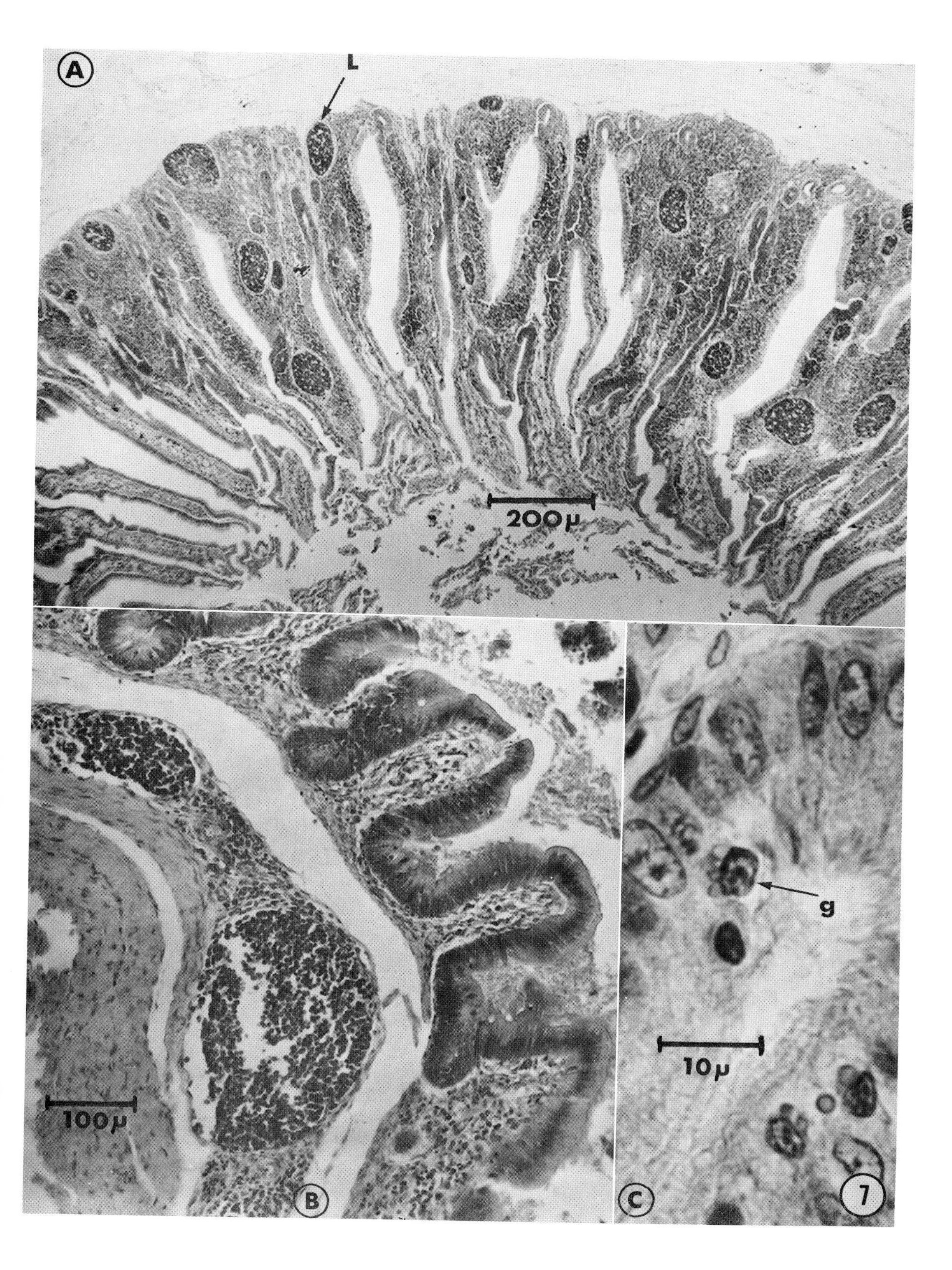
A
L
200μ
100μ
B
10μ
g
C
7

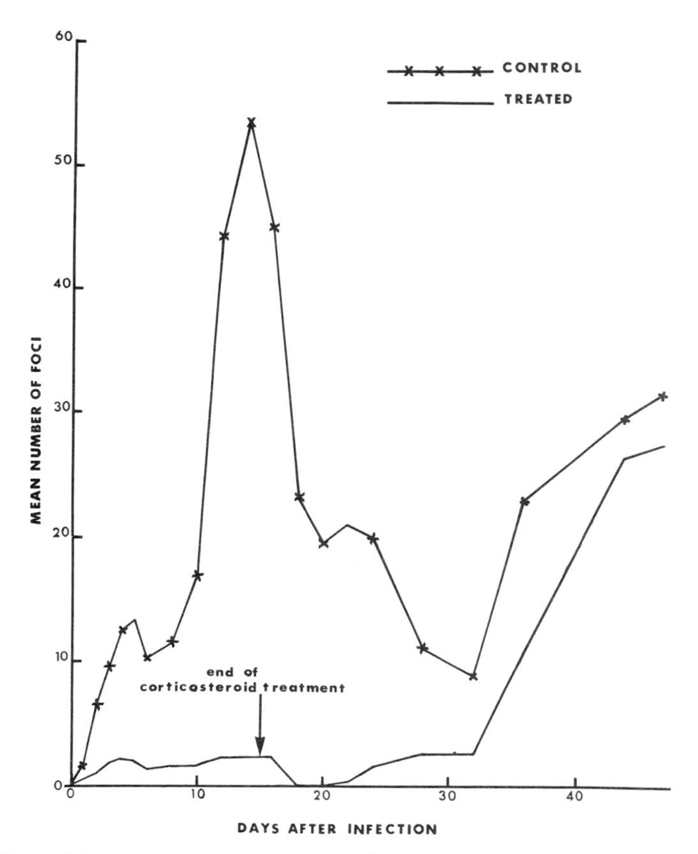

Fig. 8. The effect of betamethasone treatment on the numbers of lymphoid foci in the intestine of *E. mivati* infected chickens. The chickens were five weeks of age at the time of oocyst inoculation. Note peak in numbers of foci occurring two weeks after inoculation of the controls with oocysts. Note also the effect of betamethasone in preventing the formation of lymphoid foci. (Data from Long and Rose, 1970, Parasitology, 60: 147-155.)

The schizonts developed in both nonreaction and reaction centers and proliferating reticuloendothelial cells were present.

With *E. stiedai* infection of the rabbit, the large numbers of gametocytes and oocysts which are produced block the bile ducts and the inflammatory reaction and tissue damage cause jaundice and liver dysfunction (Bachman and Menedez, 1930). Kotlán and Pellérdy (1937) reported that proliferation of the capillary bile ducts was disproportional to the number of coccidia present; cell infiltrations and giant cell formation are very marked.

Fig. 9. Photomicrographs of transverse paraffin sections of the small and large intestine of calves after inoculation of *E. bovis* oocysts. Tissues fixed in Zenker's solution and stained with iron hematoxylin. *A*. Eighteen days after inoculation. Note large first-generation schizont in central lacteal, which has caused gross distention of the villus. × 350. (Unpublished photograph kindly supplied by D. M. Hammond.) *B*. Sixteen days after inoculation. Note young gametocytes (g) in the epithelial cells of a crypt in the large intestine and the effect on the parasitized cells which have enlarged nuclei and which have lost much cytoplasm and no longer have a columnar arrangement. × 750. *C*. 18 days after inoculation. Note large number of gametocytes and oocysts which have destroyed the epithelium lining the crypts. × 400. (*B*. and *C*. Kindly supplied by D. M. Hammond and published in the Faculty Honor Lecture, Utah State University, 1964.)

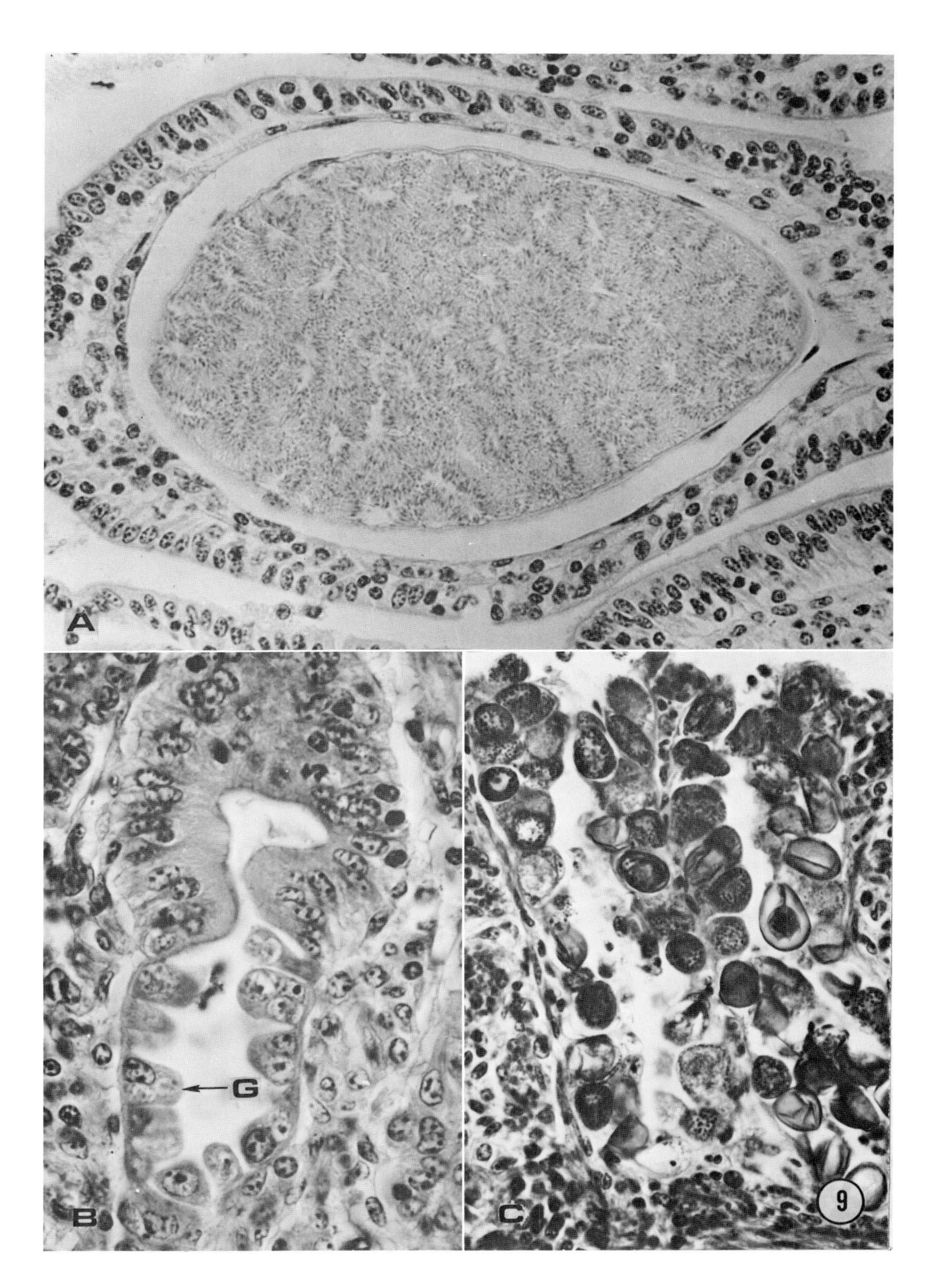
A
G
B
C
9

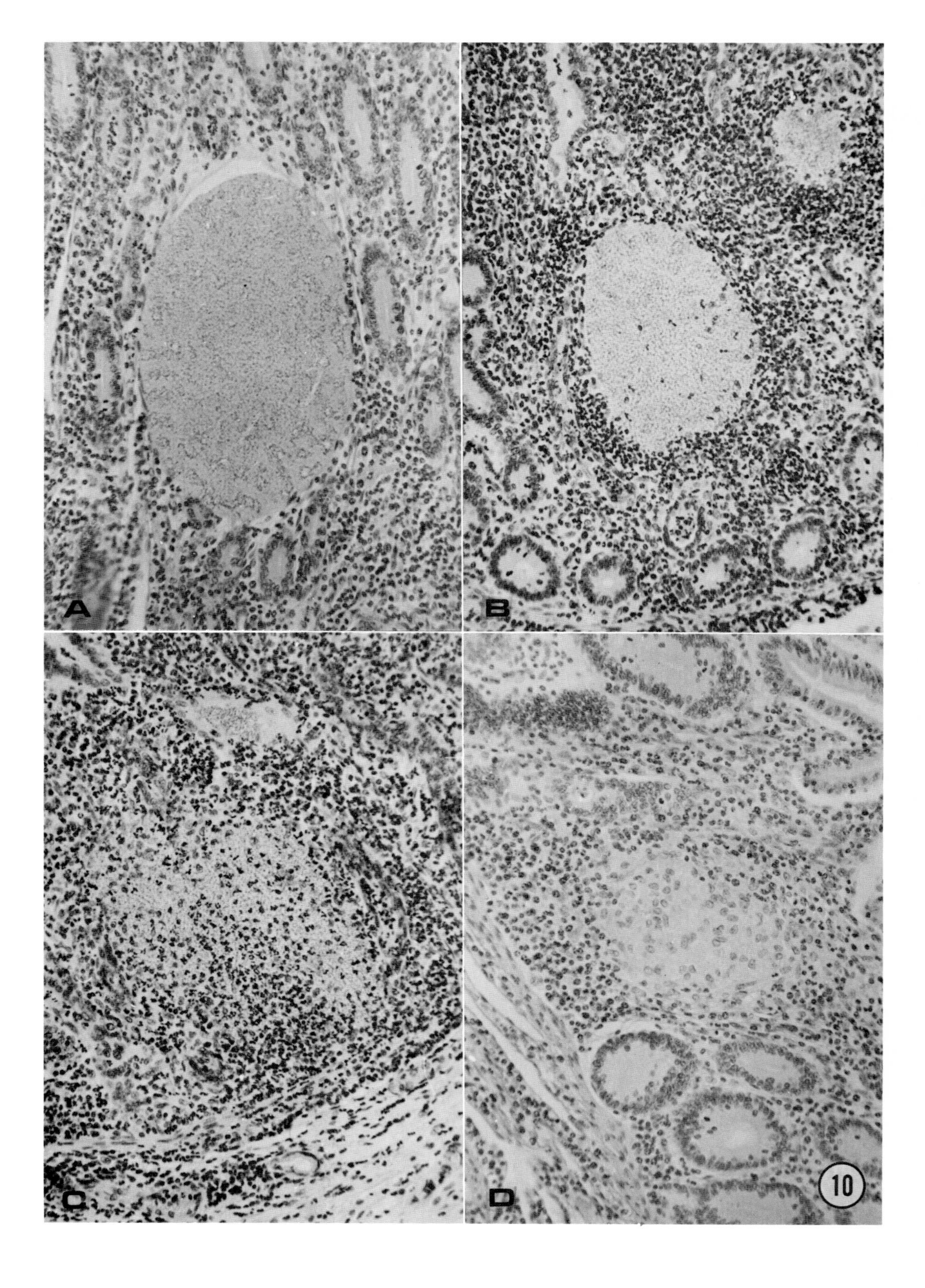
A
B
C
D
10

In the pig, Biester and Murray (1929) noted thickening of the intestinal wall and stated that an interstitial inflammatory reaction occurred, with infiltration by eosinophils. Most of the epithelial destruction occurred at the tips of the villi.

Brandborg, Goldberg, and Breidenbach (1970) demonstrated the presence of both asexual and sexual stages of *I. belli* in biopsy specimens of human intestinal mucosa. In one patient, there was a clear deposition of collagen in the lamina propria, similar to that found in collagenous sprue. Lymphocytes, plasma cells, and eosinophils were found in large numbers in the lamina propria. The capillaries were more prominent, but the lacteals were not dilated.

V. Experimental Pathology

A. EFFECT OF COCCIDIOSIS ON FOOD AND WATER CONSUMPTION, BODY WEIGHT GAIN, AND EGG PRODUCTION

With coccidial infections of chickens, Hilbrich (1963), Siegmann (1960), and Reid and Pitois (1965) have investigated the effect of coccidiosis on the ability of affected birds to eat and drink. Reid and Pitois (1965) studied the effect of six species of *Eimeria* of chickens and found that in infections of each of the species, food and water consumption was depressed on the 4th and 5th day after infection. These authors thought that sick birds were more likely to eat than to drink, and medication programs should be modified accordingly. Long (1968b) found that both food and water consumption were significantly reduced in from four to six days after infection with *E. acervulina* and *E. praecox*. Shumard (1957) reported that the food consumption of lambs infected with *E. ninakohlyakimovae* and *E. faurei* was reduced; but no significant decrease in water consumption occurred.

Reduction in body weight gain or even body weight loss occurs commonly with coccidial infection. Indeed, Tyzzer (1929) considered the body weight changes to be one of the most obvious effects of coccidiosis and many workers consider changes in body weight the only way to detect subacute coccidiosis, because a lower body weight gain often occurs in the absence of clinical signs and changes in feed and water consumption. *E. praecox* infection in the fowl is considered to be nonpathogenic, since affected birds show no clinical signs, yet heavy infections still depress body weight gain (Long, 1968b). Better performances, judged by body weight gain, are produced by using low-level coccidiostatic drugs in groups of birds where coccidiosis is not detectable clinically. A similar effect has been noted in calves by Fitzgerald and Mansfield (1969). In chickens, the maximum depression in body weight usually occurs just after the end of the prepatent period (from four to seven days). After this, weight gain appears to be similar to that of uninfected birds, but there is little indication that affected birds attain their expected weight after 28 days (Long, 1967b). Michael and Hodges (1971) showed that repeated daily oocyst inoculations extended the period of body weight depression beyond that produced by single infections, and the anorexia induced by *E. acervulina* infection is not the only factor responsible for body weight depression because uninfected birds, artificially starved to

Fig. 10. Photomicrographs of macroschizonts of *E. ninakohlyakimovae* in the small intestine of experimentally infected lambs. × 275. Transverse paraffin sections stained with Feulgen method by Mrs. Linda Bennett; photography by Clarence A. Speer. *A*. Schizont nine days after oocyst inoculation. *B*. Ten days after inoculation. Note increase in leukocytes and lymphocytes in the tunica propria and some infiltration of the schizont by leukocytes. *C*. Ten days after inoculation; note advanced stage of leukocytic infiltration of the schizont. *D*. Thirteen days after inoculation. Schizont now completely destroyed and the site of the schizont now replaced by fibrocytes and macrophages. (From Wacha, Hammond, and Miner, 1971, (Proceedings of the Helminthological Society of Washington. Reproduced by kind permission of the authors and publishers.)

the same extent as those of the infected groups, did not lose as much weight and recovered better when returned to normal *ad libitum* feeding.

Johnson (1931) noted the effect of several species of *Eimeria* on egg production of the fowl. Dickinson (1941) studied the effect of *E. acervulina* infection on egg production in some detail and found that egg production ceased entirely in from 8 to 19 days after infection. Berg, Hamilton, and Bearse (1951) found that a dose of approximately 8,000 *E. maxima* oocysts given to White Leghorn pullets resulted in cessation of laying within one week of inoculation; the effects lasted for from seven to ten days. They also observed that the eggs laid by infected birds had smoother shells and that the egg shells were thinner.

It is possible that female birds at the time of egg production are more susceptible than male birds of the same age. It has also been shown that female embryos are significantly more susceptible to *E. tenella* infection than male embryos (Jeffers and Wagenbach, 1969).

B. EFFECTS OF COCCIDIOSIS ON CARBOHYDRATE METABOLISM AND BLOOD LOSS OF THE HOST

Parasitic diseases appear to cause gross alteration in the physiology of the host, but these physiological effects are not yet fully understood. A physiological disturbance frequently observed in parasitic infections of poultry is muscular weakness. Ackert and Herrick (1928) described leg weakness in *Ascaridia galli* infection and Stafseth (1931) in intestinal coccidiosis. Bayon (1930) and Stafseth (1931) reported that teniasis caused muscular paralysis in poultry. Levine and Herrick (1954 and 1957) showed that chickens affected by *E. tenella* had a reduced ability to perform muscular work. This myasthenia was observable on the third day after infection, and became progressively more severe during the hemorrhagic phase of the infection.

Schildt and Herrick (1955) found that the muscular activity of the crop and ceca was greatly reduced during the acute phases of *E. tenella* infection. Waxler (1941) and Pratt (1944) showed that the muscle of affected birds contains appreciably less glycogen than normal birds and that affected birds cannot compensate for lowered environmental temperatures by increasing the muscular activity.

Daugherty and Herrick (1952) showed that the voluntary muscle of *E. tenella*-infected birds has a lower respiratory quotient than that of normal birds. Freeman (1970) suggested that measurements made by Pratt and Waxler were of doubtful value, since they used whole blood rather than plasma to measure glycogen and pointed out that Bell and Culbert (1967) had shown that if substances of small molecular weight are unequally distributed between plasma and blood cells it is not correct to measure changes in whole blood when changes in the number of blood cells are expected. Freeman (1970) therefore considered that the hyperglycemia shown by Pratt and by Waxler was caused by a loss of erythrocytes. Freeman noted a transient rise in cardiac glycogen stores and a depletion in the glycogen content of voluntary muscle. A reduction in plasma lactate occurred during the first two days of infection. He considered the depletion of glycogen stores in voluntary muscle the result of a reduction in muscular activity rather than from an impairment of glycogenesis. The argument that the glycogen depletion is the result of hypoxia is not proved since bleeding alone was not found by Pratt (1944) to induce a reduction in glycogen stores of voluntary muscle. Freeman (1970) further suggested that the reduced muscular activity of affected birds was the result of an impairment of nervous conduction at the neuromuscular junction caused by a substance released by the developing coccidia or by damaged host cells.

The author has frequently noted that the hemorrhagic cecal contents of birds which have died as a result of infection often remain fluid or semifluid. This indicates that the clotting is impaired by some substance produced by the parasite or as a result of the interaction of the parasite and the host. Such a substance is not thought to arise by a change in the bacterial flora of the ceca, since blood released into the allantoic cavity of bacteria-free, *E. tenella*-infected embryos frequently fails to clot. It is significant that groups of chickens fed a vitamin K-deficient diet and given *E. tenella* oocysts have greater mortality rates than infected chickens given a diet containing vitamin K (Harms and Tugwell, 1956).

The effects of *E. tenella* on the host are probably quite different from the effects of other *Eimeria* infections in which little or no blood loss occurs. With *E. tenella* infection, erythrocyte numbers are frequently depressed by more than 50%, but with *E. necatrix* infection, the packed erythrocyte volume is not depressed to such an extent (Natt and Herrick, 1955; Joyner and Davis, 1960). Whereas in other *Eimeria* infections, hemorrhage is often only slight, Moynihan (1950), working with *E. acervulina*, found no changes in total leukocytes or erythrocytes of peripheral blood. Stephens, Kowalski, and Borst (1967) found no significant differences in the packed erythrocyte volume of chickens infected with *E. maxima* during the acute phase of the disease, but found a significant increase in erythrocytes at 6, 10, and 14 days after inoculation and also slight increases in blood leukocyte numbers. Thus with most *Eimeria* infections of the fowl, the pathogenic effects do not include a substantial effect on the blood cell numbers. These authors found that weight gain and a decrease in pH of the intestinal contents occurred on the 6th day after inoculation.

C. EFFECT OF COCCIDIOSIS ON GUT PERMEABILITY AND THE UPTAKE OF NUTRIENTS

Preston-Mafham and Sykes (1967) were the first to show that serum proteins were passed into the lumen of the intestine during the inflammatory phases of *E. acervulina* infection. They demonstrated this change in permeability by inoculating pontamine sky blue intravenously shortly before killing the birds and observing the dye in the lumen of the gut and also by staining of the gut wall. This dye and Evans blue are known to combine with serum albumen. This study marked the start of their work on the effect of *E. acervulina* coccidiosis on the absorption of nutrients from the intestinal tract. These findings indicate that some changes occur in the mucosal blood capillaries as well as the epithelium during the period of infection when massive proliferation of the parasites occurs. In *E. acervulina*, this occurred after the third day of infection, when serum proteins escaped from the submucosa into the mucosa. The affected parts of the intestinal wall become edematous and remain so until about the 7th day of infection. Long (1968b) demonstrated that a similar change in gut permeability occurred between 3½ and 7 hr after inoculating birds with oocysts of *E. praecox*. This reaction occurred shortly after sporozoites had begun to invade the epithelium. However, within 24 hr, the permeability was normal but became abnormal again after 72 hr, when gross epithelial cell destruction occurred as a result of schizogony and gametogony. Rose and Long (1969) found that a similar permeability change occurred a few hours after inoculating birds with *E. acervulina* or *E. maxima* and were able to "block" the permeability change by inoculating the birds with the corticosteroid betamethasone.

Turk and Stephens (1967a,b) showed that in *E. acervulina*-infected chickens the absorption of Zn^{65}-labeled and I^{131}-labeled oleic acid is impaired. Preston-Mafham and Sykes (1967 and 1970), working with the same parasite, showed that the uptake from the gut of L-histidine, D-glucose, and fluids was also reduced. By giving uninfected birds similar amounts of food as consumed by *E. acervulina*-infected birds and studying their body weight gain, they were able to conclude that the body weight loss in infected birds was attributable partly to a reduction in feed intake and partly to a decrease in the absorption of nutrients. Kouwenhoven and Van der Horst (1969) in their study on *E. acervulina*-infections found that serum protein levels were reduced to less than half the normal concentrations from the 4th to the 7th day after infection. The vitamin A and carotene levels in the plasma were greatly reduced over the same period and the intestinal contents were strongly acid in reaction. Several workers have indicated that birds fed diets low in vitamin A suffer more from the effects of this parasite (Coles, Biely and March, 1970). In this connection, it was interesting that Kriz and Klimés (1969) obtained some beneficial effect of vitamin A administered orally shortly before birds were infected with the parasite. Kouwenhoven (1970) concluded that the fall in serum protein concentration during infection with *E. acervulina* occurs as a result of leakage into the intestinal lumen; this type of coccidiosis could be regarded as a protein-losing enteropathy and the strongly acidic intestinal contents impair the absorption of nutrients. Stephens (1965), working with *E. necatrix*, observed that intestinal contents had a low pH from six to nine days after infection and the plasma had an increased light transmittance during the hemorrhagic phase of the disease. The increased light transmittance of the plasma indicated that carotene pigments were not absorbed normally from the gut. Quite similar results were observed by Stephens, Kowalski, and Borst (1967) in their study on *E. maxima* infection of chickens. Shumard (1957), in his study of lamb coccidiosis, observed a fall in serum albumen during the peak of infection and also a reduction in the digestion of proteins by infected lambs. Svanbaev and Gorbunova (1968) produced clinical disease in lambs with *E. ninakohlyakimovae* and found that there was a fall in the alkaline reserves, chlorine, and glucose content of the blood as well as lowered erythrocyte counts and hemoglobin concentration. Deviations in the serum protein of calves during coccidial infection was observed by Fitzgerald (1962b) and, later, Fitzgerald (1964) reported that blood serum sodium and potassium levels were affected by coccidiosis.

D. THE POSSIBLE ROLE OF TOXINS IN COCCIDIAL INFECTIONS

Bertke (1963) studied renal clearance in chickens infected with *E. tenella*, using uric acid as a marker. He observed that the rate of clearance was greatest from two to four days after infection. He concluded that the reduction in uric acid rates was indicative of systemic changes caused by the parasite and infected chickens were in a state of shock. He postulated that the changes were caused by toxic material produced by or in response to the presence of the parasites and death from coccidiosis may not be caused solely by cecal damage or hemorrhage.

Burns (1959), Rikimura, Galysh, and Shumard (1951), and Sharma and Foster (1964) demonstrated toxic effects in rabbits inoculated with disrupted oocysts of chicken origin. Sharma and Foster thought that the lethal toxin was not present in the sporocysts, sporozoites or oocyst walls, but in the cytoplasm of the oocysts. Rikimura *et al.* used filtered ground-up sporulated oocyst material in their studies and obtained elevated body temperatures, hyperglycemia followed by hypoglycemia and blood-sugar depletion at death. These results are difficult to interpret, since the toxic effects were produced only by inoculating rabbits with the oocyst material of chicken origin. No effects could be produced in chickens or in mice or guinea pigs. Furthermore, the contents of oocyst material is not usually available to the infected host, since it is surrounded by a relatively impermeable membrane and the oocysts are discharged relatively quickly from the host via the feces. The material examined appears to be different from toxotoxin, recovered from the peritoneal exudate of mice infected with *Toxoplasma*, which caused deaths in mice inoculated intraperitoneally with the material (Weinman and Klatchko, 1949). Ryley (Chapter 5) has given more information on the toxins associated with *Toxoplasma gondii* and *E. tenella*.

Beyer (1963), as a result of her work on *E. intestinalis* of the rabbit, suggested two stages in the development of resistance to reinfection. In the first stage the host can withstand the toxic effects of the parasite but is unable to inhibit the development of the parasite. In the second stage the host can inhibit development of the parasite. Some support for this view was given by Hammond, Andersen, and Miner (1963b), working with *E. bovis* infections of calves. Unfortunately, the two-stage concept is not supported by the demonstration of toxins and antitoxins in response to infection. However, if the concept is true, it would mean that lesions associated with a primary infection produce pathogenic effects on the host but lesions produced by subsequent infections do not. The theory could explain the observed situation in the field when parasitism occurs in the absence of pathological effects on the host.

VI. Pathogenesis of Coccidial Infections

Pout (1967a) made the interesting observation that the height of the villi became progressively shorter during infection with *E. acervulina* in chickens and *E. crandallis* infection of lambs. The maximum depression in villus height occurred at the time when the maximum number of host cells were parasitized. Furthermore, villus atrophy (flat mucosa) occurred only in the parasitized portions of the intestine. This phenomenon had been observed in human malabsorption syndromes, but differed, in that with chickens the mucosal tissue recovered quickly from the effects of *E. acervulina* coccidiosis. With *E. mivati* infection, the villi were infiltrated rapidly by lymphocytes and some shortening of the villi occurred (Figs. 11b and 11c). *E. maxima* infection caused a more dramatic change; the villi took on a more "clubbed" appearance (compare Fig. 11d with Fig. 11a). In another paper Pout (1967b) pointed out that the villus atrophy results in a reduction of the surface area for absorption and a reduction in the number of functional epithelial cells; there is also a reduction in alkaline phosphatase activity in the villi. These changes were most severe at the time of maximum parasitism by the different species of *Eimeria* studied. Pout discussed the possible causes of these reactions, including the views of those working on human malabsorption syndromes that proliferation of the progenitor cells in the crypts of Lieberkühn is influenced by enzyme activity, and made some suggestions as to how coccidia might influence intestinal function. The coccidia could inhibit the division of the progenitor cells they invade or affect the rate of cell division in the crypts as a whole. Some support for the view that division of parasitized cells in the crypt is inhibited by the presence of parasites was given by Long and Millard (1968), who found that sporozoites were able to survive in crypts for up to 60 days in birds treated with some anticoccidial drugs. Gresham and Cruikshank (1959) thought that the second-generation schizonts of *E. tenella* developed within macrophages because they observed the grossly enlarged nuclei and the high nucleic acid content of the host cells. However, Long (1970c), indicated that division of epithelial cells parasitized by *E. tenella* in tissue cultures was inhibited and changes occurred in the host cells, which made them similar in appearance to macrophages. Bergmann (1970) made an electron microscope study of host cells parasitized by *E. tenella* and found regressive changes in host cells shortly after invasion. Later, the parasites caused atrophy of the cell components. The affected cells acquired the functional and structural characteristics of macrophages. This view that coccidia directly or indirectly alter the replication of cells in the villi is supported by the fact that they often take long periods of time to complete their development within host cells and, if cell turnover was as rapid in the parasitized intestine as in the normal intestine (Imondi and Bird, 1966), many species of *Eimeria* would have difficulty in completing their cycle.

It has been shown recently (Long, 1970d, 1971b) that development of *Eimeria tenella* can occur in the liver of chickens given corticosteroids. This suggests that in normal birds, sporozoites (and perhaps merozoites) may migrate away from the intestine and invade cells in other tissues. Further support for this view is provided by the observations of Owen (1970), who demonstrated sporozoites of *E. stiedai* in the bone marrow. Work on *Isospora* in birds by Box (1970) indicates that sporozoites and/or merozoites migrate and develop in sites other than the intestine. This work provides evidence that migration of sporozoites or merozoites of coccidia is a common occurrence. The coccidia do show marked site specificity, judged by their inability to develop in unusual sites. Even when parasites are inoculated parenterally they develop in the usual site (intestine), although there may be some delay in the completion of the life cycle (Landers, 1960; Sharma and Reid, 1962; Davies and Joyner, 1962).

The author has observed inflammatory reactions in the liver of chickens given *E. tenella* oocysts by mouth. Greven (1953) observed the infiltration of the pancreas by granulocyte leukocytes (presumably heterophil leukocytes) on the fifth day after *E. tenella* infection. She also noted congestion of the liver blood capillaries, erythrocyte pigment in the Kupffer cells and infiltration of the heart pericardium by granulocytes, mast cells, and plasma cells. Bertke and Herrick (1954) observed hemorrhagic lesions in the kidney of *E. tenella*-infected chickens; these were most obvious on the fourth day of infection. Such reactions may be produced by migration of sporozoites or merozoites, or by toxic products released by the parasite.

The manner in which coccidia produce effects on the host's intestine is still not understood, but the anatomical effects observed by Pout (1967 a,b, 1970) are similar to those of nontropical sprue (Padykula, Strauss, Ladman, and Gardner, 1961), hookworm disease (Sheehy, Meroney, Cox, and Soler, 1962), and *Nippostrongylus* in the rat (Symons and Fairbairn, 1962). In the recent report on *I. belli* infection in man, Brandborg, Goldberg, and Breidenbach (1970) indicate that flat mucosa occurred in some cases and state that "one had tall villi with 'clubbed' tips, marked dilation of the vascular spaces and excessive collagen in the lamina propria." This indicates that *I. belli* infection may cause human intestinal malabsorption.

Fig. 11. Photomicrographs of transverse sections of duodenum of chickens. Tissues were fixed in formol-sublimate and stained with Ehrlich's hematoxylin and eosin. *A*. Duodenum of uninfected bird. Note long slender villi. *B*. Four days after infection with *E. mivati*. Note gametocytes (g) and young schizonts (s) in the villi and infiltration of the lamina propria by lymphoid cells (L). *C*. One day after infection with *E. maxima*. Note infiltration of the lamina propria and tunica propria by lymphoid cells (L) but little or no shortening of the villi at this stage. *D*. Six days after infection with *E. maxima*. Note large number of gametocytes (g) in the villi. Infected villi have become short and broad.

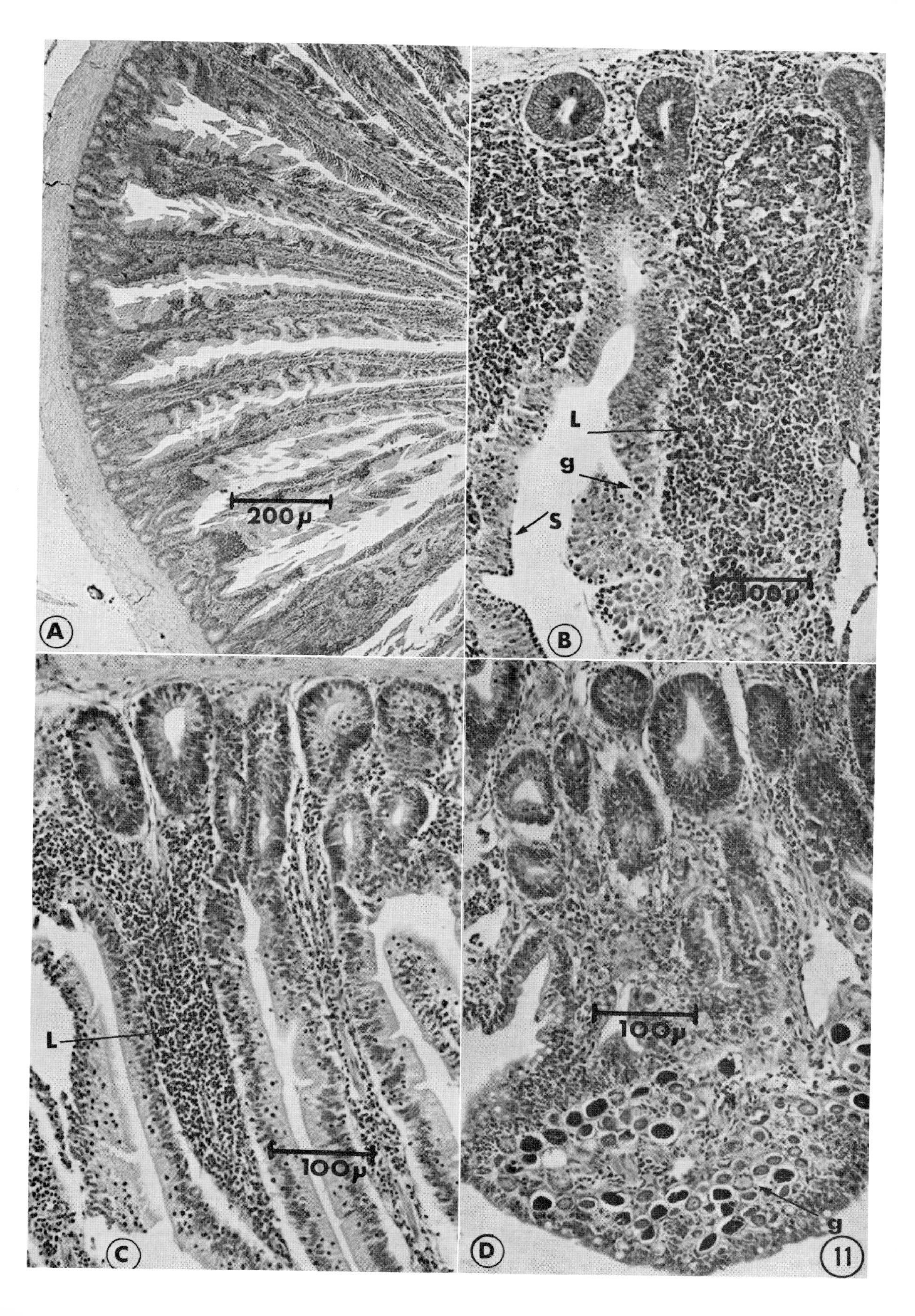
200µ
A
L
g
S
100µ
B
L
100µ
C
100µ
g
D
11

The anatomical changes in the intestine in response to *Eimeria* infection may also contribute to the malabsorption effects. The changes in gut permeability and the associated serum protein loss observed by Preston-Mafham and Sykes (1967) affect the efficacy of the digestive tract. Leakage of protein into the gut lumen could contribute to the occurrence of highly acidic intestinal contents noted by Kouwenhoven (1970), which could affect digestion and absorption of nutrients over a long period of time. The degree to which the different species of *Eimeria* cause these effects differs, and this may be related to their ability to produce toxins or induce the production of toxins by host cells parasitized by different stages in the life cycle. The studies made so far on toxin production by coccidia have been limited to the toxic effects of oocyst extracts and infected cecal tissue extracts. The oocyst had been chosen for study because oocysts can be cleaned and freed from extraneous matter, but it is a stage which is not wholly available to the host, since after the liberation of sporozoites in the intestine, the oocyst debris is discharged in the feces of the host. It is possible that toxins are produced as the result of the interaction of the parasite and host cell. Some progress might be made in the study of the pathogenesis of *E. tenella* in chicken embryos, since there should be no complication by the presence of bacterial endotoxins. In the embryo allantois, colonies of second-generation schizonts appear to have an affinity for blood capillaries, the permeability of which are probably increased, causing severe hemorrhage and death of the embryos. It should be noted that these effects are produced in the absence of bacteria. If toxins are responsible for the deaths in chickens infected with *E. tenella*, it seems remarkable that removal of one cecum (half of the available area for this parasite) should reduce mortality to exactly half of the expected level (Horton-Smith, Beattie, and Long, 1961). The addition of grass meal (a source of vitamin K) caused a similar reduction in mortality caused by this parasite (Long, 1968a).

Sporozoites of *Eimeria* do not appear to produce hyaluronidase to aid the invasion of cells (Fayer, Romanowski, and Vetterling, 1970), as in the case of another coccidium, *Toxoplasma gondii* (Lycke, Lund, and Strannegard, 1965). However, no evidence for toxin production by *Toxoplasma* could be found in experiments using chick embryos (Iygiste, Koplus, and Nutzoya, 1970). Stages other than oocysts can now be purified in glass bead columns (Wagenbach, 1969) and the possible toxin production induced by invading sporozoites and merozoites might be profitably studied.

Coccidiosis as a disease of chickens appears to occur in the absence of other organisms, although Johansson and Sarles (1948) considered that *Clostridium perfringens* and coliform bacteria might be involved in the pathology of *E. tenella* infection.

Clark, Smith, and Dardas (1962) showed that the gross pathology and life cycle stages of *E. tenella* were similar in gnotobiotic and conventionally reared chickens, although there was a delayed release of second-generation schizonts in the gnotobiotic chickens.

A necrotic enteritis often associated with *E. brunetti* infection has been considered to be caused by *Clostridium welchii*, which multiplies well in the mucoid discharges caused by *E. brunetti*. However, Hegde, Reid, Johnson, and Womack (1969) showed that the pathogenicity of *E. brunetti* was similar in bacteria-free and conventional chickens and the necrotic condition occurred in the gnotobiotic chickens.

Parish (1961) described both acute and chronic forms of necrotic enteritis in chickens and found these conditions to be caused by *Clostridium welchii* despite the fact that this organism is present in the intestine of most "normal" birds. He succeeded in reproducing the disease experimentally by feeding

C. welchii cultures. However, he found it easier to produce infections when various inorganic alkalis were given at the time of infection. He suggested that some dysfunction of the alimentary tract predisposed for this condition.

It is probable that *E. brunetti* does produce one form of necrotic enteritis not associated with *C. welchii*, but that the most common form of necrotic enteritis may be caused by the interaction of a number of factors. It is also possible that changes in the gut contents brought about by infection with intestinal coccidia may predispose to *C. welchii*-caused necrotic enteritis. Of particular interest is the possible effect on the intestinal bacterial flora of the serum proteins which enter the intestine during coccidial infection. In this connection, it is interesting that Stephens and Barnett (1964) found that *Salmonella typhimurium* produces greater effects in the host when introduced into chickens infected with *E. necatrix*. They suggested that the coccidial infection produced changes in the intestine which enable greater proliferation of *S. typhimurium* within the intestine and invasion of other internal organs.

It is also possible that coccidial infections may be potentiated by the presence of other disease agents. Biggs, Long, Kenzy, and Rootes (1968) demonstrated that infection of chickens with Marek's disease, a lymphoid proliferative disease caused by a herpes virus, increased their susceptibility to *Eimeria* infections and also interfered with the development of resistance to reinfection with the coccidia. Marek's disease is a common and severe disease, and field evidence has supported the results of this observation. It is known that Marek's disease impairs the immune response to a variety of antigens (Purchase, Chubb, and Biggs, 1968) and that this may be the reason for the increased susceptibility to coccidiosis. It was interesting to note that the life cycle in primary infections of *E. maxima* and *E. mivati* was extended in the Marek's disease-infected chickens. These results indicated that resistance to *Eimeria* infection can commence during the primary infection. Further support for this view was provided by Long and Rose (1970) when they demonstrated that corticosteroid treatment of chickens prior to and during the primary infection substantially increased the patent period. Oocyst production occurred for up to 50 days and late developing schizonts were seen in the intestinal tissues 24 days after infection. It therefore seems clear that infection of animals which reduces their immune competence can affect the severity of coccidial infection. In this connection it was interesting that Randall, Grant, and Sutherland (1971) thought that continous output of *E. acervulina* oocysts occurred in chickens affected by Marek's disease and housed in a way in which reinfection by *Eimeria* would be both difficult and unlikely. Niilo (1970c) gave dexamethasone treatment to calves infected with *E. zuernii* and *E. bovis* and found that the treatment caused increased oocyst production and aggravated the clinical signs of the disease.

Davis, Herlich, and Bowman (1959a,b; 1960a,b) have demonstrated that in calves infected with coccidia and intestinal nematodes of the genera *Cooperia, Strongyloides*, and *Trichostrongylus*, the severity of coccidiosis is increased, as evidenced by numbers of parasites and associated tissue damage. No such changes were observed when calves were concurrently infected with coccidiosis and the stomach worm *Ostertagia*. These results suggest that the enhancement in the development of the coccidia by nematodes may not occur as a result of reduced immunological competence of the host but rather as a result of physiological changes in the intestine.

Pout (1969) suggested that coccidia are only part of a hostile environment which produces the disease situation in the animal and that the estimation of numbers of oocysts passed in the feces of the host is not a good criterion of the severity of disease. However, the large numbers of oocysts which are discharged may be a significant challenge to susceptible contact animals and monitoring their numbers over a period of time may provide useful information.

VII. Concluding Remarks

It is probable that all coccidia are potentially capable of inflicting damage on their hosts. Determined attempts to evaluate the pathogenicity of most species of coccidia have not yet been made. Most of the work has been carried on with farm and laboratory animals.

The severity of coccidial infection occurring in animals is dependent upon the dose of sporulated oocysts ingested; the rate of reproduction within the host is, within limits, fixed by the number of oocysts ingested. Thus, it is possible with coccidial infections to obtain a graded series of infections in groups of animals which have ingested different numbers of oocysts. The numbers of oocysts necessary to produce different pathological effects varies according to the species of *Eimeria* given, the age and condition of the host, and whether the host has already experienced infection by the parasite. For example, *E. necatrix* is a poor oocyst producer, but only small numbers of oocysts are necessary to cause disease in a previously uninfected host. On the other hand, chickens infected with *E. acervulina* produce very large numbers of oocysts and large numbers of oocysts are needed to produce disease. There are other examples of this within the species of *Eimeria* affecting the various hosts. It is therefore incorrect to grade the pathogenicity of a species of *Eimeria* according to the number of oocysts required to cause disease.

An obstacle to our understanding of the disease process is our use of single massive doses of oocysts in experimental infections. This procedure is followed for convenience in measuring pathogenic effects, but it is not likely to reflect the kinds of host-parasite relationship operative in nature. We are still unaware of the kind of oocyst dose and frequency of dose occurring under natural conditions. With highly immunogenic species (e.g., *E. maxima* and *E. praecox*), an intake of a small number of oocysts may induce a substantial immunity, but with other species, regular doses of oocysts may produce an additive effect on the host. Turk and Stephens (1970) and Michael and Hodges (1971) found that the period of body weight gain depression was increased in birds given daily doses of *E. acervulina* and *E. necatrix* compared with those birds which received a single oocyst dose. Allowing animals to ingest oocysts in their food may be a better way of producing experimental infections than the feeding of single doses of oocysts in water. Fitzgerald (1967) found that it was possible to infect calves by giving oocysts in dry grain and Reid *et al.* (1969) used methods of infecting chickens which simulated natural conditions.

The discharge of oocysts by animals is no guide to the effect the coccidia have on the host. It is possible that such occurrences of oocyst discharge encountered in the field, reflect the process of immunization against the coccidia in the animals. With most species of *Eimeria*, immunity does not follow a single exposure to the parasite and partially resistant animals discharge large numbers of oocysts. It is probable that these animals also have lesions, caused by the parasite in the gut wall, but do not exhibit clinical signs of the disease. This view is suggested by the work of Reid and Johnson (1970), who found slight and moderate lesion scores in some birds which did not show body weight gain depression. Thus it may prove difficult to relate the presence of lesions in naturally infected animals to other changes, e.g., food and water consumption and body weight changes.

It is clear from the preceding discussion that the pathogenic effects caused by coccidia on their hosts are not restricted to the loss of epithelial cells damaged by the parasites. This factor alone probably causes only a minor effect, since epithelial cell regeneration is rapid. Nor are the most pathogenic effects caused by the most highly immunogenic species since one of the most immunogenic species (*E. praecox*) is relatively nonpathogenic. The size of the parasites and their position in the tissues is an important factor in producing damage as well as the rate of reproduction. Host effects in response to the parasites include the proliferation of defense cells, changes in tissue permeability, loss of proteins, impairment in the uptake of nutrients and damage to the intestine. In some species, there is considerable hemorrhage and an interference with carbohydrate metabolism. More evidence is needed in order to show that toxins are produced in response to coccidial infections and that toxins are involved in producing effects on the host. Little progress has been made in our knowledge of the pathological changes caused by coccidia and more information is required, in particular the influence of climate conditions and concurrent infections on the disease-producing agents, effect of infection with multiple species of coccidia, and the effect of diet.

Literature Cited

Ackert, J.E. and C.A. Herrick. 1928. Effects of the nematode *Ascaridia lineata* (Schneider) on growing chickens. J. Parasitol. 15:1-13.

Allen, E.A. 1936. *Tyzzeria perniciosa* gen. et sp. nov., a coccidium from the small intestine of the "Peking" duck, *Anas domesticus L.* Arch. Protistenk. 87 : 262-267.

Bachman, G.W. and P.E. Menendez. 1930. Jaundice in experimental coccidiosis of rabbits. Am. J. Hyg. 12 : 650-656.

Barksdale, W.L. and C.F. Routh. 1948. *Isospora hominis* infections among American personnel in the Southwest Pacific. Am. J. Trop. Med. 18 : 639-644.

Bayon, H.P. 1930. Common and preventable parasitic ailments of the domestic fowl. A contribution to the study of the hygiene of poultry. Vet. Rec. 10 : 1129-1142.

Becker, E.R. 1934. Coccidia and coccidiosis of domesticated, game and laboratory animals and of man. Iowa State College. Division of Industrial Science Monograph 2.

Bell, D.J. and J. Culbert. 1967. Oxo acids in the plasma of two lines of Brown Leghorn hens; the presence of an unidentified compound. Comp. Biochem. Physiol. 22: 121-129.

Berg, I.R., C.M. Hamilton, and G.E. Bearse. 1951. The effect of coccidiosis (caused by *Eimeria maxima*) on egg quality. Poult. Sci. 30 : 298-301

Bergmann, V. 1970. Elektronmikroskopische Untersuchungen zur Pathogenese der Blinddarmkokzidiose der Hühnerküken. Arch. Exp. Vet. Med. 24 : 1169-1184.

Bertke, E.M. 1963. Renal function during the course of *Eimeria tenella* infection. J. Parasitol. 49 : 937-942.

Bertke, E.M. and C.A. Herrick. 1954. The pathology of the kidney of the chicken produced by cecal coccidiosis. J. Parasitol. 40 : (No.5, Section 2), 130.

Beyer, T. 1963. Immunity in experimental coccidiosis of rabbits caused by heavy infective doses of *Eimeria intestinalis*. Progress in Protozoology (Proc. 1st Int. Conf. Protozool.), p. 448. Publishing House of the Czechoslovak Academy of Sciences, Prague.

Biester, H.E. and C. Murray. 1929. Studies on infectious enteritis of swine. IV. Intestinal coccidiosis. J. Amer. Vet. Med. Ass. 75 : 705.

Biggs, P.M., P.L. Long, S.G. Kenzy, and D.G. Rootes. 1968. Relationship between Marek's disease and coccidiosis. II. The effect of Marek's disease on the susceptibility of chickens to coccidial infection. Vet. Rec. 83 : 284-289.

Boughton, D.C. 1942. An overlooked macroscopic intestinal lesion of value in diagnosing bovine coccidiosis. N. Am. Vet. 23 : 173-175.

Boughton, D.C. 1943. Sulphaguanidine therapy in experimental bovine coccidiosis. Am. J. Vet. Res. 4 : 66-72.

Boughton D.C. 1945. Bovine coccidiosis. From

carrier to clinical case. N. Am. Vet. 26 : 147-153.

Box, Edith D. 1970. *Atoxoplasma* associated with an Isosporan oocyst in canaries. J. Protozool. 17 : 391-396.

Brandborg, L.L., S.B. Goldberg, and W.C. Breidenbach. 1970. Human coccidiosis – a possible cause of malabsorption. The life cycle in small bowel mucosal biopsies as a diagnostic feature. New Engl. J. Med. 283 : 1306-1313.

Burns, W.C. 1959. The lethal effect of *Eimeria tenella* extracts on rabbits. J. Parasitol. 45 : 38-46.

Challey, J.R., and W.C. Burns. 1959. The invasion of the cecal mucosa by *Eimeria tenella* sporozoites and their transport by macrophages. J. Protozool. 6 : 238-241.

Champion, L.R. 1954. The inheritance of resistance to cecal coccidiosis in the domestic fowl. Poult. Sci. 33 : 670-681.

Christensen, J.F. 1941. Experimental production of coccidiosis in silage fed feeder lambs with observations on oocyst discharge. N. Am. Vet. 22: 606-610.

Clark, D.T., C.K Smith, and R.B. Dardas. 1962. Pathological and immunological changes in gnotobiotic chickens due to *Eimeria tenella*. Poult. Sci. 41 : 1635-1636.

Clarkson, M.J. 1958. Life history and pathogenicity of *Eimeria adenoeides*, Moore and Brown, 1951 in the turkey poult. Parasitology. 48 : 70-88.

Clarkson, M.J. 1959. The life history and pathogenicity of *Eimeria meleagrimitis*, Tyzzer 1929, in the turkey poult. Parasitology. 49 : 70-82.

Clarkson, M.J. 1960. The coccidia of the turkey. Ann. Trop. Med. Parasitol. 54 : 253-257.

Clarkson, M.J. and M.A. Gentles. 1958. Coccidiosis in turkeys. Vet. Rec. 70 : 211-214.

Coles, B., J. Biely, and B.E. March. 1970. Vitamin A deficiency and *Eimeria acervulina* infection in the chick. Poult. Sci. 49 : 1295-1301.

Daugherty, J.W. and C.A. Herrick. 1952. Cecal coccidiosis and carbohydrate metabolism in chickens. J. Parasitol. 38 : 298-204.

Davies, S.F.M. 1956. Intestinal coccidiosis in chickens caused by *Eimeria necatrix*. Vet. Rec. 68 : 853-857.

Davies, S.F.M. 1957. An outbreak of duck coccidiosis in Britain. Vet. Rec. 69 : 1051-1052.

Davies, S.F.M. 1963. *Eimeria brunetti*, an additional cause of intestinal coccidiosis in the domestic fowl in Britain. Vet. Rec. 75 : 1-4.

Davies, S.F.M. and L.P. Joyner. 1962. Infection of the fowl by the parenteral inoculation of oocysts of *Eimeria*. Nature. 194: 996-997.

Davies, S.F.M., L.P. Joyner, and S.B. Kendall. 1963. Coccidiosis, p. 257. Oliver and Boyd, Edinburgh.

Davis, L.R., D.C. Boughton, and G.W. Bowman. 1955. Biology and pathogenicity of *Eimeria alabamensis* Christensen, 1941, an intranuclear coccidium of cattle. Am. J. Vet. Res. 16 274-281.

Davis, L.R. and G.W. Bowman. 1952. Coccidiosis in cattle. Proc. U.S. Livestock Sanitary. Ass. 55 : 39-50.

Davis, L.R., H. Herlich, and G.W. Bowman. 1959a. Studies on experimental concurrent infections of dairy calves with coccidia and nematodes. I. *Eimeria* spp. and the small intestinal worm *Cooperia punctata*. Am. J. Vet. Res. 20 : 281-286.

Davis, L.R., H. Herlich, and G.W. Bowman. 1959b. Studies on experimental concurrent infections of dairy calves with coccidia and nematodes. II. *Eimeria* spp. and the medium stomach worm *Ostertagia ostertagia*. Am. J. Vet. Res. 20 : 487-489.

Davis, L.R., H. Herlich, and G.W. Bowman. 1960a. Studies on experimental concurrent infections of dairy calves with coccidia and nematodes. III. *Eimeria* spp. and the threadworm *Strongyloides papillosus*. Am. J. Vet. Res. 21 : 181-187.

Davis, L.R., H. Herlich, and G.W. Bowman. 1960b. Studies on experimental concurrent infections of dairy calves with coccidia and nematodes. IV. *Eimeria* spp. and the small hairworm *Trichostrongylus colubriformis*. Am. J. Vet. Res. 21 : 188-194.

Dickinson, E.M. 1941. The effects of variable dosages of sporulated *Eimeria acervulina* oocysts on chickens. Poult. Sci. 20 : 413-424.

Dickinson, E.M. and R.M. Schofield. 1939. The effect of sulphur against artificial infection with *Eimeria acervulina* and *Eimeria tenella*. Poult. Sci. 18 : 419-431.

Dobson, E.L. 1966. Studies on the immunity of sheep to *Oesophagostomum columbianum* : The nature and fate of the globule leucocyte. Aust. J. Agr. Res. 17 : 955-956.

Dubey, J.P. and B.P. Pande. 1963. On a coccidian schizont in the Indian domestic duck (*Anas platyrhynchos domesticus*). J. Parasitol. 49 : 770.

Edgar, S.A. 1954. Effect of temperature on the

sporulation of oocysts of the protozoan, *Eimeria tenella.* Trans. Am. Microsc. Soc. 63 : 237-242.

Edgar, S.A. and C.T. Seibold. 1964. A new coccidium of chickens, *Eimeria mivati* sp n. (Protozoa : Eimeriidae) with details of its life history. J. Parasitol. 50 : 193-204.

Farr, M.M. and E.E. Wehr. 1952. *Eimeria truncata* associated with morbidity and death of domestic goslings. Cornell Vet. 42 : 185-187.

Farr, M.M., E.E. Wehr, and W.T. Shalkop. 1962. Studies on the pathogenicity of *Eimeria gallopavonis* to turkeys. J. Protozool. 9: (Suppl.) 8-9.

Fayer, R. and S. Baron. 1971. Activity of interferon and an inducer against development of *Eimeria tenella* in cell culture. J. Protozool. 18(Suppl.), Abst. 21.

Fayer, R., R.D. Romanowski, and J.M. Vetterling. 1970. The influence of hyaluronidase and hyraluronidase substrates on penetration of cultured cells by *Eimerian* sporozoites. J. Protozool. 17 : 432-436.

Fitzgerald, P.R. 1962a. Coccidia in Hereford calves on summer and winter ranges and in feedlots in Utah. J. Parasitol. 48 : 347-351.

Fitzgerald, P.R. 1962b. Deviations in serum proteins associated with *Eimeria bovis* infections in calves and some related immune reactions. J. Parasitol. 48 : 38-39.

Fitzgerald, P.R. 1964. Effect of bovine coccidiosis on the blood serum sodium and potassium levels. J. Parasitol. 50 (Suppl.): 52-53.

Fitzgerald, P.R. 1967. Results of continuous low level inoculations with *Eimeria bovis* in calves. Am. J. Vet. Res. 28 : 659-670.

Fitzgerald, P.R. and M.E. Mansfield. 1969. Economic significance of coccidiosis in calves. Progr. and Abstr., 44th Ann. Meeting Am. Soc. Parasitol., p. 39.

Ford, E.J.H. 1956. A survey of turkey mortality. Brit. Vet. J. 112 : 3-13.

Freeman, B.M. 1970. Carbohydrate stores in chickens infected with *Eimeria tenella*. Parasitology, 61 : 245-251.

Gill, B.S. and R.D. Katiyar. 1961. An acute outbreak of coccidiosis (*Eimeria arloingi*, Marotel, 1950) in kids of Orai sheep breeding farm. Ind. J. Vet. Sci. 31 : 212-216.

Gresham, G.A. and J.G. Cruickshank. 1959. Protein synthesis in macrophages containing *Eimeria tenella*. Nature, 184 : 1153.

Greven, Ursula. 1953. Zur Pathologie der Geflügelcoccidiose. Arch. Protistenk. 98 : 342-414.

Hammond, D.M. 1964. "Coccidiosis of cattle, some unsolved problems." Thirtieth faculty Honor lecture. The Faculty Association, Utah State University.

Hammond, D.M. 1965. Cellular interactions between bovine coccidia and their hosts. Progress in Protozoology, 2nd Int. Conference on Protozool. London, pp. 53-54.

Hammond, D.M., F.L. Andersen, and M.L. Miner. 1963a. The occurrence of a second generation schizont in the life cycle of *Eimeria bovis* in calves. J. Parasitol. 49 : 428-434.

Hammond, D.M., F.L. Andersen, and M.L. Miner. 1963b. The site of the immune reaction against *Eimeria bovis* in calves. J. Parasitol. 49 : 415-424.

Hammond, D.M., G.W. Bowman, L.R. Davis, and B.T. Simms. 1946. The endogenous phase of the life cycle of *Eimeria bovis*. J. Parasitol. 32 : 409-427.

Hammond, D.M., L.R. Davis, and G.W. Bowman. 1944. Experimental infections with *Eimeria bovis* in calves. Am. J. Vet. Res. 5 : 303-311.

Hammond, D.M., J.V. Ernst, and M.L. Miner. 1966. The development of first generation schizonts in *Eimeria bovis*. J. Protozool. 18 : 559-564.

Hammond, D.M., J.E. Kuta, and M.L. Miner. 1967. Amprolium for control of experimental coccidiosis in lambs. Cornell Vet. 57 : 611-623.

Hawkins, P.A. 1952. Coccidiosis of turkeys. Michigan State College Agricultural Experimental Station Technical Bulletin, Number 226.

Harms, R.H. and R.L. Tugwell. 1956. The effect of experimentally induced prolonged blood clotting time on cecal coccidiosis of chicks. Poult. Sci. 35 : 937-938.

Hegde, K.S., W.M. Reid, J. Johnson, and H.E. Womack, 1969. Pathogenicity of *Eimeria brunetti* in bacteria free and conventional chickens. J. Parasitol. 55 : 402-405.

Hein, H. 1968. Pathogenic effects of *Eimeria acervulina* in young chicks. Exp. Parasitol. 22 : 1-11.

Hein, H. 1969. *Eimeria adenoeides* and *E. meleagrimitis*; Pathogenic effect in turkey poults. Exp. Parasitol. 124 : 163-170.

Hill, K.J. 1971. The structure of the alimentary tract. *In* Physiology and biochemistry of the

domestic fowl, p. 1-23. D.J. Bell and B.M. Freeman (eds.). Academic Press, London.
Hillbrich, P. 1963. Krankheiten des Geflügels. Verlag Hermann Kuhn, Schwenningen am Necker.
Hitchcock, D.J. 1955. The life cycle of *Isospora felis* in the kitten. J. Parasitol. 41 : 383-393.
Horton-Smith, C. 1947. Coccidiosis — some factors influencing its epidemiology. Vet. Rec. 56 : 645-646.
Horton-Smith, C. 1957. Factors affecting the transmission of coccidia and the development of disease in fowls. *In* Biological aspects of the transmission of disease. C. Horton-Smith (ed.). Oliver and Boyd, Edinburgh.
Horton-Smith, C., J. Beattie, and P.L. Long. 1961. Resistance to *Eimeria tenella* and its transference from one caecum to the other in individual fowls. Immunology 4 : 111-121.
Horton-Smith, C. and P.L. Long. 1954. Preliminary observations on the physical conditions of built up litter and their possible effects on the parasite populations. Proc. 10th Worlds Poult. Sci. Congr. Edinburgh, pp. 266-272.
Horton-Smith, C. and P.L. Long. 1965. The development of *Eimeria necatrix*, Johnson, 1930 and *Eimeria brunetti*, Levine, 1942 in the caeca of the domestic fowl (*Gallus domesticus*). Parasitology 55 : 401-405.
Imondi, A.R. and F.H. Bird. 1966. The turnover of intestinal epithelium in the chick. Poult. Sci. 45 : 142-147.
Iygiste, A.K., M.O. Koplus, and Kh.I. Nutzoya. 1970. Effect of *Toxoplasma* toxin on the development of chick embryos. Medskoya Parasitol. 39 : 282-284.
Jeffers, T.K. and G.E. Wagenbach. 1969. Sex differences in embryonic response to *Eimeria tenella* infection. J. Parasitol. 55 : 949-951.
Jeffers, T.K. and G.E. Wagenbach. 1970. Embryonic response to *Eimeria tenella* infection. J. Parasitol. 56 : 656-662.
Johansson, K.R. and W.B. Sarles. 1948. Bacterial population changes in the ceca of young chickens infected with *Eimeria tenella*. J. Bacteriol. 56 : 635
Johnson, W.T. 1931. Effect of five species of *Eimeria* upon egg production of single comb White Leghorns. J. Parasitol. 18 : 122.
Jones, E.E. 1932. Size as a species characteristic in coccidia. Arch. Protistenk. 77 : 130-170.
Joyner, L.P. 1969. Immunological variation between two strains of *Eimeria acervulina*. Parasitology. 59 : 725-732.
Joyner, L.P. and S.F.M. Davies. 1960. Detection and assessment of sublethal infections of *Eimeria tenella* and *Eimeria necatrix*. Exp. Parasitol. 9 : 243-249.
Joyner, L.P. and C.C. Norton. 1969. A comparison of two laboratory strains of *Eimeria tenella*. Parasitology. 59 : 907-913.
Kheysin, E.M. (Cheissin, E.M.) 1959. Cytochemical investigations of different stages of the life cycle of coccidia of the rabbit. XVth Int. Zool. Congr. London. Section 9, paper 25.
Kotlán, S. and L. Pellérdy. 1937. Experimental studies of hepatic coccidiosis in domestic rabbits. II. Közl. Összehas. Kortan. 28 : 105-120.
Kouwenhoven, B. 1970. *Eimeria acervulina* infectie bij de kip een parasitologisch en biochemisch onderzoek. Doctoral Thesis. Rijks Universiteit Utrecht. June 1970.
Kouwenhoven, B. and C.J.G. van Der Horst. 1969. Strongly acid intestinal content and lowered protein, carotene and Vitamin A blood levels in *Eimeria acervulina* infected chickens. Z. Parasitenk. 32 : 347-353.
Krassner, S.M. 1963. Factors in host susceptibility and oocyst infectivity in *Eimeria acervulina* infections. J. Protozool. 10 : 327-332.
Kriz, H. and B. Klimés. 1969. The effect of single vitamin A application to chickens on the course of experimental *Eimeria acervulina* infection in the following four weeks and its influence on oocyst elimination. Acta. Vet. (Brno). 38 : 77-86.
Landers, E.J. 1960. Studies on excystation of coccidial oocysts. J. Parasitol. 46 : 195-200.
Leathem, W.D. and W.C. Burns. 1968. Duration of acquired immunity of the chicken to *Eimeria tenella* infection. J. Parasitol. 54 : 227-232.
Lee, C.D. 1934. The pathology of coccidiosis in the dog. J. Am. Vet. Med. Ass. 85 : 760-781.
Levine, L. and C.A. Herrick. 1954. The effect of the protozoan *Eimeria tenella* on the ability of the chicken to do muscular work when its muscles are stimulated directly or indirectly. J. Parasitol. 40 : 525-531.
Levine, L. and C.A. Herrick. 1957. Effects of cecal coccidia on the ability of Plymouth Rocks to do muscular work. Poult. Sci. 36 : 24-29.
Levine, N.D. 1961. Protozoan parasites of domestic animals and of man. Burgess Publishing Co., Minneapolis.
Levine, N.D. and V. Ivens. 1965. The coccidian parasites (Protozoa, Sporozoa) of rodents. Il-

linois Biological Monographs number 33. The University of Illinois Press, Chicago.

Levine, N.D. and V. Ivens. 1970. The coccidian parasites (Protozoa, Sporozoa) of ruminants. Illinois Biological Monographs number 44. University of Illinois Press, Chicago.

Levine, N.D., V. Ivens, and T.E. Fritz. 1962. *Eimeria christenseni* sp.n. and other coccidia (Protozoa : Eimeriidae) of the goat. J. Parasitol. 48 : 255-269.

Levine, P.P. 1942. A new coccidium pathogenic for chickens, *Eimeria brunetti* n. sp. (Protozoa : Eimeriidae). Cornell Vet. 32 : 430-439.

Long, P.L. 1959. Study of *Eimeria maxima*, Tyzzer, 1929, a coccidium of the fowl. Ann. Trop. Med. Parasitol. 53 : 325-333.

Long, P.L. 1967a. Studies on *Eimeria praecox*, Johnson, 1930, in the chicken. Parasitology. 57 : 351-361.

Long, P.L. 1967b. Studies on *Eimeria mivati* in chickens and a comparison with *Eimeria acervulina*. J. Comp. Pathol. 77 : 315-325.

Long, P.L. 1968a. The effect of breed of chickens on resistance to *Eimeria* infections. Brit. Poult. Sci. 9 : 71-78.

Long, P.L. 1968b. The pathogenic effects of *Eimeria praecox* and *E. acervulina* in the chicken. Parasitology. 58 : 691-700.

Long, P.L. 1970a. Studies on the viability of sporozoites of *Eimeria tenella*. Z. Parasitenk. 35: 1-6.

Long, P.L. 1970b. Some factors affecting the severity of infection with *Eimeria tenella* in chicken embryos. Parasitology. 60 : 435-447.

Long, P.L. 1970c. *In vitro* culture of *Eimeria tenella*. J. Parasitol. 56 (No. 4, Sect. 2) : 214-215.

Long, P.L. 1970d. Development (schizogony) of *Eimeria tenella* in the liver of chickens treated with corticosteroid. Nature. 225 : 290-291.

Long, P.L. 1971a. Maternal transfer of anticoccidial drugs in the chicken. J. Comp. Pathol. 81 : 373-382.

Long, P.L. 1971b. Schizogony and gametogony of *Eimeria tenella* in the liver of chick embryos. J. Protozool. 18 : 17-20.

Long, P.L. and C. Horton-Smith. 1968. Coccidia and coccidiosis in the domestic fowl. *In* Advances in parasitology. Number 6, 313-325. Academic Press, New York and London.

Long, P.L. and B.J. Millard. 1968. *Eimeria* : effect of meticlorpindol and methyl benzoquate on endogenous stages in the chicken. Exp. Parasitol. 23 : 331-338.

Long, P.L. and B.S. Milne. 1971. The effect of an artificial interferon inducer on *Eimeria maxima* in the chicken. Parasitology. 62 : 295-302.

Long, P.L. and M.E. Rose. 1970. Extended schizogony of *Eimeria mivati* in betamethasone-treated chickens. Parasitology. 60 : 147-155.

Lotze, J.C. 1952. The pathogenicity of the coccidian parasite, *Eimeria arloingi* in domestic sheep. Cornell. Vet. 42 : 510-517.

Lotze, J.C. 1953. Life history of the coccidian parasite, *Eimeria arloingi*, in domestic sheep. Am. J. Vet. Res. 14 : 86-95.

Lotze, J.C. 1954. The pathogenicity of the coccidian parasite, *Eimeria ninae-kohl-yakimovi* (Yakimov and Rastegaieff, 1930) in domestic sheep. Proc. Amer. Vet. Med. Ass. 19 : 141-146.

Lotze, J.C. and R.G. Leek. 1970. Failure of development of the sexual phase of *Eimeria intricata* in heavily inoculated sheep. J. Protozool. 17 : 414-417.

Lotze, J.C., W.T. Shalkop, R.G. Leek, R. Behin. 1964. Coccidial schizonts in mesenteric lymph nodes of sheep and goats. J. Parasitol. 50 : 205-208.

Lund, E.E. 1954. Estimating relative pollution of the environment with oocysts of *Eimeria stiedae* J. Parasitol. 40 : 663-667.

Lycke, E., E. Lund, and O. Strannegard. 1965. Enhancement by lysozyme and hyaluronidase of the penetration by *Toxoplasma gondii* into cultured host cells. Brit. J. Exp. Pathol. 46 : 189-199.

Mahrt, J.L. 1966. Life cycle of *Isospora rivolta* (Grassi, 1879) Wenyon, 1923. J. Protozool. 13 (Suppl.) : 18.

Matsubayashi, H., and T. Nozawa. 1948. Experimental infection of *Isospora hominis* in man. Am. J. Trop. Med. 28 : 633-637.

Mayhew, R.L. 1934. Studies on coccidiosis. VIII. Immunity or resistance to infection in chickens. J. Vet. Med. Ass. 85 : 729-734.

Michael, E., and R.D. Hodges. 1971. The pathogenic effects of *Eimeria acervulina* : a comparison of single and repeated infections. Vet. Rec. 89 : 329-333.

Moore, E.N., and J.A. Brown. 1951. A new coccidium pathogenic for turkeys, *Eimeria adenoeides* n. sp. (Protozoa : Eimeriidae). Cornell Vet. 41 : 125-136.

Morehouse, N.F., and W.C. McGuire. 1958. The pathogenicity of *Eimeria acervulina*. Poult. Sci. 37 : 665-672.

Moynihan, I.W. 1950. The role of the protozoan

parasite *Eimeria acervulina* in disease of the domestic chicken. Can. J. Comp. Med. 14 : 74-82.

Natt, M.P. and C.A. Herrick. 1955. The effect of cecal coccidiosis on the blood cells of the domestic fowl. 1. A comparison of the changes in the erythrocyte count resulting from hemorrhage in infected and mechanically bled birds. The use of the hematocrit value as an index of the severity of the hemorrhage resulting from the infection. Poult. Sci. 34 : 1100-1106.

Nemeséri, L. 1960. Beiträge zur Äetiologie der Coccidiose der Hunde. 1. *Isospora canis* sp. n. Acta Vet. Acad. Sci. Hung. 10 : 95-99.

Niilo, L. 1970a. Bovine coccidiosis in Canada. Can. Vet. J. 11 : 91- 98.

Niilo, L. 1970b. Experimental winter coccidiosis in sheltered and unsheltered calves. Can. J. Comp. Med. 34 : 20-25.

Niilo, L. 1970c. The effect of dexamethasone on bovine coccidiosis. Can. J. Comp. Med. 34 : 325-328.

Owen, D. 1970. Life cycle of *Eimeria stiedae*. Nature. 227 : 304.

Padykula, H.A., E.W. Strauss, A.J. Ladman, and F.H. Gardner. 1961. A morphologic and histochemical analysis of the human jejunal epithelium in non-tropical sprue. Gastroenterology. 40 : 735-765.

Parish, W.E. 1961. Necrotic enteritis in the fowl. III. The experimental disease. J. Comp. Pathol. 71 : 405-413.

Pattillo, W.H. 1959. Invasion of the cecal mucosa of the chicken by sporozoites of *Eimeria tenella*. J. Parasitol. 45 : 253-258.

Pellérdy, L. 1965. Coccidia and coccidiosis. Akademiai Kiado. Budapest.

Pierce, A.E., P.L. Long, and C. Horton-Smith. 1962. Immunity of *Eimeria tenella* in young fowls. Immunology. 5 : 129-152.

Pout, D.D. 1967a. Villous atrophy and coccidiosis. Nature. 213 : 306-307.

Pout, D.D. 1967b. The reaction of the small intestine of the chicken to infection with *Eimeria* spp. *In* The reaction of the host to parasitism. Veterinary Medical Review, N.G. Elwert Universitäts-und Verlagsbuchhandlung. Marburg-Köln. p. 28-38.

Pout, D.D. 1969. Coccidiosis of sheep. A critical review of the disease. Vet. Bull. 39 : 609-618.

Pout, D.D. 1970. The mucosal surface patterns of the small intestine in grazing lambs. Brit. Vet. J. 126 : 357-363.

Pout, D.D., D.C. Ostler, L.P. Joyner, and C.C. Norton. 1966. The coccidial population in clinically normal sheep. Vet. Rec. 78 : 455-460.

Pratt, I. 1944. The effect of *Eimeria tenella* (Coccidia) upon the blood sugar of the chicken. Trans. Am. Microsc. Soc. 59 : 31-37.

Preston-Mafham, R.A. and A.H. Sykes. 1967. Changes in permeability of the mucosa during intestinal coccidiosis infections in the fowl. Experientia (Basel). 23 : 972-973.

Preston-Mafham, R.A. and A.H. Sykes. 1970. Changes in body weight and intestinal absorption during infections with *Eimeria acervulina* in the chicken. Parasitology. 61 : 417-424.

Purchase, H.G., R.C. Chubb, and P.M. Biggs. 1968. Effect of lymphoid leukosis and Marek's disease on the immunological responsiveness of the chicken. J. Nat. Cancer Inst. 40 : 583-592.

Randall, C.J., G. Grant, and I.M. Sutherland. 1971. Coccidiosis with concurrent Marek's disease. Vet. Rec. 88 : 618.

Reid, W.M., R.N. Brewer, J. Johnson, E.M. Taylor, K.S. Hegde, and L.M. Kowalski. 1969. Evaluation of techniques used in studies on efficacy of anticoccidial drugs in chickens. Am. J. Vet. Res. 30 : 447-459.

Reid, W.M. and J. Johnson. 1970. Pathogenicity of *Eimeria acervulina* in light and heavy coccidial infections. Avian Dis. 14 : 166-171.

Reid, W.M. and M. Pitois. 1965. The influence of coccidiosis on feed and water intake of chickens. Avian Dis. 9 : 343-348.

Rikimari, M.T., F.T. Galysh, and R.F. Shummard. 1961. Some pharmacological aspects of a toxic substance from oocysts of the coccidium *Eimeria tenella*. J. Parasitol. 47 : 407-412.

Rommel, M. 1970. Verlauf der *Eimeria scabra*- und *E. polita*-Infektion in vollempfänglichen Ferkeln und Läuferschweinen. Tierärztliche Wochenschrift. 83 : 181-186.

Rose, M.E. 1967. The influence of age of host on infection with *Eimeria tenella*. J. Parasitol. 53: 924-929.

Rose, M.E. and P.L. Long. 1969. Immunity to coccidiosis: Gut permeability changes in response to sporozoite invasion. Experientia 25 : 183-184.

Rose, M.E. and P.L. Long. 1971. Immunity to coccidiosis : protective effects of transferred serum and cells investigated in chick embryos infected with *Eimeria tenella*. Parasitology 63 : 299-313.

Rosenberg, M.M. 1941. A study of the inheritance of resistance to *Eimeria tenella* in the domestic fowl. Poult. Sci. 20 : 472.

Rosenberg, M.M., J.E. Alicata, and A.L. Palafox. 1954. Further evidence of hereditary resistance and susceptibility to cecal coccidiosis in chickens. Poult. Sci. 33 : 972-980.

Schildt, C.D. and C.A. Herrick. 1955. The effect of cecal coccidiosis on the mobility of the digestive tract of the domestic fowl. J. Parasitol. 41 (No. 6, Sect. 2) : 18.

Sharma, N.N. and J.W. Foster. 1964. Toxic substance in various constituents of *Eimeria tenella* oocysts. Am. J. Vet. Res. 25 : 211-216.

Sharma, N.N. and W.M. Reid. 1962. Successful infection of chickens after parenteral inoculation of oocysts of *Eimeria* spp. J. Parasitol. 48 (Suppl.) : 33.

Sheehy, T.W., W.H. Meroney, R.S. Cox, and J.E. Soler. 1962. Hookworm disease and malabsorption. Gastroenterology, 42 : 148-156.

Shumard, R.F. 1957. Studies on ovine coccidiosis. 1. Some physiological changes taking place in experimental infections with *Eimeria ninae - kohl - yakimovi* (Yakimov and Rastegaeva, 1930) and *Eimeria faurei* (Moussu and Marotel, 1901). J. Parasitol. 43 : 548-554.

Siegmann, O. 1960. Die Beeinflussung des Futter- und Trinkwasserverbrauchs wachsender Küken durch Coccidieninfektionen. Arch. Geflügelk. 24 : 442-450.

Smetana, H. 1933. Coccidiosis in the liver of rabbits. III. Experimental study of the histogenesis of coccidiosis of the liver. Arch. Pathol. 15 : 516-536.

Smith, W.N., and L.R. Davis. 1965. A comparison of dry and liquid feed as vehicles for coccidial infection of cattle and sheep. Am. J. Vet. Res. 26 : 273-279.

Smith, W.N., L.R. Davis, and G.W. Bowman. 1960. The pathogenicity of *Eimeria ah - sa - ta*, a coccidium of sheep. J. Protozool. 7 (Supp.) : 8.

Smith, T. and H.W. Graybill. 1918. Coccidiosis in young calves. J. Exp. Med. 28 : 89-108.

Stafseth, H.J. 1931. Fowl paralysis and the "roup complex". J. Amer. Vet. Med. Ass. 78 : 423-429.

Stephens, J.F. 1965. Some physiological effects of coccidiosis caused by *Eimeria necatrix* in the chicken. J. Parasitol. 51 : 331-335.

Stephens, J.F. and B.D. Barnett. 1964. Concurrent *Salmonella typhimurium* and *Eimeria necatrix* infections in chicks. Poult. Sci. 45 : 353-356.

Stephens, J.F., L.M. Kowalski, and W.J. Borst. 1967. Some physiological effects of coccidiosis caused by *Eimeria maxima* in young chickens. J. Parasitol. 53 : 176-179.

Svanbaev, S.K. 1967. Patogennost, razlichnykh vidov koktsidiĭ ovets. Trudy Institute Zoology Akademie Nauk Kazakh. SSR 28 : 57-61.

Svanbaev, S.K. and Z.I. Gorbunova. 1969. Coccidiosis in lambs : biochemical and morphological changes in the blood. Vet. Bull. 39 : 331.

Swanson, L.E. and K.C. Kates. 1940. Coccidiosis in a litter of pigs. Proc. Helm. Soc. Wash. 7 : 29-30.

Symons, L.E.A. and D. Fairbairn. 1962. Pathology, absorption, transport, and activity of digestive enzymes in rat jejunum parasitized by the nematode, *Nippostrongylus brasiliensis*. Fed. Proc. 21 : 913-918.

Turk, D.E. and J.F. Stephens. 1967a. *Eimeria necatrix* infections and oleic acid absorption in broilers. Poult. Sci. 46 : 775-777.

Turk, D.E. and J.F. Stephens. 1967b. Upper intestinal tract infection produced by *E. acervulina* and absorption of Z^{65} and ^{131} I-labelled oleic acid. J. Nutrition. 93 : 161-165.

Turk, D.E. and J.F. Stephens. 1970. Effects of serial inoculations with *Eimeria acervulina* or *Eimeria necatrix* upon zinc and oleic acid absorption in chicks. Poult. Sci. 49 : 523-526.

Tyzzer, E.E. 1929. Coccidiosis in gallinaceous birds. Am. J. Hyg. 10 : 269-383.

Tyzzer, E.E., H. Theiler, and E.E. Jones. 1932. Coccidiosis in gallinaceous birds. II. A comparative study of species of *Eimeria* of the chicken. Am. J. Hyg. 15 : 319-393.

Van Doorninck, W.M. and E.R. Becker. 1957. Transport of sporozoites of *Eimeria necatrix* in macrophages. J. Parasitol. 43 : 40-44.

Versényi, L. and L. Pellérdy. 1970. Pathological and immunological investigations of the anseris – coccidiosis of the domestic goose (*Anser anser dom*). Acta. Vet. Acad. Sci. Hung. 20 : 103-107.

Vetterling, J.M. and D.J. Doran. 1969. Storage polysaccharide in coccidial sporozoites after excystation and penetration of cells. J. Protozool. 16 : 772-775.

Wacha, R.S., D.M. Hammond, and M.L. Miner. 1971. The development of the endogenous stages of *Eimeria ninakohlyakimovae* (Yakimoff and Rastegaieff, 1930) in domestic sheep. Proc. Helm. Soc. Wash. 38 : 167-180.

Wagenbach, G.E. 1969. Purification of *Eimeria tenella* sporozoites with glass bead columns. J. Parasitol. 55 : 833-838.

Warren, E.W., S.J. Ball, and J.R. Fagg. 1963. Age resistance by turkeys to *Eimeria meleagrimitis* Tyzzer, 1929. Nature 200 : 238-240.

Waxler, S.H. 1941. Changes occuring in the blood and tissue of chickens during coccidiosis and artificial hemorrhage. Am. J. Physiol. 134 : 19-26.

Weinman, D. and H.J. Klatchko. 1949. Description of a toxin in toxoplasmosis. Yale J. Biol. Med. 22 : 323-326.

Whur, P. 1967. Globule leucocyte response in hyperimmune rats infected with *Nippostrongylus brasiliensis*. J. Comp. Pathol. 77 : 271-277.

8 Immunity

M. ELAINE ROSE

Houghton Poultry Research Station, Houghton, Huntingdon, England

---

Contents

I. Introduction. . . *296*
II. Innate Immunity. . . *296*
  A. Host specificity. . . *296*
  B. Age of Host. . . *297*
  C. Genetic Constitution. . . *297*
III. Acquired Immunity. . . *297*
  A. Immunogenicity of Different Species. . . *297*
  B. Mode of Immunization. . . *298*
  C. Host Factors. . . *299*
  D. Duration of Acquired Immunity. . . *299*
  E. Species Specificity of Immunity. . . *302*
  F. Stage Stimulating Immunity and Stage Specificity of the Response. . . *302*
  G. Stage Affected by the Immune Response. . . *304*
IV. Antibody Responses. . . *306*
  A. Coccidial Antigens. . . *306*
  B. Reactions with Particulate Antigens. . . *307*
  C. Reactions with Soluble Antigens. . . *307*
  D. Opsonizing Antibodies. . . *312*
  E. Naturally Occurring Antibodies. . . *312*
  F. Autoantibodies. . . *313*
  G. Enhancing Antibodies. . . *313*
  H. Locally Produced Antibodies. . . *313*
V. Protective Effects of Coccidial Antibodies. . . *315*
  A. Passive Transfer of Immunity. . . *315*
  B. Maternal Transfer of Immunity. . . *317*
  C. Possible Mode of Action of Protective Antibodies. . . *318*

VI. Cellular Responses. . . *319*
A. Role of the Macrophage. . . *320*
B. Adoptive Transfer of Immunity with Cells. . . *321*
VII. Interferon. . . *322*
VIII. Experimental Modifications of The Immune Response. . . *322*
A. Splenectomy. . . *323*
B. Treatment with Corticosteroids. . . *323*
C. Treatment with Ionizing Radiation. . . *325*
D. Bursectomy and Thymectomy in Avian Hosts. . . *325*
E. Treatment with Antilymphocytic Serum. . . *326*
F. Tolerance. . . *327*
IX. Active Immunization. . . *330*
A. Immunization by Infection . . . *330*
B. Immunization with Nonviable Material . . . *332*
X. Concluding Remarks. . . *333*
Literature Cited. . . *333*

I. Introduction

The aim of this review is to discuss the more recent work on immunity to the coccidia, to speculate on the nature of the mechanisms involved, and to indicate possible future lines of approach to the subject. An historical review of the literature is not included and much of the earlier work is not discussed; these may be found in previous reviews (Augustin and Ridges, 1963; Horton-Smith *et al.*, 1963a). Other reviews include those by Horton-Smith, 1963; Horton-Smith and Long, 1963; and Rose, 1963, 1968a, and 1971a.

Published observations on immunity to the coccidia are practically confined to those made on species of the genus *Eimeria* and the majority of these relate to species parasitic in the domesticated animals – poultry and the larger farm mammals, with the greater emphasis on poultry. This is because of the economic importance of coccidiosis in these animals. However, the difficulties involved in maintaining adequate numbers of the larger species free from extraneous infection, of adequate experimentation on this scale, and the lack of precise knowledge (available, for instance, for the laboratory mouse or rat) of their immune responses have contributed to the imperfect state of our understanding of immunity to the coccidia.

II. Innate Immunity

Hosts show varying degrees of innate immunity to the coccidia ranging from complete species immunity to "foreign" coccidia (host specificity) to the partial resistance caused by genetic constitution and age.

A. HOST SPECIFICITY

The coccidia, in common with other members of the Sporozoa, exhibit a remarkable degree of host specificity, particularly in the genus *Eimeria* (see Pellérdy, 1965 and Chapter 2). It has been generally assumed that this specificity is the result of physiological factors but it is possible that host immune responses may also be involved (see Section VIII, B).

B. AGE OF HOST

Earlier workers, noting the resistance of older animals to experimental infection with the coccidia, assumed that this was caused by "age resistance." It is now known that older animals provide a more favorable environment for parasite multiplication and that the "age resistance" mentioned above is probably attributable to acquired immunity since keeping animals coccidia-free for any length of time is extremely difficult.

Very young animals are more resistant to infection than older ones. Kotlán and Pellérdy (1949) and Rose (1959a) found that very young suckling rabbits were resistant to infection with *E. stiedai*, but this resistance could be overcome by dosing with very large numbers of oocysts (Dürr and Pellérdy, 1969). This relative insusceptibility of the very young applies to other mammalian host species, e.g., rats, mice, and dogs (see Schrecke, 1969; Schrecke and Dürr, 1970) and also to the fowl (Krassner, 1963; Rose, 1967a). The comparative inefficiency of the excystation mechanism in the very young is partly responsible, but the trend continues after the excystation rate ceases to change, and, therefore, other host factors, possibly biochemical or physiological, must be responsible. In mammals, the deficiency of *para*-aminobenzoic acid consequent to a milk diet may affect the growth of the parasite adversely, as shown for plasmodia (see Hawking, 1955). The influence of age of host on pathogenesis is also discussed in Chapter 7.

C. GENETIC CONSTITUTION

Breeds of chickens and lines within breeds differ in susceptibility to coccidial infection (Rosenberg, 1941; Rosenberg, Alicata, and Palafox, 1954; Champion, 1954; Jeffers, Challey, and McGibbon, 1967; Long, 1968; Johnson and Edgar, 1969) which seems to be controlled by multiple genetic factors which are not sex linked. The mechanism underlying these differences is not known and the subject is further discussed in Chapter 7.

III. Acquired Immunity

The acquisition of immunity to the coccidia as a result of active infection is well documented. An historical review of the literature has been given by Horton-Smith *et al.* (1963a).

A. IMMUNOGENICITY OF DIFFERENT SPECIES

Species of coccidia differ in the ease with which they elicit immunity, within the same host species and in different hosts. It is, however, very difficult to compare the innate immunogenicity of different species, especially when experiments have been conducted by different authors, since host factors and mode of immunization are also important.

There are few published data available on comparisons of different species within the same experimental design. Rose and Long (1962) investigated the immunizing abilities of four different species in the fowl and used the reproductive indices after successive infections as a basis for comparison. They placed the test species in the following order of descending immunogenicity: *E. maxima*, *E. acervulina*, *E. tenella*, and *E. necatrix*. The immunizing schedules for three species were identical and the comparisons for these are therefore valid but the doses of oocysts of *E. acervulina* were tenfold greater. Subsequent observations have however confirmed the general conclusions, especially in regard to *E. maxima*, of which a single infection may immunize completely to challenge given within from three to six weeks. Rommel (1970a) stated that single inoculations with 200 oocysts of *E. scabra* or 20,000 oocysts of *E. polita* produced complete immunity in pigs, presumably indicating that *E. scabra* is more immunogenic than *E. polita*. Smith and Davis (1961), comparing several species of *Eimeria* of sheep, concluded that infection with *E. ahsata* produced a stronger resistance (which was species specific) to reinfection than the other species examined, but few experimental details were given.

Among the *Eimeria* of cattle, immunity to *E. alabamensis, E. bovis*, and *E. zürnii* has been examined but, unfortunately, by different authors and in different ways. Immunity to *E. zürnii* induced by three infections was found to last at least five months (Wilson and Morley, 1933) and *E. bovis* is also fairly immunogenic (Senger *et al.*, 1959) but four infections with *E. alabamensis* are said to be necessary to stimulate immunity (Davis, Boughton, and Bowman, 1955). However, the methods used by these workers are not comparable and therefore direct comparisons cannot be made.

There may be differences in immunogenicity among the species parasitic in the rabbit. A single inoculum with moderate numbers of *E. stiedai* leads to substantial immunity (Rose, 1959a) and *E. magna* also is fairly immunogenic (Niilo, 1967) whereas *E. intestinalis* may be less so; a single inoculum with 10,000 oocysts of this species failed to protect against a challenge infection but larger doses did so, and immunity became complete after three infections (Beyer, 1963).

In most species, an immune response occurs, but there have been reports of only a very slight or no immunity; for example, those of Todd and Hammond (1968) for *E. callospermophili* in the ground squirrel; Ernst, Hammond, and Chobotar (1968) for *E. utahensis* in kangaroo rats; and Versényi and Pellérdy (1970) for *E. anseris* infection of the goose.

B. MODE OF IMMUNIZATION

Again, few data are available. Senger *et al.* (1959) found that a single inoculum of 10,000 oocysts of *E. bovis* resulted in less immunity to challenge than doses of 50,000 or 100,000 between which there was little difference except in the greater severity of the immunizing infection. Equal numbers of oocysts given as single inocula or on five successive days yielded similar results. Fitzgerald (1967) compared the immunizing effects of consecutive daily dosing for 47 to 62 days with 10, 100, 500, 1,000, 5,000, or 15,000 oocysts of *E. bovis*; challenge was with 300,000 or 500,000 oocysts given as a single inoculum. All the animals resisted the challenge infection (judged by clinical signs) but the calves given the two lower dose rates were somewhat less immune than the others.

Babcock and Dickinson (1954), working with *E. tenella*, reported that with a standard total of 1,600 oocysts, variations in the number of individual doses given did not influence immunity. A substantial amount of the experimentation on immunity to *E. tenella* has followed the immunization schedule of Pierce, Long, and Horton-Smith (1962); i.e., three doses comprising 500, 5,000, and 50,000 oocysts given at two-week intervals to fairly young birds. This system has proved fairly satisfactory for short-term experiments. Hein (1968), working with *E. acervulina*, obtained a satisfactory immune response to challenge at six weeks of age by giving inoculations of 80,000 and 160,000 oocysts at two and four weeks of age, respectively. A fourfold increase in the numbers of oocysts given did not immunize the birds so effectively and possibly caused a permanent impairment in intestinal function.

Clearly, satisfactory immunity can be achieved by giving moderate numbers of oocysts, especially with some species, and it is not necessary to give doses so high as to cause severe disease. However, the survivors of single large infections with *E. tenella* are immune to short-term challenge (see Horton-Smith, Beattie, and Long, 1961). Dosing with excessive numbers of oocysts may result in over-crowding with consequent reduction of oocyst yield. Comparative investigations of different immunizing schedules using well-defined criteria (i.e., resistance to disease or to parasitism) for assessing immunity should be profitable.

No definitive information on optimum time intervals between immunizing doses exists. They probably differ for different host species and may even differ for the different species of coccidia, possibly depending upon the length of the life cycle. The time interval between antigenic stimuli necessary for the induction of an anamnestic response in any given host, if known, could be taken as a guideline. However, this does not necessarily apply to an immune response to coccidial infection.

C. HOST FACTORS

Apart from host factors responsible for innate immunity, there are others which may influence immunological responses.

1. *Age*

Immunological responses mature with age; thus older animals should respond more readily than younger ones. In addition, parasite multiplication is usually greater for moderate doses in the more mature host, thus increasing the antigenic stimulus.

2. *Intercurrent infections*

Outbreaks of coccidiosis have often been associated with other disease conditions. This is especially true of Marek's disease in chickens, which has now been shown experimentally to result in increased oocyst production in primary and subsequent infections (Biggs *et al.*, 1968; Randall, Grant, and Sutherland, 1971).

Infection with Marek's disease virus impairs the immune response to a variety of antigens (Purchase, Chubb, and Biggs, 1968), and this may be the reason for the increased susceptibility to coccidial infections.

3. *Experimental procedures*

Experimental procedures which affect the immunity of the host are described in Section VIII.

D. DURATION OF ACQUIRED IMMUNITY

Precise information on the duration of immunity is not easily obtainable. Data are meaningful only if strict precautions have been taken to avoid accidental infection throughout the period of observation, a difficult task in view of the ubiquitous nature of the coccidia, and particularly so when the large animal hosts are under consideration.

Different investigators have used different criteria for the immune state; in some cases this has been assessed on clinical grounds, in others on oocyst production. Variations in immunization procedures will also influence the results. Direct comparisons of results are, therefore, impossible. Publications relating to the duration of immunity are summarized in Table 1. In none of these is it explicitly stated that precautions against reinfection were taken after the immunization period was complete. However, the state of immunity achieved after this procedure was probably such that very few oocysts were passed by the test animals, thus reducing the chances of reinfection, unless oocysts gained access from extraneous sources.

These reservations only serve to strengthen the conclusion to be drawn from Table 1, i.e.,

TABLE 1. Summary of findings in experiments on the duration of immunity

Species	Mode of immunization	Challenge inoculations
E. tenella	Single inoculum of 60,000 oocysts given to 2-week-old chicks	120,000 oocysts given at 14, 21, or 28 days after initial inoculum
E. maxima	Single inoculum of 5,000 oocysts at 1 week of age	100,000 oocysts 3 weeks after initial inoculum
	As above + 2nd inoculum of 100,000 at 4 weeks of age	200,000 oocysts 10 weeks after 2nd inoculum
	As above + 3rd inoculum of 200,000 oocysts at 14 weeks of age	1,200,000 oocysts 26 weeks after 3rd inoculum
E. bovis	Single inocula of 10,000, 50,000 or 100,000 oocysts at 8-12 days	500,000 oocysts given 2 weeks after immunizing inoculum
	50,000 or 100,000 oocysts given as single inocula or as 5 equal doses on consecutive days at 2 weeks of age	1,000,000 oocysts given 28 days after 1st inoculum
	As above + 1,000,000 oocysts at 6 weeks of age	3,000,000 oocysts given 14 weeks after 2nd inoculum
	Single inoculum of 50,000 oocysts at 2 weeks of age	1,000,000 oocysts given 35 days after 1st inoculum
	As above + 1,000,000 oocysts at 7 weeks of age	3,000,000 oocysts given 57 days after 2nd inoculum
E. stiedai	Initial inoculum of 2,500-5,000 oocysts	500,000-1,000,000 oocysts given at 80, 115, or 290 days after immunizing inoculum
E. tenella	500 oocysts at 1 week of age 5,000 oocysts at 4 weeks of age 50,000 oocysts at 7 weeks of age 100,000 oocysts at 10 weeks of age	10^6 oocysts given at 21, 42, 63, 84 or 105 days after last immunizing inoculum
E. scabra *E. polita*	Three consecutive inocula of 20,000 oocysts at intervals of 14 days	20,000 oocysts at 1, 2, 3, 4, or 5 months after last immunizing inoculum

Assessment of immunity	Findings	Reference
Macroscopical and microscopical examination of ceca 6 days after challenge inoculum	Slight parasitism at 14 days Moderate parasitism at 21 days Fairly severe parasitism at 28 days	Horton-Smith, Beattie, and Long, 1961
Oocyst production	Resistant; oocyst production of immunized : oocyst production of control = 1 : 55	Long, 1962
Oocyst production, clinical signs	High oocyst yield (R.I. = 347) Clinical signs of coccidiosis	
Oocyst production, clinical signs	Low yield of oocysts (R.I. = 11) Clinical signs of coccidiosis	
Oocyst production, clinical signs	Moderate production and mild clinical signs	Senger *et al.*, 1959
Oocyst production, clinical signs	Moderate oocyst production, mild clinical signs	
Oocyst production, clinical signs	Variable, but infections less severe than in controls	
Oocyst production, clinical signs	Variable, but less severe than in controls	
Oocyst production, clinical signs	Mild infections in immunized, severe in controls	
Clinical signs, tests for liver function, oocyst production	No clinical signs, normal liver function. Very few, if any, oocysts in the feces	Rose, 1961
Oocyst production, histological examination and mortality	By mortality, immunity still effective at 105 days. By oocyst production, complete immunity never achieved but oocyst production negligible at 1st challenge, gradually increasing with each succeeding one. Histological examination revealed an increasing number and further development of stages as time interval to challenge increased.	Leathem and Burns, 1968
Oocyst production	Immunity began to lapse after 4 months in the case of *E. scabra* and 3 months in the case of *E. polita*.	Rommel, 1970a

that immunity wanes with time. For any given species, the duration of immunity is probably dependent on the method of immunization used, age of host, etc. For example, chicks given a single inoculum of *E. tenella* at two weeks of age were fairly susceptible to challenge four weeks later (Horton-Smith *et al.*, 1961), whereas those given four immunizing inocula with this species, the first at one week and the fourth at ten weeks of age, were fairly resistant to challenge 15 weeks after the last immunizing inoculum (Leathem and Burns, 1968).

Comparisons between the results of different experimenters and between species are impossible to make because of the variations mentioned and the design of some of the long-term experiments. Only in the work of Leathem and Burns was response to challenge at different time intervals after a single immunizing schedule carefully examined over a long period of time.

E. SPECIES SPECIFICITY OF IMMUNITY

The general consensus does indicate that immunity to the *Eimeria* is species specific, although there may be some doubt about it with species which appear to be very closely related and where oocyst production methods have been used to assess infections.

Tyzzer, Theiler, and Jones (1932) and Johnson (1938) found that immunity to *E. tenella* and *E. necatrix* was species specific.

Species specificity of immunity to the *Eimeria* of other host species has also been described: between *E. perforans* and *E. stiedai* in the rabbit (Bachman, 1930a), *E. myairii* and *E. separata* in the rat (Becker and Hall, 1933), *E. meleagridis* and *E. adenoeides* in the turkey (Clarkson, 1959). The principle has been used to establish new species (Levine, 1938 and 1942; Moore and Brown, 1951) and also to determine the prevalence of various species in the field (Reid, 1964).

Most of these studies were made using the presence of oocysts in the feces, rate of weight gain, or other clinical signs as criteria of cross-infection, although Becker and Hall (1933) did measure oocyst production.

Rose and Long (1962) found that *E. tenella* infections (judged by lesion scoring on the fifth day) in birds which had been immunized with *E. necatrix* were not as severe as might have been expected from the inoculum given but, in the reverse procedure, very severe intestinal infections with *E. necatrix* were produced in birds which had been immunized with *E. tenella*.

Some cross-protection, judged by oocyst output, between *E. necatrix* and *E. tenella*, particularly when the entire life cycle of *E. necatrix* occurred in the ceca, and between *E. maxima* and *E. brunetti*, was described by Rose (1967b,c), but Hein (1971) has produced convincing evidence of a complete lack of any cross-immunity between *E. maxima* and *E. brunetti*. Lee (1967) found partial cross-immunity between *E. mitis* and *E. maxima* and Rommel (1970a) obtained similar results for infections with *E. scabra* and *E. polita* in pigs. However, Joyner (1969) has found that there may even be immunogenic variation between two strains of the same species (*E. acervulina*).

In general, immunity to the *Eimeria* is remarkably species specific, and it is probable that any cross-immunity which might exist is confined to very closely related species and is only partial.

When conducting experiments on cross-immunity, very great care must be taken to maintain absolute purity of the cultures used, as well as complete isolation of the infections.

F. STAGE STIMULATING IMMUNITY AND STAGE SPECIFICITY OF THE RESPONSE

The complexity of the coccidial life cycle stimulates speculation concerning the roles of the different stages in the induction of immunity and in the stage specificity of the immune response. These problems can be properly investigated only in the following circumstances:

a. in species in which there is little overlap between the different stages in the life cycle so that each stage may be recovered in isolation from the others. In this way infections in which gametogony stages predominate may be obtained by recovering merozoites from the last schizont generation and inoculating them into susceptible hosts.
b. where a coccidiostat is available which can be administered so as to terminate an infection at the desired stage in the life cycle. This may enable the immunizing effects of schizogony stages only to be assessed. Little overlap in the timing of developmental stages is a necessity in these experiments also.

Even when the above criteria are fulfilled, it is still very difficult to compare the immunizing abilities of, for example, the gametogony stages with the entire life cycle of any given species, since it will be almost impossible to ensure that the same amount of parasite material is offered to the host in each case. However, some indication of relative immunogenicity may be obtained, especially if a combination of methods is used.

Some of the species parasitizing the domestic fowl, and *E. bovis* of cattle, are the only ones in which this type of work has been attempted.

Horton-Smith (1947, 1949), in experiments on the chemotherapy of coccidiosis, found that chicks treated with sulfonamide coccidiostats were immune to subsequent challenges with *E. tenella* (after withdrawal of drug) provided that the treatment had been such as to allow the original infection to proceed to the second-generation schizont stage. Ideally this stage should be reached but segmentation of the schizont with merozoite formation should be avoided to minimize the pathogenic effect.

This work was confirmed by Kendall and McCullough (1952) who also showed that chicks challenged with large numbers of *E. tenella* oocysts at 96 hr after an initial inoculum of 2,000 oocysts were fairly resistant; there was less, but still some, resistance to challenge at 72 hr and none to earlier challenges. These results imply that the partial development of the second-generation schizont is responsible for immunity. However, it must be borne in mind that an effective immune response is not instantaneous and enough time must elapse between stimulation and challenge for the immune response to develop.

Horton-Smith *et al.* (1963a) inoculated suspensions of second-generation merozoites of *E. tenella* or *E. necatrix* intrarectally. Oocyst production of *E. tenella* declined after each inoculum until none was passed. There was no diminution in the numbers of oocysts of *E. necatrix* after succeeding inocula. In neither species was there immunity to later oral challenge with oocysts. Similar experiments were carried out in more detail by Rose (1967b), who confirmed inability to immunize against gametogony stages of *E. necatrix* by intracecal inoculation of second-generation merozoites but found lowered oocyst production resulting from oral challenge with oocysts, i.e., entire life cycle challenge. This suggests that the gametogony stages of these two species have little immunizing ability, even against the homologous stage in *E. necatrix*. In the case of *E. necatrix*, though, a combination of several immunizing gametocyte infections with the schizogony stages of the challenge infection caused a reduction in oocyst output.

There are no published observations on similar experiments with other avian species of *Eimeria*. Some preliminary work in this laboratory with *E. maxima* has indicated that the sexual stages of this species may be immunogenic; this would be in accord with the relatively large size of the gametocytes of this species but further investigation is necessary.

Hammond, Andersen, and Miner (1964) inoculated first-generation merozoites of *E. bovis* into the ceca of calves and thereby successfully immunized them against subsequent oral challenge with oocysts, thus showing that development of the first-generation schizont is not essential for the induction of immunity.

The results of preliminary work with *E. maxima* implicate gametocyte stages, but it must be remembered that this species is very immunogenic and a very light contamination with other stages may well influence the outcome of challenge inoculations.

Because of the limited number of species investigated it is difficult to draw any general conclusions as to the relative values of different life cycle stages in generating an immune response.

G. STAGE AFFECTED BY THE IMMUNE RESPONSE

In partially immune hosts, reduced numbers of oocysts are produced and most authors agree that those parasites which complete their life cycles are normal in every way and the length of the prepatent period is unaffected; the patent period, however, is usually shortened. Extensions of the length of the prepatent period in semi-immune hosts have been reported by Biester and Schwarte (1932), Becker, Hall, and Hager (1932) and, more recently, by Rommel (1969), who found the prepatent period of *E. polita* and *E. scabra* in partly immune pigs to be one to two days longer than in the susceptible hosts, but found no differences in the time of peak oocyst production. Rommel (1970b) also reported that the oocysts which had been formed in partly immune pigs differed in morphology from and did not sporulate as well as those formed in control hosts. Similar accounts of reduced sporulation have been given by Henry (1931) and Niilo (1967), whereas Beyer (1963), working with *E. intestinalis* of the rabbit, found that sporulation was normal but that the oocysts lost infectivity more rapidly than those formed in susceptible hosts. Differences in morphology and sporulation have been noted in oocysts passed in infections resulting from very high doses (Jones, 1932; Kheysin, 1947; Pout, 1965; Rommel, 1970c), but whether this could be a reflection of the effect of an immune response becoming active towards the end of an initial infection is not known. Other factors such as the "overcrowding" phenomenon must also be considered.

In thoroughly immune hosts a challenge infection does not become patent, and it is thought that the parasite is halted at a very early stage. Excystation of sporozoites occurs normally and at least a proportion of the sporozoites invades the epithelial cells of the host. There is some evidence to suggest that entry into the host cells may be blocked; Morehouse (1938) and Edgar (1944 – quoted by Leathem and Burns, 1967) considered this blockage to be complete in rats immune to *E. nieschulzi* and in chickens immune to *E. tenella* respectively. A partial inhibition of sporozoite or merozoite invasion has been reported for different species (Augustin and Ridges, 1963; Hammond *et al.*, 1964; Leathem and Burns, 1967). The manner in which this inhibition might be effected is not known; the invasiveness and viability of sporozoites of *E. tenella* recovered from the lumina of immune ceca are unimpaired (Horton-Smith *et al.*, 1963b; Leathem and Burns, 1967) and extracts of mucosa or of the intestinal contents of immune animals have not been shown to affect sporozoites adversely (Augustin and Ridges, 1963; Herlich, 1965), but such preparations are very labile, because of the presence of proteolytic enzymes. (For further discussion see Sections IV, G and V, C.)

Most workers have stated that, in completely immune animals, the sporozoites which have penetrated do not undergo further development. Tyzzer, Theiler, and Jones (1932) found that at 24 hr after oocyst inoculation the sporozoites of *E. necatrix* in the cells of immunized chickens were shrunken, with little evidence of structure apart from the refractile globule, and there were no further developmental stages. Horton-Smith *et al.* (1963b) reported similar findings in infections with *E. tenella*, as did Leathem and Burns (1967), also working with *E. tenella*, but these authors did find some abnormal schizonts at 72 hr, and 96 hr. In a later publication (1968) they reported finding that in "relatively immune" birds (clinically completely immune) the predominant stages were sporozoites and trophozoites, but there were also second-generation schizonts at 96 hr, especially in groups in which immunity was waning (challenged 12 weeks after the last immunizing inoculum). These were in pockets between the submucosa and fundi of the cecal glands. In addition, these authors recovered parasites from the tissues of immunized birds and tested their infectivity for susceptible animals. At 12 hr and 24 hr after oocyst inoculation the transferred parasites produced infections in susceptible hosts but at 48 hr they were not able to do so (Leathem and Burns, 1967).

Suppression at the sporozoite stage has also been described in *E. stiedai*-immune rabbits; sporozoites are found in the epithelial cells of the duodenum but no developmental stages occur in the bile duct epithelium (Rose, 1959a, quoted in Horton-Smith *et al.*, 1963a) and the development of *E. magna* in immunized rabbits is also halted at a very early stage (Niilo, 1967).

Early experiments on the fate of *E. bovis* in immunized calves (Hammond *et al.*, 1959) indicated that the early asexual stages were not affected by the immune reaction which was, instead, directed against the sexual stages. Later work (Hammond, Anderson, and Miner, 1963), however, showed that both asexual and sexual stages were affected. An effect on the sexual stages of other species of *Eimeria* has also been described; Roudabush (1935) was not able to induce infections by the oral inoculation of merozoites of *E. nieschulzi* into immunized rats and Rose (1967b) found no oocyst output resulting from the intracecal inoculation of second generation merozoites of *E. tenella* into immunized chickens. Horton-Smith *et al.* (1963b) examined histologically the sequence of events following this procedure; merozoites invaded the cecal tissue but did not develop into gametocytes.

The immune response obviously affects all stages and its effects in any one host will depend upon the immune status of that host. In the very thoroughly immune animal it is possible that a considerable proportion of sporozoites may not even be able to invade the epithelial cells; in the not so immune, sporozoites may invade but the greater part of them will not be able to grow; thus there will be little or no schizogony. In hosts with lower degrees of immunity, some schizogony with or without gametogony may occur; in partially immunized hosts, oocysts may be produced. It is obvious that, as the cycle proceeds, albeit on a smaller scale, in the partially immunized host, it acts as a "boosting" antigenic stimulus and will elicit a response from the host which affects the terminal stages of its own life cycle, hence the shorter patent period of a challenge infection, almost invariably found in the most lightly immunized host.

It is tempting to speculate that the life cycles of even initial infections, especially of the more immunogenic species, are affected to some degree by an immune host reaction. Long and Rose (1970) have shown that the life cycle of *E. mivati* in chickens whose immune reactions have been suppressed by a corticosteroid is considerably prolonged; schizogony was seen up to the 24th day after oocyst inoculation and oocyst production continued up to the 50th day (8th and 18th days, respectively, in the controls). Corticosteroid injection was discontinued on the 15th day (see Section VIII, B).

Yet another stage in immunity can be considered in which the life cycle of the parasite may be completely unaffected but its pathogenic effects on the host may be moderated to such an extent that no clinical signs may be elicited, even by the inoculation of large numbers of oocysts. This has been suggested by Beyer (1963) in *E. intestinalis* infections in rabbits and also by Hammond *et al.* (1963); Beyer described this as an "antitoxic" immunity, but there has not yet been an unequivocal demonstration of the production by the *Eimeria* of a substance toxic to the specific host.

IV. Antibody Responses

Early attempts to demonstrate humoral antibodies in coccidial infections gave rather inconclusive results (see reviews by Horton-Smith *et al.*, 1963a, and Augustin and Ridges, 1963), but agglutinating antibodies for second-generation merozoites of *E. tenella* were reported in the serum of infected chickens by McDermott and Stauber in 1954. Since then the presence of serum antibodies has been unequivocally demonstrated by agglutinating, lytic, precipitation, and complement-fixation tests and by indirect fluorescent staining. (Heist and Moore, 1959; Rose, 1959b, 1961, 1963; Pierce, Long, and Horton-Smith, 1962, 1963; Rose and Long, 1962; Augustin and Ridges, 1963; Long, Rose, and Pierce, 1963; Andersen *et al.*, 1965; Burns and Challey, 1965; Herlich, 1965; Černá, 1966a,b, 1967, 1969, 1970).

A. COCCIDIAL ANTIGENS

Particulate and soluble antigens have been used for the detection of antibodies in coccidial infections.

The particulate antigens consist of suspensions of the free, invasive stages, i.e., sporozoites or merozoites and, depending upon the conditions in which the test is performed, antibody may be demonstrated by agglutination or lysis of the test organisms; by a reduction in infectivity after incubation in antiserum (neutralization), and by the attachment of antibody to the surface of the antigen, shown by indirect fluorescent labeling. Combination of antibody with these antigens in tissue sections may also be shown by fluorescent labeling methods.

Soluble antigens have been prepared from various stages by extraction of the free parasites or infected tissues, and used to precipitate antibody (in liquid and in agar gel) and, in combination with antibody, to take up complement.

A systematic analysis of the antigens has not been carried out but they are numerous and probably differ at different stages of the life cycle. Rose (1959b) examined some soluble antigens of *E. stiedai* consisting of bile duct exudates taken at different times after infection. At least six different antigens were present, the greatest number being obtained with exudate taken from 17 to 19 days after infection, a time when both schizogony and gametogony stages are found in the bile duct.

The antigens of different species have not been compared extensively but there are indications of shared antigens (Rose, 1959a; Herlich, 1965; Pierce *et al.*, 1962; Rose and Long, 1962) especially among closely related species.

There has been conjecture that the "functional" antigens are metabolic products of the parasite or enzymes, from evidence of a reduction in numbers of sporozoites which penetrate the intestine of immunized hosts (Augustin and Ridges, 1963; Rose, 1971a).

It has also been suggested that the parasites possess antigens with determinants similar to those of the host tissues (Augustin and Ridges, 1963), in which case they may be able to evade, to a certain extent, the response of the host. Augustin and Ridges (1963) were however unable to detect any reactions with antigens prepared from *E. meleagrimitis* in the serum of a rabbit immunized with normal turkey serum. Rose and Long (unpublished observations) have found that rabbits immunized with a highly purified preparation of *E. tenella* sporozoites form antibodies which precipitate with normal fowl serum, indicating that the advantageous situation of shared host and parasite antigens may well exist in coccidiosis.

The assumption of a "host disguise" by some helminth parasites is now well established (Damian, 1964, 1967; Capron *et al.*, 1968; Smithers, Terry, and Hockley, 1969).

B. REACTIONS WITH PARTICULATE ANTIGENS

Agglutination of sporozoites or merozoites can be satisfactorily demonstrated only when high concentrations of clean organisms are available; in low concentrations, contact between organisms is reduced. Suspensions of sporozoites excysted by *in vitro* methods necessitating the use of fresh bile tend to agglutinate spontaneously, possibly because of a coating of bile salts. The presence of debris in the suspensions is a further complication.

Pure suspensions of first-generation merozoites of *E. bovis* are fairly readily obtainable from the macroscopic schizonts, and Andersen *et al.* (1965) have studied agglutinin titers in the sera of infected calves (see Table 2).

Lytic reactions may be observed with lower concentrations of organisms in the presence of complement, but the titers are generally fairly low.

Where frank agglutination or lysis does not occur there may be morphological changes and usually a loss in infectivity. A summary of some of the more recently published work is given in Table 2.

Animals which have been infected produce circulating antibodies which, depending on the conditions in which the test is carried out, agglutinate or lyse the invasive stages of the parasite. Less severe effects such as slight changes in morphology or immobilization may be observed and reduced infectivity is a constant feature. These antibodies therefore have definite coccidiocidal or coccidiostatic properties and may be termed neutralizing antibodies. Some are clearly directed against surface antigens of the organisms as shown by agglutination, lysis, and the indirect fluorescent test. Others may be directed against metabolic products or the cell contents (Augustin and Ridges, 1963); it is also possible that some of the antibodies may be directed against some of the vital processes of the cells, e.g., enzyme systems, resulting in reduced infectivity.

The adverse effects of incubation in high concentrations of normal serum, particularly in the case of sporozoites noted by some authors, are probably the result of the presence of naturally occurring antibodies (see Section IV, D).

A possible role for antibodies in the expression of immunity will be discussed in Section V.

C. REACTION WITH SOLUBLE ANTIGENS

A summary of the more recent work on the demonstration of circulating antibodies to soluble antigens is given in Table 3. The results of precipitation and complement fixation tests show that antibodies are demonstrable from three to seven days after the initial oocyst inoculation and persist for at least several months. The possibility of autoreinfection during the test period cannot be excluded

TABLE 2. Summary of published observations on antibody reactions with particulate cocci-

Species	Antigen	Type of reaction
E. tenella	Formalinized 2nd-generation merozoites	Agglutination
E. meleagrimitis	Live 3rd-generation merozoites	Immobilization; precipitate formation due to extrusion of body contents. Reduced infectivity
E. tenella	Live sporozoites, Live 2nd-generation merozoites	Lysis; changes in morphology; reduction in infectivity
	Sporocysts	
E. bovis	1st-generation merozoites	Agglutination. Lysis. Changes in morphology, Indirect fluorescent labeling
E. tenella	2nd-generation merozoites	Lysis; agglutination changes in morphology; immobilization
	Sporozoites	
E. tenella	Sporozoites; 2nd-generation merozoites	Immobilization reduced infectivity
	Oocysts	
E. acervulina	Sporozoites	
	Oocysts	
E. magna *E. stiedai* *E. tenella* *E. acervulina* *E. pragensis*	Asexual and sexual stages in tissues; free 2nd-generation merozoites (*E. tenella*)	Indirect fluorescent labeling
E. tenella	2nd-generation merozoites	Lysis; changes in morphology

dial antigens

Observations	Reference
Maximum titers of 1 : 320 at 10-15 days post infection. Positive reactions for at least 30 days post infection	McDermott and Stauber, 1954
Sera from completely immune turkeys. Lysis in fresh immune serum, loss of activity and precipitation in heated immune sera and reduced infectivity. Merozoites unaffected by incubation in normal serum	Augustin and Ridges, 1963
Maximum lytic titers of 1 : 24 with sera from completely immune fowls. Heated and fresh sera from immune fowls reduced infectivity; morphological changes noted after incubation in heated immune sera. Fresh serum from normal birds reduced the infectivity of sporozoite suspensions; merozoites unaffected	Long, Rose and Pierce, 1963
Unaffected by incubation in serum	
Agglutinating antibody first detected 7-17 days after inoculation, maximum titers (1 : 640) at 23-41 days followed by gradual decline. Positive titers present up to 65 days. Lysis in low dilutions (1 : 20) of immune serum. Changes in morphology noted. Neat normal serum caused slight changes in morphology. Indirect fluorescent labeling first seen with serum taken 14-15 days after oocyst inoculation, with maximum titers (up to 1 : 800) at 19-34 days. Positive results obtained up to 177 days	Andersen *et al.* 1965
Lysins from 8 days after inoculation, positive up to 3 months, negative at 4 months. Anamnestic response after reinfection. Lytic response weaker in bursectomized birds; obtainable in cecectomized fowls. Changes in morphology and immobility in heated immune serum. Unaffected by normal serum	Burns and Challey, 1965
Lysed in undiluted normal serum	
Immobilization and reduced infectivity of sporozoites and merozoites by both immune and hyperimmune sera. Unaffected by incubation in *E. acervulina* immune sera	Herlich, 1965
Unaffected by incubation in serum	
Sporozoites immobilized by incubation in hyperimmune serum. Unaffected by *E. tenella* antisera	
Unaffected by incubation in serum	
Sera obtained after single oocyst inoculum. Use of asexual stages as antigens resulted in higher antibody titers. Higher antibody titers persisting for a longer time obtained in rabbits than in chickens. Cross reactions between the asexual stages of *E. stiedai* and *E. magna*	Černá, 1966, 1966a, 1967, 1969, 1970
Sera obtained 1, 3, 6, 9, 12 and 15 weeks after the last of three immunizing inocula given at 3 weekly intervals and were positive throughout	Leathem and Burns, 1968

TABLE 3. Summary of published observations on antibody reactions with soluble coccidial antigens

Species	Antigen	Type of reaction
E. stiedai	Extract of infected liver	Complement fixation
E. stiedai	Exudate from infected bile duct and saline extract of crushed oocysts	Precipitation in gel and liquid media; complement fixation
E. stiedai	Exudate from infected bile duct and saline extract of crushed oocysts	Complement fixation
E. tenella	Saline extract of infected ceca (2nd-generation schizogony) and saline extract of ultrasonically disintegrated oocysts	Precipitation in agar gel
E. tenella *E. maxima* *E. necatrix* *E. acervulina*	Saline extracts of infected tissues	Precipitation in agar gel
E. meleagrimitis	Saline extracts of infected tissues, oocysts, sporozoites, and merozoites	Precipitation in agar gel
E. tenella	Polysaccharide preparation of oocysts	Precipitation
E. tenella	Saline extract of ceca containing 2nd-generation schizonts	Complement fixation
E. bovis	Saline extract of disintegrated oocysts	Precipitin ring test in liquid medium

Observations	Reference
Antibodies detected from 9 to 16 days after inoculation with oocysts. Maximum reached at 1-2 months, titers persisted for more than 123 days	Heist and Moore, 1959
Antibodies detected by precipitation and complement fixation from 7 days after initial oocyst inoculum. At least seven different antibody/antigen systems detected in serum of a rabbit recovering from a heavy infection by precipitation against a pooled antigen. Antisera taken early in infection (shortly after schizogony) reacted only with the comparable antigen	Rose, 1959
Positive titers from 3-7 days after inoculation of oocysts, peak value between 30-40 days, remaining level up to approx. 70 days then declining gradually but still detectable at 228 days. Challenge inoculations did not affect declining antibody titers.	Rose, 1961
Precipitates formed from 7 days after first oocyst inoculum, persisting up to 105 days (repeated infections)	Pierce, Long, and Horton-Smith, 1962
Precipitins found with all four species; the greatest initial response was obtained with *E. maxima*. There was very little response to reinfection except with *E. acervulina* and the reactions of *E. necatrix* infected fowls were weak throughout. Precipitins were present at 14 days after the initial inoculum, persistence in the absence of reinoculation not known	Rose and Long, 1962
No precipitins with oocyst antigen, bands obtained with sporozoite and infected tissue extract, parasite antigens thought to be metabolic products. Antisera from solidly immune turkeys	Augustin and Ridges, 1963
Precipitin titers obtained 15-20 days after oocyst inoculation, disappeared after 5 months. (Data obtained from English summary of Japanese paper)	Itagaki, K., and M. Tsubokura, 1963
Fixation of complement from 14 days after oocyst inoculum was given, maximum values between 27 and 42 days, declining to 0 at day 58	Rose, 1963
Titers of 1 : 10 - 1 : 40 obtained with pre-oocyst inoculation sera, maximum titers (up to 1 : 640) on days 23-41 after inoculation, latest positive results days 38-65	Andersen *et al.*, 1965

in any of the published reports; where multiple and challenge inocula have been given, however, antibody levels have been little affected. This may be due to the prompt immune response which reduces parasite invasion and prevents multiplication in the sensitized host (Rose, 1961; Rose and Long, 1962; Pierce *et al.*, 1962; Andersen *et al.*, 1965). The persistence of antibody titers may be caused by autoreinfection or by retained antigen in the form of oocysts trapped in tissues of the host (Rose, 1961; Pierce *et al.*, 1962). Maximum titers were recorded at approximately four to eight weeks after inoculation.

D. OPSONIZING ANTIBODIES

Recent work on the phagocytosis (further discussed in Section VI, A) of coccidial stages by macrophages has raised the question of the participation of opsonizing antibodies in immunity. Huff and Clark (1970) investigated the destruction of sporozoites of *E. tenella* by macrophages in the celomic cavities of chickens. They found this to be greater in immunized birds and the effect was further enhanced by serum from immune birds. Patton (1970) carried out similar work with sporocysts *in vitro* with macrophages derived from heparinized blood. He found variations in phagocytosis during infection but did not mention any effects of serum, although the macrophages were presumably already sensitized. Further work on phagocytosis by macrophages from immunized and normal animals and the effects of sera from such animals should provide useful information.

E. NATURALLY OCCURRING ANTIBODIES

Naturally occurring antibodies to coccidial antigens have not been systematically studied, but there are many indications that they exist and they are probably of similar origin and may even be identical to those directed against the erythrocytes of various animal species and some bacterial antigens, e.g., those of *Escherichia coli.*

Serum from normal, coccidia-free animals, neat or in low dilution, has adverse effects on the infectivity or morphology of some of the invasive stages of their respective coccidia (see Table 2). In general, sporozoites appear to be more susceptible than merozoites, although opinions differ.

The effect of globulin fractions of serum from coccidia-free chickens on *E. tenella* infections in embryos was shown by Rose and Long (1971). Infections in embryos inoculated with such serum fractions were lower than in those which received saline, and this effect was greater in the serum from older donors. The serum of heterologous hosts lyses or agglutinates sporozoites at a higher dilution than that of the homologous host – an indication of the possible existence of "host antigens" discussed above. Titers of both homologous and heterologous hosts increase with age until approximately adult status is reached (Long, 1971a) in a similar manner to erythrocyte agglutinins (e.g., see Janković and Isaković, 1960, for the development of naturally occurring agglutinins in fowl sera).

It seems likely that the coccidia have antigenic determinants in common with other microorganisms and tissue components of the host or other macroorganisms. Germ-free chickens have much reduced antibody titers for human Group B erythrocytes and these levels become elevated on experimental infection with *E. coli* (Springer, Horton, and Forbes, 1959). A study of the naturally occurring antibodies to the coccidia in germ-free hosts might yield similar results.

Work on the effect of coccidial infection on the development of heterophile antibodies (known to increase in pigs infected with helminths, Soulsby, 1958) in chickens is being pursued in this laboratory.

The relevance of these antibodies to resistance to infection is not known, and for most species, increasing age results in increased susceptibility to infection (see Chapter 7). Work on age resistance or susceptibility is complicated by many factors, e.g., choice of correct dose of oocysts, maintenance of animals coccidia-free, increased length of gut providing more *lebensraum* for the parasites in older hosts, maturation of enzyme systems leading to more efficient excystation, etc., and such experiments need to be very carefully controlled. Natural antibodies and their effect on host-parasite relationships in other protozoal infections have been discussed by Desowitz (1970). Heterophile antigens and their significance in a wide range of host-parasite relationships have been reviewed by Jenkins (1963).

F. AUTOANTIBODIES

Autoimmunization undoubtedly occurs in protozoal infections and may be caused by sensitization of the target host cell with either parasite antigen or antiparasitic antibody, by alteration of the host cell by the infectious agent or its products so that it is no longer recognized as "self" by the host, or by the sharing of antigens by the parasite and host (see reviews by Zuckerman, 1964; Asherson, 1968).

Asherson and Rose (1963) have shown that infection with *E. stiedai* caused a significant rise in titer of complement-fixing autoantibodies of the macroglobulin class to rabbit kidney, liver, lung, brain, spleen, heart, muscle, and placenta; serum titers of rabbits kept coccidia-free were very low. Augustin and Ridges (1963) also found that turkeys infected with *E. meleagrimitis* developed autoantibodies to turkey pancreas, duodenum, ceca, and liver. It is not known whether these antibodies cause disease.

G. ENHANCING ANTIBODIES

Enhancing antibodies, described by Kaliss (1958), facilitate the growth of antigenically foreign cells which would otherwise be rejected. Their presence in hosts infected with coccidia has not been demonstrated with certainty but there are indications of their existence in the results of attempts to immunize birds passively with antiserum (Pierce *et al.*, 1963). Recently we have obtained enhanced oocyst production in *E. maxima* infections of birds which had been actively immunized with the homologous parasite antigens. Humoral precipitating antibody levels were high and it is possible that these "nonfunctional" antibodies interfered with the host's own active immune response either by "blocking" or by modifying cell-mediated immunity (CMI) (see Mackaness and Blanden, 1967; Gray and Cheers, 1969).

H. LOCALLY PRODUCED ANTIBODIES

Antibodies in the alimentary tract, "copro-antibodies," are undoubtedly involved in the immunological defense of the digestive mucosa against microorganisms, and it is highly likely that these antibodies are locally formed and not derived from the serum since there is little correlation between serum and copro-antibody titer, or serum titer and protection, but good correlation between copro-antibody and protection (see reviews by Pierce, 1959; Tomasi and Bienenstock, 1968). The copro-antibodies, together with the antibodies found in secretions and at mucosal surfaces, are now generally regarded as belonging to the IgA class of immunoglobulins, synthesized locally and involved in the immune defense of mucous surfaces.

A possible role for copro-antibodies in immunity to the coccidia is an attractive hypothesis for the following reasons:

a. these parasites are, in the normal host, confined to mucous membranes, and such infections with other microorganisms are now known to involve this type of defense mechanism of the host;
b. it would account for resistance in the absence of serum antibodies; and
c. a booster response to attempted reinfection is lacking due to rapid sequestration of antigen; also
d. passive protection with serum is brief and unreliable since the presence of protective humoral factors may result from a "spillover" of local antibody at the time of its peak production. (In the dog and mouse a large part of serum IgA is derived from the intestinal mucosa; see Bazin, 1970.)

Attempts to detect copro-antibodies in animals infected with or resistant to coccidia have met with little success until recently.

Augustin and Ridges (1963) examined saline extracts of feces from infected turkeys for antibodies to *E. meleagrimitis* with negative results; as demonstrated by immunoelectrophoresis, the extracts contained only traces of γ-globulins. Horton-Smith *et al.* (1963a) harvested sporozoites of *E. tenella* from the cecal lumina of resistant fowls and found their infectivity for susceptible fowls to be unimpaired. A comparison of the infections induced by merozoites which had been incubated in saline extracts of the ceca of resistant and susceptible fowls gave equivocal results. Herlich (1965) also incubated merozoites and sporozoites of *E. tenella* in saline extracts of the ceca of immune and hyperimmune birds and found their morphology and infectivity to be unaffected; similarly, sporozoites of *E. acervulina* were not immobilized in saline extracts of the intestines of immunized birds. From these results Herlich concluded that antibodies to *E. tenella* and *E. acervulina* are "not localised or stored in the intestinal or cecal epithelium."

Leathem and Burns (1967) showed that sporozoites harvested from the cecal mucosa of resistant animals later than 24 hr after oral inoculation of oocysts had lost their infectivity, but there is no indication of the nature of the destructive effect; it might be caused by copro-antibody either passing into the cell or being carried in on the surface of the invading sporozoite, by serum antibody penetrating the cell, or by an entirely different type of immune mechanism.

The importance of locally produced antibodies in gut infections with other organisms and the growing interest in the secretory antibodies stimulated more attempts in this laboratory to demonstrate a similar mechanism in immunity to coccidiosis. The chicken is an attractive host for these studies because of the abundance of lymphoid tissue in the chicken gut (there are no lymph nodes as such in this animal), its enormous increase in response to infection, and the known immunological competence of cells of the cecal tonsil, an aggregate of lymphoid tissue at the neck of the cecum (Janković and Mitrović, 1967; Orlans and Rose, 1970). In addition, lymphoid cell suspensions prepared from the cecal tonsils and transferred to chick embryos have been shown to cause a reduction in coccidial infection of the chorioallantoic membranes (Rose and Long, 1971).

Ammonium sulfate fractionation of saline washings of gut and of saline extracts of feces resulted in very dilute globulin preparations. Fractionation of saline extracts of cecal contents, however, yielded solutions containing IgG and an immunoglobulin which is possibly the analogue of mammalian IgA (Orlans and Rose, 1972) in reasonably high concentration. These extracts have been used in protection tests *in vivo* and *in embryo* with some success. It therefore seems likely that, as in *Vibrio cholera, Escherichia coli*, and poliomyelitis infection, antibodies present in the gut are active in immunity. It has not yet been shown that the antibodies and protective activity of the cecal extracts in coccidiosis are of local origin; sporozoite invasion is accompanied by increased gut capillary permeability and an extravasation of serum proteins (Rose and Long, 1969); thus at least some of the immunoglobulins may be of serum origin. However, the cecal contents were not collected during infection and they have been shown to be protective at a time when no such activity can be demonstrated in serum.

The reduction in *E. tenella* infections of embryos by the transfer of cecal tonsil cell suspensions from immunized donors (Rose and Long, 1971) would provide strong evidence for the role of locally produced antibody in coccidiosis immunity if it were certain that the cell transfer effect is obtained by antibody production rather than by a cell-mediated response (see Section VI, B).

Further work is necessary to determine the origin and exact nature of the copro-immunoglobulin tentatively identified above.

V. Protective Effects of Coccidial Antibodies

Antibodies may be shown to have deleterious effects on various stages of the parasite *in vitro* including neutralization but, in the case of an intracellular organism, this is not proof of their effectiveness *in vivo*. Passive protection by cell-free immunoglobulins (including those maternally transmitted) is the most valid criterion.

A. PASSIVE TRANSFER OF IMMUNITY

All the early attempts to demonstrate passive transfer of resistance to coccidiosis with serum gave negative results (Tyzzer, 1929; Bachman, 1930b; Becker, Hall, and Madden, 1935). Interest in this aspect of immunity then lapsed until it was shown that humoral antibodies were present in infected and convalescent animals and that immunity was not confined to the site of previous parasitization.

Heist and Moore (1959) briefly reported a modification of the gross pathology of the liver in rabbits inoculated with oocysts of *E. stiedai* by intraperitoneal injections of immune serum or whole blood, whereas intravenous injections had no effect. Rose (1959a and 1963) found, in a limited series of experiments, that the pathology of *E. stiedai* infection in rabbits was not, or only very slightly, modified by the intravenous or intraperitoneal injection of a globulin fraction of serum from resistant hosts. Similarly, Senger *et al.* (1959) reported no reduction in infection with *E. bovis* in calves given transfusions of plasma or leukocytes from immunized hosts. Augustin and Ridges (1963) obtained similar oocyst counts in turkeys given *E. meleagrimitis* oocysts and injected intravenously with normal turkey serum or serum from immunized turkeys. Pierce *et al.* (1963) injected chickens intravenously or intraperitoneally with globulin fractions of serum obtained from immunized or susceptible fowls and tested the effects on infection with *E. tenella* induced by oral inoculation with oocysts or by the intrarectal inoculation of second-generation merozoites. The infections in birds which received immune globulin were, if anything, rather greater than in the controls (see Section IV, G).

Fitzgerald (1964), working with *E. bovis* infections in calves, found no detectable differences in oocyst production or clinical signs in calves given concentrated serum or globulin fractions of serum from immune hosts compared with control animals.

Until recently the only experiments in which incontrovertible evidence of passive immunization against coccidiosis was provided were those of Long and Rose (1965) in which the challenge infections were induced by the intravenous inoculation of sporozoites of *E. tenella*. In these conditions a globulin fraction of serum from birds made resistant by repeated oral dosing with oocysts, given either intravenously or intraperitoneally, protected against the challenge inoculum. Protection was greatest when the interval between globulin and sporozoite inoculation was four to eight hours and least when this was extended to 19 hr. The conclusions of the authors were that the transferred antibodies were effective only so long as they were in direct contact with the invasive stages; if injection of globulin was deferred until the sporozoites had entered the mucosal cells, no protection ensued. Rose (1971b) examined serum samples taken at different time intervals during immunization with *E. maxima* and found that serum taken between two and three weeks after a primary infection and inoculated daily into susceptible recipients afforded considerable protection against oral challenge infection with oocysts of the homologous species. The protection was dose-dependent, varying directly with the volume of serum inoculated and inversely with the size of the challenge oocyst inoculum. Treatment early in the course of infection was more effective than similar volumes of serum given in the middle or late portion. The effect was species-specific; serum from birds infected with *E. maxima* was ineffective against challenge with oocysts of *E. acervulina* or *E. praecox*.

Not all pools of serum taken two to three weeks after oocyst inoculation were protective even though the donors were resistant to challenge, and pools of serum obtained after three weeks, when donors were completely recovered (and resistant) or after multiple inoculations of oocysts when immunity was complete, were not as effective as the samples taken at two to three weeks. It is possible that, in these instances, local immunity is at a high level, but serum antibody is low; challenge of these animals might lead to a rapid anamnestic response, either of a secretory immunoglobulin kind (Section IV, H), or of a cellular kind, by the sensitized gut. This would explain the inactivity of serum antibodies in these donors, which contrasts with their resistant state. This local response is probably so effective that it greatly limits the antigenic challenge, since sporozoites either would be prevented from invading or would be destroyed very soon after invasion, thus provoking little serological response. Serum precipitins are very rarely found in completely resistant animals after challenge (see Rose and Long, 1962; Long and Rose, 1962). The limited time in the host-parasite relationship during which serum can be shown to be protective is probably related to the peak of immunity. If this is the result of high local production of immunoglobulin (IgA), it may result in "spillover" into the serum.

Similar experiments, i.e., with serial bleedings during immunization, have now been carried out with *E. tenella* infections using the chick embryo as a test system (Rose and Long, 1971). No protection was afforded by globulin fractions of serum samples taken 1, 2, 3, or 4 weeks after the first oocyst inoculum or one week after the second; fractions of serum obtained two weeks after a second inoculum of oocysts caused some reduction in infection of the embryos and this was greatly increased when serum taken one or two weeks after a third inoculum was used.

The times at which the serum contained protective antibodies correspond with the development of immunity by the donors, as shown by the resulting oocyst output from challenge infections. The different time scales of onset of protection by serum in the two species is reflected in their immunizing abilities; a single infection with *E. maxima* usually results in a good immunity, whereas three or more infections with *E. tenella* are required.

Serum from solidly immune donors which have received numerous oocyst inocula and pass no oocysts in the feces have not been examined in the *E. tenella* test system but in previous *in vivo* work, such sera have not protected. An examination of serum throughout the different stages of immunity should be made.

B. MATERNAL TRANSFER OF IMMUNITY

The results of earlier work (Long and Rose, 1962; Heckmann, Gansen, and Hom, 1967) indicated that immunity was not transferred from mother to offspring. Long and Rose (1962) compared infections with *E. tenella* in the progeny of immune and susceptible hens at 4, 7, or 14 days of age. The immune hens received their original immunizing schedule at four weeks of age and were boosted at two-week intervals with one million oocysts throughout the progeny-testing period. They were completely immune and neither their pooled sera nor extracts of yolk precipitated with an antigen prepared from schizont stages of *E. tenella*; serum samples obtained at 14 and 21 days after the initial oocyst inoculum (at four weeks of age) contained precipitins. No differences were found in the susceptibilities of the chicks from the two groups of hens to infection with *E. tenella*, judged by mortality. Oocyst production resulting from a smaller test inoculum was measured in a small number of chicks from one hatch only; chicks from immune hens produced slightly fewer oocysts than those of the control hens, but this was probably not significant.

In the light of present knowledge, it is not surprising that no maternal transfer was found in these experiments. The immune status of the hens was such that it is highly unlikely that any protective antibodies would have been present in the serum; therefore none could be expected in the yolks of the eggs or in the progeny. The lack of serological response by these "ultra-immune" hens is illustrated by the absence of precipitins.

Heckmann *et al.* (1967) examined *E. nieschulzi* infections in young rats born to mothers which had received one, two, or three inocula of oocysts. They found the progeny of all three groups of parents to be susceptible to infection whereas the parents themselves were completely immune. No comparisons were made with the offspring of susceptible rats and the young were challenged at approximately six or seven weeks of age, when any maternally transmitted immunity would surely have waned.

Successful passive transfer of immunity with serum has led to a reinvestigation of maternal transmission of immunity to the coccidia. This work has been done using the fowl as host and, as the presence of protective serum antibody has been shown to have a critical temporal relationship with infection and to depend upon the stage of immunity, attention has been paid to this. Infections with *E. tenella* in embryonated eggs and with *E. tenella* or *E. maxima* in hatched chicks of susceptible and immune hens were compared; the yolks of eggs from hens infected with *E. maxima* were fractionated and the fractions tested for passive transfer of immunity in the same way as serum.

The work with *E. tenella* infections has been published (Rose and Long, 1971) as well as that concerning *E. maxima* infections (Rose, 1972).

Embryo infections with *E. tenella*, judged by the numbers of foci which developed on the CAM, were smaller in embryos derived from hens in the second week after each of the second, third, fourth, and fifth oocyst inocula than in the embryos from control hens, but the reduction was statistically significant only in the batch obtained in the second week after the fourth oocyst inoculum. Oocyst production in some groups of chicks hatched from eggs laid by infected hens was lower than in controls, and this was particularly marked (73% reduction) in the hatch corresponding to the eggs in which statistically significant lowered CAM foci numbers were found. The CAM foci method of measuring infection is probably not a suitable one for these experiments since antibody is confined to the yolk for the first two-thirds of incubation and does not appear in the allantois in any great amount until after the 14th day (Kramer and Cho, 1970), by which time the parasites are already well established.

E. maxima infections in chicks hatched from the eggs of hens laid at different times in relation to the infection of the hens were compared with those in control chicks from uninfected hens; the chicks were given oocysts at seven days of age. There was a reduction in oocyst output in chicks from eggs laid in the third, fourth, and fifth weeks after the inoculation of hens but not in the preinoculation, first, second, and sixth weeks. The greatest reduction (80%) was present in the third week (from 16 to 22 days) batch; allowing for the delay in appearance of antibody in yolk, relative to serum (Patterson *et al.*, 1962), this corresponds very well to the protective antibody activity of serum in this infection.

Immunoglobulin preparations made from yolks of eggs laid from 16 to 19 days after a single inoculation with oocysts of *E. maxima* and injected subcutaneously or intraperitoneally into 14-day-old susceptible chicks resulted in a lower fecal oocyst output from an oral inoculum with oocysts of *E. maxima*, compared with controls injected with a control preparation, or untreated.

Thus maternally transmitted protection can be demonstrated in coccidiosis provided that the correct timing in relation to infection of the dams is determined. This will be similar to that for the demonstration of serum protective antibody, allowing for the time lapse ensuing between peak serum titer and yolk titer and, possibly, in the case of mammals, placental or colostral titers and challenge inoculation of the young.

C. POSSIBLE MODE OF ACTION OF PROTECTIVE ANTIBODIES

Antibodies, both humoral and enteric are effective against the coccidia; by various *in vitro* tests they can be shown to reduce infectivity of the various stages and to reduce infections *in vivo* and *in embryo* by passive transfer experiments.

The relevance of antibodies which are destructive *in vitro* in immunity to infection with an organism which does not normally inhabit the blood stream but is generally found within epithelial cells, may be questioned. Their undoubted effectiveness in conferring protection passively in the *in vivo* tests, however, indicates that they can play a part in host resistance.

The protective action of muco- or copro-antibodies in infection may perhaps be visualized more readily than that of humoral antibodies. Muco-antibody may immobilize or destroy sporozoites on the mucosa and thus prevent or hinder cell penetration. This process might be extended so as to affect metabolism or growth of the "neutralized" sporozoite which may have succeeded in penetrating cells; antibody already attached to the parasite may be active within the cell, although the experiments of Leathem and Burns (1967) in which sporozoites transferred from "immune" cells to "susceptible" cells recover their infectivity would argue against this. This hypothesis could explain the histological findings on the reinfection of immunized animals. Immobilization or destruction or both may equally well apply to other invasive stages, resulting in a presumably increasing response as infection proceeds and accounting for the various stages in immunity discussed in Section III, G.

Humoral antibody may also participate in this type of reaction, possibly at a slightly later stage. Cell penetration by sporozoites is accompanied by a local increase in capillary permeability, enabling plasma proteins to pass into the gut lumen (Rose and Long, 1969). This occurs in normal susceptible animals but is greatly increased in the immunized host, and may be suppressed by corticosteroid treatment, which also suppresses immunity (see Section VIII, B). Increased permeability may also accompany merozoite invasion, although this has not been shown, and it is, of course, known to occur at the height of infection when the integrity of the epithelium is affected (Preston-Mafham and Sykes, 1967; see Chapter 7, this book).

The passage of serum proteins into the gut lumen during infection or cell invasion throws doubt on the origin of the immunoglobulins (protective) which have been demonstrated in fowl cecal contents. However, these preparations were made after recovery from infection and it is highly likely that the antibodies were of local rather than humoral origin. Further corroboration is given by the predominant type of immunoglobulin present; although this has not been identified with certainty as an IgA, its electrophoretic mobility and low concentration in serum suggests that it may be (Orlans and Rose, 1972).

Thus, the action of antibody on the parasite during its extracellular phases may be visualized; whether it can affect the intracellular stages, other than as an extension of the extracellular reaction, is unknown. There is controversy over whether antibody can penetrate in an effective form into living cells; the permeability of the parasitized cell may, of course, be altered and Preston-Mafham and Sykes (1967) noted that the mucosa and submucosa of infected birds were colored blue by dye leakage from the circulation, whereas in the uninfected controls, color was restricted to the submucosa. These authors associated the change in permeability with rapid growth of the parasite within the epithelial cells, in addition to the sloughing of the epithelium. The effects of parasitization on the host cells are discussed in Chapters 3 and 7. Protective humoral antibody to *Plasmodium knowlesi* is thought to have little effect upon the growth of intracellular parasites, but is believed to prevent reinvasion of red cells and thus inhibit the succeeding cycle of parasite development (Cohen and Butcher, 1970).

VI. Cellular Responses

Infection with coccidia is a stimulus to a massive cellular reaction at and infiltration of the site of parasitization, and a generalized humoral response may also be noted. This topic is discussed in detail in Chapters 7 and 9. Among the cells commonly found in proxmity to the infection are those which are generally considered to be concerned with the immune response : plasma cells, lymphocytes, macrophages, and cells resembling globule leukocytes.

Since antibodies are formed and have been shown to be capable of modifying coccidial infections, at least during some stages of the infection, some of the intense cellular activity seen at the site of infection is probably the result of the stimulation of antibody-producing cells. Immune mechanisms of the cell-mediated type are probably active in immunity to coccidiosis, as they are in infections with other intracellular organisms.

They may act in conjunction with antibody-mediated responses or may represent the predominant mechanism at a certain stage of infection. At a time when evidence for the effectiveness of antibodies in coccidiosis immunity was not available, it was suggested that cell-mediated responses might be solely responsible for immunity. However, there is little evidence to confirm this hypothesis, probably because of the greater difficulties of work in this field and the less understood nature of the mechanisms involved.

A. ROLE OF THE MACROPHAGE

Macrophages play a prominent part in cellular immunity to infections with some intracellular bacteria (see Mackaness and Blanden, 1967). In one form this immunity, transferable with cells but not serum, is induced by a specific stimulus but, once established, becomes nonspecific in its effects which are caused by an enhancement of the microbicidal activity of the macrophages. There have been reports of cross-immunity between intracellular bacteria and protozoa (Ruskin and Remington, 1968; Gentry and Remington, 1971; Barrett *et al.*, 1971), and between unrelated protozoa (Cox, 1968) but there are also conflicting reports (Frenkel, 1967). As immunity to the coccidia is remarkably specific, extending even to strains (see Section III, E), it seems unlikely that such a mechanism is part of it.

Macrophages may act in conjunction with humoral antibodies, either opsonins which promote phagocytosis (Nelson, 1969) or cytophilic antibodies which are absorbed onto the surface of macrophages and change their reactivity (Mackaness, 1964), and this might be a mechanism active in immunity to coccidiosis as has been suggested by Burns and Challey (1959). However, there have been no experimental investigations until recently. Macrophages are thought to transport at least a proportion of the sporozoites of *E. necatrix* and *E. tenella* (Van Doorninck and Becker, 1957; Challey and Burns, 1959; Pattillo, 1959) through the lamina propria to epithelial cells and this might be an opportunity, early in the life cycle, for destruction of the parasite in the sensitized host.

Huff (1966) and Huff and Clark (1970) injected sporozoites into the peritoneal cavities of birds in which cellular exudates had been stimulated and observed the "infection rates" of the cells subsequently recovered. Macrophages and cells described as degranulated granulocytes contained sporozoites; the numbers found in cells from immune birds were significantly fewer than in cells from susceptible birds and lower infection rates were associated with damaged sporozoites. When the incubation time was reduced, infection rates in cells of normal and immune birds were similar except when the macrophages were treated with immune serum, in which case they were lowered. These data were interpreted as indicating that sporozoite destruction occurred in immune macrophages and this was enhanced by treatment with immune serum; treatment of normal macrophages with immune serum was without effect. Rather different results were obtained by Patton (1970), who examined the phagocytic activity of macrophages in heparinized blood for sporocysts of *E. tenella*. Phagocytic activity of macrophages from infected chickens rose from ten days after oocyst inoculation to a peak between 12 and 19 days and declined to the level found in control, uninfected chickens at 27 days post inoculation. Thus he found greater numbers of parasites in macrophages from infected chickens and, as the stage used was the sporocyst, this must represent engulfment of the parasite by the macrophage; in the experiments of Huff and Clark, invasion of macrophages by sporozoites cannot be excluded.

Damaged merozoites of *E. tenella* in cells thought to be macrophages have been described by Scholtyseck, Strout, and Haberkorn (1969) in an electron microscope study of infected tissues.

Caution is necessary in identifying parasitized host cells as macrophages since changes may be induced in parasitized epithelial cells which cause them to resemble macrophages (see Bergmann, 1970; Long, 1970a). The interactions of lymphocytes from immune animals with infected macrophages have not yet been investigated.

B. ADOPTIVE TRANSFER OF IMMUNITY WITH CELLS

The transfer of immunity from sensitized immune donors to recipients by means of lymphoid cells is not necessarily proof of the cell-mediated nature of the immune response, even in the absence of transfer with serum. The transferred lymphoid cells may secrete antibody which may be difficult to detect in the circulation, as the transferred cells may "home" to the site of parasitization so that the antibody secretion may be very localized, and the methods of detection of antibody used may not be very suitable. With these reservations in mind, adoptive transfer of immunity with lymphoid cells will be discussed in this section (cell-mediated immunity).

For successful transfer of adoptive immunity with cells a high degree of inbreeding between donor and recipient animals is essential. As most of the recent work on immunity to coccidiosis has been done in the domestic animals, in which it is an important cause of economic loss, the conditions for histocompatibility have not been satisfactorily fulfilled. Frenkel (1970) has stressed the importance of a very high degree of inbreeding in hamsters for lymphoid cell transplants, the success of which cannot be predicted by skin graft acceptance.

Many cell transfer experiments in chickens involving infections with *E. tenella* and *E. maxima* have been carried out in this laboratory over several years (Rose and Long, unpublished observations), but the results have been equivocal and have not been published in detail. Animals were inbred Reaseheath line. In some experiments donors and recipients were paired by skin grafting before cell transfer; in others, recipients were irradiated; and, in some, cells from many donors were pooled and aliquots injected into recipients in an attempt to promote acceptance. In some experiments oocyst production was lower in recipients of cells from immunized donors; in others there were no differences.

More success has been obtained with cell transfer experiments in chick embryos infected with *E. tenella*. The advantages of this system are:

a. The host is relatively incompetent from an immunoogical aspect and does not reject donor cells, thus obviating the problems of inbreeding and cell acceptance in more mature hosts.
b. The quantitation of infection with a species such as *E. tenalla* can more accurately be made by counting the foci which develop on the CAM (Long, 1970b) than by fecal oocyst counts.

Cecal tonsil cells from both immune and susceptible donors moderated infection but a greater effect was obtained with cells from immune donors (Rose and Long, 1971). Some of the schizonts which had grown in the midst of the lymphoid cells appeared to be very degenerate. The means by which the transferred cells affected parasite development is not known; in some of the experiments serum and cells obtained from the same donor were transferred and both reduced infection. The protection afforded the immunologically incompetent embryos by cells from susceptible donors was considered to be caused by a primary response by the cells, whereas the greater effect of the cells from immunized animals was caused by an anamnestic response.

Cell transfer experiments have also been carried out in pigs (Rommel, 1969) and rats (Heydorn, 1970; Liburd and Mahrt, 1971a). Rommel (1969) injected lymphocytes from the mesenteric lymph nodes of immunized pigs intraperitoneally into nonisogeneic susceptible recipients which he then challenged with oocysts of *E. scabra*. In one experiment he attempted to make the recipient pigs tolerant of the donor cells by injecting them with donor spleen cells. Transfer of cells did not influence the extent of the infection (see Rommel, 1970c).

Inbred isogeneic rats were used by Heydorn (1970) (see also Rommel and Heydorn, 1971) in experiments on adoptive transfer of immunity to *E. nieschulzi* with cells derived from the mesenteric lymph nodes or from Peyer's patches. A statistically significant ($P < 0.01$) reduction in oocyst output was obtained in rats inoculated with cells from mesenteric lymph nodes, compared with controls. Cells from Peyer's patches from the same donors had no effect on oocyst output. From these results Heydorn concludes that immunity in coccidiosis is cell-mediated, at least in part, but he does not mention the possibility of antibody formation by the transferred cells. Liburd and Mahrt (1971a) also transferred resistance to *E. nieschulzi* in inbred rats with cells; thoracic duct lymphocytes were used and the degree of resistance obtained could be directly correlated with the numbers of cells transferred.

Thus, immunity can be transferred with cells, but as this is also possible with serum in certain circumstances, no conclusions can be drawn as to the nature of the immune response.

VII. Interferon

The antimicrobial effect of interferon, once considered to be limited to viruses, is now known to extend to other intracellular organisms, including protozoa. Interferon induced by viruses or nonviable inducers, such as statolon, protects mice against *Plasmodium berghei* malaria (Jahiel *et al.* 1968) and chick and mouse monolayers against infection with *Toxoplasma gondii* (Remington and Merigan, 1968). In addition, infection of mice with *P. berghei* induces the production of a virus inhibitor which has been characterized as an interferon (Huang, Schultz, and Gordon, 1968). Long and Milne (1971) examined the possibility of a similar phenomenon in coccidial infections. They found that statolon injected intraperitoneally induced a virus interference factor in the serum of chicks, and that treated chicks produced fewer oocysts of *E. maxima* ($P < 0.05$) than controls. A small proportion of serum samples from birds infected with *E. tenella* or *E. maxima* had slight but significant antiviral effects.

More recently, Fayer and Baron (1971) have demonstrated the anticoccidial effects of chick interferon and an interferon inducer (poly I-poly C) in an *in vitro* tissue culture system. Neither treatment affected cell penetration by sporozoites of *E. tenella* (measured at two hours), but the numbers of developmental stages present at 48 hr were reduced.

Thus, interferon induction was shown to depress oocyst production, and coccidial infection to be capable of inducing an antiviral substance. Whether interferon, or something similar, plays an important part in immunity to the coccidia is not known. It may be implicated in the "overcrowding effect" in which an immunological mechanism is also suspected (Rommel, 1970c).

VIII. Experimental Modifications of the Immune Response

Immunity can be modified by various experimental techniques and the results obtained may indicate the nature of the immune response; e.g., if "blocking" of the reticuloendothelial system results in enhanced infections, then clearly the functions of this system are of importance in the development of immunity in the normal animal.

Some of the methods which have been

used to modify immunity in coccidial infections have been chosen in an attempt to determine the relative importance of antibody and cell-mediated immune responses, but few methods are sufficiently specific to provide irrefutable evidence. Perhaps the most specific are bursectomy and thymectomy in avian hosts; treatment with antilymphocytic serum is said to affect primarily cell-mediated responses, whereas other methods such as splenectomy, treatment with corticosteroids and ionizing radiation have fairly wide-reaching effects and thus can contribute little to this problem. However, immunosuppression by various means may provide useful information on the normal immune response.

A. SPLENECTOMY

The effects of splenectomy on infections with the *Eimeria* have been investigated in rats (Becker *et al.*, 1935), chickens (Rose, 1968b; Rouse, 1967), pigs (Rommel, 1969), and mice (Haberkorn, 1970). In contrast to infections with some other species of protozoa, e.g., plasmodia (see Brown, 1969), splenectomy has little effect on infections with the *Eimeria*. The authors cited above, with the exception of Rouse (1967), considered that splenectomy did not affect the acquisition or the maintenance of immunity. Rouse found that more neonatally splenectomized cockerels died from a primary infection with *E. tenella* given at from five to six weeks of age than controls, but they became immune after reinfection, and splenectomy after the acquisition of resistance had no effect. It seems unlikely that the spleen plays a major role in immunity to the coccidia. The differences in this respect between the coccidia and the bloodborne parasites are probably due to the limitation of the coccidia to epithelial cells. The spleen plays a minor role in antibody formation to antigens given by routes other than the intravenous one (Draper and Süssdorf, 1957), and is relatively unimportant in prolonged immunization (see Taliaferro, 1956). In addition, the role of the spleen in the clearance of parasitized red cells is obviously of major importance in blood-borne infections.

B. TREATMENT WITH CORTICOSTEROIDS

Treatment with corticosteroids has a profound depressing effect on immunity to the coccidia. The acquisition of immunity can be prolonged, existing acquired immunity can be broken down and, perhaps more interestingly, the site and species specificity of the coccidia may be affected, as can be the life cycle. These effects have been investigated in avian species, pigs, and cattle. The effects of treatment with paramethasone and dexamethasone on the results of challenge inoculations given to pigs immunized with *E. scabra* were examined by Rommel (1970d). Patent infections were obtained in the treated pigs but the untreated control pigs were completely resistant; the greatest effects were found in pigs given paramethasone intramuscularly together with dexamethasone orally. Rose (1970) treated chickens with betamethasone during primary and challenge infections with *E. mivati* and found that oocyst output during primary infections was increased and that acquired immunity in prolonged treatment was abolished, since oocyst output in these cases was the same or even greater than that resulting from primary infections in untreated controls. These results differed from earlier work by Rose and Long (1970) in which cortisone acetate was used; it is likely that insufficient amounts of this compound were given. Work on the effect of treatment with the corticosteroid betamethasone on primary infections was extended by Long and Rose (1970). Peak oocyst production in treated groups was lower than in controls but the length of the patent period was extended from 12 to 45 days, and the total oocyst production increased. This was shown histologically to be the result of the continuation of schizogony (up to 24 days in treated groups) during gametogony and after this had ceased in the controls. Extended oocyst production of *E. zürnii* and *E. bovis* has also been noted in dexamethasone-treated calves (Niilo, 1970). In the absence of reinfection these results suggest that the life cycle of even primary infections may be modified by the immune response of the host, and this may be the stimulus for large scale transition from asexual to sexual reproduction as suggested by Wilson and Morley (1933). This may be one of the factors contributing to the difficulties of obtaining gametogony and oocyst production in tissue culture and embryo infections. Schizogony continues in embryos and in cultured cells for longer than in mature animals. Gametocyte formation of *E. tenella* may be more readily attained in tissue cultured CAM cells which have not been removed from the embryo before 94 hr after infection than in such cells removed before this time (Long, 1969). Although the embryo is not very immunologically competent in comparison with the hatched chick, it is probably more so than cells in cultures. Other factors, such as nutrients, may be of even more importance in affecting the life cycle when the parasite is grown outside the normal host.

Treatment with corticosteroids can also affect site and host specificity, indicating that these features also may be under immunological control. The distribution of *E. mivati* in the intestines of treated birds was much wider than in controls (Long and Rose, 1970). Other tissues were not examined. More interestingly, *E. tenella* will grow in the liver of dexamethasone-treated chickens and chick embryos (Long, 1970c and 1971b). Similarly, a strain of *Trichomonas gallinae* which is normally confined to the upper digestive tract produces visceral lesions in dexamethasone-treated pigeons (Kocan, 1971).

McLoughlin (1969) overcame the specificity of *E. meleagrimitis* for the turkey and obtained its complete development in chickens which had been treated intramuscularly with dexamethasone. No oocysts of *E. tenella* were found in the feces of dexamethasone-treated turkeys; the turkey may not be so sensitive to corticosteroid treatment, or specificity of *E. tenella* may be more difficult to overcome. As no histological observations were made, it is not known whether any partial development occurred. Long (1971b) recorded the growth of schizonts of *E. tenella* in the chorioallantoic membrane of dexamethasone-treated goose embryos, whereas there was none in control eggs.

Corticosteroid treatment, therefore, effectively depresses or even, in some instances, abolishes the acquisition of immunity and already existing immunity; it also affects the life cycle and reduces the specificity for site of development and even host species. These latest findings provide most interesting data for speculation on the factors controlling the life cycle and site and species specificity of the coccidia. They suggest that these may be governed, at least in part, by the immune response of the host. The depression of acquired immunity does not provide much information about the nature of the immune response to the coccidia since cortisone and its derivatives have such a diversity of effects on resistance to infection (see Kass and Finland, 1953; Gabrielsen and Good, 1967). Cellular infiltration of the gut, seen as diffuse lymphoid tissue and in the form of foci, was very much reduced in betamethasone-treated chickens infected with *E. mivati*, but as cells associated with both antibody formation and cell-mediated responses were affected, i.e., plasma, Russell body, pyroninophilic and globule leukocyte cells, no conclusions can be drawn.

Corticosteroid treatment does inhibit the increased capillary permeability which occurs when chickens immune to *E. maxima*, *E. praecox*, and *E. acervulina* are given challenge inocula of oocysts (Rose and Long, 1969) and, in this way, might prevent contact between the parasites and any antiparasitic factors which might be present in serum. The reticuloendothelial system is also affected by treatment (see Vernon-Roberts, 1969) and, if the macrophage does play a role in the expression of immunity to the coccidia, this may account for some of the effects observed.

C. TREATMENT WITH IONIZING RADIATION

The effects of x-irradiation on infection and immunity have not been very fully investigated. Neonatal x-irradiation of chicks has been used as an adjunct to bursectomy and thymectomy, and chicks which had had one infection of *E. tenella* have been x-irradiated shortly before receiving a challenge inoculation.

1. *Neonatal x-irradiation*

Neonatal x-irradiation with 700 r or 750 r given 2, 3, 6½, or 8½ weeks before inoculation with oocysts of *E. maxima* or *E. brunetti* produced equivocal effects on the resulting oocyst production. An increase was found in only two of the six trials; in the remainder, production was less than or similar to that found in controls.

The results of challenge infections were similar; increased oocyst production was found in only one of the three trials. Irradiated chicks had depressed body and thymus weights and lower numbers of circulating lymphocytes during the first three weeks after irradiation (Rose and Long, 1970).

2. *X-irradiation before challenge inoculations*

The effects of x-irradiation of previously infected chickens with 500 r for 25 or 50 minutes on a challenge inoculation of *E. tenella* oocysts given five days later were described by Grosjean (1969). Untreated birds had neither cecal lesions nor parasites, whereas the ceca of irradiated birds contained lesions and stages of the parasite. Their thymuses and spleens were smaller than those of the controls. These results indicate that x-irradiation shortly before challenge inoculation depressed immunity acquired as the result of an earlier infection.

D. BURSECTOMY AND THYMECTOMY IN AVIAN HOSTS

A great deal of discussion on immunity to coccidiosis has centered around the question of whether it is essentially of a humoral or cell-mediated nature. It is likely that both systems are involved, interacting and perhaps changing in emphasis at different stages in immunization (see Section X). However, for experimental purposes, it is convenient to consider them separately, and the fowl is the host *par excellence* in which to do so, because of the dissociation of its immune responses, the bursa of Fabricius largely governing humoral immune responses and the thymus those of a cell-mediated nature (Warner, Szenberg, and Burnet, 1962; Warner, 1967). A number of experiments on immunity to coccidiosis, utilizing this concept, have been described but with fairly inconclusive results. It has proved very difficult to remove completely these organs surgically, and it is likely that even very small residua are sufficient to repopulate the body with the relevant "T" or "B" cells. Warner *et al.* (1969), by hormonal bursectomy, obtained birds 40% of which did not react to *Brucella* antigen and 64% of which did not produce antibody to human γ-globulin after two injections; some were agammaglobulinemic. However, most workers have experienced difficulty in producing more than a very small proportion of such birds. Unresponsiveness to single intravenous injections of soluble antigens is more readily attained than to particulate antigens or to repeated antigenic stimuli Janković and Isaković, 1966; Rose and Orlans, 1968). Complete abrogation of thymus-dependent responses has proved even more difficult to achieve. It is therefore hardly surprising that immunity to coccidiosis has been but little affected by these procedures.

A summary of published work on the effects of thymectomy and bursectomy is presented in Table 4. In four of the five publications, bursectomy was found to have some enhancing effect on infections, judged on oocyst production, histological examination, or, in one experiment only, mortality. This latter finding (Challey, 1962) was somewhat surprising as bursectomy was done at two weeks of age, a time normally considered to be too late to have a profound effect. Rouse and Burns (1971) were unable to confirm this result and found that growth and hematocrit values were also unaffected by bursectomy, even when this was done within 48 hr after hatching. Hormonal bursectomy is more effective than surgical bursectomy (Rose and Orlans, 1968), and the authors who used this method (Long and Pierce, 1963; Pierce and Long, 1965) were the only ones to find that resistance to a challenge infection was affected. Rose and Long (1970), working with two highly immunogenic species (*E. brunetti* and *E. maxima*) found that surgical bursectomy on hatching, coupled with irradiation, consistently caused some enhancement of oocyst production during initial infections. Surgical bursectomy alone had no effect.

The partial depression of immunity by hormonal bursectomy, described by Long and Pierce, occurred in birds in which serum γ-globulin levels were much reduced and in the virtually complete absence of serum lysins to *E. tenella*. Reduction in serum lysins to *E. tenella* in bursectomized chickens was also noted by Burns and Challey (1965). This imbalance led the authors to conclude that immunity was probably mediated not by humoral antibody but by cellular factors, and that the effects observed by them were probably caused by the effects of the hormone on thymus development. However, attempts to implicate cell-mediated, thymus-controlled immunity by surgical thymectomy have not been very successful. Slight increases in oocyst output have been noted (Pierce and Long, 1965; Rose and Long, 1970), as have "subtle differences" in mortality and growth (Rouse and Burns, 1971); but it must be remembered that it is very difficult to remove the thymus in its entirety and, as with bursectomy, removal on hatching is probably too late to produce a complete effect (see results of tests of tuberculin hypersensitivity by Rose and Long, 1970). Although it is highly doubtful whether the circulating antibodies demonstrated by means other than passive protection have any relevance to protective immunity, the local effects, i.e., depletion of germinal centers and lymphoid cells of the plasmacytic series noted by Pierce and Long (1965), would be expected to have produced rather greater effects on the results of challenge infection than those observed if free-antibody-mediated systems were of paramount importance in immunity to coccidiosis.

E. TREATMENT WITH ANTILYMPHOCYTIC SERUM

Antilymphocytic serum (ALS) is widely used as an immunosuppressive agent. It can affect the recognition of antigen, immunological memory (induction of a secondary response), expression of delayed hypersensitivity, the production of circulating antibody, and can prolong the retention of skin grafts (see review by Sell, 1969); but it is generally considered to be particularly effective in suppressing cell-mediated immunity (Levey and Medawar, 1967). In view of this, its effect on immunity to coccidiosis has been tested but on a limited scale to date. Euzeby, Garcin, and Grosjean (1969) immunized chickens with *E. tenella* and then treated them with antibursa or antithymus serum (produced in rabbits) shortly before challenge with *E. tenella*. Results were judged on macroscopic appearance of the ceca and microscopic examination for parasites. The extent of infection in the group treated with antibursa serum was almost as great as in the control susceptible group and, in the group treated with antithymus serum, it was intermediate between this and the completely immune control group. Despite these results the authors concluded that immunity was of the "cell-mediated (delayed hypersensitivity) type". No attempt was made to determine the effects of the treatment on antibody formation or delayed hypersensitivity. There are few reports of the use of ALS in fowls. Tucker (1968) prepared it in rabbits and found prolonged skin graft retention but no impairment of the graft versus host response. Janković *et al.* (1970), investigated the effects of ALS prepared in rabbits on antibody production to bovine globulin and the development of experimental allergic encephalomyelitis (EAE). The authors also injected guinea pig complement into the recipients of antisera since rabbit antibody is not cytolytic in the presence of fowl complement (Rose and Orlans, 1962). They found that both antithymus globulin and antibursa globulin depressed precipitating antibody formation but that only antithymus globulin suppressed EAE (an immune reaction of the cell-mediated type). These findings cast even greater doubt on the interpretation by Euzeby *et al.* (1969) of their data.

A limited amount of work on the effects of antilymphocyte serum has been done in this laboratory using antisera produced in ducks, rabbits, or pigs, with varying results. In most experiments skin grafting was done simultaneously with the coccidial infection (*E. maxima*) to provide a measure of the effectiveness of ALS treatment. Enhanced oocyst production, which was noted particularly when the recipient chicks were of the same breed as that of the donors used to provide thymocytes for pig immunization, was obtained in three of five experiments with pig ALS. In one of these experiments, skin graft retention was also increased. A slightly increased oocyst production was noted in the two experiments in which duck ALS was used but the rabbit ALS was ineffective.

Rommel (1970d) tested the effects of sheep and rabbit ALS on *E. scabra* infections in pigs. In primary infections approximately twice as many oocysts were discharged by recipients of ALS than by controls which received normal sheep serum. Injections of ALS given to immunized pigs did not affect the outcome of secondary infections. The ALS was not tested for its action on cell-mediated immune responses such as contact hypersensitivity, tuberculin sensitivity, or skin graft rejection, as a control. Liburd and Mahrt (1971b) concluded, from studies on rats thymectomized as neonates and as adults, that thymus-dependent mechanisms were not involved in immunity to *E. nieschulzi*.

Thus not enough evidence is available for an assessment of the effects of ALS on immunity to coccidiosis; in most of the experiments described, vital controls to demonstrate the efficacy of the ALS used are lacking. Further work, including these controls, should prove rewarding.

F. TOLERANCE

Tolerance is a state of immune unresponsiveness, i.e., when animals do not respond immunologically to a substance which is normally antigenic. This occurs when the animal is exposed to antigen in fetal or early neonatal life, before the maturation of the immune system. It is thought that this is why animals do not normally respond to components of their own tissues. Tolerance or immunological paralysis may also be induced in adult animals by the administration of

TABLE 4. Summary of published work on effects of bursectomy and thymectomy on immunity to coccidiosis in the fowl[a]

Species	Experimental procedure	Findings	Reference
E. tenella	SBX at 2 weeks, one infection with 35,000 oocysts at 4 - 4½ weeks	Mortality in SBX groups significantly ($P < 0.01$) higher than in controls	Challey, 1962
E. tenella	HBX *in ovo* (9th or 12th day of incubation). Immunization with three doses of oocysts at 7, 14 and 21 days. Results of challenge infection with 50,000 oocysts at 30th day assessed on 36th day	All birds resistant to challenge, judged macroscopically. Histologically, mild coccidial infection in 6/15 completely BX and 5/9 incompletely BX compared with 1/7 intact controls. Completeness of bursectomy judged by immuno-electrophoretic detection of serum γ-globulin, presence of lysins, histological examination for lymphoid tissue and examination *postmortem* for bursal remnants	Long and Pierce, 1963
E. tenella	HBX *in ovo* (6th-9th day of incubation). Three immunizing doses given. Effects of challenge infections assessed	Slightly increased oocyst production in HBX birds during 2nd and 3rd immunizing infections. Most birds resistant to challenge but parasites detected histologically in 15/15 completely BX and 10/25 partially BX compared with 0/8 controls. Serum lysins present in only one bird in the BX groups. Secondary lymphoid foci and pyroninophilic cells reduced in HBX birds	Pierce and Long, 1965
	TX within 95 min of hatching	Slightly increased oocyst production during immunization. All birds completely resistant to challenge infections macroscopically and histologically. Serum lysins equally present in TX and controls. No differences in splenic secondary lymphoid foci, pyroninophilic cells or tissue lymphocytes. Reduced numbers of large and small blood lymphocytes	
E. tenella	HBX *in ovo* on 6th day of incubation, with or without splenectomy at 4 or 5 weeks	Neither treatment alone nor in combination caused greater oocyst production or affected the acquisition of resistance	Rose, 1968b

E. brunetti *E. maxima*	SBX at hatching or on following day, with or without X-irradiation. TX at hatching, with or without irradiation. Oocyst output resulting from the single immunizing infection measured. Results of challenge infections assessed	Only consistent finding was slightly increased oocyst production during immunizing infection in SBXI groups compared with I controls. All birds immunized with *E. maxima* completely resistant to challenge with exception of very small numbers of oocysts produced by TXI groups in one experiment. No differences in resistance to challenge with *E. brunetti* between operated and controls. TXI groups had lowered body weights and circulating lymphocytes, but reduced tuberculin hypersensitivity present in only a small proportion	Rose and Long, 1970
E. tenella	SBX at 2 weeks of age, inoculated with 10,000 oocysts at 4 weeks	No significant difference in mortality at five days after oocyst inoculation, or in total mortality	Rouse and Burns, 1971
	SBX or sham SBX within 48 hr of hatching. Inoculated with 2,000 oocysts at 5 weeks	No significant differences in mortality, weight or hematocrit values	
	SBX or sham SBX at 24-48 hr. Inoculated with 500 oocysts at 2 weeks. Challenged with 200,000 oocysts at 5 weeks	No significant differences in mortality, weight or hematocrit values at challenge	
	TX or sham TX at 12-40 hr. Inoculated with 40,000 sporulated oocysts at 36 days	No significant differences in mortality, weight or hematocrit values	
	TX or sham TX at 24-72 hr. Inoculated with 4,000 oocysts at 6 weeks. Challenged with 400,000 oocysts at 9 weeks	No differences in resistance to the challenge with 400,000 oocysts but "subtle differences" in mortality, growth and hematocrit values between TX and sham TX birds given 400,000 oocysts as an initial infection at 9 weeks	

[a]Abbreviations: SBX = surgical bursectomy, HBX = hormonal bursectomy, TX = thymectomy, I = x-irradiation.

very large amounts of a potential immunogen especially at a time when general immune responsiveness has been depressed, e.g., by thymectomy, x-irradiation, or cortisone treatment.

The ability to grow some species of *Eimeria* in the developing chick embryo has raised the question of the possible induction of tolerance to the coccidia, since apparently healthy chicks can hatch from eggs whose CAMs have supported the growth of considerable numbers of parasites. Such chicks would presumably have been in contact with parasite antigen in increasing amounts from the tenth day approximately of incubation.

Jeffers and Wagenbach (1970) compared the mortality resulting from *E. tenella* infection in chicks hatched from eggs which had harbored the parasite in the CAM from 10 to 12 days incubation with that found in control chicks. They found that this was higher in the chicks which had been exposed to the developing parasite during the incubation period, although in terms of general thriftiness (e.g., body weight) there were no differences between the two groups. This suggests that a state of partial tolerance may have been induced, although, as the authors point out, no measurements of immune response to coccidia, other than the outcome of approximately LD_{50} infections, were made.

Niilo (1969) inoculated calves with oocysts of species of *Eimeria* parasitic in cattle three to 40 hr after birth and four times thereafter, at intervals ranging from three to 11 days. The effects of challenge at five, seven, and nine months were compared with infections in calves which had not been neonatally exposed. The neonatally exposed calves passed fewer oocysts and suffered less disease than the controls at the time of the first challenge but at the second and third challenges they passed more oocysts than the controls and suffered mild diarrhea, whereas the controls did not. It is difficult to interpret these results in terms of tolerance since the neonatally exposed animals had obviously developed some resistance as shown by the results of the first challenge inoculum. From the figures given for oocyst production during the period of neonatal inoculation, it is clear that none resulted from the very early oocyst doses since, in three of the four calves, oocyst discharge began only at 32 to 35 days. It is thus possible that infection occurred only toward the end of the neonatal exposure period, perhaps due to the relative insusceptibility of very young animals, discussed in Section II, B.

IX. Active Immunization

The economic importance of coccidiosis, especially in the poultry industry, is such that effective immunization has long been the goal of many research workers. A thoroughly satisfactory method however has yet to be evolved, and it possibly may not be achieved until more is known of the nature of the immune responses and the antigens which elicit protective immunity.

A. IMMUNIZATION BY INFECTION

1. *With fully viable oocysts*

Since subclinical infection with small or moderate numbers of oocysts will result in immunity to massive challenge doses, a system of planned immunization can be evolved which works well in laboratory conditions. In practice, however, there can be no control over reinfection, but the method has been tried out in conjunction with medication. Early workers (Johnson, 1927, 1932, and 1938; Farr, 1943; Dickinson, Babcock, and Osebold, 1951) showed that it was important that the coccidiostat be not too effective or immunity would not develop. The ideal coccidiostat should allow enough multiplication of the parasite to provide a good immunogenic stimulus to the host but limit it so that the rate of weight gain and feed conversion is not adversely affected. Immunity must also develop sufficiently quickly to enable the coccidiostat to be withdrawn early or the cost becomes too great.

Edgar (1954) has developed a method which is in commercial use in the U.S.A. This system is effective where the management is such as to allow a steady, low, but adequate level of exposure to reinfection after cessation of the immunization program. It suffers from the disadvantage, common to all methods in which medication is given in the food, that reduced food intake necessarily reduces drug intake and thus loss of control over the infection. Some of the factors involved in the immunization of commercial flocks by this method have been discussed by Dorsman (1956) and Reid (1960).

2. With attenuated oocysts

Methods used for attenuation of oocysts as a means of immunization have included heat treatment, freezing, ultrasonic vibrations and, notably, ionizing radiation.

Jankiewicz and Scofield (1934) did not protect chickens against *E. tenella* by feeding heated oocysts, but others have claimed some success for this method (Uricchio, 1953; Lowder, 1966) and for freezing (Uricchio, 1953).

Early work on the x-irradiation of oocysts showed that the pathogenic effects of a suspension of treated oocysts could be reduced proportionately to the amount of radiation given (Albanese and Smetana, 1937; Waxler, 1941), and Waxler stated that chicks which received irradiated oocysts were almost as resistant to reinfection as those given normal oocysts. These chicks, however, did suffer to some degree from coccidiosis and it is possible that the effects seen were caused by destruction of some of the oocysts administered rather than to attenuation of the suspension as a whole.

The successful development of a vaccine, consisting of irradiated larvae, against the cattle lungworm *Dictyocaulus viviparus* Jarrett *et al.*, 1958, 1960), stimulated fresh interest in the effects of ionizing radiation on *Eimeria* oocysts and Hein (1963) reported that two doses of oocysts of *E. tenella* x-irradiated with 11,000 r, administered either orally or in the drinking water with a two-week interval between doses, provided considerable protection against a challenge inoculum given two weeks later. Weight gains during immunization were satisfactory; there was no decrease in hemoglobin concentration, but there was slight hemorrhage on the fifth day after inoculation and the infections became patent but with reduced numbers of oocysts. Oocyst production after the challenge inoculum was low. Fitzgerald (1965, 1968), working with *E. bovis*, concluded that inoculation with irradiated oocysts resulted in immunity to challenge only when the level of radiation was such as to enable the parasite to give rise to a moderate infection with completion of its life cycle. Similar results were obtained by Baldelli *et al.*, (1966) who exposed oocysts of *E. tenella* to γ-radiation.

The results obtained with ionizing radiation suggest that, in all cases where immunity resulted, the immunizing infections were patent ones and thus the aim of immunization by means of a halted life cycle with no oocyst production was not achieved. There was no conclusive evidence that a high dose of irradiated oocysts differed in its effects, either pathogenic or immunogenic, from a lower dose of fully viable oocysts.

In other protozoal infections the inoculation of irradiated infective forms has resulted in nonpatent infections with subsequent immunity (Sanders and Wallace, 1966; Nussenzweig *et al.*, 1967; Phillips, 1970, 1971). If it were possible to irradiate coccidial oocysts so that the life cycle were halted at or beyond the immunogenic stages, it is quite likely that the conditions would have to be carefully investigated for each species since the immunizing stages may differ in the different species (see Section III, F).

3. *With viable parasites inoculated parenterally*

The presence of the living parasite, metabolizing and possibly passing through several stages of its life cycle, is probably more likely to stimulate active immunity than the injection of dead organisms or antigens prepared from them, and hence the attempts at attenuation as a means of immunization (discussed above).

A parenteral infection with no contamination of the environment with infective material has obvious advantages. Fitzgerald (1965) injected viable sporulated oocysts or merozoites of *E. bovis* by various parenteral routes but produced immunity to challenge only when a patent infection resulted from the parenteral inoculum, i.e., by intraperitoneal inoculation only, in his experiments. It is of course possible that insufficient antigen was available in the parenteral site as there is, presumably, no multiplication of the parasite outside the intestine by those species normally parasitic in the intestinal epithelium, except in infections of the chick embryo and in immunosuppressed hosts. Infection of part of the alimentary tract does confer protection on the previously unparasitized remainder, and the idea of setting up an immunizing, parenteral infection which, of necessity, is nonpatent, is still an attractive one. Many workers have found that parenteral inoculation in many different sites often results in an intestinal infection.

B. IMMUNIZATION WITH NONVIABLE MATERIAL

Attempts to immunize against coccidiosis by the injection of nonviable parasite material have, so far, met with little success.

The early workers, as a result of some evidence for the development of circulating antibodies in infected animals, injected ground oocysts or infected intestine parenterally (Bachman, 1930a; Becker *et al.*, 1935). After the conclusive demonstration of serum antibodies in infected animals, further attempts were made to immunize with antigens prepared from schizonts of *E. tenella* (infected tissue) or the exudate present in bile ducts of rabbit livers infected with *E. stiedai.* These antigens were precipitated with aluminum hydroxide (*E. stiedai*) or emulsified with Freund's complete adjuvant (*E. tenella*) before injection, but no protection from subsequent infection produced by oral inoculation of oocysts resulted (Rose, 1961, 1963; Horton-Smith *et al.*, 1963a) although antibodies indistinguishable, by gel precipitation or complement-fixation tests, from those found in infected animals were present. However, chickens injected with the *E. tenella* antigen were protected from challenge inoculations of sporozoites given intravenously (Long and Rose, 1965); presumably the sporozoites were affected by direct contact with circulating antibodies which have been shown to have deleterious effects *in vitro* on invasive stages (see Section IV, B).

Lowder (1966), however, claimed to have immunized calves against *E. bovis* infections by oral inoculation of a soluble extract of disrupted oocysts; the oocyst extract given intravenously failed to protect.

The results of these immunization trials with nonviable parasite material have not been very encouraging, but recent work showing partial protection with transferred antiserum (see Section V) suggests that active immunization in the absence of patent infection may yet be possible. Immunization against plasmodia, in which antigenic variation occurs (Brown and Brown, 1965), has been achieved with dead parasites in Freund's adjuvant (Freund *et al.*, 1948; Targett and Fulton, 1965; Brown, Brown, and Hills, 1970). Tissue and embryo culture of some of the *Eimeria* should provide material, hitherto unavailable, which might profitably be used in immunization experiments.

X. Concluding Remarks

The nature of protective immunity to the coccidia is little understood despite the considerable amount of experimentation on the subject. As with other parasitic infections, the efforts of many research workers have been directed toward determining whether immunity is antibody- or cell-mediated. The dangers of dividing the mechanisms of immunity to parasites into these two categories, brought about largely by the limitations of experimentation, have been pointed out by Ogilvie (1970) and Soulsby (1970), who stressed the necessity for considering changes in the different components of the immune response throughout infection.

The results obtained in the work discussed in the previous sections support this thesis and indicate that probably both types of immune response are active, their relative importance perhaps varying with the stage of infection (see Rose, 1971a).

Humoral antibodies have been shown to be protective and maternal transmission of immunity to occur at certain stages in the immune response, but not at others; at these times the predominant mechanism must change to mediation by locally produced antibody, or by cellular mechanisms, or a combination of these. These may also be operating in conjunction with humoral antibody at the time when this can be shown to be effective, since its demonstrated activity does not match the immunity of the host.

There is some evidence for the participation of locally produced antibodies, but much more information is required. The work on isolation and protective activities of immunoglobulins in secretions should be extended and efforts made to demonstrate the local production of antibodies, possibly by fluorescent labeling.

Investigations on cell-mediated responses have been few; most of the assumptions as to their importance in immunity have been based on the absence of evidence for protective antibodies and the results of experiments on cell transfer and on immunosuppression. Protection with transferred cell suspensions does not necessarily indicate that this is caused by a cell-mediated mechanism (discussed in Section VI, B), and most immunosuppressive methods are not completely specific. Attention should therefore be given to experimentation designed to indicate more specifically the functioning of cell-mediated immune responses. The role of the macrophage should be further investigated, including the phagocytosis and intracellular destruction of coccidia and interactions with sensitized lymphocytes.

Where immunosuppressive methods are used in an attempt to determine coccidial immunity mechanisms, controls for well-established immune responses should be included to corroborate the effectiveness and selectivity of immunosuppression.

There are indications that host immune responses influence parasite behavior, for example, life cycle limitation, site- and species-specificity. This is an aspect of the coccidia which merits thorough investigation; it is unlikely that immune responses are entirely or even largely responsible for parasite behavior, but they are probably contributory to other (physiological and biochemical) factors.

The development of *in vitro* methods for culturing the coccidia should provide an opportunity for more thorough investigation of antigens, and this may lead to a more rewarding approach to active immunization in the absence of infection.

Literature Cited

Albanese, A.A. and H. Smetana. 1937. Studies on the effects of X-rays on the pathogenicity of *Eimeria tenella*. Am. J. Hyg. 26: 27-39.

Andersen, F.L., L.J. Lowder, D.M. Hammond, and P.B. Carter. 1965. Antibody production in experimental *Eimeria bovis* infections in calves. Exp. Parasitol. 16: 23-35.

Asherson, G.L. 1968. The role of micro-organisms in autoimmune responses. Progr. Allergy 12: 192-245.

Asherson, G.L. and M.E. Rose. 1963. Autoantibody production in rabbits. III. The effects of infection with *Eimeria stiedae* and its relation to natural antibody. Immunology 6: 207-216.

Augustin, R. and A.P. Ridges. 1963. Immunity mechanisms in *Eimeria meleagrimitis*. *In* Immunity to protozoa. P.C.C. Garnham, A.E. Pierce and I. Roitt (eds.), pp. 296-335. Blackwell Scientific Publications, Oxford.

Babcock, W.E. and E.M. Dickinson. 1954. Coccidial immunity studies in chickens. 2. The dosage of *Eimeria tenella* and the time required for immunity to develop in chickens. Poult. Sci. 33: 596-601.

Bachman, G.W. 1930a. Immunity in experimental coccidiosis of rabbits. Am. J. Hyg. 12: 641-649.

Bachman, G.W. 1930b. Serological studies in coccidiosis of rabbits. Am. J. Hyg. 12: 624-640.

Baldelli, B., G. Asdrubali, A. Begliomini, T. Frescura, and D. Massa. 1969. Studio degli effetti delle radiazioni gamme sui coccidia dei polli. IV. Potere immunizzante di oocisti di *Eimeria tenella* irradiate dopo la sporulazione. Atti. Soc. Ital. Sci. Vet. 20: 712-716.

Barrett, J.T., M.M. Rigney, and R.P. Breitenbach. 1971. The influence of endotoxin on the responses of chickens to *Plasmodium lophurae* malaria. Avian Dis. 15: 7-13.

Bazin, H. 1970. L'immunité intestinale. Pathol. Biol. 18, 1101-1106.

Becker, E.R. and P.R. Hall. 1933. Cross immunity and correlation of oocyst production during immunization between *Eimeria miyairii* and *Eimeria separata* in the rat. Am. J. Hyg. 18: 220-223.

Becker, E.R., P.R. Hall, and A. Hager. 1932. Quantitative, biometric and host-parasite studies on *Eimeria miyairii* and *Eimeria separata* in rats. Iowa State College J. Sci. 6: 299-316.

Becker, E.R., P.R. Hall, and R. Madden. 1935. The mechanism of immunity in murine coccidiosis. Am. J. Hyg. 21: 389-404.

Bergmann, V. 1970. Elektronenmikroskopische Untersuchungen zur Pathogenese der Blinddarmkokzidiose der Hühnerküken. Arch. Exp. Vet. 24: 1169-1184.

Beyer, T.V. 1963. Immunity in experimental coccidiosis of rabbit caused by heavy infective doses of *Eimeria intestinalis*. *In* Progress in protozoology. J. Ludvík, J. Lom and J. Vávra (eds.), p. 448. Academic Press, New York.

Biester, H.E. and L.H. Schwarte. 1932. Studies in infectious enteritis of swine, VI. Immunity in swine coccidiosis. J. Am. Vet. Med. Ass. 34: 358-375.

Biggs, P.M., P.L. Long, S.G. Kenzy, and D.G. Rootes. 1968. Relationship between Marek's disease and coccidiosis. II. The effect of Marek's disease on the susceptibility of chickens to coccidial infection. Vet. Rec. 83: 284-289.

Brown, I.N. 1969. Immunological aspects of malaria infection, Adv. Immunol. 11: 267-349.

Brown, K.N. and I.N. Brown. 1965. Immunity to malaria: Antigenic variation in chronic infections of *Plasmodium knowlesi*. Nature 208: 1286-1288.

Brown, K.N., I.N. Brown, and L.A. Hills. 1970. Immunity to malaria. 1. Protection against *Plasmodium knowlesi* shown by monkeys sensitized with drug-suppressed infections or by dead parasites in Freund's adjuvant. Exp. Parasitol. 28: 304-317.

Burns, W.C. and J.R. Challey. 1959. Resistance of birds to challenge with *Eimeria tenella*. Exp. Parasitol. 8: 515-526.

Burns, W.C. and J.R. Challey. 1965. Serum lysins in chickens infected with *Eimeria tenella*. J. Parasitol. 51: 660-668.

Capron, A., J. Biguet, A. Vernes, and D. Afchain. 1968. Structure antigenique des helminthes. Aspects immunologiques des relations hôte-parasite. Pathol. Biol. 16: 121-138.

Černá, Ž. 1966a. Anwendung der indirekten Fluoreszenzantikörperreaktion zum Nachweis der Antikörper bei Kaninchenkokzidiose. Zentralbl. Bakteriol. Parasitenk. Abt. I. Orig. 199: 264-267.

Černá, Ž. 1966b. Studies on the occurrence and dynamics of coccidial antibodies in *Eimeria stiedae* and *E. magna* by the indirect fluorescent antibody test. Fol. Parasitol. Prague 13: 332-342.

Černá, Ž. 1967. The dynamics of antibody against *Eimeria tenella* under the fluorescent microscope. Fol. Parasitol. Prague 14: 13-18.

Černá, Ž. 1969. Antibodies in chicks infected by *E. tenella* and *E. acervulina*, detected by an indirect fluorescent reaction. Acta Vet. Brno 38: 37-41.

Černá, Ž. 1970. The specificity of serous antibodies in coccidioses. Fol. Parasitol. Prague 17: 135-140.

Challey, J.R. 1962. The role of the bursa of Fabricius in adrenal response and mortality due to *Eimeria tenella* infections in the chicken. J. Parasitol. 48: 352-357.

Challey, J.R. and W.C. Burns. 1959. The invasion of the cecal mucosa by *Eimeria tenella* sporozoites and their transport by macrophages. J. Protozool. 6: 238-241.

Champion, L.R. 1954. The inheritance of resistance to cecal coccidiosis in the domestic fowl. Poult. Sci. 33; 670-681.

Clarkson, M.J. 1959. The life history and pathogenicity of *Eimeria meleagridis* Tyzzer, 1927, in the turkey poult. Parasitology 49: 519-528.

Cohen, S. and G.A. Butcher. 1970. Properties of protective malaria antibody. Immunology 19: 369-383.

Cox, F.E.G. 1968. Immunity to malaria after recovery from piroplasmosis in mice. Nature 219: 646.

Damian, R.T. 1964. Molecular mimicry: antigen sharing by parasite and host and its consequences. Amer. Natur. 98: 129-149.

Damian, R.T. 1967. Common antigens between adult *Schistosoma mansoni* and the laboratory mouse. J. Parasitol. 53: 60-64.

Davis, L.R., D.C. Boughton, and G.W. Bowman. 1955. Biology and pathogenicity of *Eimeria alabamensis* Christensen, 1941, an intranuclear coccidium of cattle. Am. J. Vet. Res. 16: 274-281.

Desowitz, R.S. 1970. Antiparasitic mechanisms in parasitic infections. J. Parasitol. 56 (No. 4, Sect. 2): 521-525.

Dickinson, E.M., W.E. Babcock, and J.W. Osebold. 1951. Coccidial immunity studies in chickens 1. Poult. Sci. 30: 76-80.

Dorsman, W. 1956. De enting van kuikens tegens coccidiose. Tijdschr. Diergeneesk. 81: 783-790.

Draper, L.R. and D.M. Süssdorf. 1957. The serum hemolysin response in intact and splenectomized rabbits following immunization by various routes. J. Infect. Dis. 100: 147-161.

Dürr, U. and L. Pellérdy. 1969. The susceptibility of suckling rabbits to infection with coccidia. Acta Vet. Acad. Sci. Hung. 19: 453-462.

Edgar, S.A. 1954. Control of cecal coccidiosis by active immunization. Auburn Vet. 10: 79-81, 116.

Ernst, J. V., D.M. Hammond, and B. Chobotar. 1968. *Eimeria utahensis* sp.n. from kangaroo rats (*Dipodomys ordii* and *D. microps*) in northwestern Utah. J. Protozool. 15: 430-432.

Euzeby, J., C. Garcin, and N. Grosjean. 1969. Physio-pathologie de la coccidiose caecale du poulet; étude expérimentalle de l'immunogenèse, action de sérums antilymphocytaires spécifiques. Compt. Rend. Acad. Sci. Paris, Série D. 268: 1616-1618.

Farr, M.M. 1943. Resistance of chickens to cecal coccidiosis. Poult. Sci. 22: 277-286.

Fayer, R. and S. Baron. 1971. Activity of interferon and an inducer against development of *Eimeria tenella* in cell culture. *J. Protozool.* 18 (Suppl.): 12.

Fitzgerald, P.R. 1964. Attempted passive immunization of young calves against *Eimeria bovis.* J. Protozool. 11: 46-51.

Fitzgerald, P.R. 1965. The results of parenteral injections of sporulated or unsporulated oocysts of *Eimeria bovis* in calves. J. Protozool. 12: 215-221.

Fitzgerald, P.R. 1967. Results of continuous low-level inoculations with *Eimeria bovis* in calves. Am. J. Vet. Res. 28: 659-665.

Fitzgerald, P.R. 1968. Effects of ionizing radiation from cobalt-60 on oocysts of *Eimeria bovis.* J. Parasitol. 54: 233-240.

Frenkel, J.K. 1967. Adoptive immunity to intracellular infection. J. Immunol. 98: 1309-1319.

Frenkel, J.K. 1970. Cells as principal effectors of immunity. J. Parasitol. 56 (No. 4, Sect. 2): 107.

Freund, J., W.J. Thomson, H.E. Sommer, A.W. Walter and T.M. Pisani. 1948. Immunization of monkeys against malaria by means of killed parasites with adjuvant. Am. J. Trop. Med. 28: 1-22.

Gabrielsen, A.E. and R.A. Good. 1967. Chemical suppression of adoptive immunity. Adv. Immunol. 6: 91-229.

Gentry, L.E. and J.S. Remington. 1971. Resistance against *Cryptococcus* conferred by intracellular bacteria and protozoa. J. Infect. Dis. 123: 22-31.

Gray, D.F. and C. Cheers. 1969. The sequence of enhanced cellular activity and protective humoral factors in murine pertussis. Immunology 17: 889-896.

Grosjean, N. 1969. Immunologie de la coccidiose caecale du poulet. Thesis, École Nationale Vetérinaire de Lyon, Lyon, France.

Haberkorn, A. 1970. Die Entwicklung von *Eimeria falciformis* (Eimer 1870) in der weissen Maus (*Mus musculus*). Z. Parasitenk. 34: 49-67.

Hammond, D.M., F.L. Andersen, and M.L. Miner. 1963. The site of the immune reaction against *Eimeria bovis* in calves. J. Parasitol. 49: 415-424.

Hammond, D.M., F.L. Andersen, and M.L. Miner. 1964. Response of immunized and non-immunized calves to cecal inoculation of first-generation merozoites of *Eimeria bovis.* J.Parasitol. 50: 209-213.

Hammond, D.M., R.A. Heckmann, M.L. Miner, C.M. Senger, and P.R. Fitzgerald. 1959. The life cycle stages of *Eimeria bovis* affected by the immune reaction in calves. Exp. Parasitol. 8: 574-580.

Hawking, F. 1955. The pathogenicity of protozoal and other parasites: general considerations. *In* Mechanisms of microbial pathogenicity, Fifth Symposium of the Society for General Microbiology. J. W. Howie and A.G. O'Hea (eds.), pp. 176-190. Cambridge University Press, Cambridge.

Heckmann, R., B. Gansen, and M. Hom. 1967. Maternal transfer of immunity to rat coccidiosis; *Eimeria nieschulzi* (Dieben: 1924). J. Protozool. 14: (Suppl.): 35.

Hein, H. 1963. Vaccination against infection with *Eimeria tenella* in broiler chickens. Proc. 17th World Vet. Congr. 2: 1443-1452.

Hein, H. 1968. Resistance in young chicks to reinfection by immunization with two doses of oocysts of *Eimeria acervulina*. Exp. Parasitol. 22: 12-18.

Hein, H. 1971. The effects of cross infections with *Eimeria brunetti* in chickens immunized with multiple doses of *Eimeria maxima*. Exp. Parasitol. 29: 367-374.

Heist, C.E. and T.D. Moore. 1959. Serological and immunological studies of coccidiosis of rabbits. J. Protozool. 6 (Suppl.): 7.

Henry, D.P. 1931. Allergy and immunity in coccidial infections. Proc. Soc. Exp. Biol. Med. 28: 831-832.

Herlich, H. 1965. Effect of chicken antiserum and tissue extracts on the oocysts, sporozoites and merozoites of *Eimeria tenella* and *Eimeria acervulina*. J. Parasitol. 51: 847-857.

Heydorn, A.O. 1970. Versuche zur adoptiven Übertragung der Immunität gegen *Eimeria nieschulzi* (Dieben, 1924) durch Lymphozyten. Inaugural Dissertation, Free University, Berlin.

Horton-Smith, C. 1947. Coccidiosis—some factors influencing its epidemiology. Vet. Rec. 59: 645-646.

Horton-Smith, C. 1949. The acquisition of resistance to coccidiosis by chickens during treatment with sulphonamides. Vet. Rec. 61: 237-238.

Horton-Smith, C. 1963. Immunity to avian coccidiosis. Brit. Vet. J. 119: 99-109.

Horton-Smith, C., J. Beattie, and P.L. Long. 1961. Resistance to *Eimeria tenella* and its transference from one caecum to the other in individual fowls. Immunology 4: 111-121.

Horton-Smith, C. and P.L. Long. 1963. Coccidia and coccidiosis in the domestic fowl and turkey. Adv. Parasitol. 1:68-107.

Horton-Smith, C., P.L. Long, A.E. Pierce, and M.E. Rose. 1963a. Immunity to coccidia in domestic animals. *In* Immunity to Protozoa. P.C.C. Garnham, A.E. Pierce and I. Roitt (eds.), pp. 273-295. Blackwell Scientific Publications, Oxford.

Horton-Smith, C., P.L. Long, and A.E. Pierce. 1963b. Behavior of invasive stages of *Eimeria tenella* in the immune fowl (*Gallus domesticus*). Exp. Parasitol. 14: 66-74.

Huang, K-Y, W.W. Schultz, and F.B. Gordon. 1968. Interferon induced by *Plasmodium berghei*. Science 162: 123-124.

Huff, D.K. 1966. Cellular aspects of the resistance of chickens to *Eimeria tenella* infections. Ph.D. Thesis, Michigan State University. University Microfilms Inc., Ann Arbor.

Huff, D. and D.T. Clark. 1970. Cellular aspects of the resistance of chickens to *Eimeria tenella* infections. J. Protozool. 17: 35-39.

Itagaki, K. and M. Tsubokura. 1963. Serological studies on coccidium in the fowl II. Precipitation in infected chicken serum. Jap. J. Vet. Sci. 25: 187-192.

Jahiel, R., J. Vilcek, R. Nussenzweig, and J. Vanderberg. 1968. Interferon inducers protect mice against *Plasmodium berghei* malaria. Science 161: 802-804.

Jankiewicz, H.A. and R.H. Scofield. 1934. The administration of heated oocysts of *Eimeria tenella* as a means of establishing resistance and immunity to cecal coccidiosis. J. Am. Vet. Med. Ass. 37: 507-526.

Janković, B.D. and K. Isaković. 1960. Haemagglutinins in chicken. 1. The rate of formation of naturally occurring haemagglutinins. Acta Med. Iugoslavi 14: 246-255.

Janković, B.D. and K. Isaković. 1966. Antibody production in bursectomized chickens given repeated injections of antigen. Nature 211: 202-203.

Janković, B.D., K. Isaković, S. Petrović, D. Vujić, and J. Horvat. 1970. Studies on antilymphocyte antibody in the chicken. II. Effect of rabbit antithymus and antibursa globulin on immune reactions in young chickens. Clin. Exp. Immunol. 7: 709-722.

Janković, B.D. and W. Mitrović. 1967. Antibody producing cells in the chicken as observed by fluorescent antibody technique. Fol. Biol. Prague 13: 406-410.

Jarrett, W.F.H., F.W. Jennings, B. Martin,

W.I.M. McIntyre, W. Mulligan, N.C.C. Sharp, and G.M. Urquhart. 1958. A field trial of a parasitic bronchitis vaccine. Vet. Rec. 70: 451-454.

Jarrett, W.F.H., F.W. Jennings, W.I.M. McIntyre, W. Mulligan and G.M. Urquhart. 1960. Immunological studies on *Dictyocaulus viviparus* infection. Immunity produced by the administration of irradiated larvae. Immunology 3: 145-151.

Jeffers, T.K., J.R. Challey, and W.H. McGibbon. 1967. Response of inbred lines and their F_1 progeny to experimental infection with *Eimeria tenella*. Poult. Sci. 46: 1276.

Jeffers, T.K. and G.E. Wagenbach. 1970. Embryonic response to *Eimeria tenella* infection. J. Parasitol. 56: 656-662.

Jenkins, C. R. 1963. Heterophile antigens and their significance in the host-parasite relationship. Adv. Immunol. 3: 351-376.

Johnson, L.W. and S.A. Edgar, 1969. Effects of B blood group genotypes in Leghorn lines selected for resistance and susceptibility to cecal coccidiosis and their reciprocal crosses. Poult. Sci. 48: 1827.

Johnson, W.T. 1927. Immunity or resistance of the chicken to coccidial infection. Oregon Agr. Exp. Sta. Bull. 230: 1-16.

Johnson, W.T. 1932. Immunity to coccidiosis in chickens produced by inoculation through the ration. J. Parasitol. 19: 160-161.

Johnson, W.T. 1938. Coccidiosis of the chicken with special reference to species. Oregon Agr. Exp. Sta. Bull. 358: 3-33.

Jones, E.E. 1932. Size as a species characteristic in coccidia: variations under diverse conditions of infection. Arch. Protistenk. 76: 130-170.

Joyner, L.P. 1969. Immunological variation between two strains of *Eimeria acervulina*. Parasitology 59: 725-732.

Kaliss, N. 1958. Immunological enhancement of tumour homografts in mice: a review. Cancer Res. 18: 992-1003.

Kass, E.H. and M. Finland. 1953. Adrenocortical hormones in infection and immunity. Ann. Rev. Microbiol. 7: 361-387.

Kheysin, E.M. 1947. Variability of the oocysts of *Eimeria magna*, Pérard. Zool. J. Moscow. 26: 17-30. (cited by Horton-Smith and Long, 1963).

Kendall, S.B. and F.S. McCullough. 1952. Relationships between sulphamezathine therapy and the acquisition of immunity to *Eimeria tenella*. J. Comp. Pathol. Therap. 62: 116-124.

Kocan, R.M. 1971. The effect of dexamethasone on immune pigeons infected with various strains of *Trichomonas gallinae*. J. Protozool. 18 (Suppl.): 30.

Kotlán, A. and L. Pellérdy. 1949. A survey of the species of *Eimeria* occurring in the domestic rabbit. Acta Vet. Acad. Sci. Hung. 1: 93-97.

Kramer, T.C. and H.C. Cho. 1970. Transfer of immunoglobulins and antibodies in the hen's egg. Immunology 19: 157-167.

Krassner, S.M. 1963. Factors in host susceptibility and oocyst infectivity in *Eimeria acervulina* infections. J. Protozool. 10: 327-333.

Leathem, W.D. and W.C. Burns. 1967. Effects of the immune chicken on the endogenous stages of *Eimeria tenella*. J. Parasitol. 53: 180-185.

Leathem, W.D. and W.C. Burns. 1968. Duration of acquired immunity of the chicken to *Eimeria tenella* infection. J. Parasitol. 54: 227-232.

Lee, H.H. 1967. Oocyst production of two *Eimeria* species in mixed infections. Diss. Abstr. 28, 5: 2052-B.

Levey, R.M. and P.B. Medawar. 1967. The mode of action of antilymphocytic serum. *In* Antilymphocytic serum. Ciba Foundation Study Group No.29. G.E.W. Wolstenholme and M. O'Connor (eds.), pp. 72-90. J. and A. Churchill, London.

Levine, P.P. 1938. *Eimeria hagani* n.sp. (Protozoa: *Eimeriidae*) a new coccidium of the chicken. Cornell Vet. 28: 263-266.

Levine, P.P. 1942. A new coccidium pathogenic for chickens. *Eimeria brunetti* n. sp. (Protozoa: *Eimeriidae*). Cornell Vet. 32: 430-439.

Liburd, E. M. and J. L. Mahrt. 1971a. Adoptive immunity to *Eimeria nieschulzi* in rats. J. Protozool. 18 (Suppl.): 13.

Liburd, E. M. and J. L. Mahrt. 1971b. Acquired immunity to *Eimeria nieschulzi* in thymectomized rats. J. Protozool. 18 (Suppl.): 30.

Long, P.L. 1962. Observations on the duration of the acquired immunity of chickens to *Eimeria maxima* Tyzzer, 1929. Parasitology 52: 89-93.

Long, P.L. 1968. The effect of breed of chickens on resistance to *Eimeria* infections Brit. Poult. Sci. 9: 71-78.

Long, P.L. 1969. Observations on the growth of *Eimeria tenella* in cultured cells from the parasitized chorioallantoic membranes of the developing chick embryo. Parasitology. 59: 757-765.

Long, P.L. 1970a. *In vitro* culture of *Eimeria tenella*. J. Parasitol. 56 (No. 4, Sect. 2): 214-215.

Long, P.L. 1970b. *Eimeria tenella*: Chemotherapeutic studies in chick embryos with a description of a new method (chorioallantoic membrane foci counts) for evaluating infections. Z. Parasitenk. 33: 329-338.

Long, P.L. 1970c. Development (schizogony) of *Eimeria tenella* in the liver of chickens treated with corticosteroid. Nature 225: 290-291.

Long, P.L. 1971a. Studies on the biology of *Eimeria tenella* Railliet and Lucet, 1891 (Sporozoa : Protozoa) with special reference to its culture *in vitro*. Dissertation for the Degree of Doctor of Philosophy, Brunel University.

Long, P.L. 1971b. Schizogony and gametogony of *Eimeria tenella* in the liver of chick embryos. J. Protozool. 18: 17-20.

Long, P.L. and B.S. Milne. 1971. The effect of an interferon inducer on *Eimeria maxima* in the chicken. Parasitology. 62: 295-302.

Long, P.L. and A.E. Pierce. 1963. Role of cellular factors in the mediation of immunity to avian coccidiosis (*Eimeria tenella*). Nature 200: 426-427.

Long, P.L. and M.E. Rose. 1962. Attempted transfer of resistance to *Eimeria tenella* infections from domestic hens to their progeny. Exp. Parasitol. 12: 75-81.

Long, P.L. and M.E. Rose. 1965. Active and passive immunization of chickens against intravenously induced infections of *Eimeria tenella*. Exp. Parasitol. 16: 1-7.

Long, P.L. and M.E. Rose. 1970. Extended schizogony of *Eimeria mivati* in betamethasone-treated chickens. Parasitology 60: 147-155.

Long, P.L., M.E. Rose, and A.E. Pierce. 1963. Effects of fowl sera on some stages in the life cycle of *Eimeria tenella*. Exp. Parasitol. 14: 210-217.

Lowder, L.J. 1966. Artificial acquired immunity to *Eimeria bovis* infections in cattle. Proc. 1st. Int. Congr. Parasitol. 1: 106-107.

Mackaness, G.B. 1964. The immunological basis of acquired cellular immunity. J. Exp. Med. 120: 105-120.

Mackaness, G.B. and R.V. Blanden. 1967. Cellular immunity. Progr. Allergy 11: 89-140.

Moore, E.N. and J.A. Brown. 1951. A new coccidium pathogenic for turkeys, *Eimeria adenoeides* n.sp. (Protozoa : Eimeriidae). Cornell Vet. 41: 124-135.

Morehouse, N.F. 1938. The reaction of the immune intestinal epithelium of the rat to reinfection with *Eimeria nieschulzi*. J. Parasitol. 24: 311-317.

McDermott, J.J. and L.A. Stauber. 1954. Preparation and agglutination of merozoite suspensions of the chicken coccidium, *Eimeria tenella*. J. Parasitol. 40 (Suppl.): 23-24.

McLoughlin, D.K. 1969. The influence of dexamethasone on attempts to transmit *Eimeria meleagrimitis* to chickens and *E. tenella* to turkeys. J. Protozool. 16: 145-148.

Nelson, D.S. 1969. Macrophages and immunity. Frontiers of Biology 11, North-Holland Research Monographs, North-Holland Publishing Company, Amsterdam, London.

Niilo, L. 1967. Acquired resistance to reinfection of rabbits with *Eimeria magna*. Canad. Vet. J. 8: 201-208.

Niilo, L. 1969. Experimental infection of newborn calves with coccidia and reinfection after weaning. Canad. J. Comp. Med. Vet. Sci. 33: 287-291.

Niilo, L. 1970. The effect of dexamethasone on bovine coccidiosis. Canad J. Comp. Med. Vet. Sci. 34: 325-328.

Nussenzweig, R., J. Vanderberg, H. Mast, and C. Orton. 1967. Protective immunity produced by the injection of X-irradiated sporozoites of *Plasmodium berghei*. Nature 216: 160-162.

Ogilvie, B.M. 1970. Immunoglobulin responses in parasitic infections. J. Parasitol. 56 (No. 4, Sect. 2): 525-534.

Orlans, E. and M.E. Rose. 1970. Antibody formation by transferred cells in inbred fowls. Immunology 18: 473-482.

Orlans, E. and M.E. Rose. 1972. An IgA-like immunoglobulin in the fowl. Immunochemistry 9: 833-838.

Patterson, R., J.S. Youngner, W.O. Weigle, and F.J. Dixon. 1962. Antibody production and transfer to egg yolk in chickens. J. Immunol. 89: 272-278.

Pattillo, W.H. 1959. Invasion of the cecal mucosa of the chicken by sporozoites of *Eimeria tenella*. J. Parasitol. 45: 253-258.

Patton, W.H. 1970. *In vitro* phagocytosis of coccidia by macrophages from the blood of infected chickens. J. Parasitol. 56 (No. 4, Sect. 2): 260.

Pellérdy, L.P. 1965. Coccidia and coccidiosis. Publishing House of the Hungarian Academy of Sciences, Budapest.

Phillips, R.S. 1970. Resistance of mice and rats to challenge with *Babesia rodhaini* after inocula-

tion with irradiated red cells infected with *Babesia rodhaini*. Nature 227: 1255.
Phillips, R.S. 1971. Immunity of rats and mice following infection with ^{60}Co-irradiated *Babesia rodhaini* infected red cells. Parasitology 62: 221-231.
Pierce, A.E. 1959. Specific antibodies at mucous surfaces. Vet. Rev. Annot. 5: 17-36.
Pierce, A.E. and P.L. Long. 1965. Studies on acquired immunity to coccidiosis in bursaless and thymectomized fowls. Immunology 9: 427-439.
Pierce, A.E., P.L. Long, and C. Horton-Smith. 1962. Immunity to *Eimeria tenella* in young fowls (*Gallus domesticus*). Immunology 5: 129-152.
Pierce, A.E., P.L. Long, and C. Horton-Smith. 1963. Attempts to induce a passive immunity to *Eimeria tenella* in young fowls (*Gallus domesticus*). Immunology 6: 37-47.
Pout, D.D. 1965. Coccidiosis in lambs. Vet. Rec. 77: 887-888.
Preston-Mafham, R.A. and A.M. Sykes. 1967. Changes in permeability of the mucosa during intestinal coccidiosis infections in the fowl. Experientia 23: 972-973.
Purchase, H.G., R.C. Chubb. and P.M. Biggs. 1968. Effect of lymphoid leukosis and Marek's disease on the immunological responsiveness of the chicken. J. Nat. Cancer Inst. 40: 583-592.
Randall, C.J., G. Grant, and I.H. Sutherland. 1971. Coccidiosis with concurrent Marek's disease. Vet. Rec. 88: 618.
Reid, W.M. 1960. The relationship between coccidiostats and poultry flock immunity in coccidiosis control programs. Poult. Sci. 39: 1431-1437.
Reid, W.M. 1964. *Eimeria brunetti* — studies on incidence and geographical distribution. Am. J. Vet. Res. 25: 224-229.
Remington, J.S. and T.C. Merigan. 1968. Interferon : Protection of cells infected with an intracellular protozoan (*Toxoplasma gondii*). Science 161: 804-806.
Rommel, M. 1969. Untersuchungen über Infektionsverlauf sowie Ausbildung und Natur der Immunität an experimentell mit *Eimeria scabra* (Henry, 1931) und *E. polita* (Pellérdy, 1949) infizierten Schweinen. Habilitationsschrift der Veterinärmedizinischen Fakultät der Freien Universität Berlin, 1969.
Rommel, M. 1970a. Ausbildung und Dauer der Immunität gegen *Eimeria scabra* (Henry, 1931) und *E. polita* (Pellérdy, 1949). Berlin. Münch. Tierärztl. Wochschr. 83: 236-240.
Rommel, M. 1970b. Verlauf der *Eimeria scabra* - und *E. polita*-Infektion in vollempfänglichen Ferkeln und Läuferschweinen. Berl. Münch. Tierärztl. Wochenschr. 83: 181-186.
Rommel, M. 1970c. Studies on the nature of the crowding effect and of the immunity to coccidiosis. J. Parasitol. 56 (No. 4, Sect. 2): 846.
Rommel, M. 1970d. Die Wirkung von Antilymphozytenserum und Kortikosteroiden auf den Übervölkerungseffekt und die Immunität bei der *Eimeria scabra* - Infektion des Schweines. Zentrbl. Veterinärmed B. 17: 797-805.
Rommel, M. and A. O. Heydorn, 1971. Versuche zur Übertragung der Immunität gegen *Eimeria* - Infektionen durch Lymphozyten. Z. Parasitenk. 36: 242-250.
Rose, M.E. 1959a. A study of the life cycle of *Eimeria stiedae* (Lindemann, 1865) and the immunological response of the host. Dissertation for the Degree of Doctor of Philosophy, University of Cambridge, England.
Rose, M.E. 1959b. Serological reactions in *Eimeria stiedae* infection of the rabbit. Immunology 2: 112-122.
Rose, M.E. 1961. The complement-fixation test in hepatic coccidiosis of rabbits. Immunology 4: 346-353.
Rose, M.E. 1963. Some aspects of immunity to *Eimeria* infections. Ann. N.Y.Acad. Sci. 113: 383-399.
Rose, M.E. 1967a. The influence of age of host on infection with *Eimeria tenella*. J. Parasitol. 53: 924-929.
Rose, M.E. 1967b. Immunity to *Eimeria tenella* and *Eimeria necatrix* infections in the fowl. I. Influence of the site of infection and the stage of the parasite. II. Cross-protection. Parasitology 57: 567-583.
Rose, M.E. 1967c. Immunity to *Eimeria brunetti* and *Eimeria maxima* infections in the fowl. Parasitology 57: 363-370.
Rose, M.E. 1968a. Immunity to the *Eimeria*. *In* Immunity to parasites. A.E.R. Taylor, ed., pp. 43-50. Sixth Symposium of the British Society for Parasitology. Blackwell Scientific Publications, Oxford.
Rose, M.E. 1968b. The effect of splenectomy upon infection with *Eimeria tenella*. Parasitology 58: 481-487.
Rose, M.E. 1970. Immunity to coccidiosis; effect of betamethasone treatment of fowls on *Eimeria mivati* infections. Parasitology 60: 137-146.

Rose, M.E. 1971a. Immunity to infection; coccidiosis as an example. *In* Poultry disease and world economy. R.F. Gordon and B.M. Freeman (eds.), pp. 93-108. British Egg Marketing Board Symposium No. 7. Longman, Edinburgh.

Rose, M.E. 1971b. Immunity to coccidiosis; protective effect of transferred serum in *Eimeria maxima* infections. Parasitology 62: 11-25.

Rose, M.E. 1972. Immunity to coccidiosis: maternal transfer in *Eimeria maxima* infections. Parasitology 65: 273-282.

Rose, M.E. and P.L. Long. 1962. Immunity to four species of *Eimeria* in fowls. Immunology 5: 79-92.

Rose, M.E. and P.L. Long. 1969. Immunity to coccidiosis; gut permeability changes in response to sporozoite invasion. Experientia 25: 183-184.

Rose, M.E. and P.L. Long. 1970. Resistance to *Eimeria* infection in the chicken; the effects of thymectomy, bursectomy, whole body irradiation and cortisone treatment. Parasitology 60: 291-299.

Rose, M.E. and P.L. Long. 1971. Immunity to coccidiosis; protective effects of transferred serum and cells investigated in chick embryos infected with *Eimeria tenella*. Parasitology 63: 299-313.

Rose, M.E. and E. Orlans. 1962. Fowl antibody, III. Its haemolytic activity with complements of various species and some properties of fowl complement. Immunology 5: 633-641.

Rose, M.E. and E. Orlans. 1968. Normal immune responses of bursaless chickens to a secondary antigenic stimulus. Nature 217, 231-234.

Rosenberg, M.M. 1941. A study of the inheritance of resistance to *Eimeria tenella* in the domestic fowl. Poult. Sci. 20: 472.

Rosenberg, M.M., J.E. Alicata, and A.L. Palafox. 1954. Further evidence of hereditary resistance and susceptibility to cecal coccidiosis in chickens. Poult. Sci. 33: 972-980.

Roudabush, R.L. 1935. Merozoite infection in coccidiosis. J. Parasitol. 21: 453-454.

Rouse, T.C. 1967. The effect of neonatal bursectomy, thymectomy and splenectomy on resistance to cecal coccidiosis (*Eimeria tenella*) in White Leghorn cockerels. Diss. Abstr. 28: 3.

Rouse, T.C. and W.C. Burns. 1971. Hematocrits, growth, and mortality in surgically bursectomized and thymectomized chickens infected with *Eimeria tenella*. J. Parasitol. 57: 40-48.

Ruskin, J. and J.S. Remington. 1968. Immunity and intracellular infection; resistance to bacteria in mice infected with a protozoon. Science 160: 72-74.

Sanders, A. and F.G. Wallace. 1966. Immunization of rats with irradiated *Trypanosoma lewisi*. Exp. Parasitol. 18: 301-304.

Scholtyseck, E., R.G. Strout, and A. Haberkorn. 1969. Schizonten und Merozoiten von *Eimeria tenella* in Makrophagen. Z. Parasitenk. 32: 284-296.

Schrecke, W. 1969. Experimentelle Coccidieninfektionen bei neugeborenen Tieren. Inaugural Dissertation, Veterinary Faculty of the Justus Liebig University, Giessen.

Schrecke, W. and U. Dürr. 1970. Excystations- und Infektionsversuche mit Kokzidienoocysten bei neugeborenen Tieren. Zentralbl. Bakteriol. Parasitenk. Infekt. Hyg. Abt. I. Orig. 215: 252-258.

Sell, S. 1969. Antilymphocytic antibody: effects in experimental animals and problems in human use. Ann. Intern. Med. 71: 177-196.

Senger, C.M., D.M. Hammond, J.L. Thorne, A.E. Johnson, and G.M. Wells. 1959. Resistance of calves to reinfection with *Eimeria bovis*. J. Protozool. 6: 51-58.

Smith, W.N. and L.R. Davis. 1961. Studies on resistance of sheep to reinfections by coccidia. J. Protozool. 8 (Suppl.):9.

Smithers, R.S., R.J. Terry, and D.J. Hockley. 1969. Host antigens in schistosomiasis. Proc. Roy. Soc. Ser. B. 171: 483-494.

Soulsby, E.J.L. 1958. Studies on the heterophile antibodies associated with helminth infections. I. Heterophile antibody in *Ascaris lumbricoides* infection in rabbits. II. Heterophile antibody in *Ascaris lumbricoides* infection in pigs. III. Heterophile antibody in *Oesophagostomum dentatum* infection in pigs. J. Comp. Pathol. 68: 71-81; 345-351; 380-387.

Soulsby, E.J.L. 1970. Cell mediated immunity in parasitic infections. J. Parasitol. 56 (No. 4, Sect. 2): 534-547.

Springer, G.F., R.E. Horton, and M. Forbes. 1959. Origin of antihuman blood group B agglutinins in White Leghorn chicks. J. Exp. Med. 110: 221-244.

Taliaferro, W.H. 1956. Functions of the spleen in immunity. Am. J. Trop. Med. Hyg. 5: 391-410.

Targett, G.A.T. and J.D. Fulton. 1965. Immunization of rhesus monkeys against *Plasmodium knowlesi* malaria. Exp. Parasitol. 17: 180-193.

Todd, K.S. and D.M. Hammond. 1968. Life cycle and host specificity of *Eimeria callospermophili* Henry, 1932 from the Uinta ground squirrel *Spermophilus armatus*. J. Protozool. 15:1-8.

Tomasi, T.B. and J. Bienenstock. 1968. Secretory immunoglobulins. Adv. Immunol. 9: 1-96.

Tucker, D.F. 1968. Effect of anti-spleen serum on skin allograft survival and blood lymphocyte reactivity in the chicken. Nature 218: 1259-1261.

Tyzzer, E.E. 1929. Coccidiosis in gallinaceous birds. Am. J. Hyg. 10: 269-383.

Tyzzer, E.E., H. Theiler, and E.E. Jones. 1932. Coccidiosis in gallinaceous birds. II. A comparative study of species of *Eimeria* of the chicken. Am. J. Hyg. 15: 319-393.

Uricchio, W.A. 1953. The feeding of artificially altered oocysts of *Eimeria tenella* as a means of establishing immunity to cecal coccidiosis in chickens. Proc. Helminthol. Soc. Washington 20: 77-83.

Van Doorninck, W.M. and E.R. Becker. 1957. Transport of sporozoites of *Eimeria necatrix* in macrophages. J. Parasitol. 43: 40-44.

Vernon-Roberts, B. 1969. The effect of steroid hormones on macrophage activity. Intern. Rev. Cytol. 25: 131-159.

Versényi, L. and L. Pellérdy. 1970. Pathological and immunological investigations of the anseris-coccidiosis of the domestic goose. (*Anser anser dom.*) Acta Vet. Acad. Sci. Hung. 20: 103-107.

Warner, N.L. 1967. The immunological role of the avian thymus and bursa of Fabricius. Fol. Biol. Prague 13: 1-17.

Warner, N.L., A. Szenberg, and F.M. Burnet. 1962. The immunological role of different lymphoid organs in the chicken. I. Dissociation of immunological responsiveness. Austral. J. Exp. Biol. Med. Sci. 40: 373-388.

Warner, N.L., J.W. Uhr, G.J. Thorbecke, and Z. Ovary. 1969. Immunoglobulins, antibodies and the bursa of Fabricius; induction of agammaglobulinemia and the loss of all antibody-forming capacity by hormonal bursectomy. J. Immunol. 103: 1317-1330.

Waxler, S.H. 1941. Immunization against cecal coccidiosis in chickens by the use of X-ray attenuated oocysts. J. Am. Vet. Med. Ass. 99: 481-485.

Wilson, I.D. and L.C. Morley. 1933. A study of bovine coccidiosis, II. J. Am. Vet. Med. Ass. 35: 826-850.

Zuckerman, A. 1964. Autoimmunization and other types of indirect damage to host cells as factors in certain protozoan diseases. Exp. Parasitol. 15: 138-183.

9 Toxoplasmosis: Parasite Life Cycle, Pathology, and Immunology

J. K. FRENKEL
Department of Pathology and Oncology, School of Medicine, The University of Kansas Medical Center, Kansas City, Kansas 66103

Contents

I. Introduction. . . *344*
II. Life Cycle of *Toxoplasma*. . . *345*
  A. Historical Review . . . *345*
  B. Stages of *Toxoplasma* . . . *346*
    1. Introduction. . . *346*
    2. Enteroepithelial Cycle. . . *348*
    3. Extra-intestinal (Tissue) Cycle of Infection. . . *350*
III. Infection. . . *353*
  A. Comparison of Virulence and Infectivity According to Route of Inoculation of Tachyzoites, Bradyzoites, and Sporozoites in Mice . . . *353*
  B. Dynamics of Infection. . . *354*
IV. Pathogenetic Considerations. . . *355*
  A. Microbial Effects. . . *355*
  B. Natural Resistance. . . *356*
  C. Acquired Immunity. . . *357*
  D. Acquired Resistance. . . *357*
  E. Hypersensitivity. . . *358*
V. Pathology. . . *358*
  A. Asymptomatic Infection. . . *358*
  B. Lesions of Symptomatic Infections. . . *359*
    1. Acute Infection. . . *359*
    2. Subacute Infection. . . *361*
    3. Chronic Infection. . . *362*

4. Relapsing Toxoplasmosis. . . *364*
5. Lesions Observed in Individual Species. . . *364*
VI. Immunology. . . *365*
A. Introduction. . . *365*
B. Acquisition of Immunity. . . *365*
C. Recovery from Infection. . . *365*
D. Immunity to Reinfection. . . *366*
E. Premunition (Infection Immunity). . . *368*
F. Immunization with Live and Killed Vaccines. . . *369*
G. Cells, Antibody, and Immunity. . . *370*
H. Antibody Response and *In Vivo* Effects. . . *371*
I. Immunodeficient Hosts. . . *372*
J. Hypersensitivity. . . *375*
K. Acquired Resistance. . . *376*
VII. Transmission. . . *377*
A. Biology of Transmission. . . *377*
1. Transplacental Transmission. . . *377*
2. Transmission by Carnivorism. . . *378*
3. Fecal Transmission. . . *378*
4. Possible Other Means of Transmission. . . *384*
B. Epidemiology and Prevalence. . . *387*
C. Prevention. . . *388*
VIII. Comparison of *Toxoplasma* with Other Coccidia. . . *390*
A. Apparently Related Organisms. . . *390*
1. *Besnoitia*. . . *390*
2. *Sarcocystis*. . . *395*
3. *Frenkelia*. . . *397*
B. *Isospora* and *Eimeria* . . . *399*
Literature Cited. . . *400*

I. Introduction

Toxoplasma has been known since 1908 as a tissue parasite of many species of mammals and birds. However, in 1970 it was shown to have coccidian affinities with an enteroepithelial* cycle and oocyst production in cats and other felines. The studies leading to this discovery will be reviewed below. Since then, the two intestinal cat coccidia, *Isospora felis* and *I. rivolta*, were recognized to invade other cat tissues (Dubey and Frenkel, 1972a), and their oocysts were found infectious to rodents, parasitizing extra-intestinal tissues

* I am using "enteroepithelial" and "enteric" to describe the classically coccidian cycle in the intestinal epithelium. "Extra-intestinal" and "tissue" refer to the location of the long-known *Toxoplasma* stages in cats and of the entire cycle in other mammals. These descriptive terms are scientifically accurate with two reservations, that tissue stages also occur in the lamina propria of the gut, and that enteroepithelial forms of *Toxoplasma* have occasionally been found in the bile duct epithelium. I am introducing two other terms: "tachyzoites" for the rapidly multiplying forms of the acute infection, previously called trophozoites, aggregations, and proliferative forms; and "bradyzoites" for the slowly multiplying encysted forms characteristic of chronic infection, which have been variously called merozoites or just zoites. This avoids the use of conventional terms no longer sufficiently descriptive: "trophozoites," referring to feeding forms, might be applicable to many stages, "proliferative forms" could apply to enterozoites and cyst forms as well; and "merozoites" being part of a whole, might be more descriptive for the products of schizogony than for the cyst organisms which result from endodyogeny. The term "group" is used to designate intracellular tachyzoites within a vacuole and "cyst" for the complex of bradyzoites (Table 1). This avoids the imprecise term "pseudocyst" which has been used both for noncysts and for cysts, even by the same author (Lainson, 1955a, b, 1958).

(Frenkel and Dubey, 1972a). After this relationship had been recognized, knowledge of *Toxoplasma* and the coccidia mutually complemented each other. However, although our concepts broadened recently, many of our present views may still be found too narrow, transitional, speculative, or misleading. Still, a review at this juncture will hopefully enhance progress in the two areas which coexisted independently, but can now be cross-fertilizing.

Certain difficulties in terminology can be partially resolved at this time even though the life cycle of *Toxoplasma* is still incompletely known, and the life cycle of representative coccidia will have to be reinvestigated. Lacking host specificity and significant serologic characteristics for differentiation of species, *Toxoplasma* (*gondii*) is regarded as a monospecific genus with the following characteristics: endodyogeny, endopolygeny, schizogony, and gametogony in the small intestinal epithelium of cats, oocysts with two sporocysts, and four sporozoites developing outside of the host; tachyzoites multiplying by endodyogeny in many types of cells, leading to the production of cysts with many bradyzoites, mainly in brain and muscles; facultatively heteroxenous in many mammals and birds in which only the asexual, extraintestinal cycle has been observed (Frenkel, Dubey, and Miller, 1970). Transmissible forms are tachyzoites, bradyzoites, and sporozoites, which are infectious to complete hosts (cats) and to incomplete hosts (nonfelines).

II. Life Cycle of *Toxoplasma*

A. HISTORICAL REVIEW

Toxoplasma was first recognized in acute infections of gondis (rodents) and rabbits (for a brief review of the early history, see Frenkel, 1970, 1971a). The name was derived from the Greek *toxon*, bow or arc, alluding to their lunate shape; they are not toxic, and their harmful effects depend mainly on intracellular parasitism. The tachyzoites, rapidly dividing tissue forms, infect and destroy the cells in many tissues and organs. After about seven to 12 days, cysts with slowly multiplying bradyzoites make their appearance in brain, eye, skeletal muscle, heart muscle, and elsewhere. Cysts are characterized by an argyrophilic cyst wall, by from tens to hundreds of merozoites with prominent PAS-positive granules, and by an increased resistance to peptic digestion (Jacobs, Remington, and Melton, 1960). Cysts are infectious when ingested, and this readily explains the acquisition of toxoplasmosis in obligatory or facultative carnivores. Desmonts *et al.* (1965) described transmission to humans who ate raw or undercooked meat. But how do herbivores, the usual sources of meat, become infected? Investigations of transmission by arthropods did not yield any significant clues (Woke *et al.*, 1953; Frenkel, 1965a). Hutchison (1965), however, discovered that *Toxoplasma* was fecally discharged by a cat which had eaten infected mice. At first, *Toxoplasma* was believed to be contained in eggs of the nematode *Toxocara cati* (Hutchison, 1967) and then in a "new cyst" (Work and Hutchison, 1969a, b). This will be discussed under fecal transmission and other means of transmission (pp. 378–384). The association of *Toxoplasma* with *Toxocara* was shown to be circumstantial (Sheffield and Melton, 1969; Frenkel, Dubey, and Miller, 1969) and so was its association with the "new cyst" (Dubey, Miller, and Frenkel, 1970b). However, the oocyst and its preceding enteroepithelial cycle found in cats was linked to the long-known *Toxoplasma* stages by means of separate experiments and criteria which furnished consistent evidence (Frenkel, Dubey, and Miller, 1970; Dubey, Miller, and Frenkel, 1970b). Three types of prepatent periods were observed: from three to five days after feeding cysts, from five to ten days after feeding tachyzoites, and from 20-24 days, extending occasionally to 35 days, after feeding oocysts. Of these infections with different prepatent periods, only the cyst-induced infection in domestic cats has been studied in detail (Dubey and Frenkel, 1972b).

B. STAGES OF *TOXOPLASMA*

1. Introduction

Two cycles in separate biotopes and five stages of *Toxoplasma* are known and can be linked into a life cycle (Table 1). An "enteroepithelial" cycle and an "extra-intestinal" cycle can be conveniently distinguished. In cats, the (1) enteroepithelial multiplicative stage is generally similar to that in other coccidia, and leads to (2) gametogony, and (3) oocyst production, with sporogony. Two additional stages occur in the extra-intestinal tissues of cats, and appear to constitute the entire cycle in other mammalian and avian hosts: (4) tachyzoites forming "groups" occur during the acute infection, and (5) bradyzoites within "cysts" are found during chronic infection.

TABLE 1. Classification of cycles, stages, and types of *Toxoplasma*[a]

A. Enteroepithelial cycle

1. Enteroepithelial *multiplicative stages* (after cyst infection[b]) in domestic cats.

	Division type	Occurrence after cyst feeding
a. Type A	Endodyogeny	12-18 hr
b. Type B	Endodyogeny + endopolygeny	12-54 hr
c. Type C	Schizogony	28-54 hr
d. Type D	Schizogony, endopolygeny, splitting	32 hr - 15 days
e. Type E	Schizogony	3-15 days

2. Gametocyte stage
3. Oocyst stage
   a. Sporont
   b. Sporoblast
   c. Sporozoites (mode of division unknown)

B. Extra-intestinal or tissue cycle

4. Tissue *group stage* in cats and in intermediate hosts (synonyms: "terminal colonies," aggregates, pseudocysts — in part)
   a. *Tachyzoites* (trophozoites, proliferative forms) divide by endodyogeny
5. Tissue *cyst stage* in cats and intermediate hosts (pseudocysts — in part)
   a. *Bradyzoites* (merozoites, cyst forms) divide by endodyogeny

[a]Glossary: *aggregatus* (L), added to; *bradys* (Gr), slow; *tachos* (Gr), speed; *trophicos* (Gr), nursing, feeding; *meros* (Gr), part or segment; *zoon* (Gr), animal.
[b]The types of *Toxoplasma* found in cats infected with tachyzoites and oocysts have not yet been clarified.

TABLE 2. Characterization of *Toxoplasma* types in the small intestine of kittens fed cysts[a]

	Type designation				
	A	B	C	D	E
1. Duration of occurrence	12-18 hr	24-54 hr	24-54 hr	32 hr-15 days	3-15 days
2. Number of merozoites in a group					
Common	3	5-10	32	5-10	12-16
Range	2-3	2-30	16-40	2-35	4-24
3. Size in sections (μ)					
Group	4-4.5 × 2-3	3-17 × 2-11	9-17 × 9-11	4.5-8 × 3-8	4.5-9 × 3-8
Individual	1.7-2.3 × 1.1-1.7	1.5-2 × 2.5-3	2.5-4 × 1-1.1	2.8-4.5 × 1-1.1	3-4.5 × 1-1.1
4. (a) Pattern of arrangement of merozoites in the group	No fixed pattern	No fixed pattern	Rosette	Rosette or irregular	Rosette
(b) Probable mode of division	?	Endodyogeny Endopolygeny	Schizogony	Schizogony Endopolygeny "Splitting"	Schizogony
5. Shape					
Group	Round-ovoid	Irregular	Round	Round	Round
Individual	Round	Ovoid	Elongated	Elongated	Elongated
6. Residual body	−	−	+	−	+
7. Cytoplasmic staining in sections					
Giemsa	Light	Deep[b]	Light	Deep	Deep
PAS	Negative	Light	Heavy	Light	Light
8. Location					
Surface epithelium	Common	Common[c]	Common[c]	Common[d]	Common[d]
Lamina propria[e]	Common	Common	Common	Rare	Rare
Glandular epithelium	Not seen	Rare	None	Rare	Rare
9. Effect on host cell					
(a) Host cell nuclei displaced	Yes	Yes	Yes	Yes	Yes
(b) Host cell nuclei hypertrophied	No	Yes	Yes	No	No
10. Level of small intestine where found	Jejunum	Jejunum and ileum	Jejunum and ileum	Jejunum, ileum, and colon	Jejunum, ileum, and colon

[a]Based on tissues fixed in Zenker-formol, from Dubey and Frenkel, 1972b.
[b]Not apparent after Carnoy fixation.
[c]Near base of villi.
[d]Near tips of villi.
[e]Epithelial cell displaced.

2. *Enteroepithelial cycle*

This has been studied in the intestine of newborn and weanling kittens and adult cats infected with cysts derived from the brain and skeletal muscle of mice. Although for certain experiments infections with *Isospora felis* and *I. rivolta* were rigidly excluded, either by the use of newborn kittens, or by raising kittens in isolation to weaning, it was noted that the presence of these other coccidia did not interfere with cyst-produced toxoplasmosis. Piekarski and Witte (1971), using adult cats, observed that infections with *I. bigemina (cati), I. felis* or *I. rivolta*, or with *I. felis* and *I. bigemina (cati)* did not interfere with subsequent *Toxoplasma* infection. Cystic bradyzoites are relatively resistant to peptic digestion (Jacobs, Remington, and Melton, 1960). In the small intestine, these bradyzoites still have a terminal nucleus. After they enter the intestinal epithelial cells, their PAS-positive storage material, probably amylopectin, gradually disappears and a number of morphologic types of multiplicative stages have been observed. Like typical coccidia, they undergo sequential development stages, but atypically some of these stages pass through several multiplicative generations, and hence, they are designated as "types." After cysts are fed, the prepatent period to oocyst production is from three to five days. During this time, the multiplicative types occur, which are designated types A through E; they are followed by gametocytes and oocysts (Table 2).

Type A, which is the smallest of the five intestinal types, consists of collections of two or three organisms occurring in the jejunum from 12 to 18 hr after infection (Fig. 1). Multiplication is indicated by the fact that multiple organisms are found more often than single ones. Type B is characterized by a centrally located nucleus, a prominent nucleolus and dark blue cytoplasm, giving rise to the appearance of bipolar staining with Giemsa. Type B occurs from 12 to 54 hr after infection. It appears to divide by simple endodyogeny and by endopolygeny (Piekarski, Pelster, and Witte, 1971). Type C organisms are elongate, with a subterminal nucleus and strongly PAS-positive cytoplasm. They occur frequently from 24 to 54 hr and divide by schizogony. Type D organisms are smaller than Type C and contain only a few PAS-positive granules. They occur from 32 hr to 15 days and account for over 90% of all *Toxoplasma* found in the small intestine during this period of time. Organisms grouped under Type D appear to divide by endodyogeny, by schizogony, and by splitting of single merozoites from the main nucleated mass. No residual body is left. It is not clear whether Type D constitutes a sequential group, since the three divisional processes go on simultaneously for several days. Type E resembles the type D which divides by schizogony, but has a residual body. It occurs from 3 to 15 days after infection (Figs. 1-3).

Gametocytes occur throughout the small intestine but more commonly in the ileum

Fig. 1. Comparison of asexual stages of *Toxoplasma* in kittens. Tissues fixed in Zenker-formol, stained with Giemsa (*left*), hematoxylin and eosin (*center*) and periodic acid-Schiff, hematoxylin (*right*). × 1600. *CY*, Cyst from brain of a cat infected for 57 days. The PAS stain distinguishes this stage, although the staining of amylopectin and nuclei obscures individual bradyzoites. *T*, Tachyzoites from the liver, 7 days. Larger groups are commonly found. *A*, Enteroepithelial type A, from jejunum, 12 hr; this type is best demonstrated with Giemsa stain. *B*, Enteroepithelial type B, from jejunum, 24 hr; the heavy cytoplasmic staining with Giemsa, and division by endodyogeny or endopolygeny are distinctive. *C*, Enteroepithelial type C, from jejunum, 32 and 48 hr; PAS-positivity, schizogony, and a densely staining residual body are distinctive. *D*, Enteroepithelial type D, from jejunum, 48 and 72 hr; although there are a few small PAS-positive granules, staining with Giemsa is most distinct. Schizogony is shown here; division by endopolygeny and splitting are not shown. *E*, Enteroepithelial type E, from ileum, 8 days; the residual body is distinctive and is most prominent after staining with Giemsa. (From Dubey and Frenkel, 1972b.)

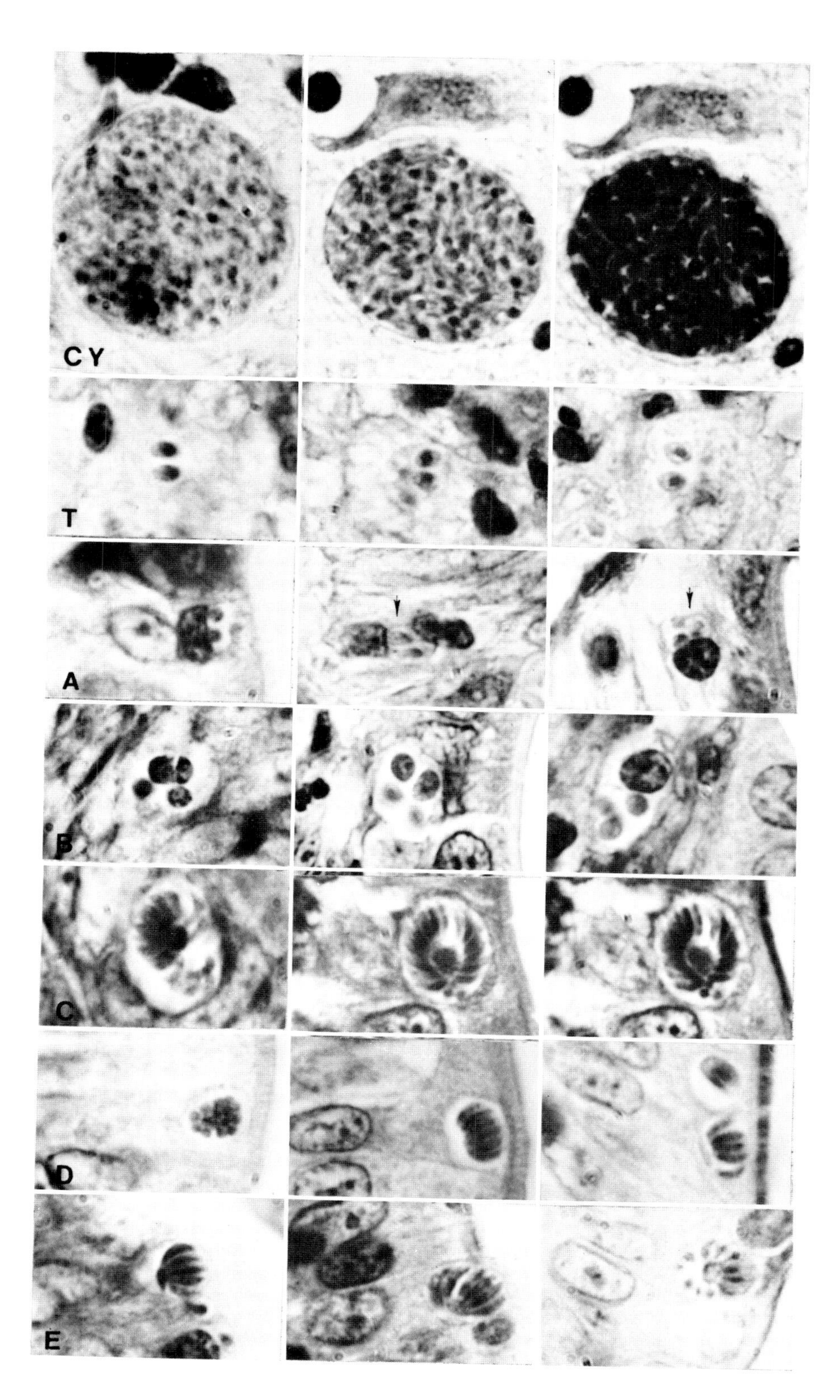
CY
T
A
B
C
D
E

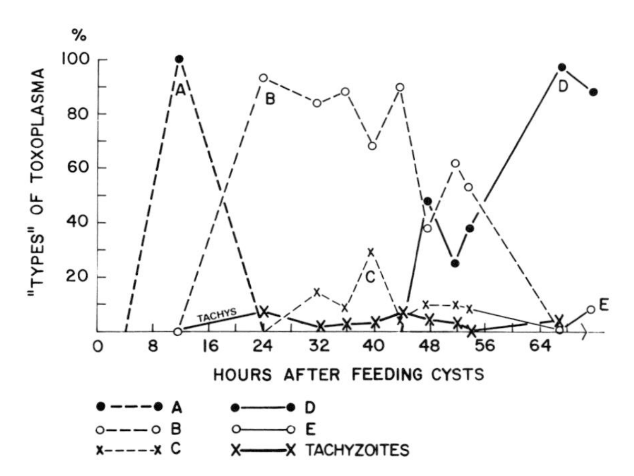

Fig. 2. Enteroepithelial types and tachyzoites found in the small intestine of kittens 12 to 67 hr after the ingestion of *Toxoplasma* cysts. Mature forms were counted at × 1000 from tissues fixed in Zenker-formol and stained with Giemsa. (From Dubey and Frenkel, 1972b.)

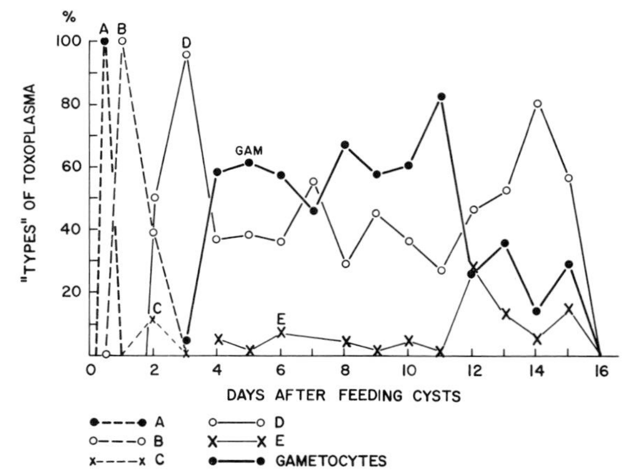

Fig. 3. Mature enteroepithelial types and gametocytes found in the small intestine of kittens 1 to 16 days after the ingestion of *Toxoplasma* cysts. Counts at × 1000 from tissues fixed in Zenker-formol and stained with Giemsa. (From Dubey and Frenkel, 1972b.)

from 3 to 15 days after infection. Male gametocytes produce on the average 12 microgametes and represent from 2% to 4% of the gametocyte population. The prepatent period after cyst-induced infection is from three to five days, with peak oocyst production occurring between five and eight days, and a patent period varying from 7 to 20 days (Dubey and Frenkel, 1972b) (Figs. 4, 21).

Oocysts become surrounded by argyrophilic membranes and the plastic granules in the cytoplasm of the macrogametes disappear. The oocysts are detached from the intestinal epithelium and are discharged. Sporulation occurs in from one to five days depending on temperature and availability of oxygen (Dubey, Miller, and Frenkel, 1970a,b). Two sporocysts are formed, each of which contains four sporozoites (Fig. 5).

3. Extra-intestinal (tissue) cycle of infection

These stages, which occur also in cats, appear to constitute the entire cycle in nonfeline. Even in cats infected orally with cysts, the enteroepithelial cycle, described above, and the extra-intestinal cycles start almost simultaneously. Tachyzoites develop in the lamina propria, mesenteric lymph nodes, and in distant organs, as shown by titrations (Dubey and Frenkel, 1972b). In other mammals infected orally, which, as far as we know lack an enteroepithelial cycle, tachyzoite-forming groups are the first stages of the infection. Cysts start to develop after an interval of from about one to two weeks in all animals which have been studied. They may coexist with the enteroepithelial cycle in cats.

a. Tachyzoite-forming groups

In the acute visceral infection as seen in many hosts, the tachyzoites develop within a vacuole in a multitude of cell types, fibroblasts, reticular cells, hepatic parenchymal cells, pneumocytes, myocardial cells, and neurons, and they have even been seen in nucleated red blood cells of birds (Fig. 10). They may occur briefly after oral infection in

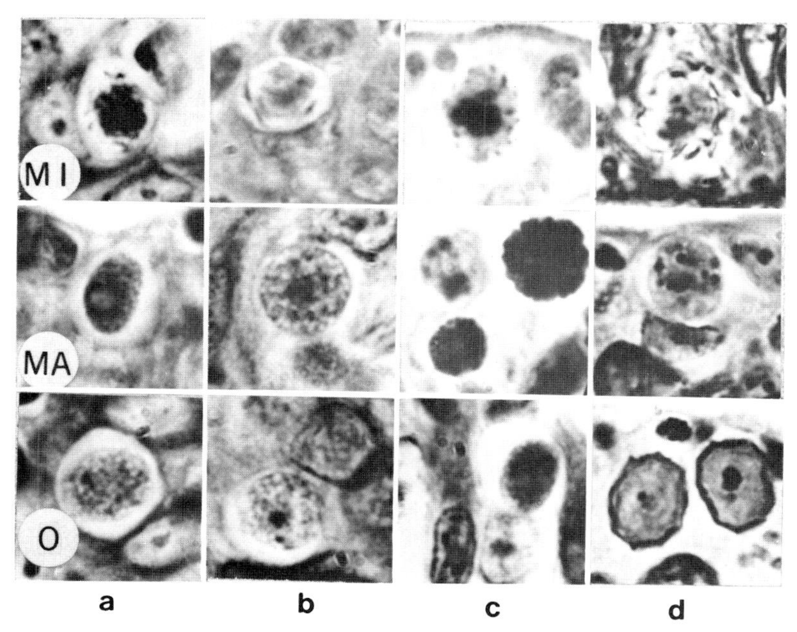

Fig. 4. Comparison of gamonts and oocysts of *Toxoplasma* in the small intestine of kittens. Tissues fixed in Zenker-formol stained either with (*a*) Giemsa, (*b*) hematoxylin and eosin, (*c*) periodic acid-Schiff, hematoxylin, or (*d*) Wilder's ammoniacal silver method and neutral-red. × 1600. MI, Microgamonts. The microgametes are best shown with Giemsa and Wilder's; the residual body is PAS-positive. Ma, Macrogamonts. Giemsa and H&E stains the nucleus. The PAS-positivity of the cytoplasm obscures the nuclei of the two organisms shown. Wilder's shows the plastic granules. O, Oocysts. The oocyst wall is best demonstrated with Wilder's; it stains faintly (not visible here) with other methods. Nuclei are shown with Giemsa, H&E and Wilder's; the PAS-positivity of the oocyst is reduced compared to that of the macrogamonts. (From Dubey and Frenkel, 1972b.)

the gut epithelium of mice. Multiplication is by endodyogeny (Goldman, Carver, and Sulzer, 1958), an internal budding with two organisms arising from the development of two new conoids and related organelles associated with nuclear division; eventually the two organisms escape from the parent (Gavin, Wanko, and Jacobs, 1962; Sénaud, 1967; Sheffield and Melton, 1968). The pellicle is formed in part by an infolding from that of the parent membranes and on liberation (for more details, see chapter 3); an additional membrane is derived from the outer membrane of the parent organism (Vivier and Petitprez, 1969). From about 8 to 16 or more organisms accumulate in the host cell before it disintegrates and new cells are infected. These accumulations of tachyzoites have been called terminal colonies, aggregates, and pseudocysts. These terms are objectionable, as they are not sufficiently descriptive (see footnote on p. 344).

b. Bradyzoite-forming cysts

Cysts are characteristic of the chronic infection and occur mainly in the brain, heart, and skeletal muscle, but also elsewhere (Figs. 1, 12, and 22). The bradyzoites are slowly multiplying, resting stages, which divide mainly by endodyogeny (Wanko, Jacobs, and Gavin, 1962; Wildfuhr, 1966; Zypen and Piekarski, 1966, 1967). Cysts may persist for months and have been recorded five years after experimental infection of guinea pigs (Lainson, 1959). They have been isolated from the brain and skeletal muscle of humans with low antibody titers, suggesting the passage of many years since primary infection (Remington and Cavanaugh, 1965; Robertson, 1966). As they may persist for the life of the host, released cyst organisms stimulate persistence of immunity and antibody. However, if immunity wanes, they are capable of initiating renewed proliferation of tachyzoites, causing a localized or generalized relapse (Frenkel, 1956b). Additional cysts may be formed from these tachyzoites if immunity returns.

Appropriate to their transmission by ingestion, bradyzoites may resist peptic and tryptic digestion (Jacobs, Remington, and Melton, 1960). This has led to a definition of cysts based on resistance of the contained organisms to enzymatic digestion.

Although cyst formation coincides usually with the development of immunity, and the proliferation of tachyzoites with its absence, other circumstances may lead to cyst formation. The occurrence of tightly packed *Toxoplasma* in older cell cultures has been observed by Hogan *et al.* (1960) and was shown by Jacobs and Melton (1965) and Jacobs (1967) to be related to an increased resistance to peptic digestion as compared with the forms in younger cultures. According to Matsubayashi and Akao (1963), who studied cyst development in cultures, chick embryos, and mice, cysts are produced whenever multiplication is retarded.

Another definition of cyst is based on the three-to-five-day prepatent period to oocyst

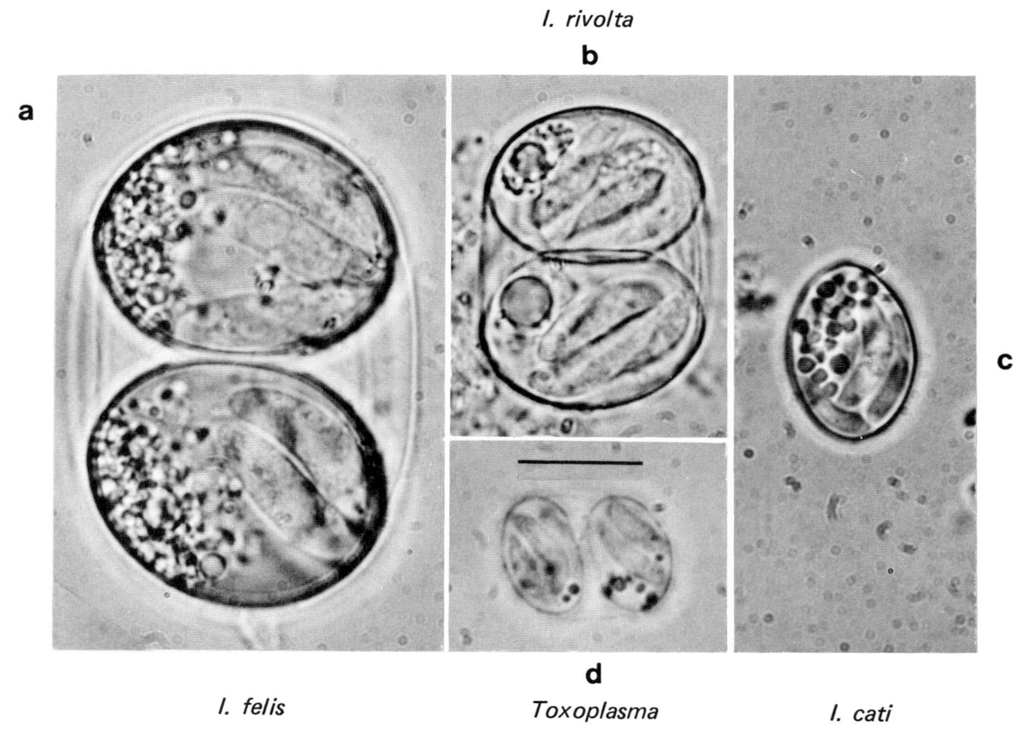

Fig. 5. Comparison of coccidia from feces of cats. × 1600. Flattened oocysts of (*a*) *I. felis*, (*b*) *I. rivolta*, and (*d*) *Toxoplasma gondii*, all of which are discharged in the unsporulated form. Sporocysts of *I. cati* (*c*) were discharged already sporulated from a cat in Costa Rica, which after being fed *Toxoplasma* cysts also developed *Toxoplasma* oocysts. Scale = 10 μ.

production in cats, with trophozoites having a 5-to10-day prepatent period, and oocysts one of more than 20 days (Frenkel, Dubey, and Miller, 1970). This observation suggests an additional significance of the cyst in the life cycle of *Toxoplasma*, beyond its function as a resting stage. Candidate cysts from cell culture should be tested for the prepatent period.

Cysts are surrounded by a tough, argyrophilic cyst wall, which isolates them from the host, so that they do not provoke an inflammatory reaction when intact (Frenkel, 1956b). Antibody levels may decline and disappear although the cysts persist (Eyles, 1954; Jacobs, 1964). Cyst organisms store amylopectin, which stains vividly with the PAS technique, and aids the histologic identification of cysts even by low power objectives. The host cell and its nucleus are not modified.

The cyst, its wall, and the bradyzoites develop intracellularly, and the cyst wall does not include the host-cell nucleus. The nature of the cyst wall has been analyzed ultrastructurally, and consists of a from 15nm-20nm-thick membrane with an irregular surface. It is lined by a granular layer, which fills the space between the bradyzoites, together with a canalicular system which communicates with the cyst wall (Zypen and Piekarski, 1966,1967).

III. Infection

A. COMPARISON OF VIRULENCE AND INFECTIVITY ACCORDING TO ROUTE OF INOCULATION OF TACHYZOITES, BRADYZOITES, AND SPOROZOITES IN MICE

Toxoplasma isolates show marked differences of virulence ("relative pathogenicity") when compared in a single host species (Table 3). Although on primary isolation such a difference may be merely related to the *size* of the inoculum, virulence was compared by titration in a number of studies. Sometimes virulence undergoes a change after repeated passages in a new host ("adaptation"). However, virulence may be fixed, as observed in old laboratory strains. Sometimes, for ease of experimentation, strains in which the minimal infectious dose (MID) equals a

TABLE 3. Comparison of the infectivity and pathogenicity of cysts, and of oocysts and tachyzoites derived from these cysts, by feeding and after subcutaneous and intraperitoneal infection (M-7741 strain)[a]

	Oocysts			Cysts			Tachyzoites		
TID[b]	p.o.	s.c.	i.p.	p.o.	s.c.	i.p.	p.o.	s.c.	i.p.
10^5	5.2/-/-	9.0/-/-	10/-/-	11/-/-	12/-/-	12/-/-	-/1/5	17/1/-	18/3/-
10^4	7.5/-/-	19/-/-	18/-/-	12/-/-	18/-/-	12/-/-	-/-/6	26/4/-	24/3/-
10^3	8.0/-/-	13/2/-	13/2/-	20/-/-	16/-/-	13/-/-	-/-/6	15/5/-	19/1/-
10^2	12/-/-	13/2/-	13/3/-	17/-/5	18/-/-	16/2/-	-/-/6	17/3/-	21/5/-
10^1	17/1/1	14/3/1	17/2/-	-/-/6	22/-/1	18/4/-	-/-/6	18/2/2	13/4/-
10^0	14/1/4	-/1/5	16/1/3	-/-/6	-/-/6	48/-/5	-/-/6	-/1/5	-/2/4
10^{-1}	-/-/6	-/-/6	-/-/6	-/-/6	-/-/6	-/-/6	-/-/6	-/-/6	-/-/6
LD_{50}	5.35	4.25	4.55	3.49	5.39	4.61	0	2.79	3.19
ID_{50}	5.62	5.49	6.0	3.49	5.39	5.59	0	5.35	5.75

[a]The original ID_{50} and LD_{50} values of the titration are noted.
[b]Six mice per group. First figure: mean day of death; second figure: number of mice with chronic infection as shown by the examination of their brains for cysts; third figure: number remaining uninfected. The highest dilution of each material giving rise to infection by any route was taken as "threshold infectious dose" (TID) and used as base for tabulation. A threshold infectious dose is equal to, or slightly greater than, a minimal infectious dose, but it is usually smaller than a 50% infectious dose. (Dubey and Frenkel, unpublished.)

lethal dose (LD) have been used. The endpoint is death. However, because of the greater biologic and epidemiologic importance of strains that do not kill their hosts, many studies have been conducted with such low virulence strains where death results only from large inocula, or from certain routes of administration, and all survivors must be examined serologically to determine whether they were infected. Sometimes, this is coupled with a search for cysts in the brain of these animals.

A comparison of infectivity and pathogenicity of oocysts, cysts, and tachyzoites administered by three different routes is shown in Table 3. The threshold infectious dose (TID) for each inoculum has been designated from the highest titer obtained for each inoculum, irrespective of route. Comparing *infectivity* via the oral route, that of oocysts is greater than that of tachyzoites and cysts; oocysts are similarly infectious via the subcutaneous and intraperitoneal routes. However, it takes almost 10^2 cysts and 10^5 tachyzoites to produce infection by feeding. Comparing the *sensitivity* of routes to detect infectivity of an inoculum, the intraperitoneal route is most sensitive, followed closely by the subcutaneous route. Time to death is usually shorter after intraperitoneal than after subcutaneous inoculation. To detect threshold infectious doses, a minimal examination of dying animals should include peritoneal exudate (if any), or the subcutaneous infection site, and lungs, lymph nodes, and brain; the survivors should also be examined for antibody development (20 days after inoculation) and for the development of *Toxoplasma* cysts in the brain. The pathogenicity can be compared by the number of ten-fold dilutions in which chronic infections were observed, excluding those dilutions where some of the six mice remained uninfected. Chronicity was observed in from zero to three dilutions of oocyst infections, in from zero to two dilutions of cyst infections, but in five dilutions after tachyzoite infections. In general, the time to death was shorter for cysts than for tachyzoite infections. However, oocysts were most rapidly fatal when fed, giving rise to an intense enteritis. In such mice, the primary site of infection, the intestinal mucosa, may by autolyzed, so that infection must be diagnosed from the mesenteric lymph nodes.

B. DYNAMICS OF INFECTION

From the inoculation site, or from the sites of primary infection in the gut, *Toxoplasma* spreads from cell to cell and is disseminated by the lymphatics to the lymph nodes, and by the bloodstream. Under special circumstances, such as after intraperitoneal inoculation, intracavitary spread is also important. Dissemination occurs with *Toxoplasma* in macrophages, lymphocytes, and granulocytes, as well as free in the plasma. After ingestion of *Toxoplasma* cysts or oocysts, the enteric infection spreads to the regional lymph nodes and via the portal circulation to the liver. From the lymph nodes, *Toxoplasma* reaches the thoracic duct and the lungs, and from the liver the organisms extend via the venous circulation also to the lungs. From there, organisms are disseminated via the arterial circulation. From studies of animals killed at intervals, it is clear that the majority of organisms reach first the lymph nodes, the liver, or the lungs, and are arrested there, multiplying in these sites; subsequently, a variable proportion of their progeny are swept into the lymphatics and into the bloodstream to be further disseminated.

The infection builds up if the strain is virulent to the particular species. Multiplication takes place and lesions are produced, depending on several factors. Important are the accessibility of a tissue via lymphatics and bloodstream, which is high for lymphoid tissues and low for the brain, and the suitability of a tissue as a substrate for *Toxoplasma*, which is high for lymphoid tissues and low for the kidney. Production of lesions depends on the vulnerability of the tissue, which is high for the brain and retina, whose function is concentrated in cells that do not regenerate, but is low for the intestinal epithelium and lymphoid tissues which normally turn over quickly. Hypersensitivity adds to these lesions, and immunity diminishes their magnitude.

The presence of *Toxoplasma* in the bloodstream has been studied in an attempt to find an explanation of (1) dissemination, (2) appearance of immunity as shown by declining parasitemia, (3) infection of fetuses in successive pregnancies, and (4) possible transmission of the infection by blood-sucking arthropods. Such studies have involved mice (Jacobs and Jones, 1950; Remington, Melton, and Jacobs, 1961), rats (Jacobs and Jones, 1950), rabbits (Jacobs and Jones, 1950; Remington, Melton, and Jacobs, 1961; Huldt, 1966a), guinea pigs (Remington, Melton, and Jacobs, 1961; Huldt, 1963), cats (Dubey and Frenkel, 1972b), pigeons (Jacobs and Jones, 1950; Jacobs, Melton, and Cook, 1953), chickens (Jacobs and Melton, 1966), and, sporadically, human beings (Huldt, 1963; Frenkel, Weber, and Lunde, 1960). During the acute infection, *Toxoplasma* is frequently isolated from the blood and the titers observed are high, whereas during the chronic infection, parasitemia is sporadic and the titers are low. Only a few studies encompass both the acute and the chronic infection with a given strain and species (Remington, Melton, and Jacobs, 1961). Often, the animals died from acute infection, and either chemotherapy or a strain of low virulence has been used to study chronic infection.

IV. Pathogenetic Considerations

The term "infection" is used in the scientific meaning; infection may be asymptomatic or accompanied by disease. From a consideration of the prevalence of *Toxoplasma* antibody and of the rarity of clinical disease, it is clear that the majority of infections are asymptomatic, both in animals and humans. There is a spectrum of virulence of strains in reference to a given host, a spectrum of susceptibility among species, and of susceptibility within each species, especially that related to the age of the host, and the degree of hypersensitivity acquired. As there is some variation in how immunologic terms are used, their intended meanings are briefly defined here.

A. MICROBIAL EFFECTS

Tachyzoites proliferate in many tissues, and by diverting cell metabolites to themselves, tend to kill cells at a rate faster than their natural turnover (Fig. 10). Hence, necrosis of individual cells results. Enteroepithelial stages are, on the whole, less injurious, since they live in cells with a normally fast turnover rate; in fact, the infection is concentrated in the cells of the villous tips, which are shed normally. It is less common, therefore, to find an ulceration (surface defect) in the intestine than a focus of necrosis (cell death) in tissues that turn over more slowly such as liver, lymph node, brain, and eye.

Microbial factors contributing to the pathogenicity of acute infection include the ability to invade cells, to utilize available substrates, to multiply at a rate faster than the host cell cycle, and to persist in cells without being exposed to the immunologic effector or expressor mechanisms.

Multiplication can be studied in immunosuppressed animals or in cell culture independently of host immunity. A variety of cells are infected, including even nucleated red blood cells of birds. There is no predilection for the cells of the reticuloendothelial system as is sometimes believed.

Tachyzoites enter cells actively within a few seconds (Lund, Lycke, and Sourander, 1961; Bommer, Hofling, and Keunert, 1968), and have also been seen to multiply after having been phagocytized (Pulvertaft, Valentine, and Lane, 1954). A penetration-enhancing factor, probably an enzyme located in the paired organelle, facilitates penetration of tachyzoites (Norrby, 1970;1971). Dissemination from cell to cell in the tissues, and via the bloodstream, lymphatics, and serosa has been observed.

The intracellular growth requirements for *Toxoplasma* have not been defined. Although it possesses respiratory and glycolytic enzymes, a cytochrome system (Lund *et al.*, 1966), and mitochondria, a Golgi apparatus with ribosomes and lysosomes, and a unique RNA and DNA (Neimark and Blake, 1967; Remington, Bloomfield, *et al.*, 1970; Perrotto, Keister, and Gelderman, 1971), *Toxoplasma* has not so far been grown outside of living cells.

A toxic factor has been suspected to account for certain lesions observed and for death after intravenous inoculation of exudates from infected animals. An erroneous etymological interpretation of *Toxoplasma*, which refers to the lunate arciform shape (toxon-bow, arc) but not to a poisonous nature (toxicon), may have contributed to the latter interpretation (Hogan *et al.*, 1971). The "toxin" was effective in immune animals and it was not neutralized by antibody, hence it is not likely to be a true toxin. The lethal intravenous effect observed could be traced to the thromboplastin and fibrin co-precipitation activities of proteinaceous material injected (Pettersen, 1971). The effects on cells which could not be explained by direct parasitization can generally be traced to hypersensitivity (see below).

Cysts develop intracellularly, especially in neurons, retinal cells, myocardium, and skeletal muscle cells. They persist in tissues throughout a period of immunity of the host. There is some evidence of their preferentially developing in certain tissues, and they appear to persist in immunologically isolated tissues. In most experimental infections, the number of cysts found appeared to depend on the degree of dissemination of tachyzoites. Cysts coexist with their host cells for a prolonged period of time. Eventually the host cell may die, as indicated by the disappearance of the nucleus, and ultrastructurally, host cell elements are no longer associated with certain cysts (Zypen, 1966). It is believed that cysts can persist in the brain without host cells, at least for some time.

When a cyst wall breaks, the contained organisms generally are killed by immunity and inflammatory cells; however, a large lesion may result from hypersensitivity to the antigen released. There are differences in the rate of decay of cysts in different hosts, which together with the different degrees of hypersensitivity developed by the host, influence the number and the extent of lesions produced from cyst rupture.

B. NATURAL RESISTANCE

This is a relative quality since most mammals and birds are susceptible to infection with certain strains of *Toxoplasma*. However, certain host species are genetically more resistant to the effects of primary infection than are other hosts, irrespective of the strain of *Toxoplasma* involved. Natural resistance is an important consideration in explaining why members of a susceptible species do not develop clinical disease. For example, rats are highly resistant genetically, but white mice, cotton rats (*Sigmodon hispidus*) and multimammate rats (*Mastomys natalensis*) are generally susceptible (Jacobs, 1956); chickens are more resistant than canaries (Harboe and Erichsen, 1954; Lainson, 1955c). Differences in natural resistance of hosts can be compared only with virulent strains of *Toxoplasma*, just as differences in virulence can be studied only in species where disease generally develops. A "natural" or peripheral route of infection should be used, i.e., oral or subcutaneous, since the results of intracerebral inoculation, for example, may be misleading. Maturity is important for resistance in rats and chickens, the young of which are fully susceptible (Lewis and Markell, 1958; Harboe and Erichsen, 1954). This situation finds a parallel in humans, where asymptomatic infection is the rule in mothers who have a higher degree of natural resistance than their infants.

The reasons for natural resistance or susceptibility are probably many, and are largely unknown. Interferon has been shown to partially protect cell cultures against *Toxoplasma*, and the number of organisms produced was from one to two logs lower in interferon-treated cultures than in controls (Remington and Merigan, 1968). However, in hamsters, Newcastle disease virus as interferon inducer which was active against three virus infections, showed no effect against toxoplasmosis (Frenkel, unpublished data).

C. ACQUIRED IMMUNITY

This term characterizes the acquired capacity of the host to *specifically* control numbers of *Toxoplasma*; specific immunity begins to develop a few days after infection, thereby restricting and occasionally terminating infection. Acquired immunity is of major concern in trying to analyze recovery from infection or to prevent disease in a susceptible host. Acquired immunity depends on an interplay of cellular factors and antibodies. Transfer of immune spleen and lymph node cells conferred more passive immunity than did serum from the same donors (Frenkel, 1967). Relapse and death from pneumonia or encephalitis occurred in hamsters given cortisone, although antibody levels remained high (Frenkel and Lunde, 1966). However, antibody does prepare extracellular *Toxoplasma* for lysis by the complement-accessory factor system (Desmonts, 1960), intracellular organisms not being affected (Sabin and Feldman, 1948). Most of the pathogenic effects of acute toxoplasmosis are terminated by acquired immunity.

Although transferrable immunity is linked to lymphoid cell populations in hamsters, there is evidence of immunity being expressed or developing in some other cells. For example, macrophages destroy *Toxoplasma* in cultures of peritoneal exudate cells, and by intracerebral challenge, immunity can be shown to be expressed in the brain.

Immunity, however, does not generally terminate infection. Infection immunity (premunition) is the rule, and *Toxoplasma* persists in the cyst stage in brain, skeletal muscle, and elsewhere following the acute infection. Cyst organisms multiply slowly, suggesting the presence of some intracellular immunity. The labile nature of this immunity, sensitive to the effect of hypercorticism, total body radiation, cyclophosphamide and some other agents, accounts for the renewed active proliferation leading to clinical relapse (Frenkel, 1957,1960; Frenkel and Lunde, 1966). This will be discussed in greater detail below.

D. ACQUIRED RESISTANCE

This term refers to the *nonspecifically increased* capacity to handle *Toxoplasma*, induced by immunologically unrelated agents. This phenomenon recently has been studied by Ruskin and Remington (1968b) and Ruskin, McIntosh, and Remington (1969) using relatively avirulent strains and carefully adjusted challenge inocula. The degree of protection is several orders of magnitude lower than that afforded by acquired specific immunity; sometimes, death is merely delayed (Krahenbuhl and Remington, 1971).

There is no evidence that the mechanisms involved in nonspecific acquired resistance are comparable to those mediating specific acquired immunity. While puzzling and intriguing, these two acquired conditions should be studied comparatively, to determine the relative roles of nonspecific resistance and of specific immunity in handling challenge inocula of *Toxoplasma*. The effect of interferon is discussed under "Natural resistance" (see above).

E. HYPERSENSITIVITY

This concept characterizes the cell death, tissue necrosis, and intense inflammatory reaction with harmful effects on the sensitized host which are observed after exposure to the specific antigen. Delayed hypersensitivity is an important pathogenic mechanism, especially in chronic toxoplasmosis. In such chronically infected, hypersensitive animals, a rupture of a *Toxoplasma* cyst kills adjacent uninfected cells, whereas, during the acute infection, only the infected cells themselves are visibly damaged. Delayed hypersensitivity in humans and guinea pigs is illustrated by the toxoplasmin skin test (Frenkel, 1948; Krahenbuhl, Blazkovek, and Lysenko, 1971a). Since they occur together, cellular immunity and delayed hypersensitivity are often considered related. However, their separation is clinically desirable and scientifically necessary in analyzing pathogenesis, as previously reviewed (Frenkel, 1969a). Whereas hypersensitivity is measured by its deleterious effect on the host, immunity is measured by its effect on the microbe.

A reaction of immediate hypersensitivity of the Arthus type appears to give rise to the area of periventricular necrosis observed in neonatal toxoplasmosis encephalitis in humans (Frenkel and Friedlander, 1951).

V. Pathology

The wide host range of *Toxoplasma* and the great variability in capacity to handle the infection makes a general description necessarily inexact. A distinction should be drawn also between natural infection via the oral route by which toxoplasmosis is generally acquired via cysts or oocysts, and artificial avenues used in experimental studies employing tachyzoite inocula. A considerable portion of the latter deal with intraperitoneal inoculation, which is practical, but almost as artificial as the intracerebral route. In cats, the natural host, even subcutaneous inoculation may lead to a more generalized and frequently fatal infection, involving liver and lungs, whereas after oral inoculation, the parasite spends much of the initial period of infection, during which immunity is acquired, in the essentially expendable intestinal epithelium. The general aspects of lesions found in various stages of infection are briefly described as seen in animals and man as well as certain syndromes occurring typically in several species. Unless specifically mentioned, reference is made to natural infections or experimental ones acquired by mouth. The lesions in humans and their clinical significance have recently been reviewed elsewhere (Frenkel, 1971c).

A. ASYMPTOMATIC INFECTION

Most infections in most species are probably asymptomatic, judging from the large number of animals and humans with antibody (Feldman and Miller, 1956; Remington, Efron *et al.*, 1970). In adult cats, and kittens after weaning, toxoplasmosis is generally asymptomatic (Dubey and Frenkel, 1972b). The active infection with proliferating tachyzoites may be brief, with immunity developing before any significant lesions are produced.

Asymptomatic acute toxoplasmosis may result in infection of the fetus *in utero*. Parasitemia has been observed in experimentally infected mice, rabbits, and guinea pigs for as long as a year (Remington, Melton, and Jacobs, 1961). Parasitemia or chronic infection of the uterus could result in placental and fetal infection and lead to abortion, fetal disease, or death. Placental and fetal infections have not been studied in detail in cats. However, in sheep this occurs epidemically and is said to account for about half of the ovine abortions from all causes in England and New Zealand (Hartley and Moyle, 1968; Beverley, Watson, and Spence, 1971). In humans, infection of the fetus and newborn is seen sporadically. In both sheep and humans, this is associated with the acquisition of toxoplasmosis during or close to the pregnancy in question (Eichenwald, 1960; Couvreur, 1971). In chronically infected mice, repeated transmission to the fetus has been briefly reported (Beverley, 1959; Nakayama, 1968), but a detailed study is not available.

Acute infection is followed by chronic infection, which generally is also asymptomatic. Symptomatic chronic toxoplasmosis is discussed below. Asymptomatic chronic toxoplasmosis may reactivate following immunodepression with corticosteroids and other agents (Frenkel, 1956b;1960).

B. LESIONS OF SYMPTOMATIC INFECTIONS

1. Acute infection (Figs. 6 through 11)

In most acute infections the first site of infection is the gut. In cats, there may be intense parasitization of the epithelial cells with stages described in section II,B,2. There is little evidence of necrosis of individual cells, since regenerative capacity can compensate, or since the infected cells are not destroyed before they are normally sloughed off. Developmental stages are generally found in the cells covering the upper half of the villi (Fig. 6), which have a high natural turnover, and ulceration rarely occurs. The actively dividing epithelium of the crypts and of the intestinal glands are rarely parasitized. Infected young kittens often develop diarrhea with foul-smelling stools. They fail to gain weight normally. *Toxoplasma* is disseminated via the lymphatics, the portal vein, and the general blood circulation. If such animals are examined after a week, lesions of the lymph nodes and liver are usually prominent. At death, after from two to three weeks of infection, widely disseminated lesions are found, with those in the lymph nodes, heart, and brain most severe (Dubey and Frenkel, 1972b). In mice fed oocysts, enteritis with tachyzoites is present for up to a week, often with ulceration, and sometimes fatal with minimal lesions elsewhere (Figs. 7 and 8). Hepatitis is present for from three to six days, and later, pneumonia and encephalitis are clinically most important (unpublished). In mice and cats, the oral infection tends to produce fewer lesions than subcutaneous inoculation. Even after extending beyond the intestinal epithelium, infection is in cells capable of regenerating as in lymph nodes and liver, which serve as filters before the infection disseminates (Figs. 9 and 10). Thus, after oral infection, immunity can begin to develop, together with a few serious lesions, whereas after subcutaneous inoculation infection spreads to the lungs and frequently results in significant pneumonia and general dissemination, which is often fatal (Fig. 11). However, with high doses, fatal toxoplasmosis has been observed even after oral inoculation; in mice, much of the small bowel becomes denuded of mucosa (Fig. 7). Such infection may be difficult to diagnose, since, after death, autolysis destroys the main lesion in the gut, and liver lesions are still few. However, the diagnosis can be made by identifying *Toxoplasma* in the mesenteric lymph nodes.

Acute symptomatic infection terminates either with death of the host, or with the development of immunity and chronic infection,

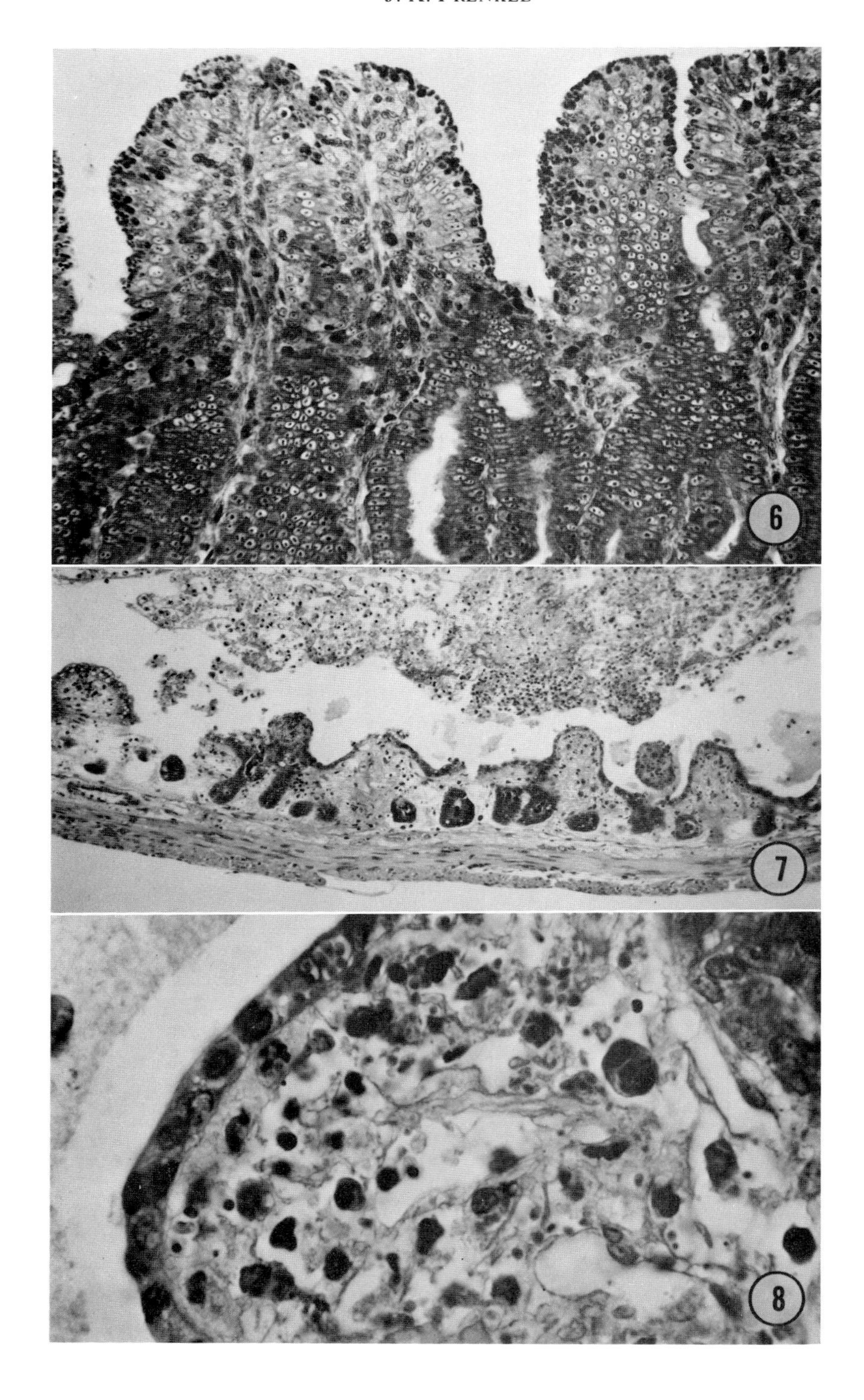
6
7
8

to be discussed below. A prolonged activity of infection can be called subacute infection.

2. Subacute infection

Human babies infected *in utero* often show a protracted clinical course; they are partially protected by antibody transferred *in utero* and with milk, but they are developmentally too immature to acquire sufficient active immunity to contain the infection. Thus, they may show healing lesions in the lung, liver, heart, and lymph nodes, while lesions progress in the brain and eyes, where immunity is acquired most slowly (Frenkel and Friedlander, 1951; Frenkel, 1971c, 1972b).

A similarly protracted course of disease has been described in some animals in which immunity develops quite slowly in the central nervous system. Lesions result either from multiplying tachyzoites, leading to individual cell necrosis, or from cysts, some of which rupture from time to time, leading to tissue necrosis from hypersensitivity. Such a course of infection has been described from chin-

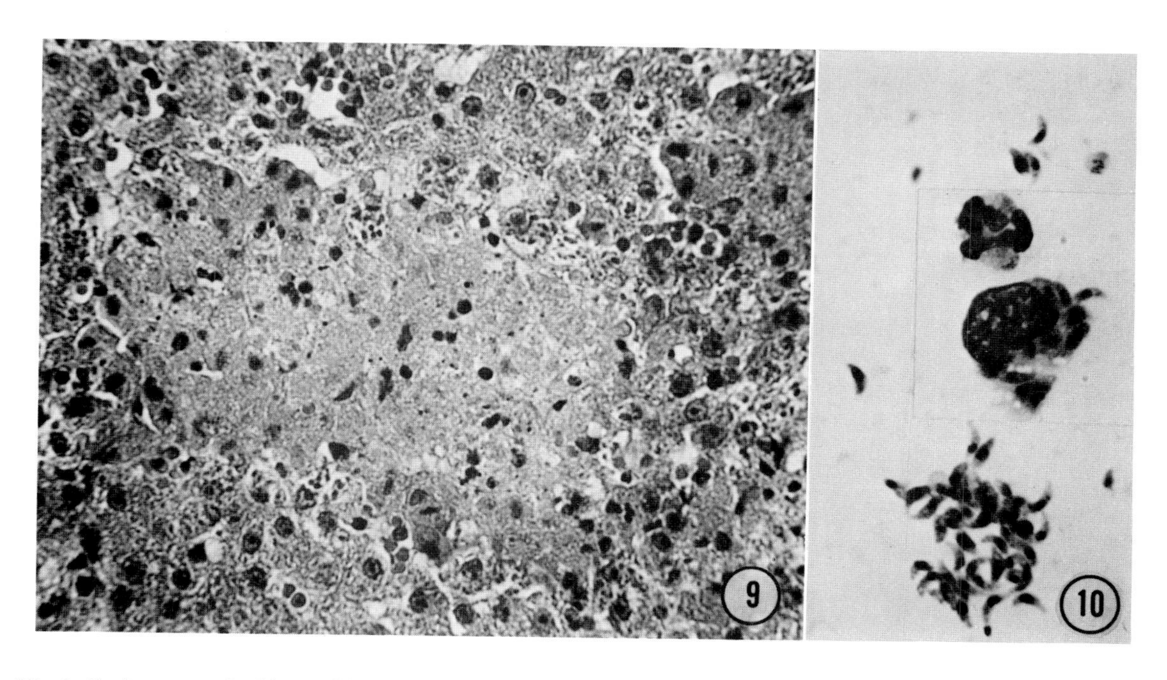

Fig. 6. Ileal mucosa of a kitten with schizonts and gametocytes of *Toxoplasma*, seven days after the ingestion of cysts. The superficial portion of the epithelium lining the villous segment is parasitized (dark bodies); the glands of Lieberkühn are not. Giemsa, × 240. (from Dubey and Frenkel, 1972b, Fig. 110.)

Fig. 7. Toxoplasmic enteritis in mouse due to tachyzoites six days after the ingestion of oocysts. The epithelium and the villous stroma are parasitized; there is ulceration, the villi are blunted and covered with a reduced number of flattened epithelial cells. Ulcerated tissue and inflammatory cells are visible in the lumen. PASH, × 120. (See Fig. 8 for detail.)

Fig. 8. Toxoplasmic enteritis in same mouse as in Fig. 7. Tachyzoites are present in epithelial and stromal cells of villus. PASH, × 800.

Fig. 9. Focus of toxoplasmic hepatitis with necrosis and tachyzoites in newborn kitten, nine days after the ingestion of cysts. H & E, × 625. (From Dubey and Frenkel, 1972b, Fig. 119.)

Fig. 10. Tachyzoites of *Toxoplasma*, free and intracellular, from the peritoneal exudate of a mouse three days after intraperitoneal injection of tachyzoites. Dried imprint, methanol fixation, Giemsa, × 1000. (From Frenkel, 1971a.)

chillas (Keagy, 1949) and from hamsters (Frenkel, 1951). Sporadic instances should be found in other species, wherever immunity is sufficiently rapid in development to prevent death from acute infection, but not sufficient to contain infection in brain and eye. Often these are younger animals of a given species, such as kittens, small dogs, and lambs, whereas adults develop an adequate immunity. Chemotherapy may have been given during the acute infection, but was terminated before immunity had been acquired in the brain, cord, and retina.

3. *Chronic infection*

Toxoplasmosis becomes chronic in most animals that can develop immunity. The tough, argyrophilic, cyst wall apparently protects the organisms from antibody and cellular immunity. Multiplication of the cystic bradyzoites is slow or is negligible. Intact cysts may remain for months or years without invoking an inflammatory reaction (Figs. 1, 12). However, where a cyst wall breaks down, a more or less intense hypersensitive inflammatory reaction develops, generally associated with death of the bradyzoites. Only rarely will the liberated organisms enter new cells, and if immunity has waned, give rise to a few generations of tachyzoites before cysts are again formed. With approximately similar numbers of cysts, lesions are larger in hamsters than in mice, which is probably related to the degree of hypersensitivity developed. The lesions also appear to be more numerous in hamsters than in mice; and judging from the great sensitivity of the cyst wall to pepsin and trypsin (Jacobs, Remington, and Melton, 1960), one might suspect that spontaneous cyst rupture is brought about by local increases in tissue enzyme levels.

The necrotic lesions in the brain are followed by development of microglial nodules without organisms (Fig. 13). Occasionally, the lesions are associated with a number of cysts which developed from liberated organisms that were not killed. If many such lesions are formed, animals may develop signs of chronic encephalitis. Hamsters become irritable, run in circles, hold their heads at an angle, and may develop spastic paralysis, usually first of the hind quarters (Frenkel, 1951;1953a;1956b). Certain strains (RH, LRH) give rise to progressive lesions more frequently than others. A similar syndrome has been seen in chinchillas (Keagy, 1949).

Ocular lesions arise in the retina and involve the choroid secondarily, so that the lesion is properly called retinochoroiditis, although chorioretinitis is also used. Small or confluent self-limited inflammatory foci follow cyst rupture and lead to glial scarring. Sometimes progressive lesions with tachyzoites are seen which gradually destroy the retina. Such lesions produced by *Toxoplasma* and the related *Besnoitia* have been studied in hamsters where ocular lesions developed following generalized infection (Frenkel, 1955;1961). Clinically such lesions have been seen in cats (Vainisi and Campbell, 1969;

Fig. 11. Toxoplasmic encephalitis with glial nodule and two cysts from brain of patient after an illness of over nine weeks. The two intact cysts do not elicit inflammation. The glial nodule probably originated from the rupture of a cyst; similar glial nodules are produced by tachyzoites. PASH, × 225. [Slide courtesy of Dr. Edward H. Kass (Kass *et al.*, 1952).] (From Frenkel, 1971b.)

Fig. 12. Intracellular *Toxoplasma* cyst in brain of hamster infected for six months. The nucleus of the neuronal host cell is visible below the PAS-positive, deeply staining granules of the bradyzoites. There is no inflammatory reaction. PASH, × 400.

Fig. 13. Necrotic focus (arrows) in cerebellum involving the granular and Purkinje layers, resulting from rupture of a cyst in hamster infected for one year. No viable *Toxoplasma* were present in the necrotic cell debris. An older glial nodule (*bottom, left*) in the molecular layer represents the permanent scar following a cyst rupture. PASH, × 135. (From Frenkel, 1956b.)

Fig. 14. Relapsing toxoplasmic encephalitis, with numerous tachyzoites and deficient inflammatory reaction in a chronically infected hamster that had been receiving injections of cortisone for the preceding 13 days. H & E. × 475.

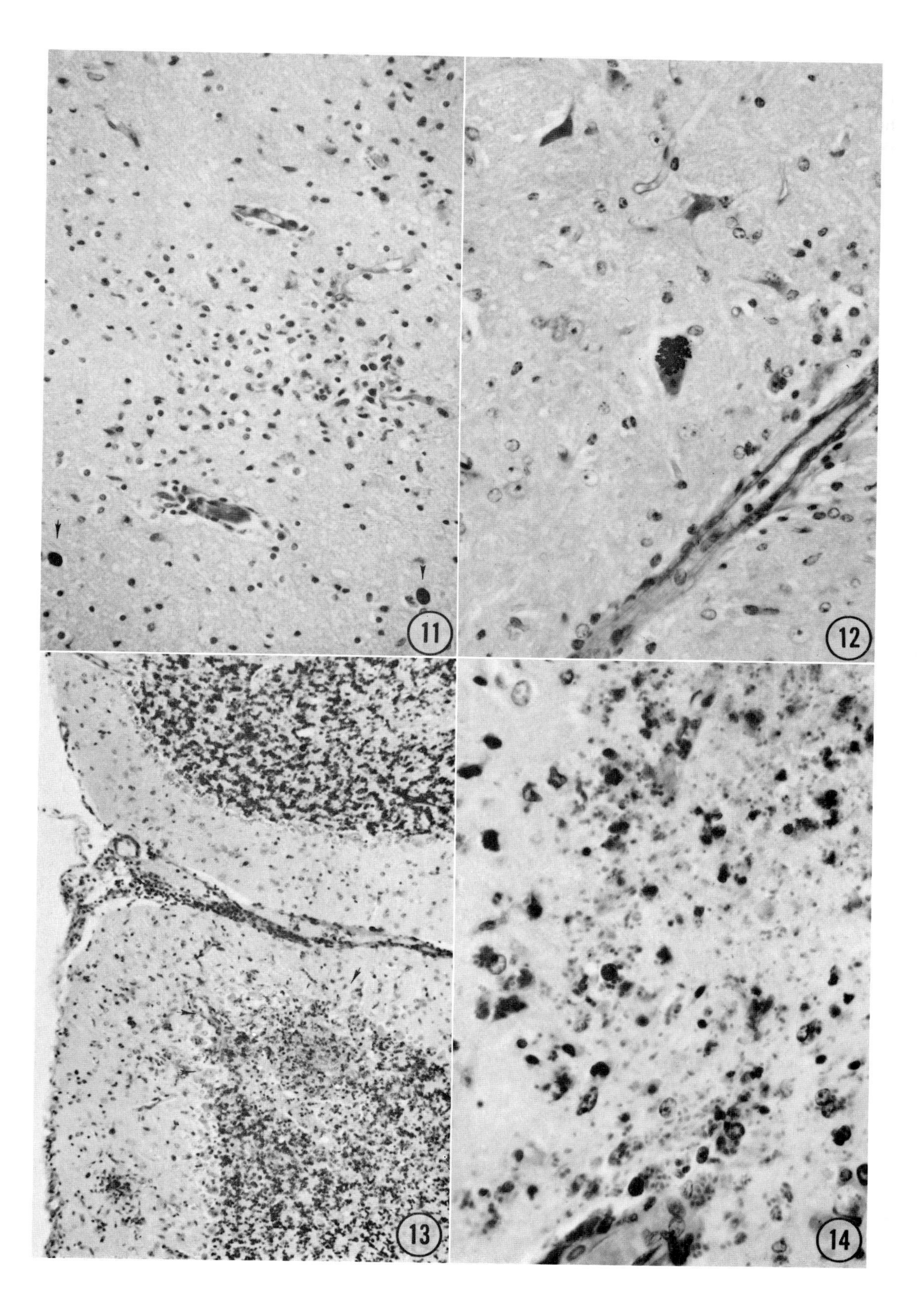
11
12
13
14

Piper, Cole, and Shadduck, 1970) and in humans (Wilder, 1952; Zimmerman, 1961); this is discussed in detail elsewhere (Frenkel 1961,1971c).

Myocardial infection develops in cats (Fig. 22) (Dubey and Beverley, 1967), guinea pigs (Kean and Grocott, 1945), as well as in humans (Kass *et al.*, 1952; Hooper, 1957) and is generally mild or subclinical. However, the patient reported by Kass *et al.*, (1952) showed widespread involvement of the heart and skeletal muscle. In addition, protracted myocarditis has been observed in one patient, confirmed by isolation of parasites at autopsy (Durge, Baqai, and Ward, 1967).

Organizing pneumonia with *Toxoplasma* cysts and a few tachyzoites has recently been observed in Costa Rica in a seven-month-old child with a five-month history of illness (Piza and Troper, 1971).

4. *Relapsing toxoplasmosis*

When immunity is impaired or abolished by other diseases or by administration of corticosteroids, cytotoxic agents, such as cyclophosphamide, or total body irradiation, toxoplasmosis may relapse. This probably originates from an accidental cyst rupture after which the liberated bradyzoites are not destroyed, but can enter new cells and multiply as tachyzoites (Frenkel, 1957; Frenkel and Wilson, 1972). This is further discussed in section VI, I.

Extensive necrotic lesions with large numbers of *Toxoplasma* and little inflammatory reaction are characteristically found. In humans, they are seen mainly in the brain (Vietzke *et al.*, 1968; Frenkel, 1971c), in hamsters mainly in the lungs and brain (Fig 14) (Frenkel, 1956b; 1957), and in cats in the liver and lymph nodes.

5. *Lesions observed in individual species*

Toxoplasmosis has been reported from numerous species either naturally or experimentally infected. Infection in the domestic animals has been reviewed by Siim, Biering, Sorensen, and Møller (1963). The publications concerning toxoplasmosis in various species have been analyzed in the bibliography of Jira and Kozojed (1970). Attention will be drawn here to the lesions of special interest.

In ocular toxoplasmosis, the retina is usually involved (Frenkel and Jacobs, 1958; Frenkel, 1961); however, in pigeons the choroid might be destroyed while the retina is left intact (Frenkel, 1953a). Ocular toxoplasmosis occurs in humans, hamsters, and cats during natural infection (Vainisi and Campbell, 1969; Piper, Cole, and Shadduck, 1970). In rabbits it can be studied after intraocular injection (Nozik and O'Connor, 1971).

Adrenal involvement with toxoplasmosis is uncommon, but has been observed in mice (Cowen and Wolf, 1951), hamsters (Frenkel, 1956a), humans (Cespedes and Morera, 1955; Corpening, Stembridge, and Rigdon, 1952), in several species of monkeys (Benirschke and Richart, 1960; McKissick, Ratcliffe and Koestner, 1968; Seibold and Wolf, 1971), cats (Dubey and Frenkel, 1972b), and kangaroos (Hackel, Kinney, and Wendt, 1953). At least in hamsters and humans, which are known to secrete cortisol (Frenkel *et al.*, 1965), adrenal infection has been related to endogenous hypercorticism which produces a localized immune defect (Frenkel, 1956a, 1972a).

Monkeys vary greatly in their susceptibility to toxoplasmosis. Rhesus monkeys often survive infection with the mouse-virulent RH strain of *Toxoplasma* organisms. Marmoset (*Oedipomides*) monkeys develop a fulminant disease fatal in from five to six days, with intense parasitism of cells of liver, spleen, and gut (Benirschke and Richart, 1960); also squirrel monkeys (*Saimiri*) often develop clinical illness (McKissick, Ratcliffe, and Koestner, 1968), and the disease has been reported from a wooly monkey, *Lagothrix* sp. (Benirschke and Low, 1970), and others

(Seibold and Wolf, 1971; Ruch, 1959). Marmosets have been infected with cysts and with oocysts (Miller, Frenkel, and Dubey, 1972).

VI. Immunology

A. INTRODUCTION

The emphasis here will be on the acquired specific immunity, accounting for recovery from infection and for immunity to reinfection. Natural resistance comprises mainly genetic, fixed host factors and is therefore discussed in the section on pathogenesis. Acquired nonspecific resistance is discussed here since it evolves simultaneously with acquired specific immunity, but is seen in its pure form only when nonspecific immunization is used.

Hypersensitivity will be touched only briefly to delineate it from immunity, since it is considered principally a factor of pathogenesis; it is discussed in section IV,E. Diagnostic uses of hypersensitivity and of serology are discussed in Chapter 10. Analytic and other studies of antigens and antibodies have been reviewed by Jacobs (1967) and Remington (1970). As observed by Piekarski and Witte (1971), previous or concomitant infections with *Isospora bigemina* (*cati*), *I. felis* and *I. rivolta* do not interfere or induce cross immunity against toxoplasmosis. Dubey and Frenkel (unpublished observations) made similar observations in respect to the latter two species. *Besnoitia* infection will be frequently mentioned for comparative purposes; it is briefly discussed in Section VIII,A,1 of this chapter.

B. ACQUISITION OF IMMUNITY

With the CJ strain of *Toxoplasma* controlled by 15mg% of sulfonamide therapy, only from four to six days of treatment were necessary to produce immunity on the seventh day after intraperitoneal infection. As little as three days of treatment assured immunity on the tenth day after subcutaneous injection. The dye test titers were between 1:500 and 1:1000 (Frenkel, 1956b).

Besnoitia injected subcutaneously in hamsters usually is fatal in from nine to ten days. However, hamsters treated with 60 mg of sodium sulfadiazine per 100 ml of drinking water for from seven to ten days usually survived (Frenkel, 1972a). A few animals die late, suggesting partial immunity. The number of *Besnoitia* in the hamster organs showed an increase or levelling off into the third week, declining and reaching a minimum between 30 and 90 days after infection (Frenkel and Lunde, 1966).

C. RECOVERY FROM INFECTION

The clearance of organisms proceeds at different rates in the several organs. When *Besnoitia*-infected female hamsters were treated with 60 mg of sodium sulfadiazine per 100 ml of drinking water, the infection progressed at a controlled rate without fatalities. Titrating of the organs of such hamsters at approximately two-day intervals and pooling of the data for weekly periods indicated delayed clearing from the eyes (retina) as compared with brain, liver, spleen, lung, adrenals, and ovaries. Evaluation of the number of organisms per 20 mg of tissue between one and four months after infection indicated that the spleen maintained parasite numbers from 100 to 1000-fold greater than the other organs mentioned (Frenkel and Lunde, 1966).

Clearance of organisms from the adrenal cortex takes several days longer than for other organs, based on a study of individual hamsters by titration and sections.

The clearance of *Toxoplasma* from the brain is delayed. This is found in mice treated with graded doses of sulfonamides; in these mice the last deaths are seen to occur with encephalitis, at a time when the extraneural viscera are essentially cleared of *Toxoplasma*. In such brains, both tachyzoites and developing cysts are found (Frenkel, 1961). Also, if mice are treated with a given dose of sulfonamide for four weeks, a certain number of them will die from active toxoplasmosis following the cessation of chemotherapy, indicating that acquired immunity had not fully developed and that tachyzoites persisted. The greater vulnerability of the brain to injury is, of course, also a factor in mortality. However, the persistence of tachyzoites and the lesions in the brain, but not in other organs, suggests differences in the acquisition of immunity. Also, in human infants infected *in utero* while partially protected by antibody transferred from the mother, active infection persists longer in the brain than in other organs, leading to advanced encephalitis (Frenkel and Friedlander, 1951).

D. IMMUNITY TO REINFECTION (TABLE 4)

Mice infected with the CJ strain of *Toxoplasma* usually die, whereas the BDA strain causes only a 10% mortality in 30 days. When three groups of 30 mice were vaccinated subcutaneously with living BDA organisms, their acquisition of immunity could be studied by daily challenge. Three mice were challenged daily with the virulent CJ strain, one intracerebrally, one intraperitoneally, and one subcutaneously. Unvaccinated mice were used as controls for slight variations in the challenge numbers. In mice challenged *intracerebrally*, on the eighth day after inoculation, the expected death on the 11th day was delayed, and when the mice were challenged on the tenth day, indefinite survival occurred instead of death on the 13th day. In mice challenged *intraperitoneally* on the fifth day, the expected death on the 11th day was delayed, and when the mice were challenged on the sixth day, indefinite survival occurred instead of death on the 12th day. In mice challenged *subcutaneously* only one day after vaccination, the expected death on the eighth day was delayed and when the mice were challenged on the fifth day, indefinite survival occurred instead of death on the 12th day as seen in the control (Table 4). In all three challenge groups, immunity began to develop around the 8th day and was essentially complete by the 12th or 13th day (Frenkel, 1956b). This coincided with the appearance of antibodies, as measured in the dye test, which showed a titer of 1:16 on the 9th day, 1:256 on the 11th day, and 1:8000 on the 13th day (Frenkel, 1956b).

Infection with a given strain of low virulence does not always cause effective immunity to a more virulent strain. Although mice vaccinated with BDA resisted challenge with CJ, most died after challenge with the RH strain even though their period of illness was prolonged (Frenkel, 1956b: Table 3). As this phenomenon is limited to certain hosts, factors other than differences among strains of parasites must be involved; such factors may include inability of the host to acquire an adequate level of immunity. For example, it was difficult for mice to acquire immunity to the RH strain; even prolonged or repeated sulfonamide treatment did not result in development of immunity in all of the mice (Frenkel, 1961: Figs. 81,82). Hamsters and guinea pigs, however, could easily be immunized so as to resist RH challenge either by using a strain with low virulence as vaccine, or chemotherapy (Krahenbuhl, Blazkovek, and Lysenko, 1971b; Frenkel, unpublished data).

Roever-Bonnet (1963,1964) investigated partial immunity in the survivors of vaccination with a strain of low virulence, and found that many of the mice had become infected with the virulent challenge strain also. However, when testing individual organs she sometimes isolated cysts derived from the avirulent primary infection, and at other times cysts from the infection resulting from challenge with the virulent strain. Although the groups were small, from one to ten per organ, each organ suspension was tested in several mice, and one can get an idea of the degree of parasitism by the number of mice that showed infection with either or neither of the strains. In the unchallenged mice the avirulent (vaccine) strain persisted in brain, lung,

TABLE 4. Development of immunity to toxoplasma in mice. Vaccination with BDA strain, challenge with the virulent CJ strain[a]

Group:	1	2	3	4	5	6	7		8	1a	3a	5a
Vaccination on 0-day with	BDA	None	BDA	None	BDA	None	BDA		BDA	BDA	BDA	BDA
CJ strain injected	Intracerebrally		Intraperitoneally		Subcutaneously		None		None	IC	IP	SC
Day of CJ injection	Day of death after CJ injection						Reciprocal titers dye-test-CF		Vaccine deaths 40 mice	Day of death after vaccination over expected day of death		
1	4	3	6	5	18	7	0	0		5/4	7/6	19/8
2	3	2	5	3	8	7				5/4	7/5	10/9
3	4	4	4	4	14	6	0	0		7/7	8/7	17/9
4	3	3	5	5	10	7				7/7	9/9	14/11
5	4	3	8	6		7	0	0		9/8	13/11	/12
6	3	3		6		7				9/9	/12	/13
7	3	2	5	5		6	2	0		10/9	12/12	/13
8	24	3	16	4	11	6				32/11	24/12	18/14
9	7	3		6	7	6	16	0		16/12	/15	16/15
10		3	26	6		7				/13	36/16	/17
11		3		6		7	256	0		/14	/17	/18
12		3		5		6				/15	/17	/18
13	5	3		6		7	8000	8		18/16	/19	/20
14		4		5		7				/18	/19	/21
15		3		5		6	1024	8		/18	/20	/21
16		4		6		7				/18	/22	/23
17		3		5		6	4096	8	+	/20	/22	/23
18		3		5		7						
19		4		5		7	4096	24				
20		2		5		7						
21		3		5		8	2048	16	+			
22		3		5		7						
23		3		7		7	2048	40				
24		3		5		7						
25		3		9		7	8000	12	+			
26		4		6		8			+			
27		3		6		6	4096	20				
28		4		6		7						
29		3		6		7	2048	32				
30		3		5		6						

[a]Blank spaces indicate survival for 60+ days. Reciprocal values of serologic titers are noted. (From Frenkel, 1956b.)

liver, heart, spleen, stomach, small and large intestines, mesentery, kidneys, bladder, uterus, testis, vesicular gland, skeletal muscles, tongue, eye, and bloodstream. The virulent challenge strain was found in the liver and in the lung and spleen of most mice. Both of the strains were found in brain, stomach, large intestine, and kidney of hamsters and mice. Avirulent vaccine *Toxoplasma* alone remained in the small intestine, mesentery, uterus, vesicular gland, heart, and skeletal muscle, and although only a few mice were tested, these organs can tentatively be regarded as immune to superinfection. No organisms were cultured from the tongue, testis, ovary, and bloodstream. A curious fact was that chronic infections were more common in mice infected with the avirulent strain only. Apparently the virulent organisms did not become established in certain organs; chronicity was evidently diminished in the challenged animals. A host difference was evident in that a greater percentage of mice than hamsters remained infected (De Roever-Bonnet, 1964).

Huldt (1967b) studied immunity to superinfection in rabbits, and showed that with subcutaneous challenge on the fifth day after vaccination all the rabbits survived, whereas the controls died after an average of 8.2 days. One could reasonably conclude, therefore, that most of the rabbits acquired an effective immunity between 9 and 14 days after immunization with a strain of low virulence. The latest deaths occurred on the 13th and 14th days.

E. PREMUNITION (INFECTION IMMUNITY)

Following the acute infection, the numbers of *Toxoplasma* are reduced, as discussed in Sections C and D, but some organisms persist in several tissues during a prolonged chronic infection, generally with little proliferation. This has been characterized as infection immunity or premunition. As reviewed in previous sections, chronic infection occurs characteristically in mice and hamsters (Roever-Bonnet, 1963,1964; Frenkel, 1953a) and in cats (Dubey, 1967; Katsube *et al.*, 1969), with brain as well as skeletal and cardiac muscle most consistently infected. Infection for five years has been documented in a guinea pig (Lainson, 1959), for more than a year in mice (Pope, Derrick and Cook, 1957), for three years in rats and mice (Jacobs, 1957; Nakayama and Matsubayashi, 1961), for 33 months in pigeons (Jacobs, Melton and Cook, 1953) and for ten months in chickens (Jacobs and Melton, 1966). At least the mice were shown to be immune (Frenkel, 1956b), whereas in rats this could not have been tested since they normally develop an age resistance. Likewise, the persistence of *Toxoplasma* antibody titers in humans for many years is believed to be based on the persistence of chronic infection with cysts. I have traced a woman who had given birth to a *Toxoplasma*-infected baby, and who 25 years later still had a dye test titer of 1:32. Of course, we know nothing about the immune status, as opposed to antibody titers, in these people. It is likely that chronically infected animals and people would show at least a moderate degree of immunity. When the latter is impaired, such as after the administration of corticosteroids, relapse occurs. The percentage of animals and humans in whom premunition persists for one or two years, and the rate of clearance of infection, has not been studied longitudinally. It is possible that some species can eradicate the infection better than others. What happens to their immunity has not been studied, but we have some idea about this from the study of antibody titers and immunity after the use of killed vaccine, as discussed below.

Premunition is a less effective immunity than "sterile" immunity, since the host cannot eradicate infection. In the presence of "sterile" immunity, we cannot detect persistent infection; however, the existence of the infection in an unrecognized form cannot be excluded. For example, immunity to vaccinia is not accompanied by continuing infection, but immunity to herpes is often characterized by premunition. Although it is a less effective immunity, conceptually speaking, premunition may confer some biologic advantage to the host and be of survival value. From time to time, the *Toxoplasma* cyst ruptures, or a herpetic fever blister develops, and immunity is probably boosted and maintained at a high level. Anti-vaccinial immunity wanes and after an interval of perhaps from 10 to 20 years, revaccination is followed by a second infection with primary characteristics (maximal lesion about the 12th day). In other infections with an apparently sterile immunity, this double jeopardy may be avoided by the chance for frequent reinfection, thus maintaining an effective immunity, as is believed to occur in chickenpox and poliomyelitis. How often, under natural circumstances, man and several species of animals are reinfected with *Toxoplasma* is not known. But it is likely to vary with locality and would be a function parallel to the rate of antibody acquisition. But even in areas of low incidence of infection, premunition would tend to maintain an effective immunity.

Premunition is of practical importance in preventing infection of fetuses during successive pregnancies in humans, a species in which the fetus is infected generally only during primary infection of the mother, or shortly thereafter (Desmonts, 1971; Couvreur, 1971). In following 216 mothers through 380 pregnancies subsequent to their giving birth to a toxoplasmic baby, Dr. Heinz Eichenwald could not find an infected infant or fetus, and the incidence of miscarriage and of congenital defects was similar to that of a matched control group (personal communication, enlarging on statistics published by Sabin *et al.*, 1952).

In species where chronic infection is accompanied by repeated infection of fetuses during successive pregnancies, the incomplete, premunition type of immunity probably works at a disadvantage to the host. Mice infected *in utero* were reported to have a reduced life span, number of litters and litter size (Beverley, 1959).

The question of the disappearance of immunity when chronic infection is ended has not been studied experimentally. One would surmise that it will diminish, but whether quickly or at a slow rate would be of interest.

F. IMMUNIZATION WITH LIVE AND KILLED VACCINES

Guinea pigs vaccinated with killed vaccine and Freund's adjuvant developed CF antibody and were immune to intraperitoneal and intradermal challenge after four weeks. However, guinea pigs with infection immunity were immune even to intracerebral challenge (Cutchins and Warren, 1956).

Rabbits were studied for immunity to challenge after immunization with from four to six weekly doses of killed vaccine. They had developed antibody titers similar to those of rabbits that had undergone "natural infection." The course of infection in the vaccinated, sero-positive rabbits was similar, however, to that in nonimmune rabbits as measured by parasitemia. Titers of *Toxoplasma* isolated from the blood and lymph stream, and the drop in the lymphocyte numbers in the peripheral blood paralleled that of control rabbits. Meanwhile the naturally infected rabbits showed no or only a few *Toxoplasma* in the blood and lymph stream (Huldt, 1966b,c). All the animals were killed on from the sixth through the eighth day so that no survival data were obtained.

When mice were injected with killed vaccine, 70% developed antibody after five doses and about 90% developed antibody after ten doses (Nakayama, 1969). On intraperitoneal challenge with 3000 RH strain organisms, performed between two and six weeks after first vaccination, from 30% to 50% of mice survived. However, when challenged between 8 and 14 weeks, only from 0% to 3% of the mice survived, although the antibody titers remained similarly high.

All the challenge experiments so far mentioned utilized virulent *Toxoplasma*, usually the RH strain. Ruskin and Remington, however, used strains of low virulence and small inocula, and demonstrated that even killed vaccine can produce some degree of immunity against *Toxoplasma* (Remington, 1970). It is not clear, however, whether this immunity is significantly greater than the nonspecific resistance which they produced by activating macrophages as discussed below.

G. CELLS, ANTIBODY, AND IMMUNITY

Antibody inactivates or lyses *Toxoplasma* in the presence of accessory factor, which is the basis of the dye test (Sabin and Feldman, 1948; Desmonts, 1955). It was shown also that antibody was effective only against free *Toxoplasma* and that intracellular forms were protected. It is clear, therefore, that one can ascribe to antibody a major role in immunity only against free *Toxoplasma*. As most of the organism's life cycle takes place within cells, additional immune mechanisms should also operate.

Since the presence of antibody which developed after exposure of the host to killed vaccine conferred little immunity, it was thought that some essential antigenic component might have been destroyed (Huldt, 1966b,c). However, the transfer of high titer antibody from animals with infection immunity persisting for the duration of infection failed to impart significant immunity in *Besnoitia* infection (Frenkel and Lunde, 1966). Also, the infection immunity of hamsters with toxoplasmosis and besnoitiosis was abolished by hypercorticism, although levels of antibody and its *in vitro* activity in the dye test were unaffected (Frenkel, 1957,1960; Frenkel and Lunde, 1966). This suggested the important participation of cells in immunity.

Cell-mediated immunity was demonstrated in the following experiment. Spleen and lymph node cells and sera from immune hamsters were separately transferred to different groups of nonimmune hamsters. The cell recipients showed more immunity than the serum recipients as indicated by the finding that they frequently survived challenge, whereas the hamsters which received antiserum always died, although usually after a prolonged illness. The fact that the transplantation of cells required isogenicity, neonatal tolerance, or pre-irradiation of the recipients indicated that the transferred cells must survive in order to express immunity. The immunity transferred was specifically directed either against *Toxoplasma* or *Besnoitia*, and there was no cross-immunity. Neither did cells from BCG-infected donors confer any nonspecific resistance against *Besnoitia* (Frenkel, 1967).

Lymphoid cells which were immunocompetent on transfer were found in lymph nodes and spleens. Cells from the thoracic duct, the peritoneum, the thymus or bone marrow did not transfer immunity (Frenkel, 1967). However, in cell culture, peritoneal cells from immune donors exhibited immune behavior (Vischer and Suter, 1954; Frenkel, unpublished). These peritoneal lymphoid cells may be too short-lived to demonstrate their immunocompetence on passive transfer, judging from observations on lymphocytes in the inflamed peritoneum of rats (Koster and McGregor, 1971).

Although lymphoid cells play the important role in *Toxoplasma* and *Besnoitia* immunity of hamsters, antibodies augment cell effects. Their relative role is not easy to assess, since cellular and antibody immunity generally occur together. Although one can abolish cellular immunity with corticosteroids, leaving antibody, which does not prevent relapse, it has not been possible so far to abolish antibody immunity in order to study cellular immunity alone. It is probable that antibody is decisive against extracellular *Toxoplasma*, resulting in lysis, at least in those animals that have accessory factor. As antibody is not active against intracellular *Toxoplasma* (Sabin and Feldman, 1948), it is reasonable to try to account for intracellular immunity by the participation of lymphoid cells. However, we are unable, so far, to account for the immunity in parenchymal cells, such as those of liver, adrenal, and brain. Also the hamster model, although apparently representative of what is seen in other animals, does not help in understanding immunity in the intestinal epithelium of cats. Here coccidian models will be useful, as are discussed in Chapter 8.

H. ANTIBODY RESPONSE AND *IN VIVO* EFFECTS

Numerous antibodies have been described in animals and humans infected with *Toxoplasma*. Antibodies have generally been studied in serologic tests for diagnostic purposes, which are reviewed in Chapter 10. Dye test antibody is most closely linked to *in vivo* effects. It appears earliest during infection, being measurable in from three to five days depending on the host, and it persists longest. Dye test antibody can account for neutralization of *Toxoplasma* in certain hosts possessing accessory factor, since after the attachment of antibody, the complement-like accessory factor brings about lysis (Sabin and Feldman, 1948; Desmonts, 1955). Ultrastructural studies of the effect of antibody-accessory factor mixtures have demonstrated defects in the pellicle of *Toxoplasma* tachyzoites, and histochemical studies have shown a loss of cytoplasm and of ribonucleic acid (Kulasiri and Das Gupta, 1959; Strannegard, 1967b). The dye serves as a convenient indicator that cytoplasmic loss has occurred by seepage through perforations in the pellicle of *Toxoplasma*; however, the action of antibody-accessory factor can also be observed by phase microscopy without dye (Desmonts, 1955).

The antibody which attaches to the surface of *Toxoplasma* and is detected with fluorescein-labeled anti-species globulin is probably identical to the dye test antibody (Suggs, Walls, and Kagan, 1968). It appears at the same time, attains similar titers and persists for the same length of time. With sera that have been stored or heat inactivated, or in which any associated accessory factor has been diluted, lysis of organisms is not seen. Whether and how this antibody immunity is effective in hosts such as mice, that do not possess accessory factor, has been debated (Feldman, 1956). The transfer of antiserum does prolong the life of *Toxoplasma* and *Besnoitia*-infected hamsters (Frenkel and Lunde, 1966; Frenkel, unpublished) in whose serum little or no accessory factor has been detected. This passive immunity was not augmented by the injection of human accessory factor (Frenkel, unpublished). In addition, such antibody is active in a cell culture test in the absence of demonstrable accessory factor activity (Lycke and Lund, 1964a,b). However, such antibody alone did not affect the respiration *in vitro* of *Toxoplasma* (Strannegard, Lund, and Lycke, 1967), and either accessory factor or properdin had to be present also for the occurrence of respiratory inhibition. This suggests that either trace amounts of normal serum factors, or cells, are necessary for the biologic activity of *Toxoplasma* antibody *in vivo*.

The other antibodies, demonstrated by complement-fixation, hemagglutination of antigen-coated red blood cells, agglutination of *Toxoplasma*, flocculation, and precipitation have not been studied sufficiently to characterize their participation in immunologic *in vivo* events. Complement-fixing antibodies appear from two to three weeks after infection and persist only for months or a few years in man, so that it is doubtful whether they participate in immunity. Although the hemagglutination antibody titers usually parallel the titers of dye test and fluorescent antibodies, some notable exceptions have been cited and discussed in respect to their serologic implications (Jacobs, 1967). It is conceivable that development of dye test antibodies and nonappearance of hemagglutinating antibodies in congenitally infected babies indicates an immunologic defect.

IgG and IgM globulins have been described as participating in the dye test and fluorescent antibody test reactions. IgM appears earlier and disappears as IgG increases in quantity (Remington, Miller, and Brownlee, 1968a,b). In human infants congenitally infected, both are present; in uninfected infants born to a mother with *Toxoplasma* antibody, only the IgG fraction is generally present. IgM is held back by an intact placenta. Anamnestic IgM responses have been observed.

A heat labile *Toxoplasma* "hostile factor" has been observed *in vitro* in several species, including man. It is believed to represent a "natural antibody," possibly an IgA globulin in rabbits (Strannegard, 1967a). However, rabbits with "hostile factor" are not immune. The nature of "hostile factor" in human serum has not been identified. It interferes with the use of human sera as accessory factor, but its activity is markedly decreased by citrate ions (Wallace, 1969a).

Antibody responses vary among animal species. High antibody titers (from 10^3-10^5) are produced during acute infection in humans, dogs, mice, rats, hamsters, guinea pigs, and rabbits. Cats, however, develop lower antibody titers, generally below 10^3 (Dubey, Miller, and Frenkel, 1970a). Chickens, Japanese quail, blue jays, and crows failed to develop antibody measurable in the dye test, even after repeated infection (Miller, Frenkel, and Dubey, 1972); however, complement-fixing antibodies have been demonstrated in chickens (Harboe and Reenaas, 1957). Pigeons generally show an excellent antibody response.

The persistence of antibody titers varies. It is especially interesting that chronic infections may continue even though dye test titers decline below measurable threshold. This has been found in rats, pigeons, raccoons, squirrels, and humans (see the section on Epidemiology).

With the dye test, no cross-reactions between *Besnoitia, Toxoplasma, Isospora felis, Eimeria nieschulzi, E. scabra, E. polita,* and *Sarcocystis tenella* have been found (Andrade and Weiland, 1971). Minor cross-reactions have been seen in the fluorescent antibody test, the hemagglutination test, and the agar precipitation test (Andrade and Weiland, 1971; Lunde and Jacobs, 1965; Suggs, Walls, and Kagan, 1968).

I. IMMUNODEFICIENT HOSTS

Immunodeficiency is a state in which a subnormal quality of immunity is generated as compared with most "immunocompetent" members of a species. Immunodeficiency may be the result of immaturity, of maldevelopment of the lymphoreticular system, or the result of disease. Apart from hypogammaglobulinemia and defects or absence of the thymic anlage, the functional lymphoreticular system may be "preoccupied" with another antigenic stimulus, replaced by metastatic tumors, or it may itself have undergone malignant transformation. It may be depressed by malnutrition or drugs.

Certain tissues can be considered immunodeficient. The brain and eye are poorly penetrated by immunocompetent cells and antibody due to a "blood-neuroectodermal barrier," and the development of immunity is often delayed in comparison with the extraneural tissues. The adrenals contain a pharmacologic concentration of corticosteroids as compared with the rest of the body, and the expression of immunity is interfered with.

Experimentally, hypercorticism delays or prevents the acquisition of immunity (Fig. 15). In hamsters, whose immunity mechanism is highly corticoid-sensitive, established immunity is reversed and relapsing toxoplasmosis and besnoitiosis are fatal (Fig. 16) (Frenkel, 1960; Frenkel and Lunde, 1966). As antibody levels remain stable (Fig. 17), it is likely that the hypercorticism acts on immunocompetent cells or expressor mechanisms.

Ever since the first recognition of human toxoplasmosis, the majority of cases were observed in immunodeficient hosts: fetuses and newborns. Each of these patients was accompanied by an immunocompetent control, the mother, who generally showed no, or occasionally, only slight symptoms of disease. Several cases were reported in immunodeficient adults. The first recorded fatal toxoplasmosis in an adult was observed in a malnourished Peruvian who also suffered from bartonellosis (Pinkerton and Weinman, 1940); other patients had myeloid metaplasia and cytomegalovirus infection (Hemsath and Pinkerton, 1956), or a lymphoma (O'Reilly, 1954). In some patients, disease manifestations were confined to an immunodeficient organ such as the brain (Sabin, 1941; Pinkerton and Henderson, 1941; Guimarães, 1943; Sexton, Eyles, and Dillman, 1953; Bobowski and Reed, 1958). Ocular toxoplasmosis is remarkable for its chronicity in most humans and animals (Wilder, 1952).

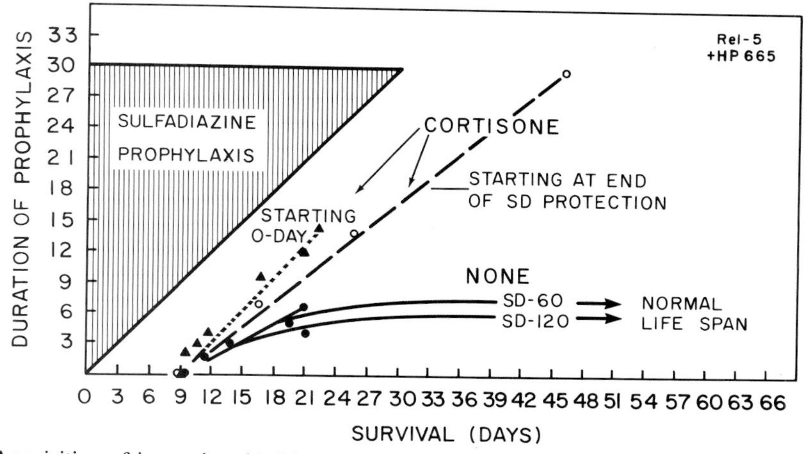

Fig. 15. Acquisition of immunity aided by sulfadiazine prophylaxis and its impairment by the administration of cortisone. Prophylaxis with sulfadiazine (60-120 mg% in drinking water for 6-10 days) after *Besnoitia* infection resulted in indefinite survival of hamsters. When 2.5 mg of cortisone acetate was injected, twice weekly, following sulfadiazine prophylaxis (from 1-4 weeks), death from relapse resulted. The hamsters died after 8-16 days, with the later deaths occurring in animals having the longer immunogenic periods (*broken line*). However, when cortisone was started on the day of infection, death resulted after a fixed interval following the discontinuation of sulfadiazine protection (*dotted line*). Apparently no immunity had been acquired, although antibody was formed. (From Frenkel and Lunde, 1966.)

Also in animals, clinical toxoplasmosis has been recorded more often from fetal and newborn sheep, from young kittens, and from dogs that suffered from distemper and other infections (Watson and Beverley, 1971; Dubey and Frenkel, 1972b; Siim, Biering-Sorensen, and Møller, 1963).

Apart from such spontaneous instances of immunodeficiency and *Toxoplasma* infection in immunodeficient hosts or organs, modern medicine has added iatrogenous (physician-generated) instances of clinical toxoplasmosis. Probably the first observed was a patient with Hodgkin's disease treated with cortisone, radiation, and nitrogen mustard, and who died with toxoplasmic encephalitis and cytomegalovirus pneumonia (Arias-Stella, 1956). This observation led to experimental studies in hamsters with asymptomatic chronic toxoplasmosis in which a fatal relapse could be produced by cortisone treatment, potentiated by radiation, but not by nitrogen mustard therapy (Frenkel, 1957).

Although Hodgkin's disease alone produces an immune defect, so that these patients sometimes develop tuberculosis or cryptococcosis, no instance of toxoplasmosis has been recorded in a patient with untreated Hodgkin's disease. The treatment for Hodgkin's disease and other lymphomas with pharmacologic doses of corticosteroids, with radiation, cytostatic drugs, and antimetabolites can transform a small immune defect to a major immunodeficiency. Frequently, *Toxoplasma* and *Pneumocystis*, cytomegalovirus and herpes virus, and the fungi, *Aspergillus* and *Candida* complicate such treatment in patients with leukemia, lymphoreticular and

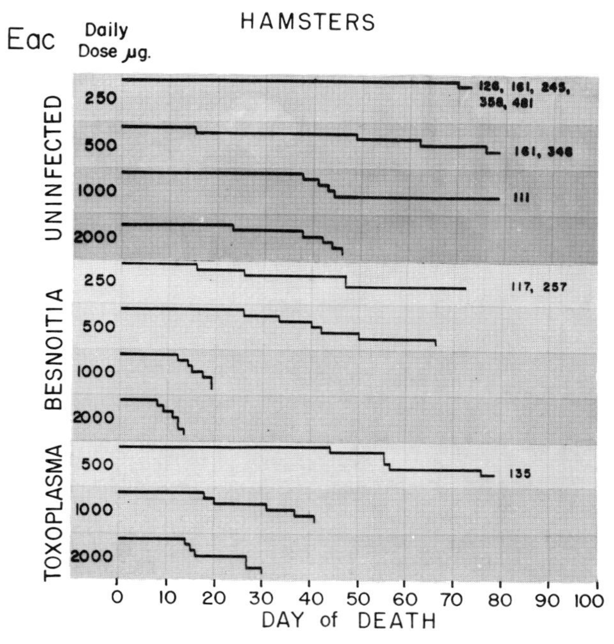

Fig. 16. Effects of cortisone acetate, given subcutaneously, on uninfected hamsters and on hamsters with chronic *Besnoitia* and *Toxoplasma* infections. Ordinates represent the number of groups of six hamsters surviving at each cortisone level. Note the rapid and uniform mortality in hamsters with *Besnoitia* infection and treated with the two largest doses of cortisone. Chronic *Toxoplasma* infection is intermediate in corticoid sensitivity between those of normal and *Besnoitia*-infected hamsters. (From Frenkel, 1960.)

occasional other neoplasms (Cheever, Valsamis, and Rabson, 1965; Vietzke *et al.*, 1968).

In addition, toxoplasmosis and the other infections are seen in patients immunodepressed deliberately in order to overcome immunologic rejection of their kidney or heart transplants (Reynolds, Walls, and Pfeiffe, 1966; Flament-Durand *et al.*, 1967; Cohen, 1970; Ghatak, Poon and Zimmerman, 1970; Stinson *et al.*, 1971).

A double jeopardy situation exists when blood from a donor with chronic myelogenous leukemia is transfused into patients with acute leukemia. Although his granulocytes are sufficiently mature to benefit patients without granulocytes, the donor may be immunodeficient from his leukemia or from treatment. The recipient is usually markedly immunosuppressed. Four cases of leukocyte-transmitted toxoplasmosis have been described (Siegel *et al.*, 1971).

Toxoplasmosis in the immunodepressed host may be a primary generalized infection, more prolonged and severe than usual, since immunogenesis is impaired. Toxoplasmosis may also result from the reactivation of a chronic infection in brain or lungs. Both types may be fatal (Fig. 14).

J. HYPERSENSITIVITY

Hypersensitivity characterizes the damaging effect of antigen on specifically sensitized cells and tissues. It is distinct from toxicity,

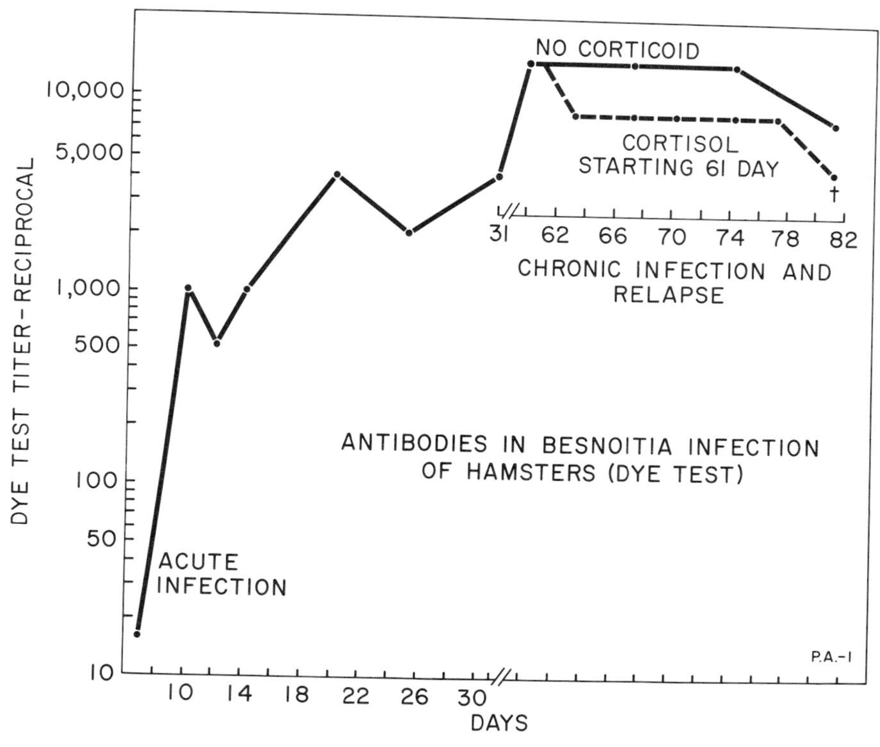

Fig. 17. Development of dye test antibody during acute *Besnoitia* infection of hamsters, its persistence during chronic infection, and after cortisol was administered (2.5 mg subcutaneously, twice weekly) even though all the animals died from relapse, similar to the results shown in Fig. 16. (From Frenkel, 1967.)

since nonsensitized cells are not similarly affected. It is part of immunology in the broad sense, dealing with all modified host reactions following sensitization. Hypersensitivity is different from immunity, which is defined as the altered humoral and cellular reactions resulting from infection or vaccination, specifically potentiating microbial defense. Immunity and hypersensitivity are medical concepts, deduced from the observations that following prior contact, a host may be either protected or develop active disease; in either case hypersensitivity may or may not be present.

In toxoplasmosis, both immediate hypersensitivity (transferred by serum) and delayed hypersensitivity (transferred by cells) can be found. Both are important pathogenically and the latter diagnostically.

Delayed hypersensitivity may be observed irrespective of the state of immunity. Guinea pigs were immune when challenged 14 days after infection with a strain of low virulence. However, delayed hypersensitivity was present after seven days (Krahenbuhl, Blazkovek, and Lysenko, 1971a,b). With *Besnoitia* infection, delayed hypersensitivity appears after from four to five days in hamsters, immunity after three weeks (Frenkel, 1967).

When both immunity and hypersensitivity are present, the latter may be diminished by desensitization without affecting the former. This has been used in patients with retinochoroiditis, usually a result of cyst rupture in the retina, before corticosteroid therapy was available (Frenkel, 1949). Nowadays, one would depress hypersensitivity with corticosteroids and guard against simultaneous depression of immunity with anti-toxoplasmic chemotherapy. The suppression of toxoplasmin reactions by multiple simultaneous injection of homologous antigens has been studied by Fuller, Chaparas, and Kolb (1967).

Transfers of cells or serum in many infections transfers hypersensitivity without immunity. In besnoitiosis of hamsters, one can transfer immunity without hypersensitivity (Frenkel, 1967; unpublished data).

When dealing with antigen-antibody relationships or antigen-cell relationships *in vitro*, it may not be clear whether we are dealing with a process analogous to either immunity or hypersensitivity. Hence, in a system of inductive immunology based on *in vitro* phenomena, it would not be possible to distinguish between immunity and hypersensitivity. The lymphocyte transformation observed when buffy coat leukocytes of toxoplasmic patients are exposed to antigen (Huldt, 1967a; Tremonti and Walton, 1970), and the inhibition of macrophage migration with peritoneal exudate of toxoplasmic guinea pigs in the presence of toxoplasmin (Krahenbuhl, Blazkovek, and Lysenko, 1971a) are not clearly assignable to one or the other. Macrophage inhibition was not closely correlated with delayed hypersensitivity in guinea pigs; it reached a maximum in one week in half of the animals, hypersensitivity was maximally developed between three and ten weeks, and clinical immunity as measured by challenge with the virulent RH strain was well established after two weeks; it was not specifically investigated at the one-week time interval (Krahenbuhl, Blazkovek, and Lysenko, 1971b). Lymphocyte transformation needs to be correlated with immunity and hypersensitivity in the course of experimental and clinical toxoplasmosis. A factor inhibiting macrophage migration effective on normal cells is released by the interaction of sensitized cells and antigen in guinea pigs (Krahenbuhl, Blazkovek, and Lysenko, 1971a); such soluble factors, which have been well studied in the tuberculin system (David, 1968), were not active in imparting immunity to *Besnotia*-infected hamsters (Frenkel, unpublished data).

K. ACQUIRED RESISTANCE

Mice infected with *Besnoitia jellisoni* were reported to be resistant to *Toxoplasma* and vice versa (Ruskin and Remington, 1968a), as well as to a great variety of phylogenetically unrelated intracellular bacteria, such as *Listeria monocytogenes* and *Salmonella typhimurium* (Ruskin and Remington, 1968b, 1971), and to Mengo virus and *Brucella melitensis* (Remington and Merigan, 1969; Ruskin and Remington, 1968a). In the instance of *Toxoplasma* challenge, strains of low virulence were used and small challenge doses, such as from one to ten doses, lethal to 50% of the mice. Although it has been demonstrated that heterospecific infection can slightly modify certain fatal infections, this modification was minor compared to the effect of specific serum and cell-mediated immunities, protective against a thousand-to million-fold 50% lethal dose (Frenkel, 1973). Whereas it had been hoped that common mechanisms of immunity underlie intracellular infections of all types (Ruskin, McIntosh, and Remington, 1969), the minor degree of this nonspecific cross-resistance observed against phylogenetically unrelated organisms does little to sustain this hope. The effect of interferon and interferon inducers is discussed in the section on natural resistance.

VII. Transmission

A. BIOLOGY OF TRANSMISSION

1. Transplacental transmission

Toxoplasmosis was first described to be transmitted transplacentally by Cowen, Wolf, and Paige (1942). It was soon recognized as a disease of the fetus and newborn, whereas the mother, although showing serologic evidence of a recent infection, generally remained asymptomatic. Gradually the concept developed that a woman usually transmitted the infection to her unborn child only if she contracted primary infection during the pregnancy in question (Eichenwald, 1960); subsequently born children did not become infected *in utero*. All of this, however, was based on retrospective studies.

Prospective studies on the transmission of toxoplasmosis in man are those of Desmonts, Couvreur, and Ben Rachid in Paris (1965). Although only about 16% of women in the child-bearing age were found to be without antibody and at risk, still, this is the largest study series available, with 118 cases of toxoplasmosis that were acquired during pregnancy (Couvreur, 1971). The study of Alford *et al.* (1969) reports on six cases from Alabama, and Kimball, Kean, and Fuchs (1971a) present three cases from New York.

A suspicion developed that chronic toxoplasmosis might give rise to habitual abortion (Langer, 1966). However, these reports were generally based on faulty techniques, either the examination of fresh tissues, or the examination of such tissues with fluorescent antibody technique. Langer's observations on fresh specimens were invalidated by his own statement that he found pollen grains, which were morphologically indistinguishable from *Toxoplasma* cysts, and that it was not possible to decide retrospectively in which of his cases pollen grains or *Toxoplasma* cysts had been seen (Langer, 1966). Other observations based on staining of suspected tissues with fluorescein-labeled *Toxoplasma* antibody (Thomascheck, Werner, and Schmidtke, 1966) were not supplemented by an examination of the fluorescent bodies in permanent preparations, or by attempts to isolate *Toxoplasma* from such tissues. Although occasionally two successive conceptuses may be infected, resulting in abortion (Remington, Newell, and Cavanaugh, 1964; Garcia, 1968), there is no valid evidence to indicate that it is either common or habitual (Janssen, Piekarski, and Korte, 1970; Kimball, Kean, and Fuchs, 1971b).

Transplacental toxoplasmosis has been observed also in animals (Siim, Biering-Sorensen, and Møller, 1963). In sheep, the available evidence links this transmission to the acquisition of primary maternal infection during the period of pregnancy (Hartley and Moyle, 1968;1973), and infection before pregnancy prevents fetal infection (Beverley and Watson, 1971). However, transmission of *Toxoplasma* does appear to have been observed in repeated pregnancies in mice, where indeed vertical transmission for several generations has been reported (Beverley, 1959). This observation deserves careful study, which has not been forthcoming in the ten years following publication of the preliminary note. However, additional studies in rats, mice, guinea pigs, and hamsters are available, confirming transmission of *Toxoplasma* during chronic infection in certain hosts (Remington, Jacobs, and Melton, 1961; Giraud *et al.*, 1965; Roever-Bonnet, 1969; Nakayama, 1968)

2. Transmission by carnivorism

The existence of cysts as an apparent chronic stage in skeletal and cardiac muscle prompted Jacobs, Remington, and Melton (1960) to investigate the transmission of *Toxoplasma* by carnivorism. They found that cyst organisms were more resistant to peptic digestion than were tachyzoites. Utilizing peptic digestion techniques, they showed that a variable percentage of samples of meat for human consumption contained *Toxoplasma*. The gourmet custom of eating of raw or undercooked meat, such as steak tartare, favors transmission of toxoplasmosis to humans. The same applies when raw meat is eaten since "it is healthy," as in France (Desmonts *et al.*, 1965), or when undercooked meat may be eaten on account of "being in a rush," as in the United States (Kean, Kimball, and Christenson, 1969). Mutton and pork are most commonly infected (Work, 1967) but beef has also been found to contain cysts (Catar, Bergendi, and Holkova, 1969). As indicated in the next section, the manner in which the animals are maintained, on the range or in feed lots or barns, and what and how they are fed, is likely to influence the incidence of infection. As *Toxoplasma* are not resistant to heat, such as frying, roasting or other means of cooking, these measures are generally adequate to destroy the infectivity of meat (Work, 1968). Freezing of meat is injurious to *Toxoplasma* bradyzoites, but to eliminate infectivity, freezing is not a dependable measure.

Mere contact with raw meat should also be considered as a mode of transmission. For example, Price (1969) found that both dog and cat owners who fed raw meat showed a 90% incidence of antibody, whereas in those who did not, the incidence was only about 65%.

Biologically determined carnivorism, as in dogs and cats, would explain the acquisition of infection by carnivores. But frequently the animals eaten are herbivores (sheep) and it was impossible to explain how these became infected.

3. Fecal transmission

A Scottish parasitologist, W.M. Hutchison, first discovered in 1965 that *Toxoplasma* could be transmitted fecally. He fed *Toxoplasma* cysts from chronically infected mice to a cat and demonstrated that its feces were capable of infecting mice. This *Toxoplasma* infectivity survived for a year in the shed feces, which were said to contain bacteria, oocysts of *Isospora*, and the ova of *Toxocara cati* (Hutchison, 1965). He then fed *Toxoplasma* cysts to two cats, one of which carried *Toxocara* and the other did not. *Toxoplasma* was fecally discharged by the *Toxocara*-infested cat and not by the other. He then eliminated *Toxocara* from the first cat with an anthelmintic, and infected the second cat with *Toxocara*. When *Toxoplasma* was refed to both cats, only the second cat discharged fecal *Toxoplasma* (Hutchison, 1967). These feces were filtered through a 36 μ sieve and the *Toxoplasma* infectivity remained with the nematode eggs. Based on these simple correlations but without statistical considerations, Hutchison postulated that the infectious *Toxoplasma* was protected within the eggs of *Toxocara cati* (Hutchison, 1967). Several other instances of the transmission of *Toxoplasma* in nematode and cestode eggs were reported in rats (Machado *et al.*, 1967), cats (Dubey, 1968; Rommel *et al.*, 1968), from dogs (Tishovskaya, 1969), and from guinea pigs (Kheysin *et al.*, 1969). However, Rommel and collaborators (1968), who studied 15 species of nematodes in six hosts, linked the transmission of *Toxoplasma* only to *Toxocara cati* in cats. Vermeil and Marguet (1967) obtained negative results with *Ascaridia galli* from chickens.

The nematode transmission hypothesis motivated me to closely examine the role of *Toxocara* in the transmission of *Toxoplasma*. This resulted first, in "deworming" the fecal transmission of *Toxoplasma* (Frenkel, Dubey, and Miller, 1969) and second, in the identification of fecal infectivity within an isosporan oocyst (Frenkel, Dubey, and Miller, 1970). The identification of *Toxoplasma* as a coccidian was based on at least 20 mutually independent criteria, contributed by several investigators, which are discussed together with the "new cyst" on pp. 382-383. A number of divergent and congruent observations have been analyzed by Dubey, Miller, and Frenkel (1970b) and have been partially resolved (Hutchison *et al.*, 1971).

Toxoplasma oocysts are produced regularly by sero-negative cats or kittens after the ingestion of cysts of *Toxoplasma*; the prepatent period is from three to five days (Fig. 18). The oocyst production was less efficient, four times in nine attempts, after the feeding of tachyzoites, with a prepatent period of from five to ten days. It was still less efficient after the feeding of oocysts, after which only eight out of 17 cats developed oocysts within a prepatent period of from 20 to 24 days (Dubey, Miller, and Frenkel, 1970b). The low efficacy of tachyzoite transmission was not surprising in view of the high susceptibility of these forms to peptic and tryptic digestion. However, the low incidence of oocyst development by cats fed oocysts was unexpected, especially since most of the cats developed *Toxoplasma* antibody, and since oocysts are highly infectious to various mammals and birds (Miller, Frenkel, and Dubey, 1972). Whereas cysts generally give rise to oocyst production in newborn kittens (in 23 out of 24 trials), oocyst feeding led to oocyst production only in weaned kittens (8 out of 17), although 37 of 44 nursling kittens in which it had been tried became infected. This phase of infection is currently under investigation.

The immunologic status, as reflected by serum antibodies, influences the production of oocysts. Whereas 23 out of 24 cats free of antibody discharged oocysts after cyst feeding, this occurred in only seven out of ten cats with *Toxoplasma* antibody on arrival in the laboratory, and in none of nine sero-positive cats that had been infected recently in the laboratory (Frenkel, 1970). This suggests that recent infection is associated with solid immunity, and remote infection with partial immunity. However, in more recent studies, an occasional cat in the recently infected sero-positive group has been found to shed a few oocysts after a delay of several days. Similar findings have been reported by Kühn and Weiland (1969) and by Piekarski and Witte (1971).

Toxoplasma oocyst production was observed in domestic cats, one out of seven bobcats (*Lynx rufus*), one mountain lion (*Felis concolor*), two out of five ocelots (*Felis pardalis*), one out of two jaguarundis (*F. yagouaroundi*), but in none of 23 species of wild nonfeline mammals and birds from 12 orders (Jewell *et al.*, 1972; Miller, Frenkel, and Dubey, 1972). All this suggests that *Toxoplasma* may have arisen as an enteric coccidian of domestic cats or other felines, transmitted only by oocysts, and later developed the ability to parasitize many tissues and hosts. The formation of a persistent cyst in the brain, skeletal muscle, and cardiac muscle has made it possible for carnivorism to become an additional means of transmission. As feeding cysts gives rise to oocyst production more efficiently than feeding oocysts, we may be witnessing transition to a two-host cycle, which is more common than the one-host cycle at present, although the intermediate host with cysts is not obligatory. This idea is supported by observations of Wallace (1971b) who noted that kittens began to acquire *Toxoplasma* antibodies when they were more than six months old, suggesting that they became infected in Hawaii mainly when they started to hunt. Although nursling and weanling kittens can be experimentally infected with *Toxoplasma* oocysts, we have not observed them to shed oocysts before weaning, as is common with *Isospora felis* (Dubey and Frenkel, 1972a).

Infection with oocysts (sporozoites) has been achieved in all 13 species of animals

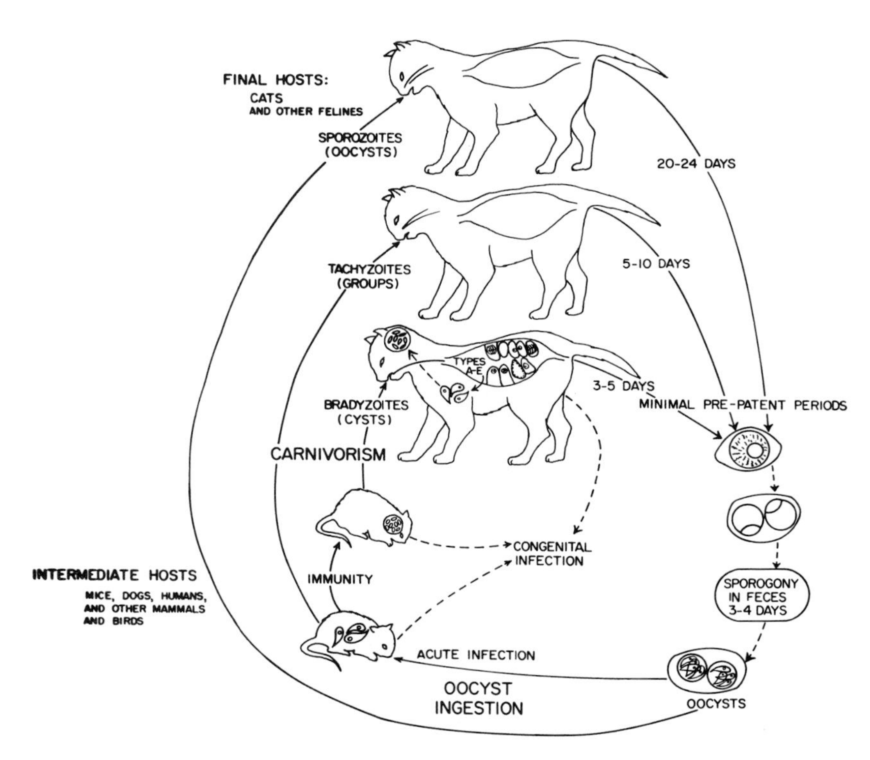

Fig. 18. Life cycle of *Toxoplasma*. Three cats are used to indicate infections with different prepatent periods to oocyst appearance. After the ingestion of bradyzoites (cysts) the minimal prepatent period is 3-5 days. After the ingestion of tachyzoites of the acute infection, the minimal prepatent period is 5-10 days. After the ingestion of sporozoites (oocysts), the minimal prepatent period is 20 to 24 days. Only the endogenous cycle in cats following cyst ingestion has been described (Dubey and Frenkel, 1972b). Mice are shown to represent the many intermediate, or nonspecific hosts in which only the extra-intestinal, or tissue cycle of *Toxoplasma* is found. In cats this extra-intestinal cycle occurs together with the enteroepithelial cycle (types A- E and gametocytes and oocysts). Congenital infection has been observed in some animals in the course of acute or chronic infection.

where it has been tried (Miller, Frenkel, and Dubey, 1972). Oocyst transmission serves to explain natural infection in herbivores, such as in sheep, which so far have not been studied in nature (Fig. 19). As the outbreaks in sheep are epidemic, followed sometimes by several years without evidence of further dissemination (Hartley and Moyle, 1973), they could well be caused by cats that contaminate the watering places of such sheep with oocysts. The findings on islands and atolls where *Toxoplasma* antibody in sheep, pigs, as well as humans coincided with the presence of cats, lend support to this thesis (Munday, 1972; Wallace, Marshall, and Marshall, 1972; Wallace, 1969b). Additional dissemination by transport hosts such as filth flies explored by Wallace (1971a) and the potential role of earthworms in distributing *Toxoplasma* through the soil should also be considered (Dubey, Miller, and Frenkel, 1970a).

Helminths and other fecal constituents. Recovery of the *Toxoplasma* oocyst from worm-free cats does not by itself invalidate the hypothesis that *Toxoplasma* can be transmitted inside the worm egg. However, the role of *Toxocara cati* was excluded by separating the *Toxoplasma* infectivity from the worm eggs by thorough washing. Since, however, occasionally some *Toxoplasma* infectivity remained with the worm eggs, particu-

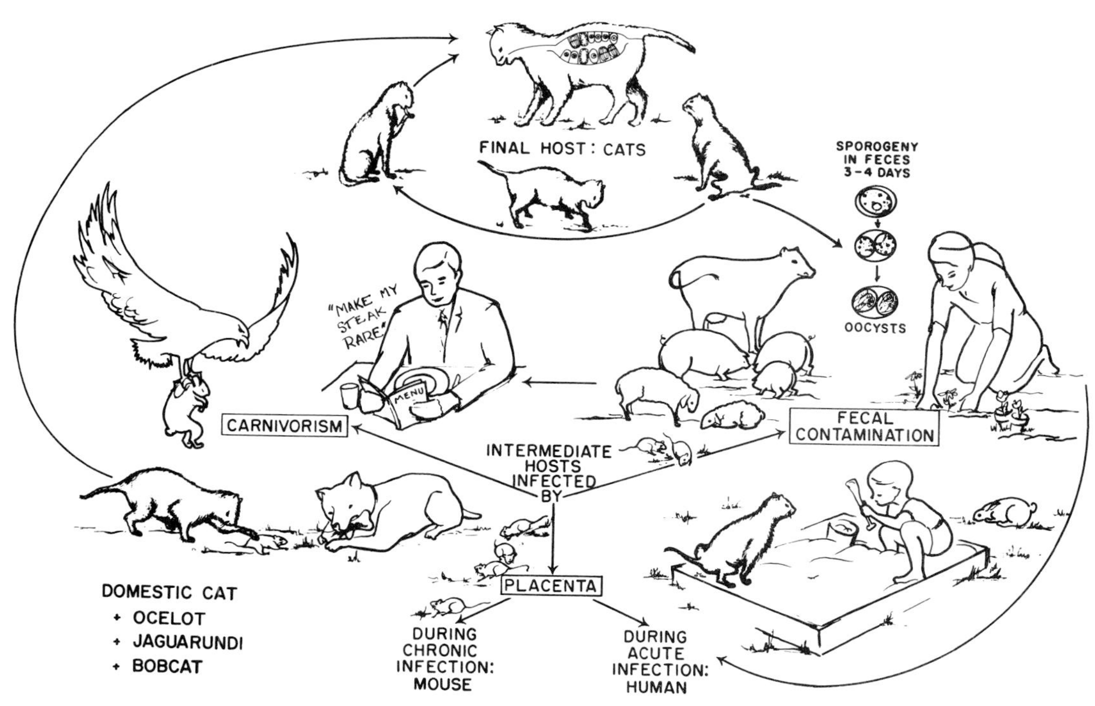

Fig. 19. Postulated transmission of toxoplasmosis. Oocysts are shed in the feces of cats, and after sporulation are infectious to a great variety of hosts by fecal contamination, probably directly and via transport hosts (*right*). Carnivorism is the other important means of transmission (*left*). Cats probably contract infection more often by carnivorism than by the ingestion of oocysts. Transplacental transmission is indicated below. (By permission of Williams and Wilkins; from Frenkel, 1971a.)

larly when fungal hyphae, to which the oocysts became attached, were present in the fecal suspension, a method of separating the worms from the infectivity had to be devised. This was accomplished by hatching the eggs and depositing the suspension of larvae in the center of an agar plate. The larvae which migrated away from the site of deposition were shown to be free from *Toxoplasma* infectivity, which remained in the center of the plate. Hence, by completely separating worm eggs from *Toxoplasma* infectivity on a sieve and in the filtrate, obtaining pure *Toxoplasma* infectivity free of eggs, as well as obtaining 54% of *Toxocara cati* larvae completely free of any *Toxoplasma*, we could prove that *Toxocara cati* eggs were not the vehicle in which *Toxoplasma* was conveyed (Frenkel, Dubey, and Miller, 1969).

The following nematodes were excluded from playing a role in the transmission of *Toxoplasma* by Rommel *et al.*, (1968): *Haemonchus contortus, Ostertagia circumcincta, Chabertia ovina,* and *Oesophagostomum venulosum* from eight sheep, *Ascaris suum, O. quadrispinulatum*, and *dentatum* from six pigs, *Toxocara canis, Toxascaris leonina, Ancylostoma caninum*, and *Uncinaria stenocephala* from eleven dogs, *Ascaridia galli* and *Heterakis gallinae* from two chickens, and *Syphacia obvelata* from 30 mice. Although positive results were observed in cats infected with *Toxocara cati*, these results can be ascribed to *Toxoplasma* oocysts.

Hymenolepis nana from rats infected with sublethal doses of *Toxoplasma* by mouth and intraperitoneally were said to transmit toxoplasmosis. After washing the feces in 10% formalin and 0.1% merthiolate for ten minutes, the ova and the adult worm were triturated separately; infection resulted in 7 of 30 mice given the ova and in a slightly larger number of mice given adult worms (Machado *et al.*, 1967). This interesting observation has not yet been confirmed to my knowledge.

A "new cyst" was described by Work and Hutchison (1969a). It measured 9μ x 14μ and was surrounded by a wall of uniform thickness, with the interior occupied by one slightly granular mass; after three weeks, two separate organisms had developed, measuring about $3\mu \times 7\mu$, and no definite structures except for some granules could be seen inside them. Based on these observations and the finding that ten of these organisms infected four of six mice, and that four single micro-isolated new cysts infected all four mice injected, the authors concluded that, "it seems most likely that a new cystic form of *Toxoplasma gondii* has been observed."

Several inconsistent observations indicate that these "new cysts," observed in unpreserved fecal suspensions, in which fungi are common, are unrelated to *Toxoplasma*: (a) the inner cyst measured $3\mu \times 7\mu$, about twice as long as broad, and not ovoid as are the sporocysts ($6\mu \times 8\mu$); (b) the easily visible sporozoites were not seen in these inner cysts by investigators who were well familiar with the coccidia (Hutchison, 1965); (c) the relationship of infectivity with single micro-isolated new cysts is not valid without inoculation of control mice with a similar sample which did not include a new cyst; (d) the observation that four micro-isolated new cysts would each infect a mouse would be unusual when in a titration of the same materials, a calculated dose of ten new cysts produced infection in only four out of six mice, and a dose of one new cyst produced no infection at all (Work and Hutchison, 1969a); (e) in another titration in which new cysts were compared with infectivity, a calculated dose of a single new cyst produced infection in only one of six mice, and 1000 new cysts were necessary to infect all in a group of mice, equivalent to what had been reported from one micro-isolated new cyst (Work and Hutchison, 1969b).

In summarizing these observations which attempt to relate the new cyst to *Toxoplasma* infectivity, Siim, Hutchison, and Work (1969) shift to the term oocyst to characterize what had been described previously as the "new cyst." In addition, they report finding four sporozoites in sporocysts stained with hematoxylin, stating that they were remarkably similar to *Isospora bigemina*. They also refer to an observation of profuse schizogony and gametogony which was communicated in a letter to the editor (Hutchison *et al.*, 1969). The identification of the newly found oocyst as belonging to *Toxoplasma* was based on an incomplete *post hoc ergo propter hoc* relationship of finding profuse enteric stages in one test cat, a single schizont in another, none in two test cats, and none in one control (Hutchison *et al.*, 1970; Hutchison *et al.*, 1971). The role of intuition in the research which resulted in these findings is well described by Garnham (1971). After the size differences between the separate inner organisms of the "new cysts" and the sporocysts, and the discrepancies between numbers of "new cysts" and *Toxoplasma* infectivity were pointed out (Dubey, Miller, and Frenkel, 1970b), only the size differences were explained by error in measurements (Hutchison *et al.*, 1971).

Although there is now general agreement that oocysts containing sporocysts and sporozoites are identical with the *Toxoplasma* infectivity in feces of cats which had fed on *Toxoplasma*-infected mice, it is important to examine the criteria on which this assertion was based. The use of so-called specific pathogen-free (not germ-free) cats by some investigators, and of newborn kittens used in our investigations, merely exclude certain other known infections, such as *Toxocara*, which had already been excluded, and other cat coccidia, which might be confused histologically with stages of *Toxoplasma*. However, the crucial problem of linking the observed oocysts with *Toxoplasma* infectivity had to be based on more than ten qualitative and quantitative correlations which exceeds a figure which could likely be obtained by chance alone. They were the following: (1) oocysts appeared and disappeared simultaneously with infectivity; (2) numbers of oocysts were correlated with infectivity titers; (3) when oocysts were unsporulated, infectivity was absent; and the appearance of sporozoites was related with respect to time with the appearance of infectivity; (4) oxygen was necessary for both sporulation and for the development of infectivity, and anaerobiasis inhibited both; (5) lowered temperature similarly delayed sporulation and development of infectivity; (6) both were similarly inactivated by heating between 45°C and 50°C; (7) by freezing; (8) by chemicals such as formalin, ammonia, tincture of iodine; and (9) by drying; (10) oocysts measuring about $10\mu \times 12\mu$ and infectivity passed the same filters, but both were retained whenever filter pores were less than 9 μ; (11) oocysts and infectivity showed the same density characteristics, with a peak at sp. gr 1.11; (12) both moved similarly in an electric field; (13) antigenic similarities were found between *Toxoplasma* tachyzoites, oocysts, and sporocyst walls, and smears of sporozoites and tissue stages using fluorescent antibody techniques. More importantly, the capacity of *Toxoplasma* antiserum to stain oocyst walls could be absorbed with tachyzoites of the standard RH strain of *Toxoplasma*; (14) the feeding of mouse brains harboring *Toxoplasma* cysts to one-to two-day old kittens gave rise to the discharge of oocysts just as in adults, excluding possible activation by toxoplasmosis of a latent infection with another coccidium; (15) examination of the intestinal lining of such infected newborn kittens revealed coccidia-type schizogony, gametogony, and oocysts, whereas control littermate kittens that were uninfected or received normal mouse brains remained negative; (16) feeding each of the three stages of *Toxoplasma* to cats resulted in identical prepatent periods to oocyst production and to the production of fecal infectivity: from three to five days after cyst feeding, from five to ten days after feeding of tachyzoites, and from 20 to 24 days after feeding of mouse-infectious cat feces containing oocysts; (17) the serologic status of cats with respect to *Toxoplasma*, measured with the standard RH strain, affected the production of oocysts and fecal infectivity. The latter occurred in 23 of 24 sero-negative cats, but in only seven of ten cats which had *Toxoplasma* antibody on arrival in the laboratory, probably acquired a long time before, but in none of nine sero-positive cats that had recently been infected in the laboratory; (18) the infectivity of a given number of oocysts was increased four-fold after some of the enclosed sporozoites had been liberated. Identical qualitative or quantitative behavior in these correlates proved conclusively that the *Toxoplasma* infectivity coincided with the candidate oocyst, which could be designated the *Toxoplasma* oocyst (Frenkel, Dubey, and Miller, 1970; Dubey, Miller, and Frenkel, 1970b; Frenkel, 1970; (19) microdroplets with oocysts were shown to be infectious whereas control droplets were not (Overdulve, 1970); (20) oocysts became infectious to cell cultures only after the contained sporozoites were liberated (Sheffield and Melton, 1970); (21) time and temperature required for the sporulation of oocysts were shown to be identical with those for the development of *Toxoplasma* infectivity (Weiland and Kühn, 1970); (22) in cats with a *Toxoplasma* antibody titer in excess of 1:256, no discharge of oocysts or fecal infectivity occurred after inoculation with either cysts or oocysts; and (23) two strains of *Toxoplasma* isolated in Japan, one from a human, also gave rise to infectious oocysts in cats (Walton and Werner, 1970). Prior infection with coccidia with the morphology of *I. rivolta*, *I. felis* and *I. bigemina* (*cati*) did not interfere with the excretion of candidate oocysts and occurrence of *Toxoplasma* infectivity (Piekarski and Witte, 1971).

If one wished to relate oocysts of other organisms such as *Besnoitia*, *Sarcocystis* and *Frenkelia* to a coccidian cycle, similar evidence, exceeding that likely to accumulate by chance alone would need to be accumulated. In a sequence of an infectious feeding and oocyst discharge, significance at the 5% level is attained after seven correlations and at the 1% level after nine such sequences, in the absence of any divergent results (Dixon and Massey, 1969). Statistically significant numbers of correlations would be especially important if the postulated final host were an animal harboring several species of coccidia, and if it were difficult to obtain coccidia-free animals for experimentation. Establishing a relationship between infectivity and oocyst, and between oocyst and the preceding multiplicative and gametogonic stages, would involve separate considerations.

4. Possible other means of transmission

Venereal. Although sexual and osculatory transmission of toxoplasmosis have been suspected in humans, an extraneous source of infection for both partners is difficult to exclude in individual cases. The isolation of *Toxoplasma* from the saliva of 12 of 20 patients, mostly with lymph node enlargement, reported by Levi *et al.* (1968) is noteworthy and requires confirmation. Relevant information can be obtained from a comparison of antibody presence in married couples. If venereal infection were common, there should be a greater percentage of conforming than of divergent serologic findings. In a recent study (Price, 1969) involving 43 couples, both partners were sero-positive in 20 instances, both were negative in three, and only one partner had antibody in 20. This distribution is close to what could be expected by chance alone, and argues against the probable occurrence of conjugal infection. *Toxoplasma* does not commonly invade the oral mucosa or that of the respiratory or reproductive tracts; however, pulmonary macrophages are parasitized during the acute infection so that positive sputum might be obtained. Although orchitis has been observed in animals, not many *Toxoplasma* organisms would be likely to appear in the semen. While this route of infection cannot be entirely excluded, it appears improbable.

Lactation. Eichenwald (1948) described transmission through the milk of lactating mice. *Toxoplasma* was isolated from the colostrum of a cow (Sanger *et al.*, 1953) and from asymptomatic carrier pigs (Sanger and Cole, 1955). Milk from experimentally infected dogs, cats, and goats transmitted infection (Mayer, 1965). Rommel and Breuning (1967) observed *Toxoplasma* in the milk of acutely infected mice, guinea pigs, rabbits, sheep, and a cow (Rommel, 1969). Although infected milk might occasionally transmit toxoplasmosis in animals, pasteurization of milk would destroy all forms of *Toxoplasma* for man.

Eggs. Chicken eggs have rarely been found infected with *Toxoplasma.* Jacobs and Melton (1966) found one egg that was infected out of 327 coming from chickens with chronic toxoplasmosis. They found none of 108 commercial eggs infected. However, 12 of 62 pools of ovary and oviduct were found positive; each pool contained tissues of ten birds. The report by Pande, Shukla, and Sekariah (1961) is questionable as evidence, because it was fraudulently illustrated (DuShane *et al.*, 1961).

Blood transfusions. As parasitemia has been observed in animals and humans (Miller, Aronson, and Remington, 1969), blood transfusions can be considered as a possible mode of transmission. However, the risk from normal donors is apparently very small, since Kimball, Kean, and Kellner (1965) found no increased incidence of *Toxoplasma* antibody in 43 patients with thalassemia major who had received 4,805 transfusions. The incidence was 7%, which is in the range of that which is expected for this age group. However, the likelihood of parasitemia is significant in donors with chronic myelogenous leukemia, who may be immunologically impaired because of their disease, their treatment, or both. The recipients, usually patients with acute leukemia, are deficient in granulocytes and are more susceptible than normal to infection, since they are also immunodeficient. Two donors with chronic myelogenous leukemia and high antibody titers each gave rise to two infections in patients with acute leukemia (Siegel *et al.*, 1971).

Organ transplants. Patients receiving transplants may develop disseminated toxoplasmosis either when infected via the organ transplanted, from the blood transfusions, or from conventional outside sources (oocysts or raw meat). The patient of Reynolds, Walls, and Pfeiffe (1966) appears to belong to one of these categories. Since these patients are immunosuppressed to facilitate "take" of the transplanted tissue, *Toxoplasma* multiplication would likely persist longer and accumulate more lesions than in normal individuals. They may be already infected, and their chronic toxoplasmosis may relapse because of immunosuppression (Stinson *et al.*, 1971), as discussed in Section V,B,4. Such relapses are usually localized in the brain, in contrast to the generalized distribution of acute infections.

Laboratory and autopsy. Infection from accidental aspiration, inhalation of spray, needle puncture, and cuts have been reported sporadically (Rawal, 1959; Neu, 1967). This danger is slight. If the accident is noticed and infection ensues, prompt treatment can be instituted. Indeed, it is probable that the patient treated soonest for toxoplasmosis was one with a laboratory infection (Frenkel, Weber, and Lunde, 1960). Since *Toxoplasma* is killed by drying, airborne infection other than by the inhalation of a direct spray does not appear to be a danger. Piekarski and Witte (1971) have found the dust of animal rooms containing cats to be noninfectious, and we have examined dry cat feces and found them to remain infectious for about two weeks, although the original one million were soon reduced to 100 oocysts (Frenkel and Dubey, 1972b).

Arthropods. Toxoplasma was discovered in 1908 shortly after the recognition of the mosquito transmission of malaria. Investigators attempting to classify *Toxoplasma* compared it to *Leishmania*, piroplasms, and hemogregarines, all of which were suspected or known to be transmitted by arthropods. The stage was set, therefore, to investigate the ectoparasites of gondis, in which *Toxoplasma* was first found, for their possible role as vectors of *Toxoplasma*. Eduard Chatton and Georges Blanc (1917) so examined two species of ticks, one of chiggers, one of fleas, and two species of diptera, a *Simulium* and a chironomid. Some of these were examined microscopically, others were tested for infectivity in gondis. However, *Toxoplasma* was not found nor did an analysis of the biology of these arthropods appear consistent with their role as a vector. Chatton and Blanc (1917) observed that the gondis were not infected in nature and that they became infected only after coming to the Pasteur Institute in Tunis. Therefore, they directed their attention to parasitic mites, *Dermanyssus*, fleas, bed bugs, and ticks found in and near the animal quarters and an adjacent dog kennel, however, without coming to any conclusions.

This was followed by several reports of the survival of *Toxoplasma* in one or another arthropod. A systematic study was performed in 17 species from eight groups (Woke *et al.*, 1953) and another in 26 species from eight groups (Frenkel, 1965a). It was possible to show that *Toxoplasma* was ingested by practically all of the species examined except the chiggers (Frenkel, 1965a). Both studies showed occasionally longer survival of *Toxoplasma* in certain groups. In *Dermacentor* and *Haemaphysalis*, *Toxoplasma* was recovered as late as 48 hr after feeding; and in an injected *Haemaphysalis* in 19 days. In *Aedes varipalpus* fed on infected peritoneal fluid, *Toxoplasma* could be recovered by subinoculation after three days. Similar findings were made by Chapman and Ormerod (1966) who observed survival in *Aedes aegypti* for three days, and in two species of *Anopheles* for two days. The arthropods were refed after appropriate intervals (after molting in the *Ixodidae*, and after rearing through the adult to the succeeding larval stage in *Trombidiformes*), but in no instance was *Toxoplasma* transmitted by bite. When, after this feeding, the arthropods were macerated and injected into mice, *Toxoplasma* was never recovered.

Shortcomings of these negative observations include the use of old laboratory strains. Only in work with chiggers, and in certain experiments with *Dermacentor andersoni* and *Haemophysalis*, recently isolated *Toxoplasma* strains were used and the inocula were within ten intraperitoneal passages from the cyst (Frenkel and Loomis, unpublished data). In some of the sporadically positive results, the recipient rabbits and guinea pigs were not shown to be free of *Toxoplasma* before the experiment. As these hosts are sometimes spontaneously infected, and since some of the strains recovered from the transmission experiments were different from those used, the possibility of chance infection must be considered.

In reviewing these studies, one is struck with the negative and inconclusive results. However, these do not exclude the possibility that *Toxoplasma* may sometimes be transmitted by bloodsucking arthropods, and of course, flies and cockroaches may serve as carriers of oocysts as shown by Wallace (1971a; Wallace, Marshall, and Marshall, 1972). The sampling of such arthropods in endemic foci might provide the best clues for further investigation.

B. EPIDEMIOLOGY AND PREVALENCE

Surveys for *Toxoplasma* antibody in animals and humans from many areas show that incidence of infection increases with age (Feldman and Miller, 1956). The rate of increase differs widely in different localities (Fig 20). It is highest in hot and humid climates, where in the lowlands of Guatemala, El Salvador, the Easter Island, and Tahiti, infection rates approximating 100% have been found in young adults (Gibson and Coleman, 1958; Remington, Efron *et al.*, 1970; Morales *et al.*, 1961). The lowest incidence of *Toxoplasma* antibody has been found in inhabitants of dry and cold climates, such as in the Navajo and the Eskimos (Feldman and Miller, 1956). Whereas moisture favors transmission, it is diminished by dry or cold climates, conditions which parallel viability of the *Toxoplasma* oocyst, as observed in the laboratory (Dubey, Miller, and Frenkel, 1970a,b). The lesser incidence of *Toxoplasma* antibody at high altitudes may be related to such climatic conditions (Walton, Arjona, and Benchoff, 1966). The differences in incidence observed in humans and animals on three Pacific atolls studied by Wallace (1969b) suggested a relationship to the animal fauna, and indicated that transmission of toxoplasmosis did not take place in the absence of cats (Wallace, Marshall, and Marshall, 1972; Munday, 1972). In the United States, a yearly antibody acquisition rate of from 1/2% to 1% is common in humans (Feldman, 1965). In El Salvador, the average yearly rate was 5 1/2% and even higher in children between six months and five years, which is quite compatible with cat and soil contacts, but in view of the food habits, not related to the ingestion of raw meat (Remington, Efron *et al.*, 1970) (Fig. 20). In United States cities, the antibody acquisition rate in children is lower than in adults (Kimball, Kean, and Fuchs, 1971b), whereas in Paris, where raw meat is sometimes given

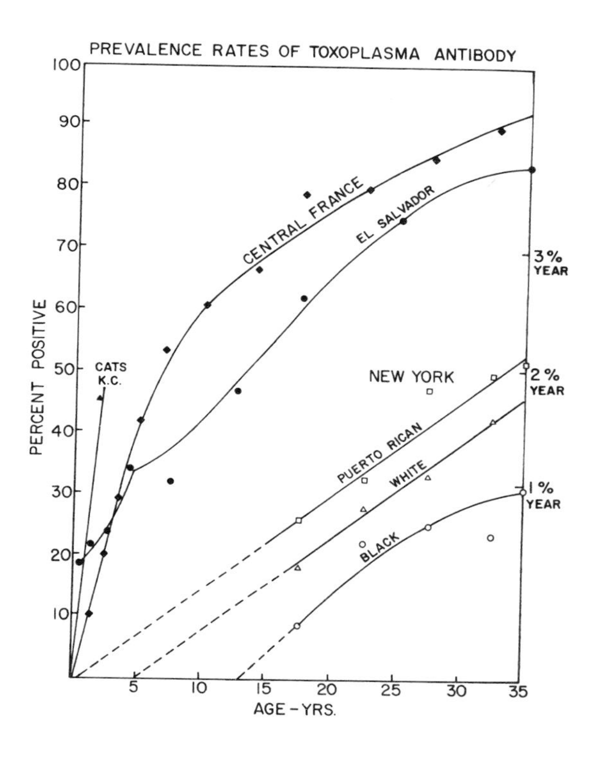

Fig. 20. Prevalence rates of *Toxoplasma* antibody and speculations about the relative importance of meat and cats in transmission to several population groups. Forty-five per cent of cats from Kansas City, with an approximate mean age of two years, had antibody (Dubey, Miller, and Frenkel, 1970a). Similarly high prevalence rates have been found in other populations of cats. In New York, the antibody rates of black, white and Puerto Rican populations have been tabulated by Kimball, Kean, and Fuchs (1971b). The prevalence is lower in the black and white populations most likely to be autochthonous to New York. The rate of infection either is higher between 15 and 35 years than during childhood, or the manner of infection commences only in late childhood or adolescence. This transmission is most likely to result from culturally acquired carnivorism; cats are considered of subsidiary importance. In El Salvador a higher rate of antibody acquisition has been found in children than in adults (Remington *et al.*, 1970). Since little meat is eaten, and all well cooked, this can be taken to reflect infection by contact with cats and cat-contaminated soil. The climate appropriately is moist and without frost. In central France, antibody acquisition has been studied by Desmonts, Couvreur, and Ben Rachid (1965). Although eating undercooked and raw meat is a known cultural habit in France (Desmonts *et al.*, 1965), transmission via cats may also be important.

to children, antibody is acquired early in life (Fig. 20).

Serologic surveys show antibody to be common in some species and rare in others. This is not necessarily a true indication of the frequency with which animals become infected, because members of highly vulnerable species such as mice, multimammate rats, canaries, and marmoset monkeys in the laboratory rarely survive infection. If such animals became infected in nature, one would rarely encounter antibody. Other species, such as chickens, crows, and probably other birds, do not develop dye test antibody although they develop infection and survive it (Miller, Frenkel, and Dubey, 1972). Some animals do not maintain antibody titers for more than a few months, although they remain infected; so, in pigeons, rats, squirrels, raccoons, humans, and others, the index of infection may be higher than revealed by antibody (Jacobs, Melton, and Cook, 1953; Eyles, 1954; Walton and Walls, 1964; Jacobs, 1967). These observations also suggest that serologic surveys based on titers of 1:16 or 1:64, or higher, may significantly underestimate the prevalence of infection.

C. PREVENTION

Toxoplasmosis has been found in numerous animal species (Jira and Kozojed, 1970), and I am not aware of any species of mammal or bird found to be resistant. Indeed, *Toxoplasma* was first described from laboratory infections of rabbits and gondis, whereas wild gondis in Tunisia were free of infection (Chatton and Blanc, 1917). It is likely that these animals became infected from cats that used to be kept in animal colonies to reduce the number of escaped or stray rodents. Direct contamination of some of the foodstuffs or bedding appears possible, and transport hosts may also have been involved. Laboratory animals continued to be infected with *Toxoplasma* in the 1940's when Perrin (1943) found it in one out of 21 rabbits and 15 of 292 guinea pigs; it is now rare in the laboratory animals in the United States; however, infections of pet and zoo animals continue to be reported.

Stray and "outdoor" cats that hunt may well be the key source of infection (Wallace, 1971b) and their control appears essential if one wishes to curb the spread of toxoplasmosis to other animals and man (Fig. 19).

Cats defecate in loose soil and sand, so that children playing in unprotected sand piles, and anyone working in soil contaminated with cat feces is exposed. Children's sand piles should be covered. Wearing work gloves and hand washing will diminish the risk.

Young domestic cats that are fed raw meat are endangered themselves, and can be a source of infection to humans. Owners of dogs fed raw meat showed as high an incidence of antibody as owners of cats (Price, 1969), so that the handling of raw meat may actually be more important for pet owners than the oocysts, which are discharged only by cats.

"Indoor" cats fed only canned, cooked or dry food have little chance to become infected. To be safe, cat feces should be disposed of every day, before the oocysts have had time to sporulate. They should be flushed down the toilet or burned. Disposable gloves can be used by vulnerable individuals, such as pregnant women without antibody. Oocysts remaining in litterpans can be killed with near boiling water or with dry heat of over 140°F (Frenkel and Dubey, 1972b).

Cats with antibody are safer pets than cats without, since the presence of antibody is usually accompanied by some immunity, which prevents, or markedly decreases, repetition of oocyst discharge. A vaccine would be useful and is being investigated. Antibody determinations are not practical for the determination of infectivity of cats, since the infectious oocysts are generally discharged before cats have developed antibody (Fig. 21) (Dubey and Frenkel, 1972b). If unsporulated *Toxoplasma* oocysts (Fig. 5) are found in cat feces, the animal should be isolated for a couple of weeks until oocyst discharge stops. The finding of sporulated oocysts and free sporocysts in fresh cat feces indicates infection with *I. cati*, not *Toxoplasma*.

Flies, cockroaches, earthworms, and probably other coprophagic animals can serve as transport hosts for *Toxoplasma* (Wallace, 1971a; Frenkel, Dubey, and Miller, 1970). They should be controlled, and their access to food prevented. In the presence of cockroaches (*Blatella germanica*), *Isospora felis* spreads easily under laboratory conditions (unpublished) and this may also apply to *Toxoplasma*. Control of cockroaches and tightening of cleanliness greatly reduced the rate of spread of *I. felis*.

In the laboratory, toxoplasmosis is communicable by means of oocysts, to a greater extent than via tachyzoites or cysts. Several investigators who had worked safely with the latter two stages for many years, became infected soon after beginning to work with infected cat feces. In part, this occurred before the highly resistant nature of the oocyst was recognized, such as while examining candidate forms under the fluorescent microscope, and when testing large numbers of fecal suspensions for infectivity by feeding and injection of mice (Miller, Frenkel, and Dubey, 1972). The oocysts resist 2% sulfuric acid (which is a good preservative), and 24 hr of undiluted commercial sodium hypochlorite, or concentrated sulfuric acid cleaning solution, and six hours of 10% formalin.

Disinfection is easy after contamination with tissue forms but difficult after exposure to oocysts. Tachyzoites from the peritoneal exudate of acutely infected mice, such as are used for the dye test, are readily killed by

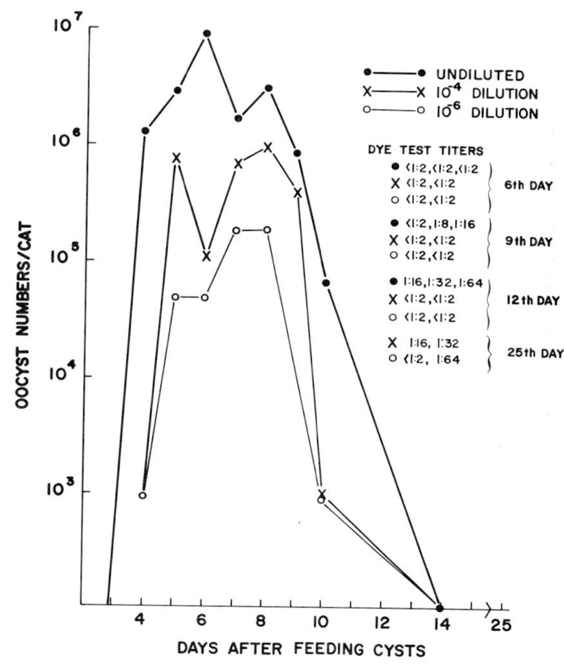

Fig. 21. Correlation of oocyst shedding and antibody development in kittens. An emulsion of ten brains containing numerous cysts, a 10^{-4} and a 10^{-6} dilution thereof, were fed to three groups of from two to three mice. Most of the oocysts were shed between the 5th and 10th days. By the 12th day only the most heavily infected kittens had developed antibody. The antibody levels reached after acute infections are similar to those of old infections (Dubey, Miller, and Frenkel, 1970a), and are not readily useful diagnostically. (From Dubey and Frenkel, 1972b.)

drying, and by tap water, 70% ethanol and common disinfectants. *Toxoplasma* cysts are killed in cooked meat (over 140°F throughout), but freezing is not an altogether certain method for their destruction. Oocysts are destroyed in 24 hr by 10% formalin, in ten minutes by 10% ammonium hydroxide, 7% tr. iodine and by heating to 60°C (140°F) (Dubey, Miller, and Frenkel, 1970a,b). Rapid disinfection of contaminated hands is not easily accomplished. Washing removes most contamination, and tincture of iodine could be used in the hope of reaching oocysts that were missed, such as under the fingernails.

The hazard of working with oocysts may be reduced by carrying out flotation procedures prior to sporulation of the oocysts. Oocysts should be injected via the intraperitoneal route, since after oral administration, some oocysts pass through the gut of mice, and survive in the moist feces and cage bedding for days, endangering animal caretakers and uninfected cagemates. Leakage of injected oocysts from the site of inoculation can also give rise to infection of cagemates. The use of 7% tincture of iodine at the inoculation site reduces this possibility.

Concerning the possibility of airborne infection, it is well to realize that oocysts may survive even in "dry feces" for two weeks or more (depending on relative humidity), although their number decreases sharply. In view of the cohesiveness of dry feces, airborne dissemination appears unlikely. In the same laboratory where four individuals who handled sporulated oocysts became infected, others remained sero-negative although working with infected cats or where infected feces were concentrated by centrifugation. To date, we have more than six contact-years of observation on a total of seven people (Miller, Frenkel, and Dubey, 1972). Piekarksi and Witte (1971) found dust from animal rooms in which infected cats were kept to be noninfectious to mice.

Infected meat or eggs are made safe for human consumption by cooking, frying, broiling, and roasting. Heating to 140°F throughout will kill bradyzoites in cysts. Steak or hamburgers would generally still be pink after attaining such a temperature. It is important to wash hands after handling raw meat and to avoid contact of potentially contaminated hands with the mucous membranes of mouth and eyes. Mutton and pork are more frequently infected than beef. Smoking, brine curing, and freezing at -40°C (-40°F) for nine days have been shown to kill bradyzoites in infected pork (Work, 1968).

VIII. Comparison of *Toxoplasma* with Other Coccidia

A. APPARENTLY RELATED ORGANISMS

Conjecturally, we can best include here *Besnoitia*, since it is biologically similar to *Toxoplasma*. After the injection of cysts, *Besnoitia* rapidly develops tachyzoites and can multiply indefinitely, similar to *Toxoplasma*. Cysts form, as with *Toxoplasma*, and develop during the second week of infection.

Often, *Sarcocystis* and *Frenkelia* are included here, likewise, because of their morphologic similarity. However, after the subinoculation of cyst organisms of *Sarcocystis* and *Frenkelia* into animals, tachyzoites are not formed readily. Their life cycles have not yet been clarified.

The principal characteristics of *Isospora*, *Eimeria*, *Toxoplasma*, and the apparently related organisms are shown synoptically in Table 5.

1. Besnoitia

This protozoan is closely related to *Toxoplasma* biologically, but it is serologically and immunologically distinct (Frenkel, 1953b, 1971b). A sexual cycle has not yet been described. The extra-intestinal infection with *Besnoitia* is widely disseminated in the tissues, first forming groups of tachyzoites, and later cysts, which develop probably in connective tissue cells (Fig. 23). *Besnoitia besnoiti* occurs in cattle, wildebeest, and impala in South Africa and the Mediterranean countries (Pols, 1960; Basson, McCully and Bigalke, 1970), *B. tarandi* in reindeer and caribou in Alaska (Choquette *et al.*, 1967), *B. jellisoni* in white-footed mice (*Peromyscus maniculatus*) from Idaho, with similar organisms having been observed in kangaroo rats (*Dipodomys*) from Utah, and in opossums from Utah, California, Missouri and Kentucky (Frenkel, 1953b; Ernst *et al.*, 1968; Conti-Diaz *et al.*, 1970; Flatt, Nelson and Patton, 1971), and *B. darlingi* has been described from lizards and opossums in Panama and Belize (Schneider, 1965, 1967; Garnham, 1966).

The specific names are largely based on different origins and hosts. *B. jellisoni* and *B. besnoiti* are serologically different. *B. jellisoni* protects against challenge with *B. darlingi*, but the latter protects incompletely against the former (Schneider, 1967,1968). A serological comparison of *Besnoitia jellisoni*, *B. darlingi* and several *Toxoplasma* isolates showed variable relationships among and even within each species, with the IFA, CF and IHA tests (Suggs, Walls, and Kagan, 1968). The morphology of the cyst wall is an undependable characteristic since it is of host origin. The several species need to be compared in one host. *B. besnoiti* has been transferred only to rabbits; *B. tarandi* has not yet been reported from an experimental animal. However, *B. darlingi* from lizards produces cysts also in mice (Schneider, 1967).

Acute *Besnoitia* infection is accompanied by a tachyzoite stage, which is almost indistinguishable from that of *Toxoplasma* in size, shape, and internal organization, as shown by studies of Sheffield (1966).

Besnoitia tachyzoites are motile at room temperature whereas those of *Toxoplasma* acquire active motility between 30°C and 37°C. The acute infections in mice, hamsters, guinea pigs, and rabbits are similar in besnoitiosis and toxoplasmosis. Tachyzoites of *Besnoitia* can be grown in the peritoneal cavity of mice for use in serologic tests. Acute infection lasts from seven to ten days and can be fatal. Like *Toxoplasma*, the organism is sensitive to sulfadiazine, pyrimethamine, and chlortetracycline, all of which can be used to suppress infection sufficiently for the host to acquire immunity. Also, many fresh isolates of *Besnoitia* fail to kill mice and they produce chronic infection without chemotherapy. In one of the natural hosts of *B. jellisoni*, the white-footed or deer mouse (*Peromyscus maniculatus*), survival appears to be the rule after acute infection (Frenkel, 1953b). Chobotar *et al.* (1970) found evidence of pathogenicity of *B. jellisoni* in naturally infected kangaroo rats (*D. ordii*) collected in the field in northwestern Utah.

Chronic *Besnoitia* infection is accompanied by cysts. The cyst of *Besnoitia* encloses the entire host cell, including nuclei; the wall is from 1μ to 10μ thick and is PAS-positive. In contrast, the cyst of *Toxoplasma* encloses only the cytoplasmic vacuole containing the parasites and excludes the nucleus; the cyst wall is thin and stains faintly with PAS. The earliest *Besnoitia* cysts appear after from 10 to 12 days infection in mice and are recognized by enlarged and multinucleate host cells and by a thin capsule (Frenkel, 1965b,1970). Within a vacuole of the multinucleate host cell, the crescentic bradyzoites develop. The cyst enlarges and may measure up to almost 2 mm in size, with the bradyzoites numbering into the hundreds or thousands. In the large cysts, the cyst wall nuclei are compressed between the vacuole and the cyst wall (Fig. 23). The cysts persist for many months. The ultrastructure of the cyst and the contained bradyzoites have been studied by Sheffield (1968) and Sénaud (1969). The cyst wall was shown to consist of fine fibrils and small dense granules embedded in an electron-lucid matrix. The plasma membrane of the host cell had a microvillar or pseudopodial configuration and the cytoplasm contained vesicles, probably of pinocytotic origin. The parasitophorous vacuole is irregular, forming small blebs. The bradyzoites are similar to the tachyzoites and appear to divide by endodyogeny and binary fission.

After several months, some of the cysts degenerate, beginning with partial lysis of organisms next to the break in the cyst wall, followed by granulocytic and monocytic infiltration. The cyst wall remains. This sequence has been studied in the general viscera of

TABLE 5. Comparison of classical coccidia with *Toxoplasma*, *Besnoitia*, *Sarcocystis* and *Frenkelia*

	Coccidia	*Toxoplasma*	*Prepatent period to oocyst*	*Besnoitia*	*Sarcocystis*	*Frenkelia*
Enteroepithelial cycle						
Modes of division	Schizogony + other	Schizogony endopolygeny, splitting, endodyogeny		?	?	?
Pregametogonic multiplication	Variable numbers of generations (2-5); in approximately 10 coccidia studied, each generation differs from the next in morphology. (Exception: *Isospora felis*)	At least 5 morphologic types, with multiplication likely within each type = 5+ generations				
Gametogony	Similar	Similar			In lamina propria of dog jejunum and ileum	?
Gametocytes	Similar	Similar				
Oocyst + sporogony	Similar	Similar	20-24-40 days	?	Cat, dog, man generally shed as sporulated sporocysts, occas. isosporan oocysts.	?
Extra-intestinal (tissue) cycle						
Mode of division	Largely unknown	Endodyogeny, and possibly binary and multiple division		Endodyogeny	Endodyogeny and binary division	Endodyogeny
Group stage with tachyzoites	Extensive in *I. lacazei*; limited in *I. felis*; little in *I. rivolta*	Extensive, multiplication destroys host cell, many tissues involved	5-10-21 days	Like *Toxoplasma*	?	?
Cyst stage with bradyzoites	Not generally recognized. Possibly the "ensheathed" forms of *I.*	Regularly formed, host cell persists *outside* of cyst wall especially in brain, stri-	3-5 days	Regularly formed, host cell nuclei persist *inside* cyst	Thick-walled (radial spines), septate cysts in striated	Thin-walled septate cysts in brain (homology

	felis and *I. rivolta*		wall; in connective tissue; cyst formation may be lost after 30 passages as tachyzoites	muscle (homology uncertain)	uncertain)
Complete, final, host	Usually assumed to be host specific, at least as measured by oocyst development; with some exceptions.	Members of family *Felidae* are complete hosts.	?	*S. tenella*—cat *S. fusiformis*—cat, dog, man *S. miescheriana*—man	?
Facultative, incomplete, *intermediary hosts*	Largely unknown; *I. felis* and *I. rivolta*: mice, rats, and hamsters	Widespread, epidemiologically important, efficiently transmitted	Moderately widespread (Mammals, reptiles)	? (Mammals, birds reptiles)	? (Mammals)
Immunity	Usually immunogenic but this varies interspecifically; *Eimeria bilamellata* in *Spermophilus armatus* – highly immunogenic; *E. callospermophili*—little or no immunity even after 5 infections (Anderson, 1971); *I. felis, I. rivolta* intermediate	Premunition, moderately firm against tissue forms; less in gut and affecting oocyst production	Premunition, moderately firm against tissue stages	Repeated and prolonged oocyst shedding observed (Rommel et al., 1972; Heydorn et al., 1972)	?
Transmission					
Transmissible stages	Sporozoites	Sporozoites, tachyzoites, bradyzoites	Tachyzoites, bradyzoites	Cyclic; sporozoites to herbivores; bradyzoites to carnivores	?
To final host	Oocyst + ? tissue stages	Cyst more efficient than oocyst	?	Cyst	?
To intermediate hosts	?	Cyst – carnivorism; oocyst	Cyst, carnivorism	Probably sporocyst	?
Growth in cell culture	Cyclic, partial or complete	Continuous – tachyzoites and cyst formed	Continuous – tachyzoites; no cysts formed	Cyclic, partial to oocyst?	?

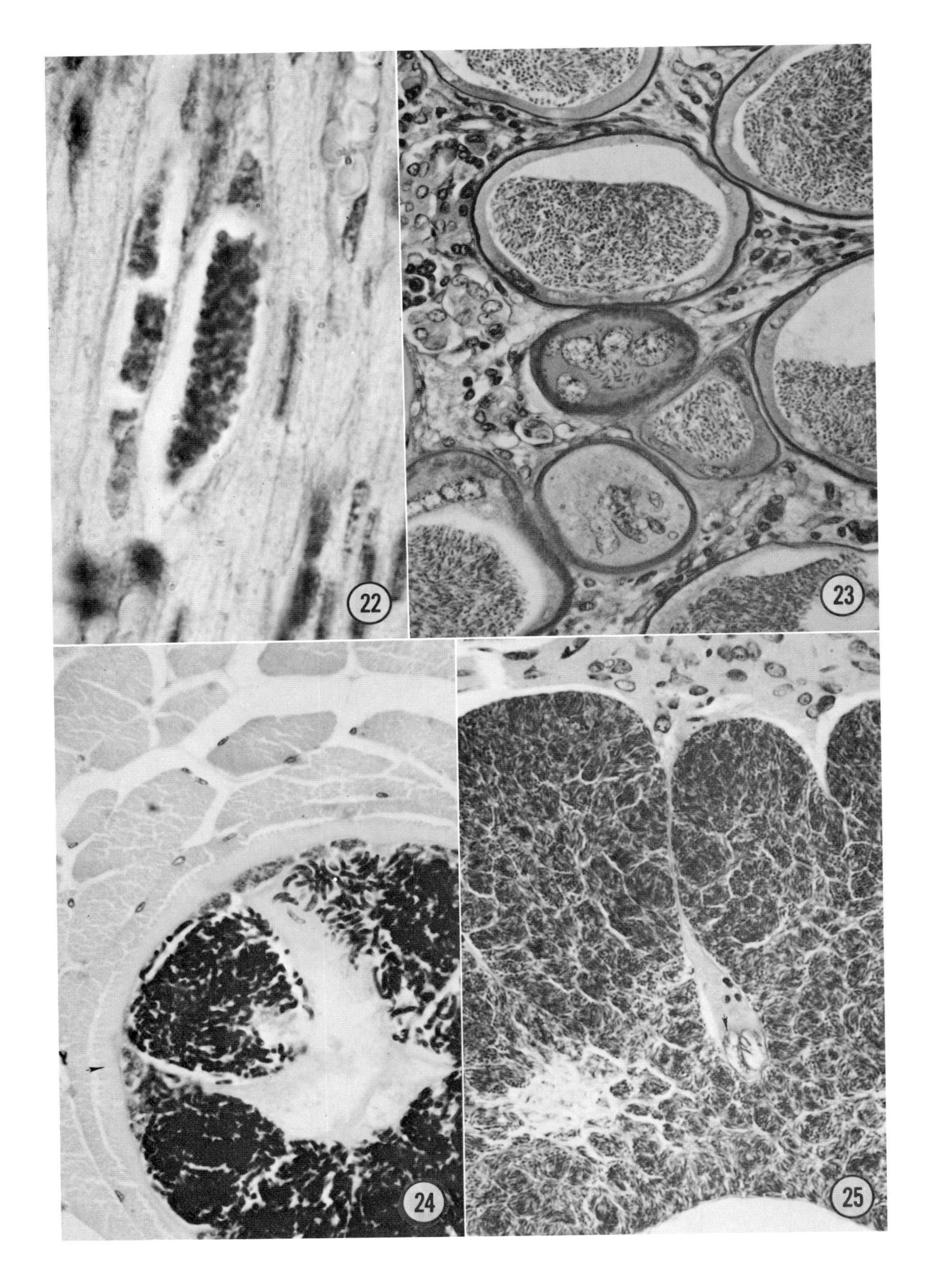
22
23
24
25

hamsters and in the retina, where the beginning of cyst rupture can be determined ophthalmologically (Frenkel, 1961; Basson, McCully and Bigalke, 1970; Bigalke *et al.*, 1967).

Cyst-less strains of *Besnoitia* develop after about 30 twice-weekly passages as tachyzoites in the peritoneum of mice. That chronic infection occurs also in such cyst-less strains can be shown in mice and hamsters by subinoculation. However, the organisms are few and have not been studied morphologically. Cyst-less strains have generally been used for immunologic studies to avoid the unpredictable events following cyst rupture (Frenkel and Lunde, 1966; Frenkel, 1967). In contrast to *Besnoitia*, *Toxoplasma* continues to form cysts even when passed in the tachyzoite stage for many years. However, *Toxoplasma* tends to lose the capacity to form oocysts after about 30-40 passages as tachyzoites (unpublished).

In hamsters and mice, toxoplasmosis and besnoitiosis present parallel infectious and immunologic behavior, although, as previously mentioned, they are serologically and immunologically distinct. Study of *Besnoitia* infection aided the development of the hypothesis that retinochoroiditis may result from cyst rupture in toxoplasmosis; the small *Toxoplasma* cyst was not found in the lesions, whereas the thick-walled *Besnoitia* cyst wall remained as a marker (Frenkel, 1961). Subsequently, the pathogenesis of adrenal necrosis leading to Addison's disease was worked out in hamsters with besnoitiosis, showing that the local hypercorticism in the adrenal was so immunosuppressive that microbial proliferation destroyed the adrenal. Finally, *Besnoitia* infection served to demonstrate the importance of cellular immunity, and, with toxoplasmosis, to characterize its specificity (Frenkel, 1967). Both infections have been discussed in Section VI, Immunology, of this chapter.

Fig. 22. Two small elongate *Toxoplasma* cysts from the myocardium of a kitten, ten days after subcutaneous inoculation with cysts. PASH, × 1600. (From Dubey and Frenkel, 1972b.)

Fig. 23. *Besnoitia jellisoni*, developmental stages of cysts. The dark staining, PAS-positive cyst wall encloses a host cell with many "cyst wall nuclei." The crescentic bradyzoites develop within a vacuole of the cell. In pancreas of hamster infected 37 days. Zenker-formol, PASH, × 400. (From Frenkel, 1971b.)

Fig. 24. *Sarcocystis* from skeletal muscle of cottontail rabbit. The thick cyst wall with radial spines is visible, although lightly stained. Septa extend from the cyst wall to form compartments. So-called sporoblasts or metrocytes are visible just under the cyst wall (*arrows*). The elongated bradyzoites are seen to be slightly curved and have blunt ends. The central part of the cyst is free of organisms. Zenker-formol, hematoxylin, azure II, eosin, × 500. (From Frenkel, 1971b.)

Fig. 25. *Frenkelia microti*, chronic infection from the brain of *Microtus modestus*. The cysts are lobate, the cyst wall is thin and argyrophilic (not shown). What may be the cyst's host cell shows hypertrophied and hyperplastic nuclei (*arrow*) compared to the normal (*top*). Septa and compartments are best seen in the center of the cyst (lower *left*), where the bradyzoites have degenerated. Hematoxylin, azure II, eosin, × 400. (From Frenkel, 1971b.)

2. *Sarcocystis*

As implied by its name, *Sarcocystis* forms cysts in muscle (Fig. 24). These cysts may be visible grossly as whitish streaks corresponding to the direction of muscle fibers. They frequently have been reported to be 100 μ in length, and occasionally may reach a length of 5 mm. The cysts are limited by membranes, sometimes smooth, as in the form found in house mice or with radial spines (cytophaneres), as in the parasites of sheep and cottontail rabbits (Frenkel, 1956b). In some hosts, the cysts are subdivided by septa (Fig. 24). Cysts are filled with bradyzoites (so-called spores), which appear to have arisen from endodyogeny and binary fission (Sénaud, 1967). In mature cysts, the central zoites sometimes degenerate. Rounded cells ("sporoblasts," "metrocytes," or "cytomeres") may be present peripherally in mature cysts; they divide by binary fission, and may fill young cysts entirely (Levine,

1961; Sénaud, 1967). The mature bradyzoites are sausage-shaped, generally with two rounded ends, although one end which has a conoid is slightly pointed. Bradyzoites of most known species measure from 5 μ to 12 μ in length and from 1 μ to 4 μ in width. In Giemsa-stained smears, the nucleus can be seen near the blunt end; it is surrounded by polysaccharide granules, next to which are some vacuoles; the anterior end has fairly clear cytoplasm (Sénaud, 1967; Frenkel, 1956b). When liberated from the cysts, the bradyzoites are motile, especially when warmed to body temperature.

Sarcocystis species have been described from mammals, birds, and reptiles. Lainson and Shaw (1971) described *S. kinosterni* from a Brazilian tortoise and reviewed other reptilian forms.

By electron microscopy, a conoid and other structures resembling those of *Toxoplasma* have been seen. Chapter 4 should be consulted for a detailed comparison. The species from sheep (Sénaud, 1967), from the grackle bird (Zeve, Price, and Herman, 1966), and from cattle (Simpson, 1966) have been examined for their ultrastructure.

Much that has been written concerning the life cycle is of doubtful accuracy. Scott (1943), who studied sheep in Wyoming, 100% of which were infected, lists some pitfalls characterizing previous attempts at defining the life cycle, such as use of "many scattered observations, the drawing of tentative, frequently erroneous conclusions from incomplete data or inaccurate observations and without the critical analysis of all the pertinent facts available." Some of the detailed descriptions from sheep must be considered a mixture of stages from *Sarcocystis, Toxoplasma,* and *Aspergillus* compounded by error and imagination. Transmission experiments are of doubtful accuracy, because even the cycle in house mice, which could be studied under fairly controlled conditions, includes a silent period of 45 days, as described by Theobald Smith (1901). The large size of sheep and pigs, as well as the difficulty of keeping them isolated and of sampling their tissues, make them unsuited for critical experiments; indeed, the conclusion that *Sarcocystis* is a stage of the fungus *Aspergillus* is illustrative of this (Spindler, 1947).

Smith (1901) believed that he could transmit *Sarcocystis* in mice by feeding infected muscle. For donors he used mostly gray house mice with a spontaneous infection rate of 6% that had been captured near the animal rooms. In white mice, also showing a spontaneous infection rate of about 6%, the infection rate rose to 63% after they had been fed carcasses of infected mice. The infection rate in a "permanently infected cage," to which bodies of infected mice were added from time to time was 50%. The white recipient mice had been in contact with unprocessed plant material; oats, hay, and straw were sterilized only in the late stages of the experiment.

Smith allowed that this cycle was not necessarily applicable to herbivores, which, curiously, are infected more commonly in nature than carnivores. In some areas, all adult sheep are infected (Scott, 1943), as well as a large percentage of cattle and ducks, in all of which infection may be strikingly evident in the muscle.

Since the introduction of modern methods for the production of laboratory rodents, including a rapid turnover rate for animals, processed food and wood pulp for bedding, *Sarcocystis* has practically disappeared from mouse colonies. The same applies to guinea pigs and rabbits, in which *Sarcocystis* was reported in the early 1900's. Twort and Twort (1932) observed *Sarcocystis* in 14% of laboratory mice used for cancer research, and Theobald Smith (1901) reported an incidence varying between 6% and 60%. It has been mentioned that in the 1920-30's, it was rare to find an uninfected laboratory mouse or rat (Tuffery and Innes, 1963); now it is rare to find an infected animal. Hence, *Sarcocystis* was still discussed by Heston (1941) in the first edition of *Biology of the Laboratory Mouse*, which appeared in 1941, but it is not even mentioned by Hoag and Meyer (1966) in the second edition of 1966. Recent observations of *Sarcocystis* concern a colony of guinea pigs in Singapore, Malaysia (Muir, 1968), which were fed bread, milk, and fresh greens, and also a colony of mice in Mexico City (Martinez-Baez, Roch, and Varela, 1967).

Most experimenters agree that the injection of *Sarcocystis*-infected muscle does not transmit infection. Macro- and microgametes and cyst-like bodies have been described by Fayer (1972), who inoculated *Sarcocystis* from grackles into cell cultures. Many microgametes are formed (more than 10 are visible in some of the illustrations), placing *Sarcocystis* with the *Eimeriina* rather than the *Adeleina* (Baker, 1965). In an exciting series of papers, Rommel, Heydorn, and Gruber (1972), who fed *Sarcocystis tenella* from sheep to cats, reported shedding of sporulated sporocysts and occasional isosporan oocysts after a 12-day prepatent period. The average sporocyst measurement was 12.4 × 8.1 μ. Dogs did not shed coccidian stages from sheep sarcocysts. *Sarcocystis fusiformis* from cattle gave rise to sporocysts and occasional oocysts in cats, dogs, and man (Heydorn and Rommel, 1972). In cats the sporocysts appeared after a prepatent period of from 7 to 9 days and measured 12.5 × 7.8 μ. In dogs they appeared after a prepatent period of from 9 to 10 days and measured 16 × 8 μ. In humans the prepatent period was 9 days and the sporocysts measured 14.7 × 9.3 μ on the average. In one dog, oocysts were observed in the lamina propria of the villar tips in the posterior two-thirds of the jejunum and the ileum. *Sarcocystis miescheriana* from pigs gave rise to sporocysts and occasionally isosporan oocysts in man (Rommel and Heydorn, 1972). The prepatent period was 10, 13, and 17 days each in three human subjects, and the sporulated sporocysts measured 12.6 × 9.3 μ on the average. Evidence was presented that each of the three *Sarcocystis* species could repeatedly infect cats, dogs and presumably man as indicated. The authors also pointed out that *I. hominis* may comprise the sporocysts of two species, *S. fusiformis* and *S. miescheriana*. The same applies to cats, where *Isospora cati*, at least the form shed as sporulated sporocysts, may represent *Sarcocystis tenella* or *S. fusiformis*. So far, these observations have not been confirmed.

In the absence of experimental evidence, it is not certain whether there are many host-specific species or only a few species infecting many hosts. *S. muris* is considered by Levine (1961) to be the species present in the mouse, the Norway rat, and the black rat, and *S. cuniculi* is that in domestic and cottontail rabbits. A guinea pig species has been mentioned repeatedly, but in most instances, the infective species appears to have been *Toxoplasma*. The parasite of man has been called *S. lindemanni* but is suspected by some to be an infection with a form usually found in another animal (Levine, 1961).

Lesions result from degeneration of cysts, which is associated with an inflammatory reaction, usually with the participation of lymphocytes, plasma cells, and eosinophils. This is later followed by fibrosis. Myositis is usually focal and widely separated; however, in the heart muscle, heavy infections in rabbits, pigs, and cattle may be associated with significant lesions (Köberle, 1958). Human infection has been described occasionally, as discussed by Frenkel (1971b). Some of the reports on humans, however, are now regarded as having concerned toxoplasmosis (Chaves-Carballo, 1970).

3. *Frenkelia*

These protozoa form compartmented, thin-walled cysts in the brain (Fig. 25). The cysts may be grossly visible as whitish bodies on the surface of the brain, even through the skull (Bell, Jellison, and Glesne, 1964). Cysts measure up to 1000 μ in diameter and probably form in the cytoplasm of cells whose nuclei undergo hypertrophy and sometimes hyperplasia. In the periphery of the cyst, and in young cysts, the organisms are spherical (metrocytes). Mature organisms become crescentic in the interior of the cyst and measure up to from 8 μ to 10 μ in length and from 1 μ to 3 μ in width; they may degenerate in the center of the cyst, leaving compartments filled with debris. The cysts are lobate in field voles (*Microtus agrestris*), meadow mice (*M. modestus*) and muskrats (*Ondatra zibethica*); cysts remain spherical in bank voles (Rötelmäuse) (*Clethrionomys glareolus*). Unlike *Toxoplasma, Frenkelia* cannot be transmitted by subinoculation, suggesting a complex life cycle as in *Sarcocystis* (Frenkel, 1956b; Bell, Jellison and Glesne, 1964; Sebek, 1962; Tadros, 1970). The organism is differentiated from other cyst-producing organisms in Table 4. Type species is *F. microti* Biocca, 1968 (synonym: *Toxoplasma microti* Findlay and Middleton, 1934). The genus *Frenkelia* was proposed by Biocca (1968) to replace the term "M-organism," which was used noncommittally by myself previously (Frenkel, 1953b, 1956b).

Developmental stages have been studied by Enemar (1963) in *Clethrionomys glareolus* from Lund. The early intracellular parasites are spherical, with a central nucleus, and cause enlargement of host cell nucleus and cytoplasm. When the diameter of the mass of parasites has reached approximately 100 μ, some elongated organisms develop, which later become preponderant. The cysts enlarge and septa develop, but the cyst shape remains essentially spherical. When the cyst exceeds 300 μ, the central organisms may degenerate. In the equivalent parasites of *M. agrestris* from England (Findlay and Middleton, 1934), *M. modestus* from Hamilton, Montana (Frenkel, 1956b), *M. arvalis* from Czechoslovakia (Sebek, 1962) and of muskrats from Canada (Karstad, 1963), the cysts develop marked lobulations. At least the parasites from *M. modestus* and *Clethrionomys* stimulate hypertrophy and hyperplasia of host cell nuclei.

Findlay and Middleton (1934) called this organism "*Toxoplasma microti*," and suggested that it might cause mortality in voles, resulting in a downward swing in their population cycles. No significant encephalitis was present, transmission experiments in retrospect appear negative or at least inconclusive, and the cause of vole cycles is now sought elsewhere (Christian and Davis, 1964).

The ultrastructure of *Frenkelia* from *C. glareolus* has been studied by Scholtyseck, Kepka, and Piekarski (1970) and by Kepka and Scholtyseck (1970).Metrocytes, intermediate cells, and bradyzoites divide by endodyogeny. Even in the metrocytes a conoid has been found.

In the manner of cyst production, the degeneration of trophozoites in central compartments and the inability to multiply on subinoculation, these organisms resemble *Sarcocystis*. Ludvik (1960) described ultrastructural similarities with *Sarcocystis* and the trophozoites of the two species could be distinguished on a morphologic basis only with difficulty (Sebek, 1962; 1970; Tadros, 1970). According to Sénaud (1967), the metrocytes of *Sarcocystis* are nonpolar, lack a conoid, and divide by binary fission. Associated with the three vole species, *M. modestus, M. arvalis*, and *M. agrestris* infected with *Frenkelia*, nonseptate cysts, which resemble those of *Sarcocystis,* have been found in the muscle, with the contained bradyzoites, however, very similar to those found in the brain (Frenkel, 1953b; Sebek, 1962). Whether double infection with *Sarcocystis* and *Frenkelia* was present, or whether one or the other gives rise to cysts both in brain and muscle requires clarification.

Although the amount of brain involved may be considerable, up to 3.6% (Enemar, 1963), there are usually no symptoms other than diuresis. Diagnosis can be made grossly, sometimes even through the calvarium (Bell, Jellison, and Glesne, 1964), or by the increased urine output.

B. *ISOSPORA* AND *EIMERIA*

Considering the classical single-host coccidian life cycle of *Isospora* and *Eimeria* in the gut wall or its derivatives, *Toxoplasma* may be regarded as having evolved to multiply extensively in other tissues and hosts. With the development of a persisting cyst in brain and muscle, carnivorism could become another mode of transmission. As sexual development was not found, and apparently is absent in nonfeline hosts, they are regarded as facultative, intermediate hosts (Frenkel, Dubey, and Miller, 1970; Miller, Frenkel, and Dubey, 1972) or nonspecific hosts (Piekarski and Witte, 1971). In fact, as cysts give rise consistently to oocyst production in cats, but oocyst ingestion results in discharge of oocysts only irregularly, we may be witnessing transition to a two-host cycle, although at present the intermediate host is not obligatory (Frenkel, 1970).

However, it was apparent that the one-host hypothesis applied customarily to *Isospora* and *Eimeria* had never been tested critically. In tests with the two common cat coccidia, mice, rats, and hamsters were found capable of serving as intermediary hosts for *Isospora felis*, and as transport hosts for *I. rivolta* (Frenkel and Dubey, 1972a). Furthermore, extra-intestinal tissue stages were found even in cats (Dubey and Frenkel, 1972a).

In mice fed with oocysts, motile forms appeared in the mesenteric lymph nodes and in other organs. The zoites of *I. felis* were seen to undergo binary fission; when infected mouse tissues were fed to newborn kittens, the prepatent period was from five to six days compared with from eight to ten days after the feeding of oocysts. The zoites of *I. rivolta* were not observed to multiply actively, and when infected mouse tissues were fed to newborn kittens, the prepatent period was from five to six days, identical with that following oocyst ingestion. After four or five days the zoites of both *Isospora* species contracted, becoming ensheathed in what previously was their pellicle; as these ensheathed forms become predominant after a week and since they may survive for from three months to a year in mice, they might be regarded as the morphologic and biologic equivalent of a cyst.

In cats, the presence of extra-intestinal forms of *I. felis* and *I. rivolta* was shown by subinoculation of cat liver, spleen, and mesenteric lymph nodes into newborn kittens. The prepatent period after feeding tissue forms of *I. felis* was from four to eight days, which was shorter than after oocyst ingestion; active multiplication, with up to 15 identifiable zoites in one vacuole, was observed. The prepatent period after the ingestion of tissue forms of *I. rivolta* was from five to six days, identical to that after the infection with oocysts; and no more than two zoites of *I. rivolta* were seen in a single vacuole in the mesenteric lymph nodes (Dubey and Frenkel, 1972a).

The finding of both intermediate and transport hosts, and of extra-intestinal tissue stages of classical coccidia thought to be monoxenous and essentially limited to the intestine, raises the question: are these findings unique for the two *Isospora* species studied in this respect, or do other coccidia show similar stages? Did we discover an exception or a rule?

Optional intermediary hosts could extend the range of the coccidia by serving as reservoirs in climates where the oocysts are killed by cold, heat, or dryness. As in *Toxoplasma*, the narrow host specificity of other coccidia may apply only to the presexual multiplication and to the gametogony. Some intermediate hosts, infected by oocysts or by carnivorism, if unusually susceptible or immunosuppressed, might develop clinical infection or disease. If they occur, such infections might be difficult to diagnose, since the binary divisional stages are more cryptic in the tissues than the conspicuous schizogonic and gametogonic forms located in the intestinal epithelium.

The occurrence of extra-intestinal tissue stages in the specific or final host could provide an endogenous source of reinfection of the intestine, resulting in oocyst shedding. Possible occurrence of transplacental and fetal infection must be considered. Also, the mechanisms of maintenance of immunity in coccidia should be reexamined, since conceivably those species with persistent extra-intestinal tissue infection, or with a high degree of extra-intestinal proliferation might exhibit a different quality of immunity to reinfection than those without either of these.

As already suggested by the studies on *Atoxoplasma* in canaries, which is associated with oocysts resembling *I. lacazei* (Box, 1970), the concepts of the life cycles in the *Eimeriina* may be more complex and varied than realized. A reappraisal is needed.

Literature Cited

Alford, C.A., J.W. Foft, W.J. Blankenship, G. Cassady, and J.W. Benton. 1969. Subclinical central nervous system disease of neonates: A prospective study of infants born with increased levels of IgM. J. Pediat. 75 (Sect. 6, Part 2):1167-1178.

Anderson, L.C. 1971. Experimental coccidian infections in captive hibernating and non-hibernating Uinta ground squirrels, *Spermophilus armatus*. Doctoral dissertation. Dept. of Zoology, Utah State Univ., Logan.

Andrade, C.M. de and G. Weiland. 1971. Serologische Untersuchungen zur Feststellung gemeinsamer Antigene von Toxoplasmen, Sarkosporidien und Kokzidien. Berlin and Münch. Tierärztl. Wochenschr. 84:61-64.

Arias-Stella, J. 1956. Personal communication.

Baker, J.R. 1965. The evolution of parasitic protozoa. *In* A.E.R. Taylor, ed., Evolution of parasites, pp. 1-27. Oxford: Blackwell Scientific Publishers.

Basson, P.A., R.M. McCully, and R.D. Bigalke. 1970. Observations on the pathogenesis of bovine and antelope strains of *Besnoitia besnoiti* (Marotel, 1912) infection in cattle and rabbits. Onderstepoort J. Vet. Res. 37:105-126.

Bell, J.F., W.L. Jellison, and L. Glesne. 1964. Detection of *Toxoplasma microti* in living voles. Exp. Parasitol. 15:335-339.

Benirschke, K. and R.J. Low. 1970. Acute toxoplasmosis. Comp. Pathol. Bull. 2:3-4.

Benirschke, K. and R. Richart. 1960. Spontaneous acute toxoplasmosis in a marmoset monkey. Am. J. Trop. Med. Hyg. 9:269-273.

Beverley, J.K.A. 1959. Congenital transmission of toxoplasmosis through successive generations of mice. Nature 183:1348-1349.

Beverley, J.K.A. and W.A. Watson. 1971. Prevention of experimental and of naturally occurring ovine abortion due to toxoplasmosis. Vet. Rec. 88:39-41.

Beverley, J.K.A., W.A. Watson, and J.B. Spence. 1971. The pathology of the foetus in ovine abortion due to toxoplasmosis. Vet. Rec. 88:174-178.

Bigalke, R.D., J.W. Niekerk, P.A. Basson, and R.M. McCully. 1967. Studies on the relationship between *Besnoitia* of blue wildebeest and impala, and *Besnoitia besnoiti* of cattle. Onderstepoort J. Vet. Res. 34:7-28.

Biocca, E. 1968. Class Toxoplasmatea: critical review and proposal of the new name *Frenkelia* gen.n. for M-organism. Parassitologia 10:89-98.

Bobowski, S.J. and W.G. Reed. 1958. Toxoplasmosis in an adult, presenting as a space-occupying cerebral lesion. AMA Arch. Pathol. 65:460-464.

Bommer, W., K.H. Hofling, and H.H. Keunert. 1968. Lebendbeobachtungen über das Eindringen von Toxoplasmen in die Wirtszelle. Dtsch. Med. Wochenschr. 93:2365-2367.

Box, E.D. 1970. *Atoxoplasma* associated with an isosporan oocyst in canaries. J. Protozool. 17:391-396.

Catar, G., L. Bergendi, and R. Holkova. 1969. Isolation of *Toxoplasma gondii* from swine and cattle. J. Parasitol. 55:952-955.

Cespedes, R. and P. Morera. 1955. Toxoplasmosis humana generalizada. Rev. Biol. Trop. 3:183-202.

Chapman, J. and W.E. Ormerod. 1966. The survival of *Toxoplasma* in infected mosquitoes. J. Hyg. Camb. 64:347-355.

Chatton, E. and G. Blanc. 1917. Notes et réflexions sur le toxoplasme et la toxoplasmose du gondi (*Toxoplasma gondii* Ch. Nicolle et Manceaux, 1909). Arch. Inst. Pasteur, Tunis 10:1-41.

Chaves-Carballo, E. 1970. Samuel T. Darling and human sarcosporidiosis or toxoplasmosis in Panama. J. Am. Med. Ass. 211:1687-1689.

Cheever, A.W., M.P. Valsamis, and A.S. Rabson. 1965. Necrotizing toxoplasmic encephalitis and herpetic pneumonia complicating treated Hodgkin's disease. N.E. J. Med. 272:26-29.

Chobotar, B., L.C. Anderson, J.V. Ernst, and D.M. Hammond. 1970. Pathogenicity of *Besnoitia jellisoni* in naturally infected kangaroo rats (*Didodomys ordii*) in northwestern Utah. J. Parasitol. 56:192-193.

Choquette, L.P.E., E. Broughton, F.L. Miller, H.C. Gibbs, and J.G. Cousineau. 1967. Besnoitiosis in barren-ground caribou in northern Canada. Can. Vet. J. 8:282-287.

Christian, J.J. and D.E. Davis. 1964. Endocrines, behavior and population. Science 148:1550.

Cohen, S.N. 1970. Toxoplasmosis in patients receiving immunosuppressive therapy. J. Am. Med. Ass. 211:657-660.

Conti-Diaz, I.A., C. Turner, D.T. Tweeddale, and M.L. Furcolow. 1970. Besnoitiosis in the opossum (*Didelphis marsupialis*). J. Parasitol. 56:457-460.

Corpening, T.N., V.A. Stembridge, and R.H. Rigdon. 1952. Toxoplasmosis in Texas. Report of a case with autopsy. Texas St. J. Med. 48:469-471.

Couvreur, J. 1971. Prospective study of acquired toxoplasmosis in pregnant women with a special reference to the outcome of the foetus. *In* D. Hentsch, ed., Toxoplasmosis, pp. 119-135. Bern, Hans Huber.

Cowen, D., and A. Wolf, 1942. Experimental congenital toxoplasmosis. IV. Genital and secondary lesions in the mouse infected with *Toxoplasma* by the vaginal route. J. Neuropathol. Exp. Neurol. 10:1-15.

Cowen, D., A. Wolf, and B.H. Paige, 1942. Toxoplasmic encephalomyelitis. VI. Clinical diagnosis of infantile or congenital toxoplasmosis; survival beyond infancy. Arch. Neurol. Psychol. 48:689-739.

Cutchins, E. and J. Warren. 1956. Immunity patterns in the guinea pig following *Toxoplasma* infection and vaccination with killed *Toxoplasma*. Am. J. Trop. Med. Hyg. 5:197-209.

David, J.R. Macrophage migration. 1968. Fed. Proc. 27:6-12.

Desmonts, G. 1955. Sur la technique de l'épreuve de lyse des toxoplasmes. La Semaine des Hôpitaux de Paris 31:1-6.

Desmonts, G. 1960. Diagnostic serologique de la toxoplasmose. Pathol. Biol. 8:109-125.

Desmonts, G. 1971. Congenital toxoplasmosis: problems in early diagnosis. *In* D. Hentsch, ed., Toxoplasmosis, pp. 137-149.

Desmonts, G., J. Couvreur, F. Alison, J. Baudelot, J. Gerbeaux, and M. Lelong. 1965. Etude épidémiologique sur la toxoplasmose: de l'-influence de la cuisson des viandes de boucherie sur la fréquence de l'infection humaine. Rev. Franc. Etudes Clin. Biol. 10:952-958.

Desmonts, G., J. Couvreur, and M.S. Ben-Rachid. 1965. Le toxoplasme, la mere, et l'enfant. Arch. Franc. Ped. 22:1183-1200.

Dixon, W. and F. Massey. 1969. Introduction to Statistical Analysis, 3rd ed., p. 522, Table A12-F.

Dubey, J.P. 1967. Distribution of *Toxoplasma gondii* in the tissues of infected cats. I. Isolation in mice. Trop. Geogr. Med. 19:199-205.

Dubey, J.P. 1968. Feline toxoplasmosis and its nematode transmission. Vet. Bull. 38:495-499.

Dubey, J.P. and J.K.A. Beverley. 1967. Distribution of *Toxoplasma gondii* in the tissues of infected cats. II. Histopathology. Trop. Geogr. Med. 19: 206-212.

Dubey, J.P. and J.K. Frenkel. 1972a. Extra-intestinal stages of *Isospora felis* and *I. rivolta* (Protozoa:Eimeriidae) in cats. J. Protozool. 19:89-92.

Dubey, J.P. and J.K. Frenkel. 1972b. Cyst-induced toxoplasmosis in cats. J. Protozool. 19:155-177.

Dubey, J.P., N.L. Miller, and J.K. Frenkel. 1970a. Characterization of the new fecal form of *Toxoplasma gondii*. J. Parasitol. 56:447-456.

Dubey, J.P., N.L. Miller, and J.K. Frenkel. 1970b. The *Toxoplasma gondii* oocyst from cat feces. J. Exp. Med. 132:636-662.

Durge, N.G., M.U. Baqai, and R. Ward. 1967. Myocardial toxoplasmosis. Lancet 2:155.

DuShane, G., K.B. Krauskopf, E.M. Lerner, P.M. Morse, H.B. Steinbach, W.L. Straus, and E.L. Tatum. 1961. An unfortunate event. Science 134:945-946.

Eichenwald, H. 1948. Experimental toxoplasmosis. I. Transmission of the infection *in utero* and through the milk of lactating female mice. Am. J. Dis. Child. 76:307-315.

Eichenwald, H. 1960. A study of congenital toxoplasmosis. *In* J.C. Siim, ed., Human toxoplasmosis, pp. 42-49.

Enemar, A. 1963. Studies on a parasite resembling *Toxoplasma* in the brain of the bank-vole, *Clethrionomys glareolus*. Ark. Zool. 15:381-392.

Ernst, J.V., B. Chobotar, E.C. Oaks, and D.M. Hammond. 1968. *Besnoitia jellisoni* (Sporozoa:Toxoplasmea) in rodents from Utah and California. J. Parasitol. 54:545-549.

Eyles, D.E. 1954. Serologic response of white rats to *Toxoplasma* infection. J. Parasitol. 40:77-83.

Fayer, R. 1972. Gametogony of *Sarcocystis* sp. in cell culture. Science 175:65-67.

Feldman, H.A. 1956. The relationship of *Toxoplasma* antibody activator to the serum-properdin system. Ann. N.Y. Acad. Sci. 66:263-267.

Feldman, H.A. 1965. A nationwide serum survey of United States military recruits, 1962. VI. *Toxoplasma* antibodies. Am. J. Epidemiol. 81:385-391.

Feldman, H.A. and L.T. Miller. 1956. Serological study of toxoplasmosis prevalence. Am. J. Hyg. 64:320-335.

Findlay, G.M. and A.D. Middleton. 1934. Epidemic disease among voles (*Microtus*) with special reference to *Toxoplasma*. J. Anim. Ecol. 3:150-160.

Flament-Durand, J., C. Coers, C. Waelbroeck, J. Van Geertruyden, and C. Toussaint. 1967. Encephalite et myosite a toxoplasmes au cours d'un traitement immuno-depresseur. Acta Clin. Belg. 22:44-54.

Flatt, R.E., L.R. Nelson, and N.M. Patton. 1971. *Besnoitia darlingi* in the opossum (*Didelphis marsupialis*). Lab. Anim. Sci. 21:106-109.

Frenkel, J.K. 1948. Dermal hypersensitivity to *Toxoplasma* antigens (toxoplasmins). Proc. Soc. Exp. Biol. Med. 68:634-639.

Frenkel, J.K. 1949. Uveitis and toxoplasmin sensitivity. Am. J. Ophthal. 32 (Sect. 6, Part 2): 127-135.

Frenkel, J.K. 1951. Pathology of chronic toxoplasmosis in the golden hamster. Am. J. Pathol. 27:746-747.

Frenkel, J.K. 1953a. Host, strain and treatment variation as factors in the pathogenesis of toxoplasmosis. Am. J. Trop. Med. Hyg. 2:390-416.

Frenkel, J.K. 1953b. Infections with organisms resembling *Toxoplasma*, together with the description of a new organism *Besnoitia jellisoni*. Atti. del VI Cong. Int. Microbiol. 5:426-434.

Frenkel, J.K. 1955. Ocular lesions in hamsters with chronic *Toxoplasma* and *Besnoitia* infections. Am. J. Ophthal. 39:203-225.

Frenkel, J.K. 1956a. Effects of hormones on the adrenal necrosis produced by *Besnoitia jellisoni* in golden hamsters. J. Exp. Med. 103: 103:375-398.

Frenkel, J.K. 1956b. Pathogenesis of toxoplasmosis and of infections with organisms resembling *Toxoplasma*. Ann. N.Y. Acad. Sci. 64:215-251.

Frenkel, J.K. 1957. Effects of cortisone, total body irradiation, and nitrogen mustard on chronic, latent toxoplasmosis. Am. J. Pathol. 33:618-619.

Frenkel, J.K. 1960. Evaluation of infection-enhancing activity of modified corticoids. Proc. Soc. Exp. Biol. Med. 103:552-555.

Frenkel, J.K. 1961. Pathogenesis of toxoplasmosis with a consideration of cyst rupture in *Besnoitia* infection. Surv. Ophthal. 6 (Sect. 6, Part 2):799-825.

Frenkel, J.K. 1965a. Attempted transmission of *Toxoplasma* with arthropods. 2nd Int. Conf. on Protozoology; London 1965. Excerpta Med. Int. Cong. Series. 91, 190.

Frenkel, J.K. 1965b. The development of the cyst of *Besnoitia jellisoni*; usefulness of this infection as a biologic model. Prog. Protozool., Excerpta Med. Int. Cong. Series, 91, 122-124.

Frenkel, J.K. 1967. Adoptive immunity to intracellular infection. J. Immunol. 98:1309-1319.

Frenkel, J.K. 1969. Models for infectious diseases. Fed. Proc. 28: 179-190.

Frenkel, J.K. 1970. Pursuing *Toxoplasma*. J. Inf. Dis. 122:553-559.

Frenkel, J.K. 1971a. Toxoplasmosis. *In* R. Marcial-Rojas, ed., Pathology of protozoal and helminthic diseases, pp. 254-290.

Frenkel, J.K. 1971b. Protozoal diseases of laboratory animals. *In* R. Marcial-Rojas, ed. Pathology of protozoal and helminthic diseases, 318-369. Williams and Wilkins, Baltimore.

Frenkel, J.K. 1971c. Toxoplasmosis. Mechanisms of infection, laboratory diagnosis and management. Curr. Topics Pathol. 54:29-75.

Frenkel, J.K. 1972a. Infection and immunity in hamsters. Progr. Exp. Tumor Res. 16:326-367.

Frenkel, J.K. 1972b. Neuropathology of toxoplasmosis. *In* J. Minckler, ed., Pathology of the nervous system, vol. 3. pp. 2521-2538. McGraw Hill, New York.

Frenkel, J.K. 1973. Specific immunity and nonspecific resistance to viral and protozoan infections in hamsters. In preparation.

Frenkel, J.K., K. Cook, H.J. Grady, and S.K. Pendleton. 1965. Effects of hormones on adrenocortical secretion of golden hamsters. Lab. Invest. 14:142-156.

Frenkel, J.K. and J.P. Dubey. 1972a. Rodents as vectors for feline coccidia, *Isospora felis* and *Isospora rivolta*. J. Inf. Dis. 125:69-72.

Frenkel, J.K. and J.P. Dubey. 1972b. Toxoplasmosis and its prevention in cats and man. J. Inf. Dis. 126:664-673.

Frenkel, J.K., J.P. Dubey, and N.L. Miller. 1969. *Toxoplasma gondii*: fecal forms separated from eggs of the nematode *Toxocara cati*. Science 164:432-433.

Frenkel, J.K., J.P. Dubey, and N.L. Miller. 1970. *Toxoplasma gondii*: fecal stages identified as coccidian oocysts. Science 167:893-896.

Frenkel, J.K. and S. Friedlander. 1951. Toxoplasmosis. Pathogenesis of neonatal disease. Pathogenesis, diagnosis, and treatment. Pub. Health Serv. Publ. 141, pp. 1-105. U.S. Govt. Printing Office, Washington.

Frenkel, J.K. and L. Jacobs. 1958. Ocular toxoplasmosis. AMA Arch. Ophthal. 59:260-279.

Frenkel, J.K. and M.N. Lunde. 1966. Effects of corticosteroids on antibody and immunity in *Besnoitia* infection of hamsters. J. Inf. Dis. 116:414-424.

Frenkel, J.K., R.W. Weber, and M.N. Lunde. 1960. Acute toxoplasmosis. Effective treatment with pyrimethamine, sulfadiazine, leucovorin calcium and yeast. J. Am. Med. Ass. 173:1471-1476.

Frenkel, J.K. and H.R. Wilson. 1972. Effects of radiation of specific cellular immunities: besnoitiosis and a herpes-virus infection of hamsters. J. Inf. Dis. 125:216-230.

Fuller, V.J., S.D. Chaparas, and R.W. Kolb. 1967. Suppression of toxoplasmin skin reactivity in sensitized guinea pigs by multiple inoculation. Proc. Soc. Exp. Biol. Med. 124:603-607.

Garcia, A.G.P. 1968. Congenital toxoplasmosis in two successive sibs. Arch. Dis. Child. 43:705-710.

Garnham, P.C.C. 1966. *Besnoitia* (Protozoa: Toxoplasmea) in lizards. Parasitology 56:329-334.

Garnham, P.C.C. 1971. Progress in parasitology, pp. 116-210. Athlone Press, London.

Gavin, M.A., T. Wanko and L. Jacobs. 1962. Electron microscope studies of reproducing and interkinetic *Toxoplasma*. J. Protozool. 9:222-234.

Ghatak, N.R., T.P. Poon, and H.M. Zimmerman. 1970. Toxoplasmosis of the central nervous system in the adult. Arch. Pathol. 89:337-348.

Gibson, C.L. and N. Coleman. 1958. The prevalence of *Toxoplasma* antibodies in Guatemala and Costa Rica. Am. J. Trop. Med. Hyg. 7:334-338.

Giraud, P., H. Payan, M. Toga, M. Berard, D. Dubois, and M. Laugier. 1965. Toxoplasmose congénitale aiguë expérimentale du cobaye. Corrélations cliniques, biologiques et anatomiques. Ann. Anat. Pathol. (Paris) 10: 337-350.

Goldman, M., R.K. Carver, and A.J. Sulzer. 1958. Reproduction of *Toxoplasma gondii* by internal budding. J. Parasitol. 44:161-171.

Guimarães, F.N. 1943. Toxoplasmose humana. Meningoencefalomielite toxoplasmica: ocorrência em adulto e em recemnascido. Memorias do Inst. Oswaldo Cruz 38:257-320.

Hackel, D.B., T.D. Kinney, and W. Wendt. 1953. Pathologic lesions in captive wild animals. Lab. Invest. 2:154-163.

Harboe, A. and S. Erichsen. 1954. Toxoplasmosis in chickens. 3. Attempts to provoke a systemic disease in chickens by infection with a chicken strain and a human strain of *Toxoplasma*. Acta Pathol. Microbiol. Scand. 35: 495-502.

Harboe, A. and R. Reenaas. 1957. The complement fixation inhibition test with sera from chickens experimentally infected with toxoplasms. Acta Pathol. Microbiol. Scand. 41:511-516.

Hartley, W.J. and G. Moyle. 1968. Observations on an outbreak of ovine congenital toxoplasmosis. Aust. Vet. J. 44:105-107.

Hartley, W.J. and G. Moyle. 1973. Long term observations on the natural development of *Toxoplasma* antibodies in sheep. In Press.

Hemsath, F.A. and H. Pinkerton. 1956. Disseminated cytomegalic inclusion disease and disseminated toxoplasmosis in an adult with myeloid metaplasia. Am. J. Clin. Pathol. 26:36-41.

Heston, W.E. 1941. Parasites. *In* G.E. Snell, ed. Biology of the laboratory mouse, p. 349. Dover Publications, New York.

Heydorn, A.O. and M. Rommel. 1972. Beiträge zum Lebenszyklus der Sarkosporidien. II. Hund und Katze als Überträger der Sarkosporidien des Rindes. Berl. Münch. Tierärztl. Woch. 85:121-123.

Hoag, W.G. and H. Meier. 1966. Infectious diseases. *In* E.L. Green, ed. Biology of the laboratory mouse, p. 589. McGraw Hill, New York.

Hogan, M.J., G.B. Moschini, and O. Zardi. 1971. Effects of *Toxoplasma gondii* toxin on the rabbit eye. Am. J. Ophthal. 72:733-742.

Hogan, M.J., C. Yoneda, L. Feeney, P. Zweigart, and A. Lewis. 1960. Morphology and culture of *Toxoplasma*. Arch. Ophthal. 64:665-667.

Hooper, A.D. 1957. Acquired toxoplasmosis. AMA Arch. Pathol. 64:1-9.

Huldt, G. 1963. Experimental toxoplasmosis. Parasitemia in guinea pigs. Acta Pathol. Microbiol. Scand. 58:457-470.

Huldt, G. 1966a. Experimental toxoplasmosis. Studies of the multiplication and spread of *Toxoplasma* in experimentally infected rabbits. Acta Pathol. Microbiol. Scand. 67:401-423.

Huldt, G. 1966b. Experimental toxoplasmosis. Effect of inoculation of *Toxoplasma* in seropositive rabbits. Acta Pathol. Microbiol. Scand. 68:592-604.

Huldt, G. 1966c. Experimental toxoplasmosis. Effect of corticosteroids on rabbits with varying degree of immunity. Acta Pathol. Microbiol. Scand. 68:605-621.

Huldt, G. 1967a. In vitro studies of some immunological phenomena in experimental rabbit toxoplasmosis. Acta Pathol. Microbiol. Scand. 70:129-146.

Huldt, G. 1967b. Studies on pathogenesis in experimental toxoplasmosis. Doctoral dissertation, Karolinska Institute, Stockholm.

Hutchison, W.M. 1965. Experimental transmission of *Toxoplasma gondii*. Nature 206:961-962.

Hutchison, W.M. 1967. The nematode transmission of *Toxoplasma gondii*. Trans. Roy. Soc. Trop. Med. Hyg. 61:80-89.

Hutchison, W.M., J.F. Dunachie, J.C. Siim, and K. Work. 1969. Life cycle of *Toxoplasma gondii*. Brit. Med. J. 4(5686):806.

Hutchison, W.M., J.F. Dunachie, J.C. Siim, and K. Work. 1970. Coccidian-like nature of *Toxoplasma gondii*. Brit. Med. J. 1(5689): 142-144.

Hutchison, W.M., J.F. Dunachie, K. Work, and J.C. Siim. 1971. The life cycle of the coccidian parasite, *Toxoplasma gondii*, in the domestic cat. Trans. Roy. Soc. Trop. Med. Hyg. 65:380-399.

Jacobs, L. 1956. Propagation, morphology, and biology of *Toxoplasma*. Ann. N.Y. Acad. Sci. 64:154-179.

Jacobs, L. 1957. The distribution of *Toxoplasma gondii* in muscles of rats with chronic infections. J. Parasitol. 43:41-42.

Jacobs, L. 1964. The occurrence of *Toxoplasma* infection in the absence of demonstrable antibodies. Proc. 1st Int. Cong. Parasitol. 1:176-177.

Jacobs, L. 1967. *Toxoplasma* and toxoplasmosis. Adv. Parasitol. 5:1-45.

Jacobs, L. and F.E. Jones. 1950. The parasitemia in experimental toxoplasmosis. J. Inf. Dis. 87:78-89.

Jacobs, L. and M.L. Melton. 1965. *Toxoplasma* cysts in tissue culture. Progr. in Protozool. Excerpta Med. Int. Cong. Series. 91. 187-188.

Jacobs, L. and M.L. Melton. 1966. Toxoplasmosis in chickens. J. Parasitol. 52:1158-1162.

Jacobs, L., M.L. Melton, and M.K. Cook. 1953. Experimental toxoplasmosis in pigeons. Exp. Parasitol. 2:403-416.

Jacobs, L., J.S. Remington, and M.L. Melton. 1960. The resistance of the encysted form of *Toxoplasma gondii*. J. Parasitol. 46:11-21.

Janssen, P., G. Piekarski, and W. Korte. 1970. Zum Problem des Abortes bei latenter *Toxoplasma*-Infektion der Frau. Klin. Wochenschr. 48:25-30.

Jewell, M.L., J.K. Frenkel, K.M. Johnson, V. Reed, and A. Ruiz. 1972. Development of *Toxoplasma* oocysts in neotropical *Felidae*. Am. J. Trop. Med. Hyg. 21:512-517.

Jira, J. and V. Kozojed. 1970. Toxoplasmosis 1908-1967, 2 vols. Gustav Fischer, Stuttgart.

Karstad, L. 1963. *Toxoplasma microti* (the M-organism) in the muskrat (*Ondatra zibethica*). Can. Vet. J. 4:249-251.

Kass, E.H., S.B. Andrus, R.D. Adams, F.C. Turner, and H.A. Feldman. 1952. Toxoplasmosis in the human adult. AMA Arch. Int. Med. 89:759-782.

Katsube, Y., T. Hagiwara, H. Miyakawa, T. Muto, and K. Imaizumi. 1969. Studies on toxoplasmosis. 2. Distribution of *Toxoplasma* in the organs of cat and dog cases of latent infection occurring naturally. Jap. J. Med. Sci. Biol. 22:319-326.

Keagy, H.F. 1949. Toxoplasma in the chinchilla. J. Am. Vet. Med. Ass. 114:15.

Kean, B.H. and R.G. Grocott. 1945. Sarcosporidiosis or toxoplasmosis in man and guinea pig. Am. J. Pathol. 21:467-483.

Kean, B.H., A.C. Kimball, and W.N. Christenson. 1969. An epidemic of acute toxoplasmosis. J. Am. Med. Ass. 108:1002-1004.

Kepka, O. and E. Scholtyseck. 1970. Weitere Untersuchungen der Feinstruktur von *Frenkelia* spec. (=M-organismus, Sporozoa). Protistologica 6:249-266.

Kheysin, E.M., V. Krylov, A.N. Sokolov, and A.J. Kirilov. 1969. The role of the poultry and mammalian roundworms (Nematode) in the transmission of toxoplasmosis. 3rd Int. Congr. Protozool.; Leningrad, pp. 227-228.

Kimball, A.C., B.H. Kean, and F. Fuchs. 1971a. Congenital toxoplasmosis: a prospective study of 4,048 obstetric patients. Am. J. Obstet. Gynec. 111:211-218.

Kimball, A.C., B.H. Kean, and F. Fuchs. 1971b. The role of toxoplasmosis in abortion. Am. J. Obstet. Gynecol. 111:219-226.

Kimball, A.C., B.H. Kean, and A. Kellner. 1965. The risk of transmitting toxoplasmosis by blood transfusion. Transfusion 5:447-451.

Köberle, F. 1958. Über Sarkosporidiose beim Menschen. Z. Tropenmed. Parasitol. 9:1-6.

Koster, F.T. and D.D. McGregor. 1971. The mediator of cellular immunity. III. Lymphocyte traffic from the blood into the inflamed peritoneal cavity. J. Exp. Med. 133:864-876.

Krahenbuhl, J.L., A.A. Blazkovek, and M.G. Lysenko. 1971a. *In vivo* and *in vitro* studies of delayed-type hypersensitivity to *Toxoplasma gondii* in guinea pigs. Inf. Imm. 3:260-267.

Krahenbuhl, J.L., A.A. Blazkovek, and M.G. Lysenko. 1971b. The use of tissue culture-grown trophozoites of an avirulent strain of *Toxoplasma gondii* for the immunization of mice and guinea pigs. J. Parasitol. 57:386-390.

Krahenbuhl, J.L. and J.S. Remington. 1971. *In vitro* induction of nonspecific resistance in macrophages by specifically sensitized lymphocytes. Inf. Imm. 4:337-343.

Kühn, D. and G. Weiland. 1969. Experimentelle *Toxoplasma*-Infektionen bei der Katze. 1. Wiederholte Übertragung von *Toxoplasma gondii* durch Kot von mit Nematoden infizierten Katzen. Berl. Münch. Tierärztl. Wochenschr. 82:401-404.

Kulasiri, C. and B. Das Gupta. 1959. A cytochemical investigation of the Sabin-Feldman phenomenon in *Toxoplasma gondii* and an explanation of its mechanism on this basis. Parasitology 49:586-593.

Lainson, R. 1955a. Isolation of *Toxoplasma gondii* from domestic rabbits in England. Trans. Roy. Soc. Trop. Med. Hyg. 49:10-11.

Lainson, R. 1955b. Development and persistence of "pseudocysts" of *Toxoplasma gondii* (rabbit strain) in the lungs of experimentally infected animals. Trans. Roy. Soc. Trop. Med. Hyg. 49:296-297.

Lainson, R. 1955c. Toxoplasmosis in England. II. Variation factors in the pathogenesis of *Toxoplasma* infections: the sudden increase in virulence of a strain after passage in multimammate rats and canaries. Ann. Trop. Med. Parasitol. 49:397-416.

Lainson, R. 1958. Observations on the development and nature of pseudocysts and cysts of *Toxoplasma gondii*. Trans. Roy. Soc. Trop. Med. Hyg. 52:396-407.

Lainson, R. 1959. A note of the duration of *Toxoplasma* infection in the guinea pig. Ann. Trop. Med. Parasitol. 53:120-121.

Lainson, R. and J.J. Shaw. 1971. *Sarcocystis gracilis* n.sp. from the Brazilian tortoise, *Kinosternon scorpioides*. J. Protozool. 18:365-372.

Lainson, R. and J.J. Shaw. 1972. *Sarcocystis* in tortoises: A replacement name, *Sarcocystis kinosterni*, for the homonym *Sarcocystis gracilis* Lainson and Shaw, 1971. J. Protozool. 19:212.

Langer, H. Die Bedeutung der latenten mütterlichen *Toxoplasma*-Infektion für die Gestation. 1966. *In* H. Kirchhoff and H. Kräubig, eds., Toxoplasmose, pp. 123-138. Georg Thieme Verlag, Stuttgart.

Levi, G.C., S. Hyakutake, V. Amato, and M.O.A. Correa. 1968. Observacões complementares sôbre a presença do *Toxoplasma gondii* na saliva de pacientes com toxoplasmose. Rev. Soc. Bras. Med. Trop. 2:275-278.

Levine, N.D. 1961. Protozoan parasites of domestic animals and of man, 1st ed., p. 412. Burgess Publishing Co., Minneapolis.

Lewis, W.P. and E.K. Markell. 1958. Acquisition of immunity to toxoplasmosis by the newborn rat. Exp. Parasitol. 7:463-467.

Ludvik, J. 1960. The electron microscopy of *Sarcocystis miescheriana* Kuhn 1865. J. Protozool. 7:128-135.

Lund, E., H.A. Hanson, E. Lycke, and P. Sourander. 1966. Enzymatic activities of *Toxoplasma gondii*. Acta Pathol. Microbiol. Scand. 68:59-67.

Lund, E., E. Lycke, and P. Sourander. 1961. A cinematographic study of *Toxoplasma gondii* in cell cultures. Brit. J. Exp. Pathol. 42:357-362.

Lunde, M.N. and L. Jacobs. 1965. Antigenic relationship of *Toxoplasma gondii* and *Besnoitia jellisoni*. J. Parasitol. 51:273-276.

Lycke, E. and E. Lund. 1964a. A tissue culture method for titration of infectivity and determination of growth rate of *Toxoplasma gondii*. 1. Acta Pathol. Microbiol. Scand. 60:209-220.

Lycke, E. and E. Lund. 1964b. A tissue culture method for titration of infectivity and determination of growth rate of *Toxoplasma gondii*. 2. Acta Pathol. Microbiol. Scand. 60:221-233.

McKissick, G.E., H.L. Ratcliffe, and A. Koestner. 1968. Enzootic toxoplasmosis in caged squirrel monkeys *Saimiri sciureus*. Pathol. Vet. 5:538-560.

Machado, J.O.L., S. Silva, A.L. De Pinho, and F.J.R. Gomes. 1967. Forésia no mecanismo de transmissão da Toxoplasmose adquirida. O Hospital (Rio de Janeiro) 72:1161-1165.

Martinez-Baez, M., E. Roch, and G. Varela. 1967. Sarcosporidiosis en ratones inoculados experimentalmente con *Toxoplasma gondii*. Rev. Invest. Salud Publ. (Mexico) 27:23-28.

Matsubayashi, H. and S. Akao. 1963. Morphological studies on the development of the *Toxoplasma* cyst. Am. J. Trop. Med. Hyg. 12:321-333.

Mayer, H. 1965. Investigaciones sobre toxoplasmosis. Bol. Oficina Sanit. Panamer. 58:485-497.

Miller, M.J., W.J. Aronson, and J.S. Remington. 1969. Late parasitemia in asymptomatic acquired toxoplasmosis. Ann. Int. Med. 71:139-145.

Miller, N.L., J.K. Frenkel, and J.P. Dubey. 1972. Oral infections with *Toxoplasma* cysts and oocysts in felines, other mammals, and in birds. J. Parasitol. 58:928-937.

Morales, A., A. Mosca, S. Silva, A. Sims, E. Thiermann, F. Knierim, and A. Atias. 1961. Estudio serologico sobre toxoplasmosis y otras parasitosis en Isla de Pascua. Bol. Chil. Parasitol. 16:82-87.

Muir, C.S. 1968. Personal communication.

Munday, B.L. 1972. Serological evidence of *Toxoplasma* infection in isolated groups of sheep. Res. Vet. Sci. 13:100-103.

Nakayama, I. 1968. Investigation on the congenital transmission of toxoplasmosis in chronically infected mice which were reinoculated during pregnancy. Jap. J. Parasitol. 17:128-138.

Nakayama, I. 1969. Persistence of positive hemagglutination titer in animals vaccinated with killed toxoplasmas and their resistance to challenge with virulent strain. Jap. J. Parasitol. 18:539-549.

Nakayama, I. and H. Matsubayashi. 1961. Experimental transmission of *Toxoplasma* gondii in mice. Keio J. Med. 10:63-79.

Neimark, H. and R.G. Blake. 1967. DNA base composition of *Toxoplasma gondii* grown *in vivo*. Nature 216:600.

Neu, H.C. 1967. Toxoplasmosis transmitted at autopsy. J. Am. Med. Ass. 202:844-845.

Norrby, R. 1970. Host cell penetration of *Toxoplasma gondii*. Inf. Imm. 2:250-255.

Norrby, R. 1971. Immunological study on the host-cell penetration factor of *Toxoplasma gondii*. Inf. Imm. 3:378-386.

Nozik, R.A. and G.R. O'Connor. 1971. Studies on experimental ocular toxoplasmosis in the rabbit. III. Recurrent inflammation stimulated by systemic administration of antilymphocyte serum and normal horse serum. Arch. Ophthal. 85:718-722.

O'Reilly, M.J.J. 1954. Acquired toxoplasmosis: an acute fatal case in a young girl. Med. J. Aust. 2:968-970.

Overdulve, J.P. 1970. The identity of *Toxoplasma* Nicolle and Manceaux, 1909 with *Isospora* Schneider, 1881 (I). Koninkl. Nederl. Akad. van Wetenschappen (Amsterdam) Proc., Series C 73:129-151.

Pande, P.G., R.R. Shukla, and P.C. Sekariah. 1961. *Toxoplasma* from the eggs of the domestic fowl (*Gallus gallus*). Science 133:648.

Perrin, T.L. 1943. *Toxoplasma* and *Encephalitozoon* in spontaneous and in experimental infections of animals. A comparative study. Arch. Pathol. 36:568-578.

Perrotto, J., D.B. Keister, and A.H. Gelderman. 1971. Incorporation of precursors into *Toxoplasma* DNA. J. Protozool. 18:470-473.

Pettersen, E.K. 1971. An explanation of the biological action of toxotoxin based on some *in vitro* experiments. Acta Pathol. Microbiol. Scand. B 79:33-36.

Piekarski, G., B. Pelster, and H.M. Witte. 1971. Endopolygeny in *Toxoplasma gondii*. Z. Parasitenk. 36:122-130.

Piekarski, G. and H.M. Witte. 1971. Experimental and histological studies on *Toxoplasma* infections in cats. Z. Parasitenk. 36:95-121.

Pinkerton, H. and R.G. Henderson. 1941. Adult toxoplasmosis. A previously unrecognized disease entity simulating the typhus-spotted fever group. J. Am. Med. Ass. 116:807-814.

Pinkerton, H. and D. Weinman. 1940. *Toxoplasma* infection in man. Arch. Pathol. 30:374-392.

Piper, R.C., C.R. Cole, and J.A. Shadduck. 1970. Natural and experimental ocular toxoplasmosis in animals. Am. J. Ophthal. 69:662-668.

Piza, J. and L. Troper. 1971. Personal communication.

Pols, J.W. 1960. Studies on bovine besnoitiosis with special reference to the aetiology. Onderstepoort J. Vet. Res. 28:265-356.

Pope, J.H., E.H. Derrick, and I. Cook. 1957. *Toxoplasma* in Queensland. I. Observations on a strain of *Toxoplasma gondii* isolated from a bandicoot, *Thylacis obesulus*. Aust. J. Exp. Biol. Med. Sci. 35:467-479.

Price, J.H. 1969. *Toxoplasma* infection in an urban community. Brit. Med. J. 4:141-143.

Pulvertaft, R.J.V., J.C. Valentine, and W.F. Lane. 1954. The behaviour of *Toxoplasma* gondii on serum-agar culture. Parasitology 44:478-484.

Rawal, B.D. 1959. Laboratory infection with *Toxoplasma*. J. Clin. Pathol. 12:59-61.

Remington, J.S. 1970. Toxoplasmosis: recent developments. Ann. Rev. Med. 21:201-218.

Remington, J.S. 1969. The present status of the IgM fluorescent antibody technique in the diagnosis of congenital toxoplasmosis. J. Ped. 75(Sect. 6, Part 2):1116-1124.

Remington, J.S., M.M. Bloomfield, E. Russell, and W.S. Robinson. 1970. The RNA of *Toxoplasma gondii*. Proc. Soc. Exp. Biol. Med. 133:623-626.

Remington, J.S. and E.N. Cavanaugh. 1965. Isolation of the encysted form of *Toxoplasma gondii* from human skeletal muscle and brain. N.E. J. Med. 273:1308-1310.

Remington, J.S., B. Efron, E. Cavanaugh, H.J. Simon, and A. Trejos. 1970. Studies on toxoplasmosis in El Salvador. Prevalence and incidence of toxoplasmosis as measured by the Sabin-Feldman dye test. Trans. Roy. Soc. Trop. Med. Hyg. 64:252-267.

Remington, J.S., L. Jacobs, and M.L. Melton. 1961. Congenital transmission of toxoplasmosis from mother animals with acute and chronic infections. J. Inf. Dis. 108:163-173.

Remington, J.S., M.L. Melton, and L. Jacobs. 1961. Induced and spontaneous recurrent parasitemia in chronic infections with avirulent strains of *Toxoplasma gondii*. J. Immunol. 87:578-581.

Remington, J.S. and T.C. Merigan. 1968. Interferon: protection of cells infected with an intracellular protozoan (*Toxoplasma gondii*). Science 161:804-806.

Remington, J.S. and T.C. Merigan. 1969. Resistance to virus challenge in mice infected with protozoa or bacteria. Proc. Soc. Exp. Biol. Med. 131:1184-1188.

Remington, J.S., M.J. Miller, and I. Brownlee. 1968a. IgM antibodies in acute toxoplasmosis. II. Prevalence and significance in acquired cases. J. Lab. Clin. Med. 71:855-866.

Remington, J.S., M.J. Miller, and I. Brownlee. 1968b. IgM antibodies in acute toxoplasmosis. I.Diagnostic significance in congenital cases and a method for their rapid demonstration. Pediatrics 41:1082-1091.

Remington, J.S., J.W. Newell, and E. Cavanaugh. 1964. Spontaneous abortion and chronic toxoplasmosis. Obstet. Gynec. 24:25-31.

Reynolds, E.S., K.W. Walls, and R.I. Pfeiffe. 1966. Generalized toxoplasmosis following renal transplantation. Arch. Int. Med. 118:401-405.

Robertson, J.S. 1966. Chronic toxoplasmosis with negative dye test? Postgrad. Med. J. 42:61-64.

Roever-Bonnet, H. de. 1963. Mice and golden hamsters infected with an avirulent and a virulent *Toxoplasma* strain. Trop. Geogr. Med. 15:45-60.

Roever-Bonnet, H. de. 1964. *Toxoplasma* parasites in different organs of mice and hamsters infected with avirulent and virulent strains. Trop. Geogr. Med. 16:337-345.

Roever-Bonnet, H. de. 1969. Congenital *Toxoplasma* infections in mice and hamsters infected with avirulent and virulent strains. Trop. Geogr. Med. 21:443-450.

Rommel, M. 1969. Möglichkeiten der Übertragung von *Toxoplasma gondii* bei Haustieren durch Nematoden und durch die Milch. Schlacht und Viehhof-Zeitung 7:268-269.

Rommel, M. and J. Breuning. 1967. Untersuchungen über das Vorkommen von *Toxoplasma gondii* in der Milch einiger Tierarten und die Möglichkeit der laktogenen Infektion. Tierärztl. Wochenschr. 80:365-384.

Rommel, M. and A.O. Heydorn. 1972. Beiträge zum Lebenszyklus der Sarkosporidien. III. *Isospora hominis* (Railliet und Lucet, 1891) Wenyon, 1923, eine Dauerform der Sarkosporidien des Rindes und des Schweins. Berl. Münch. Tierärztl. Woch. 85:143-145.

Rommel, M., A.O. Heydorn, and F. Gruber. 1972. Beiträge zum Lebenszyklus der Sarkosporidien. I. Die Sporozyste von *S. tenella* in den Fäzes der Katze. Berl. Münch. Tierärztl. Woch. 85:101-105.

Rommel, M., K. Janitschke, W. Dalchow, H-P. Schulz, J. Breuning, and E. Schein. 1968. Versuche zur Übertragung von *Toxoplasma gondii* durch parasitische Nematoden bei Schafen, Schweinen, Hunden, Katzen, Hühnern und Mäusen. Tierärztl. Wochenschr. 81:309-313.

Roth, J.A., S.E. Siegel, A.S. Levine, and C.W. Berard. 1971. Fatal recurrent toxoplasmosis in a patient initially infected via a leukocyte transfusion. Am. J. Clin. Pathol. 56:601-605.

Ruch, T.C. 1959. Diseases of laboratory primates, 1st ed. W.B. Saunders Co., Philadelphia.

Ruskin, J., J. McIntosh, and J.S. Remington. 1969. Studies on the mechanisms of resistance to phylogenetically diverse intracellular organisms. J. Immunol. 103:252-259.

Ruskin, J. and J.S. Remington. 1968a. Role for the macrophage in acquired immunity to phylogenetically unrelated intracellular organisms. Antimicrob. Agents Chemoth. 474-477.

Ruskin, J. and J.S. Remington. 1968b. Immunity and intracellular infection: resistance to bacteria in mice infected with a protozoan. Science 160:72-74.

Ruskin, J. and J.S. Remington. 1971. Resistance to intracellular infection in mice immunized with *Toxoplasma* vaccine and adjuvant. J. Reticulo. Soc. 9:465-479.

Sabin, A.B. 1941. Toxoplasmic encephalitis in children. J. Am. Med. Ass. 116:801-807.

Sabin, A.B., H. Eichenwald, H.A. Feldman, and L. Jacobs. 1952. Present status of clinical manifestations of toxoplasmosis in man. Indications and provisions for routine serologic diagnosis. J. Am. Med. Ass. 150:1063-1069.

Sabin, A.B. and H.A. Feldman. 1948. Dye as microchemical indicators of a new immunity phenomenon affecting a protozoon parasite (*Toxoplasma*). Science 108:660-663.

Sanger, V.L., D.M. Chamberlain, C.R. Cole, and R.L. Farrell. 1953. Toxoplasmosis. V. Isolation of *Toxoplasma* from cattle. J. Am. Vet. Med. Ass. 123:87-91.

Sanger, V.L. and C.R. Cole. 1955. Toxoplasmosis. VI. Isolation of *Toxoplasma* from milk, placentas, and newborn pigs of asymptomatic carrier sows. Am. J. Vet. Res. 16:536-539.

Schneider, C.R. 1965. *Besnoitia panamensis*, sp.n. (Protozoa: Toxoplasmatidae) from Panamanian lizards. J. Parasitol. 51:340-344.

Schneider, C.R. 1967. *Besnoitia darlingi* (Brumpt, 1913) in Panama. J. Protozool. 14:78-82.

Schneider, C.R. 1968. Personal communication.

Scholtyseck, E., O. Kepka, and G. Piekarski. 1970. Die Feinstruktur der Zoiten aus reifen Cysten des sog. M-organismus (=*Frenkelia* spec.). Z. Parasitenk. 33:252-261.

Scott, J.W. 1943. Life history of sarcosporidia, with particular reference to *Sarcocystis tenella*. U. Wyo. Agric. Exp. Station Bull. 259.

Sebek, Z. 1962. *Sarcocystis* und M-Organismen bei Insektenfressern und Nagetieren. Zool. Listy, Folia Zool. 11:355-366.

Sebek, Z. 1970. Lebenszyklus der *Frenkelia* (M-Organismus). J. Parasitol. 56 (4, Sec. II, Part 1):310-311.

Seibold, H.R. and R.H. Wolf. 1971. Toxoplasmosis in *Aotus trivirgatus* and *Callicebus moloch*. Lab. Anim. Sci. 21:118.

Sénaud, J. 1967. Contribution a l'étude des sarcosporidies et des Toxoplasmes (Toxoplasmea). Protistologica 3:167-242.

Sénaud, J. 1969. Ultrastructure des formations kystiques de *Besnoitia jellisoni* (Frenkel, 1953) protozoaire, *Toxoplasmea*, parasite de la souris (*Mus musculus*). Protistologica 5:413-430.

Sexton, R.C., D.E. Eyles, and R.E. Dillman. 1953. Adult toxoplasmosis. Am. J. Med. 14:366-377.

Sheffield, H.G. 1966. Electron microscope study of the proliferative form of *Besnoitia jellisoni*. J. Parasitol. 52:583-594.

Sheffield, H.G. 1968. Observations on the fine structure of the "cyst stage" of *Besnoitia jellisoni*. J. Protozool. 15:685-693.

Sheffield, H.G. and M.L. Melton. 1968. The fine structure and reproduction of *Toxoplasma gondii*. J. Parasitol. 54:209-226.

Sheffield, H.G. and M.L. Melton. 1969. *Toxoplasma gondii*: transmission through feces in absence of *Toxocara cati* eggs. Science 164: 431-432.

Sheffield, H.G. and M.L. Melton. 1970. *Toxoplasma gondii*: the oocyst, sporozoite, and infection of cultured cells. Science 167:892-893.

Siegel, S.E., M.N. Lunde, A.H. Gelderman, R.H. Halterman, J.A. Brown, A.S. Levine, and R.G. Graw. 1971. Transmission of toxoplasmosis by leukocyte transfusion. Blood 37:388-394.

Siim, J.C., U. Biering-Sorensen, and T. Møller. 1963. Toxoplasmosis in domestic animals. Adv. Vet. Sci. 8:335-429.

Siim, J.C., W.M. Hutchison, and K. Work. 1969. Transmission of *Toxoplasma gondii*. Further studies on the morphology of the cystic form in cat faeces. Acta Pathol. Microbiol. Scand. 77:756-757.

Simpson, C.F. 1966. Electron microscopy of *Sarcocystis fusiformis*. J. Parasitol. 52:607-613.

Smith, T. 1901-1905. The production of sarcosporidiosis in the mouse by feeding infected muscular tissue. J. Exp. Med. 6:1-22.

Spindler, L.A. 1947. A note on the fungoid nature of certain internal structures of Miescher's sacs (*Sarcocystis*) from a naturally infected sheep and a naturally infected duck. Proc. Helminth. Soc. Wash. 14:28-30.

Stinson, E.B., C.P. Bieber, R.B. Griepp, D.A. Clark, N.E. Shumway, and J.S. Remington. 1971. Infectious complications after cardiac transplantation in man. Ann. Int. Med. 74:22-36.

Strannegard, O. 1967a. The so-called *Toxoplasma*-hostile factor and its relation to antibody. Acta Pathol. Microbiol. Scand. 69:465-476.

Strannegard, O. 1967b. An electron microscopic study on the immunoinactivation of *Toxoplasma gondii*. Acta Pathol. Microbiol. Scand. 71:463-470.

Strannegard, O., E. Lund, and E. Lycke. 1967. The influence of antibody and normal serum factors on the respiration of *Toxoplasma gondii*. Am. J. Trop. Med. Hyg. 16:273-277.

Suggs, M.T., K.W. Walls, and I.G. Kagan. 1968. Comparative antigenic study of *Besnoitia jellisoni*, *B. panamensis* and five *Toxoplasma gondii* isolates. J. Immunol. 101:166-175.

Tadros, W.A. 1970. Speculations on the life-cycle of *Frenkelia*. J. Parasitol. 56 (4, Sec. II, Part 1):338-339.

Thomascheck, G., H. Werner, and L. Schmidtke. 1966. *Toxoplasma*-Infektion und Schwangerschaft. Klinische Symptomatologie, serologische und histologische Untersuchungsergebnisse. Klin. Wochenschr. 44:921-928.

Tishovskaya, T.M. 1969. Transmission of *Toxoplasma* by eggs of helminthes. 3rd Int. Congr. Protozool.; Leningrad, pp. 246-247.

Tremonti, L. and B.C. Walton. 1970. Blast formation and migration-inhibition in toxoplasmosis and leishmaniasis. Am. J. Trop. Med. Hyg. 19:49-56.

Tuffery, A.A. and Innes, J.R.M. 1963. Diseases of laboratory mice and rats. *In* W. Lane-Petter, ed., Animals for research, p. 48. Academic Press, New York.

Twort, J.M. and C.C. Twort. 1932. Disease in relation to carcinogenic agents among 60,000 experimental mice. J. Pathol. Bact. 35:219-242.

Vainisi, S.J. and L.H. Campbell. 1969. Ocular toxoplasmosis in cats. J. Am. Vet. Med. Ass. 154:141-152.

Vermeil, C. and S. Marguet. 1967. Sur la transmission de la toxoplasmose par les helminthes et leurs oeufs. Ann. Parasitol. 42:283-284.

Vietzke, W.M., A.H. Gelderman, P.M. Grimley, and M.P. Valsamis. 1968. Toxoplasmosis complicating malignancy. Cancer 21:816-827.

Vischer, W.A. and E. Suter. 1954. Intracellular multiplication of *Toxoplasma gondii* in adult mammalian macrophages cultivated *in vitro*. Proc. Soc. Exp. Biol. Med. 86:413-419.

Vivier, E. and A. Petitprez, A. 1969. Le complexe membranaire superficiel et son evolution lors de l'elaboration des individus-fils chez *Toxoplasma gondii*. J. Cell Biol. 43:329-342.

Wallace, G.D. 1969a. Sabin-Feldman dye test for toxoplasmosis. Am. J. Trop. Med. Hyg. 18: 395-398.

Wallace, G.D. 1969b. Serologic and epidemiologic observations on toxoplasmosis on three Pacific atolls. Am. J. Epidemiol. 90:103-111.

Wallace, G.D. 1971a. Experimental transmission of *Toxoplasma gondii* by filth-flies. Am. J. Trop. Med. Hyg. 20:411-413.

Wallace, G.D. 1971b. Isolation of *Toxoplasma gondii* from the feces of naturally infected cats. J. Inf. Dis. 124:227-228.

Wallace, G.D. 1972. Experimental transmission of *Toxoplasma gondii* by cockroaches. J. Inf. Dis. 126:545-547.

Wallace, G.D., L. Marshall, and M. Marshall. 1972. Cats, rats, and toxoplasmosis on a small Pacific island. Am. J. Epidemiol. 95: 475-482.

Walton, B.C., I. de Arjona, and B.M. Benchoff. 1966. Relationship of *Toxoplasma* antibodies to altitude. Am. J. Trop. Med. Hyg. 15:492-495.

Walton, B.C. and K.W. Walls. 1964. Prevalence of toxoplasmosis in wild animals from Fort Stewart, Georgia as indicated by serological tests and mouse inoculation. Am. J. Trop. Med. Hyg. 13:530-533.

Walton, B.C. and J.K. Werner. 1970. Schizont, gamete, and oocyst production in the cat by human and porcine strains of *Toxoplasma gondii* from Japan. Jap. J. Parasitol. 19:628-634.

Wanko, T., L. Jacobs, and M.A. Gavin. 1962. Electron microscope study of *Toxoplasma* cysts in mouse brain. J. Protozool. 9:235-242.

Watson, W.A. and J.K.A. Beverley. 1971. Epizootics of toxoplasmosis causing ovine abortion. Vet. Rec. 88:120-124.

Weiland, G. and D. Kühn. 1970. Experimentelle *Toxoplasma*-Infektionen bei der Katze. II. Entwicklungstadien des Parasiten im Darm. Tierarztl. Wochenschr. 83:128-132.

Wilder, H.C. 1952. *Toxoplasma* chorioretinitis in adults. AMA Arch. Ophthal. 48:127-137.

Wildführ, W. 1966. Elektronenmikroskopische Untersuchungen zur Morphologie und Reproduktion von *Toxoplasma gondii*. II. Beobachtungen zur Reproduktion von *Toxoplasma gondii* (Endodyogenie). Zbl. Bakt. Parasitenk. (I.Abt.) 201:110-130.

Witte, H.M., and Piekarski, G. 1970. Die Oocysten-Ausscheidung bei experimentell infizierten Katzen in Abhängigkeit vom *Toxoplasma*-Stamm. Z. Parasitenk. 33:358-360.

Woke, P.A., L. Jacobs, F.E. Jones, and M.L. Melton. 1953. Experimental results on possible arthropod transmission of toxoplasmosis. J. Parasitol. 39:523-532.

Work, K. 1967. Isolation of *Toxoplasma gondii* from the flesh of sheep, swine and cattle. Acta Pathol. Microbiol. Scand. 71:296-306.

Work, K. 1968. Resistance of *Toxoplasma gondii* encysted in pork. Acta Pathol. Microbiol. Scand. 73:85-92.

Work, K. and W.M. Hutchison. 1969a. A new cystic form of *Toxoplasma gondii*. Acta Pathol. Microbiol. Scand. 75:191-192.

Work, K. and W.M. Hutchison. 1969b. The new cyst of *Toxoplasma gondii*. Acta Pathol. Microbiol. Scand. 77:414-424.

Zeve, V.H., D.L. Price, and C.M. Herman. 1966. Electron microscope study of *Sarcocystis* sp. Exp. Parasitol. 18:338-346.

Zimmerman, L.E. 1961. Ocular pathology of toxoplasmosis. Surv. Ophthal. 6 (6, Part 2):832-838.

Zypen, E. van der. 1966. Licht- und elektronenmikroskopische Studien zur Frage der Entwicklung von *Toxoplasma*-cysten im Gehirn der weissen Maus. Z. Parasitenk. 28:31-44.

Zypen, E. van der and G. Piekarski. 1966. Zur Ultrastruktur der Cystenwand von *Toxoplasma gondii* im Gehirn der weissen Maus. Z. Parasitenk. 28:45-59.

Zypen, E. van der and G. Piekarski. 1967. Ultrastrukturelle Unterschiede zwischen der sog. Proliferationsform (RH-Stamm, BK-Stamm) und dem sog. Cysten Stadium (DX-Stamm) von *Toxoplasma gondii*. Zentralbl. Bakteriol. Parasitenk. 203:495-517.

10 Techniques

L. R. DAVIS

United States Department of Agriculture, Agricultural Research Service, Veterinary Sciences Research Division, Regional Parasite Research Laboratory, Post Office Drawer 952, Auburn, Alabama

---

Contents

I. Introduction . . . *413*
II. *Eimeria* . . . *414*
  A. Establishing Infections of Coccidia . . . *414*
    1. Establishing Cultures from Single Oocysts . . . *414*
    2. Dry Feed Inoculations of Coccidia of Ruminants . . . *414*
    3. Tracing Inoculations of Coccidia in Animals . . . *414*
  B. Obtaining Oocysts of Coccidia . . . *415*
    1. Mechanical Device for Screening Oocysts from the Feces . . . *415*
    2. Isolation of Oocysts . . . *415*
    3. Isolation of Coccidia by Micromanipulators . . . *416*
    4. Isolation of Pure Cultures of Coccidia with a Laser Microscope . . . *416*
  C. Isolating Experimental Calves . . . *417*
    1. Calf Muzzle . . . *417*
    2. Portable Pens . . . *417*
  D. Collecting Samples and Counting Oocysts of Coccidia . . . *417*
    1. Collecting Fecal Samples from Domestic Animals . . . *417*
    2. Counting Oocysts . . . *418*
  E. Sporulating Oocysts of Coccidia . . . *419*
    1. Sporulating Oocysts of Coccidia with Aeration . . . *419*
    2. Sporulating in Tap Water . . . *419*
    3. Temperature and Sporulation of Oocysts . . . *420*
    4. Sporulation of Oocysts in Large Volumes . . . *420*
    5. Testing for Sporulation . . . *420*
    6. Testing for Survival of Oocysts in the Field . . . *421*
  F. Preparation of Pure Oocyst Suspensions . . . *421*
    1. Preparing Bacterially Sterile Oocysts of Coccidia . . . *421*
    2. Cleaning Coccidial Oocysts by Density-Gradient Sedimentation . . . *421*

3. Purifying Coccidian Oocysts . . . *422*
4. Obtaining Pure Oocyst Suspensions . . . *422*
5. Oocyst Isolation Techniques . . . *424*
6. Trypsinization of Tissues to Release Oocysts . . . *428*
G. Obtaining Freed Sporocysts of Coccidia . . . *429*
1. Freeing Sporocysts from Oocysts by Crushing . . . *429*
2. Releasing Sporocysts from Oocysts by Grinding . . . *429*
H. Preparation of Sporozoites of Coccidia . . . *429*
1. Excystation from Oocysts . . . *429*
2. Purifying Sporozoites of *Eimeria tenella* . . . *432*
3. Excysting Sporozoites . . . *432*
4. Obtaining Pure Sporozoite Suspensions . . . *432*
5. Obtaining Sporozoites for Use in Cell Cultures . . . *433*
6. Strout's Method of Obtaining Sporocysts of Coccidia . . . *434*
I. Isolation of Merozoites of Coccidia . . . *434*
1. Obtaining Merozoites for Inoculations . . . *434*
2. Stimulating Motility of Merozoites . . . *434*
J. Cultivating Coccidia in Chick Embryos . . . *435*
1. Cultivation of *Eimeria stiedai* in Avian Embryos . . . *435*
2. Chick Embryo Infections . . . *435*
K. Low Temperature Storage of Coccidia . . . *436*
1. Cold Storage Preservation of Excysting Sporozoites . . . *436*
2. Freezing Coccidia . . . *436*
3. Determining Infectivity of Frozen Sporozoites . . . *437*
L. Some Techniques Used in Studies on Life Cycles of Coccidia . . . *438*
1. Surgery Used in Life Cycle Studies on Coccidia of Cattle . . . *438*
2. Rapid Preservation of Tissues for Sectioning . . . *438*
3. Counting Schizonts of Coccidia in Intestines . . . *439*
4. Examining Sections of Intestine for Life Cycle Stages of Coccidia . . . *440*
5. Flattening Intestinal Samples for Fixation . . . *440*
6. Triple Staining for Parasites in Tissue . . . *440*
7. Staining Oocysts of Coccidia . . . *441*
8. An All-Glass Staining Rack for Coverslips . . . *441*
9. Phase-Contrast and Interference Microscopy . . . *441*
10. Fluorescence Microscopy . . . *442*
11. Fixation of Merozoites for Electron Microscopy . . . *442*
12. Locating Parasites by Light Microscopy before Using Electron Microscopy . . . *442*
M. Techniques for Immunological Studies . . . *443*
1. Particulate Antigens . . . *443*
2. Soluble Antigens . . . *443*
N. Plastic Film Isolators for Producing Gnotobiotic Poultry . . . *444*
1. Plastic Film Isolators . . . *444*
2. Equipment . . . *444*
3. Sterilization of Unit and Accessories . . . *444*
4. Sterilization of Feed and Water . . . *445*
5. Testing for Sterility . . . *445*

6. Production of Gnotobiotic Birds . . . *445*
O. Other Techniques for Investigating Coccidia . . . *445*
III. *Toxoplasma* . . . *445*
A. Surveying Methods for *Toxoplasma* in Domestic Animals . . . *445*
B. Laboratory Diagnosis of Toxoplasmosis . . . *446*
1. The Indirect Fluorescent Antibody Test for *Toxoplasma* . . . *446*
2. Dye Test (DT) . . . *447*
3. Isolation of *Toxoplasma* from Blood or Tissues . . . *447*
4. The Indirect Fluorescent Antibody Test for Toxoplasmosis with Filter Paper Blood Specimens . . . *448*
C. Propagation of *Toxoplasma* . . . *448*
1. Maintenance in Laboratory Animals . . . *448*
2. Preservation . . . *449*
3. Obtaining *Toxoplasma* for Use in Tissue Cultures . . . *449*
4. Obtaining Proliferative *Toxoplasma* Free of Host Cells . . . *449*
5. The Plaque Method for Studying *Toxoplasma* . . . *449*
D. Techniques for Handling *Toxoplasma* . . . *450*
IV. *Sarcocystis* . . . *451*
A. Examination of Birds for Presence of *Sarcocystis* . . . *451*
B. Culturing *Sarcocystis* in Avian and Mammalian Cells . . . *451*
V. *Besnoitia* . . . *452*
A. Examination for *Besnoitia jellisoni* in Rodents . . . *452*
B. Transmission of *Besnoitia* . . . *452*
1. Transmission of *Besnoitia jellisoni* . . . *452*
2. Transmission of *Besnoitia besnoiti* . . . *452*
C. Isolation of *Besnoitia* . . . *453*
Literature Cited . . . *453*

I. Introduction

There are many techniques that are useful in research on coccidia. The literature is voluminous, and complete coverage would fill more than an entire book. This chapter is devoted to selected methods that have been helpful in research on certain coccidia or other parasites and should be of benefit to those working with these or related organisms.*

No attempt has been made to list techniques of electron microscopy or of histochemistry because excellent books are already available on these subjects.

Some specialists have developed or modified techniques that they have found to be very productive in their own laboratories and have contributed these procedures for use in this chapter. Their particular contributions are acknowledged in appropriate places.

Acknowledgments. I wish to express my appreciation for the assistance of the following: David J. Doran and Ronald Fayer of the National Animal Parasite Research Laboratory, Veterinary Sciences Research Division, ARS, U. S. Department of Agriculture, Beltsville, Maryland 20705; J. P. Dubey and J. K. Frenkel of the Department of Pathology and Oncology, School of Medicine, Univer-

*Mention of a trademark name or proprietary product does not constitute a guarantee or warranty of the U. S. Department of Agriculture and does not imply its approval to the exclusion of other products that may also be suitable.

sity of Kansas Medical Center, Kansas City, Kansas 66103; P. L. Long and Elaine Rose of the Houghton Poultry Research Station, Houghton, Huntington, England; John F. Ryley of the Imperial Chemical Industries Ltd., Pharmaceutical Division, Alderly Park, Macclesfield, Cheshire, England; and to John V. Ernst of this laboratory for the loan of numerous reprints.

II. *Eimeria*

A. ESTABLISHING INFECTIONS OF COCCIDIA

1. Establishing cultures from single oocysts

Edgar and Seibold (1964) outlined a method of obtaining oocysts from isolation of single oocysts. Single cell isolates were given separately to from one- to five-day-old coccidia-free chicks isolated in rearing cans. The chicks were killed on the sixth or seventh day and intestinal contents were aerated. The preparation from each chick was passed "blindly" through a second coccidia-free chick, and the oocysts were harvested from each. These oocysts were passed through different groups of clean chicks and the resultant oocysts furnished the cultures for further studies.

2. Dry feed inoculations of coccidia of ruminants

There are indications that the method of inoculating ruminant animals is important. Smith and Davis (1965) reported that oocysts mixed with dry feed produced greater infections than those given in liquid feed and suggested that the dry feed mix introduced the additional factors of mechanical mixing and grinding and a more prolonged action of salivary juices on the oocysts. Lotze (1953) reported that oocysts administered by drench may pass readily into the small intestine and those given in a gelatin capsule are more likely to be retained in the rumen, and he suggested that oocysts should remain in the upper portion of the digestive tract as long as six hours in order to produce adequate infections in experimental work. The work of Smith and Davis showed that challenge inoculations given in liquid to previously infected lambs usually did not produce reinfections, whereas challenge inoculations given in dry feed resulted in the deaths of 13 of 16 lambs. Also, their work showed that far fewer oocysts than are usually used in experimental infections are necessary when the oocysts are given in dry feed. Only 31,000 oocysts of *E. ahsata* in dry feed caused death as compared with 100,000 oocysts required when administered in liquid. When given in dry feed the minimum number of oocysts of *Eimeria ninakohlyakimovae* resulting in death or producing clinical signs was 20,000 or 10,000, respectively, far fewer than had been reported previously. Fitzgerald (1967) stated that the administration of oocysts in dry or liquid medium to calves seemingly did not influence the "take," but he suggested that the dry inoculum, fed immediately after the evening feeding of milk, may have "equalized" the effects of the two methods. Also, all his challenge inoculations consisted of oocysts suspended in water and administered by dose syringe rather than in dry feed.

3. Tracing inoculations of coccidia in animals

Inoculated oocysts given in drench or dry feed can be traced in the early portion of the life cycle of coccidia by mixing inert fluorescent powders and then, at postmortem, using an ultraviolet light on the opened intestine or smears made from the intestine. Davis and Smith (1959) showed that such pigments with extremely brilliant hues even in daylight

(Switzer Brothers, Cleveland, Ohio) fluoresced with even brighter colors under ultraviolet light. They mixed the powder 2 : 1 with bentonite as a carrier for oocysts suspended in liquids or mixed in dry feed. The particles (average size 4.5 μ) are reported to be relatively harmless, even when fed in greater amounts than the trace quantities used in the studies. When as little as 3 g were mixed with oocysts given orally to a yearling calf, the inoculum could be located in the digestive tract by using a hand-held ultraviolet light. Oocysts and early tissue stages were found adjacent to the brilliantly glowing pigment particles. The system is simpler, safer, and cheaper than using radioisotopes and is far more efficient than the old method of using carbon particles and microscopy to locate an inoculum in early stages of the life cycles.

The fluorescing particles in the feces made it possible to determine which feces in the feedlot came from inoculated animals. The authors suggested that the inert pigments could be used to determine sites of injection of organisms as well as drugs, chemicals, and other substances in experimental animals.

B. OBTAINING OOCYSTS OF COCCIDIA

1. Mechanical device for screening oocysts from the feces

Hill and Zimmerman (1961) produced a unit that was designed for screening worm eggs from feces in a very rapid manner and found that it also removed oocysts of *Eimeria debliecki* from pig feces by screening the specimen through a 400-mesh screen. They used a planetary-type electric mixer made by the Hobart Manufacturing Company of Troy, Ohio (Model K5-A). They made special blades to be used to agitate the material just above a screened container. A copper bucket was designed to serve as an intermediate catch basin and to support the screened cylindrical basket which had connections for supplying running water. Critical tests on 300-g samples of sheep feces and 500-ml samples of swine and cattle feces revealed that more than 99% of nematode eggs were removed in approximately five minutes running time when a 200-mesh screen was used. The authors gave a complete description for modifying the planetary mixer without destroying its usefulness as a regular laboratory mixer.

2. Isolation of oocysts

Oocysts can be isolated by the method of Wilson and Fairbairn (1961). Washings from contents of small intestines of chickens were made at temperatures below 10°C. The suspension was sedimented overnight and the sediment treated by a series of centrifugal flotations in a continuous flow chemical centrifuge (International Centrifuge Corporation, Model No. 450) using an unperforated basket-rotor, in lieu of the standard perforated basket. The sediment was suspended in zinc sulfate solution (sp. gr. 1.181) and recentrifuged. The oocysts were collected in the effluent and the majority of the contaminants remained in the basket. The effluent was diluted with six volumes of water and recentrifuged; the oocysts were collected in the basket. Using a refrigerated centrifuge, subsequent flotations were carried out in decreasing concentrations of zinc sulfate to a specific gravity of 1.156. Some large particles that floated on the zinc sulfate solutions were removed by screening. The oocysts were washed several times in water by centrifugation and stored in 0.1 N H_2SO_4 at 4°C. The oocysts contained virtually no visible contaminants and were isolated in a few hours in temperatures low enough to stop development. The yield was from 30% to 50% of the original number and the viability after treatment was about 90%. [Note Vetterling's (1969) method, p. 422.]

3. *Isolation of coccidia by micromanipulators*

Micromanipulators have been used in isolating stages of coccidia. El-Badry (1963) covered many useful techniques; however, most laboratories do not have an instrument available. Davis and Bowman (1966) described for use with coccidia a unique micromanipulator that can be made by unskilled workers in less than 30 min at a cost of less than a dollar. They utilized a microscope that has the usual focusing stage to produce the three-dimensional movements necessary for a micromanipulator. A rubber stopper was cemented to a 3 in by 1 in microscope slide. The top of the stopper was notched and a handmade micropipette was placed in the slot. A two-foot length of flexible tubing enabled the operator to use his mouth on the micropipette to pick up oocysts or other stages of coccidia and to expel them when desired. The microscope with the simplified micromanipulator was placed in front of another viewing microscope in such a manner that the tip of the micropipette was in the center of the field of the viewing microscope. To prevent breakage of the micropipettes the viewing microscope should be one in which the stage does not focus. Assorted handmade microtools can be used in lieu of the micropipette. The authors used the micromanipulator for isolating pure cultures of oocysts from mixed species of coccidia. Coating the microtools and micropipettes with silicones (El-Badry, 1963) reduces surface tension that builds up in the small orifices when positive or negative pressure is used for picking up or expelling oocysts. This pressure causes violent movements of the fluids used for suspension of the parasites, making micromanipulators difficult to maneuver unless silicone coatings are used.

4. *Isolation of pure cultures of coccidia with a laser microscope*

A laser beam through a microscope was used by Davis (1969) to purify mixed cultures of oocysts of coccidia of cattle, sheep, and rabbits. A TRG laser that was custom made for the Leitz Ortholux microscope was adjusted by means of the apertures in the laser head and by changing the magnification of the microscope objectives to obtain approximately a 10-μ diameter beam on the microscope slide. Oocysts were exposed with and without a coverslip, but the coverslip offered some protection for the objective of the microscope. It was found that some of the colorless oocysts showed very little damage until a small amount of methylene blue dye in aqueous solution was added to color the suspension. When the beam ruptured oocyst walls, the dye stained the inner contents and the oocyst walls, and enabled one to distinguish exposed and control oocysts. The energy could be adjusted between 150 and 230 joules in the power unit to vary the extent of oocyst damage from the formation of a tiny bubble of steam inside the oocyst to complete disruption of oocyst walls and contents. Mixed cultures of oocysts were exposed to the laser beam to destroy unwanted species and to produce pure cultures for experimental use.

Because dust particles on the lenses of the microscope can act as a "heat sink," resulting in damage to the objective lenses, inexpensive objectives were used instead of the Leitz objectives.

Operators of a laser microscope should be warned that there is a possibility of damage to the operator's eyes from stray light sources in the microscope and also from specular reflection from the microscope slide and coverslip or from shiny surfaces in the room. Davis (1969) devised a special enclosure of aluminum and plywood to reduce or eliminate such danger. The microscope was placed inside the enclosure with the power supply adjoining. When the sliding door was closed the safety case prevented the operator from looking into the microscope while the power supply unit was fired. When the sliding door was moved to cover the power supply, the operator could look through the microscope without any possibility of being able to fire the laser accidentally.

C. ISOLATING EXPERIMENTAL CALVES

1. Calf muzzle

Hammond *et al.* (1944) placed calves in outdoor clean pens, and in the first experiments a special muzzle was placed on the calf as an additional precaution against infection. The muzzle (not described in the publication) was made of a tin can flared at one end in order to fit the head of the calf, lined with foam rubber to prevent abrasion, and was fitted at the opposite end with an ointment can which was perforated and filled with fiberglass to act as a strainer. The muzzle was held in place by means of a collar with two leather straps. It was replaced each day with a freshly autoclaved muzzle. The calves were fed milk by means of sterilized nippled pails. The muzzle was very effective in prevention of ingestion of extraneous species not desired in the experiments, but when the muzzles were removed at six weeks, the calves were apparently more susceptible to respiratory infections than normal.

2. Portable pens

Davis (1949) described a portable pen system developed at the USDA Regional Parasite Research Laboratory at Auburn, Alabama, for prevention of clinical coccidiosis as well as other infections. Calves were removed from the dam within 24 hr, after receiving colostrum, and were placed in movable pens 5 ft × 10 ft in size, where they remained until from three to four months of age, depending on the size of the calves. The pens were placed in rows and were moved zigzag uphill and were not placed on the same ground for one year or for six months if the surface had been plowed under. A complete description of the calf pens and a comparison with other enclosures for raising calves were given by Davis *et al.* (1952).

D. COLLECTING SAMPLES AND COUNTING OOCYSTS OF COCCIDIA

1. Collecting fecal samples from domestic animals

Swan (1970) devised a mechanical aid to speed up collection of rectal samples from sheep. He developed an extractor made of a hollow clear plastic cylinder, 2 cm outside diameter and 20.5 cm long, bevelled at an angle of 45° at the tip and with smooth edges. The inner plunger was a solid clear plastic rod of a diameter to fit snugly into the outer component. A similar extractor with an outside diameter of 1.6 cm was made to use with lambs. In use, the plunger was withdrawn as the outer shell was placed gently through the anus. The extractor was then removed, the fecal sample expelled, and the extractor washed thoroughly between collections.

A more simplified method (unpublished) was developed at the USDA Regional Parasite Research Laboratory many years ago for collecting samples from calves. Instead of using a glass test tube, which could break easily, a one-half inch clear plastic rod was used. These were obtained in from 10 in to 12 in lengths from a radio supply mail order company where they are sold as electrical insulators. A grinding wheel was used to round off one end and then three grades of jewellers' abrasive compound on a cloth polishing disc were used to polish the rounded end to the same degree of clarity as the rest of the rod.

Inexperienced animal caretakers are issued one with an encircling ring of red nail polish three inches from the rounded tip. The tail of the calf is raised vertically and the plastic rod is inserted to the mark and pressed downward gently to allow air to enter the rectum. This usually stimulates defecation. The polished plastic enables one to see that the rod has been washed thoroughly between samplings. Tapering toothbrush handles were modified as above when working with lambs. Over the years this method proved more satisfactory than a commonly used glass test tube or a gloved finger tip for obtaining samples.

2. *Counting oocysts*

Boughton (1943) strained a level tablespoon of feces through a 60-mesh sieve with a liter of water under pressure, allowed it to sediment overnight, poured off all but 150 ml, after stirring removed a 15-ml sample, centrifuged, poured off the water, added concentrated sugar solution, centrifuged again, removed a sample from the surface with a wire loop 5 mm in diameter, and examined it as a hanging drop. The resultant count represented an estimated number of oocysts in a loopful from a flotation of 0.1 tablespoon of fecal sample. This gave a rough approximation of the number of oocysts present, and could be used to compare samples.

The original McMaster helminth egg counting method (Gordon and Whitlock, 1939) has been modified by several researchers. Whitlock (1948) made the following changes to increase the speed and maintain the accuracy of the original technique:

1. Weigh a 2-g sample into a 2 fl oz jar and add 20 ml of water. Allow to stand from one to two hours or overnight.
2. Add saturated NaCl solution to make up to a volume of 60 ml. Stir with an electric mixer.
3. Use a sieve pipette to add the mixed sample to the McMaster counting chamber by allowing the suspension to run between the base plate and the top piece. Allow the oocysts to rise to the under surface of the top piece of glass.
4. Adjust the draw tube of the microscope to give a 4-mm diameter field of view. Count all oocysts observed in two traverses between the parallel lines made on the under surface of the top piece. Multiply the number counted by a factor of 200 to obtain the number of oocysts per gram in the original sample.

If commercially made McMaster counting chambers are not available, the researcher can make his own by referring to the two original references listed above. Whitlock (1948) gave details for making a glass cement that allows the finished counting chamber to be washed in hot water and does not become brittle on aging. The McMaster method of counting is commonly used by many workers. Long and Rowell (1958) evaluated it statistically. Peters and Leiper (1940) found less variance in the McMaster method than in the Stoll dilution method (Stoll, 1923), for which the coefficient of variation ranged from 11% to 81%. Ryley (1971, personal communication)* uses a 12-lane McMaster slide to coincide with the field of view of the microscope objective used in counting.

Levine (1939) developed a modification of the commonly used direct centrifugation-flotation method. Instead of lifting the coverslip for examination after centrifugation, he examined the coverslip without removing it from the flat-bottomed shell vial developed for this purpose. Ray (1953) reported that 30% of the oocysts remained behind after the first coverslip was removed. Levine *et al.* (1960) found that the McMaster method was faster and more accurate than the direct centrifugation-flotation method.

*John F. Ryley, Imperial Chemical Industries Ltd., Pharmaceutical Division, Alderly Park, Macclesfield, Cheshire, England.

Dorney (1964) developed a modification of the direct centrifugation-flotation (Levine vial) method for use in small mammal research, in which only small fecal quantities of less than 0.1 g are available. This method is capable of detecting low numbers of oocysts and of providing a reasonably accurate estimate of total numbers. Dorney made 5.5-ml flat-bottomed vials 34 mm in height with an inside diameter of 14.2 mm from standard 16-mm test tubes. Later, a commercially manufactured vial 30 mm in height (Kontes Glass Company, Vineland, New Jersey) proved satisfactory and had greater stability. Dorney used Levine vials cemented to microscope slides for stability and after centrifuging and adding sugar solution, the material was resuspended with an applicator stick. More sugar solution was added until the meniscus was just slightly above the rim level and a 22 mm × 22 mm coverslip was placed on top. The sugar solution was permitted to seal the coverslip to the top of the vial by allowing it to stand motionless for ten minutes. Centrifugation was done at 1,250 rpm for ten minutes. The entire vial with coverslip was placed on a microscope slide with a drop of sugar solution to help cement it more firmly to the slide. Two vials were placed on one slide and the coverslips were examined microscopically. The light source had to be increased, so a 35-mm slide projector with a 300-watt lamp was used. Removal of the substage condenser allowed the focal point of the light to be adjusted to a higher level. In a comparison of different methods, Dorney found that with small volumes of suspension the coefficient of variation was higher for hemacytometer counts than for counts made with the modified Levine vial, Stoll, or McMaster methods.

E. SPORULATING OOCYSTS OF COCCIDIA

1. *Sporulating oocysts of coccidia with aeration*

Goff (1942) reported the use of fine sieves to reduce the amount of fecal matter in isolating oocysts from chickens. Smetana (1933) aerated suspensions of oocysts by bubbling air through material to bring about sporulation in work on coccidia of rabbits. Boughton (1943) used compressed air for sporulation. Hammond and Davis (1944) reported a modification of both of these systems and found increased percentages of sporulation and more uniform experimental infection resulting from inoculation of oocysts sporulated by their method. Samples with oocysts of the species desired were washed through a 100-mesh sieve, followed by a 200-mesh and then a 325-mesh sieve (44-μ openings). The suspensions were allowed to sediment between the sievings. The suspensions in 2% potassium dichromate were placed in round-bottomed tubes 5 cm × 60 cm with a capacity of 675 ml. A small aquarium aerating pump with connecting tubing carried air to the bottom of each tube, where a stone or porous air breaker was used. When the pump was operating, streams of fine air bubbles produced adequate agitation and aeration. One pump could aerate three or four tubes. Paper was fastened over the top to prevent spattering. Small squares of tinfoil were later found to be superior to paper. Siliconizing the inner surfaces of the tubes would prevent some of the suspension from adhering to the sides while bubbling.

2. *Sporulating in tap water*

Lotze and Leek (1961) used tap water for sporulating oocysts of coccidia. Fecal material was washed and mixed in tap water. Solid particles were screened out with 200- and 400-mesh screens until the sieved liquid became nearly clear. An electric drill with a special stirring blade was used to break up the suspended material above the sieves. Centrifugation was used. The oocysts were freed of water-soluble substance by continuous centrifugation at from 1,000 rpm to 2,000 rpm. The concentrated oocysts were transferred to centrifuge tubes or bottles for washing and were stored in tap water or were poured into glass or stainless steel containers to a depth of not greater than ⅜ in of liquid. The cultures were sporulated in tap water at room temperature, being careful to keep the samples moist. Since no antiputrefactive agents were used, free-living protozoa, rotifers, and algae sometimes developed.

3. Temperature and sporulation of oocysts

Edgar (1954) described simple methods of determining the time required for sporulation of *Eimeria tenella* oocysts at various temperatures. Infected chicks were killed on the eighth day to obtain unsporulated oocysts from the ceca. The suspensions were placed in 2.5% potassium dichromate, stirred for one minute in a Waring Blendor, strained to remove large particles, and divided into several test lots. Each lot was poured into a sterile petri dish and sporulated at a specific temperature, ranging from 8°C to 41°C (±1° C). Results showed that the optimum temperatures in the preliminary tests were between 20° C and 32° C, so 20° C, 24° C, 28° C, and 32° C were used in the final tests. Infectivity of the oocysts was determined by feeding a portion of each lot to three coccidia-free chickens of from two to three weeks of age. Edgar found that 29° C was optimum for most rapid sporulation; the oocysts reached the infective stage between 15 and 18 hr at that temperature. Edgar pointed out the necessity of sporulation under regulated conditions in the production of consistently potent oocyst suspensions. He emphasized the necessity of controlling the range of temperature used for sporulation, rather than conducting sporulation at variable room temperature in producing oocyst cultures for critical tests.

4. Sporulation of oocysts in large volumes (David J. Doran)

Sporulation of oocysts is sometimes necessary in order to identify a species and is essential in order to obtain sporozoites for experimental work. The process is most frequently carried out by placing feces suspended in from 2% to 4% potassium dichromate or other antimicrobial agents into petri dishes or other flat containers to a depth of a few millimeters. Oxygen, which is required for sporulation, reaches the oocysts by diffusion. This procedure is satisfactory for small quantities of feces, but becomes impractical when quantities of fecal material larger than one liter are to be handled. For large amounts, columns through which air is forced to achieve aeration have been used (Boughton, 1943; Hammond and Davis, 1944). Although the columns can handle from 800 ml to 1,000 ml of fecal suspension, the procedure has limitations because (1) large particles in the suspension must be removed by sieving beforehand, and (2) the diameter of the column has to be small so that the stream of bubbles can keep the fecal suspension agitated. Vetterling (1969) described a method for sporulating poultry coccidia in large volumes of fecal material. During the first three days of the patent period, droppings from 5 to 15 birds are collected in pans containing 2.5% potassium dichromate. These are diluted to nine liters with dichromate and placed in a nine-liter spinner-flask. The suspension is then constantly stirred and aerated for three days at 21° ± 2° C. It is not necessary to remove the large fecal particles by sieving because stirring circulates the air to all parts of the container. The method routinely results in better than 85% sporulation.

5. Testing for sporulation

Oocysts in feces were prepared for sporulation studies (Davis *et al.*, 1955) by using water pressure to force them through a 60-mesh sieve, washing three times in tap water and centrifuging to remove floating material. A siphon was used to remove oocysts floated to the surface of saturated sugar solutions. Instead of using a straight glass tube at the end of rubber tubing for the siphon, the glass tube was bent to form a J-shape, so as to reduce the amount of sugar solution siphoned with the oocysts from the surface layer. Sugar was removed by washing and centrifugation, and the oocysts were then placed in 2% potassium dichromate solution. A drop of the resulting suspension was placed on each of 12 coverslips which were supported on rubber stoppers above a layer of distilled water in a glass dish, which was then covered to reduce evaporation. A coverslip was removed each day to determine the progress of sporulation. The tests were conducted at room temperature or at specific temperatures in incubators.

6. *Testing for survival of oocysts in the field*

Farr and Wehr (1949) summarized longevity studies made on oocysts exposed on soil under various field conditions. The authors obtained droppings from chickens known to be passing oocysts in great numbers. The droppings were mixed thoroughly, weighed into equal parts, and distributed over six different plots. At various intervals, infectivity of the oocysts in the soil was tested by obtaining soil samples down to a depth of one-half inch obtained from three or four places in each plot and feeding to coccidia-free chickens. The chickens were killed at the end of the seventh day after feeding the samples and the small intestines and ceca were carefully examined for coccidia. When no lesions were noted, the intestinal and cecal contents were examined by sugar flotation to recover oocysts. Infective oocysts of *E. acervulina* were recovered from the plots as long as 86 weeks after deposition. *Eimeria tenella* and *E. maxima* disappeared from all plots in less than a year's time. No severe infection with any of the three species was produced after 34 weeks.

F. PREPARATION OF PURE OOCYST SUSPENSIONS

1. *Preparing bacterially sterile oocysts of coccidia*

Smith and Herrick (1944) described methods of preparation of unsporulated oocysts by removing cecal pouches from chickens killed on the ninth day after infection. The cores and scrapings of the pouches containing the oocysts were placed in the Waring Blendor for from three to five minutes to separate the oocysts from the cecal material. The volume was increased with water three times, and the suspension was placed in a separatory funnel. After 30 min, the sediment was transferred to 50-ml centrifuge tubes and washed free of extraneous material by centrifugation with tap water. Saturated sodium chloride solution was added to the sediment, and the floating oocysts were removed with glass tubing. The oocysts were centrifuged again with sterile distilled water to free them of the concentrated salt solution. Only a few bacteria remained, and these were removed by 5% antiformin made up in 10% formalin and allowed to stand for five minutes. The antiformin-formalin solution was removed by centrifugation and the sterility of the suspension was tested by placing 0.1 ml in tubes of nutrient agar.

2. *Cleaning coccidial oocysts by density-gradient sedimentation*

Sharma and Reid (1963) produced pure suspensions of oocysts free of foreign material. Two Sedgwick-Rafter funnels were connected with tubing; 100 ml of saturated sucrose solution was placed in one funnel and the other contained distilled water. A jet of air was introduced into the bottom of one funnel to mix the two solutions. Near the bottom of the funnel containing sugar solution, an outlet with a stopcock was used to drain off fluid into a 250-ml centrifuge bottle. The first layer drawn off contained saturated sugar solution. The other layers showed a decreasing density as the water from the second funnel diluted the sugar solution. After draining off 220 ml, the bottle contained a layered continuous gradient in decreasing density from the bottom to the top. Oocysts were concentrated by washing them through different sieves. With intestinal scrapings, the tissues were removed by the peptic digestion method of Rikimaru *et al.* (1961). Twenty milliliters of oocyst suspensions were placed on top of a layered continuous gradient of sugar solutions. Bottles were handled carefully to prevent mixing and were centrifuged for six minutes at 1,100 *g* to produce layering and concentration of the oocysts. A syringe can be used to remove the second layer from the top, which contains concentrated pure oocysts. If necessary, the suspensions can be purified even more by overlayering on a second fluid gradient.

3. *Purifying coccidian oocysts*

Wagenbach *et al.* (1966) devised a method of producing purified oocysts of coccidia. To tissue homogenates containing oocysts, from 20 to 30 ml of 1:1 Clorox (5.25% sodium hypochlorite) were added per 10 ml of suspension, and the mixture was stirred in an ice bath for from 10 to 20 min. The mixture was passed through a 140-mesh screen and centrifuged at 200 *g*. The oocysts were resuspended in 20% (w/v) sodium chloride, then removed with a spatula and the suspension was washed by centrifugation to remove the Clorox. The suspensions were bacteriologically sterile and almost free of debris. Additional debris was removed by adding from 80 to 100 ml of cold acid-dichromate solution to concentrated suspensions of oocysts. (The solution was made by adding 174 ml concentrated sulfuric acid to 100 ml of 20% (w/v) sodium dichromate and filtering through glass wool.) While adding the acid, the suspension was stirred continuously in Erlenmeyer flasks in ice. The authors centrifuged at 250 *g*, removed oocysts with a spatula as before, and resuspended quickly in a mixture of ice and water. Washing in 10% sodium carbonate was used to remove traces of dichromate. The oocysts were left in contact with the acid less than 30 min.

Hammond *et al.* (1968) modified Jackson's (1964) method of cleaning debris from oocysts. Twenty-five milliliters of oocyst suspension were mixed with 25 ml of Sheather's sugar solution in a glass petri dish, 90 mm in diameter. After from five to ten minutes, the bottom portion of a plastic petri dish that would fit inside the glass petri dish was placed with the bottom downward on the surface of the suspension and allowed to stand for one hour. The plastic dish was lifted gently and the adhering oocysts and liquid were washed into a beaker. The process was repeated once. The authors centrifuged the mixture several times to remove the sugar and the suspension was placed in Ringer's solution.

4. *Obtaining pure oocyst suspensions* (David J. Doran)

In order to carry out many types of research on coccidia, oocyst suspensions must be free of fecal debris and bacteria. Several flotation techniques, i.e., coverslip flotation (Lane, 1924), gravity pan flotation (Jackson, 1964), linear gradients (Sharma and Reid, 1963), and zonal gradients (Marquardt, 1961; Patnaik, 1966), can be used for separating oocysts from fecal debris. All of these techniques are satisfactory in varying degrees for small quantities of fecal material, but require sieving of the material and considerable manual dexterity. Vetterling (1969) reviewed these methods and compared them with his modification of the zonal gradient technique of Marquardt (1961). The modification was found better for recovering oocysts from quantities of fecal material less than one liter. The technique is as follows:

a. Mix sieved fecal material with an equal amount (v/v) of 2 M sucrose.
b. Place 200 ml of this mixture into a 250-ml centrifuge bottle.
c. Layer with 25 ml of 0.5 M and 0.2 M sucrose and pass a wire spring quickly down through the layers.
d. Centrifuge at 1,000 *g* for ten minutes at 20° C.

After recovering oocysts by any of these methods, they can be rendered bacteria-free by the following method, which is essentially that of Jackson (1964).

a. Sediment oocysts in a centrifuge tube (preferably screw-top) having a volume of 40 ml or greater. More than 1 ml of packed oocysts is not recommended.
b. Decant and add 20 ml of 5.25% sodium hypochlorite (undiluted Clorox).
c. Resuspend oocysts and allow them to stand for from 20 to 40 min. Shake the tube to resuspend the oocysts every ten minutes or so.
d. Add 20 ml or more of sterile water.
e. After sedimentation, pour off the supernatant and wash with sterile water until the supernatant appears clear.

Vetterling (1969) described a continual-flow differential density flotation (CFDDF) method for recovering oocysts in a bacteria-free state from large volumes of fecal material. No sieving is required and approximately 85% of the oocysts can be recovered. In this method, the fecal suspension is processed through the following steps using a chemical centrifuge having an unperforated basket with an influent reservoir and an effluent reservoir (Fig. 1).

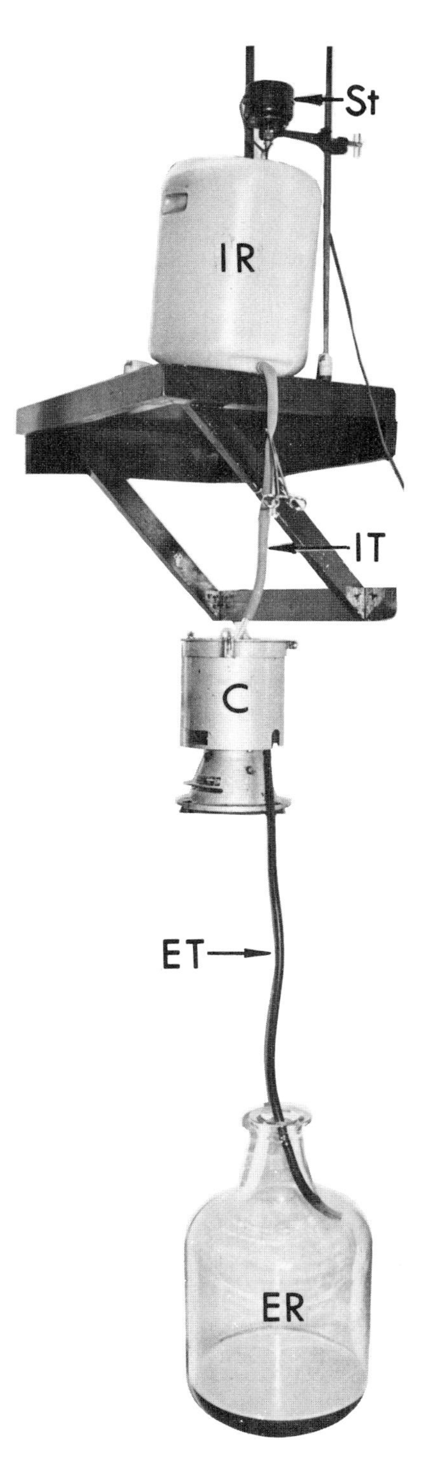

Fig. 1. Chemical centrifuge set up for continuous-flow differential density flotation of oocysts of coccidia. St, stirrer; IR, influent reservoir; IT, influent tube; C, centrifuge; ET, effluent tube; ER, effluent reservoir (Vetterling, original.) (Modified from Vetterling (1969), J. Parasitol. 55, 414, Fig. 2. Appreciation is expressed to the editor, Justus F. Mueller, for permission to use the modification, and to J. M. Vetterling for furnishing the illustration.)

a. Mix four liters of fecal suspension taken from a nine-liter sporulation flask with four liters of 2 M sucrose solution (D^{20}_4 1.130) and pour it into the influent reservoir. Start the stirrer and centrifuge. When the centrifugal rate has reached 1575 *g*, open the tube leading to the centrifuge and allow the suspension to run through the basket at a rate of one liter/min. The supernatant fluid is collected in the effluent reservoir. Upon completion of the run, the contents of the basket are discarded.
b. Add two liters of water to the effluent reservoir to dilute the supernatant from step 1 to the desired density (D^{20}_4 1.106). Pour this into the influent reservoir and centrifuge as above. The sediment in the basket is discarded.
c. Mix supernatant with six liters of water (D^{20}_4 1.067), centrifuge as above, and discard sediment.
d. Mix supernatant with 24 liters of water (D^{20}_4 1.030), centrifuge as above, and discard supernatant. Set sediment aside and repeat steps 1-4 until all fecal material is processed.
e. Take sediment from several runs (step 4) and add water to make four liters, mix with two liters of 2 M sucrose solution (D^{20}_4 1.067), centrifuge as above, and discard sediment.
f. Mix supernatant with 40 liters of water (D^{20}_4 1.01), centrifuge as above, and discard supernatant. Set sediment aside and repeat steps 5 and 6 until all material from step 4 is processed. Combine sediments from step 6 and repeat steps 5 and 6.
g. Suspend sediments in undiluted Clorox at 4° C (D^4_4 1.085), incubate for five minutes at 4° C, centrifuge as above, and discard sediment.
h. Mix supernatant 1/6 (v/v) with sterile water, centrifuge as above, and discard supernatant.
i. Resuspend sediment in sterile water and place in sterile centrifuge tubes with caps, centrifuge at 500 *g* for 15 min at 4° C, and discard supernatant. Repeat until odor of chlorine is absent.

The only drawback to using Clorox is that it alters the oocyst wall. Oocysts treated with Clorox are satisfactory for most experimental work, but should not be used for studying the chemical or physical nature of the oocyst wall.

5. *Oocyst isolation techniques* (John F. Ryley)

In any program of coccidiosis research, it is necessary to isolate oocysts from tissues or feces as a source of infective material or an object of study; the degree of purity required will depend on the use to which the material will be put. In a large-scale drug screening program, or with biochemical investigations of oocysts or sporozoites, where large amounts of material are required, the efficiency of the oocyst production method is of paramount importance. The following notes, based on experience with *E. tenella*, indicate some of the procedures which can be used to obtain the maximum number of oocysts of a required degree of purity and viability with the expenditure of the minimum amount of physical effort, time, animals, and chemicals. The various stages in a flotational isolation from feces will be discussed and attention drawn to some of the factors which influence oocyst yield at each stage; the optimal procedure will then be described. The methods can be adapted for use with other species if individual species characteristics are investigated at the several stages.

a. Age of bird and size of inoculum

The larger the bird, the greater is the area of intestinal epithelium available to support coccidial development, and presumably the greater the output of oocysts per bird. Older birds, too, are thought to be more susceptible to infection than younger ones. The older the bird, however, at the time of infection, the greater will have been the effort in raising it coccidia free to that age, and the greater the risk of adventitious coccidial infection, and therefore contamination of the strain being multiplied. We find the optimal compromise to be from three to four weeks at the time of infection. Above this age we found a decrease in oocyst yield from any inoculum when measured in terms of the weight of chicken infected. Inoculation of birds with small numbers of oocysts allows a relatively large production of oocysts in terms of numbers inoculated; large inocula may result in the death of birds before oocyst production starts, or at least sufficient parasitization and destruction of intestinal epithelium by the earlier asexual stages to prevent optimal sexual development. As the actual numbers of oocysts used for inoculation of passage birds are insignificant, we seek to define the inoculum which gives the biggest absolute production of oocysts. For birds from three to four weeks of age, experiment showed that total oocyst production was approximately constant over the inoculum range from 1,000 to 5,000 oocysts; above this, mortality resulted in chicken wastage and a decrease in absolute oocyst production caused by tissue damage, while below this level the full productive capacity of the epithelium was not realized.

b. Feeding status at infection and during collection

It has been suggested that a more uniform infection with a larger percentage "take" results following starvation of chicks for several hours prior to infection. With less material in the intestine to traverse, it might be easier for the sporozoite to gain access to the gut wall; on the other hand, gut digestive activity necessary for oocyst excystation will be at a low ebb. Our experiments have shown that starvation for up to 24 hr before infection can boost oocyst production up to 90%, but effects were extremely variable and inconsistent. Starvation before infection may be recommended, but is not critical.

By the time oocyst production is starting, withholding food might not be expected to influence the total oocysts shed. It might, however, influence the oocysts recovered, as starved birds should produce less feces containing a greater concentration of oocysts, which might in turn lead to more efficient isolation, and would certainly cut down the physical effort involved. Unless a very efficient type of hopper is used, food readily gets into the collection trays; this both increases the bulk of material to be processed and often causes problems in getting pure preparations of oocysts because of difficulties in removing starch grains. Under conditions where feces were collected over a three day period, and food was given during the collection period for one-half hour per day rather than absolute starvation, experiments have shown that the absolute production of oocysts was 20% greater under conditions of continuous feeding compared with production during restricted feeding. In contrast, restricted feeding during collection reduced the total bulk of material to be processed from one-fifth to one-quarter that produced with continuous feeding, so that obviously the small loss in oocyst production per bird was more than compensated for by the marked reduction in effort during isolation.

c. Mode and period of collection

Although oocyst production from infected birds may continue for a prolonged period, levels for much of this time are so low that the effort involved in extraction is not justified by the numbers of oocysts obtained. Experiments showed that when birds were inoculated on day zero, and oocyst production was followed over days 5 through 10, 32.8% of the total output was recovered in the days 5 through 6 collection, 41.6% of the total over days 6 through 7 and a further 12.9% over days 7 through 8. Thus three-quarters of the total output would be recovered over days 5 through 7. We currently consider it worthwhile to process feces excreted over days 5 through 8.

It is thought preferable to collect feces for oocyst isolation into liquid, rather than collecting dry, since drying out may impair oocyst viability, and certainly makes it more difficult to prepare the homogeneous suspension necessary for subsequent isolation procedures. Following collection of feces into water, unprecedented difficulties may be experienced at the salt-flotation stage because of a mechanical trapping of oocysts with unwanted fecal debris. This is apparently the result of mucilaginous material produced by bacterial growth during fecal decomposition in water, binding oocysts to grosser debris. Although the binding can be overcome by acidification at the flotation stage, the smell generated is so revolting that the method is unworkable. If feces are collected into dichromate solution (from 0.5% to 2.5%), this trouble does not arise, and a three-day collection can be made and processed all at once instead of processing daily.

d. To sieve or not to sieve, that is the question

In processing feces obtained over days 5 through 8 from birds on restricted feeding it was found that passing the fecal homogenate once through a 40-mesh sieve followed by once through a 100-mesh sieve eliminated 67.5% of the solids requiring subsequent processing, coupled with a loss of 25% of the oocysts initially present. A further 8% of the oocysts could be recovered by resuspending the material on the sieves in water and resieving. This, however, almost doubled the volume of material to be processed subsequently. The use of a continuous flow centrifuge and a massive dilution of feces has been advocated as a method of avoiding sieving with its consequent loss of oocysts. We consider that the time involved in centrifuging the vast quantities of dilute feces and the amount of dichromate used is not justified by the increased recovery of oocysts. We consider the more profitable method to be the use of a concentrated fecal homogenate which is passed once through a 40-mesh sieve and once through a 100-mesh sieve.

e. Flotation

Although oocysts can be isolated by repeated washing and sedimentation, by far the most rapid way of producing oocysts in a reasonable degree of purity and in good yield is by flotation. Saturated sodium chloride is much cheaper than either sucrose or zinc sulfate, and does not have the stickiness associated with sucrose. Although plasmolysis and collapse of the inner oocyst wall can occur on prolonged contact with saturated salt, the effect is reversible in water and does not necessarily result in permanent damage or loss in viability. By centrifuging at the salt flotation stage and by washing the separated oocysts in the centrifuge, freeing them from salt, duration of contact with salt is very limited, and no damage is likely to result. (It should be remembered that other species may be more sensitive to saturated salt than *E. tenella*.) Although oocysts float in saturated salt solution, and indeed in salt solutions above 66% saturation, some invariably get trapped mechanically in the unwanted debris which sediments. The extent of this trapping is less at acidic than at alkaline pH and, further, oocysts can be recovered by resuspending the deposit and recentrifuging it.

f. Sporulation

Sporulation is a strictly aerobic process, and it can be carried out prior to or after oocyst isolation. It can be achieved under static conditions provided adequate control over microbial growth, which would result in oxygen depletion, is exercised, and surface area/depth conditions are suitable to allow adequate diffusion. Alternatively, forced aeration may be used. Sporulated oocysts are somewhat less dense than unsporulated ones; neither, however, has any difficulty in floating in salt solutions above 66% saturation. Some workers feel that sporulation should be carried out as soon as possible if maximum viability is to be achieved. While sporulation prior to isolation may be in order when the isolation methods used involve long washing and sedimentation procedures, isolation by sieving and salt flotation is a matter of but a few hours, and it is far simpler to sporulate a small volume of clean oocyst preparation than to forcibly aerate vast amounts of fecal homogenate for several days.

g. Recommended procedure

Groups of 15 chicks from three to four weeks of age are caged, and after withholding food overnight, are infected with 5,000 sporulated oocysts of *E. tenella* (day zero). On day 5 of infection, food is withdrawn, the trays under the cages are cleaned out, and 1.5 liters of 2% aqueous potassium dichromate solution are added. The birds are subsequently provided with food for one-half hour each day. On day 8, the birds are killed, and the three-day fecal collection in dichromate is homogenized for three minutes with a Vortex mixer. The resulting slurry is passed once through a 40-mesh and then once through a 100-mesh sieve with the help of a 3-in paint scraper. The slurry is centrifuged for three minutes at 1500 rpm in 250-ml glasses or 700-ml plastic bottles, and the supernatant discarded. The sediment is resuspended in saturated NaCl so that the bottles are about two-thirds full. After centrifuging again for three minutes at 1000 rpm, the oocyst-containing scum is removed from the surface by means of a syringe fitted with a wide-bore cannula (1.5 mm ID), and is squirted into water. The residue in the centrifuge bottle is resuspended and more saturated salt added to restore the volume. A second collection of oocysts is recovered following centrifugation. The two collections of oocysts are resuspended with water and centrifuged. The sediment is subsequently made up in 2% potassium dichromate to a concentration not exceeding 5×10^6 oocysts/ml. The suspension is aerated at 30°C for three days in a measuring cylinder by means of an air line fitted with a sintered glass disc. Alternatively, sporulation may be carried out under static conditions in open dishes or flat-bottomed culture vessels, provided the oocyst concentration does not exceed 10^6/ml and the depth of liquid does not exceed 5 mm; if a greater depth of liquid is used, the oocyst concentration must be reduced correspondingly.

Oocyst preparations isolated by this method will still be contaminated by some fecal material, but at least 50% of the solid material will be oocysts. Such preparations are eminently suitable for animal infection, but not for culture inoculation or biochemical studies. Although further purification can be achieved by repeated salt flotation, this is not the most convenient or efficient method. Two methods are available which will give completely pure preparations of oocysts; the method used depends on the use to which the oocysts are to be put.

h. Hypochlorite method

When oocysts are required for *in vitro* production of sporozoites suitable for inoculation of tissue cultures or embryos, or for biochemical studies, this is the preferred method, since it is the easiest and quickest, and may be carried out under aseptic conditions. It does, however, result in oocysts with the outer coat of the shell stripped off, and so is unsuitable for studies involving the oocyst wall. Oocysts are treated with hypochlorite at room temperature. This dissolves the outer layer of the oocyst wall, but does not harm the contents in any way. Bacteria and other fecal debris are destroyed and partially degraded and dissolved. Provided the sodium hypochlorite concentration is high enough, centrifugation after treatment results in a pure layer of floating oocysts, the degraded unwanted debris being sedimented.

Crude oocyst material is suspended in 0.8% saline and mixed with half its volume of 30% sodium hypochlorite solution. After standing for from 20 to 30 min at room temperature, the digest is briefly centrifuged. The pure white oocyst scum is removed with a syringe and wide-bore cannula and washed by centrifugation with water until free from the smell of hypochlorite.

i. Glass bead column and gradient centrifugation

For studies on the oocyst wall where stripping with hypochlorite is not acceptable, a somewhat more tedious method is required. In our hands, oocyst purification with continuous or discontinuous density gradients is not satisfactory unless a preliminary washing stage is used. A column of glass beads (Superbrite, type 100-5005; Minnesota Mining and Manufacturing Co.) 10 cm in diameter and 8 cm deep is supported on a sintered glass disc, porosity 3. A crude concentrated oocyst preparation is shaken briefly by hand with 7 mm glass beads in 5% Tween 80 to disperse thoroughly, and is then loaded into the column. The oocysts are washed through the column with two liters of 5% Tween 80 and the effluent concentrated by brief centrifugation. Twenty-milliliter amounts of this are then layered onto continuous sucrose gradients (SG1-1.15) in 250-ml centrifuge bottles and spun at 1500 rpm for eight minutes. The oocyst band which forms partway down the bottle is recovered with a syringe and diluted with five volumes of water. Final cleaning of the oocysts is achieved by three washings with large volumes of water in the centrifuge, centrifugation being carried out at the slowest speed attainable – sufficient to sediment most of the oocysts, but not the residual finely particulate debris. An overall yield of 25% of the original oocysts may be expected from combined column and gradient purification.

6. *Trypsinization of tissues to release oocysts* (P. L. Long)

a. Harvesting of oocysts from infected ceca

The infected ceca were removed from the bird and chopped into small pieces into an Atomix tissue homogenizer filled to the cutting blades with phosphate buffer at pH 8.0. The pieces were then homogenized for from three to four minutes with repeated treatments if needed, until most of the tissue was macerated. This coarse material was then filtered through cheesecloth to remove any large pieces of tissue and the filtrate treated with from 0.5% to 1% trypsin (Difco 1:250) for about 20 min at 39°C in a shaking water bath.

When most of the tissue had been degraded and the oocysts released, the mixture was removed from the water bath and washed in water by repeated centrifugation. A short hypochlorite treatment followed by salt flotation (using saturated NaCl solution) was suggested, as the culture by this time was not sterile and still contained cell debris. If hypochlorite was used, the oocysts were subjected to a 1% final strength solution of sodium hypochlorite in an ice water bath for not more than 15 min, as any longer treatment was likely to cause adverse effects on the oocyst wall. Repeated centrifugation in sterile water washed the oocysts clean of residual chlorine.

Subsequent salt flotation gave a very clean culture and for this the oocyst deposit was suspended in a saturated solution and slowly centrifuged in 15-ml glass centrifuge tubes with an 18 mm × 18 mm cover glass on top. The oocysts were removed from the cover glass and the top of the salt solution. Repeated centrifugation in water gave a salt-free culture, which at this stage was sterile. The oocysts were sporulated by making up into a 2% potassium dichromate solution and incubating at 28° C for 48 hr (during the first 24 hr the oocysts were aerated with a simple air pump) after which they were stored at 4° C.

b. Harvesting of oocysts from embryos

Infected embryos were cut longitudinally with scissors and the urate deposits and desquamated epithelium removed with forceps, and placed in phosphate buffer, pH 8.0. Before treatment with trypsin, the material was broken with a magnetic stirrer to produce a larger surface area on which the enzyme could work.

Trypsin (Difco 1 : 250) was added at a rate of 0.5% w/v and the mixture incubated in a shaking water bath at 39° C for about 20 min or until most of the oocysts were released. The suspension was filtered through cheesecloth to remove any large particles unaffected by trypsin, and the filtrate washed twice in water by centrifugation.

At this stage, oocysts from embryos were sporulated as before (*see* Harvesting of oocysts from infected ceca). If any subsequent treatment proved necessary, the sporulated oocysts, unchanged by most laboratory procedures, may be given hypochlorite treatment followed by salt flotation (both as before). Salt flotation in 15-ml glass centrifuge tubes with cover glasses on top was a means of separating sporulated oocysts from unsporulated oocysts. The sporulated oocysts rose to the cover glasses (Fig. 2), while the damaged and unsporulated oocysts remained as a deposit at the bottom of the tube (Fig. 3). The sporulated oocysts were collected from the cover glass and top centimeter of the salt solution, washed several times in water to remove the salt and stored in 2% potassium dichromate at 4° C.

G. OBTAINING FREED SPOROCYSTS OF COCCIDIA

1. Freeing sporocysts from oocysts by crushing

Davis and Smith (1960) obtained freed sporocysts from oocysts of *Eimeria ahsata*, *E. auburnensis* and *E. ellipsoidalis* by placing water suspensions of oocysts on a 3 in × 1 in microscope slide, placing another slide on top and using finger pressure on the top slide with a sliding motion until oocysts were crushed and sporocysts were freed. The sporocysts were then rinsed into small vials for staining. The crushing process can be observed under a microscope to determine when all oocysts have been released.

2. Releasing sporocysts from oocysts by grinding

Oocysts were suspended in distilled water and ground in a Teflon-coated tissue grinder for from five to ten minutes, releasing intact sporocysts (Patton, 1965). Fayer and Hammond (1967) used this method and then centrifuged the suspension at 290 *g* for five minutes. The sporocysts and broken oocyst walls were washed once with saline A, resuspended in excysting medium, and incubated at 37° C for from 2 to 2.5 hr.

H. PREPARATION OF SPOROZOITES OF COCCIDIA

1. Excystation from oocysts

Jackson (1962) used *in vitro* methods similar to those described by Rogers (1960). Jackson found that carbon dioxide is the essential part of the first stimulus required in excystation. Anaerobic conditions and reducing agents, such as sodium dithionite, cysteine and ascorbic acid, assisted in the stimulation. Oocysts incubated in isotonic buffers could be washed in water prior to incubation in bile-trypsin solutions to produce complete excystation. The optimal temperature was between 37° C and 41° C. Oocysts sporulated in potassium dichromate had a lower excystation rate than oocysts sporulated in distilled water. Jackson used an excysting fluid of 0.25% trypsin, 1:250, and 10% sheep bile in a phosphate-buffered balanced salt solution at pH 7.4. The first free sporozoites were found after about 0.5 hr. The process continued for several hours. Both trypsin and bile had to be present for the second stimulus to be effective. The stimulus was effective if the oocysts were placed in bile solution first, washed, and then placed in a trypsin solution. Reversing the procedure was not effective, however. He found that papain could be used to replace trypsin. Sodium taurocholate and Tween 80 reacted similarly to bile. Unless the osmotic pressure and pH of the fluid were kept within physiological limits, the sporozoites were not viable. Jackson placed oocysts in a cellophane sack in the rumen and confirmed Hammond *et al.*'s (1946) observations that oocysts in a cellophane sack in the large intestine of the calf began to break up after a few hours. Jackson observed excystation of sporozoites from oocysts of *E. parva, E. intricata, E. faurei*, and *E. ninakohlyakimovae* from sheep.

Nyberg *et al.* (1968) developed a modification of Jackson's (1962) method for stimulating excystation. Feces of infected chickens were homogenized in a blender for from one to two minutes and suspended in 2.5% $K_2Cr_2O_7$ solution. Sporulation was accomplished at 20° C with continuous stirring. Sporulated oocysts were passed through a series of sieves, concentrated by sedimentation, resuspended in 2.5% $K_2Cr_2O_7$ and stored at 5° C. Oocysts were freed of dichromate by washing via centrifugation and were stored in distilled water. Wagenbach *et al.'s* (1966) method was modified to separate the oocysts from debris. Clorox 1:1-water solution (2.6% sodium hypochlorite) was used with the oocyst suspension in a proportion of 1:10 (v/v) and the mixture was chilled in ice with frequent stirring for a 15 min period. Water was used to remove Clorox and the oocysts were concentrated in a small volume of distilled water. Nyberg and Hammond's (1964) pretreatment method was used with slight modification. Oocyst suspensions were made 0.02 M with cysteine hydrochloride and placed in desiccators containing water-moistened filter paper. A vacuum pump was used to remove air and replace it with controlled atmosphere. At appropriate temperatures the pH was adjusted to 7.0 with 0.2 M Na_2HPO_4 and then to 7.5 with 1.0 N KOH. The oocysts were exposed to 1% trypsin-10% bovine bile made up with 0.85% NaCl and buffered to pH 7.5 with 0.2 M Na_2HPO_4 added to an equal volume of buffered suspension of oocysts.

2. *Purifying sporozoites of* Eimeria tenella

Sporozoites of *Eimeria tenella* can be purified with glass bead columns (Wagenbach, 1969). Oocysts were suspended in Ringer's solution buffered with 0.025 M tris at pH 8.0. Sporocysts were obtained from oocysts by agitating the oocyst suspension with 4-mm glass beads in a Stevens Pipet Shaker (W. M. Welch), or by handshaking with beads. Sporocysts were placed in an excysting medium modified from Doran and Farr (1962), consisting of 0.25% trypsin and 5.0% chicken bile in buffered Ringer's solution and incubated for from 2 to 4 hr at 41.5° C. Washing

Fig. 2. *Eimeria tenella*, mostly sporulated oocysts removed from the top of saturated NaCl tube after centrifugation. × 800. (Long, original.)

Fig. 3. *Eimeria tenella*, mostly damaged, unsporulated oocysts and debris removed from the bottom of saturated NaCl tube after centrifugation. × 800. (Long, original.)

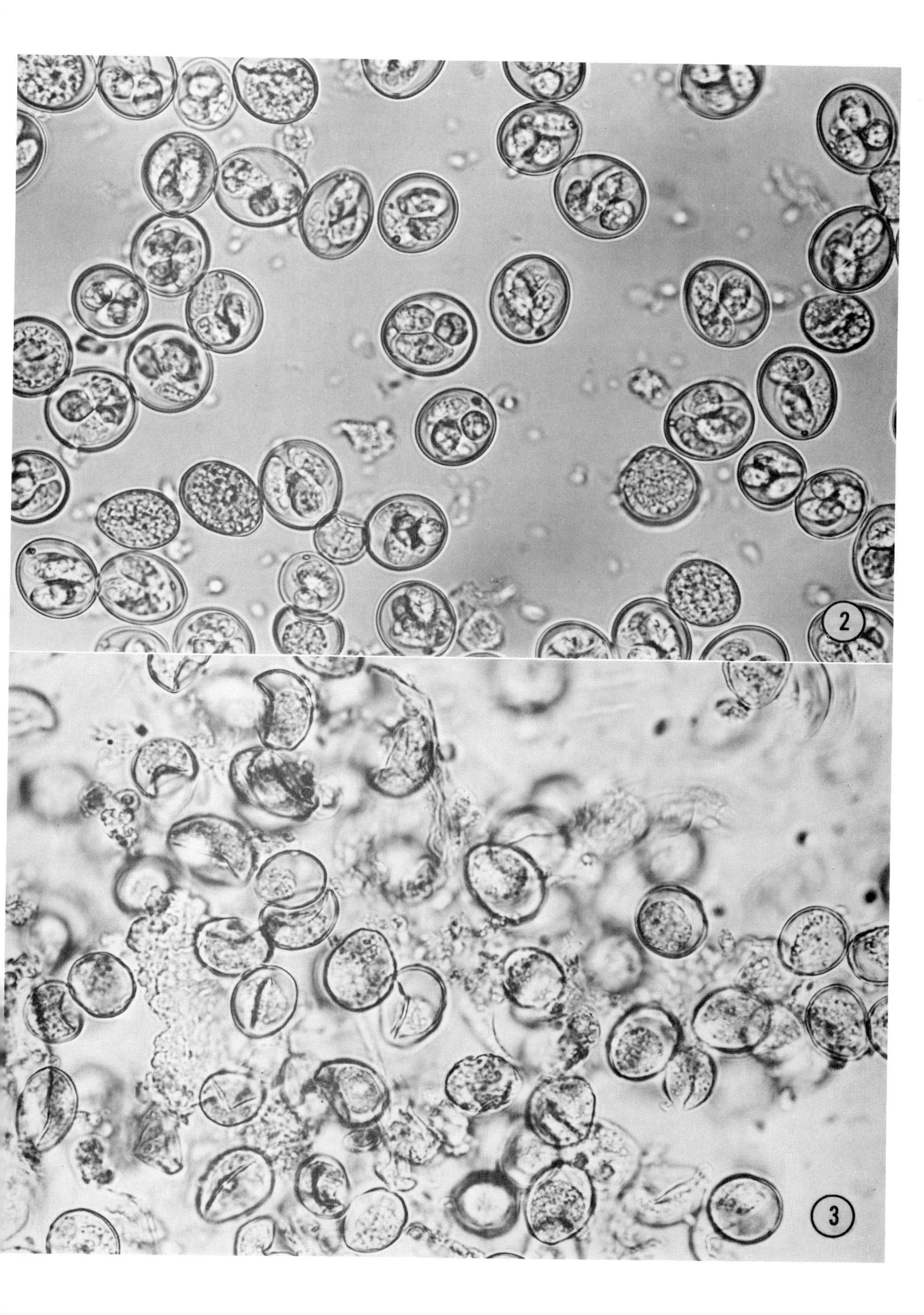
2
3

three times with Ringer's solution removed the excystation medium.

The glass columns were made from Pyrex tubing 2.5 cm in diameter and 16 cm in length and were enclosed in a water jacket to control temperature. Glass beads 200 μ (No. 100 Superbrite, Minnesota Mining and Manufacturing Co.) were placed in the columns at 5-cm, 10-cm, or 15-cm depths. Glass wool formed the bottom layer to retain beads. Concentrated nitric acid was used to wash the beads, followed by tap and distilled water rinses. After the column of beads was washed with Ringer's solution, the cell suspension was poured onto the beads. Ringer's solution was then passed through the column at from 2 to 25 ml/min until elution was completed. Concentration of sporozoites in stock preparation and in the effluent was determined with a cell-counting chamber. Viability of excysted sporozoites was checked in trypan blue 1% solution and by orally inoculating coccidia-free cockerels with 8×10^5 sporozoites each. Doran's and Vetterling's (1968) method was used to preserve sporozoites for long periods by freezing in a solution of dimethyl sulfoxide.

3. Excysting sporozoites (David J. Doran)

In order to obtain large numbers of sporozoites sufficient for experimental work in a short interval of time, investigators have, with few exceptions, mechanically released sporocysts from oocysts before treating with excystation fluid. A good procedure for obtaining *in vitro* excystation is as follows:

a. Pellet the oocysts in a 12-ml conical tissue grinder (Bellco, No. 1977). More than 0.5 ml of packed oocysts is not recommended.
b. After decanting all but a small amount (0.1 ml) of the supernatant, grind the oocysts for about one minute at from 2,500 to 3,000 rpm. If all the supernatant is removed, grinding releases and destroys many of the sporozoites.
c. Add from 5 to 8 ml of distilled water, sediment the ground material, and decant all of the supernatant.
d. Add at least 2 ml of excystation fluid at pH of from 7.3 to 7.8; this consists of:

trypsin 1-300*	0.25 g
NaCl	0.65 g
KCl	0.014 g
$CaCl_2$	0.012 g
bile (undiluted)	5.0 ml
water (distilled)	95.0 ml

e. Resuspend the ground material and incubate at the temperature of the natural host.

With poultry coccidia, the method as outlined is most reliable and can result in as high as 50 million sporozoites/0.5 ml of packed oocysts. To obtain this level, however, the oocyst sample must contain a high percentage (from 90% to 95%) of sporulated oocysts and little or no extraneous fecal debris. The time necessary for a high percentage of excystation depends on the species.

Sodium taurocholate, sodium glycocholate, or ox bile (ox gall) can be substituted for bile, and enzyme preparations other than the one specified can be used. However, a decreased yield can be expected (Doran and Farr, 1962; Farr and Doran, 1962).

4. Obtaining pure sporozoite suspensions (David J. Doran)

After excystation, suspensions contain unwanted debris (oocyst and sporocyst walls). Wagenbach (1969) described a method involving passage over glass beads for separating *E. tenella* sporozoites from this debris. The method is excellent and results in pure sporozoite suspensions. However, the glassware is quite costly. The following procedure, which is essentially that of Wagenbach, is less expensive and also results in high yields of pure sporozoites.

Preparation of bead column (Fig. 4):

a. Treat glass beads with nitric acid and dis-

*Nutritional Biochemicals Corporation

tilled water as prescribed by Wagenbach.

b. With a thin strip of masking tape, mark a point (A) on the glass column 200 mm from the glass wool. Add Ringer's solution (pH 7.0) to the column to a level above (A).
c. Add dried glass beads until they reach (A).
d. Draw off until the liquid and bead levels coincide.
e. Add 15 ml Ringer's. Mark this as (C) and a point halfway between (A) and (C) as (B).

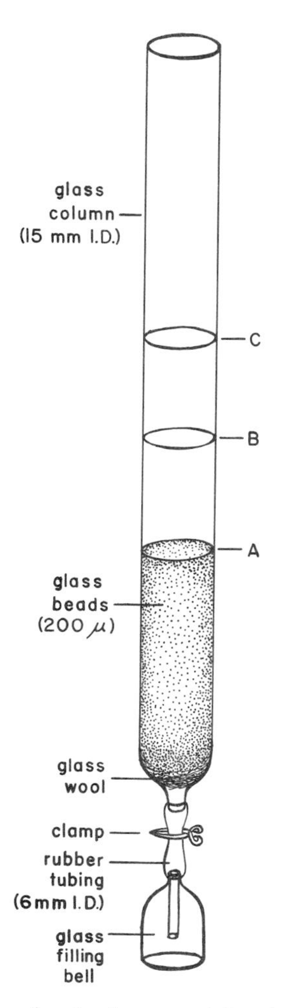

Fig. 4. Glass bead column used for cleaning sporozoite suspensions from debris. (Doran, original.)

Purification technique:

a. Dilute excystation fluid containing oocysts to 15 ml with Ringer's solution.
b. Slowly pour this into the column and draw off and discard fluid until the level in the column reaches (B).
c. Add Ringer's to (C).
d. Draw off to (A) and retain (first half of sporozoite sample).
e. Add 15 ml Ringer's to column.
f. Draw off to (A) again and retain (second half of sporozoite sample).
g. Repeat steps 2 through 6 in two other columns with each of the two halves of the sporozoite sample.

The column can be autoclaved if bacteria-free sporozoites are desired. After cooling, however, the beads must be repacked to mark (A).

The amount of material that can be used without hindering or stopping flow through the column is critical. One should not use material from more than 0.5 ml of packed oocysts; less if the suspension also contains fecal debris.

5. *Obtaining sporozoites for use in cell cultures*

Clark and Hammond (1969) increased the yield of sporozoites from ground oocysts by pretreatment of cleaned oocysts for 24 hr at 37° C in cysteine hydrochloride solution with an atmosphere of carbon dioxide and air 50:50 before grinding and treating with bile-trypsin mixture of Hibbert and Hammond (1968), but there was some indication of a decrease in developmental capability of sporozoites obtained in this way. Clark and Hammond cleaned oocysts by sugar flotations and then subjected them to Wagenbach *et al.*'s (1966) salt flotation method modified by centrifugation at from 300 to 400 *g* for five minutes. They found they could substitute taurocholic acid at 0.5% or 0.75% or glycotaurocholic acid at a concentration of 0.75% for fresh whole bile. Hibbert *et al.* (1969) found that excystation of oocysts of *Eimeria bovis* occurred at pH's of 6.0 through 10.0 with tris-maleate buffer, with the highest levels at from pH 7.5 to pH 8.5. The use of boric acid-borax, ammediol, glycine-sodium hydroxide, and phosphate buffers inhibited excystation or caused sporozoites to disintegrate. Excystation occurred in dilutions of taurocholic, glycocholic, glycotaurocholic, and cholic acids, as well as in dilutions of fresh and lyophilized bile. Unless trypsin was used in conjunction with bile acids, and fresh or bovine bile, no excystation occurred except in fresh bile with a heavy suspension of bacteria and fungi. Satisfactory excystation resulted from treatment of oocysts with a medium containing tris-maleate at from pH 7.5 to pH 8.5, 0.25% trypsin, and 1% of one of the bile acids.

6. *Strout's method of obtaining sporocysts of coccidia*

Strout *et al.* (1965) isolated oocysts from the upper small intestines of chicks killed 106 hr after infection. The oocysts were washed with tap water and agitated by shaking, concentrated by low-speed centrifugation, 175 *g* for ten minutes, and were sporulated in 2.5% potassium dichromate solution for 27.5 hr at 30° C (Krassner, 1963). Dichromate was washed from oocysts, which were then subjected to pepsin digestion (Rikimaru *et al.*, 1961) for 24 hr at 37° C. The solution was removed by washing and the oocysts were bacterially sterilized by six daily changes of 50 ppm of chlorine solution; they were then washed free of chlorinated water with sterile saline. Bacterial sterility was verified in thioglycollate broth at 37° C for seven days. Aseptic *in vitro* excystation of the oocysts was accomplished with trypsin and bile (Doran and Farr, 1962, and Farr and Doran, 1962). The oocysts were exposed to the excystation medium for 40 min at 37° C, and the trypsin and bile were removed from motile sporozoites by centrifugation, and the sporozoites were then suspended in cell culture nutrient. The suspension of sterile motile sporozoites was then used to inoculate cell culture systems.

I. ISOLATION OF MEROZOITES OF COCCIDIA

1. *Obtaining merozoites for inoculations*

Ceca were obtained from chicks from two to three weeks old that were killed from five to six days after inoculation *per os* with from 50,000 to 150,000 sporulated oocysts of *E. tenella* (Bedrník, 1969). Merozoites were released from the ceca in two ways:

a. The ceca were cut longitudinally and their contents, consisting mostly of blood and merozoites, were agitated in the tissue culture medium used as inoculum.
b. The merozoites enclosed in old schizonts and in host cells were released by the use of trypsin. The cut ceca were rinsed with PBS and agitated at intervals in 1.5% trypsin at 39° C for one hour. Trypsin was removed by centrifugation or filtration through membrane filters. Merozoite suspensions were resuspended in tissue culture medium. The second-generation merozoites developed third-generation merozoites, sexual stages, and oocysts in tissue cultures.

Bedrník (1969) suggested that information obtained from the cultivation of merozoites could save time and effort in learning to cultivate coccidia in tissue cultures.

2. *Stimulating motility of merozoites*

Speer *et al.* (1970) examined extracellular merozoites and mature schizonts with merozoites in monolayers, using a phase-contrast microscope. After checking for degree of motility, a solution of trypsin and/or bile salts or bovine bile in minimal essential medium (MEM) was added to the edge of the double coverslip. The solutions included trypsin 0.2% and 4% bovine bile, 0.2% trypsin and 0.5% sodium taurocholate (bile salt), 0.2% trypsin, 0.5% sodium taurocholate, 0.5% sodium glycotaurocholate, 4% bovine bile, and fresh MEM as a control. The pH was adjusted to 7.4 with sodium bicarbonate solution. A warm stage was used to maintain 37° C, or the preparations were examined at room temperature. Three *Eimeria* species from the Uinta ground squirrel, *E. nieschulzi* from the rat, and *E. ninakohlyakimovae* from the sheep were examined. In all preparations to which solutions containing bile or bile salts were added, the merozoites, except for *E. ninakohlyakimovae,* moved within the schizonts and left the host cells. There was a marked increase in motility, and merozoites of *E. callospermophili* penetrated new host cells only when treated with solutions containing bile or bile salts. The authors emphasized that the use of bile salts may play an important role in obtaining further development beyond the first generation of schizonts in cell cultures when researchers work with *Eimeria* species in mammals.

J. CULTIVATING COCCIDIA IN CHICK EMBRYOS

1. Cultivation of Eimeria *stiedai in avian embryos*

Fitzgerald (1970) treated freed sporocysts with a mixture of trypsin and bile to release sporozoites which were then washed and inoculated into the chorioallantoic cavity of ten-day-old chick embryos. Parasites were observed in chorioallantoic membranes of 15-day-old chick embryos five days after inoculation. The inoculations did not produce infections consistently, however.

2. Chick embryo infections (P. L. Long)

a. Embryo maintenance

From the onset of incubation to ten days hence, the eggs were maintained at 100° F (38° C) in an egg incubator in which the eggs were turned automatically. On the tenth day, the embryos were candled with a Melbourne egg candler and sites for inoculation marked on the egg shell.

For intra-allantoic injections a place on the chorioallantoic membrane (CAM), free from any large blood vessels, was chosen and a small hole was made into the shell at that point with a fine pointed dissecting needle. Two thousand i/u of penicillin and streptomycin in 0.05-ml volume were given as a precaution against any possible bacterial contamination in the infective material (sporozoites) and the hole sealed with methylated collodin flexile.

The embryos were then transferred to an incubator at 105° F (41° C) where they were maintained in an upright position and were inoculated with sporozoites on the same or following day (11th day).

Embryos for intravenous injections were given penicillin and streptomycin and sealed as before. On the day of infection (day 10 or 11), they were held upright under a candler and a triangle was marked on the shell above a large vein; the apex of the triangle being directly above the center of the vein. The embryos were then laid horizontally and the sides of the triangle cut using a dental drill fitted with a carborundum wheel, care being taken to avoid cutting the shell membrane. Blunt forceps were used to remove the cut shell, and prior to injection the shell membrane was "cleared" with liquid paraffin. Sporozoites, free from oocyst debris, were inoculated in from 0.05 ml to 0.1 ml volumes in a syringe fitted with a fine hypodermic needle.

K. LOW TEMPERATURE STORAGE OF COCCIDIA

1. *Cold storage preservation of excysting sporozoites*

Activities of coccidia in stages of *in vitro* excystation were arrested by lowering the temperature of the parasites to 40° F; at this temperature parasites remained alive for extended periods, sometimes as long as two weeks (Lotze and Leek, 1967). When placed on a warming stage, excystation proceeded. There was a gradual reduction in the number of oocysts with parasites in stages of excystation after storage of from one to two days, however.

2. *Freezing coccidia*

Freeze preservation of parasitic protozoa has been used successfully as a modern tool for prolonging the life of organisms and reducing the numbers of laboratory animals necessary for maintaining strains of various culture material. Levine and Andersen (1966), Farrant (1966), and Walker (1966) have reported the techniques used in freeze-preserving parasites. Kouwenhoven (1967) reported the freezing of sporocysts of *Eimeria tenella* using dimethyl sulfoxide as a preservative. Norton *et al.* (1968) reported that sporozoites of *E. acervulina*, excysted *in vitro*, with the techniques of Doran and Farr (1962) and Jackson (1964), could be frozen by using the following technique:

Sterile borate buffered solution of salts (BS-2, Lumsden *et al.*, 1963): The BS-2 was mixed with 10% bovine serum, and the mixture was used to suspend freshly excysted sporozoites:

Freezing:

a. To the sporozoite suspension an equal volume of 15% glycerol in BS-2+ serum was added slowly over a period of five minutes.
b. A syringe and needle were used to place 1 ml of the protected sporozoite suspension into ampules, which were heat sealed.
c. Equilibration at room temperature was allowed for 20 min, counting from the time the glycerol solution was first added.
d. The ampules were cooled to -70° C at 1° C/min in a Union Carbide L.R. 35 liquid nitrogen refrigerator, with a Union Carbide B.F. 5 biological freezer operating in the neck of the refrigerator. The ampules were clipped to pre-cooled metal spines which were stored in the liquid nitrogen refrigerator.

Thawing: The suspension of sporozoites was thawed by plunging the ampule into four liters of water at 37° C. The researchers determined that a glycerol concentration of 7.5% was the lowest satisfactory one and this was used in subsequent experiments.

The researchers tested dimethyl sulfoxide (DMSO) and found that the material was ineffective as a freeze-protectant for sporozoites of *E. acervulina*, because resulting infections were very poor. They reported that there was no evidence of decreasing viability or virulence of the sporozoites after a 6-month storage period at -196° C with glycerol. The authors tried the same technique on oocysts and sporocysts with very poor results, obtaining only light infections in three out of five chicks inoculated orally with sporocysts after storage for 80 days. The oocysts stored for the same period produced no infections.

Norton and Joyner (1968) found that their original technique with glycerol as a freeze protectant that worked satisfactorily for sporozoites of *Eimeria acervulina* was also effective with *E. tenella* and *E. maxima*. They found that it was easier to use sporocysts rather than sporozoites because oral inoculations could be used with sporocysts instead of placing them directly into the intestine as was necessary with sporozoites. Norton *et al.* (1968) found that dimethyl sulfoxide

(DMSO) gave good protection with an equilibration time of only 15 min, but glycerol required overnight equilibration. A greater number of parasites survived freezing when sporocysts were used with DMSO as compared with glycerol. The same technique produced good survival rates with *E. adenoeides* and *E. meleagrimitis* of the turkey, *E. phasiani, E. colchici*, and *E. duodenalis* from the pheasant, *E. stiedai* from the rabbit, and *E. ahsata* and *E. arloingi* from the sheep.

Doran (1970) found that when sporozoites of *E. adenoeides* and *E. tenella* which had been frozen and previously unfrozen sporozoites were inoculated into cell cultures of bovine embryonic kidney, more of the latter sporozoites than of the former were intracellular at five hours. More sporozoites that were young when frozen were intracellular at five hours than old sporozoites; however, at 48 hr the percentages of developmental stages from young, previously frozen sporozoites were similar to those from young, previously unfrozen sporozoites and greater than those from sporozoites that were old when frozen.

Doran (1969b) placed oocysts, sporocysts released from oocysts, and excysted sporozoites of *E. meleagrimitis* and *E. tenella* in media containing 7.5% dimethyl sulfoxide and froze them to -80° C at the rate of 1° C/min and stored them in liquid nitrogen vapor for as long as four months.

He found the following:

> Even though the oocysts appeared fresh when frozen, they did not produce infections when fed to their natural host. Sporocysts of both species produced infections after 30 days. The infections were less severe than those of the controls. Sporozoites thawed after four months of storage had a loss of less than 10% per vial. When sporozoites were injected into the duodenum or cecum of the natural host, infections were similar to those of the controls. When frozen sporozoites were thawed and placed in tissue cultures, the development was similar to that of fresh sporozoites. The sporozoites tested at 7½ months were similar to those stored for one month.

Doran and Vetterling (1968) freed sporozoites of *Eimeria meleagrimitis* and *E. tenella*, suspended them in media containing 7% dimethyl sulfoxide and froze them to -80° C at the rate of 1° C/min. They then stored them in a tank three inches above liquid nitrogen. After one month, sporozoites were thawed in water at 43° C and were tested for viability by inoculating into cell cultures and into the ceca or duodenum of chicks and turkey poults. There was no significant difference between frozen and nonfrozen sporozoites in producing infections and also in production of oocysts in lesions. Frozen sporocysts also produced infections, but these were less than in birds fed freshly released sporocysts.

Doran (1969a) reported on concentration of freezing protectant required, equilibration period, and cooling rates during freezing of excysted sporozoites of *E. adenoeides, E. mivati*, and *E. tenella*. When sporozoites were placed in concentrations of dimethyl sulfoxide (DMSO) between 2.5% and 15%, survival of sporozoites was better at the lower concentrations in the absence of freezing. When the sporozoites were frozen with a cooling rate of 1° C/min from 25° C - 26° C to -30° C and 10° C/min from -30° C to -80° C, survival was better at the higher concentrations. Best results (more than 70% survival) were obtained when the equilibration period was from 45 to 50 min and the cooling rate from the freezing point to -30° C was 1° C/min. Increasing the equilibration period to as long as 2.5 hr did not increase survival of sporozoites in DMSO.

3. Determining infectivity of frozen sporozoites

Doran and Vetterling (1969a,b) gave detailed information on the techniques used in determining infectivity of excysted sporozoites. Sporozoites within released sporocysts, and sporozoites within intact oocysts of *E. meleagrimitis* and *E. tenella* were placed in media containing 7% dimethyl sulfoxide, frozen to -80° C at a rate of 1° C/min and stored above liquid nitrogen for intervals up to four months; results were as follows:

> Oocysts produced no infection. After three months, sporocysts of both species produced infections but less than those produced by fresh oocysts. After storage for four months, sporozoites inoculated into the duodenum or cecum of their natural host resulted in infections comparable to those caused by fresh oocysts.

The authors suggested that freeze-storage of sporozoites or released sporocysts by their method could be used to save time, effort, and expense in maintaining cultures in the laboratory. They pointed out that this method makes possible retention of pathogenic and drug-resistant strains without genetic alteration and that it might be possible that a type specimen collection could be established.

Doran (1969b) pointed out that even though there was a slight loss of sporozoites caused by freezing, he was planning to use frozen sporozoites in future work with tissue cultures because of the saving in time and availability of sporozoites of a uniform age which are provided by this method.

L. SOME TECHNIQUES USED IN STUDIES ON LIFE CYCLES OF COCCIDIA

1. Surgery used in life cycle studies on coccidia of cattle

Hammond *et al.* (1946) reported that laparotomies and Thiry's fistulas were useful in life cycle studies of coccidia of cattle. They performed laparotomies on calves approximately one week old and placed viable schizonts and merozoites directly into the ileum with a hypodermic syringe. The Thiry's fistula was used in a calf about four months old. A segment of the ileum about one foot long located about six feet anterior to the ileocecal valve was sutured to an opening in the abdominal wall. The fistula was used in two ways: One consisted of introducing sporulated oocysts enclosed in cellophane bags into the isolated portion of the intestine, leaving them there for from 18 to 46 hr, removing them and examining the oocysts for signs of excystation. The other use consisted of introducing schizonts and merozoites directly into the segment, and making daily examinations of curetted mucosa from this segment for evidence of oocyst formation.

Hammond *et al.* (1963) used a surgical method for inoculating merozoites directly into the cecum. Mucosal scrapings were collected in saline or Ringer's solution and were concentrated by sedimentation and centrifugation, homogenized for 30 sec in a blender to free merozoites from schizonts, and the merozoites were introduced into the cecum of another calf within from three to five hours. Under local anesthesia, laparotomy was performed. The cecum was ligated and a small piece was removed from the distal end. Mucosal scrapings containing from 0.5 to 1.0 billion merozoites of *E. bovis* were introduced by syringe, and the incision in the cecum was closed with sutures. Three or four cecal biopsies were performed on each calf at various intervals to determine stages of coccidia present. Excised cecal tissues were bisected; one portion was fixed in Serra's and the other in Helly's or Zenker's solution. The former tissues were stained in periodic-acid-Schiff and with Himes and Moriber's (1956) method. Iron hematoxylin and hematoxylin-eosin were used with the other tissues.

2. Rapid preservation of tissues for sectioning

Postmortem changes occur rapidly in intestinal cells. Immediate fixation is extremely important in making parasitological examinations. Davis (1944) developed an enterotome knife that reduced the time required for slitting a bovine intestine of approximately from 50 to 75 ft to approximately 1½ min. The instrument opened the intestine in a straight line on the side opposite the attachment of mesentery. A stainless steel rod 250 mm × 10 mm was machined on a lathe to taper one end to a diameter of 6 mm, leaving a knob 8 mm in diameter at this end of the rod. The rod was bent before a longitudinal slot 1 mm wide by 6 mm deep was cut on the inside surface of the angle, extending approximately 35 mm from the angle toward each end, and then the rod was bent further to form a 90° angle. Half of a double-edged razor blade was clamped in the groove by means of a screw at each end of the blade. Sealing wax or beeswax could be used instead of screws in anchoring the blade in a more simplified knife.

In preparation for use of the enterotome, the mesentery must be cut as close as possible to the intestine. The knife can be used in the hand by pushing the bulbous end into the lumen of the intestine and moving it lengthwise toward the other end of the intestine. One can use a laboratory clamp to fasten the knife to the edge of a bucket, thus freeing both hands to pull the intestine out of the bucket over the knife edge and into another bucket. Davis (1944) made a similar enterotome knife to cut intestines of smaller animals. He added a rounded point to a Bard-Parker No. 12 hooked blade. The knob was made by holding resin-cored solder in a flame until the rounded knob was formed. The knob and the shank were cut approximately 14 mm long. The tip of the Bard-Parker blade was inserted in the open end of the shank and pliers were used to flatten the lead tubing to clamp the flat sides of the tip of the blade. The small enterotome blade was used in the same manner as the large enterotome.

Bizzell and Ciordia (1962) modified the Davis (1944) enterotome by replacing the double-edged razor blade with the stronger injector-type blade, which worked better on older cattle and sheep. After using both types, Davis had selected the double-edged razor blade for the younger animals because the blade was thinner and seemed to be sharper. Bizzell and Ciordia made the enterotome from brass rods. Instead of having a groove cut longitudinally, they made a separate plate from one of the rods to fit into and lie flush with the surface of the other rod, anchoring the razor blade between the two. In addition, they made a wooden platform to hold iron rods for clamping the enterotome blade, and instead of pulling the intestine over the enterotome by hand, the slit intestine was started by hand, fed through two parallel metal scrapers, and then wrapped around a windlass operated by a hand crank. Intestinal contents were washed off with a shower sprayer and the intestinal contents as well as the opened intestine could be examined for parasites and pathological changes.

Shumard *et al.* (1955) developed a somewhat similar motor-driven enterotome. Detailed plans of this apparatus have not been published but are available from the North Dakota Agricultural Experiment Station.

3. Counting schizonts of coccidia in intestines

Hammond *et al.* (1959) reported a unique method of counting schizonts in the intestine of calves used to determine immune reactions in the experimental animals. Starting at the ileocecal valve, examinations were made at each two-foot interval along the length of the small intestine. At each site, a 3 in × 1 in glass slide, which had been marked with a square inch and a square 0.4 in area, was laid over the intestine and the schizonts visible under a 10× dissecting microscope were counted. The 0.4 in square was used when the schizont numbers were large.

4. Examining sections of intestine for life cycle stages of coccidia

Edgar *et al.* (1944) fixed intestinal tissue in 95% alcohol, Carnoy's or Bouin's fixatives. After sectioning in paraffin, they used iodine staining for glycogen, with Galigher and Kozloff's method (1971). They found that oocysts stained with iodine were readily seen, even in lightly infected tissues, because host tissue did not take the iodine stain.

5. Flattening intestinal samples for fixation

Hammond *et al.* (1946) used a method to prevent curling of intestinal tissues of calves when placed in fixing fluid. Approximately one-half inch lengths of intestine were excised from 20 regions along the entire intestinal tract. The pieces were placed in saline solution and were immediately inverted over sections of tongue depressors to prevent curling in fixatives. One must be very careful in using this method to prevent damaging the cells on the surface of the villi.

6. Triple staining for parasites in tissue

Himes and Moriber (1956) developed a triple histochemical stain that Davis *et al.* (1961), and Davis and Bowman (1963) found to be very useful in studying life cycle stages of the coccidia. Several types of fixatives, such as Carnoy, formalin, or Helly and Smith, can be used. The exposure time in each solution can be varied from 30 sec to 30 min, with the exception of step 2, in which acid hydrolysis occurs.

Reagent I (Azure A-Schiff Reagent). Dissolve 0.5 g of Azure A in 100 ml of the bleach (Reagent II); add a few drops of 10% K-metabisulfite before use. Keeps for several weeks.

Reagent II. Bleach, 1 N HCl, 5 ml; K-metabisulfite, 5 ml; distilled water, 90 ml; use fresh.

Reagent III. Periodic acid, 0.8 g; water, 90 ml; 0.2 M sodium acetate solution, 10 ml; use fresh.

Reagent IV (Basic Fuchsin-Schiff reagent). Follow Stowell's (1945) method. Stores indefinitely if refrigerated in glass-stoppered bottle.

Reagent V. Naphthol yellow S, 0.02% stock solution Naphthol yellow S, 1 g; 1% acetic acid, 100 ml; dilute 2:100 with 1% acetic acid. Lasts indefinitely.

Staining Procedure:

(Rinse in water after steps 2 through 8)

1. Remove paraffin. Pass sections through graded alcohols to water.
2. Hydrolyze 12 min in 1 N HCl at 60°C.
3. Stain 5 min in Reagent I.
4. Bleach 4 min, two washes of 2 min each in Reagent II.
5. Place in Reagent III for 2 min.
6. Stain 2 min in Reagent IV.
7. Repeat bleach method of step 4.
8. Stain 2 min in Reagent V.
9. Dehydrate with tertiary butyl alcohol, two changes, 2 min each or longer.
10. Clear in xylene, two changes, and affix coverglass with balsam or synthetic resin mounting medium.

Results (Davis *et al.*, 1961): Nuclei stained blue to green; polysaccharides red; proteins yellow. Various combinations of these colors produce very striking colors in stages of coccidia. Merozoites in developing schizonts may be light green to blue, depending on maturity. Residual matter is usually yellow and, as merozoites reach final stages of maturation, pink color appears. Male and female gametocytes are easily distinguished by color as well as morphology. Immature macrogametes have a yellowish tint, and as maturity is attained the color changes from pink to a bright red. Microgametes are green while immature and change to a deep blue or nearly black in the mature forms.

7. Staining oocysts of coccidia

Staining of oocysts is extremely difficult because of the nearly impervious outer wall of the oocysts. Monné and Hönig (1954) reported a method of removal of the outer wall of the oocyst by using hypochlorite. Davis and Smith (1960) used antiformin to remove the outer wall of oocysts, after which there was much greater penetration of fluorescent stains such as acridine orange, auramine O, acridine yellow, neutral red, primuline yellow, and rivanol. Acridine orange used at pH 4.7 or unbuffered at a concentration of 0.2% proved to be the best of the stains listed. The authors were able to distinguish between heat-killed and live oocysts using this method and fluorescence microscopy with a high pressure mercury arc lamp. After releasing sporocysts from oocysts by crushing and rolling them between two glass slides, the best differentiation between living and dead sporocysts resulted when these were stained with acridine orange solutions of 0.002 and 0.0002%.

8. An all-glass staining rack for coverslips (David J. Doran)

Cells grown on coverslips in cell culture are frequently stained by passing the coverslip from one container to another with tweezers or after placing the coverslips in wire mesh baskets attached to slides. A much simpler way to do this that permits the use of certain metallic stains is by using a rack described as follows:

a. Place a microscope slide in slots 1, 4, 7, and 10 of a rectangular glass slide rack that will accommodate ten slides. Place thin (from 2 mm to 3 mm) glass rods in the slots between the slides.
b. Construct 12 glass coverslip holders shown as in Figure 5. These can be made by a glass blower or by using the top 25 mm of a test tube having less than a 10-mm opening. The holder shown is for a 10 mm × 35 mm coverslip. For a 10 mm × 22 mm coverslip, the height of the holder should be from 15 mm to 16 mm. The distance between the indentation at the upper end and the back wall should be no more than 8 mm.

Coverslips are placed in the holders with the cells away from the indentation. The indentation prevents the coverslip from turning in the holder. The rack accommodates three rows of four holders. It can be run through the staining process in a manner similar to that used for slide preparations.

9. Phase contrast and interference microscopy

Ordinary bright field microscopy has its drawbacks when examining unstained, living oocysts or endogenous stages. Phase contrast and interference microscopy have been used extensively in studying such stages. Mellors (1955) summarizes the histories and methods of use for both phase contrast and interference microscopy. Phase contrast produces greater detail than does bright field. With phase contrast one usually does not need intravital and other stains.

The interference microscope (Baker, 1955) takes longer to set up and adjust than does the phase microscope. Color fringes have to be spread to have a uniform color filling the aperture of the objective. This requires the

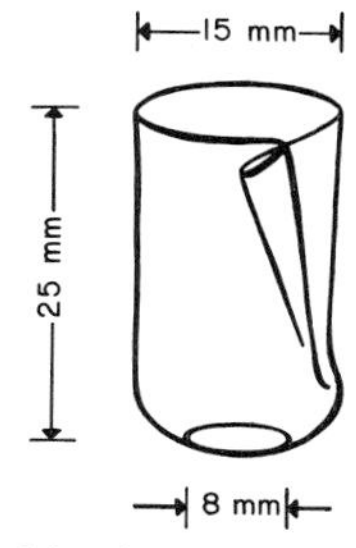

Fig. 5. Glass holder for coverslips. (Doran, original.)

use of the telescope on nearly every slide because the thickness of the coverslips or the object being examined may cause a prismatic effect. The adjustments usually have to be made when one changes from one objective to another. Once the microscope has been adjusted one can move a goniometer scale to produce colors in the object being examined and in the background or in tissues. These colors can be used to differentiate morphological structures in stages of coccidia because of their variations in density.

10. Fluorescence microscopy

Hicks and Matthaei (1955) gave excellent information as to use of fluorescence in microscopy. They described a method of using blue light in a range of from 400 mμ to 500 mμ wavelength to excite fluorescence of tissues for observation with an ordinary microscope, using a 12-volt, 100-watt projection lamp to provide sufficient light for blue fluorescence. They gave information on the optical principles involved, apparatus to be used, and the results obtained with different types of fluorescent dyes.

11. Fixation of merozoites for electron microscopy

Sheffield and Hammond (1966) described effective methods of fixing merozoites of *E. bovis* for electron microscopy. Merozoites of *E. bovis* were obtained from calves by the method of Hammond *et al.* (1965). Scrapings of the intestinal mucosa were placed in a large volume of fixative without washing. One method used for fixation involved 3% glutaraldehyde in Sorensen's phosphate buffer (Sabatini *et al.*, 1963). The tissues were fixed for two hours, then rinsed in buffer with sucrose and post fixed in veronal-acetate-buffered OsO_4 for two hours. Other specimens were fixed in veronal-acetate-buffered OsO_4 (Caulfield, 1957). A third fixative was Dalton's (1955) chrome-osmium fixative, used for the same period of time. Fixatives were buffered at pH 7.4 and used at from 0° C to 4° C. In order to store or mail specimens from one laboratory to another they were left in 70% alcohol for several days or the glutaraldehyde-fixed material was left in the rinsing buffer. They were dehydrated in ethyl alcohol and passed through propylene oxide to remove the alcohol before embedding in Epon (Sporn *et al.*, 1962).

12. Locating parasites by light microscopy before using electron microscopy

When the numbers of parasites are very low, one may waste much time in searching with electron microscopy. If one makes thick and ultrathin sections from the same tissue, the thick sections can be examined first so as to select the ultrathin sections that are likely to contain the parasitic stages desired.

The usual method is to make thick sections and then remove the embedding material before staining. Leeson and Leeson (1970) described a method for staining Epon 812-embedded material (Luft, 1961) without having to remove the plastic before staining. They used the technique for correlating electron microscopy findings with observations made by light microscopy as well as in preparing slides for histology courses.

The sections were cut at from 0.5 μ to 2 μ with glass knives and were floated onto a water bath containing 15 % acetone. The sections were removed with a wire loop and transferred to a drop of distilled water on a clean slide. They were dried at from 50° C to 60° C for at least two hours before staining.

The authors used their own modifications of toluidine blue and safranin O, methylene blue, azure-methylene blue, van Gieson's, and periodic acid-Schiff staining methods. They substituted safranin O for picrofuchsin in the van Gieson's stain in order to achieve results similar to hematoxylin-eosin methods.

They air-dried the stained sections, immersed them in xylene for from one to two minutes and then mounted them in a synthetic resin under a cover slip. They decided that the toluidine blue and safranin O method was the simplest and most useful of the stains listed.

M. TECHNIQUES FOR IMMUNOLOGICAL STUDIES (Elaine Rose)

Except for the preparation of antigens, methods for immunological investigations are largely similar, irrespective of the nature of the organism involved, and may be found in standard textbooks. There are differences in the properties of antibodies of animals of different species, and the relevant textbooks should be consulted for modifications of techniques.

Possibly the only methods which are specific for a study of the coccidia are those for the preparation of antigens.

1. Particulate antigens

Methods for the preparation of particulate antigens are those used to prepare clean suspensions of the various stages.

2. Soluble antigens

a. From free stages

Soluble antigens from pure suspensions of oocysts, sporocysts, sporozoites, or merozoites may be obtained by disrupting the cell walls, thus liberating the contents; cell wall debris may then be removed by centrifugation. Methods that have been used for disintegration include the following:

1. Mechanical fracture of oocysts. This may be done by hand with glass tissue homogenizers, but can be done efficiently with a mechanical grinding apparatus such as the Tri-R Stir-R (Tri-R Instruments, Rockville Center, N. Y.), which has a Teflon plunger.
2. Use of ultrasound. Oocysts and other stages may be disrupted by exposure to ultrasound. The suspensions should be examined at intervals to determine progress; excess treatment may be deleterious.
3. Alternate freezing and thawing. This is a fairly efficient method of rupturing organisms, especially suspensions of sporozoites or merozoites. The freezing may be carried out in liquid nitrogen or in a freezing mixture of acetone and solid CO_2, and thawing in a water bath maintained at approximately 37°C. Agitation will accelerate both procedures. Care should be taken to ensure that the containers used will withstand the treatment.

b. From infected tissues

It is advisable to starve the host before killing, to reduce contamination with food particles. The presence of parasites should first be confirmed and attention concentrated on heavily parasitized regions. After the removal of lumen contents, the mucosa of infected intestine may be stripped off with a glass microscope slide, leaving the muscle layers. The stripped-off mucosa is then mixed with a little phosphate buffered saline (PBS) and the mixture subjected to the disintegration methods described above. Debris is removed by centrifugation.

c. Collection of soluble parasite products

This collection is not feasible for those species parasitizing the intestine except in those instances where the parasites can be grown *in vitro*.

The bile ducts and gall bladders of rabbits infected with *E. stiedai* become distended with fluid, which contains soluble parasite products. Any particulate matter, stages of the parasite, host cells, and debris may be removed by centrifugation.

The allantoic fluids in infected embryos

contain soluble material of parasite origin, which can be harvested in the following way: The eggs should be candled and the extent of the air space marked on the shell; the eggs may then be refrigerated overnight, air space uppermost to reduce contamination of the allantoic fluid with embryo blood during collection. The shell is then cut off at the air-space mark and the membranes carefully peeled back, avoiding damage to blood vessels. The fluid can then be aspirated and centrifuged before storage. Centrifugation will remove any contaminating erythrocytes.

The fluid overlay in tissue culture preparations can also be harvested and stored until required for use.

All antigen preparations should be carried out as quickly as possible and precautions taken to keep all reagents cold. This applies especially to preparations made from tissues as these will probably contain enzymes which may degrade the antigens. Ice-water baths should be used throughout and the diluents kept cold. Thawing in warm water should be closely supervised and the time kept to a minimum.

Antigens should be stored in the frozen state, preferably at temperatures lower than −10° C; freeze-drying may be useful but has not been extensively used.

N. PLASTIC FILM ISOLATORS FOR PRODUCING GNOTOBIOTIC POULTRY

Trexler and Reynolds (1957) were the first to report on the development and use of environmental chambers constructed of plastic film for gnotobiotic experimental animals. Bradley *et al.* (1967) gave complete details of a technique for establishing and using the system for producing bacteria-free poultry and its application in parasitological research. They used commercially available isolators, accessories, and supplies from the same source.* Their technique is described briefly as follows:

1. Plastic film isolators

The environmental chambers were made from 1.2-mm thick polyvinyl chloride film, as described by Trexler and Reynolds (1957) and were 60 × 24 × 24 in size. Each was furnished with an air filter, blower, and outlet trap and was placed on a plywood base to allow it to be used on shelves. The air filter was made of layers of glass fibers and was changed frequently. Air flow through the filter was approximately 2.5 cu ft/min.

The outlet trap was constructed of Lucite or molded glass fiber and had a float to prevent reverse flow of air into the isolator.

After each use, the isolator was washed with a detergent and was filled with freon gas; a gas detector was used to make sure that all surface areas were free of leaks.

2. Equipment

For use with poultry, a plastic hatching tray for eggs, glass feeding dishes, sterile water, sterile vitamin solutions in glass ampules, sterile bacteriological media in screw-capped test tubes, beakers, and other containers were placed inside the isolator prior to chemical sterilization. Plastic laundry baskets fitted with stainless steel mesh covers and floors were used for animal cages.

3. Sterilization of unit and accessories

Air filters were sterilized with dry heat. The outlet trap was filled with light mineral oil containing a germicide. The interior of the isolator and all the equipment were sterilized by using air pressure to produce an aerosol spray of 2% peracetic acid containing Nacconal.† After complete saturation of the inner contents, the isolator was inflated for one hour, the outlet seal of the air filter was broken from the inside and the blower was activated for 24 hr to evacuate the acid fumes.

*G-F Supply Division, 431 N. Quentin Road, Palatine, Ill. 60067.

†Monsanto Chemical Co., St. Louis, Mo.

4. *Sterilization of feed and water*

Autoclaving was used to sterilize drinking water and commercially mixed mash and also linens, swabs, glassware, and equipment that could not be chemically sterilized. A plastic sleeve was used to transfer the sterilized material into a pre-sterilized isolator, using peracetic acid aerosol spray for complete sterilization of the sleeve.

5. *Testing for sterility*

When feed was added, samples were placed in various bacteriological media and were removed from the isolator for incubation. When chicks were hatched and twice weekly thereafter, cloacal swabs were made and cultured for bacteria.

6. *Production of gnotobiotic birds*

Embryonated chicken or turkey eggs were washed in detergent solution at 37° C and were immersed in 1% mercuric chloride solution at 37° C for eight minutes to sterilize the shells. The eggs were then passed into the isolator through a special egg chute. The entire isolator was placed in an incubator room adjusted to 37° C and a relative humidity of 80% or higher. Conventionally raised controls were produced in battery brooders in the laboratory and were fed sterile feed and water.

In the first tests, the major difficulty was inadequately sterilized feed. By driving out the heated air before closing the autoclave a sterilizing temperature was reached in the autoclave. Repeating the autoclaving cycle after 72 hr destroyed heat-resistant spore-forming bacteria and fungi.

Bradley *et al.* (1967) stated that the plastic film isolator technique was relatively simple for producing gnotobiotic birds for experiments with protozoa or cestodes, provided a rigid protocol is followed. They recommended that a complete log be kept on individual isolators in order to determine the cause of accidental contamination.

O. OTHER TECHNIQUES FOR INVESTIGATING COCCIDIA

Several symposia on investigations of coccidia have been held in the United States. These resulted in publications of many kinds of techniques that have proved to be very useful. The investigator should refer to these for details (Brackett, 1949; Edgar, 1958; Hunter, 1959; Merck, Sharp, and Dohme, 1959). The latest symposium, "Methodology for the Development, Selection, and Testing of Anticoccidial Drugs for Use in Controlling Coccidiosis," was held on the campus of the University of Georgia in 1969, with W. Malcolm Reid as chairman and organizer. Twenty-seven articles resulted.*

III. *Toxoplasma*

A. SURVEYING METHODS FOR *TOXOPLASMA* IN DOMESTIC ANIMALS

Jacobs *et al.* (1960) outlined methods of testing for *Toxoplasma* in diaphragms of swine, cattle, and sheep. They used fresh muscles removed from the diaphragm, packed in plastic bags, and kept in ice for up to three days. A household-type meat grinder was used to grind the tissues. The ground meat was placed in a two-liter flask covered with a volume of peptic digestive solution of about ten times its weight. A magnetic stirrer was used at 37° C with digestion taking place for two hours. The suspension was filtered through layers of coarse cheesecloth and centrifuged in 250-ml bottles for 15 min at 2,000 rpm. The fluid was decanted and the sediment pooled into a bottle and resuspended in saline wash and recentrifuged in

*Published in Exp. Parasitol., 1970, 28(1): entire issue. Information on battery trials, screening methods, field-collecting of oocysts, evaluation trials, floor-pen trials, drug mixtures, oocyst counting, drug resistance, and useful parameters for drug testing was included.

the same way. The sediment was suspended in from 10 ml to 20 ml of saline and inoculated intraperitoneally into ten mice, each receiving one milliliter of the inoculum. On the following day each of the same mice received additional injections of the suspension, which had been refrigerated overnight. The mice were observed for from three to four weeks, and deceased ones were examined by means of fresh smears of peritoneal exudate. Giemsa-stained impression smears were made of brain, liver, spleen, and lung. At the end of one month half the survivors were bled from the orbital sinus and the serum specimens were tested at dilutions of 1:16 and 1:64, using the dye test for toxoplasmosis. When positive titers resulted, tissues from the remaining mice were subinoculated into fresh mice. The estimate of virulence of various strains was made by counting parasites from peritoneal exudates in a Neubauer-Levy counting chamber, and suspension in 10% serum-saline solution gave the desired number of parasites per milliliter. Standard amounts were inoculated intraperitoneally into mice and the number of mice and the number of deaths and survival times were noted.

Infected cat intestines were examined by flushing segments of the intestines with ice-cold normal saline for visual examination and for isolation of *Toxoplasma* in mice (Dubey *et al.*, 1970a). Impression smears were fixed in acetone in 5% acetic acid in 95% ethanol for fluorescent antibody staining or in methyl alcohol for staining with Giemsa. Washed gut segments were flushed in ice-cold fixative and preserved with other organs in acetic alcohol for fluorescent antibody studies and in Zenker-formol or 10% formalin for histologic studies. Free sporozoites obtained from sporulated oocysts were treated with 6% sodium hypochlorite solution for one-half hour in an ice bath, washed with water and crushed between a coverglass and a slide coated with inactivated serum free of toxoplasmic antibody. Smears were fixed with methyl alcohol and stained with Giemsa, or with acetone to be used for the fluorescent antibody test. Shaking sodium hypochlorite-treated oocysts with alundum (mesh No. 60) released sporocysts, which were excysted in 0.5 trypsin and 5% cat or dog bile in Melnick's A medium with 0.5% lactalbumin hydrolysate pH 7.5 for from one-half to two hours at 37° C. Flotations containing oocysts were mixed with 10% normal cat serum or 10% egg albumin and centrifuged to form a pellet to be fixed in acetic alcohol for 24 hr and then embedded in paraffin blocks.

B. LABORATORY DIAGNOSIS OF TOXOPLASMOSIS

Eichenwald (1956) outlined the general methods for laboratory diagnoses of toxoplasmosis. His publication should be used for details. Briefly, his laboratory methods fall into three classes: (1) isolation of the organism, (2) serologic tests of various types, and (3) cutaneous tests with "toxoplasmins." He described the dosing of test animals for from three to five days with cortisone before inoculation to increase the chances of recovering *Toxoplasma*. He used three types of serologic tests: (1) the dye test, (2) the complement fixation test, and (3) the neutralization test, and stated that the dye and complement fixation tests are the most useful for diagnostic purposes.

1. The indirect fluorescent antibody test for Toxoplasma

This test is very useful but requires a fluorescent microscope. Frenkel (1971) stated that a positive serologic test indicates past or present infection but not necessarily presence of the disease, as *Toxoplasma* antibody is found in the normal population. Of the various serologic tests, the indirect fluorescent antibody test (IFAT) is an ideal method and can be used in any clinical laboratory. According to Frenkel (1971), antigens may be obtained either as formalinized suspensions (Walton *et al.*, 1966), lyophilized (Takumi *et al.*, 1966) or as direct smears of *Toxoplasma* from mouse peritoneal exudate or cell culture. He fixed the slides with 5% trichloracetic acid in 100% ethyl alcohol, then washed and stored them at -20° C. The fluorescent antispecies globulin must be free from antibody and should not stain *Toxoplasma* without added specific antiserum. The procedures are described by Walton *et al.* (1966), Suggs *et al.* (1968), and Archer *et al.* (1971). The IFAT can be applied to tissue sections in order to detect small numbers of *Toxoplasma*. Frenkel prefers the alcohol-acetic fixation of Carver and Goldman (1959). Specificity of staining should be controlled by its extinction after absorbing the antiserum with *Toxoplasma* (Carver and Goldman, 1959). Frenkel confirmed identification of the species by marking the location of the fluorescent groups, removing glycerol mounting medium, and staining the slide with hematoxylin or Giemsa, which differentiate the nuclei of *Toxoplasma*, while the cytoplasm is lysed.

2. *Dye test (DT)*

This is the most reliable test for *Toxoplasma* but requires maintenance of live organisms in the laboratory. Lysis of living *Toxoplasma* after exposure to antibody with a complement-like "accessory factor" present in normal human serum, is the basis of this test (Sabin and Feldman, 1948). The lysis can be observed directly by adding alkaline methylene-blue as an indicator. "Accessory factor" can be obtained from human serum or citrated plasma (Wallace, 1969) free of *Toxoplasma* antibody. Fresh blood-bank blood in citrate-dextrose solution was tested for suitability. Separated plasma was frozen at -70° C or -20° C in from 5-ml to 10-ml amounts for a day's run. The patients' sera are inactivated at 56° C for 30 min. Frenkel and Jacobs (1958) have outlined the test procedures. Sera were tested undiluted or at 1 : 2 dilution, progressing in two-or four-fold dilutions, depending on accuracy needed. Titrations should extend to a dilution of 1 : 1,024 to avoid falsely negative tests, as a prozone phenomenon is occasionally observed. One should compare two serum specimens taken at least a week apart, except in ocular parasitism, when one specimen will suffice.

3. *Isolation of* Toxoplasma *from blood or tissues*

Frenkel (1971) centrifuged heparinized blood obtained from infected animals, removed the buffy coat, along with some red blood cells, and gave injections of this suspension intraperitoneally into mice. There is the possibility of passive protection if antibody is present. Minced tissues from similar sources were ground in a mortar with sand or alundum and suspensions in saline were injected intraperitoneally or subcutaneously into mice or hamsters.

After about five days, peritoneal fluid should be examined for organisms. If negative, peritoneal fluid, liver, spleen, and lung suspensions should be subinoculated intraperitoneally, repeating every five to six days to build up the number of organisms. Smears of peritoneal fluid should be dried, fixed with methanol, and stained with Giemsa's or Wright's. Brain sections obtained from survivors usually contain *Toxoplasma* after 30 days. Brain suspensions from normal animals should be used as controls. Frenkel (1971) isolated *Toxoplasma* oocysts from cats by suspending fecal material in water, filtering through gauze, and concentrating by centrifugation and sucrose (40% weight/volume, specific gravity 1.15), washing the supernatant in water, and sporulating in shallow dishes of 2% sulfuric acid for from three to five days. The suspension was washed twice in saline and injected intraperitoneally into mice, using care to avoid leakage of the inoculum (Frenkel *et al.*, 1970, and Dubey *et al.*, 1970a,b). To reduce the risk of infection of personnel, it was advised that the initial washing and flotation should take place on the day the feces are shed, prior to their becoming infectious.

4. *The indirect fluorescent antibody test for toxoplasmosis with filter paper blood specimens*

The filter paper blood specimen technique of Anderson *et al.* (1961) proved an ideal way to conduct epidemiologic investigations of human schistosomiasis without the problems of venipunctures, as well as the aseptic processing, cold storage, and shipment of serum. Walton and Arjona (1971) worked out a method for adapting the test for use in investigations of toxoplasmosis. Whatman No. 2 filter paper was stamped to outline a 2.3-cm diameter circle, which was touched to welling drops of blood to saturate the marked area. The blood was air dried and was stored at ambient temperature in the Panama Canal Zone for from 2 to 15 days or was stored at 4° C. A matching set of samples was mailed to the United States and returned; the two sets were then examined together in the same test.

Veronal-buffered saline (0.1 ml) was placed with the dried blood on the filter paper in a two-inch length of one-eighth-inch diameter tygon tubing plugged with a glass bead at one end. The tubing was flexed and rolled and allowed to extract for 30 min and then a small vise was used to squeeze out the eluate into a 12-mm by 75-mm test tube in which 0.1 ml of the saline had been added. This reconstituted blood was found to be equivalent to a 1:8 dilution of serum. Two-fold serial dilutions of serum and of reconstituted blood were made starting at 1:8 and both preparations were tested simultaneously with the IFA test.

The results obtained from the two kinds of test specimens were in close agreement; with similar volumes there was little difference between the dried blood and the fresh serum preparations. The mailed specimens were satisfactory but those refrigerated for longer than three months were found to have some antibody deterioration.

C. PROPAGATION OF *TOXOPLASMA*

Jacobs (1956) gave several examples of methods of propagation of *Toxoplasma*. He reviewed the method of tissue culturing and reported that various authors were able to use chick embryonic tissue, such as heart muscle, leg muscle, or liver epithelium, mouse embryonic tissues, human epithelium, myometrium, and embryonic skin muscle, embryonic rat heart muscle, monkey kidney epithelium, and macrophages from laboratory rodents, and described the use of tissue cultures as a method for primary isolation of parasites from suspected cases. He mentioned that the use of tissue culture cell lines removed all doubt of the source of the *Toxoplasma* as compared with laboratory inoculations of animals that have to be proved to be free of spontaneous infections. Tissue cultures were also used for isolation without having to maintain laboratory host animals.

1. *Maintenance in laboratory animals*

Jacobs (1956) mentioned that laboratory strains of *Toxoplasma* are easily passed by any route of inoculation in mice, rats, hamsters, rabbits, guinea pigs, chick embryos, and pigeons. He recommended the mouse as the most convenient animal; it has the advantage that it is probably not likely to have spontaneous infections. Peritoneal fluid passages are useful routinely. After from one to two weeks, the organisms can be isolated from liver, spleen, and lung. Laboratory rats are convenient for maintenance because the animal suffers very little disease despite large numbers of the parasites inoculated. Later in the infection, the parasites are found easily in the brain. He was able to find *Toxoplasma* from brains of rats infected two years previously, thus making rats good reservoirs for demonstration purposes in teaching.

Jacobs (1956) listed reports of the use of chick embryonic tissue cultures for long-term preservation. He described chick embryos as a useful tool for primary isolation and maintenance by weekly passage by inoculation into the yolk sac.

2. *Preservation*

Eyles *et al.* (1956) developed the techniques of preserving the *Toxoplasma* by freezing in glycerine solutions, a most convenient method of preserving the original characteristics of strains.

3. *Obtaining* Toxoplasma *for use in tissue cultures*

Cook and Jacobs (1958) included a summary of the history of *in vitro* cultivation of *Toxoplasma gondii* from 1929 to 1956. The authors used trypsinized monkey kidney tissues grown in roller tubes or on coverslips for their experiments. The initial medium was made up of Hanks' balanced salt solution, pH 6.8, containing 0.5% lactalbumin hydrolysate, 2% calf serum inactivated at 56° C for 35 min, with 100 units of penicillin and 100 μg of streptomycin per milliliter. After four to five days of growth at 37° C, they used Earl's balanced salt solution, pH 7.6, containing 0.5% lactalbumin hydrolysate, 1% inactivated calf serum with 100 units of penicillin and 100 μg of streptomycin per milliliter. Suspensions of *Toxoplasma* were obtained by harvesting mouse peritoneal exudates into 10% calf serum-Hanks' solution. They used the inoculated material within one hour after preparation. Numbers of parasites were determined by hemacytometer counts.

Fulton and Spooner (1957, 1960) described use of cultured *Toxoplasma* for the dye test and for storage. They found that the parasites could be stored for as long as 120 days in a refrigerator and for 360 days while frozen.

4. *Obtaining proliferative* Toxoplasma *free of host cells*

Fulton and Spooner (1957, 1960) described a method for isolating *T. gondii* parasites, free of host cells, from peritoneal exudate of infected animals. Although the strain was maintained in mice by serial passage of peritoneal exudate, cotton rats were found to be by far the most satisfactory animals for bulk parasite production. Animals from 100 g to 200 g in weight were caged singly and killed usually on the third day after IP inoculation with 2×10^6 parasites. Peritoneal exudate was removed with a blunt Pasteur pipette, and further parasites were recovered by washing out the peritoneal cavity with a little citrated saline. The pooled material was strained through a thin layer of gauze and centrifuged. The deposit was suspended in buffered saline and shaken gently by hand for five minutes with glass beads to liberate toxoplasms from white cells. Many white cells agglutinated and could be removed by allowing the suspension to stand for a few minutes; the rest were removed by filtering through a sintered glass filter, pore size from 15 μ to 35 μ, with the aid of gentle suction. Any red cells present were removed by agglutination with a specific antiserum or by hemolysis with antiserum and guinea pig serum.

5. *The plaque method for studying* Toxoplasma

Akinshina and Zasukhina (1966) used the plaque method to isolate pure lines of *Toxoplasma*, to study genetic features, and to attempt to correlate the sizes of the plaques and the pathogenicity of the strains in cultures. They found differences in pathogenicity between different clones of the same strain but no correlation between pathogenicity and plaque sizes. They found that the parasites were able to reproduce intensively under the agar layer with the plaque technique at 42° C. Remington (1970) suggested that genetic differences can be used instead of virulence as markers for studies with this species.

D. TECHNIQUES OF HANDLING *TOXOPLASMA*

Frenkel (personal communication, 1971)* cautions against the use of animals for storage of *Toxoplasma* unless one knows the source of the animals. Ideally, the individual animals or the stock should be tested serologically to be sure that *Toxoplasma* antibody is not present. Rabbits, guinea pigs, and rats are occasionally spontaneously infected, so that a laboratory strain to be stored could be contaminated with a natural infection already present. He thinks that some of the natural infections may be produced by contamination of feed by cat feces but thinks such contamination occurs only occasionally.

Dubey (personal communication, 1971)† refers to Dubey *et al.* (1970a,b) and lists the following suggested variations of methods they use with *Toxoplasma* as compared with techniques routinely used with other coccidia:

1. We use sugar solution (40% w/v of sucrose solution of 1.15 sp.gr. with 0.8% phenol) to float oocysts of *Toxoplasma* from cat feces—Sheather's sucrose solution floats too many fecal particles.
2. Some cats excrete very few oocysts, which are generally missed in direct or flotation examinations of cat feces. Fecal suspensions, concentrated by flotation, are fed or are injected into mice for confirmation of infections.
3. *Toxoplasma* oocysts are sporulated and preserved better in 2% H_2SO_4 as compared with the routinely used 2.5% $K_2CR_2O_7$ solution, as infectivity of chromated oocysts is lower than those preserved in H_2SO_4. Oocysts should be sporulated in capped bottles, and to minimize risk to humans, oocysts should be separated from cat feces before sporulation.
4. In mice, a higher infectivity is achieved by the intraperitoneal route rather than by the oral route. Oocysts freed of sulfuric acid are injected intraperitoneally, being careful to avoid backflow.
5. Oocysts from constipated cats do not sporulate; unsporulated oocysts are killed very rapidly at 37° C.
6. Sporocysts can be released easily from Clorox-treated oocysts by shaking with the help of alundum or sand grinders. This reduces the danger of scattering contaminative material by tissue grinders.

Items No. 3,4, and 6 are important because of the possibility of infection of researchers with the infective *Toxoplasma* organisms (Hutchison *et al.*, 1969). Dr. Dubey said that almost invariably those who experiment with oocysts of this organism soon become infected, as shown by laboratory tests for *Toxoplasma* titers in human beings.*

---

*J. K. Frenkel, Department of Pathology and Oncology, School of Medicine, University of Kansas Medical Center, Kansas City, Kansas 66103.

†J.P. Dubey, Department of Pathology and Oncology, School of Medicine, University of Kansas Medical Center, Kansas City, Kansas 66103.

*For further details about toxoplasmosis, the reader is urged to read Hutchison *et al.* (1971), and to review the conference reports of the Annals of the New York Academy of Sciences, Volume 64, Article 2, pages 25-277, edited by Kenneth T. Morse, 1956, entitled "Some Protozoan Diseases of Man and Animals: Anaplasmosis, Babesiosis, and Toxoplasmosis," with Clarence R. Cole as chairman of a committee of specialists. Additional details can be found in "Toxoplasmosis," Survey of Ophthalmology, Volume 6(6), part 2, December, 1961, edited by A.E. Maumenee. These are reports from a symposium by the Council for Research in Glaucoma and Allied Diseases. Also, techniques are listed in numerous papers in the 1971 issues of Vet. Rec. on ovine abortion due to toxoplasmosis, reported by Archer, Beverly, Payne, Spence, and Watson.

IV. *Sarcocystis*

A. EXAMINATION OF BIRDS FOR PRESENCE OF *SARCOCYSTIS*

Sarcocystis develops in the muscles of reptiles, birds, and mammals, including man. Fayer and Kocan (1971) described a method of examining grackles (*Quiscalus quiscula*) for the intramuscular cysts. They examined macroscopically for gross lesions on the cutaneous surfaces of the right breast, right leg and thigh, tongue, and heart. The muscles of both obviously parasitized birds and those without visible lesions were excised, crushed with a tissue grinder, and the expressed fluid was examined microscopically for motile *Sarcocystis*. Fayer and Kocan found that 93% of 98 mature birds were parasitized. Microscopy verified occurrence of infection in all with gross lesions with the exception of five. One bird without macroscopic lesions was found to be positive on microscopic examination. Immature birds and those hatched in the laboratory did not harbor *Sarcocystis*. The parasites were not found in tongues or hearts of grackles, contrary to the findings in some other hosts.

B. CULTURING *SARCOCYSTIS* IN AVIAN AND MAMMALIAN CELLS

Fayer (1970) was the first to culture *Sarcocystis* in avian and mammalian cells. He obtained motile banana-shaped organisms by crushing cysts obtained from aseptically excised muscles from grackles. A glass tissue grinder was used. The parasites were suspended in from 2 ml to 3 ml of Eagle's basal medium with Hanks' balanced salt solution, plus 10% fetal calf serum, strained through layers of sterile gauze to remove coarse tissue particles, and then centrifuged at 290 *g* for five minutes. The pelleted parasites were resuspended in tissue culture medium and inoculated into cell line cultures of (1) embryonic bovine kidney, (2) embryonic bovine trachea, (3) Madin-Darby canine kidney, (4) primary cultures of embryonic chicken kidney, (5) embryonic chicken muscle, and (6) embryonic turkey kidney. Cinemicrography was used to film the parasites developing into enlarged ellipsoid stages and then into multinucleate cyst-like stages.

Fayer (personal communication, 1971)* outlined his technique of preparing *Sarcocystis* organisms for cell cultures as follows:

1. Kill bird by blow to the head.
2. Clip off right wing at upper humerus.
3. Wet plumage with 95% ethanol.
4. Pluck feathers of right leg and thigh and right side of breast.

In all the following steps sterile instruments are used.

5. Remove skin from leg, thigh, and breast.
6. Examine muscle surfaces for cysts and excise muscles where the greatest numbers of cysts are visible.
7. Mince excised muscles with sharp scissors.
8. Place minced pieces into ground-glass or Teflon-coated tissue grinder and add from 3 ml to 5 ml of culture medium.
9. Press the pestle firmly on the tissue and twist slightly to express fluid from the minced tissue pieces. Repeat several times.
10. Decant culture medium through several folds of sterile gauze. Rinse grinding tube with from 3 ml to 5 ml of culture medium and decant again.
11. Pellet zoite suspension by centrifugation at approximately 300 *g* for 5 min.
12. Resuspend organisms in culture medium.
13. Determine number of organisms with aid of hemacytometer.

---

*Ronald Fayer, National Animal Parasite Research Laboratory, Veterinary Sciences Research Division, ARS, U. S. Department of Agriculture, Beltsville, Maryland 20705.

14. Dilute suspensions to desired number of parasites per milliliter and inoculate into Leighton tubes.

V. *Besnoitia*

A. EXAMINATION FOR *BESNOITIA JELLISONI* IN RODENTS

In examining rodents for *Besnoitia* (Ernst *et al.*, 1968), each animal was skinned and subcutaneous tissue of the back and skull was carefully observed for cysts. Perforations of the skulls were associated with the presence of the cysts. For detailed examination of the skull, the heads were removed and placed in dermestid colonies for cleaning or were preserved in 10% formalin. Clinical signs did not appear in any of the animals kept in captivity. Bigalke and Naude (1962) found that examination of the scleral conjunctiva of cattle for cysts of *B. besnoiti* was a reliable diagnostic method, but this did not prove true for kangaroo rats, as cysts could be seen in this location in only a few of the living kangaroo rats. At necropsy, the cysts could be found in abdominal mesenteries, serosa of the cecum and pancreas, and inside these organs. Cysts also occurred in the connective tissue and at the surface of almost all the sexual organs of males and females, in the dorsal bones of the skull and in the peripheral areas of the middle ear. The cysts averaged 448.2 μ with a range of from 329 μ to 586 μ. They had a wall consisting of three layers.

B. TRANSMISSION OF *BESNOITIA*

1. *Transmission of* Besnoitia jellisoni

Jellison *et al.* (1956) described transmission experiments with ingestion methods. They obtained cysts filled with crescent shaped organisms and administered these orally to baby mice with a capillary pipette. Thirty-seven of 60 young experimental mice became infected by ingestion in this manner. These authors were also able to transmit the parasites by feeding organisms from peritoneal fluid. In both experiments the maternal females in the litters were exposed to infective suspensions unintentionally; they experienced the exposure in grooming the young and by eating the young that died. Eight of ten females in the first experiment became infected but none in the second were known to have become infected. The authors pointed out that *B. jellisoni* can be transmitted to mice by ingestion and suggested the possibility that this is a mode of transmission in nature, since mice are cannibalistic.

2. *Transmission of* Besnoitia besnoiti

Tissues containing numerous cysts of *B. besnoiti* were obtained from cattle with chronic besnoitiosis in natural infections in the Transvaal, Bushveld, and Northwestern K Province in Africa and were used in artificial transmission studies by Bigalke (1967). Citrated blood was drawn and inoculated into rabbits by the intraperitoneal or intravenous routes. Cyst-bearing tissues were minced and ground in a mortar after normal saline or Hanks' solution had been added. The normal saline contained 500 units of penicillin and 500 μg of streptomycin per milliliter; the Hanks' solution contained 200 units of penicillin, 200 μg of streptomycin, and 2 μg of fungizone per milliliter. Suspensions were strained through gauze to remove large particles. When cysts were scarce, they were dissected out of the dermis to reduce host tissue before crushing as above. A hemacytometer was used to determine the number of cyst forms in the inoculum. Smears from the inoculum were stained with Giemsa to be sure the organisms were present. Suspensions were injected into rabbits or cattle. Newly isolated strains of *B. besnoiti* were passaged in rabbits for one or more generations by subinoculations of blood at the height of the febrile reaction. A splenectomized calf was injected with one strain after two rabbit passages. After the calf became febrile, 500 ml of blood were transfused into a cow. Two rabbits were given injections with blood from this animal to verify the infection after the febrile reaction. The only sign shown in cattle was an increased temperature. The rabbit proved to be highly susceptible. Innumerable cysts were found in chronically infected cattle in the dermis, subcutis, and veins. Cyst-bearing tissues were still infective nine years after purchase of two of the cattle used as donors.

C. ISOLATION OF *BESNOITIA*

Bigalke (1962) described the following methods of isolating *Besnoitia*:

Preparation of inoculum

Rabbits with scrotal edema were killed at the height of the acute phase of the disease. Testes were removed aseptically, ground in a mortar, and from 5 ml to 10 ml of medium were added. The suspension was filtered through sterile gauze. Cyst organisms were isolated from a piece of skin from a naturally infected bull. Thin slices with cysts were placed for 4 days in Hanks' solution containing 2,000 units of penicillin, 2,000 μg of streptomycin, 1,500 μg of neomycin, and 40 μg of fungizone per milliliter. Cysts were crushed in 5 ml of medium containing 500 μg of neomycin per milliliter in addition to the usual antibiotics.

Besnoitia besnoiti of rabbit testis origin was maintained in lamb kidney tissue culture for four generations by subinoculations.

Use of embryonated eggs for culturing

The organisms were injected into the chorioallantoic sac or into the yolk sac and the membranes of these were harvested for serial passage.

Seven-day-old embryos proved to be better than nine-day-old embryos. The use of Madin's and Darby's (1958) method of removing cells from the walls of culture tubes with saline containing trypsin and versene (ATV) proved to be very helpful.

Literature Cited

Akinshina, G. T., and G. D. Zasukhina. 1966. Method of investigation of mutations in Protozoa (*Toxoplasma gondii*). Genetika, Moscow 8:71-75.

Anderson, R. I., E. H. Sadun, and J. S. Williams. 1961. A technique for the use of minute amounts of dried blood in the fluorescent antibody test for schistosomiasis. Exp. Parasitol. 11: 111-116.

Archer, J. F., J. K. A. Beverley, W. A. Watson, and D. Hunter. 1971. Further field studies on the fluorescent antibody test in the diagnosis of ovine abortion due to toxoplasmosis. Vet. Rec. 88: 178-180.

Baker, C. 1955. The Baker interference microscope, 2nd ed., Mainsail Press, Ltd., London.

Bedrník, P. 1969. Cultivation of *Eimeria tenella* in tissue cultures. I. Further development of second generation merozoites in tissue cultures. Acta Protozool. 7: 87-98.

Bigalke, R. D. 1962. Preliminary communication on the cultivation of *Besnoitia besnoiti* (Morotel [sic], 1912) in tissue culture and embryonated eggs. J. S. Afr. Vet. Med. Ass. 33: 523-532.

Bigalke, R. D. 1967. The artificial transmission of *Besnoitia besnoiti* (Marotel, 1912) from chronically infected to susceptible cattle and rabbits. Onderstepoort J. Vet. Res. 34: 303-316.

Bigalke, R. D., and T. W. Naude. 1962. The diagnostic value of cysts in the scleral conjunctiva in bovine besnoitiosis. J. S. Afr. Vet. Med. Ass. 33: 1-7.

Bizzell, W. E., and H. Ciordia. 1962. An apparatus for collecting helminth parasites of ruminants. J. Parasitol. 48: 490-491.

Boughton, D. C. 1943. Sulfaguanadine therapy in experimental bovine coccidiosis. Am. J. Vet. Res. 4: 66-72.

Brackett, S. (ed.). 1949. Coccidiosis. Ann. N.Y. Acad. Sci. 52: 431-623.

Bradley, R. E., H. Botero, J. Johnson, and W. M. Reid. 1967. Techniques in parasitology. I. Gnotobiotic poultry in plastic film isolators and some applications to parasitological research. Exp. Parasitol. 21: 403-413.

Carver, R. K., and M. Goldman. 1959. Staining *Toxoplasma gondii* with fluorescent-labelled antibody. III. Reaction in frozen and paraffin sections. Am. J. Clin. Pathol. 32: 159-164.

Caulfield, J. B. 1957. Effects of varying the vehicle for OsO_4 in tissue fixation. J. Biophys. Biochem. Cytol. 3: 827-830.

Clark, W. N., and D. M. Hammond. 1969. Development of *Eimeria auburnensis* in cell cultures. J. Protozool. 16: 646-654.

Cook, M. K., and L. Jacobs. 1958. Cultivation of *Toxoplasma gondii* in tissue cultures of various derivations. J. Parasitol. 44: 172-182.

Dalton, A. J. 1955. A chrome-osmium fixative for electron microscopy. Anat. Rec. 121: 281.

Davis, L. R. 1944. Two new enterotome devices. Am. J. Vet. Res. 5: 60-61.

Davis, L. R. 1949. Portable pens for raising healthy dairy calves. USDA Agricultural Research Administration, Bureau of Animal Industry, Zoological Division, Mimeo. Jan. 1949, Rev. Sept. 1949.

Davis, L. R. 1969. Laser microscope used in purifying mixed cultures of oocysts of coccidia of domestic animals. Progress in Protozool., Third Int. Cong. Protozool., p. 382. Swets & Zeitlinger N.V. Keizersgracht 471 & 487, Amsterdam.

Davis, L. R., D. C. Boughton, and G. W. Bowman. 1955. Biology and pathogenicity of *Eimeria alabamensis* Christensen, 1941, an intranuclear coccidium of cattle. Am. J. Vet. Res. 16: 274-281.

Davis, L. R., and G. W. Bowman. 1963. Diagnosis of coccidiosis of cattle and sheep by histochemical and other techniques, 67th Ann. Proc. U.S. Livestock San. Ass. 516-522.

Davis, L. R., and G. W. Bowman. 1966. Simplified, easily constructed, inexpensive micromanipulator. J. Protozool. 13(Suppl.): 22-23.

Davis, L. R., G. W. Bowman, and D. A. Porter. 1952. Portable pens compared with other enclosures for control of diseases of dairy calves. Vet. Med. 47: 485-490, 512.

Davis, L. R., G. W. Bowman, and W. N. Smith. 1961. Himes' and Moriber's triple histochemical stain reveals endogenous stages of coccidia of sheep and cattle. J. Protozool. 8(Suppl.): 9-10.

Davis, L. R., and W. N. Smith. 1959. Fluorescent powders and ultraviolet light for tracing inoculations of oocysts. J. Protozool. 6(Suppl.): 3.

Davis, L. R., and W. N. Smith. 1960. The use of acridine orange and fluorescence microscopy for examining oocysts of cattle and sheep. J. Protozool. 7 (Suppl.): 12.

Doran, D. J. 1969a. Freezing excysted coccidial sporozoites. J. Parasitol. 55: 1229-1233.

Doran, D. J. 1969b. Cultivation and freezing of poultry coccidia. Acta Vet., Brno 38: 25-30.

Doran, D. J. 1970. Effect of age and freezing on development of *Eimeria adenoeides* and *E. tenella* sporozoites in cell culture. J. Parasitol. 56: 27-29.

Doran, D. J., and M. M. Farr. 1962. Excystation of the poultry coccidium, *Eimeria acervulina*. J. Protozool. 9:154-161.

Doran, D. J., and J. M. Vetterling. 1968. Preservation of coccidial sporozoites by freezing. Nature 217: 1262.

Doran, D. J., and J. M. Vetterling. 1969a. Infectivity of two species of poultry coccidia after freezing and storage in liquid nitrogen vapor. Proc. Helm. Soc. Wash. 36: 30-33.

Doran, D. J., and J. M. Vetterling. 1969b. Influence of storage period on excystation and development in cell cultures of sporozoites of *Eimeria meleagrimitis* Tyzzer, 1929. Proc. Helm. Soc. Wash. 36: 33-35.

Dorney, R. S. 1964. Evaluation of a microquantitative method for counting coccidial oocysts. J. Parasitol. 50: 518-522.

Dubey, J. P., N. L. Miller, and J. K. Frenkel. 1970a. The *Toxoplasma gondii* oocysts from cat feces. J. Exper. Med. 132: 636-662.

Dubey, J. P., N. L. Miller, and J. K. Frenkel. 1970b. Characterization of the new fecal form of *Toxoplasma gondii*. J. Parasitol. 56: 447-456.

Edgar, S. A. 1954. Effect of temperature on the sporulation of oocysts of the protozoan, *Eimeria tenella*. Trans. Am. Micr. Soc. 62: 237-242.

Edgar, S. A. 1958. Problems in the control of coccidiosis. Proc. Semiannual Meet. Am. Feed Manufact. Nutr. Counc. 19-26.

Edgar, S. A., C. A. Herrick, and L. A. Fraser. 1944. Glycogen and the life cycle of the coccidium, *Eimeria tenella*. Trans. Am. Micr. Soc. 63: 199-202.

Edgar, S. A., and C. T. Seibold. 1964. A new coccidium of chickens, *Eimeria mivati* sp.n. (Protozoa: Eimeriidae) with details of its life history. J. Parasitol. 50: 193-204.

Eichenwald, H. F. 1956. The laboratory diagnosis of toxoplasmosis. Ann. N. Y. Acad. Sci. 64: 207-214.

El-Badry, H. M. 1963. Micromanipulators and micromanipulation. Springer-Verlag, Vienna.

Ernst, J. V., B. Chobotar, E. C. Oaks, and D. M. Hammond. 1968. *Besnoitia jellisoni* (Sporozoa: Toxoplasmea) in rodents from Utah and California. J. Parasitol. 54: 545-549.

Eyles, D. E., N. Coleman, and B. J. Cavanaugh. 1956. The preservation of *Toxoplasma gondii* by freezing. J. Parasitol. 42: 408-413.

Farr, M. M., and D. J. Doran. 1962. Comparative excystation of four species of poultry coccidia. J. Protozool. 9: 403-407.

Farr, M. M., and E. E. Wehr. 1949. Survival of *Eimeria acervulina, E. tenella*, and *E. maxima* oocysts on soil under various field conditions. Ann. N. Y. Acad. Sci. 52: 468-472.

Farrant, J. 1966. The preservation of living cells, tissues, and organs at low temperatures: Some underlying principles. Lab. Pract. 15: 402-404, 409.

Fayer, R. 1970. *Sarcocystis*: Development in cultured avian and mammalian cells. Science 168: 1104-1105.

Fayer, R., and D. M. Hammond. 1967. Development of first generation schizonts of *Eimeria bovis* in cultured bovine cells. J. Protozool. 14: 764-772.

Fayer, R., and R. M. Kocan. 1971. Prevalence of *Sarcocystis* in grackles in Maryland. J. Protozool. 18: 547-548.

Fitzgerald, P. R. 1967. Results of continuous low-level inoculations with *Eimeria bovis* in calves. Am. J. Vet. Res. 28: 659-665.

Fitzgerald, P. R. 1970. Development of *Eimeria stiedai* in avian embryos. J. Parasitol. 56: 1252-1253.

Frenkel, J. K. 1971. Current Topics in Pathology. 54: 29-75. Springer-Verlag, New York.

Frenkel, J. K., J. P. Dubey, and N. L. Miller. 1970. *Toxoplasma gondii* in cats: Fecal stages identified as coccidian oocysts. Science 167: 893-896.

Frenkel, J. K., and L. Jacobs. 1958. Ocular toxoplasmosis. Pathogenesis, diagnosis and treatment. A. M. A. Arch. Ophth. 59: 260-279.

Fulton, J. D., and D. F. Spooner. 1957. Preliminary observations on the metabolism of *Toxoplasma gondii*. Trans. Roy. Soc. Trop. Med. Hyg. 51: 123-124.

Fulton, J. D., and D. F. Spooner. 1960. Metabolic studies on *Toxoplasma gondii*. Exp. Parasitol. 9: 293-301.

Galigher, A. E., and E. N. Kozloff. 1971. Essentials of practical microtechnique. 2nd ed. Lea and Febiger, Philadelphia.

Goff, O. E. 1942. Flowers of sulfur and charcoal in the prevention of experimentally produced coccidiosis. Poult. Sci. 21: 23-29.

Gordon, H. McL., and H. V. Whitlock. 1939. A new technique for counting nematode eggs in sheep faeces. J. Counc. Sci. Ind. Res. (Austral.) 12: 50-52.

Hammond, D. M., F. L. Andersen, and M. L. Miner. 1963. The occurrence of a second asexual generation in the life cycle of *Eimeria bovis* in calves. J. Parasitol. 49: 428-434.

Hammond, D. M., G. W. Bowman, L. R. Davis, and B. T. Simms. 1946. The endogenous phase of the life cycle of *Eimeria bovis*. J. Parasitol. 32: 409-427.

Hammond, D. M., B. Chobotar, and J. V. Ernst. 1968. Cytological observations on sporozoites of *Eimeria bovis* and *E. auburnensis*, and an *Eimeria* species from the Ord Kangaroo Rat. J. Parasitol. 54: 550-558.

Hammond, D. M., and L. R. Davis. 1944. An improved method for sporulating oocysts in bovine fecal material. Am. J. Vet. Res. 5: 70-71.

Hammond, D. M., L. R. Davis, and G. W. Bowman. 1944. Experimental infections with *Eimeria bovis* in calves. Am. J. Vet. Res. 5: 303-311.

Hammond, D. M., J. V. Ernst, and M. Goldman. 1965. Cytological observations on *Eimeria bovis* merozoites. J. Parasitol. 51: 852-858.

Hammond, D. M., R. A. Heckman, M. L. Miner, C. M. Senger, and P. R. Fitzgerald. 1959. The life cycle stages of *Eimeria bovis* affected by the immune reaction in calves. Exp. Parasitol. 8: 574-580.

Hibbert, L. E., and D. M. Hammond. 1968. Effects of temperature on *in vitro* excystation of various *Eimeria* species. Exp. Parasitol. 23: 161-170.

Hibbert, L. E., D. M. Hammond, and J. R. Simmons. 1969. The effects of pH, buffers, bile and bile acids on excystation of sporozoites of various *Eimeria* species. J. Protozool. 16: 441-444.

Hicks, J. D., and E. Matthaei. 1955. Fluorescence in histology. J. Pathol. Bacteriol. 70: 1-12.

Hill, C. H., and R. E. Zimmerman. 1961. A mechanical apparatus for screening worm eggs from feces. J. Parasitol. 47: 357-362.

Himes, M., and L. Moriber. 1956. A triple stain for deoxyribonucleic acid, polysaccharides, and protein. Stain Tech. 31: 67-70.

Hunter, J. E. 1959. Consideration for the evaluation of coccidiostats. Proc. 20th Ann. Meet. Am. Feed Manufact. Nutr. Counc. 16-19.

Hutchison, W. M., J. F. Dunachie, and K. Work. 1969. Transmissible toxoplasms. *Nippostrongylus* and *Toxoplasma*, Symposia, Brit. Soc. Parasitol. 7: 51-63.

Hutchison, W. M., J. F. Dunachie, K. Work, and J. Chr. Siim. 1971. The life cycle of the coccidian parasite, *Toxoplasma gondii*, in the domestic cat. Trans. Roy. Soc. Trop. Med. Hyg. 65: 380-399.

Jackson, A. R. B. 1962. Excystation of *Eimeria arloingi* (Marotel, 1905): Stimuli from the host sheep. Nature 194: 847-849.

Jackson, A. R. B. 1964. The isolation of viable coccidial sporozoites. Parasitology 54: 87-93.

Jacobs, L. 1956. Propagation, morphology, and biology of *Toxoplasma*. Ann. N. Y. Acad. Sci. 64: 154-179.

Jacobs, L., J. S. Remington, and M. L. Melton. 1960. A survey of meat samples from swine, cattle, and sheep for the presence of encysted *Toxoplasma*. J. Parasitol. 46: 23-28.

Jellison, W. L., W. J. Fullerton, and H. Parker. 1956. Transmission of the protozoan *Besnoitia jellisoni* by ingestion. Ann. N.Y. Acad. Sci. 64: 271-274.

Kouwenhoven, B. 1967. The possibility of low temperature freezing of *Eimeria tenella* sporocysts. Tijdschr. Diergeneesk. 92: 1639-1642.

Krassner, S. M. 1963. Factors in host susceptibility and oocyst infectivity in *Eimeria acervulina* infections. J. Protozool. 10: 327-333.

Lane, C. 1924. The mass diagnosis of ankylostome infestation. Trans. Roy. Soc. Trop. Med. Hyg. 17: 407-436.

Leeson, C. R., and T. S. Leeson. 1970. Staining method for sections of epon-embedded tissues for light microscopy. Can. J. Zool. 48: 189-191.

Levine, N. D., and F. L. Andersen. 1966. Frozen storage of *Tritrichomonas foetus* for 5-6 years. J. Protozool. 13: 199-202.

Levine, N. D., K. N. Mehra, D. T. Clark, and I. J. Aves. 1960. A comparison of nematode egg counting techniques for cattle and sheep feces. Am. J. Vet. Res. 21: 511-515.

Levine, P. P. 1939. The effect of sulfanilamide on the course of experimental avian coccidiosis. Cornell Vet. 29: 309-320.

Long, P. L., and J. G. Rowell. 1958. Counting oocysts of chicken coccidia. Lab. Pract. 7: 515-518, 534.

Lotze, J. C. 1953. Life history of the coccidian parasite, *Eimeria arloingi*, in domestic sheep. Am. J. Vet. Res., 14: 86-95.

Lotze, J. C., and R. G. Leek. 1961. A practical method for culturing coccidial oocysts in tap water. J. Parasitol. 47: 588-590.

Lotze, J. C., and R. G. Leek. 1967. A cold storage technic for studying excystation of *Eimeria tenella*. J. Protozool. 14: 231-232.

Luft, J. H. 1961. Improvements in epoxy resin embedding methods. J. Biophys. Biochem. Cytol. 6: 409-414.

Lumsden, W. H. R., M. P. Cunningham, W. A. F. Webber, K. van Hoeve, and P. J. Walker. 1963. A method for the measurement of the infectivity of trypanosome suspensions. Exp. Parasitol. 14: 269-279.

Madin, S. H., and N. B. Darby. 1958. Established kidney cell lines of normal adult bovine and ovine origin. Proc. Soc. Exp. Biol. and Med. 98: 574-576.

Marquardt, W. C. 1961. Separation of nematode eggs from fecal debris by gradient centrifugation. J. Parasitol. 47: 248-250.

Mellors, R. C. 1955. Analytical cytology, Chapt. 3, McGraw-Hill, New York.

Merck, Sharp & Dohme. 1959. Conference on methods of drug testing coccidiostats; Cheswold Farm, Dover, Del.

Monné, L., and G. Hönig. 1954. On the properties of the shells of coccidian oocysts. Ark. Zool., Stockholm, Series II, 7: 251-256.

Norton, C. C., and L. P. Joyner. 1968. The freeze preservation of coccidia. Res. Vet. Sci. 9: 598-600.

Norton, C. C., D. D. Pout, and L. P. Joyner. 1968. Freeze preservation of *Eimeria acervulina* Tyzzer, 1929. Folia Parasitol., Prague 15: 203-211.

Nyberg, P. A., D. H. Bauer, and S. E. Knapp. 1968. Carbon dioxide as the initial stimulus for excystation of *Eimeria tenella* oocysts. J. Protozool. 15: 144-148.

Nyberg, P. A., and D. M. Hammond. 1964. Excystation of *Eimeria bovis* and other species of bovine coccidia. J. Protozool. 11: 474-480.

Patnaik, B. 1966. A technique of obtaining oocysts of coccidia in pure state from chicken feces by modified Marquardt's method. Ind. Vet. J. 43: 414-422.

Patton, W. H. 1965. *Eimeria tenella*: Cultivation of the asexual stages in cultured animal cells. Science 150: 767-769.

Peters, B. G., and J. W. G. Leiper. 1940. Variation in dilution counts of helminth eggs. J. Helm. 18: 117-142.

Ray, D. K. 1953. Comparative efficiency using sulfate, sugar and salt solutions for flotation of coccidial oocysts of sheep and goats. Prac. Zool. Soc., Bengal 6: 135-138.

Remington, J. S. 1970. Toxoplasmosis: Recent developments. Ann. Rev. Med. 21: 201-218.

Rikimaru, M. T., F. T. Galysh, and R. F. Shumard. 1961. Some pharmacological aspects of a toxic substance from oocysts of the coccidium *Eimeria tenella*. J. Parasitol. 47: 407-412.

Rogers, W. P. 1960. The physiology of infective processes of nematode parasites; the stimulus from the animal host. Proc. Roy. Soc. Ser. B. Biol. Sci. 152: 367-386.

Sabatini, D. D., K. Bensch, and R. J. Barrnett. 1963. The preservation of cellular ultrastructure and enzymatic activity by aldehyde fixation. J. Cell. Biol. 17: 19-58.

Sabin, A. B., and H. A. Feldman. 1948. Dyes as microchemical indicators of a new immunity phenomenon affecting a protozoan parasite (*Toxoplasma*). Science 108: 660-663.

Sharma, M. N., and W. M. Reid. 1963. A cleaning method for coccidial oocysts using density-gradient sedimentation. J. Parasitol. 49: 159-160.

Sheffield, H. G., and D. M. Hammond. 1966. Fine structure of first-generation merozoites of *Eimeria bovis*. J. Parasitol. 52: 595-606.

Shumard, R. F., I. A. Schipper, and D. F. Eveleth. 1955. A motor-driven enterotome for the rapid removal of helminths from the gastrointestinal tracts of large mammals. J. Parasitol. 41 Sec. 2: 38.

Smetana, H. 1933. Coccidiosis of the liver in rabbits. I. Experimental study on the excystation of oocysts of *Eimeria stiedae*. Arch. Pathol. 15: 175-192.

Smith, B. F., and C. A. Herrick. 1944. The respiration of the protozoan parasite, *Eimeria tenella*. J. Parasitol. 30: 295-302.

Smith, W. N., and L. R. Davis. 1965. A comparison of dry and liquid feeds as vehicles for coccidial infection of cattle and sheep. Am. J. Vet. Res. 26: 273-279.

Speer, C. A., D. M. Hammond, and G. L. Kelley. 1970. Stimulation of motility in merozoites of five *Eimeria* species by bile salts. J. Parasitol. 56: 927-929.

Sporn, M. B., T. Wanko, and W. Dingman. 1962. The isolation of cell nuclei from rat brain. J. Cell. Biol. 15: 109-120.

Stewart, G. L., and H. A. Feldman. 1965. Use of tissue culture cultivated *Toxoplasma* in the dye test and for storage. Proc. Soc. Exp. Biol. Med. 118: 542-546.

Stoll, N. R. 1923. Investigations on the control of hookworm disease. XV. An effective method of counting hookworm eggs in feces. Am. J. Hyg. 3: 59-70.

Stowell, R. E. 1945. Feulgen reaction for thymonucleic acid. Stain Tech. 20: 45-58.

Strout, R. G., J. Solis, S. C. Smith, and W. R. Dunlop. 1965. *In vitro* cultivation of *Eimeria acervulina* (Coccidia). Exp. Parasitol. 17: 241-246.

Suggs, M., K. W. Walls, and I. G. Kagan. 1968. Comparative antigenic study of *Besnoitia jellisoni, B. panamensis* and 5 *Toxoplasma gondii* isolates. J. Immunol., 101: 166.

Swan, R. A. 1970. An improved method for the collection of faecal samples from sheep. Austral. Vet. J. 46: 25-26.

Takumi, K., I. Takebayashi, H. Takeuchi, H. Ikeda, and N. Toshioka. 1966. The use of lyophilized parasites in the indirect fluorescent antibody technique for detection of *Toxoplasma* antibody. Jap. J. Microbiol. 10: 189-191.

Trexler, P. C., and L. I. Reynolds. 1957. Flexible film apparatus for the rearing and use of germ-free animals. Appl. Microbiol. 5: 406-412.

Vetterling, J. M. 1969. Continuous-flow differential density flotation of coccidial oocysts and a comparison with other methods. J. Parasitol. 55: 412-417.

Wagenbach, G. E. 1969. Purification of *Eimeria tenella* sporozoites with glass bead columns. J. Parasitol. 55: 833-838.

Wagenbach, G. W., J. R. Challey, and W. C. Burns. 1966. A method for purifying coccidian oocysts employing clorox and sulfuric acid-dichromate solution. J. Parasitol. 52: 1222.

Walker, P. J. 1966. Freeze preservation of parasitic protozoa. Lab. Pract. 15: 423-426.

Wallace, G. D. 1969. Sabin-Feldman dye test for toxoplasmosis. The use of sodium citrate in accessory factor, and a method for collecting and storing blood on paper disc. Am. J. Trop. Med. Hyg. 18: 395-398.

Walton, B. C., and I. Arjona. 1971. Utilization of whole blood specimens on filter paper for the indirect fluorescent antibody test for toxoplasmosis. J. Parasitol. 57: 678-680.

Walton, B. C., B. M. Benchoff, and W. H. Brooks. 1966. Comparison of the indirect fluorescent antibody test and methylene blue dye test for detection of antibodies to *Toxoplasma gondii*. Amer. J. Trop. Med. Hyg. 15: 149-152.

Whitlock, H. V. 1948. Some modifications of McMaster helminth egg-counting technique and apparatus. J. Counc. Sci. Ind. Res., Austral. 21: 177-180.

Wilson, P. A. G., and D. Fairbairn. 1961. Biochemistry of sporulation in oocysts of *Eimeria acervulina*. J. Protozool. 8: 410-416.

Index

Abortion, from toxoplasmosis, 359, 377
Acetarsone, affecting cultures, 239
Acnidosporidia, 7
Actinomyxidia, 7
Adelea, 6, 8
  ovata, 2, 5
Adeleida, 7, 8
Adeleidea, 7, 8
Adeleorina, 3, 11
Adelina tribolii, macrogamete ultrastructure of, 119, 121
Adenine, affecting cultures, 237-238
Adenosine triphosphatase, in macrogametes, 133
Adrenal gland involvement, in toxoplasmosis, 364, 365
Age
  of cells, affecting cell cultures, 234
  of host
    and immunity, 297, 299
    and oocyst production, 424-425
    and reproduction of coccidia, 260
    and susceptibility to infection, 259-262, 313
  of parasite, affecting cell cultures, 229
Aggregata, 10, 11
  eberthi
    conoid of, 90
    enzyme activity in, 133
    granular bodies of, 199
    histochemical reactions of, 171
    inner membrane complex in, 66
    macrogamete ultrastructure of, 119
    merozoite formation in, 68
    microgamete ultrastructure in, 108
    preconoidal rings of, 88
    sporozoite ultrastructure in, 101
Aggregatidae, 6, 7, 9, 10, 11-12
p-Aminobenzoic acid
  deficiency, effects of, 158-159
  effects in cultures, 238
6-Aminonicotinamide, effects of, 157
Aminopterin, affecting cultures, 238
Amphotericin B, affecting cultures, 239
Amprolium
  dietary, 156-157
  effects in cultures, 239
Amylopectin, 149-150
  in excystation, 166
  in macrogametes
    of *Eimeria*, 116
    of *Toxoplasma*, 118
  in sporozoites, 101
Amylopectin, in sporozoites—*Cont'd*
    storage affecting, 57, 259
  in sporulation, 161
  in *Toxoplasma* cysts, 353
Anaplasma, 8
Angeiocystidae, 9, 10
Angeiocystis, 10, 12
  sporozoite ultrastructure, 101
Animals, experimental, care and housing of, 30
Antibiotics, affecting cultures, 238-239
Antibodies in coccidial infections, 306-315
  agglutinating, 306
  autoantibodies, 313
  in *Besnoitia* infections, 370, 371
  coccidiocidal or coccidiostatic, 307
  complement-fixing, 306
    in toxoplasmosis, 372
  copro-antibodies, 313-314, 318
  cytophilic, 320
  enhancing, 313
  fluorescent-labeled, 306
  heterophile, 312-313
  humoral, 306, 307, 315, 318-319
    and macrophage activity, 320
  indirect fluorescent antibody test for *Toxoplasma*, 446-447, 448
  locally produced, 313-315
  lytic, 306
  naturally occurring, 312-313
  neutralizing, 307
  opsonizing, 312, 320
  precipitating, 306
  protective effects of, 315-319
    and mode of action, 318-319
  reactions with particulate antigens, 307, 308-309
  reactions with soluble antigens, 307, 310-311
  in toxoplasmosis, 370, 371-372
Anticoccidial drugs, growth factors and antagonists, 155-159
Antigens, coccidial, 306-312
  functional, 306
  host antigens, 306-307, 312
  and loss of infectivity, 307, 314
  particulate, 306
    antibody reactions with, 307, 308-309
    preparation of, 443
  persistence of, 307-312
  preparation from cultures, 236-237
  shared host/parasite, 306
  soluble, 306

Antigens, coccidial, soluble—*Cont'd*
antibody reactions with, 307, 310-311
preparation of, 443-444
storage of preparations of, 444
Antilymphocyte serum, affecting immunity, 326-327
Apical complex, 2
in sporozoite transformation into trophozoite, 62
Apicomplexa, 2-4, 10, 11-14
apical complex of, 2, 62
classes of, 3
Archispores, 6
Arthrocystis, 14, 18
Arthropods, and toxoplasmosis transmission, 386
Asexual generations, 51, 54, 55, 71
and cell cultures, 215-222
ultrastructural aspects of, 72, 121
Aspartate aminotransferase, 170
Asporocystidae, 6, 7, 8
Atoxoplasma, 12, 14
See also Lankesterella
Autoantibodies, 313
Avian embryo cultures. *See* Culture studies, in avian embryos
Avian infections, *See* Birds
Axonemes, 127
in microgametes, 111
6-Azauracil, as anticoccidial, 155

Babesia
bigemina
microtubules of, 90
ultrastructure of, 99, 105
gibsoni, micronemes of, 92
ovis
development of, 121
micronemes of, 92
micropore of, 93
paired organelles in, 91
ultrastructure of, 100, 102, 105
Babesioidea, 8
Bacteria-free poultry, production of, 444-445
Bananella, 6
Barroussia, 6, 8
Barrouxia, 6, 13, 17
Barrouxiinae, 7
Bartonella, 8
Bartonellosis, and toxoplasmosis, 373
Basal bodies, 107
Benedenia, 6, 8
Benezimidizole, affecting cultures, 237
Besnoitia, 14, 18
acquired immunity to, 365
antibody response to, 370, 371
besnoiti
cultivation of, 186
in avian embryos, 187, 236
in cell cultures, 208, 209, 236
transmission of, 452-453
comparison with *Toxoplasma,* 390-395
conoid of, 88
cross-reactions with other coccidia, 372

Besnoitia—Cont'd
cultivation of, 185, 186
in avian embryos, 186, 236
in cell cultures, 207-209, 236
agents affecting, 240
aggregates and cytopathological effects, 208-209
cell types affecting, 233
cells supporting development, 207-208
extracellular movement of, 191
histopathological effects, 209
and immunity studies, 241
intracellular movement and location, 197, 198
mode and rate of multiplication, 208
reinvasion of cells, 209
release from cells, 209
time and rate of entry, 197
examination of rodents for, 452
isolation of, 453
jellisoni
conoid of, 90
cultivation of, 186
in avian embryos, 187
in cell cultures, 207-209
cyst wall of, 53
merozoite ultrastructure, 102, 104
micropore of, 93
microtubules of, 89
preconoidal rings of, 88
transmission of, 452
ultrastructure of, 98
ocular lesions from, 362
paired organelles in, 91
recovery from infection, 365
research techniques, 452-453
transmission of, 452-453
Besnoitiinae, 14
Betamethasone
affecting immunity, 323
affecting schizogony, 71
and serum leakage in chickens, 35
Bile and bile salts
and excystation, 58, 163-164
stimulation of merozoite motility, 70
stimulation of sporozoite motility, 58
See also Trypsin and bile treatment
Biotin requirements, 157
Birds
cross transmission studies, 30, 33
examination for *Sarcocystis,* 451
excystation of coccidia in, 57-58
life cycle of coccidia in, 51
parenteral inoculation studies of, 37
Blastophores, 48, 64, 66, 218
Blood loss, in coccidiosis. *See* Hemorrhagic coccidiosis
Blood stream, *Toxoplasma* in, 355
Blood transfusions, and toxoplasmosis transmission, 385
Bone marrow, sporozoites in, 282
Bradyzoites, 344, 345, 346
cyst formation by, 344, 351-353

Brain
lesions from *Toxoplasma*, 362, 365, 366, 373
as site of infection, with *Isospora*, 40
Bursa of Fabricius, as site of infection, 37
Bursectomy, affecting immunity, 325-326, 328-329

Canaries
abnormal sites of infection in, 40, 55
cross transmission studies, 33
size of oocysts in, 28
Carbohydrate metabolism, coccidiosis affecting, 278
Carbon dioxide
affecting organisms in cultures, 230
and excystation, 162-163, 430
and micropyle changes, 57
and Stieda body changes, 58
Carnivores, cross transmission studies in, 33
Carotene levels, coccidiosis affecting, 280
Caryolysus, 6
Caryophagus, 6
Caryospora, 13, 16
Caryosporidae, 9
Caryosporinae, 7
Caryotropha, 8, 10
mesnili, 10, 53
Caryotrophidae 7, 9, 10, 12
Cats
cross transmission studies, 30, 33
excystation in, 26
gross pathology of infections in, 267-268
Isospora felis transmission to dog, 30
life cycle of *Isospora* in, 54-56
toxoplasmosis in, 358, 359, 364, 368, 372, 379, 389
Cattle
gross pathology of infections in, 265-266
histopathology of infections in, 269, 270
Cecum
merozoites inoculated directly into, 438
as site of infection, 37
Cell-mediated immunity, 319-322, 325, 333
in toxoplasmosis, 370-371
Cells
alterations in host cells, 46, 48, 53-54, 281-282
See also Host cells
necrosis from tachyzoites, 355
types of, and site specificity, 38-39
Centrifuge, for isolation of oocysts, 415
Centriolar granules, on nuclear membrane, 57
Centrioles
in merozoites, 68, 102
in microgamonts, 105, 107
occurrence in nuclear division, 64
Centrocône, 69
Centromeres, in meiosis, 57
Cephaloidophorus conformis, 4
Chamois, cross transmission studies in, 32
Chemistry. *See* Cytochemistry; Histochemistry
Chemotherapy, and nutrition, 154-159
Chickens
cross transmission studies in, 30, 31-32
Eimeria infection of embryos, 312, 317, 321
Chickens—*Cont'd*
excystation in, 59
gross pathology of infections in, 263-264
histopathology of *Eimeria* infections, 268
Marek's disease in, 285, 299
parenteral inoculation studies, 37
toxoplasmosis in, 355, 357, 368, 372
Chinchilla
cross transmission studies in, 33
toxoplasmosis in, 362
Chromatin, in nuclear division, 64
Chromatin of host cell, changes in, 46, 54
Chromosomes, in meiosis, 57
Cinemicrography, 59-60
Cisternae, in merozoites, 103
Classification of Coccidiorina, 11-14
Clear globules. *See* Refractile bodies
Clinical infections, source of, 73
Clostridium welchii, necrotic enteritis from, 284-285
Club-shaped organelles, 92
See also Paired organelles and Rhoptries
Cnidospora, 10
Cnidosporidia, 7
Coccidia, 5
Coccidiasina, 1, 3, 4, 11-14
Coccididae, 7
Coccidiida, 6, 7
Coccidiidea, 6
Coccidiomorpha, 7, 8
Coccidiorina, 11-14
Coccidiosis
in cattle, 73-74
hemorrhagic. *See* Hemorrhagic coccidiosis
in man, 268
pathology of. *See* Pathology
Coccidiostats, affecting cultures, 239-240
Coccidium 5, 6, 7, 8
cuniculi, 2
falciforme, 5
lacazei, 2
oviforme, 5
schubergi, 2
Coccidium-like bodies, in human disease, 2
Coelotropha
durchoni
microgamete ultrastructure, 108
micropore of, 94
sporozoite ultrastructure, 101
Cold storage. *See* Freezing
Color of oocysts, 29
Complexes, host cell-parasite, 53, 54
Condensation of protoplasm, in oocysts, 56
Connective tissue cells, coccidia in, 51
Conoid, 2
electron microscopy of, 88-91
and penetration of host cell, 60, 166
and preconoidal rings, 88
protrusion of, 91, 166
Copro-antibodies, 313-314, 318
Corticosteroids
and gut permeability, 319

Corticosteroids—*Cont'd*
and patent period, 285
and reactivation of toxoplasmosis, 359, 364, 368, 374
and schizogony, 257
and sporozoite migration through tissues, 282
and susceptibility to infection, 34-35, 323-325
Counting
of oocysts, 418-419
of schizonts, 439
Crescent-shaped bodies, with gamonts and schizonts, 72
Crescent-shaped organelles, in macrogametes of *Eimeria*, 118
Cretya, 6
Cross immunity studies, 25, 30
and toxoplasmosis, 365
Cross transmission studies, 26, 27, 30-34, 54
and abnormal sites for organisms, 39
negative experiments in, 30
Crowding phenomenon, 304, 322
and dosage of oocysts, 256, 258, 260
Cryptosporidiidae, 7, 9, 13, 17-18
Cryptosporidium, 13, 17
wrairi, 17-18
Crypts of Lieberkuhn
merozoites in, 48
sporozoites in, 38, 268
Crystallospora, 6, 7, 8
Culture studies, 183-242
and antigen preparations, 236-237
and antimicrobial studies, 237-241
applications of, 234-242
in avian embryos, 186-190, 435
age of embryo affecting, 190
age of inoculum affecting, 189
applications of, 234-242
Besnoitia, 187
development and pathological effects in, 186-189
factors affecting, 186-190
Eimeria, 187-189
sex of embryo affecting, 190
size of inoculum affecting, 189
strain of embryo affecting, 190
strain of inoculum affecting, 189
temperature of incubation affecting, 190
Toxoplasma, 186-187
in cell culture, 191-234
and age and number of cells, 234
and age of parasite, 228-229
applications of, 234-242
and behavior of organisms prior to development, 191-199
active penetration, 191-196
entering and leaving cells, 191-196
extracellular movement, 191
intracellular movement and location, 197-198
passive ingestion, 196
phagocytosis, 196
time and rate of entry, 196-197
Besnoitia, 207-209
cell passage number affecting, 234
cell types affecting, 232-234
Culture studies, in cell culture—*Cont'd*
cellular factors in, 232-234
culture chamber affecting, 230
debris affecting, 224, 231
development of organisms, 200-228
environmental factors affecting, 230-232
factors affecting development, 228-234
and immunity and resistance studies, 240-241
and interferon protection against *Toxoplasma*, 357
and isolation of organisms, 234
and maintenance of organisms, 235-236
media affecting, 231-232
merozoites as inoculum, 209, 216-217
parasitological factors affecting, 228-230
penetration in, 191-196, 221, 232
Sarcocystis, 227-228, 451
and size of inoculum, 228
sporozoites as inoculum, 209, 210-215
obtaining of, 433-434
staining of, 441
survival of organisms in, 223-225
and age of sporozoites, 224
and cell type, 223-224
debris affecting, 224, 231
and species of parasite, 224
temperature affecting, 230-231
Toxoplasma, 200-207
treatment of parasites prior to inoculation, 229-230
in tissue cultures, 39, 156, 157, 166
for *Toxoplasma*, 448, 449
See also Research techniques *and individual species cultured*
Cyanide, affecting metabolism, 165, 167
Cyclophosphamide, and relapsing toxoplasmosis, 357, 364
Cyclospora, 6, 7, 8, 13, 16
Cyclosporidae, 9
Cyclosporinae, 7
Cysts
bradyzoite-forming, 344, 351-353
formation in cell cultures, 207
new cystic form of toxoplasmosis, 382-383
Cytamoeba, 6
Cytochemistry, 146-151
of oocyst wall, 151-154
Cytochrome, and metabolism studies, 167, 170
Cytomegalovirus infection, and toxoplasmosis, 373
Cytomere-like bodies, 121
Cytoplasm of host cell, changes in, 46, 54
Cytoplasmastränge, 92
Cytosporidia, 6

Dactylosoma, 8
Debris in cell cultures, effects of, 224, 231
Dehydrogenases, activity of, 172
Deoxyribonuclease, affecting granules in nucleolus, 72
Deoxyribonucleic acid, 147
in nucleolus, 72
Development of coccidia. *See* Life cycle
Dexamethasone
and abnormal sites of infection, 40

Dexamethasone—*Cont'd*
effects in calves, 285
and schizonts in endothelial cells of liver, 55
and susceptibility to infection, 34-35, 323, 324
2,6-Diaminopurine, affecting cultures, 237
Diaspora, 8, 17
Diaveridine, effects of, 158
Diet. *See* Nutrition
Dimethyl sulfoxide, for preservation of coccidia, 436-437
Diplospora, 6, 8, 15, 16
brumpti, 15
subgenera of, 16
Diplosporidae, 9
Disporea, 6
Disporocystidae, 6, 7, 8
Dobellidae, 8
Dogs
cross transmission studies in, 30, 33
Entamoeba histolytica, host specificity, 24
gross pathology of infections in, 267-268
Isospora felis transmission from cat, 30
life cycle of *Isospora* in, 54-56
Dorisiella, 13, 17
scolelepidis, 17
Drepanidium, 6
Ducks
cross transmission studies in, 31
gross pathology of infections in, 264-265
Duodenum, *Isospora felis* in, 55

Echinospora, 6, 8, 17
Ectomerogony, 66-68, 121
Eggs
coccidiosis affecting production of, 278
infected, and toxoplasmosis transmission, 385
Eimeria, 1, 5, 6, 7, 8, 12, 16
acervulina, 57
affecting blood cell numbers, 279
affecting egg production, 278
affecting food and water intake, 277
affecting gut permeability, 279-280
affecting weight gain, 277
amylopectin in, storage affecting, 259
antibody reactions
to particulate antigens, 308-309
to soluble antigens, 310-311
corticosteroids affecting, 324
cross transmission studies, 31
cultivation of
in avian embryos, 188
in cell cultures, 210, 224
cytochemistry of, 146
dose of oocysts affecting host, 256, 257, 298
excystation in, 163, 165, 166
gross pathology of infection with, 264
immunogenicity of, 297
variations between strains of, 302
and intestinal function of host, 174, 175
and Marek's disease, 285
oocyst production by, 286
parenteral inoculations of, 37, 38

Eimeria, acervulina—Cont'd
pathogenesis of infections with, 281
preservation by cold storage, 436
protein in, 148
resistance to
and age of host, 259
and strains of host, 262
site of development in host, 255
sporozoite escape from sporocysts in, 59
sporulation in, 161
Stieda body of, 58
vitamin requirements of, 155-159
adenoeides
cross transmission studies, 31
cultivation in cell cultures, 210, 224, 229, 239
dose of oocysts affecting host, 258
gross pathology of infection with, 265
hyaluronidase affecting penetration of, 195
morphology of oocysts, 25
penetration of, 166
preservation by cold storage, 437
protruded conoidal complex in sporozoites, 60
refractile bodies in sporozoites, 199
resistance to, and age of host, 260
size of oocysts in, 28
species specificity of immunity to, 302
ahsata
dry feed inoculations of, 414
gross pathology of infections with, 266
preservation by cold storage, 437
ahtanumensis, sporulation, 56
alabamensis
cultivation in cell cultures, 195, 196, 197, 198, 210, 217, 218, 220, 225
gross pathology of infections with, 266
immunity to, 298
inner membrane complex in, 69
merozoite formation in, 68
micropore of, 95, 100
nutrient intake in, 66
refractile bodies in sporozoites, 199
changes in, 62
resistance to, and age of host, 260
shape of oocysts in, 29
spindles in schizonts of, 64
anseris, 265
immunogenicity of, 298
arloingi
cellular reactions to, 270
dose of oocysts affecting host, 258
excystation in, 162
gross pathology of infections with, 266
large schizonts of, 270
micropyle changes in, 57
preservation by cold storage, 437
size of endogenous stages, and pathogenicity, 256
asexual reproduction in, 121
See also Asexual generations
auburnensis
cell type for development of, 38
cultivation in cell cultures, 210, 217, 218, 224, 225, 226

Eimeria, auburnensis—Cont'd
growth stages of, 64, 66
host cell changes, 54
intravacuolar tubules in macrogametes of, 112
in lamina propria cells, 51
microgamete anlage in, 72
microgamete ultrastructure, 107, 108
microgametogenesis in, 122
microgamont ultrastructure, 105
micropore of, 92, 93
microtubules of, 89
morphology of oocysts, 25
oocyst structure, 25
outer oocyst wall variability in, 28
pellicle in macrogamete of, 111
spindles in schizonts of, 64
sporozoite-shaped schizonts in, 62
sporozoite ultrastructure, 101
Stieda body of, 58
wall-forming bodies of, 115
avium, cross transmission studies of, 31
bilamellata
cross transmission studies, 32
cultivation in cell cultures, 197, 210-211, 217, 218, 221
in lamina propria, 51
refractile bodies in sporozoites, 199
sporozoite-shaped schizonts in, 62
bovis
antibody reactions
to particulate antigens, 308-309
to soluble antigens, 310-311
blastophores in, 64
conoid of, 90
cultivation in cell cultures, 194, 211, 215, 216, 218, 220, 221, 224, 226, 231
dexamethasone affecting, 285
duration of immunity to, 300-301
excystation in, 58, 163, 164
affected by pH, 434
fertilization in, 129
gross pathology of infections with, 265-266
histopathologic effects of, 269
host-cell changes from, 54
host-cell type for development of, 38
immunity to, 298
intravacuolar tubules in macrogametes of, 112
large schizonts in, 53, 269
life cycle of, 35, 46-51
merozoites in
formation of, 66
ultrastructure of, 102, 103
micropore of, 92, 95
changes in during excystation, 57
microtubules of, 89
oocyst dosage affecting immunity to, 298
patent period of, 51
preconoidal rings of, 88
radiation affecting oocysts of, 331
refractile bodies in, 101, 199
schizogenous generations of, 25
site of development in host, 255

Eimeria, bovis—Cont'd
size of oocysts in, 28
sporozoite ultrastructure, 101
stage affected by immune response, 305
stage specificity of immune response, 303, 304
Stieda body of, 58
toxins possibly associated with, 281
and transport of material from host cell into parasite, 66
trophozoites in, 62
wall-forming bodies of, 115
and winter coccidiosis, 73
brunetti
bursectomy affecting, 326
cross-immunity with *E. maxima*, 302
cultivation of
in avian embryos, 188
in cell cultures, 211, 216, 239
cytochemistry of, 146
gross pathology of infections with, 264
and intestinal function of host, 175
necrotic enteritis from, 284-285
parenteral inoculations of, 37, 38
polysaccharide inclusions in, 116
protein in, 148
resistance to, and strains of host, 262
size of endogenous stages, and pathogenicity, 256
callospermophili, 57
conoid of, 90
cross transmission studies, 32-33
cultivation in cell cultures, 197, 211-212, 217, 218, 220, 221, 226
depth in intestinal mucosa, 38
excystation in, 164
immunogenicity of, 298
merozoite formation in, 68
multinucleate merozoites in, 70
nuclear division in schizonts of, 64
nutrient intake in, 66
oocyst wall of, 152
polar rings of, 84
preconoidal rings of, 88
refractile bodies in sporozoites, 199
sporozoite-shaped schizonts in, 62
sporozoite ultrastructure, 101
Stieda body of, 58
stimulation of merozoite motility, 435
substiedal body of, 59
cameli, 53
caviae, gross pathology of infections with, 267
chinchillae, host specificity of, 33
colchici, preservation by cold storage, 437
comparison with *Toxoplasma*, 399-400
crandallis
cultivation in cell cultures, 212, 218
pathogenesis of infections from, 281
refractile bodies in sporozoites, 198
changes in, 62
transformation to trophozoite, 217
cultivation of, 185, 186
in avian embryos, 187-189

Eimeria, cultivation of, in avian embryos—*Cont'd*
agents affecting, 239
in cell culture, 209-227
activating factors in, 222
active penetration in, 192-195
age of parasite affecting, 228-229
agents affecting, 239-242
asexual development in, 215-222
cell types affecting, 233
cells used and extent of development, 209-215
comparison of development in culture and in host, 226
cytopathological effects of, 225
debris affecting, 224, 231-232
extracellular movement of, 191
first and second generations, 215-221
gametocytes, 222-223
histopathological effects of, 225
and immunity studies, 241
intracellular movement and location, 197, 198
merozoite formation in, 218
and merozoite motility, 221
merozoites as inoculum, 209, 216-217
nuclear changes in sporozoites, 215, 217
nuclear division in, 217
oocysts in, 223
penetration of merozoites, 221
sexual development in, 222-223
size of inoculum affecting, 228
sporozoite-shaped schizonts in, 217-218
sporozoites as inoculum, 209, 210-215
survival in, 223-225
third and other generations, 221-222
time and rate of entry 196-197
and treatment of parasites prior to inoculation, 229-230
trophozoite formation, 217
debliecki, 56, 57
depth in intestinal mucosa, 38
screening from pig feces, 415
development in macrophages, 281
duodenalis, preservation by cold storage, 437
ellipsoidalis
cultivation in cell cultures, 212, 226
microtubules of, 89
resistance to, and age of host, 260
size of oocysts in, 28
sporozoite ultrastructure, 101
environ, size of oocysts in, 28
falciformis, 5
gross pathology of infections, 267
intravacuolar tubules in macrogametes of, 112
microgamete ultrastructure, 107
microgametogenesis in, 122
mitochondria in, 116
partial completion of life cycle in, 27
wall-forming bodies of, 115
faurei
affecting food and water intake, 277
stimulation of excystation, 430
gallopavonis
cross transmission studies, 32
Eimeria, *gallopavonis*—*Cont'd*
cultivation in cell cultures, 212
gross pathology of infections with, 265
morphology of oocysts in, 25
gilruthi, large schizonts in, 53
histochemical reactions of, 171
honessi, size of oocysts in, 28
host specificity of, 30-33
impalae, 265
innocua
cross transmission studies, 32
size of oocysts in, 28
intestinalis
cytochemistry of, 146
DNA in, 147
immunogenicity of, 298
merozoite formation in, 68
merozoite ultrastructure, 102, 103
microgametogenesis in, 126
micropore of, 94, 100
microtubules of, 90
nuclear division in schizonts of, 64
nutrient intake in, 66
pellicle in macrogamete of, 111
stage affected by immune response, 304, 306
toxins possibly associated with, 281
trophozoites in, 62
wall-forming bodies of, 115
intravacuolar tubules in macrogametes, 112
intricata
dose of oocysts affecting host, 258
stimulation of excystation, 430
irresidua, gross pathology of infections with, 267
labbeana, sporulation in, 160
lacazei, 5
larimerensis
cinemicrography of, 59-60
cultivation in cell cultures, 197, 198, 212-213, 217, 218, 221, 226
excystation in, 164
polar rings of, 84
refractile bodies in sporozoites, 199
sporozoite escape from oocysts in, 59
sporozoite penetration of host cells, 59-60
sporozoite-shaped schizonts in, 62
sporozoite ultrastructure, 101
Stieda body of, 58
substiedal body of, 59
lemuris, 56
leuckarti
host cell changes from, 54
in lamina propria cells, 51
magna
antibody reactions to particulate antigens, 308-309
conoid of, 90
cultivation in cell cultures, 213, 216, 217, 218, 223, 226
gross pathology of infections with, 267
immunogenicity of, 298
merozoite formation in, 68
merozoite ultrastructure, 102, 103
microtubules of, 90

Eimeria, magna—Cont'd
multinucleate merozoites in, 70, 71
spindles in schizonts of, 64
sporozoite-shaped schizonts in, 62
stage affected by immune response, 305
magnalabia, transmitted from wild to domestic geese, 29
maxima
affecting blood cell numbers, 279
affecting egg production, 278
affecting gut permeability, 279
antibody reactions to soluble antigens, 310-311
antilymphocyte serum affecting, 327
bursectomy affecting, 326
corticosteroids affecting, 324
cross-immunity with other species, 302
cross transmission studies, 31
cultivation of
in avian embryos, 188
in cell cultures, 213, 224
duration of immunity to, 300-301
immunogenicity of, 297
infection of chick embryos, 317-318
and interferon production, 250-257
and intestinal function of host, 174
intravacuolar tubules in macrogametes of, 112
in lamina propria, 51
lipid in, 148
and Marek's disease, 285
meiosis in, 57
microgamete ultrastructure, 107, 108
microgametogenesis in, 122
micropores in, 112
mitochondria in, 116
parenteral inoculations of, 37, 38
passive transfer of immunity of, 316
pathogenesis of infection with, 281
preservation by cold storage, 436
resistance to
and age of host, 259
and strains of host, 262
size of endogenous stages, and pathogenicity, 256
stage specificity of immune response, 303, 304
wall-forming bodies of, 115
media, size of oocysts in, 28
meleagridis
cross transmission studies, 31
morphology of oocysts, 25
species specificity of immunity to, 302
meleagrimitis
antibody reactions
to particulate antigens, 308-309
to soluble antigens, 310-311
in chemically treated hosts, 34-35
cross transmission studies, 32
cultivation in cell cultures, 213, 221, 223, 224, 225, 226, 228, 234
dose of oocysts affecting host, 258
excystation in, 165
gross pathology of infections, 265
preservation by cold storage, 437, 438

Eimeria, meleagrimitis-Cont'd
refractile bodies in sporozoites, 199
resistance to, and age of host, 260
sporozoite penetration of host cells, 60
microgamete ultrastructure, 107-108
microgametogenesis in, 122-127
microgamont ultrastructure, 105-107
micropores of, 95
in mature macrogametes, 112
microtubules of, 88
mitis
cross-immunity with *E. maxima*, 302
cross transmission studies, 31
cultivation in avian embryos, 188, 189
mivati
cellular reactions from, 269
corticosteroids affecting, 257, 323, 324
cultivation in cell cultures, 213, 224
and Marek's disease, 285
parenteral inoculations of, 37, 38
pathogenesis of infections with, 281
preservation by cold storage, 437
resistance to, and strains of host, 262
schizogony affected by betamethasone, 71
stage affected by immune response, 305
miyairii
merozoite ultrastructure, 103
micronemes of, 92
micropore of, 93
microtubules of, 90
species specificity of immunity to, 302
mohavensis, oocyst production in, 34
necatrix
affecting blood cell numbers, 279
affecting intestinal pH, 280
antibody reactions to soluble antigens, 310-311
cellular reactions from, 269
centrocônes in, 69
in connective tissue cells, 51
cross-immunity with *E. tenella*, 302
cultivation of
in avian embryos, 188, 189
in cell cultures, 213, 221, 224, 225, 226, 234
distribution in intestine, 264
excystation in, 163, 165
hemorrhage from, 263, 264
immunogenicity of, 297
and intestinal function of host, 174, 175
and metabolism of host, 175
oocyst production by, 286
parenteral inoculations of, 37, 38
resistance to, and age of host, 260
and *S. typhimurium infection*, 285
site of development in host, 255
species specificity of immunity to, 302
stage affected by immune response, 305
stage specificity of immune response, 303
neitzi, 56
in uterus of impala, 51
neoirresidua, size of oocysts in, 28
nieschulzi
cell transfer of immunity to, 322

Eimeria, nieschulzi—Cont'd
cross-reactions with other coccidia, 372
cultivation in cell cultures, 213, 221
gross pathology of infections with, 267
intravacuolar tubules in macrogametes of, 112
merozoite formation in, 66
merozoite ultrastructure, 103
microgamete ultrastructure, 107
microgametogenesis in, 126
micropore of, 92
microtubules of, 89
parenteral inoculations of, 36
partial completion of life cycle in, 26-27
pellicle in macrogamete of, 111
refractile bodies in, 101
schizogonous generations of, 25
sporozoite motility in, 58
sporozoite ultrastructure in, 101
stage affected by immune response, 304, 305
stimulation of merozoite motility, 435
ninakohlyakimovae
affecting food and water intake, 277
affecting intestinal absorption, 280
in connective tissue of lamina propria, 51
cultivation in cell cultures, 213-214, 217, 218, 221
dry feed inoculations of, 414
gross pathology of infections with, 266
growth of, 64
host cell changes from, 54
large schizonts of, 270
microtubules of, 84, 89
nuclear division in schizonts of, 64
nutrient intake in, 66
refractile bodies in sporozoites, 199
resistance to, and age of host, 260
size of endogenous stages, and pathogenicity, 256
sporozoite ultrastructure, 101
stimulation of excystation, 430
stimulation of merozoite motility, 435
transformation to trophozoite, 217
trophozoites in, 62
ultrastructure of, 95
nova, 7
nucleic acids in, 147
nucleus in macrogametes of, 118
ondatrazibethicae, gross pathology of infections with, 267
oocysts of, 15
wall of, 153
ovina. *See Eimeria arloingi*
paired organelles in, 91
See also club-shaped organelles; Rhoptries
papillata
sporozoite escape from sporocysts in, 59
substiedal body of, 59
parva
large schizonts of, 270
stimulation of excystation, 430
pathology of infections with. *See* Pathology

Eimeria—Cont'd
pellicle of, 84
in mature macrogametes, 111
perforans
cultivation of, 186
intravacuolar tubules in macrogametes, 112
macrogamete growth and development, 127
merozoite formation in, 68
merozoite ultrastructure, 103
microgamete ultrastructure, 107
microgametogenesis in, 122
microgamont ultrastructure, 105
micropore of, 92
microtubules of, 89
mitochondria in, 116
pellicle in macrogamete of, 111
species specificity of immunity to, 302
ultrastructure of, 96
wall-forming bodies of, 115
phagocytosis of, 196
phasiani, preservation by cold storage, 437
polita
cross-immunity with *E. scabra*, 302
cross-reactions with other coccidia, 372
duration of immunity to, 300-301
gross pathology of infections from, 267
immunogenicity of, 298
stage affected by immune response, 304
ponderosa, 265
poudrei, size of oocysts in, 28
praecox
affecting food and water intake, 277
affecting gut permeability, 279
affecting weight, 277
corticosteroids affecting, 324
cultivation of
in avian embryos, 188
in cell cultures, 214, 224
parenteral inoculations of, 37, 38
resistance to, and age of host, 259
site of development in host, 255
size of endogenous stages, and pathogenicity, 256
sporozoite invasion of chicken intestine, 268
pragensis
antibody reactions to particulate antigens, 308-309
merozoite formation in, 68
merozoite ultrastructure, 103
micropore of, 94, 100
refractile bodies in sporozoites, 198-199
research techniques, 414-445
establishment of infections, 414-415
obtaining oocysts for, 415-417
See also Research techniques
robusta, outer oocyst wall variability in, 28
scabra
antilymphocyte serum affecting, 327
corticosteroids affecting immunity to, 323
cross-immunity with *E. polita*, 302
cross reactions with other coccidia, 372
duration of immunity to, 300-301

Eimeria, scabra—Cont'd
gross pathology of infections with, 267
immunogenicity of, 298
stage affected by immune response, 304
schubergi, 5
sciurorum, gross pathology of infections with, 267
separata
gross pathology of infections with, 267
size of oocysts in, 28
species specificity of immunity to, 302
site of development in host, 255-256
species of, 15
stiedai, 2, 4, 5
abnormal sites for, 40
antibody reactions
to particulate antigens, 308-309
to soluble antigens, 310-311
in chemically treated hosts, 35
conoid of, 90
crescent-shaped organelle in, 118
cultivation in avian embryos, 188, 435
dose of oocysts affecting host, 258
duration of immunity to, 300-301
ellipsoidal organelles in, 118
enzymes in, 129, 170-172
gross pathology of infections with, 267
histochemical reactions of, 171
histopathologic effects of, 274
immunogenicity of, 298
lipid in, 148
merozoite formation in, 68
merozoite ultrastructure, 103
micropore of, 92
microtubules of, 88, 89
multinucleate merozoites in, 70
pellicle in macrogamete of, 111
preconoidal rings of, 88
preservation by cold storage, 437
respiratory rate in sporulation, 160, 161-162
site of development in host, 256
species specificity of immunity to, 302
sporozoites in bone marrow, 282
stage affected by immune response, 305
two types of schizonts in, 71
wall-forming bodies of, 115
subrotunda, cross transmission studies, 32
tenella
abnormal sites for, 40
affecting egg production, 278
antibody reactions
to particulate antigens, 308-309
to soluble antigens, 310-311
antilymphocyte serum affecting, 326-327
blood loss from, 263, 279
bursectomy affecting, 326
cellular reactions from, 269
in chemically treated hosts, 35
in connective tissue cells, 51
conoid of, 90
cross-immunity with *E. necatrix,* 302
cross transmission studies, 30, 31
cultivation of, 186

Eimeria, tenella, cultivation of—*Cont'd*
in avian embryos, 188, 189, 190, 236
in cell cultures, 214-239
depth in intestinal mucosa, 38
development of, 121
duration of immunity to, 300-301
ellipsoidal organelles in, 118
enzymes in, 172
excystation in, 163, 165
hemorrhagic coccidiosis from, 263, 279
histochemical reactions of, 171
host cell type for development of, 38, 39
immunogenicity of, 297
infection of chick embryos, 312, 317, 321
and interferon production, 257
and intestinal function of host, 175
intravacuolar tubules in macrogametes of, 112
location within cells, 39
meiosis in, 57
merozoite formation in, 68
merozoite ultrastructure, 103
and metabolism of host, 175
metabolism of sporozoites, 167-169
microgamete ultrastructure, 107
microgametogenesis in, 126
micropore of, 93, 100
changes in, 58
microtubules of, 89
myasthenia from, 278
oocyst dosage affecting immunity to, 298, 299
oocyst wall of, 152, 154
parenteral inoculations of, 37, 38
partial completion of life cycle in, 27
passive transfer of immunity to, 315-316
pathogenesis of infections with, 281, 282, 284
pellicle in macrogamete of, 111
phosphatases in, 170
polysaccharide inclusions in, 116
preconoidal rings of, 88
preservation by cold storage, 436, 437, 438
protruded conoidal complex in sporozoites, 60
purification of sporozoites, 430-433
radiation affecting oocysts, 331
refractile bodies in sporozoites, 198
resistance to
and age of host, 259
genetic differences in, 262
and strains of host, 262
respiratory rate in sporulation, 160, 161-162
site of development in host, 255-256
species specificity of immunity to, 302
splenectomy affecting immunity to, 323
sporozoite ultrastructure, 101
stage affected by immune response, 304, 305
stage specificity of immune response, 303
strain differences in pathogenicity, 259
toxins associated with, 173, 281
vitamin requirements, 155-159
wall-forming bodies of, 115
truncata
gross pathology of infection with, 265
in kidney epithelial cells, 51

Eimeria, truncata—Cont'd
site of development in host, 256
ultrastructure
of merozoites, 102-103
of sporozoites, 100-101
utahensis
excystation in, 164
immunogenicity of, 298
protein in, 148
Stieda body of, 59
substiedal body of, 59
vermiformis, in chemically treated hosts, 35
wall-forming bodies in, ultrastructure of, 115
wassilewsky, 265
zuernii, 18, 265, 266
corticosteroids affecting immunity to, 324
dexamethasone affecting, 285
histopathologic effects of, 270
immunity to, 298
resistance to chemical and physical agents, 74
and winter coccidiosis, 73
Eimeridae, 7
Eimeriidae, 1, 7, 9, 12-13, 15-17, 18
Eimeriidea, 7, 8
Eimeriinae, 7
Eimeriorina, 1, 3-4, 11-14, 17
Electron microscopy, 81-139, 442
of asexual reproduction, 60-71, 121
of conoid, 88-91
of development of coccidia, 60-73, 121-129
of developmental stages, 100-121
of endodyocytes, 84
of endoplasmic reticulum in macrogametes, 116
of enzyme activities in macrogametes, 129-133
of fertilization, 129
of Golgi complex in macrogametes, 116
of host-parasite relationship, 134-137
of intravacuolar tubules in macrogametes, 112
of macrogametes, 111-121, 127-129
of *Aggregata eberthi*, 119
of *Eimeria*, 111-118
of *Klossia helicina*, 119
of *Toxoplasma gondii*, 118
of merozoites, 102-105
of *Besnoitia jellisoni*, 104
of *Eimeria*, 102-103
of *Frenkelia*, 103-104
of *Isospora*, 103
of *Sarcocystis*, 104
of *Toxoplasma gondii*, 104
of microgametes, 107-111
of *Aggregata eberthi*, 108
of *Coelotropha*, 108
of *Eimeria*, 107-108
of *Toxoplasma gondii*, 108
of microgametogenesis, 122-127
of microgamonts, 105-107
of micronemes, 92
of micropores, 58, 92-100
in mature macrogametes, 112
of microtubules, 84-88
of mitochondria in macrogametes, 116
Electron microscopy—*Cont'd*
of motile stages, 83-100
of nuclear division in schizonts, 64
of nucleus in mature macrogametes, 118
of oocyst wall formation, 129
of pellicle, 84
in mature macrogametes, 111
of polar rings, 84
of polysaccharide inclusions in macrogametes, 116
of rhoptries, 91-92
of sexual reproduction, 121-129
of sporozoites, 100-102
in *Eimeria*, 100-101
of Stieda body, 58
of wall-forming bodies, 115
Eleutheroschizonidae, 9
Ellipsoidal organelles, in macrogametes, 118
Endodyocytes, 100
electron microscopy of, 84
in *Frenkelia*, 103-104
Endodyogeny, 68-69, 104, 121, 200, 220
in *Toxoplasma*, 345, 346
Endogenesis, 70
Endomerogony, 66, 68, 121
Endoplasmic reticulum
in macrogametes, 116
of *Toxoplasma*, 118
in merozoites, 68, 103
in sporozoites, 101
Endopolygeny, 69-70, 121
in *Toxoplasma*, 345, 346, 348
Endothelial cells
development of coccidia in, 38
schizonts in, after dexamethasone, 55
Entamoeba histolytica, host specificity of, 24
Enteritis, necrotic, pathogenesis of, 284-285
Enterotome, use of, 439
Environmental chambers, use of, 444
Enzyme activity, 169-172
in macrogametes, 129-133
and penetration, 195-196
Epithelial cells
coccidia in, 38, 51, 55, 56
merozoites in, 48
Erthyrocytes, decreased number in *Eimeria* infection, 263
Erythromycin, affecting cultures, 239
Eucoccidia, 8
Eucoccidiorida, 3, 11-14
Eucoccidium
dinophili, 71
conoid of, 90
microgamete ultrastructure, 108
micropore of, 94
microtubules of, 89
tubules in macrogametes of, 112
sporozoite ultrastructure, 101
Excystation, 26, 46, 57-59, 162-166
age of host affecting, 297
energy considerations in, 164-166
and escape of sporozoites from sporocysts and oocysts, 59

Excystation—*Cont'd*
immunity affecting, 304
lack of, 25
micropyle changes in, 57-58
oocyst wall in, 57-58
pH affecting, 434
primary phase, 162-163
secondary phase, 163-164
Stieda body and substiedal body in, 58-59
stimulation of, 429-430, 432
Exogenesis, 70
Exoschizon, 8
Experimental animals, care and housing of, 30
Experimental infections. *See* Research techniques
Extra-intestinal sites of coccidia, 39-40, 55
Eye lesions
from *Besnoitia,* 362
from *Toxoplasma,* 362, 364, 373

Fabricius, bursa of
effects of bursectomy, 325-326, 328-329
as site of infection, 37
Feces
collection of samples from animals, 417-418
for oocyst isolation, 425-426
screening of oocysts from, 415
sieving of homogenates, 426
Fertilization, 72-73, 152
of macrogametes, 51
ultrastructural aspects of, 129
Fetal infections, from *Toxoplasma,* 359, 361, 369, 377
Filaria, 4
Fish, microsporidian infections of, 53
Flagella, in microgametes, 107, 111
Fluorescence microscopy, 442, 446
Fluorescent antibody test, indirect, for *Toxoplasma,* 446-447, 448
Fluorescent powders, for tracing inoculations of coccidia, 414-415
Folic acid requirements, 158-159
Food intake of host, coccidiosis affecting, 277-278
Food vacuoles, formation of, 66
Freezing
affecting oocysts, 74
affecting organisms in cultures, 229
for sporozoite storage, 436-438
for storage of antigen preparations, 444
for *Toxoplasma* storage, 449
Frenkelia, 14, 18
asexual reproduction of, 121
comparison with *Toxoplasma,* 392-393, 397-399
conoid of, 88, 90
merozoite ultrastructure, 103-104
micropore of, 93
microtubules of, 89
paired organelles in, 91
pellicle of, 84
preconoidal rings of, 88
ultrastructure of, 97
Fur-bearers, cross transmission studies in, 33

Galago, sporulation in, 56
Gametocytes, 48
in cell cultures, 222-223
effects on host cells and tissues, 48, 51, 269-270
in mammary glands, 39
of *Toxoplasma,* 348, 350
Gametogony, 25, 46, 71-72
in abnormal hosts, 25-26, 34
and abnormal sites of infection, 40
in chemically treated hosts, 35
in *Toxoplasma,* 345, 346
Gamonts, 48, 51
Gastrocystis gilruthi, 53
Geese
gross pathology of infections in, 264-265
morphology of oocysts in, 29
Genetics, and resistance to infection, 262, 297
Genotypic determination, 71
Globidium
gilruthi, 53
leuckarti, 53
Globule leukocytes, in *Eimeria* infection, 269
Globules, clear. *See* Refractile bodies
Glugea anomala, intracellular development of, 53
Glycerol solution, for preservation of coccidia, 436
Glycogen levels, 149, 170
in host, coccidia affecting, 175
in macrogametes of *Eimeria,* 116
in merozoites, 102
storage affecting, 259
Gnotobiotic poultry, production of, 444-445
Goats
cross transmission studies, 32, 39
histopathology of *Eimeria* infection in, 270
large schizonts of *E. gilruthi* in, 53
oocyst morphology in, 25
Golgi apparatus
in macrogametes, 116
in merozoites, 66, 68, 103
in schizonts, division of, 64
in sporozoites, 101
Gonobia, 6
Goussia, 6
Granular bodies, in macrogametes of *Aggregata eberthi,* 119
Granules
in nucleolus, 72
plastic. *See* Wall-forming bodies
Gray bodies, in macrogametes, 119, 127
enzyme activity affecting, 133
Gregarina, 4
falciformis, 5
Gregarinasina, 1, 3, 11
Gregarines, 5
Gregarinida, 5, 6, 7
Gregarinidea, 6
Gregarinomorpha, 8
Grouse, cross transmission studies in, 31
Growth factors, and nutrition, 155-159
Growth of schizonts, 64-66

Guinea hen, cross transmission studies in, 31, 32
Guinea pigs
  excystation in, 26
  toxoplasmosis in, 355, 358, 372
Gymnospora, 6
Gymnosporidiida, 6

Haemamoeba gallinacea
  micropore of, 100
  paired organelles in, 91
Haemogregarina, 6
  conoid of, 90
  macrogamete ultrastructure, 119
  micronemes of, 92
  micropore of, 94
Haemogregarines, 15
Haemogregarinidae, 8
Haemogregarinidea, 7
Haemoproteus, 6, 15
  columbae
    macrogamete ultrastructure, 119
    microgamete ultrastructure, 111
  passeris, 15
Haemosporidia, 7
  kochi, microgamete ultrastructure in, 111
Haemosporidiida, 6
Haemosporidiidea, 7, 8
Haemosporina, 3, 6, 14
Halteridium, 6
Hamsters, toxoplasmosis in, 362, 364, 372, 373
Haplospora, 10
Haplosporea, 10
Haplosporidia, 7, 8
Heart, lesions from *Toxoplasma,* 364
Helminths, and *Toxoplasma* transmission, 381
Hemorrhagic coccidiosis, 279
  gross pathology in, 263
  resistance to, and age of host, 260
Heredity, and resistance to infection, 262, 297
Histochemistry, 169-172
Histopathology of infections, 268-277
Historical aspects of coccidia studies, 4-10
Hoarella, 13, 17
  garnhami, 17
Hodgkin's disease, and toxoplasmosis, 374
Holoeimeriidea, 8
Horse, *Globidium leuckarti* in, 53
Host cells
  alterations in, 46, 48, 53-54, 281-282
  hypertrophy of, 53
  and parasite complexes, 53, 54
  penetration by sporozoites, 59-60
    inhibition of, 60
    membrane interrupted by, 60
  sporozoite changes in, 60-62
  transport of material from, into parasite, 66
  types of, and site specificity, 38-39
Host immunity. *See* Immunity
Host-parasite interrelationships, 173-175
  electron microscopy of, 134-137
Host-parasite interrelationships—*Cont'd*
  and intestinal function of host, 174-175
  and metabolism of host, 175
  and toxin activity, 173-174
Host specificity, 26-35
  in chemically treated hosts, 34-35
  corticosteroids affecting, 324
  cross transmission studies, 26, 27, 30-34, 54
  and excystation, 26
  and oocyst structure, 25, 27-29
  and partial completion of life cycle, 26-29
  and shape index, 28-29
  and successful completion of life cycle, 29-34
  in sympatric species, 34
Housing and care of experimental animals, 30
Human infections
  gross pathology of, 268
  toxoplasmosis, 355, 358, 359, 361, 364, 366, 368, 373, 384
Humidity, affecting oocysts, 74
Hyaloklossia, 6
Hyaluronidase
  affecting penetration, 166, 195
  production by sporozoites, 284
Hypercorticism, and toxoplasmosis, 357, 364, 370, 373
Hypersensitivity, in toxoplasmosis, 356, 358, 361, 365, 375-376
Hypertrophy of host cells, 53
Hypochlorite, for purification of oocyst preparations, 422, 427-428

Ileum, *Isospora felis* in, 55
Immunity, 295-333
  acquired, 297-306
    to *Toxoplasma,* 357, 365
  active immunization, 330-332
    by infection, 330-332
    with nonviable material, 332
    with parenteral inoculum of viable parasites, 332
    in toxoplasmosis, 369-370
    with viable oocysts, 330-331
  and age of host, 297, 299
  and antibody responses, 306-315, 333
    in chemically treated hosts, 35
  antilymphocyte serum affecting, 326-327
  antitoxic, 306
  bursectomy affecting, 325-326, 328-329
  cell-mediated, 319-322, 325, 333
    in toxoplasmosis, 370-371
  cell transfer of, 321-322
  corticosteroids affecting, 323-325
  cross-immunity studies, 30, 302
  duration of, 299-302
  experimental infections, after, 28
  experimental modifications of, 322-330
  genetic factors in, 262, 297
  host factors in, 299
  and host specificity, 296
  humoral, 325
  innate, 296-297

Immunity—*Cont'd*
 and intercurrent infections, 299
 and interferon activity, 322
 macrophage role in, 320
 maternal transfer of, 317-318
 mechanisms in resistance to infection, 257
  See also Resistance to infection
 and mode of immunization, 298-299
 oocyst dosage affecting, 286, 298-299
 passive transfer of, 315-317
 and prepatent period length, 304
 and protective effects of antibodies, 315-319
  See also Antibodies
 radiation affecting, 325
 reticuloendothelial system in, 325
 to species of coccidia, 297-298
 splenectomy affecting, 323
 and sporulation, 304
 stage affected by immune response, 304-306
 stage specificity of immune response, 302-304
 study of factors in, 240-241
  techniques for, 443-444
 thymectomy affecting, 325-326, 328-329
 and tolerance, 327-330
 in toxoplasmosis, 365-377
Immunization with vaccines, 330-332
 in toxoplasmosis, 369-370
Immunoglobulins
 and copro-antibodies, 313, 314
 and protective antibodies, 316
 in toxoplasmosis, 372
Impala, uterine coccidiosis in, 40, 51, 56, 265
Inclusions, in macrogametes, 116-118
Infections from coccidia
 pathology of. *See* Pathology
 sources of, 73
Infectivity
 of frozen sporozoites, 437-438
 loss of, 307, 314
 of oocysts in soil, 421
Inner membrane complex and merozoite formation, 66, 69
Inner membrane of oocyst shell, 152
Interference microscopy, 441-442
Interferon, 256-257, 258
 effects in cultures, 241
 and immunity, 322
Intestinal glands. *See* Crypts of Lieberkuhn
Intestine of host
 coccidia in epithelial cells of, 38, 51, 55, 56
 coccidiosis affecting, 174-175, 281-285
 counting of schizonts in, 439
 development of organisms in, 254-255
 merozoites in epithelial cells of, 48
 permeability affected by coccidiosis, 279-280, 315, 319
 sites of infection in, 36-38
Intravacuolar folds, in macrogametes, 134
Isospora, 1, 6, 8, 12, 15, 16, 18
 abnormal sites for, 39-40
 belli, 56, 268, 277
  intestinal changes from, 282

Isospora—*Cont'd*
 bigemina, 18
  gross pathology of infections with, 267
  life cycle of, 54, 55-56
  and *Toxoplasma* infection, 348
 canis
  life cycle of, 54-55
  cultivation in cell cultures, 214-215
  prepatent period in, 55
 cati, 54
 chloridis
  cultivation in cell cultures, 186, 215, 218, 224, 230, 231, 232, 233
  host specificity of, 33
 comparison with *Eimeria*, 399-400
 conoid of, 90
 cultivation in cell cultures, 185, 186, 209, 215, 218
  cell types affecting, 233
  cells used and extent of development, 209-215
  media affecting, 231-232
  survival in, 223
 excystation in, 164, 166
 felis
  cross reactions with other coccidia, 372
  cultivation in cell cultures, 215
  gross pathology of infections with, 267
  life cycle of, 54-55
  multinucleate merozoites in, 71
  prepatent period in, 55
  sporulation in, 57
  and *Toxoplasma* infection, 348
  transmission of, 73
   from cat to dog, 30
 hominis, 56, 268
 host specificity of, 30, 33
 lacazei, 14
  abnormal sites for, 40
  cultivation in cell cultures, 186, 215, 218, 224, 230, 231, 232, 233
  host specificity of, 33
  size of oocysts in, 28
 life cycle of, 54-56
  in dogs and cats, 54-56
  in man, 56
 merozoite formation in, 68
 merozoite ultrastructure, 103
 micronemes of, 92
 micropore of, 94
 microtubules of, 89
 natalensis, 56
 oocyst wall of, 153
 paired organelles in, 91
 rara, 15, 16
 rivolta, 16
  cross transmission studies, 33
  cultivation in cell cultures, 215
  endodyogeny in, 69
  gross pathology of infections with, 267
  prepatent period in, 54
  sporulation in, 57
  and *Toxoplasma* infection, 348
  transmission of, 73

Isospora—Cont'd
sporozoite migration through tissues, 282
ultrastructure of, 96
Isosporidae, 7
Isosporinae, 7

Jejunum, *Isospora felis* in, 55
Joyeuxella, 8

Karyosome, in nucleus of gamonts, 72
Kidney
coccidia in, 51
in woodcock, 40
Eimeria infection affecting, 282
Kinetosomes, 107
Klossia, 6, 8
helicina, 5
acid phosphatase in microgamonts of, 112
enzyme activity in, 133
gray bodies of, 119, 127
histochemical reactions of, 171
macrogamete growth and development, 127
macrogamete ultrastructure, 119
micropore of, 94, 95
phosphatases in, 112, 172
Klossidae, 7
Klossiella, 8
muris, gross pathology of infections with, 267
Klossiidae, 8

Lactation, and toxoplasmosis transmission, 385
Lamina propria of villi, development of coccidia in, 38, 51
Lankesterella, 6, 10, 12, 14, 15
abnormal sites for, 39-40
adiei, 14
garnhami
merozoite ultrastructure, 102
micropore of, 94, 100
hylae
conoid of, 88, 90
microtubules of, 89
polar rings of, 84
marchouxi, microgamete ultrastructure of, 108, 111
minima, 14
sporozoite ultrastructure, 101
Lankesterellidae, 7, 9
Lankesterellinae, 7
Lankesterellonemes, 92
Laparotomy, for life cycle studies, 438
Laser microscope, for isolation of pure cultures of coccidia, 416-417
Laverania, 6
Legerella nova, 7
Legerellidae, 7, 8
Leishmania, host-specificity of, 24
Leucocytozoon simondi
macrogamete ultrastructure, 119
microgamete ultrastructure, 108, 111
Leukemia, and toxoplasmosis, 375
Leukocytes, globule, in *Eimeria* infection, 269
Life cycle of coccidia, 5, 25, 46-56
and asexual generations, 51, 54, 55, 71
of *Eimeria bovis*, 46-51
and host cell alterations, 46, 48, 53-54
and host cell-parasite complexes, 53, 54
of *Isospora*, 54-56
in dogs and cats, 54-56
in man, 56
large schizonts in, 53, 54
partial completion of, 26-29
and site specificity, 35
successful completion in range of hosts, 29-34
techniques in studies on, 438-443
ultrastructural aspects of, 121-129
in various species of coccidia, 51-54
Lipid
life-cycle stages, content in, 148
inclusions in macrogametes, 116, 127
of *Toxoplasma*, 118
in oocyst wall, 151
Liver, coccidia in, 40, 55, 282
Lizards, sporulation in, 56
Lung, coccidia in, 40, 55
Lymph nodes, coccidia in, 39, 55
Toxoplasma in, 354
Lymphocytes
antilymphocyte serum affecting immunity, 326-327
as host cells, 38
Lymphoid cells, response to *Eimeria* infection, 269
Lymphomas, and toxoplasmosis, 373
Lysosomes, in merozoites, 102
Lysozyme, affecting penetration, 166, 195

M-organism. *See Frenkelia*
Macrogametes, 46
in cell cultures, 222
endoplasmic reticulum in, 116
enzyme activities in, 129-133
fertilization of, 51
Golgi complex in, 116
growth and development of, 127-129
intravacuolar tubules in, 112
lipid inclusions in, 116
micropores of, 92, 112
mitochondria in, 116
nucleus in, 118
pellicle in, 111
polysaccharide inclusions in, 116
ultrastructure of, 111-121
in *Aggregata eberthi*, 119
in *Eimeria*, 111-118
in *Klossia helicina*, 119
in *Toxoplasma gondii*, 118
unfertilized, development of, 73
wall-forming bodies in, 115
Macrogamonts, 46, 71-72
plastic granules in, 51
Macrophages
Eimeria development in, 281

Macrophages—*Cont'd*
phagocytosis by, 312
role in immunity, 320
sporozoites transported by, 268
Magnamycin, affecting cultures, 239
Malaria parasite, discovery of, 2, 6
See also Plasmodium
Mammary glands, gametocytes in, 39
Mantonella, 13, 16-17
peripati, 16
potamobii, 17
Mantonellidae, 9
Marek's disease in chickens, 285, 299
Meiosis, in sporogony, 57
Mercuric chloride, affecting oocysts, 74
Merocystidae, 9, 10
Merocystis, 12
Merogony, 46
ultrastructural aspects of, 121
Meronts, 46
Merozoites, 46
conoid of, 88
fixation for electron microscopy, 442
formation of, 64, 66-70
and ectomerogony, 66-68
and endodyogeny, 68-69
and endomerogony, 66, 68
and endopolygeny, 69-70
internal, 68
splitting in, 68
inoculated directly into cecum, 438
as inoculum in cell cultures, 209, 216-217
isolation for research studies, 434-435
mature, 70-71
micropores of, 95, 100
microtubules of, 84
motility of
in cell cultures, 221
stimulation of, 434-435
multinucleate, 70-71
paired organelles in, 91
pellicle of, 84
penetration by, 221
polar rings of, 84
preconoidal rings of, 88
second-generation, 55
soluble antigens from suspensions of, 443
third-generation, 55
ultrastructure of, 102-105
Metabolism
of coccidia, 159-173
of host, coccidia affecting, 175
Methyl benzoquate, affecting cultures, 239
Metrocytes, 103
micropores of, 95
Mice, toxoplasmosis in, 355, 356, 359, 362, 366, 368
Microgametes, 46
flagella in, 107, 111
and spindles in microgamonts, 64
ultrastructure of, 107-111
Microgametocytes, in cell cultures, 222
Microgametogenesis, ultrastructural aspects of, 105, 122-127
Microgamonts, 46, 71, 72
of *Klossia helicina,* acid phosphatase in, 112
large, growth of, 66
micropores of, 92
ultrastructure of, 105-107
Micromanipulators, for isolation of oocysts, 416
Micronemes
in apical complex, 2
electron microscopy of, 92
in sporozoite transformation into trophozoite, 62
Micropores, 2, 56, 100
electron microscopy of, 58, 92-100
in mature macrogametes, 112
in excystation, 57-58
function of, 66, 100
Micropyle
alterations in, during excystation, 57, 58, 162, 163
in fertilization, 152
as oocyst characteristic, 57
Microscopy
electron, 81-139, 442
fluorescence, 442, 446
interference, 441-442
light, for locating parasites before electron microscopy, 442
phase-contrast, 441
Microspora, 7, 10
Microsporida, 2, 5, 6, 7
infections of fishes, 53
Microtubules
electron microscopy of, 84-88
numbers of, 88, 89-90
Migration of sporozoites through tissues, 282
Minchinia, 6, 8
Mitochondria
in macrogametes, 116
of *Toxoplasma,* 118
in sporozoite cytoplasm, 101
Molybdis, 6
Monkeys, susceptibility to toxoplasmosis, 364
Monocystidea, 6
Monocystis stiedae, 5
Monocytes, and *Isospora lacazei* infection, 40
Monosporea, 6
Morphology of oocysts, and host specificity, 25, 27-29
Motility
of merozoites
in cell cultures, 221
stimulation of, 434-435
of sporozoites, 58
Mucopolysaccharide content of oocysts and other stages, 151
Muscular weakness, in coccidiosis, 278
Myocardial lesions, from toxoplasmosis, 364
Myriospora, 10, 12
Myriosporidae, 9, 10
Myriosporides, sporozoite ultrastructure, 101
Myxospora, 7
Myxosporida, 2, 5, 6, 7

Necrosis of cells, from tachyzoites, 355

Necrotic enteritis, pathogenesis of, 284-285
Nematodes
  and coccidial infections, 285
  and *Toxoplasma* transmission, 382
Neosporidia, 7
Nicotinic acid requirements, 157
Nuclear membrane, in meiosis, 57
Nucleic acids, 147-148
  synthesis of, 155
Nucleolus
  of gamonts, 72
  enlargement of, in host cells and sporozoites, 54
Nucleus
  changes in stages in cell cultures, 215, 217, 227
  division in schizonts, 62-64
  of host cell, changes in, 46, 54
  in macrogametes of *Eimeria*, 118
Nutrient intake
  by host, coccidiosis affecting, 277-278
  of schizonts, 64-66
Nutrition, 154-159
  folic acid requirements, 158-159
  and growth factors, 155-159
  and nucleic acid synthesis, 155
  protein, 154-155
  vitamin requirements, 155-159

Octosporella, 13
  mabuiae, 17
Ocular lesions
  from *Besnoitia*, 362
  from *Toxoplasma*, 362, 364, 373
Oligoplastina, 6
Oligosporea, 6
Oocysts, 46
  age of, affecting development in cultures, 228-229
  age of host affecting production of, 424-425
  attenuated, 331
  bacterially sterile, preparation of, 421
  in cell cultures, development of, 223
  chemical nature of wall, 151-154
  in chemically treated hosts, 34-35
  cleaning by density gradient, 421-422
  color of, factors affecting, 29
  counting of, 418-419
  discharge of, 51
    in sympatric species, 34
  dose of and pathogenicity, 256-258, 286, 298-299
  escape of sporozoites from, 59
  excystation of sporozoites from, 46
    See also Excystation
  freeing of sporocysts from, 429
  harvesting of
    from embryos, 429
    from infected ceca, 428-429
  infectivity in soil, 421
  isolation of, 415-417, 424-428
  morphology of, and host specificity, 25, 27-29
  production by *Toxoplasma*, 350, 379
  pure suspensions of, 422-424
  purification of, 422, 427-428
  radiation affecting, 331

Oocysts—*Cont'd*
  residuum of, 57
  resistance to chemical and physical agents, 74
  screening from feces, 415
  shape index of, 28-29
  size of, 28
  soluble antigens from suspensions of, 443
  sporulation of, 46, 51, 56-57
    techniques for research, 419-421, 426-427
  staining of, 441
  sterilization of, 151, 153
  storage affecting, 57
  survival of
    agents affecting, 74
    in soil, 421
  unsporulated, 56
  variations in wall of, 28, 56
  viability and virulence of, 258-259
  vitality of, 57
  wall changes in excystation, 57-58
  wall formation in, 51, 72, 73
    ultrastructural aspects of, 129
  wall-forming bodies in development of, 115
Opsonizing antibodies, 312, 320
Organelles, paired, electron microscopy of, 91-92
  See also Rhoptries
Ovivora, 10, 12
Ox, cross transmission studies in, 32

Paired organelles, electron microscopy of, 91-92
  See also Rhoptries
Pancreas, *Eimeria* infection affecting, 282
Parahaemoproteus, 15
  garnhami, 15
Paramethasone, affecting immunity, 323
Paranuclear bodies, in sporozoites, 101
  See also Refractile bodies
Parasitophorous vacuole, 51, 66, 112, 134, 197-198, 220
  of *Toxoplasma gondii*, 118
Partridge, cross transmission studies in, 31, 32
Patent period
  corticosteroids affecting, 285
  in *Eimeria bovis*, 51
  immunity affecting, 305
  in *Toxoplasma*, 350
Pathogenesis of coccidial infections, 281-286
  factors affecting, 254-262
Pathology of infections, 253-287
  age of host affecting, 259-262
  blood loss, 279
  breed or strain of host affecting, 262
  carbohydrate metabolism, 278
  "crowding" factor in, 256, 258, 260, 304, 322
  dose of oocysts affecting, 256-258
  egg production, 278
  experimental, 277-281
  food and water intake, 277
  gross pathology, 262-268
    in cats, 267-268
    in cattle, 265-266
    in dogs, 267-268

Pathology of infections—*Cont'd*
in domestic fowl, 263-264
in ducks, turkeys and geese, 264-265
in mammals, 265-268
in man, 268
in pigs, 267
in poultry, 262-265
in rodents, 267
in sheep, 266-267
histopathology, 268-277
and interferon activity, 256-257, 258
intestinal pH, 279, 280
nutrient uptake, 279-280
permeability of gut, 279-280, 315, 319
and site of development in host, 254-256
size of endogenous stages affecting, 256
toxin role in, 280-281
with *Toxoplasma*, 358-365
viability and virulence of oocysts affecting, 258-259
weight changes, 277
Pellicle
electron microscopy of, 84
in mature macrogametes, 111
in macrogametes of *Toxoplasma*, 118
Penetration of cells, 59-60, 166
in cultures, 191-196, 221
media affecting, 232
by merozoites, 221
by sporozoites, 221
and enzyme activity, 195-196
inhibition of, 60
Penicillin, affecting cultures, 239
Perforatorium, 107
Perinuclear space, in sporozoites, 101
Peritoneal exudates, *Toxoplasma* in, 446
Permeability of gut, coccidiosis affecting, 279-280, 315, 319
Pfeifferella, 6
Pfeifferellinae, 7
Pfeifferinella, 13
Pfeifferinellidae, 9, 13
pH
and excystation of oocysts, 434
intestinal, in coccidiosis, 279, 280
of medium, in cell cultures, 232
Phagocytosis, 196
by macrophages, 312, 320
Pheasants, cross transmission studies in, 31, 32
Phenotypic determination of sex, 71
Phosphatases, 170-172
in macrogametes, 133
in microgamonts, 112
Pig
excystation in, 26
gross pathology of infections in, 267
histopathology of *Eimeria* infection in, 277
Pigeons
cross transmission studies in, 31
toxoplasmosis in, 355, 368, 372
Pinocytosis, 66
Pinocytotic vesicles, in macrogametes, 111
Piridium, 8
Piroplasmasida, 3, 14
Piroplasmea, 10
Piroplasmidea, 7
Placental infections, from *Toxoplasma*, 359, 377
Plasmodium, 6
berghei, and interferon production, 322
chromatin distribution in, 64
host-specificity of, 24
macrogamete ultrastructure, 119, 121
microgamete ultrastructure, 108, 111
micropore of, 93, 100
microtubules of, 89
paired organelles in, 91
parasiteophorous vacuole in, 115
pellicle of, 84
penetration mechanisms of, 134
polar rings of, 84
Plastic granules. *See* Wall-forming bodies
Pneumonia, and toxoplasmosis, 364
Polar body, 57
Polar granules, formation in sporulation, 56
Polar rings, 2
electron microscopy of, 84
Poles of spindles, and merozoite formation, 68
Polymyxin, affecting cultures, 238
Polyplastina, 6
digenica, 6
monogenica, 6
Polysaccharides, 149-151
inclusions in macrogametes, 116
in oocyst wall, 151
See also Amylopectin; Glycogen
Polysporea, 6
Polysporocystidae, 6, 7, 8
Poultry
excystation in, 26
gnotobiotic, production of, 444-445
gross pathology of infections in, 262-265
Prairie dogs, excystation in, 26
Preconoidal rings, 88
Pregnancy, toxoplasmosis in, 359, 361, 369, 377
Prepatent period
immunity affecting, 304
in *Toxoplasma*, 348, 350, 352-353, 379
Preservation. *See* Storage
Prococcidia, 8
Progenitor cells in intestinal crypts, coccidia affecting, 281
Proliferative stages, of *Toxoplasma*, 166-167
Protein
content in life-cycle stages, 148
levels in host, coccidiosis affecting, 280
nutritional requirement by organisms, 154-155
in oocyst wall, 151
Protococcidiorida, 3, 11
Protoplasm condensation, in oocysts, 56
Pseudoklossia, 10, 12
Pseudoklossiidae, 9, 10
Pseudonavicellae, 4
Psorospermien, 4
Psorospermium, 4, 5
cuniculi, 4
Purification
of oocysts, 422, 427-428
of sporozoites, 430-433

Purines
  effects in cultures, 237-238
  synthesis of, 155
Pyramid stage, in sporulation, 57
Pyridoxine requirements, 157-158
Pyrimethamine, effects of, 158, 238
Pyrimidines, synthesis of, 155
Pythonella, 13
  bengalensis, 17
  scelopori, 17

Quail
  cross transmission studies in, 30, 31, 32
  toxoplasmosis in, 372
Quinine
  affecting penetration of cells by sporozoites in cultures, 60, 240
  effects in embroyonated eggs, 240

Rabbit
  abnormal sites of infection in, 40
  cross transmission studies in, 33
  histopathology of *Eimeria* infection in, 274
  size of oocysts in, 28
  toxoplasmosis in, 355, 372
Raccoons, toxoplasmosis in, 372
Radiation
  affecting oocysts, 331
  and relapsing toxoplasmosis, 357, 364, 374
  and susceptibility to infection, 35, 325
Rats, toxoplasmosis in, 355, 356, 368
Refractile bodies, 198-199
  in merozoites, 68, 101
  protein in, 148
  in sporozoites, 101
    changes in, 60, 62
Reproduction of coccidia
  age of host affecting, 260
  asexual, ultrastructural aspects of, 64-71, 121
  sexual, ultrastructural aspects of, 72, 121-129
Reptiles, morphology of oocysts in, 29
Research techniques, 411-453
  age of host and size of inoculum in, 424-425
  Besnoitia, 452-453
  cleaning oocysts by density gradient sedimentation, 421-422
  collection of feces, 417-418, 425-426
  counting of oocysts, 418-419
  counting of schizonts in intestines, 439
  cultures established from single oocysts, 414
  cultivation of coccidia in chick embryos, 435
  Eimeria, 414-445
  enterotome use in, 439
  establishment of infections, 414-415
  feeding status at infection and during collection of feces, 425
  flattening of intestinal samples for fixation, 440
  freeing of sporocysts from oocysts
    by crushing, 429
    by grinding, 429
  freezing of sporozoites for storage, 436-438
Research techniques, freezing of sporozoites for storage—*Cont'd*
    thawing of, 436-437
  harvesting of oocysts
    from embryos, 429
    from infected ceca, 428-429
  immunological study techniques, 443-444
  inoculation methods for ruminants, 414
  isolating experimental calves, 417
  isolation of *Besnoitia,* 453
  isolation of merozoites, 434-435
  isolation of oocysts, 415-417, 424-428
    by centrifuge, 415
    by flotation, 426
    by laser microscope for pure cultures, 416-417
    by micromanipulator, 416
  isolation of *Toxoplasma,* 447-448, 449
  life cycle study techniques, 438-443
    examining sections of intestine for, 440
    surgery in, 438
  merozoite inoculation directly into cecum, 438
  microscopy in, 441-443
  plastic film isolators for gnotobiotic poultry, 444-445
  pure suspensions in
    of oocysts, 421-424
    of sporozoites, 432-433
  purification of oocysts, 422, 427-428
    by glass bead column and gradient centrifugation, 428
    hypochlorite in, 422, 427-428
  purification of sporozoites, 430-433
  rapid preservation of tissues for sectioning, 438-439
  Sarcocystis, 451
  screening oocysts from feces, 415
  sieving of fecal homogenates, 426
  sporozoite preparations, 429-434
    for cell cultures, 433-434
    in vitro excystation technique, 432
    pure suspensions of, 432-433
    purification techniques, 430-433
    stimulation of excystation, 429-430, 432
    Strout's method of, 434
  sporulation of oocysts, 419-421, 426-427
    with aeration, 419, 427
    in large volumes, 420
    in tap water, 419-420
    temperature range for, 420
    testing for, 420-421
  staining methods
    for cells grown on coverslips, 441
    for parasites in tissue, 440
  sterile oocyst preparations, 421
  testing for survival of oocysts in soil, 421
  Toxoplasma, 445-450
  tracing of inoculations, 414-415
  trypsinization of tissues to release oocysts, 428-429
Resistance to infection
  and age of host, 313
  breed or strain of host affecting, 262
  cold weather affecting, 74
  genetic differences in, 262
  immune mechanisms in, 257
  in primary infection, 285

Resistance to infection—*Cont'd*
sex differences in, 278
study of factors in, 240-241
to *Toxoplasma*
acquired, 357-358, 376-377
natural, 356-357, 365
See also Immunity
Respiration, 160
and enzyme activity, 172
in excystation, 165
in proliferative *Toxoplasma*, 167
in sporozoites of *E. tenella*, 167-169
in sporulation, 161-162
Reticuloendothelial system, in immunity, 325
Rhabdospora, 6
Rhoptries, 2
electron microscopy of, 91-92
in sporozoite transformation into trophozoite, 62
Rhytidocystis, 8
Riboflavin requirements, 157
Ribonuclease, affecting granules in nucleolus, 72
Ribonucleic acid, 147
synthesis by *Trichinella spiralis* larva, 54
Ribonucleoprotein granules, in nucleolus, 72
Ribosomes, in sporozoites, 101
Rifampin, affecting cultures, 239
Robenidene, affecting cultures, 240
Rodents
cross transmission studies in, 30, 32, 33
examination for *Besnoitia*, 452
gross pathology of infections in, 267
life cycle of coccidia in, 51
Rosette formation, by *Toxoplasma* in cell cultures, 205-206
Rumen, excystation in, 57
Ruminants
cross transmission studies in, 30, 32
excystation of coccidia in, 57-58
inoculation methods for experimental infections, 414
life cycles of coccidia in, 51

Sabin-Feldman dye test, for toxoplasmosis, 147, 447
Salmonella typhimurium, concurrent coccidial infection with, 285
Sarcocystidae, 13-14, 18
Sarcocystidia, 6
Sarcocystinae, 14
Sarcocystis, 14, 18
comparison with *Toxoplasma*, 392-393, 395-397
conoid of, 88
cultivation in cell cultures, 185, 186, 208, 227-228, 451
agents affecting, 240
cells used in cultures, 227
extracellular movement of, 191
host-cell types affecting, 233-234
intracellular movement and location, 197
nuclear changes in, 227
cyst wall of, 53
examination of birds for, 451
hylae, conoid of, 90
Sarcocystis—*Cont'd*
merozoite ultrastructure, 104
micropore of, 95
paired organelles in, 91
research techniques, 451
tenella
cross-reactions with other coccidia, 372
merozoite ultrastructure, 104
micropore of, 93
microtubules of, 89
preconoidal rings of, 88
toxins in, 174
ultrastructure of, 99
Sarcodina, 10
Sarconemes, 92
Sarcosporidia, 5, 6, 7, 8
Schellackia, 10, 12
Schellackiidae, 9
Schellackinae, 7
Schizogony, 25, 62-71, 100, 218
in abnormal hosts, 25, 34
and abnormal sites of infection, 40
in chemically treated hosts, 35
and granulocyte-cell response, 269
growth and nutrient intake in, 64-66
mature schizonts and merozoites, 70-71
and merozoite formation, 64, 66-70
nuclear division in, 62-64
temperature affecting, 230
termination of, 71
in *Toxoplasma*, 345, 346
Schizonts, 46
counting of, in intestines, 439
large, 53, 54
in *Eimeria* infections, 269, 270
in endothelial cells of lacteals, 269
mature, 70-71
nuclear division in, 62-64
sporozoite-shaped, 62, 217-218
vacuole surrounding, 134
Schizozoites, 100
Selenidium
hollandei, micropore of, 94
micronemes of, 92
Selenococcidiidae, 7, 8, 11
Selenococcidium, 11
intermedium, 8-9
Sexual reproduction, ultrastructural aspects of, 72, 121-129
Shape index of oocysts, 28-29
Sheep
cross transmission studies in, 32, 39
gross pathology of infections in, 266-267
histopathology of *Eimeria* infections in, 270
large schizonts of *E. gilruthi* in, 53
oocyst morphology in, 25
toxoplasmosis in, 359
Shrew
cross transmission studies in, 33
gametocytes in mammary glands of, 39
Sieving of fecal homogenates, 426
Site specificity, 35-40

Site specificity—*Cont'd*
and abnormal sites, 39-40
and host-cell type, 38-39
and depth in mucosa, 38
and distribution of parasites along intestine, 36-38
and life cycle of coccidia, 35
and location within host cell, 39
parenteral inoculation studies, 36-38
Sivatoshellina, 13, 17
Small intestine
excystation phase in, 59
Isospora in, 55
Sparrows
abnormal sites of infection, 40
cross transmission studies in, 33
Species differences
and antibody response to *Toxoplasma,* 372
and immunogenicity, 297-298
and specificity of immunity, 302
and toxoplasmosis, 364
Specificity
of coccidia, for their hosts, 24-35
corticosteroids affecting, 324
See also Host specificity
of site, 35-40
corticosteroids affecting, 324
species specificity of immunity, 302
Spindles, intranuclear
and microgamete formation, 72
in microgamonts, 107
in schizonts, during cell division, 64
Spirocystis, 8
Spleen, coccidia in, 55
Splenectomy, affecting immunity, 323-325
Splitting, and merozoite formation, 68
Sporoblasts, 56
Sporocysts, 57
escape of sporozoites from, 59
number per oocyst, 9
preservation by cold storage, 436, 437
release from oocysts, 429
in gizzard, 58
soluble antigens from suspension of, 443
Sporogony, 46, 56-57
Sporonts, 56
Sporozoa, 1, 4, 5-10
incertae sedis, 6, 8
Sporozoasida, 3, 11-14
Sporozoite-shaped schizonts, 62, 217-218
Sporozoites, 46
activation by trypsin-bile solution, 58, 60
affecting gut permeability, 315, 319
age of, and survival in culture, 224
in chemically treated hosts, 35
conoid of, 88
distribution in intestine, 36-37
escape from sporocysts and oocysts, 59
excystation of, 26, 46
stimulation of, 429-430, 432
See also Excystation
extracellular, covering layer of, 60
freezing for storage, 436-438
Sporozoites, freezing for storage—*Cont'd*
affecting infectivity, 437-438
hyaluronidase production by, 284
immunity affecting invasion of host cells, 304
as inoculum in cell cultures, 209, 210-215
intracellular, changes in, 60-62
metabolism of, 167-169
micropores of, 95, 100
microtubules of, 84
migration through tissues, 282
motility of, 58
nucleolar enlargement in host cell, 54
number per sporocyst, 9
paired organelles in, 91
and partial completion of life cycle, 27
pellicle of, 84
penetration of cells, 59-60, 221
inhibition of, 60
polar rings of, 84
preconoidal rings of, 88
preparations for research studies, 429-434
purification techniques, 430-433
refractile bodies in, 60, 62, 198-199
changes in, 60, 62
soluble antigens from suspensions of, 443
transformation into trophozoites, 62
transported by macrophages, 268
ultrastructure of, 100-102
Sporulation, 46, 51, 56-57, 160-162
immunity affecting, 304
techniques in research studies, 419-421, 426-427
temperature affecting, 258-259
of *Toxoplasma,* 350
Squirrels, toxoplasmosis in, 372
Staining methods
for cells grown on coverslips, 441
for oocysts, 441
for parasites in tissue, 440
Statolon, effects of, 322
Sterilization
of oocysts, 151, 153
for production of gnotobiotic poultry, 444-445
Stieda body, 57, 164
composition of, 58
disappearance of, 59
electron microscopy of, 58
in excystation, 58-59
protein in, 148
Storage
of antigen preparations, 444
freezing for, 436-438
See also Freezing
of Toxoplasma, 449
Strain differences
in avian embryos, affecting cultures, 190
in hosts, and resistance to infection, 262
in organisms
and cyst formation in cultures, 207
and multiplication in cultures, 205
and pathogenicity, 259
in cultures, 189, 207
and invasiveness, 196

Strain differences, in organisms—*Cont'd*
  and resistance to pyrimethamine, 238
Streptomycin, effects in cultures, 239
Stylocephalus, microgamete ultrastructure in, 111
Substiedal body, 57, 164
  in excystation 58-59
Sulfonamides
  anticoccidial activity of, 158-159
  effects in cell cultures, 237-238
  and immunity to reinfection, 303
  in toxoplasmosis, 365
Sunlight, affecting oocysts, 74
Surgery, in life cycle studies, 438
Survival
  of oocysts, in soil, 421
  of organisms in cell cultures, 223-225
Susceptibility to infection. *See* Resistance to infection
Swans, morphology of oocysts in, 29

Tachyzoites, 344, 345, 346
  cellular necrosis from, 355
  group formation by, 344, 350-351
Taxonomy, 1-4, 10-18
Telosporea, 10
Telosporidia, 7
Temperature
  affecting organisms in cultures, 230-231
  and excystation rate, 430
  and sporulation, 258-259, 420
  and storage of antigen preparation, 444
  and storage of coccidia, 436-438
Terramycin, affecting cultures, 238
Tetrasporea, 6
Tetrasporocystidae, 6-7, 8
Theileria parva, micropore of, 93
Thiamine requirements, 156-157
Thiry's fistula, for life cycle studies, 438
Thymectomy, affecting immunity, 325-326, 328-329
Tissue culture studies, 39, 156, 157, 166
  for *Toxoplasma*, 448, 449
Tolerance, immunological, 327-330
Toxins, 173-174
  role in coccidial infections, 280-281
  of *Toxoplasma*, 356
Toxoplasma, 8, 14, 18
  abortion from, 359, 377
  acquired resistance to, 357-358, 365, 376-377
  acute infection with, 359
  adrenal lesions from, 364, 365
  air borne infection with, 390
  antibodies to, 370, 371-372
  asexual reproduction in, 68-70, 121
  asymptomatic infection with, 358-359
  in blood stream, 355
  bradyzoite-forming groups, 344, 351-353
  brain lesions from, 362, 365, 366, 373
  cell-mediated immunity to, 370-371
  cell-type, specificity for, 39
  centrocônes in, 69
  characterization of, 347
  chronic infection with, 362-364

Toxoplasma—Cont'd
  comparison with other coccidia, 390-400
    Besnoitia, 362, 365, 370, 371, 372, 390-395
    Eimeria, 399-400
    Frenkelia, 392-393, 397-399
    Isospora, 399-400
    Sarcocystis, 392-393, 395-397
  conoid of, 88, 90
  cross-reactions with other coccidia, 372
  cultivation in avian enbryos, 186-187, 189, 190
    and antigen preparations, 236
    and isolation of organism, 234
    and maintenance of organism, 235
  cultivation in cell culture, 200-207
    active penetration in, 191-192, 195-196
    age and number of cells affecting, 234
    age of parasite affecting, 229
    agents affecting, 237-241
    aggregates and cytopathological effects, 205-206
    and antigen preparations, 236-237
    cells and culture vessels used for, 201-205
    cells supporting development, 200
    culture chamber affecting, 230
    and cyst formation, 207
    debris affecting, 231-232
    development in, 200-207
    extracellular movement, 191
    histopathological effects, 206-207
    host-cell types affecting, 232-233
    and immunity studies, 240, 241
    intracellular movement and location, 197, 198
    and isolation of organism, 234
    and maintenance of organisms, 235-236
    mode and rate of multiplication, 200-205
    reinvasion of cells, 206
    release from cells, 206
    and rosette formation, 205-206
    size of inoculum affecting, 228
    strain differences in, 200, 205, 207
    temperature affecting, 230-231
    time and rate of entry of sporozoites, 196
  cyst wall of, 53
  cytochemistry of, 146
  development of, 121
  dissemination of, 354, 359
  dye test for, 147, 447
  dynamics of infection, 354-355
  effects of inoculation routes, 354, 358, 366
  endodyogeny in, 68-69
  enteroepithelial cycle, 346, 348-350, 355
  enzymes in, 169-170
  epidemiology and prevalence of infection, 387-388
  extra-intestinal or tissue cycle of, 346, 350-353
  fetal infections with, 359, 361
  filter paper blood specimens, 448
  handling of, 450
  histochemical reactions of, 171
  host specificity, 33-34
  "hostile factor" of, 372
  human infections, 355, 358, 359, 361, 364, 366, 368, 373, 384
  hypercorticism affecting, 357, 364, 370, 373

Toxoplasma—Cont'd
- hypersensitivity to, 356, 358, 361, 365, 375-376
- iatrogenic infections, 374-375
- immunity to reinfection, 366-368
- immunization with live and killed vaccines, 369-370
- immunodeficient hosts, 372-375
- immunology of, 365-377
- indirect fluorescent antibody test for, 446-447, 448
- infection with, 353-355
- interferon protection against, in cell cultures, 357
- intestinal stages of, 70
- isolation
  - from blood or tissues, 447-448
  - free of host cells, 449
- laboratory diagnosis of infection, 446-448
- lesions in individual species, 364
- life cycle of, 345-353
- lipid in, 148
- in lymph nodes, 354
- lysis with accessory factor, 447
- macrogamete ultrastructure, 118
- maintenance in laboratory animals, 448-449
- merozoite formation in, 68, 70
- merozoite ultrastructure, 104
- and metabolism of host, 175
- metabolism of sporozoites, 167
- microbial effects of, 355-356
- microgamete ultrastructure, 108
- microgametogenesis in, 122, 126
- micronemes of, 92
- micropore of, 93
- microtubules of, 89
- myocardial lesions from, 364
- natural resistance to, 356-357, 365
- new cystic form of, 382-383
- nucleic acids in, 147
- ocular lesions from, 362, 364, 373
- oocyst morphology, 267
- oocyst production, 350, 379
- paired organelles in, 91
- pathogenetic considerations, 355-358
- pathology from, 358-365
- penetration activity, 166
- in peritoneal exudates, 446
- phagocytosis of, 196
- plaque method for studies of, 449-450
- pneumonia from, 364
- preconoidal rings of, 88
- and premunition, 357, 368-369
- prepatent period, 348, 352-353, 379
- preservation of, 449
- prevention of infection, 388-390
- proliferative stages of, 166-167
- propagation of, 448-450
- recovery from infection, 365-366
- relapsing infection with, 357, 359, 364, 366, 373
- research techniques, 445-450
- rupture of cyst, 356, 358, 369
- schizogonic stages of, 69
- stages of, 346-353
- subacute infection with, 361-362
- symptomatic infections from, 359-365

Toxoplasma—Cont'd
- tachyzoite-forming groups, 344, 350-351
- testing for, 445-446
- threshold infectious dose of, 354
- in tissue cultures, 448, 449
- toxic factor of, 356
- toxins associated with, 173, 281, 284
- transmission of infection, 73, 377-386
  - by arthropods, 386
  - by blood transfusions, 385
  - by carnivorism, 378
  - by eggs, 385
  - fecal, 378-384
  - in laboratory and at autopsy, 385-386
  - by lactation, 385
  - by organ transplants, 385
  - transplacental, 359, 361, 369, 377
  - venereal, 384-385
- transport hosts for, 389
- ultrastructure of, 98
- virulence of, 353-354
- wall-forming bodies of, 115

Toxoplasmatidae, 18
Toxoplasmatinae, 14
Toxoplasmea, 10
Toxoplasmin skin test, 358
Toxoplasmosis, 343-400
Transfusions, and toxoplasmosis transmission, 385
Transmission of infection, 73-74
- *Besnoitia,* 452-453
- *Toxoplasma,* 377-386
- *See also* Cross transmission

Transplants, and toxoplasmosis transmission, 385
Trichinella spiralis, affecting muscle fibers, 54
Trisporea, 6
Trophozoites, 46, 62
- transformation to, 217

Trypsin-bile treatment
- and activation of sporozoites, 58, 60
- and excystation, 162, 163-164
- and merozoite motility, 70, 221
- for obtaining sporozoites, 433-434
- and Stieda body changes, 58

Trypsinization of tissues, to release oocysts, 428-429
Tubules
- intravacuolar, in macrogametes, 112, 134
- microtubules, 84-90
- subpellicular, in apical complex, 2

Turkey
- cross transmission studies in, 30, 31-32
- excystation in, 59
- gross pathology of infections in, 264-265
- oocyst morphology in, 25
- resistance to infection, 260

Typhlitis, 263
Tyzzeria, 13, 16
- *anseris,* host specificity of, 29
- *perniciosa,* 264

Ultracytostome, 100
Ultrasound, for preparation of soluble antigens, 443

Ultrastructure of organisms. *See* Electron microscopy
Ultraviolet light, for tracing inoculations of coccidia, 414-415
Urobarrouxia, 17
Uterine coccidiosis, in impala, 40, 51, 56, 265

Vaccines, and immunity against *Toxoplasma,* 369-370
Vacuoles
  food, formation of, 66
  parasitophorous, 51, 66, 112, 134, 197-198, 220
Vermicule, 100
Vesicles
  pinocytotic, in macrogametes, 111
  thick-walled, in merozoites, 102
Villus atrophy, in coccidiosis, 269, 281
Vitamin A
  coccidiosis affecting levels of, 280
  requirements by coccidia, 159
Vitamin K
  and mortality from *Eimeria* infection, 279
  requirements by coccidia, 159
Vitamin requirements, 155-159
  biotin, 157
  nicotinic acid, 157
  pyridoxine, 157-158
  riboflavin, 157
  thiamine, 156-157
  vitamin A, 159
  vitamin K, 159
Vitamins, affecting cultures, 238

Wall formation, oocyst, 72, 73
Wall-forming bodies, 115, 148, 152-153
  enzyme activity in, 133
  in macrogametes of *Toxoplasma,* 118
  in macrogamonts, 51
  phosphatases in, 172
  protein in, 148
  types of, 153
  ultrastructure of, 115
Water buffalo, cross transmission studies in, 32
Water intake, coccidiosis affecting, 277-278
Weakness, muscular, in coccidiosis, 278
Weight gain, coccidiosis affecting, 277
Wenyonella, 13, 16
Winter coccidiosis, 73
Woodcock, renal coccidium of, 40

Xenoma, 53
Xenon, 53
Xénoparasitome, 53

Yakimovella, 13, 17
  erinacei, 17
Yakimovellidae, 9

Zoites, 100
Zygotes, 46